T0186042

Universitext

Universitext

Universitext is a series of textbooks that presents material from a wide variety of mathematical disciplines at master's level and beyond. The books, often well classtested by their author, may have an informal, personal even experimental approach to their subject matter. Some of the most successful and established books in the series have evolved through several editions, always following the evolution of teaching curricula, into very polished texts.

Thus as research topics trickle down into graduate-level teaching, first textbooks written for new, cutting-edge courses may make their way into *Universitext*.

More information about this series at http://www.springer.com/series/223

Olivier Bordellès

Arithmetic Tales

Advanced Edition

 Springer

Olivier Bordellès
2 allée de la combe
Aiguilhe, France

Translated by
Véronique Bordellès
2 allée de la combe
Aiguilhe, France

ISSN 0172-5939   ISSN 2191-6675   (electronic)
Universitext
ISBN 978-3-030-54945-9   ISBN 978-3-030-54946-6   (eBook)
https://doi.org/10.1007/978-3-030-54946-6

Mathematics Subject Classification: 11-01, 11A, 11L07, 11M06, 11N05, 11N13, 11N25, 11N37, 11R04, 11R09, 11R11, 11R16, 11R18, 11R21, 11R27, 11R29, 11R42

This Springer imprint is published by the registered company Springer Nature Switzerland AG.
The registered company address is: Gewerbestrasse 11, 6330 Cham, Switzerland

Preface

Significant advances have been made in number theory since the book *Arithmetic Tales* was published back in 2012. The most important breakthrough is probably the landmark result that Zhang proved on the bounded gaps between prime numbers. One of the main achievements of the proof is to establish a suitable extension of the Bombieri–Vinogradov theorem.

This naturally led me to write a thoroughly revised and updated version of the book. To begin with, the chapters devoted to the knowledge required from undergraduate mathematics students have all been removed. In particular, this includes practically all of the content of my earlier book *Thèmes d'Arithmétiques*, with only the most necessary overlapping content completely rewritten in summarized form. The prerequisites for this second edition are therefore essentially all known results in classic elementary number theory. Besides, the sections previously called "Further Developments" have become full sections that have substantially been expanded. In addition, the number of exercises, all corrected as in the first edition, has increased by almost 50%.

Each chapter has been revised, updated and enhanced, and Chap. 4 in particular has undergone some drastic alterations. Dirichlet series are now to be found at the beginning of the chapter immediately after the basic theory. A quite simple asymptotic formula has been derived for multiplicative functions close to the divisor functions. Each mean value of the usual arithmetic function is given in detail, including the important Hooley's divisor function. A section, that is rarely mentioned in the literature and dealing with arithmetic functions of several variables, is provided. Ten well-known problems in analytic number theory have also been selected and given with almost self-contained proofs. The results that are beyond the scope of this book are given without proof but always with one or several references, so that the reader who is eager to learn more about the subject may go deeper into the analysis and pursue further so as to absorb the most profound results of the subject area.

This revised edition would have been impossible without the patience, availability and enthusiasm of Remi Lodh, whom I wish to thank for his interest in this project. I would also like to thank most warmly my wife, Véronique, for her rigorous and faithful translation of this work and for her unrelenting support. I am also most grateful to my colleagues Anne-Lise Camus, Florian Daval and Igor E. Shparlinski for their careful proof-reading and to Władysław Narkiewicz and Titus W. Hilberdink for their suggestions. Last but not least, I would particularly like to thank my two children, Nicolas, a graduate student in Civil and Mining Engineering at the Ecole des Mines, and Marie, who follows a 2-year undergraduate intensive course in Mathematics, Physics and Biology.

Aiguilhe, France   Olivier Bordellès
May 2020

Preliminaries

1 General Notation

Letters n, m and a, b, c, d, k, ℓ refer to integers, whereas p indicates a prime number. We also adopt Riemann's notation for complex numbers, namely $s = \sigma + it \in \mathbb{C}$ with $\sigma = \mathrm{Re}(s)$ and $t = \mathrm{Im}(s)$.

$\mathbb{Z}, \mathbb{Q}, \mathbb{R}, \mathbb{C}, \mathbb{P}, \mathbb{F}_p$ are, respectively, the sets of integers, rational numbers, real numbers, complex numbers, prime numbers and finite field with p elements. If $a \in \mathbb{Z}$, we may also adopt the notation $\mathbb{Z}_{\geqslant a}$ of all integers $n \geqslant a$.

$a \mid b$ means a divides b, and $a \mid b^\infty$ means $p \mid a \Rightarrow p \mid b$. Also, $p^k \| n$ means $p^k \mid n$ and $p^{k+1} \nmid n$.

$P^+(n)$ is the greatest prime factor of n, with the convention $P^+(1) = 1$. This symbol is sometimes abbreviated as $P(n)$. Similarly, $P^-(n)$, or sometimes $p(n)$, is the smallest prime factor of n, with the convention $P^-(1) = \infty$.

$(a_1, \ldots, a_n)$ and $[a_1, \ldots, a_n]$ are, respectively, the greatest common divisor and the least common multiple of $a_1, \ldots, a_n$.

If $\mathcal{E}$ is a finite set of integers, $|\mathcal{E}|$ is the number of elements belonging to $\mathcal{E}$.

2 Basics of Complex Analysis

We will make a constant use of the next three results. For a proof and more details, see the book *The Theory of Functions* by E. C. Titchmarsh, Oxford Science Publication, reprinted 1993, Chapter IV and Theorems 2.8 and 3.11.

Theorem 1 (Analytic Continuation) *Let f, g be two analytic functions in a connected open subset $\Omega \subset \mathbb{C}$. If there exists a non-empty open subset $U \subset \Omega$ such that, for all $z \in U$, $f(z) = g(z)$, then, for all $z \in \Omega$, we have $f(z) = g(z)$.*

Theorem 2 (Weierstrass' Differentiation Theorem) *Let $(f_n(s))_n$ be a sequence of analytic functions in a non-empty open subset $\Omega \subset \mathbb{C}$, such that the series*

$\sum_{n\geqslant 1} f_n(s)$ *is uniformly convergent in every closed disc* $\mathcal{D} \subset \Omega$. *Then, the function:*

$$f(s) = \sum_{n=1}^{\infty} f_n(s)$$

is analytic in Ω *and, for all* $s \in \Omega$ *and all* $k \in \mathbb{Z}_{\geqslant 1}$

$$f^{(k)}(s) = \sum_{n=1}^{\infty} f_n^{(k)}(s)$$

with uniform convergence in every closed disc $\mathcal{D} \subset \Omega$.

Theorem 3 (Cauchy's Residue Theorem) *Let* Ω *be an open set of* $\mathbb{C}$ *and* S *be a subset of* Ω *without cluster points of* Ω. *Let* f *be analytic in* $\Omega \setminus S$ *and* K *be a boundary-regular compact subset of* Ω, *such that* ∂K *does not contain any point of* S. *Then,* S *has only finitely many points in* K *and*

$$\frac{1}{2\pi i} \int_{\partial K} f(s)\, ds = \sum_{a \in S \cap K} \operatorname*{Res}_{s=a} (f(s)) .$$

3 Functions

3.1 The Zeta and Gamma Functions

The *Riemann zeta function*, which is the cornerstone of this book, is first defined by the series:

$$\zeta(s) = \sum_{n=1}^{\infty} \frac{1}{n^s} \quad (\operatorname{Re} s > 1) .$$

We will see in Chap. 3, Theorem 3.7, how this function can be analytically continued in the whole complex plane to a meromorphic function.

The *Euler Gamma function* can be defined as the unique finite order meromorphic function $\Gamma(s)$, without zeros or poles in the half-plane $\sigma > 0$, such that $\Gamma(1) = 1, \Gamma'(1) \in \mathbb{R}$, and which satisfies the functional equation $\Gamma(s+1) = s\Gamma(s)$.

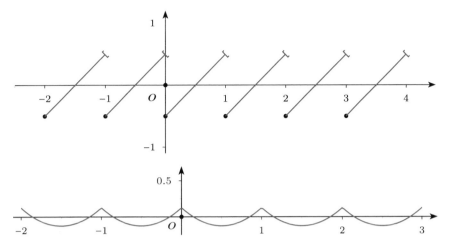

Fig. 1 Functions ψ and ψ_2

3.2 Integer Functions

For any $x \in \mathbb{R}$, $\lfloor x \rfloor$ is the integer part of x, and $\lfloor x \rceil$ is the nearest integer to x. The notation $\{x\}$ means the fractional part of x defined by $\{x\} = x - \lfloor x \rfloor$. Hence, for any $x \in \mathbb{R}$, we have $0 \leqslant \{x\} < 1$ and $\lfloor x \rfloor = x + O(1)$.[1] We will also make use of the functions:

$$\psi(x) = \{x\} - \tfrac{1}{2} \quad \text{and} \quad \psi_2(x) = \int_0^x \psi(t)\,dt = \tfrac{1}{2}\psi(x)^2 - \tfrac{1}{8}.$$

The function ψ, called the *first Bernoulli function*, is an odd, 1-periodic function and then admits a Fourier series development. Note that the function ψ_2 is also 1-periodic and then bounded. Furthermore, it is not difficult to check that (Fig. 1)

$$|\psi(x)| \leqslant \tfrac{1}{2} \quad \text{and} \quad -\tfrac{1}{8} \leqslant \psi_2(x) \leqslant 0.$$

The distance of a real number x to its nearest integer is written $\|x\|$. Hence, we have

$$\|x\| = \min\left(\tfrac{1}{2} + \psi(x),\ \tfrac{1}{2} - \psi(x)\right).$$

---

[1] It is sometimes useful to also have in mind the obvious equality $\lfloor x \rfloor = x + O(x^\alpha)$ for $\alpha \in [0, 1)$.

3.3   Sums and Products

It is important to note that, in all cases, the index p means a sum or a product running through prime numbers exclusively.

If $x \geqslant 1$ is a real number, then by convention:

$$\sum_{n \leqslant x} f(n) = \sum_{n=1}^{\lfloor x \rfloor} f(n) \quad \text{and} \quad \prod_{n \leqslant x} f(n) = \prod_{n=1}^{\lfloor x \rfloor} f(n).$$

3.4   Exponential and Logarithmic Functions

For $x > 0$, $\log x$ is the natural logarithm and, for $s \in \mathbb{C}$, e^s or $\exp s$ is the exponential function. It is convenient to define the functions:

$$e(x) = e^{2\pi i x} \quad \text{and} \quad e_q(x) = e(x/q) = e^{2\pi i x/q}.$$

For $s \in \mathbb{C}$, we also denote by $\log s$ a complex logarithm, defined as any complex number w satisfying $e^w = s$. Unless otherwise stated, we take the principal branch of the logarithm. In particular, if $f(s)$ is analytic in any connected open subset U of $\mathbb{C}$ such that $f(s) \neq 0$ and $\overline{f(s)} = f(\overline{s})$ for all $s \in U$, then $\log f(s)$ is the branch that coincides with $\ln f(s)$ for all *real* $s \in U$. As an example, let us have a brief glance at the functions $\log \zeta(s)$ and $\log \Gamma(s)$.

▷ We will see in Sect. 3.3 in Chap. 3 that $\zeta(s) \neq 0$ in the half-plane $\sigma > 1$. Hence a single-valued branch of $\log \zeta(s)$ may be defined, real and positive for real $s > 1$, and for which, inspired by the Euler product,[2] we have

$$\log \zeta(s) = -\sum_p \log \left(1 - \frac{1}{p^s}\right) \quad (\sigma > 1).$$

▷ The $\log \Gamma$ function is sometimes considered as a special function as well for which, inspired by the Weierstrass product formula, we have

$$\log \Gamma(s) = -\gamma s - \log s + \sum_{n=1}^{\infty} \left(\frac{s}{n} - \log \left(1 + \frac{s}{n}\right)\right)$$

for $s \in \mathbb{C}$, $s \neq 0, -1, -2, \ldots$

---

[2] In fact, this equality really comes from the Euler product by analytic continuation.

3.5   Comparison Relations

Let $a < b$ be real numbers and k be a non-negative integer. The notation $f \in C^k[a, b]$ means that f is a real-valued function k-times differentiable on $[a, b]$ and $f^{(k)}$ is also continuous on $[a, b]$. By convention, $f^{(0)} = f$, $f^{(1)} = f'$, $f^{(2)} = f''$ and $f^{(3)} = f'''$. Let $x \mapsto f(x)$ and $x \mapsto g(x)$ be functions defined for all sufficiently large x and $a, b > 0$.

▷ *Landau.* $f(x) = O(g(x))$, also sometimes written as $f = O(g)$, means that $g > 0$ and that there exist a real number x_0 and a constant $c_0 > 0$, such that, for all $x \geqslant x_0$, we have $|f(x)| \leqslant c_0 \, g(x)$.

▷ *Ramaré.* $f(x) = O^\star(g(x))$ means that $g > 0$ and that there exists a real number x_0, such that, for all $x \geqslant x_0$, we have $|f(x)| \leqslant g(x)$.

▷ *Vinogradov.* $f(x) \ll g(x)$ is equivalent to $f(x) = O(g(x))$.

▷ *Titchmarsh.* $a \asymp b$ means that there exist $c_2 \geqslant c_1 > 0$, such that $c_1 \, b \leqslant a \leqslant c_2 \, b$. If a, b represent two functions f and g, then $f \asymp g$ is equivalent to $f \ll g$ and $g \ll f$.

▷ *Landau.* $f(x) = o(g(x))$ for $x \to x_0$ means that $g \neq 0$ and

$$\lim_{x \to x_0} \frac{f(x)}{g(x)} = 0.$$

▷ *Landau.* $f(x) \sim g(x)$ for $x \to x_0$ means that $g \neq 0$ and

$$\lim_{x \to x_0} \frac{f(x)}{g(x)} = 1.$$

An *asymptotic estimate* is a relation of the shape $f(x) \sim g(x)$, while an *asymptotic formula* is a relation of the form $f(x) = g(x) + O(r(x))$, or equivalently $f(x) - g(x) = O(r(x))$, where $g(x)$ is called the *main term* and $O(r(x))$ is an *error term*. Obviously, such a relation is only meaningful if the error term $r(x)$ is of smaller order than $g(x)$. Otherwise, this relation is equivalent to $f(x) = O(r(x))$, so that the estimate is only an *upper bound*.

It is important to understand the difference between $f \asymp g$ and $f \sim g$. The first relation is less precise than the second one but can be used in a larger range. For instance, the Chebyshev estimates from Theorem 3.1 assert that

$$\pi(x) \asymp \frac{x}{\log x}$$

for all $x \geqslant 5$, while the Prime Number Theorem,[3] which was proved some 40 years later, implies that

$$\pi(x) \sim \frac{x}{\log x}$$

as soon as $x \to \infty$.

It should be mentioned that the constants implied in some error terms of the form $f(x) \ll g(x)$ depend sometimes on extra parameters. For instance, it is proven in Exercise 54 that

$$\tau(n) \ll n^{\varepsilon},$$

where the implied constant depends on $\varepsilon > 0$. This means that, for all $\varepsilon > 0$, there exists a constant $c(\varepsilon) > 0$ depending on $\varepsilon > 0$, such that, for all $n \geqslant 1$, we have $\tau(n) \leqslant c(\varepsilon)n^{\varepsilon}$. Such a situation[4] is sometimes denoted by

$$\tau(n) \ll_{\varepsilon} n^{\varepsilon}.$$

---

[3] See Theorem 3.27.
[4] It can be shown that $c(\varepsilon) = \exp\left(2^{1/\varepsilon} / \log 2^{\varepsilon}\right)$ is admissible.

Contents

Chapter 1
Basic Tools

1.1 The Riemann–Stieltjes Integral

1.1.1 Definition

In what follows, $a < b$ are two real numbers.

The usual Riemann integral can be generalized in the following way. Let f and g be two real-valued functions defined on the interval $[a, b]$. We denote by Δ a subdivision of $[a, b]$ by the points $a = x_0 < x_1 < \cdots < x_n = b$, and the *norm* of Δ is defined by

$$N(\Delta) = \max_{0 \leqslant k \leqslant n-1} (x_{k+1} - x_k).$$

A *tagged subdivision* of $[a, b]$ is a pair (Δ, ξ), where $\Delta = \{x_0, \ldots, x_n\}$ is a subdivision of $[a, b]$ and $\xi = \{\xi_0, \ldots, \xi_{n-1}\}$ with $\xi_k \in [x_k, x_{k+1}]$.

We call Riemann–Stieltjes sum of f with respect to g for the tagged subdivision (Δ, ξ) of $[a, b]$ the sum $S_\Delta(f, g)$ given by

$$S_\Delta(f, g) = \sum_{k=0}^{n-1} f(\xi_k)(g(x_{k+1}) - g(x_k)).$$

Definition 1.1 If the limit $\lim_{N(\Delta) \to 0} S_\Delta(f, g)$ exists independently of the manner of subdivision and of the choice of the number ξ_k, then this limit is called the *Riemann–Stieltjes integral* of f with respect to g from a to b and is denoted by

$$\int_a^b f(x)\, dg(x).$$

f is called the *integrand* and g is the *integrator*.

© Springer Nature Switzerland AG 2020
O. Bordellès, *Arithmetic Tales*, Universitext,
https://doi.org/10.1007/978-3-030-54946-6_1

Note that this integral reduces to the Riemann integral if $g(x) = x$. This definition easily extends to complex-valued functions by setting

$$\int_a^b f(x)\,dg(x) = \int_a^b \operatorname{Re} f(x)\,d(\operatorname{Re} g(x)) - \int_a^b \operatorname{Im} f(x)\,d(\operatorname{Im} g(x))$$

$$+ i\left(\int_a^b \operatorname{Re} f(x)\,d(\operatorname{Im} g(x)) + \int_a^b \operatorname{Im} f(x)\,d(\operatorname{Re} g(x))\right).$$

We do not precisely determine the pairs (f, g) for which the Riemann–Stieltjes integral exists, and the reader interested in this subject should refer to [1, 3]. However, the following class of functions plays an important role in number theory.

Definition 1.2 A function g defined on $[a, b]$ is of *bounded variation* on $[a, b]$ if there exists $M > 0$ such that, for all subdivisions $a = x_0 < x_1 < \cdots < x_n = b$, we have

$$\sum_{k=0}^{n-1} |g(x_{k+1}) - g(x_k)| \leqslant M.$$

The smallest number M satisfying this inequality is called the *total variation* of g in $[a, b]$ and is denoted by $V_{[a,b]}(g)$ or $V_a^b(g)$.

It can be shown that a real-valued function of bounded variation on $[a, b]$ is the difference of two non-decreasing bounded functions. Furthermore, if f and g are of bounded variation on $[a, b]$, then so are $f + g$ and fg, and we have

$$V_a^b(f + g) \leqslant V_a^b(f) + V_a^b(g).$$

$$V_a^b(fg) \leqslant \sup_{x \in [a,b]} |f(x)| \times V_a^b(g) + \sup_{x \in [a,b]} |g(x)| \times V_a^b(f).$$

Finally, if $g \in C^1[a, b]$, then we have

$$V_a^b(g) = \int_a^b |g'(x)|\,dx.$$

1.1.2   Basic Properties

A sufficient condition of the existence of the Riemann–Stieltjes integral is given by the following result. See [3, Chapter 1, Theorem 4a] for a proof.

Proposition 1.1 *If f is continuous on $[a, b]$ and if g is of bounded variation on $[a, b]$, then the integral $\displaystyle\int_a^b f(x)\,dg(x)$ exists.*

The usual properties of the Riemann–Stieltjes integral are similar to those of the Riemann integral. We list some of them in the following proposition [3, Theorems 5a and 5b].

Proposition 1.2 *Let f, f_1 and f_2 be continuous functions on $[a, b]$ and g, g_1 and g_2 be functions of bounded variation on $[a, b]$.*

▷ *For all constants α_1, α_2, β_1, β_2, we have*

$$\int_a^b \{\alpha_1 f_1(x) + \alpha_2 f_2(x)\} \, d\{\beta_1 g_1(x) + \beta_2 g_2(x)\} = \sum_{i,j=1}^2 \alpha_i \beta_j \int_a^b f_i(x) \, dg_j(x).$$

▷ *For any $a, b, c \in \mathbb{R}$, we have*

$$\int_a^b f(x) \, dg(x) = \int_a^c f(x) \, dg(x) + \int_c^b f(x) \, dg(x).$$

▷ *We have*

$$\left| \int_a^b f(x) \, dg(x) \right| \leqslant \int_a^b |f(x)| \, dV_a^x(g) \leqslant \sup_{x \in [a,b]} |f(x)| \times V_a^b(g).$$

▷ *Suppose f_1 is of bounded variation on $[a, b]$, f_2 and g bounded on $[a, b]$. Then*

$$\left| \int_a^b f_1(x) f_2(x) \, dg(x) \right| \leqslant \left(|f_1(b)| + V_a^b(f_1) \right) \sup_{t \in [a,b]} \left| \int_a^t f_2(x) \, dg(x) \right|$$

provided that the integrals exist.

The differential $dV_a^x(g)$ is sometimes abbreviated $|dg(x)|$, so the third result can be rewritten in the form [3, p. 9]

$$\left| \int_a^b f(x) \, dg(x) \right| \leqslant \int_a^b |f(x)| \, |dg(x)|.$$

It is quite difficult to derive general necessary and sufficient conditions for the existence of the Riemann–Stieltjes integral.[1] Nevertheless, the next inequality, due to L. C. Young [4], has received considerable attention, in particular to understand the Ito map and to develop a stochastic integration theory based on his techniques.

---

[1] See [1] for such a result.

For $\lambda > 0$, we say that $f : [a, b] \to \mathbb{R}$ has a *finite strong λ-variation* if

$$V_{[a,b]}^{\lambda}(f) := \sup_{n} \sup_{\Delta} \sum_{k=0}^{n-1} |f(x_{k+1}) - f(x_k)|^{\lambda} < \infty,$$

where the second supremum is taken over all subdivisions

$$\Delta = \{a = x_0 < x_1 < \cdots < x_n = b\}$$

of $[a, b]$. Strong λ-variation may be viewed as a measure of path irregularity since, if $V_{[a,b]}^{\lambda}(f) < \infty$ for some $\lambda \geqslant 1$, then $V_{[a,b]}^{\nu}(f) < \infty$ for all $\nu > \lambda$. Now L. C. Young's result may be stated as follows.

Theorem 1.1 *The Riemann–Stieltjes integral $\int_a^b f(x) \, dg(x)$ exists whenever f and g have no common discontinuities in (a, b), and $V_{[a,b]}^{\lambda}(f) < \infty$, $V_{[a,b]}^{\nu}(g) < \infty$, for some $\lambda, \nu > 0$ satisfying $\lambda^{-1} + \nu^{-1} > 1$. Furthermore, for any $\xi \in [a, b]$, we have*

$$\left| \int_a^b f(x) \, dg(x) - f(\xi) \, (g(b) - g(a)) \right|$$

$$\leqslant 2 \left(1 + \zeta \left(\lambda^{-1} + \nu^{-1} \right) \right) \left(V_{[a,b]}^{\lambda}(f) \right)^{1/\lambda} \left(V_{[a,b]}^{\nu}(g) \right)^{1/\nu}.$$

One of the main weaknesses of the Riemann–Stieltjes integral is that the integral

$$\int_a^b f(x) \, dg(x)$$

does not exist if f and g have a common discontinuity in (a, b). Nevertheless, if f is continuous, this integral is often used in number theory to express various sums in terms of integrals. In particular, the Riemann–Stieltjes integral provides a natural context for Abel summation seen in Sect. 1.2. More precisely, suppose $a < b$ are real numbers and f is continuous on $[a, b]$. Then we have

$$\sum_{a < n \leqslant b} f(n) g(n) = \int_a^b f(x) \, d \left(\sum_{a < n \leqslant x} g(n) \right). \tag{1.1}$$

Note that there is some freedom in the interval of integration, since the left endpoint can be any number in $\left[\lfloor a \rfloor, \lfloor a \rfloor + 1 \right)$ and the right endpoint can be chosen in

$\left[\lfloor b\rfloor, \lfloor b\rfloor + 1\right)$, without changing the value of the integral. However, one must be careful in choosing these endpoints of integration. For instance, we have

$$\sum_{k=1}^{n}\frac{1}{k} = 1 + \sum_{k=2}^{n}\frac{1}{k} = 1 + \int_{1}^{n}\frac{1}{x}\,\mathrm{d}\left(\sum_{k\leqslant x}1\right) = 1 + \int_{1}^{n}\frac{\mathrm{d}\lfloor x\rfloor}{x}.$$

1.1.3   Integration by Parts

Theorem 1.2 (Integration by Parts)   *If the integral $\displaystyle\int_{a}^{b} g(x)\,\mathrm{d}f(x)$ exists, then the integral $\displaystyle\int_{a}^{b} f(x)\,\mathrm{d}g(x)$ exists and*

$$\int_{a}^{b} f(x)\,\mathrm{d}g(x) = f(b)g(b) - f(a)g(a) - \int_{a}^{b} g(x)\,\mathrm{d}f(x).$$

Proof We choose an arbitrary subdivision Δ of $[a, b]$ and numbers $\xi_k \in [x_k, x_{k+1}]$. For convenience, we set $\xi_{-1} = a$ and $\xi_n = b$, so that we have a subdivision Ξ of $[a, b]$ such that

$$x_0 = \xi_{-1} = a \leqslant \xi_0 \leqslant x_1 \leqslant \xi_1 \leqslant x_2 \leqslant \xi_2 \leqslant \cdots \leqslant x_{n-1} \leqslant \xi_{n-1} \leqslant x_n = \xi_n = b.$$

Since

$$\xi_{k+1} - \xi_k \leqslant \begin{cases} x_1 - a, & \text{if } k = -1 \\ x_{k+2} - x_k, & \text{if } 0 \leqslant k \leqslant n - 2 \\ b - x_{n-1}, & \text{if } k = n - 1 \end{cases}$$

we have $N(\Xi) \leqslant 2N(\Delta)$. Now

$$S_\Delta(f, g) = \sum_{k=0}^{n-1} f(\xi_k)g(x_{k+1}) - \sum_{k=0}^{n-1} f(\xi_k)g(x_k)$$

$$= \sum_{k=1}^{n} f(\xi_{k-1})g(x_k) - \sum_{k=0}^{n-1} f(\xi_k)g(x_k)$$

$$= f(b)g(b) - f(a)g(a) - \sum_{k=0}^{n} g(x_k)\left(f(\xi_k) - f(\xi_{k-1})\right)$$

$$= f(b)g(b) - f(a)g(a) - S_\Xi(g, f).$$

Since $N(\Xi) \leqslant 2N(\Delta)$, the sum on the right-hand side tends to $\int_a^b g(x)\,\mathrm{d}f(x)$ as $N(\Delta)$ tends to 0, which concludes the proof.   □

The following example is very useful.

Example 1.1 (Harmonic Numbers)   For all $x \geqslant 1$, we have

$$\sum_{n \leqslant x} \frac{1}{n} = \log x + \gamma - \frac{\psi(x)}{x} + O^\star\!\left(\frac{1}{4x^2}\right).$$

Proof   Using (1.1) and Theorem 1.2, we get

$$\sum_{n \leqslant x} \frac{1}{n} = 1 + \int_1^x \frac{\mathrm{d}\lfloor t \rfloor}{t} = 1 + \int_1^x \frac{\mathrm{d}t}{t} - \int_1^x \frac{\mathrm{d}\psi(t)}{t}$$

$$= 1 + \log x - \frac{\psi(x)}{x} + \psi(1) - \int_1^x \frac{\psi(t)}{t^2}\,\mathrm{d}t$$

$$= \tfrac{1}{2} + \log x - \frac{\psi(x)}{x} - \int_1^x \frac{\mathrm{d}(\psi_2(t))}{t^2}$$

$$= \tfrac{1}{2} + \log x - \frac{\psi(x)}{x} - \frac{\psi_2(x)}{x^2} - 2\int_1^x \frac{\psi_2(t)}{t^3}\,\mathrm{d}t.$$

Now since $|\psi_2(t)| \leqslant \tfrac{1}{8}$, the last integral converges, so that we get

$$\sum_{n \leqslant x} \frac{1}{n} = \log x + \tfrac{1}{2} - 2\int_1^\infty \frac{\psi_2(t)}{t^3}\,\mathrm{d}t - \frac{\psi(x)}{x} - \frac{\psi_2(x)}{x^2} + 2\int_x^\infty \frac{\psi_2(t)}{t^3}\,\mathrm{d}t.$$

Since $\gamma = \lim_{x \to \infty}\left(\sum_{n \leqslant x} \frac{1}{n} - \log x\right)$, by letting $x \to \infty$, we get

$$\gamma = \tfrac{1}{2} - 2\int_1^\infty \frac{\psi_2(t)}{t^3}\,\mathrm{d}t,$$

and therefore

$$\sum_{n \leqslant x} \frac{1}{n} = \log x + \gamma - \frac{\psi(x)}{x} - \frac{\psi_2(x)}{x^2} + 2\int_x^\infty \frac{\psi_2(t)}{t^3}\,\mathrm{d}t,$$

and we conclude by the estimate

$$\left| -\frac{\psi_2(x)}{x^2} + 2 \int_x^\infty \frac{\psi_2(t)}{t^3} \, dt \right| \leqslant \frac{1}{8x^2} + \frac{1}{4} \int_x^\infty \frac{dt}{t^3} = \frac{1}{4x^2},$$

as required.   □

1.1.4   Sums and Integrals

One of the simplest applications of the Riemann–Stieltjes integral is the usual comparison between sums and integrals. One of the main advantages of this result lies in its minimalist assumptions.

Proposition 1.3 *Let $a < b \in \mathbb{Z}$ and f be a monotonic real-valued function on $[a, b]$. Then*

$$\sum_{a < n \leqslant b} f(n) = \int_a^b f(t) \, dt + O\left(|f(a)| + |f(b)|\right).$$

Proof Assume f is non-decreasing. Then

$$\sum_{a < n \leqslant b} f(n) = \int_{a^-}^b f(t) \, d\lfloor t \rfloor = \int_a^b f(t) \, dt - \int_{a^-}^b f(t) \, d\psi(t)$$

$$= \int_a^b f(t) \, dt - f(t)\psi(t)\big|_a^b + \int_{a^-}^b \psi(t) \, df(t)$$

$$= \int_a^b f(t) \, dt + \tfrac{1}{2} \left(f(b) - f(a)\right) + \int_{a^-}^b \psi(t) \, df(t).$$

From Proposition 1.2

$$\left| \int_a^b \psi(t) \, df(t) \right| \leqslant \tfrac{1}{2} V_a^b(f)$$

and since f is non-decreasing, $V_a^b(f) = f(b) - f(a)$, completing the proof.   □

Example 1.2 Let $\sigma \in (0, 1)$ and $n \in \mathbb{Z}_{\geqslant 1}$. Then

$$\sum_{k=1}^n \frac{1}{k^\sigma} = \frac{n^{1-\sigma}}{1 - \sigma} + O(1).$$

1.2   Partial Summation

1.2.1   Abel Transformation Formula

The following very useful result has been discovered by Abel.

Proposition 1.4 *Let $m < n$ be any non-negative integers and (a_k), (b_k) be any sequences of complex numbers. Then*

$$\sum_{k=m+1}^{n} a_k b_k = b_n \sum_{k=m}^{n} a_k - b_{m+1} a_m - \sum_{k=m+1}^{n-1} (b_{k+1} - b_k) \sum_{h=m}^{k} a_h.$$

Proof For all $m \leqslant j \leqslant n$, define $s_j = \sum_{h=m}^{j} a_h$. Then, obviously $a_k = s_k - s_{k-1}$, so that

$$\sum_{k=m+1}^{n} a_k b_k = \sum_{k=m+1}^{n} b_k (s_k - s_{k-1}) = \sum_{k=m+1}^{n} b_k s_k - \sum_{k=m}^{n-1} b_{k+1} s_k,$$

implying the asserted result.   □

Corollary 1.1 *Let $m < n$ be any non-negative integers and (a_k), (b_k) be any sequences of complex numbers. Then*

$$\left| \sum_{k=m+1}^{n} a_k b_k \right| \leqslant \left\{ 2 \max(|b_{m+1}|, |b_n|) + V_{m,n} \right\} \max_{m \leqslant k \leqslant n} \left| \sum_{h=m}^{k} a_h \right|,$$

where

$$V_{m,n} = \sum_{k=m+1}^{n-1} |b_{k+1} - b_k|.$$

In particular, if (b_k) is a monotone sequence of positive real numbers, then

$$\left| \sum_{k=m+1}^{n} a_k b_k \right| \leqslant 2 \max(b_{m+1}, b_n) \max_{m \leqslant k \leqslant n} \left| \sum_{h=m}^{k} a_h \right|.$$

1.2.2 Partial Summation Formula

This is the famous integration by parts for sums.

Theorem 1.3 *Let $x \geqslant 0$ be any real number, $a \in \mathbb{Z}_{\geqslant 0}$ and let $f : [a, x] \to \mathbb{C}$ and $g \in C^1[a, x]$. Then*

$$\sum_{a \leqslant n \leqslant x} f(n)g(n) = g(x) \sum_{a \leqslant n \leqslant x} f(n) - \int_a^x g'(t) \left(\sum_{a \leqslant n \leqslant t} f(n) \right) dt.$$

Similarly, if $x \geqslant 2$, then

$$\sum_{p \leqslant x} f(p)g(p) = g(x) \sum_{p \leqslant x} f(p) - \int_2^x g'(t) \left(\sum_{p \leqslant t} f(p) \right) dt.$$

Proof Using (1.1) and Theorem 1.2, we derive

$$\sum_{a \leqslant n \leqslant x} f(n)g(n) = f(a)g(a) + \int_a^x g(t) \, d\left(\sum_{a < n \leqslant t} f(n) \right)$$

$$= f(a)g(a) + g(t) \sum_{a < n \leqslant t} f(n) \bigg|_a^x - \int_a^x g'(t) \left(\sum_{a < n \leqslant t} f(n) \right) dt$$

$$= g(x) \sum_{a \leqslant n \leqslant x} f(n) - \int_a^x g'(t) \left(\sum_{a \leqslant n \leqslant t} f(n) \right) dt$$

$$+ f(a)(g(a) - g(x)) + f(a)(g(x) - g(a))$$

$$= g(x) \sum_{a \leqslant n \leqslant x} f(n) - \int_a^x g'(t) \left(\sum_{a \leqslant n \leqslant t} f(n) \right) dt,$$

which is the asserted result. The second one readily follows from the first by using the following function:

$$\mathbf{1}_{\mathbb{P}}(n) = \begin{cases} 1, & \text{if } n \text{ is a prime number} \\ 0, & \text{otherwise} \end{cases}$$

which yields

$$\sum_{p\leqslant x} f(p)g(p) = g(x) \sum_{2\leqslant n\leqslant x} \mathbf{1}_{\mathbb{P}}(n)f(n) - \int_2^x g'(t)\left(\sum_{2\leqslant n\leqslant t} \mathbf{1}_{\mathbb{P}}(n)f(n)\right)dt$$

$$= g(x) \sum_{p\leqslant x} f(p) - \int_2^x g'(t)\left(\sum_{p\leqslant t} f(p)\right)dt$$

as asserted.   □

As an example, let us estimate the following series related to the Riemann zeta-function.

Corollary 1.2 *Let $\sigma > 1$ be any real number. Then, for all $x \geqslant 1$*

$$\sum_{n>x} \frac{1}{n^\sigma} = \frac{1}{(\sigma-1)x^{\sigma-1}} + O^\star\left(\frac{1}{x^\sigma}\right).$$

Proof Theorem 1.3 yields for all $x \geqslant 1$

$$\sum_{n>x} \frac{1}{n^\sigma} = -\frac{\lfloor x\rfloor}{x^\sigma} + \sigma \int_x^\infty \frac{\lfloor t\rfloor}{t^{\sigma+1}}\,dt = \frac{1}{(\sigma-1)x^{\sigma-1}} + \frac{\{x\}}{x^\sigma} - \sigma\int_x^\infty \frac{\{t\}}{t^{\sigma+1}}\,dt$$

and uses $0 \leqslant \{t\} < 1$ to complete the proof.   □

1.3 The Euler–Maclaurin Summation Formula

Let f be a function of bounded variation on $[a, b]$. Then using (1.1) and Theorem 1.2, we get

$$\sum_{a<n\leqslant b} f(n) = \int_a^b f(x)\,d\lfloor x\rfloor = \int_a^b f(x)\,dx - \int_a^b f(x)\,d\psi(x)$$

$$= \int_a^b f(x)\,dx + f(a)\psi(a) - f(b)\psi(b) + \int_a^b \psi(x)\,df(x),$$

so that we have by Proposition 1.2

$$\sum_{a<n\leqslant b} f(n) = \int_a^b f(x)\,dx + f(a)\psi(a) - f(b)\psi(b) + O\left(V_a^b(f)\right), \qquad (1.2)$$

and if $f \in C^1[a, b]$, then we get the following more accurate identity:

$$\sum_{a < n \leqslant b} f(n) = \int_a^b f(x)\,\mathrm{d}x + f(a)\psi(a) - f(b)\psi(b) + \int_a^b f'(x)\psi(x)\,\mathrm{d}x,$$

(1.3)

which is usually called the *Euler summation formula*. If f has derivatives of higher orders, this can be generalized by integrating by parts repeatedly. The process makes some polynomials appear, often denoted by $B_k(x)$ and called the *Bernoulli polynomials*. These polynomials can be defined by induction by first setting $B_0(x) = 1$ and, for all $k \in \mathbb{Z}_{\geqslant 1}$, by putting

$$\frac{\mathrm{d}}{\mathrm{d}x} B_k(x) = k B_{k-1}(x).$$

Thus, apart from the constant term, $B_k(x)$ is determined by this differential equation. The constant term, still denoted by B_k and called the kth *Bernoulli number*, is given by the additional following condition:

$$\int_0^1 B_k(x)\,\mathrm{d}x = 0 \quad (k \geqslant 1),$$

which is equivalent to $B_{k+1}(1) - B_{k+1}(0) = 0$ for all positive integers k, so that we have

$$B_k(1) = B_k(0) = B_k \quad (k \geqslant 2).$$

One can prove by induction that, for all $k \geqslant 0$, we have

$$B_k(x) = \sum_{j=0}^{k} \binom{k}{j} B_j x^{k-j}.$$

Either one of these formulae can be used to discover inductively some properties of the Bernoulli numbers and polynomials. For instance, if k is odd, then we have $B_k = 0$ for all $k \geqslant 3$ and $B_k(x) = -B_k(1-x)$ for all $k \geqslant 1$. If k is even, then $B_k(x) = B_k(1-x)$ for all $k \geqslant 0$. The following table gives the first values of B_k.

k	0	1	2	4	6	8	10
B_k	1	$-\frac{1}{2}$	$\frac{1}{6}$	$-\frac{1}{30}$	$\frac{1}{42}$	$-\frac{1}{30}$	$\frac{5}{66}$

The repeated integrations by parts of the Euler summation formula yield the functions $B_k(\{x\})$, usually called the *Bernoulli functions*. These functions are periodic of period 1 and, for all $k \geqslant 2$, they are continuous since $B_k(0) = B_k(1)$. Also note that

$$B_1(\{x\}) = \{x\} - \tfrac{1}{2} = \psi(x) \quad \text{and} \quad B_2(\{x\}) = 2\psi_2(x) + \tfrac{1}{6} = \psi(t)^2 - \tfrac{1}{12}.$$

The periodicity of the functions $B_k(\{x\})$ enables us to consider their expansions in Fourier series. When $k \geqslant 2$, the series is absolutely convergent and $B_k(\{x\})$ is continuous, so that the series converges uniformly to $B_k(\{x\})$. For $k = 1$, the function $\psi(x)$ has a jump discontinuity at the integers, but is also of bounded variation in $[0, 1]$ so that the partial sums of its Fourier series converge to $\psi(x)$ when $x \notin \mathbb{Z}$. The usual computations from Fourier analysis provide the following result.

Proposition 1.5 (Fourier Series Expansions) *If $x \notin \mathbb{Z}$, we have*

$$\psi(x) = -\sum_{h=1}^{\infty} \frac{\sin(2\pi h x)}{\pi h}.$$

When $x \in \mathbb{Z}$, the series converges to 0. Furthermore, for all $k \in \mathbb{Z}_{\geqslant 1}$, we have uniformly in x

$$B_{2k+1}(\{x\}) = 2(-1)^{k+1}(2k+1)! \sum_{h=1}^{\infty} \frac{\sin(2\pi h x)}{(2\pi h)^{2k+1}}.$$

$$B_{2k}(\{x\}) = 2(-1)^{k+1}(2k)! \sum_{h=1}^{\infty} \frac{\cos(2\pi h x)}{(2\pi h)^{2k}}.$$

These formulae are of great use. For instance, taking $x = 0$ in the second identity above provides the following result discovered by Euler.

Proposition 1.6 (Zeta-Function at Even Integers) *For all integers $k \geqslant 1$, we have*

$$\zeta(2k) = (-1)^{k+1} 2^{2k-1} \pi^{2k} \frac{B_{2k}}{(2k)!},$$

where

$$\zeta(\sigma) = \sum_{n=1}^{\infty} \frac{1}{n^{\sigma}} \quad (\sigma > 1).$$

Hence, $\zeta(2k)$ is a rational multiple of π^{2k}. For example, $\zeta(2) = \frac{1}{6}\pi^2$ and $\zeta(4) = \frac{1}{90}\pi^4$. It is interesting to note that no similar formula is known for $\zeta(2k+1)$, although it has been proved that it is not a rational multiple of π^{2k+1}. Finally, we have the following bounds valid for all $k \geqslant 1$ and $x \in \mathbb{R}$ [2]

$$|B_{2k}(\{x\})| \leqslant |B_{2k}| \leqslant \frac{(2k)!}{12(2\pi)^{2k-2}} \quad \text{and} \quad |B_k(\{x\})| \leqslant \frac{k!}{12(2\pi)^{k-2}}.$$

Starting from (1.3) and using induction and the properties of the Bernoulli polynomials seen above, we get the following important result.

Theorem 1.4 (Euler–Maclaurin Summation Formula) *Let $a < b$ be real numbers, $k \in \mathbb{Z}_{\geqslant 1}$ and $f \in C^k[a, b]$. Then we have*

$$\sum_{a < n \leqslant b} f(n) = \int_a^b f(x)\,\mathrm{d}x$$

$$+ \sum_{j=1}^k \frac{(-1)^j}{j!} \left(B_j(\{b\}) f^{(j-1)}(b) - B_j(\{a\}) f^{(j-1)}(a) \right) - \mathcal{R}_f(x),$$

where

$$\mathcal{R}_f(x) = \frac{(-1)^k}{k!} \int_a^b B_k(\{x\}) f^{(k)}(x)\,\mathrm{d}x.$$

In practice, it can be interesting to have at our disposal some simpler versions. Along with identity (1.3), we will prove the following corollary.

Corollary 1.3 *Let $x \geqslant 1$ be a real number and $f \in C^2[1, +\infty)$. Then we have*

$$\sum_{n \leqslant x} f(n) = \int_1^x f(t)\,\mathrm{d}t + \frac{1}{2}f(1) - \frac{1}{8}f'(1)$$

$$- \psi(x)f(x) + \frac{1}{2}\psi(x)^2 f'(x) - \frac{1}{2}\int_1^x \psi(t)^2 f''(t)\,\mathrm{d}t.$$

(continued)

Corollary 1.3 (continued)
Furthermore, if $\lim\limits_{t\to\infty} f'(t) = 0$ *and if* $f''(t)$ *has a constant sign on* $[1, +\infty)$,
then there exists a constant γ_f *such that*

$$\sum_{n \leqslant x} f(n) = \int_1^x f(t)\, dt + \gamma_f - \psi(x) f(x)$$

$$+ \frac{1}{2}\left\{ \psi(x)^2 f'(x) + \int_x^\infty \psi(t)^2 f''(t)\, dt \right\}.$$

Proof The first identity is clearly equivalent to

$$\sum_{n \leqslant x} f(n) = \int_1^x f(t)\, dt + \frac{1}{2} f(1) - \psi(x) f(x) + \psi_2(x) f'(x) - \int_1^x \psi_2(t) f''(t)\, dt,$$

which follows from (1.3) and an integration by parts. Indeed, by (1.3), we have

$$\sum_{n \leqslant x} f(n) = f(1) + \sum_{1 < n \leqslant x} f(n)$$

$$= f(1) + \int_1^x f(t)\, dt + \psi(1) f(1) - \psi(x) f(x) + \int_1^x f'(t)\, d\psi_2(t)$$

$$= \int_1^x f(t)\, dt + \frac{1}{2} f(1) - \psi(x) f(x) + \psi_2(x) f'(x) - \psi_2(1) f'(1) - \int_1^x \psi_2(t) f''(t)\, dt,$$

which is the asserted result since $\psi_2(1) = 0$. For the second identity, suppose that
$f'' \geqslant 0$ and fix a real number $\varepsilon > 0$. Since $\lim\limits_{t\to\infty} f'(t) = 0$, there exists a real
number $A = A(\varepsilon) > 0$, such that, for all $t \geqslant A$, we have $|f'(t)| \leqslant \varepsilon$. Now let
$z > y \geqslant A$ be two real numbers. Since $f'' \geqslant 0$ and is continuous, we have

$$\left| \int_y^z \psi_2(t) f''(t)\, dt \right| \leqslant \frac{1}{8} \int_y^z f''(t)\, dt = \frac{f'(z) - f'(y)}{8} \leqslant \frac{\varepsilon}{4},$$

so that the integral

$$\int_1^\infty \psi_2(t) f''(t)\, dt$$

converges by Cauchy's theorem. Now the second formula follows by setting

$$\gamma_f = \frac{1}{2} f(1) - \int_1^\infty \psi_2(t) f''(t)\, dt$$

and using the first identity above. The case $f'' \leqslant 0$ is similar.   □

Exercises

1   Let a, q be positive integers. We denote by $\mathcal{S}_q$ the set of positive integers b such that q is the quotient of the Euclidean division of a by b. Show that

$$\sum_{q=1}^{a} |\mathcal{S}_q| = a.$$

2   Let $m, n \in \mathbb{Z} \setminus \{0\}$. Show that

(a) $\left\lfloor \dfrac{n}{m} \right\rfloor \geqslant \dfrac{n+1}{m} - 1.$

(b) If $m \nmid n$, then $\left\lfloor \dfrac{n}{m} \right\rfloor \leqslant \dfrac{n-1}{m}.$

(c) Suppose $1 \leqslant m \leqslant n$. Then

$$\left\lfloor \frac{n}{m} \right\rfloor - \left\lfloor \frac{n-1}{m} \right\rfloor = \begin{cases} 0, & \text{if } m \nmid n \\ 1, & \text{if } m \mid n. \end{cases}$$

3   Let $x \geqslant 1$ be a real number and f be any complex-valued function. Let $g \in C^1[1, x]$ be a complex-valued function. Suppose that

$$g(x) \sum_{n \leqslant x} f(n) \to 0$$

as $x \to \infty$. Using partial summation, show that

$$\sum_{n=1}^{\infty} f(n)g(n) = -\int_{1}^{\infty} g'(t) \left(\sum_{n \leqslant t} f(n) \right) dt$$

in the sense that if either side converges, then so does the other one, to the same value. Deduce that we also have

$$\sum_{n > x} f(n)g(n) = -g(x) \sum_{n \leqslant x} f(n) - \int_{x}^{\infty} g'(t) \left(\sum_{n \leqslant t} f(n) \right) dt.$$

4   Let (a_n) be a sequence of complex numbers such that $|a_n| \leqslant 1$ and suppose there exists a positive real number M such that

$$\left| \sum_{k=1}^{n} a_k \right| \leqslant M \quad (n \geqslant 1).$$

Prove that

$$\left| \sum_{n=1}^{\infty} \frac{a_n}{n} \right| \leqslant \log M + 1.$$

5 If $f \in C^1[a, b]$ where $a < b \in \mathbb{R}$, then the following inequality due to
Sobolev

$$|f(x)| \leqslant \frac{1}{b-a} \int_a^b |f(t)| \, dt + \int_a^b |f'(t)| \, dt$$

valid for any real number $x \in [a, b]$, plays an important part in analytic number
theory where it was used by Gallagher in the theory of the so-called *large sieve*.[2]
This exercise proposes a discrete version of this inequality.

Let N be a positive integer and $a_1, \ldots, a_N$ be any complex numbers. Show that
for any $n \in \{1, \ldots, N\}$ the following inequality holds

$$|a_n| \leqslant \frac{1}{N} \left| \sum_{k=1}^{N} a_k \right| + \sum_{k=1}^{N-1} |a_{k+1} - a_k|.$$

6 Prove that, for any real number $x \geqslant 0$ and any integer $n \geqslant 1$,

$$\sum_{0 \leqslant j \leqslant x} (x - j)^n = \frac{1}{n+1} x^{n+1} + \frac{1}{2} x^n + O^\star \left(\frac{1}{8} n x^{n-1} \right).$$

Deduce that

$$\sum_{0 \leqslant j \leqslant x} (x - j)^n \leqslant \frac{1}{n+1} \left(x + \frac{1}{2} \right)^{n+1}.$$

7 Let $x \geqslant 2$ be any real number, $f \in C^1[2, +\infty)$ and

$$\theta(x) = \sum_{p \leqslant x} \log p$$

be the first Chebyshev prime number function. Prove that

$$\sum_{p \leqslant x} f(p) = \frac{f(x)\theta(x)}{\log x} - \int_2^x \theta(t) \frac{d}{dt} \left(\frac{f(t)}{\log t} \right) dt.$$

---

[2] See Lemma 4.16 and Theorem 4.47.

8 | In what follows, $x \geqslant 2$ is a real number. Define the *logarithmic integral* by

$$\mathrm{Li}(x) = \int_2^x \frac{dt}{\log t}. \tag{1.4}$$

Prove that

▷ For all $x \geqslant 2$

$$\frac{x}{\log x} + \int_2^x \frac{dt}{(\log t)^2} = \mathrm{Li}(x) + \frac{2}{\log 2}. \tag{1.5}$$

▷ For all $x \geqslant 2$

$$\mathrm{Li}(x) \leqslant \frac{2x}{\log x}.$$

▷ For all $x \geqslant e^2$

$$\mathrm{Li}(x) \geqslant \frac{x}{\log x}.$$

▷ For all $x \geqslant e^2$

$$\tfrac{1}{2} x \log x \leqslant \mathrm{Li}^{-1}(x) \leqslant 2x \log x.$$

9 | Let Li be the function defined in (1.4), and

$$\pi(x) = \sum_{p \leqslant x} 1$$

be the prime number counting function we shall see in Chap. 3. Suppose that we have at our disposal the following upper bound

$$|\theta(x) - x| \leqslant R(x) \quad (x \geqslant 2),$$

where R is an integrable, positive, non-decreasing function satisfying $\sqrt{x} \log x \leqslant R(x) < x$ for all $x \geqslant 2$. Using the result of Exercise 7, show that, for $x \geqslant 2$

$$|\pi(x) - \mathrm{Li}(x)| < \frac{5R(x)}{\log x}.$$

10   Let Li be the function defined in (1.4), and assume that there exists a real number $a \geqslant 13$ such that

$$|\pi(x) - \mathrm{Li}(x)| < R(x) \quad (x \geqslant a),$$

where R is a positive, non-decreasing function satisfying $R(x) < x$ for $x \geqslant a$. Using Corollary 3.3, prove that, for $n \in \mathbb{Z}_{\geqslant 1}$, the nth prime p_n satisfies

$$\left| p_n - \mathrm{Li}^{-1}(n) \right| \leqslant 3 \log n \, R(2n \log n)$$

provided that $p_n \geqslant a$.

11   Show that, for all $x \geqslant 1$

$$\left(x - \tfrac{1}{2}\right) \log x - x + \tfrac{7}{8} \leqslant \log\left(\lfloor x \rfloor !\right) \leqslant \left(x + \tfrac{1}{2}\right) \log x - x + 1.$$

References

1. Bliss, G.A.: A necessary and sufficient condition for the existence of a Stieltjes integral. Nat. Acad. Proc. **3**, 633–637 (1917)
2. Rademacher, H.: Topics in Analytic Number Theory. Springer, Berlin (1973)
3. Widder, D.V.: The Laplace Transform. Princeton University Press, Princeton (1946)
4. Young, L.C.: An inequality of the Hölder type, connected with Stieltjes integration. Acta Math. **67**, 251–282 (1936)

Chapter 2
Linear Diophantine Equations

2.1 Basic Facts

Let k, n be positive integers and $\mathfrak{a} = \{a_1, \ldots, a_k\}$ be a finite set of integers. Any equation of the form

$$a_1 x_1 + \cdots + a_k x_k = n,$$

with $(x_1, \ldots, x_k) \in \mathbb{Z}^k$, is called a *linear Diophantine equation*. Also set $d = (a_1, \ldots, a_k)$.

2.1.1 Solutions in the Simplest Cases

The two following results are well-known.

Lemma 2.1 *Let $a, b, n \in \mathbb{Z}$ such that $ab \neq 0$ and set $d = (a, b)$. Then the Diophantine equation $ax + by = n$ has a solution in $\mathbb{Z}^2$ if and only if $d \mid n$. In this case, the solutions are given by*

$$(x, y) = \left(x_0 + \frac{kb}{d}, \ y_0 - \frac{ka}{d}\right)$$

for some $k \in \mathbb{Z}$, where (x_0, y_0) is a particular solution of $ax + by = n$.

Lemma 2.2 *Let $a, b, c, n \in \mathbb{Z}$ such that $abc \neq 0$ and set $d = (a, b, c)$. Then the Diophantine equation $ax + by + cz = n$ has a solution in $\mathbb{Z}^3$ if and only if $d \mid n$. In this case, the solutions are given by*

$$(x, y, z) = \left(x_0 + \frac{kb}{\delta} - u_0 \ell, \ y_0 - \frac{ka}{\delta} - v_0 \ell, \ z_0 + \delta \ell\right)$$

© Springer Nature Switzerland AG 2020
O. Bordellès, *Arithmetic Tales*, Universitext,
https://doi.org/10.1007/978-3-030-54946-6_2

for some $(k, \ell) \in \mathbb{Z}^2$, where $\delta = (a, b)$ and (u_0, v_0) is a particular solution of $au + bv = \delta c$, (z_0, t_0) is a particular solution of $cz + \delta t = n$ and (x_0, y_0) is a particular solution of $ax + by = \delta t_0$.

Example 2.1 The solutions of the Diophantine equation $8x + 14y + 5z = 11$ are

$$(x, y, z) = (-1 + 7k - 3\ell, \ 1 - 4k + \ell, \ 1 + 2\ell) \quad \text{with} \quad (k, \ell) \in \mathbb{Z}^2.$$

In this chapter, we will mainly focus on the *number of solutions* of such equations. If $a, b \in \mathbb{Z}$, Lemma 2.1 shows that there are infinitely many solutions. The interesting case is then when a and b are positive.

Proposition 2.1 *Let $a, b \in \mathbb{Z}_{\geq 1}$ be coprime integers and $n \in \mathbb{Z}_{\geq 0}$. The number $\mathcal{D}_2(n)$ of non-negative solutions of the equation $ax + by = n$ is*

$$\mathcal{D}_2(n) = \left\lfloor \frac{n}{ab} \right\rfloor + r$$

where $r = 0$ or $r = 1$.

Proof Since $(a, b) = 1$ and $a, b \geq 1$, there exist two non-negative integers u, v such that $-au + bv = 1$. By Lemma 2.1, the solutions of the equation $ax + by = n$ are given by the pairs $(-un + bk, vn - ak)$ with $k \in \mathbb{Z}$, and the non-negative solutions are obtained by solving the inequalities $-un + bk \geq 0$ and $vn - ak \geq 0$, so that $\mathcal{D}_2(n)$ is equal to the number of integers in the interval $\left[\frac{un}{b}, \frac{vn}{a} \right]$ and hence

$$\mathcal{D}_2(n) = \left\lfloor \frac{vn}{a} - \frac{un}{b} \right\rfloor + r = \left\lfloor \frac{n(-au + bv)}{ab} \right\rfloor + r = \left\lfloor \frac{n}{ab} \right\rfloor + r$$

as asserted.   □

Remark 2.1 We shall see in Theorem 2.4 a more precise result about the value of r. See also Exercise 23.

Lemmas 2.1 and 2.2 may be generalized by induction to any $k \in \mathbb{Z}_{\geq 2}$.

Proposition 2.2 *Given $k \in \mathbb{Z}_{\geq 2}$, $a_1, \ldots, a_k, n \in \mathbb{Z}$ with $d = (a_1, \ldots, a_k)$, the equation*

$$a_1 x_1 + \cdots + a_k x_k = n \tag{2.1}$$

has solutions if and only if $d \mid n$, and in this case, it has infinitely many solutions, which can be expressed in terms of $k - 1$ integral parameters.

Proof If $d \nmid n$, then there is no solution since the left-hand side is divisible by d. Now assume $d \mid n$. Dividing the equation by d if necessary, we can suppose $d = 1$. We prove the proposition by induction on k, the case $k = 2$ being proved in Lemma 2.1. Set $d_{k-1} = (a_1, \ldots, a_{k-1})$. Any solution of the equation satisfies the congruence

$$a_k x_k \equiv n \pmod{d_{k-1}},$$

and since $(a_k, d_{k-1}) = 1$, a_k is inversible modulo d_{k-1} by Proposition 2.6, and therefore $x_k \equiv n\overline{a_k} \pmod{d_{k-1}}$, where $\overline{a_k}$ satisfies $a_k\overline{a_k} \equiv 1 \pmod{d_{k-1}}$.[1] Writing $x_k = n\overline{a_k} + h_{k-1}d_{k-1}$ with $h_{k-1} \in \mathbb{Z}$ and substituting in (2.1) yield

$$a_1 x_1 + \cdots + a_{k-1} x_{k-1} = n - na_k\overline{a_k} - a_k h_{k-1} d_{k-1}.$$

Note that, since $na_k\overline{a_k} \equiv n \pmod{d_{k-1}}$, we infer that d_{k-1} divides the right-hand side. Hence, dividing the previous equation by d_{k-1}, we get

$$a_1' x_1 + \cdots + a_{k-1}' x_{k-1} = n',$$

with $(a_1', \ldots, a_{k-1}') = 1$ and $n' = d_{k-1}^{-1}(n - na_k\overline{a_k} - a_k h_{k-1} d_{k-1})$. Now applying the induction hypothesis to the latter equation entails that it is solvable for all $h_{k-1} \in \mathbb{Z}$, and its solutions can be written in terms of $k - 2$ integral parameters. Adding the solution $x_k = n\overline{a_k} + h_{k-1}d_{k-1}$, we obtain solutions to (2.1) in terms of $k - 1$ integral parameters. $\square$

The number of solutions of the equation $a_1 x_1 + \cdots + a_k x_k = n$ when $a_1, \ldots, a_k, n \geqslant 1$ and $k \geqslant 3$ is a very hard task. This subject will be investigated in more detail in Sect. 2.3.

2.1.2 The Frobenius Problem

Given integers $1 \leqslant a_1 < \cdots < a_k$ such that $(a_1, \ldots, a_k) = 1$ and $n \geqslant 1$, we say that n is *representable* as a non-negative integer combination of $a_1, \ldots, a_k$ if there exist integers $x_i \geqslant 0$ such that

$$n = a_1 x_1 + \cdots + a_k x_k.$$

From an algebraic point of view, the set

$$S = \{a_1 x_1 + \cdots + a_k x_k : x_1 \in \mathbb{Z}_{\geqslant 0}, \ldots, x_k \in \mathbb{Z}_{\geqslant 0}\}$$

is sometimes abbreviated in

$$S = \langle a_1, \ldots, a_k \rangle \tag{2.2}$$

---

[1] By Theorem 2.3 below, one can take $\overline{a_k} = a_k^{\varphi(d_{k-1})-1}$ for instance.

and is the subsemigroup of $\mathbb{Z}_{\geqslant 0}$ generated by $\{a_1, \ldots, a_k\}$. If $(a_1, \ldots, a_k) = 1$, then S is called a *numerical semigroup*. The starting point of this section is the following observation.

Proposition 2.3 *If* $(a_1, \ldots, a_k) = 1$, *then there exists a non-negative integer* n_0 *such that every integer* $n \geqslant n_0$ *is representable as a non-negative integer combination of* $a_1, \ldots, a_k$.

Proof Since $(a_1, \ldots, a_k) = 1$, there exist $u_1, \ldots, u_k \in \mathbb{Z}$ such that $a_1 u_1 + \cdots + a_k u_k = 1$. Define $s = a_1 + \cdots + a_k$ and

$$n_0 = s^2 \max_{1 \leqslant i \leqslant k} |u_i|.$$

Now let $n \geqslant n_0$ be any integer. There exists $h \in \mathbb{Z}_{\geqslant 0}$ such that $n = n_0 + h$, so that there exists a unique couple $(q, r) \in (\mathbb{Z}_{\geqslant 0})^2$ such that

$$n = n_0 + sq + r \quad \text{with} \quad 0 \leqslant r < s.$$

Therefore

$$n = s^2 \max_{1 \leqslant i \leqslant k} |u_i| + sq + r \sum_{i=1}^{k} a_i u_i = \sum_{i=1}^{k} \left(s \max_{1 \leqslant i \leqslant k} |u_i| + q + r u_i \right) a_i,$$

where, for all $i \in \{1, \ldots, k\}$, we have $s \max_{1 \leqslant i \leqslant k} |u_i| + q + r u_i > r(|u_i| + u_i) + q \geqslant 0$. Hence, any integer $n \geqslant n_0$ is representable as a non-negative integer combination of $a_1, \ldots, a_k$, as asserted.   □

Proposition 2.3 leads to the following definition.

Definition 2.1 Let $1 \leqslant a_1 < \cdots < a_k$ be integers such that $(a_1, \ldots, a_k) = 1$. The *Frobenius number* $g(a_1, \ldots, a_k)$ is the largest positive integer n such that the equation $a_1 x_1 + \cdots + a_k x_k = n$ has no solution in non-negative integers $x_1, \ldots, x_k$. The *Frobenius problem* is the exact determination of the Frobenius number.

The Frobenius problem has a rich and long history, with several applications and extensions, but it is not completely clear when it first appeared. Ferdinand G. Frobenius discussed the problem occasionally in his lectures in the late 1800s, although he did not publish anything on the problem. On the other hand, the origin of the problem is often attributed to Sylvester who was the first to solve the case $k = 2$.

Proposition 2.4 (Sylvester) *Let* $a_1 < a_2$ *be coprime positive integers. Then*

$$g(a_1, a_2) = a_1 a_2 - a_1 - a_2.$$

Proof Let $N := a_1 a_2 - a_1 - a_2$.

▷ We first prove that N cannot be representable as a non-negative integer combination of a_1 and a_2. Suppose on the contrary that there exist $x, y \in \mathbb{Z}_{\geqslant 0}$ such that $N = a_1 x + a_2 y$. Then

$$N \equiv a_2 y \pmod{a_1} \iff -a_2 \equiv a_2 y \pmod{a_1},$$

and hence $y \equiv -1 \pmod{a_1}$ since $(a_1, a_2) = 1$. This implies that $y \geqslant a_1 - 1$.
On the other hand, $a_2 y \leqslant N = a_2(a_1 - 1) - a_1 < a_2(a_1 - 1)$, so that $y < a_1 - 1$, giving a contradiction.

▷ We now show that any integer $n > N$ can be representable as a non-negative integer combination of a_1 and a_2. To do this, first notice that the equation $a_2 y \equiv n \pmod{a_1}$ has a unique solution $0 \leqslant y < a_1$. Hence, there exists $x \in \mathbb{Z}$ such that

$$n = a_1 x + a_2 y \quad \text{with} \quad 0 \leqslant y < a_1.$$

Furthermore, $a_1 x = n - a_2 y > N - a_2 y \geqslant N - a_2(a_1 - 1) = -a_1$, so that $x \geqslant 0$. □

The Frobenius problem is far more difficult when $k \geqslant 3$. In the case $k = 3$, Curtis [8] proved that no simple formula for the Frobenius number exists, i.e. there is no polynomial $P \in \mathbb{C}[z_1, z_2, z_3, z_4]$ such that, for any triple $\{a_1, a_2, a_3\}$ satisfying $1 \leqslant a_1 < a_2 < a_3$, $(a_1, a_2, a_3) = 1$ with a_1, a_2 primes, we have $P(a_1, a_2, a_3, g(a_1, a_2, a_3)) = 0$. However, some particular cases have been investigated over the last decades.

When $k = 3$, Brauer and Shockley [7] showed in 1962 that, if $a_1 \mid a_2 + a_3$, then

$$g(a_1, a_2, a_3) + a_1 = \max\left(a_2 \left\lfloor \frac{a_1 a_3}{a_2 + a_2} \right\rfloor, \ a_3 \left\lfloor \frac{a_1 a_2}{a_2 + a_2} \right\rfloor \right).$$

Building on Exercises 19, 20 and 21, we derive the next result that may be of some help [24, Theorem 2.6].

Theorem 2.1 *Let $a_1, a_2 > 0$ be coprime integers and $a_3 \in \mathbb{Z}_{\geqslant 1}$.*

▷ *$g(a_1, a_2, a_3) = g(a_1, a_2) \iff a_3 \in \langle a_1, a_2 \rangle$.*
▷ *$a_3 \notin \langle a_1, a_2 \rangle \iff a_3 = a_1 a_2 - a_1 x - a_2 y$ for some $x, y > 0$. In this case*

  ◇ *if $a_1 x < a_2 y$ and $0 < y \leqslant \left\lfloor \frac{1}{2} a_1 \right\rfloor$, then $g(a_1, a_2, a_3) = g(a_1, a_2) - a_1 x$;*

  ◇ *if $a_1 x > a_2 y$ and $0 < x \leqslant \left\lfloor \frac{1}{2} a_2 \right\rfloor$, then $g(a_1, a_2, a_3) = g(a_1, a_2) - a_2 y$.*

Proof The first part of the theorem follows from Exercise 20, and the first assertion of the second part follows from Exercise 19. We prove the proposition

$$a_1 x < a_2 y \text{ and } 0 < y \leqslant \left\lfloor \frac{1}{2} a_1 \right\rfloor \Rightarrow g(a_1, a_2, a_3) = g(a_1, a_2) - a_1 x,$$

the other one being similar. Recall that $a_3 = a_1 a_2 - a_1 x - a_2 y$ for some $x, y > 0$.

▷ Assume first that $g(a_1, a_2, a_3) < g(a_1, a_2) - a_1 x$. Since $g(a_1, a_2, a_3)$ is the largest positive integer not representable as a non-negative linear combination of a_1, a_2, a_3, we deduce that there exist $u, v, w \in \mathbb{Z}_{\geqslant 0}$ such that

$$g(a_1, a_2) - a_1 x = a_1 u + a_2 v + a_3 w = a_1 u + a_2 v + w(a_1 a_2 - a_1 x - a_2 y).$$

Note that we have $w > 0$, otherwise $g(a_1, a_2) = a_1(x + u) + a_2 v$ so that $g(a_1, a_2) \in \langle a_1, a_2 \rangle$, which is impossible. If w is even, then

$$g(a_1, a_2) = a_1 \left(u + x + \tfrac{1}{2}w(a_2 - 2x) \right) + a_2 \left(v + \tfrac{1}{2}w(a_1 - 2y) \right),$$

and since $a_1 x < a_2 y$ and $y \leqslant \tfrac{1}{2}a_1$, we infer $x \leqslant \tfrac{1}{2}a_2$, which implies that $g(a_1, a_2) \in \langle a_1, a_2 \rangle$, which is impossible. If w is odd, then

$$g(a_1, a_2) = a_1 \left(u + x + \tfrac{1}{2}(w - 1)(a_2 - 2x) \right) + a_2 \left(v + \tfrac{1}{2}(w - 1)(a_1 - 2y) \right) + a_3$$

$$:= k a_1 + \ell a_2 + a_3$$

say, and hence, using Proposition 2.4

$$a_3 = g(a_1, a_2) - a_1 k - a_2 \ell = a_1 a_2 - (k + 1)a_1 - (\ell + 1)a_2,$$

and since $a_3 = a_1 a_2 - a_1 x - a_2 y$, we derive $x = k + 1$, i.e.

$$u + \tfrac{1}{2}(w - 1)(a_2 - 2x) + 1 = 0,$$

which is impossible since $w > 0$ and $a_2 \geqslant 2x$.

▷ Therefore $g(a_1, a_2, a_3) \geqslant g(a_1, a_2) - a_1 x$. Assume $g(a_1, a_2, a_3) > g(a_1, a_2) - a_1 x$. There exist $m, n \geqslant 0$ such that $g(a_1, a_2, a_3) = g(a_1, a_2) - m a_1 - n a_2$ by Exercise 21, and inserting in the inequality above, we get $m a_1 + n a_2 < a_1 x$, so that $m < x$, and since $a_1 x < a_2 y$, we also get $n < y$. Now, by Proposition 2.4,

$$g(a_1, a_2, a_3) = g(a_1, a_2) - m a_1 - n a_2$$

$$= a_1 a_2 - a_1 - a_2 - m a_1 - n a_2$$

$$= (a_1 a_2 - a_1 x - a_2 y) + a_1 x + a_2 y - a_1 - a_2 - m a_1 - n a_2$$

$$= a_3 + a_1(x - m - 1) + a_2(y - n - 1),$$

with $x - m - 1 \geqslant 0$ and $y - n - 1 \geqslant 0$ since $m < x$ and $n < y$. Hence, $g(a_1, a_2, a_3) \in \langle a_1, a_2, a_3 \rangle$, which is impossible by the definition of $g(a_1, a_2, a_3)$.   □

Recently, Tripathi [26] proved formulæ that cover all cases of a_1, a_2, a_3 providing, of course, that $(a_1, a_2, a_3) = 1$. As an example, let us state here [26, Theorem 3].

Proposition 2.5 (Tripathi) *Define* $\kappa = \lfloor a_3/a_2 \rfloor$, $\rho \equiv a_3 a_2^{-1} \pmod{a_1}$, $q = \lfloor \frac{a_1}{a_1 - \rho} \rfloor$, $r = a_1 - q(a_1 - \rho)$ *and* $\lambda = \lfloor \frac{a_3 q - a_2 r}{a_2(a_1 - \rho) + a_3} \rfloor$. *Assume* $\rho \geqslant \kappa$ *and* $a_2 r < a_3 q$. *Then*

$$
g(a_1, a_2, a_3) + a_1 = \begin{cases} a_2\left((\lambda + 1)(a_1 - \rho) + r - 1\right), & \text{if } \lambda \geqslant \frac{a_3(q-1) - a_2 r}{a_2(a_1 - \rho) + a_3} \\ a_2(a_1 - \rho - 1) + a_3(q - \lambda - 1), & \text{otherwise.} \end{cases}
$$

For instance, $g(113, 127, 157) = 2554$. When $k \geqslant 4$, results concerning these problems are limited essentially to algorithms, bounds and exact formulæ in some special cases. Let us mention the very elegant upper bound discovered by Erdős and Graham in [12].

Theorem 2.2 (Erdős and Graham) *Let* $1 \leqslant a_1 < \cdots < a_k$ *be integers such that* $(a_1, \ldots, a_k) = 1$. *Then*

$$
g(a_1, \ldots, a_k) \leqslant 2a_{k-1} \left\lfloor \frac{a_k}{k} \right\rfloor - a_k.
$$

Subsequently, several results were discovered by many authors. In 1977, Selmer [22] proved that, if $a_1 \geqslant k$, then

$$
g(a_1, \ldots, a_k) \leqslant 2a_k \left\lfloor \frac{a_1}{k} \right\rfloor - a_1.
$$

In 1975, Vitek [27] showed that, for all $k \geqslant 3$,

$$
g(a_1, \ldots, a_k) \leqslant \left\lfloor \tfrac{1}{2}(a_2 - 1)(a_k - 1) \right\rfloor - 1,
$$

while, in 1982, Rödseth [20] improved Theorem 2.2 by showing that

$$
g(a_1, \ldots, a_k) \leqslant 2a_{k-1} \left\lfloor \frac{a_1 + 2}{k + 1} \right\rfloor - a_1 \tag{2.3}
$$

provided that k is odd. In 2002, Beck et al. [5] proved that

$$
g(a_1, \ldots, a_k) \leqslant \tfrac{1}{2}\left(\sqrt{a_1 a_2 a_3 (a_1 + a_2 + a_3)} - a_1 - a_2 - a_3 \right).
$$

Finally, in 2007, borrowing techniques belonging to the geometry of numbers, Fukshansky and Robins [13] obtained the following upper bound:

$$
g(a_1, \ldots, a_k) \leqslant \left\lfloor \pi^{-k/2}(k-1)^2 \Gamma\left(\tfrac{k}{2} + 1\right)^{-1} \sum_{i=1}^{k} a_i \sqrt{\|\mathbf{a}\|_2^2 - a_i^2} + 1 \right\rfloor,
$$

where $\|\mathbf{a}\|_2$ is the Euclidean norm of $\mathbf{a} = (a_1, \ldots, a_k)$. In the other direction, Davison [9] proved that, when $k = 3$

$$g(a_1, a_2, a_3) \geqslant \sqrt{3a_1 a_2 a_3} - a_1 - a_2 - a_3,$$

where it is known that the constant $\sqrt{3}$ is sharp. For $k \geqslant 4$, Killingbergtrø[15] derived the lower bound

$$g(a_1, \ldots, a_k) \geqslant \left((k-1)! \prod_{j=1}^{k} a_j \right)^{\frac{1}{k-1}} - \sum_{j=1}^{k} a_j,$$

and using the geometry of numbers, Aliev and Gruber [2] showed that the inequality is strict.

2.2 The Ring $(\mathbb{Z}/n\mathbb{Z}, +, \times)$

2.2.1 Units and Zero Divisors

For any ring R, we denote by $R^{\times}$ the group of units of R.

Let n be a positive integer. The set of residue classes modulo n, often denoted by $\bar{0}, \bar{1}, \ldots, \overline{n-1}$, is a unitary abelian ring, denoted by $\mathbb{Z}/n\mathbb{Z}$, endowed with the binary operations $+$ and $\times$ defined by $\overline{x+y} = \bar{x} + \bar{y}$ and $\overline{x \times y} = \bar{x} \times \bar{y}$, which are obviously well-defined. The basic result is then the following theorem.

Proposition 2.6 *The ring $\mathbb{Z}/n\mathbb{Z}$ is a disjoint union $\mathbb{Z}/n\mathbb{Z} = E \cup F$, where E, resp., F, is the set of units, resp., zero divisors, of $\mathbb{Z}/n\mathbb{Z}$. Furthermore, $\bar{x} \in \mathbb{Z}/n\mathbb{Z}$ is a unit if and only if $(x, n) = 1$.*

Proof Let $d = (x, n)$. Suppose first that $\bar{x}$ is a unit and define $\bar{y} = (\bar{x})^{-1}$. We then have $xy \equiv 1 \pmod{n}$, and then there exists $k \in \mathbb{Z}$ such that $xy - 1 = kn$. Thus we have $d \mid (xy - kn) = 1$ and then $d = 1$. Conversely, suppose that $d = 1$. By Bachet-Bézout's theorem, there exist $u, v \in \mathbb{Z}$ such that $xu + vn = 1$. Therefore, $xu \equiv 1 \pmod{n}$ and then $\bar{x}$ is a unit in $\mathbb{Z}/n\mathbb{Z}$. If $d > 1$, we set $x = dx', n = dn'$ with $(x', n') = 1$. Then we have

$$n'x = dn'x' = nx' \equiv 0 \pmod{n}$$

so that $\bar{x}$ is a zero divisor in $\mathbb{Z}/n\mathbb{Z}$. Conversely, a zero divisor cannot be a unit, and then $d > 1$. $\square$

2.2.2   The Euler Totient Function

Definition 2.2 *Euler's totient function* is the function φ, which counts the number of units in $\mathbb{Z}/n\mathbb{Z}$, with the convention that $\varphi(1) = 1$.

Proposition 2.6 has then the following immediate application.

Corollary 2.1 *Let n be a positive integer. Then $\varphi(1) = 1$ and, for every integer $n \geqslant 2$, $\varphi(n)$ is the number of integers $1 \leqslant m \leqslant n$ such that $(m, n) = 1$. In other words, we have*

$$\varphi(n) = \sum_{\substack{m \leqslant n \\ (m,n)=1}} 1.$$

This important function will extensively be studied in Chap. 4.

Finally, note that if $n = p$ is prime, then $\mathbb{Z}/p\mathbb{Z}$ does not have any zero divisor, and then the set of units of $\mathbb{Z}/p\mathbb{Z}$ is $(\mathbb{Z}/p\mathbb{Z})^\times = \{\overline{1}, \ldots, \overline{p-1}\}$. Thus, for every prime number p, we have $\varphi(p) = p - 1$. More generally, group theory tells us that if $n = p_1^{\alpha_1} \cdots p_r^{\alpha_r}$ is the factorization of n into prime powers, then we have the following isomorphism of cyclic groups:

$$\mathbb{Z}/n\mathbb{Z} \simeq \mathbb{Z}/p_1^{\alpha_1}\mathbb{Z} \oplus \mathbb{Z}/p_2^{\alpha_2}\mathbb{Z} \oplus \cdots \oplus \mathbb{Z}/p_r^{\alpha_r}\mathbb{Z},$$

which induces the group isomorphism

$$(\mathbb{Z}/n\mathbb{Z})^\times \simeq (\mathbb{Z}/p_1^{\alpha_1}\mathbb{Z})^\times \oplus (\mathbb{Z}/p_2^{\alpha_2}\mathbb{Z})^\times \oplus \cdots \oplus (\mathbb{Z}/p_r^{\alpha_r}\mathbb{Z})^\times,$$

whence we deduce that

$$\varphi(n) = \varphi(p_1^{\alpha_1})\varphi(p_2^{\alpha_2}) \cdots \varphi(p_r^{\alpha_r})$$

characterizing the multiplicativity of Euler's totient function. We shall prove this again in Chap. 4 by using purely arithmetic arguments.

2.2.3   The Euler–Fermat Theorem

One of the most important results in elementary number theory is certainly the next theorem.

Theorem 2.3 (Euler–Fermat) *Let* a, n *be positive integers such that* $(a, n) = 1$. *Then*

$$a^{\varphi(n)} \equiv 1 \ (\mathrm{mod}\, n).$$

Proof This result is a direct consequence of Lagrange's theorem[2], which implies that, if G is any finite group of order $|G|$ and identity element e_G, then, for all $a \in G$, we have $a^{|G|} = e_G$. Theorem 2.3 then follows at once by applying this result with $G = (\mathbb{Z}/n\mathbb{Z})^\times$ since, by Theorem 2.6, $(a, n) = 1 \Leftrightarrow \overline{a} \in (\mathbb{Z}/n\mathbb{Z})^\times$ and hence $\left|(\mathbb{Z}/n\mathbb{Z})^\times\right| = \varphi(n)$. □

2.3 Denumerants

The term "denumerant" is due to Sylvester, who seemed to have been quite inventive when it came to coining new words and phrases.

2.3.1 Definition

Definition 2.3 Let k, n be positive integers and $\mathfrak{a} = \{a_1, \ldots, a_k\}$ be a finite set of positive relatively prime integers. The *denumerant* of n with respect to k and to the set $\mathfrak{a}$ is the number $\mathcal{D}_k(n)$ of solutions of the Diophantine equation

$$a_1 x_1 + \cdots + a_k x_k = n,$$

where the unknowns are $(x_1, \ldots, x_k) \in \left(\mathbb{Z}_{\geq 0}\right)^k$. When $a_1 = \cdots = a_k = 1$, this denumerant will be denoted by $\mathcal{D}_{(1,\ldots,1)}(n)$ where the vector $(1, \ldots, 1)$ is supposed to have k components.

2.3.2 Denumerants with Two Variables

2.3.2.1 Paoli's Theorem

In this section, $\mathfrak{a} = \{a, b\}$ with $(a, b) = 1$. The next result is a refinement of Proposition 2.1.

---

[2]See Theorem 7.1.

Theorem 2.4 (Paoli) *Let $a, b \in \mathbb{Z}_{\geq 1}$ such that $(a, b) = 1$, $n \in \mathbb{Z}_{\geq 0}$, and let r be the remainder in the Euclidean division of n by ab. Then, $\mathcal{D}_2(r) = 0$ or 1, and*

$$\mathcal{D}_2(n) = \left\lfloor \frac{n}{ab} \right\rfloor + \mathcal{D}_2(r).$$

Proof From the proof of Proposition 2.1, $\mathcal{D}_2(n)$ is the number of integers belonging to the interval $\left[\frac{un}{b}, \frac{vn}{a} \right]$, where $u, v \in \mathbb{Z}_{\geq 0}$ are such that $-au + bv = 1$, so that

$$\mathcal{D}_2(n) = \left\lfloor \frac{vn}{a} \right\rfloor + \left\lfloor 1 - \frac{un}{b} \right\rfloor.$$

Write $n = abq + r$ with $q = \left\lfloor \frac{n}{ab} \right\rfloor$ et $0 \leq r < ab$. Since $bv - au = 1$, we get

$$\mathcal{D}_2(r) = \left\lfloor \frac{vr}{a} \right\rfloor + \left\lfloor 1 - \frac{ur}{b} \right\rfloor \leq \frac{vr}{a} + 1 - \frac{ur}{b} = 1 + \frac{r(bv - au)}{ab} = 1 + \frac{r}{ab} < 2,$$

and therefore $\mathcal{D}(r) = 0$ or 1.

▷ If $\mathcal{D}_2(r) = 0$, then the interval $\left[\frac{ur}{b}, \frac{vr}{a} \right]$ does not contain any integer, implying in particular that $\left\lfloor \frac{ur}{b} \right\rfloor = \left\lfloor \frac{vr}{a} \right\rfloor$, and hence

$$\mathcal{D}_2(n) = \left\lfloor \frac{vn}{a} \right\rfloor + \left\lfloor 1 - \frac{un}{b} \right\rfloor = \left\lfloor \frac{vr}{a} + bvq \right\rfloor + \left\lfloor 1 - \frac{ur}{b} - auq \right\rfloor$$

$$= \left\lfloor \frac{vr}{a} \right\rfloor + bvq + 1 + \left\lfloor -\frac{ur}{b} \right\rfloor - auq$$

$$= \left\lfloor \frac{vr}{a} \right\rfloor + bvq + 1 - 1 - \left\lfloor \frac{ur}{b} \right\rfloor - auq$$

$$= (bv - au)q = q = \left\lfloor \frac{n}{ab} \right\rfloor,$$

where we used the fact that $\frac{ur}{b} \notin \mathbb{Z}$.

▷ If $\mathcal{D}_2(r) = 1$, then the interval $\left[\frac{ur}{b}, \frac{vr}{a} \right]$ contains a sole integer, so that $\left\lfloor \frac{ur}{b} \right\rfloor = \left\lfloor \frac{vr}{a} \right\rfloor - 1$. Note that $\frac{ur}{b} \notin \mathbb{Z}$ also in this case, for $0 \leq \frac{vr}{a} - \frac{ur}{b} = \frac{r}{ab} < 1$ and thus, if $\frac{ur}{b} \in \mathbb{Z}$, then $\left\lfloor \frac{ur}{b} \right\rfloor = \left\lfloor \frac{vr}{a} \right\rfloor$, which is impossible. □

2.3.2.2 Ehrhart's Theorem

A similar result exists in the case of three variables [11, Theorem 10.5].

Theorem 2.5 (Ehrhart) *Let $a, b, c \in \mathbb{Z}_{\geq 1}$ such that $(a, b) = (b, c) = (c, a) = 1$, $n \in \mathbb{Z}_{\geq 0}$, and let r be the remainder in the Euclidean division of n by abc. Then*

$$\mathcal{D}_3(n) = \frac{1}{2} \left\lfloor \frac{n}{abc} \right\rfloor (n + r + a + b + c) + \mathcal{D}_3(r).$$

2.3.2.3 Tripathi's Theorem

By Proposition 2.11, the generating function $F(z)$ of $\mathcal{D}_2(n)$ is given by

$$F(z) = \frac{1}{(1 - z^a)\left(1 - z^b\right)}.$$

Theorem 2.7 yields[3]

$$\mathcal{D}_2(n) = \frac{n}{ab} + \frac{1}{2}\left(\frac{1}{a} + \frac{1}{b}\right) + \frac{1}{a}\sum_{h=1}^{a-1}\frac{e_a(nh)}{1 - e_a(-bh)} + \frac{1}{b}\sum_{h=1}^{b-1}\frac{e_b(nh)}{1 - e_b(-ah)},$$

(2.4)

where $e_a(x) = e^{2\pi i x/a}$. Note that the sums are periodic in n, the first one with period a and the second one with period b, and since $(a, b) = 1$, the sum of these two sums is periodic of period ab, and then the expression of $\mathcal{D}_2(n)$ is essentially determined modulo ab. Now define a' and b' to be integers satisfying

$$\begin{cases} 1 \leqslant a' \leqslant b \\ 1 \leqslant b' \leqslant a \\ aa' \equiv -n \pmod{b} \\ bb' \equiv -n \pmod{a}. \end{cases}$$

(2.5)

Note that

$$\{1 - e_a(-bh)\} \sum_{j=0}^{b'-1} e_a(-bjh) = 1 - e_a(-b'bh) = 1 - e_a(nh),$$

hence we get

$$\frac{e_a(nh)}{1 - e_a(-bh)} = \frac{1}{1 - e_a(-bh)} - \sum_{j=0}^{b'-1} e_a(-bjh),$$

and using the formulæ

$$\sum_{h=1}^{a-1} e_a(-mh) = \begin{cases} a - 1, & \text{if } a \mid m \\ -1, & \text{otherwise} \end{cases}$$

---

[3] See Exercise 24 for another proof of this identity.

valid for positive integers a, m, and

$$\sum_{h=1}^{a-1} \frac{1}{1 - e_a(-mh)} = \frac{a-1}{2},$$

valid for positive integers a, m such that $(a, m) = 1$, we derive

$$\sum_{h=1}^{a-1} \frac{e_a(nh)}{1 - e_a(-bh)} = \sum_{h=1}^{a-1} \frac{1}{1 - e_a(-bh)} - \sum_{j=0}^{b'-1} \sum_{h=1}^{a-1} e_a(-bjh)$$

$$= \frac{a-1}{2} - (a-1) + \sum_{j=1}^{b'-1} 1$$

$$= b' - \frac{a+1}{2},$$

and the same is true for the second sum, so that we finally get

$$\mathcal{D}_2(n) = \frac{n}{ab} - 1 + \frac{a'}{b} + \frac{b'}{a}.$$

Note that by (2.5) both a and b divide $n + aa' + bb'$, so that $n + aa' + bb'$ is divisible by ab. Furthermore, since $n + aa' + bb' \geqslant n + a + b$, we infer that the number on the right-hand side above is indeed a non-negative integer. We can then state the following result [25].

Theorem 2.6 (Tripathi) *Let $a, b \geqslant 1$ be coprime integers and define positive integers a', b' as in (2.5). Then*

$$\mathcal{D}_2(n) = \frac{n}{ab} - 1 + \frac{a'}{b} + \frac{b'}{a}.$$

Note that the following four properties characterize uniquely $\mathcal{D}_2(n)$.

$$\mathcal{D}_2(n + hab) = \mathcal{D}_2(n) + h \quad (h \geqslant 0).$$

$$\mathcal{D}_2(n) = 1 \quad \text{if } ab - a - b < n < ab.$$

$$\mathcal{D}_2(n) + \mathcal{D}_2(m) = 1 \quad \text{if } m + n = ab - a - b \quad (m, n \geqslant 0).$$

$$\mathcal{D}_2(n) = 1 \quad \text{if } n = ax_0 + by_0 < ab - a - b \quad (x_0, y_0 \geqslant 0).$$

See [3, 4, 23, 28] for more details.

Another formula, discovered by Popoviciu [19] in 1953, is given in Exercise 24.

2.3.3   Denumerants with k Variables

2.3.3.1   Bounds

An upper bound can be derived in an elementary way by taking into account the following observation. Since

$$a_1 x_1 + \cdots + a_k x_k = n \iff a_1 x_1 + \cdots + a_{k-1} x_{k-1} = n - a_k x_k$$

we obtain $x_k \leqslant \lfloor n/a_k \rfloor$ and thus

$$\mathcal{D}_k(n) = \sum_{j=0}^{n/a_k} \mathcal{D}_{k-1}(n - ja_k). \tag{2.6}$$

The aim is then to prove the following result.

Proposition 2.7 *Define the positive integers s_k and r_k as $s_1 = 0$, $r_1 = 1$ and, for any $k \in \mathbb{Z}_{\geqslant 2}$*

$$s_k = a_2 + \tfrac{1}{2}(a_3 + \cdots + a_k) \quad and \quad r_k = a_2 \cdots a_k,$$

with $s_2 = a_2$. Then

$$\mathcal{D}_k(n) \leqslant \frac{(n + s_k)^{k-1}}{(k - 1)! \, r_k}.$$

Furthermore, if $a_1 = \cdots = a_k = 1$, then $\mathcal{D}_{(1,\ldots,1)}(n) = \binom{k + n - 1}{n}$.

Proof By induction on k, the case $k = 1$ being immediate since

$$\mathcal{D}_1(n) \leqslant 1 = \frac{n^{1-1}}{(1 - 1)!} = \frac{(n + s_1)^{1-1}}{(1 - 1)! r_1}.$$

Now suppose $k \geqslant 2$. The case $k = 2$ follows easily from Proposition 2.1, and suppose the inequality is true for some $k \geqslant 2$. Using (2.6), induction hypothesis and Exercise 6, we derive

$$\mathcal{D}_{k+1}(n) = \sum_{j=0}^{n/a_{k+1}} \mathcal{D}_k(n - ja_{k+1})$$

$$\leqslant \frac{1}{(k - 1)! \, r_k} \sum_{j=0}^{n/a_{k+1}} (n - ja_{k+1} + s_k)^{k-1}$$

$$\leq \frac{a_{k+1}^{k-1}}{(k-1)!\, r_k} \sum_{j=0}^{(n+s_k)/a_{k+1}} \left(\frac{n+s_k}{a_{k+1}} - j\right)^{k-1}$$

$$\leq \frac{a_{k+1}^{k-1}}{k!\, r_k} \left(\frac{n+s_k}{a_{k+1}} + \frac{1}{2}\right)^{k} = \frac{(n+s_{k+1})^k}{k!\, r_{k+1}},$$

which is the desired result. The last part of the theorem may be treated similarly, using the well-known identity (see [14, identity 1.49])

$$\sum_{j=0}^{n} \binom{k+j-1}{j} = \binom{k+n}{n}$$

in the induction argument.   □

The particular case of Proposition 2.7 can be generalized as follows [16].

Proposition 2.8 *Let $c, k, n \in \mathbb{Z}_{\geq 1}$ and $c_1, \ldots, c_k \in \mathbb{Z}_{\geq 0}$. Then the number of integer solutions of the equation $x_1 + \cdots + x_k = n$ satisfying $x_1 > c_1, x_2 > c_2, \ldots, x_k > c_k$, is given by*

$$\binom{n - s_k - 1}{k - 1},$$

where $s_k := \sum_{i=1}^{k} c_i$.

In [1], Agnarsson proved by a purely combinatorial method the next estimates.

Proposition 2.9 (Agnarsson) *Let $\{a_1, \ldots, a_k\}$ be a fixed k-tuple of positive integers and define recursively the integers $A_1, \ldots, A_k$ and $B_1, \ldots, B_k$ by*

$$A_1 = 0, \quad A_j = A_{j-1} + a_j \times \frac{(a_1, \ldots, a_{j-1})}{(a_1, \ldots, a_j)} \quad (2 \leq j \leq k)$$

$$B_1 = 0, \quad B_j = B_{j-1} + a_j \times \left(\frac{(a_1, \ldots, a_{j-1})}{(a_1, \ldots, a_j)} - 1\right) - 1 \quad (2 \leq j \leq k).$$

Set $d = (a_1, \ldots, a_k)$. If $d \mid n$ and $n \geq B_k + k - 1$, then

$$\frac{d}{a_1 \cdots a_k} \binom{n - B_k}{k - 1} \leq \mathcal{D}_k(n) \leq \frac{d}{a_1 \cdots a_k} \binom{n + A_k}{k - 1}.$$

The proof uses induction and the following bound, which may have its own interest. *Let m, n, q be positive integers satisfying $n \geqslant m - 1$. Then*

$$\frac{1}{q}\binom{n+1}{m} \leqslant \sum_{k=0}^{\left\lfloor \frac{1}{2}(n-m+1) \right\rfloor} \binom{n-kq}{m-1} \leqslant \frac{1}{q}\binom{n+q}{m}.$$

The author points out that the integers A_k and B_k can be computed efficiently, essentially equally as expensive as the computation of the greatest common divisor of two numbers. In particular, when $a_1 = 1$, the bounds above have the following simpler expression.

$$\frac{1}{a_2 \cdots a_k}\binom{n+k-1}{k-1} \leqslant \mathcal{D}_k(n) \leqslant \frac{1}{a_2 \cdots a_k}\binom{n+a_2+\cdots+a_k}{k-1}.$$

2.3.3.2   Asymptotic Formula

In this section, we intend to prove the following result.

Proposition 2.10   *Let $\{a_1, \ldots, a_k\}$ be a fixed k-tuple of positive pairwise relatively prime integers. Then*

$$\mathcal{D}_k(n) = \frac{n^{k-1}}{(k-1)! \, a_1 \cdots a_k} + O\left(n^{k-2}\right).$$

Proof   We proceed by induction on k, the case $k = 1$ being obvious and the case $k = 2$ being already proved in Theorem 2.4. Now let $k \geqslant 3$ and assume the Proposition is true for $k - 1$. Applying the induction hypothesis to (2.6), we derive

$$\mathcal{D}_k(n) = \sum_{j=0}^{n/a_k} \mathcal{D}_{k-1}(n - ja_k)$$

$$= \frac{1}{(k-2)! \, a_1 \cdots a_{k-1}} \sum_{j=0}^{n/a_k} (n - ja_k)^{k-2} + O\left(\sum_{j=0}^{n/a_k} (n - ja_k)^{k-3}\right)$$

$$= \frac{a_k^{k-2}}{(k-2)! \, a_1 \cdots a_{k-1}} \sum_{j=0}^{n/a_k} \left(\frac{n}{a_k} - j\right)^{k-2} + O\left(n^{k-2}\right),$$

where we used the fact that the error term does not exceed

$$\leqslant n^{k-3} \sum_{j=0}^{n/a_k} 1 \leqslant 2n^{k-2}.$$

Now from Exercise 6, we have

$$\sum_{j=0}^{n/a_k} \left(\frac{n}{a_k} - j\right)^{k-2} = \frac{1}{k-1}\left(\frac{n}{a_k}\right)^{k-1} + \frac{1}{2}\left(\frac{n}{a_k}\right)^{k-2} + O^{\star}\left(\frac{k-2}{8}\left(\frac{n}{a_k}\right)^{k-3},\right)$$

and since $a_k \ll n$, we infer

$$\mathcal{D}_k(n) = \frac{n^{k-1}}{(k-1)! \, a_1 \cdots a_k} + O\left(n^{k-2}\right),$$

as required.   □

Note that a similar result holds assuming only $(a_1, \ldots, a_k) = 1$. The proof proceeds the same way replacing (2.6) by

$$\mathcal{D}_k(n) = \sum_{j=0}^{m/a_k} \mathcal{D}_{k-1}^{\star}(m - ja_k),$$

where $m := d^{-1}(n - ua_k)$, $d := (a_1, \ldots, a_{k-1})$, u is the unique solution in $\{0, \ldots, d-1\}$ of the equation $n \equiv ua_k \pmod{d}$ and $D_{k-1}^{\star}$ stands for the denumerant associated to the $(k-1)$-tuple $\left(a_1^{\star}, \ldots, a_{k-1}^{\star}\right)$ with $a_j^{\star} := d^{-1}a_j$. See [17], for instance.

2.3.4 Generating Functions

The *generating function* of a sequence (u_n) is the function $F(z)$ formally defined by

$$F(z) = \sum_{n \geqslant 0} u_n z^n.$$

If $F(z)$ can be written as an elementary function of z, then the comparison of the coefficients of z^n can give a closed formula for u_n. Nevertheless, we must notice that such a series could either be studied formally or must be taken as a function of the complex variable z. In the latter case, convergence problems must be studied. Generating functions are the main tools in combinatorial theory or additive number theory, where results from complex analysis are frequently used. See [18] or [28] for instance.

One of the simplest examples of generating functions is the following identity, which will prove to be of great use.

For any positive integer k and any real number $|x| < 1$, we have

$$\sum_{j=0}^{\infty} \binom{k+j-1}{j} x^j = \frac{1}{(1-x)^k}. \tag{2.7}$$

This identity allows us to derive the generating function of the denumerant $\mathcal{D}_k(n)$.

Proposition 2.11 *The generating function of $\mathcal{D}_k(n)$ is given by*

$$F(z) = \prod_{i=1}^{k} \frac{1}{1 - z^{a_i}}.$$

Proof Using (2.7) with $k = 1$, we get

$$\frac{1}{(1 - z^{a_1})(1 - z^{a_2}) \cdots (1 - z^{a_k})} = \sum_{x_1 \geqslant 0} z^{a_1 x_1} \sum_{x_2 \geqslant 0} z^{a_2 x_2} \cdots \sum_{x_k \geqslant 0} z^{a_k x_k}$$

$$= \sum_{x_1, \ldots, x_k \geqslant 0} z^{a_1 x_1 + \cdots + a_k x_k}$$

$$= \sum_{n \geqslant 0} \left(\sum_{\substack{x_1, \ldots, x_k \geqslant 0 \\ a_1 x_1 + \cdots + a_k x_k = n}} 1 \right) z^n$$

$$= \sum_{n \geqslant 0} \mathcal{D}_k(n) z^n,$$

as required.   □

Example 2.2 Take $k = 3$ and $\mathfrak{a} = \{1, 2, 3\}$ so that $\mathcal{D}_3(n)$ counts the number of non-negative integer solutions of the equation $x + 2y + 3z = n$. The generating function of $\mathcal{D}_3(n)$ is then

$$F(z) = \frac{1}{(1-z)(1-z^2)(1-z^3)},$$

and partial fractions expansion theory gives

$$\frac{1}{(1-z)(1-z^2)(1-z^3)} = \frac{1}{6(1-z)^3} + \frac{1}{4(1-z)^2} + \frac{17}{72(1-z)}$$

$$+ \frac{1}{8(1+z)} + \frac{1}{9}\left(\frac{1}{1 - z/\rho} + \frac{1}{1 - z/\rho^2} \right),$$

where $\rho = e_3(1) = e^{2\pi i/3}$. Using (2.7), we obtain

$$\sum_{n \geq 0} \mathcal{D}_3(n)z^n = \sum_{n \geq 0} z^n \left\{ \frac{1}{6}\binom{n+2}{n} + \frac{1}{4}\binom{n+1}{n} + \frac{17}{72} \right.$$

$$\left. + \frac{(-1)^n}{8} + \frac{1}{9}(\rho^{-n} + \rho^{-2n}) \right\},$$

and comparing the coefficients of z^n in each side gives

$$\mathcal{D}_3(n) = \frac{1}{6}\binom{n+2}{n} + \frac{1}{4}\binom{n+1}{n} + \frac{17}{72} + \frac{(-1)^n}{8} + \frac{1}{9}\left(\rho^{-n} + \rho^{-2n}\right)$$

$$= \frac{6n^2 + 36n + 47}{72} + \frac{(-1)^n}{8} + \frac{2}{9}\cos\left(\frac{2\pi n}{3}\right)$$

$$= \frac{(n+3)^2}{12} + O^\star\left(\frac{1}{3}\right) = \left\lfloor \frac{(n+3)^2}{12} \right\rceil,$$

where $\lfloor x \rceil$ is the nearest integer to x.

Building the general case on the previous example, Proposition 2.11 enables us to derive the next identity.

Theorem 2.7 *Let* $\{a_1, \ldots, a_k\}$ *be a fixed k-tuple of positive relatively prime integers. Then*

$$\mathcal{D}_k(n) = \sum_{j=1}^{k} c_j \binom{n+j-1}{n} + \sum_{j=1}^{k} \sum_{h=1}^{a_j-1} A_{a_j}(h)\zeta_{a_j}^{-nh},$$

where the coefficients c_j *and* $A_{a_j}(h)$ *are, respectively, given in (2.9) and (2.10) below. In particular*

$$c_k = (a_1 \cdots a_k)^{-1} \quad and \quad c_{k-1} = \frac{a_1 + \cdots + a_k - k}{2a_1 \cdots a_k}.$$

Proof The generating function $F(z)$ of $\mathcal{D}_k(n)$ has 1 as a pole of order k, and roots of unity $\zeta = e_s(r)$ for some $(r, s) = 1$, as poles whose orders are $< k$. Indeed, roughly speaking, the multiplicity with which each point ζ occurs as a pole of F is the number of a_is that are divisible by s, and hence this number is $< k$ since $(a_1, \ldots, a_k) = 1$. Let us be more precise. Expanding $F(z)$ into partial fractions yields

$$F(z) = \sum_{j=1}^{k} \frac{c_j}{(1-z)^j} + \sum_{j=1}^{k} \sum_{h=1}^{a_j-1} \frac{A_{a_j}(h)}{1 - \zeta_{a_j}^{-h}z}. \tag{2.8}$$

Multiplying both sides by $(1 - z)^k$, differentiating $k - j$ times and letting $z \to 1$ entails

$$c_j = \frac{(-1)^{k-j}}{(k-j)!} \frac{\mathrm{d}^{k-j}}{\mathrm{d}z^{k-j}} \left(\frac{(1-z)^k}{(1-z^{a_1}) \cdots (1-z^{a_k})} \right) \bigg|_{z=1}. \tag{2.9}$$

Similarly, multiplying both sides by $1 - \zeta_{a_j}^{-h} z$ and taking limits as $z \to \zeta_{a_j}^h$ yield, for all $1 \leqslant j \leqslant k$ and $1 \leqslant h \leqslant a_j - 1$,

$$A_{a_j}(h) = \frac{1}{a_j} \prod_{\substack{\ell=1 \\ \ell \neq j}}^{k} \left(1 - \zeta_{a_j}^{a_\ell h} \right)^{-1}. \tag{2.10}$$

Now using (2.7) and comparing the coefficients of z^n yield the asserted identity.   □

When $k \geqslant 4$, it seems extremely difficult to calculate the coefficients c_j for $1 \leqslant j \leqslant k-2$ using (2.9), so that Theorem 2.7 is useless for large k.[4] On the other hand, this result implies immediately the next estimate, generalizing Proposition 2.10.

Corollary 2.2 *Let* $\{a_1, \ldots, a_k\}$ *be a fixed k-tuple of positive relatively prime integers. Then*

$$\mathcal{D}_k(n) \sim \frac{n^{k-1}}{(k-1)! \, a_1 \cdots a_k} \quad (n \to \infty).$$

2.3.5 The Barnes Zeta Function

Let $k \in \mathbb{Z}_{\geqslant 2}$ and $\mathfrak{a} = \{a_1, \ldots, a_k\}$ be a fixed k-tuple of positive integers. If D is a common multiple of $a_1, \ldots, a_k$, then it has long been known [6] that the denumerant $\mathcal{D}_k(n)$ is a quasi-polynomial of degree $k-1$ with period D, i.e.

$$\mathcal{D}_k(n) = d_{k-1}(n) n^{k-1} + \cdots + d_1(n) n + d_0(n),$$

with $d_{k-1}(n) \neq 0$ and $d_j(n + D) = d_j(n)$ for all $0 \leqslant j \leqslant k-1$. The *Barnes zeta function* associated to $\mathfrak{a}$ and to a real number $x > 0$ if defined by

$$\zeta_{\mathfrak{a}}(s; x) := \sum_{n=0}^{\infty} \frac{\mathcal{D}_k(n)}{(n+x)^s} = \sum_{n_1, \ldots, n_k = 0}^{\infty} \frac{1}{(n_1 a_1 + \cdots + n_k a_k + x)^s}.$$

---

[4]See Exercise 17 for the case $k = 3$.

This function generalizes the classical Hurwitz zeta function

$$\zeta(s; x) := \sum_{n=0}^{\infty} \frac{1}{(n+x)^s} \qquad (\sigma > 1,\ x > 0).$$

From Proposition 2.11, $\zeta_a(s; x)$ converges in the half-plane $\sigma > k$. It can be shown that it can be analytically continued to a meromorphic function having poles at most in the set $\{1, \ldots, k\}$. The Dirichlet series $L(s, \mathcal{D}_k)$ of $\mathcal{D}_k(n)$, also denoted by $\zeta_a(s)$, is given by

$$\zeta_a(s) := \sum_{n=1}^{\infty} \frac{\mathcal{D}_k(n)}{n^s} = \lim_{x \searrow 0} \left(\zeta_a(s; x) - x^s\right) \qquad (\sigma > k).$$

It has been proved in [21] the following result, which relates all the parameters above.

Proposition 2.12 *With the notation above*

$$\zeta_a(s) = \sum_{j=0}^{k-1} \sum_{n=1}^{D} d_j(n) D^{j-s} \zeta\left(s - j; D^{-1}n\right).$$

Exercises

12 (Thue's Lemma). Let $a > 1$ be an integer and $p \geqslant 3$ be a prime number such that $p \nmid a$. Prove that the equation $au \equiv v \pmod{p}$ has a solution $(u, v) \in \mathbb{Z}^2$ such that $1 \leqslant |u|, |v| < \sqrt{p}$.

13 Let n be any positive integer. Compute $\mathcal{D}_{(1,1,2)}(n)$.

14 Determine the positive integers n such that the Diophantine equation

$$x + 2y + 7z = n$$

has exactly 225 solutions in non-negative integers x, y and z.

15 *Infinite sequence of coprime integers* [10]. Let u_1 be an odd integer and (u_n) be the sequence of integers defined by $u_n = u_{n-1}^2 - 2$ for any $n \geqslant 2$. Prove that, for any integers $n \geqslant 3$ and $r = 2, \ldots, n-1$, we have $(u_n, u_{n-r}) = 1$.

16   Let $a, b, c, d \in \mathbb{Z}$. Assume that, for all $m, n \in \mathbb{Z}$, there exist $x, y \in \mathbb{Z}$ such that $ax + by = m$ and $cx + dy = n$. Prove that $|ad - bc| = 1$.

17   Let $a, b, c \in \mathbb{Z}_{\geqslant 1}$ such that $(a, b) = (b, c) = (c, a) = 1$ and let $n \in \mathbb{Z}_{\geqslant 1}$. Prove that

$$\mathcal{D}_{(a,b,c)}(n) = \frac{n(n + a + b + c)}{2abc} + 1 - \left(\frac{S_a}{a} + \frac{S_b}{b} + \frac{S_c}{c} \right),$$

where

$$S_a = \sum_{h=1}^{a-1} \frac{1 - \zeta_a^{-nh}}{(1 - \zeta_a^{bh})(1 - \zeta_a^{ch})}$$

$$S_b = \sum_{h=1}^{b-1} \frac{1 - \zeta_b^{-nh}}{(1 - \zeta_b^{ch})(1 - \zeta_b^{ah})}$$

$$S_c = \sum_{h=1}^{c-1} \frac{1 - \zeta_c^{-nh}}{(1 - \zeta_c^{ah})(1 - \zeta_c^{bh})},$$

and $\zeta_m = e_m(1) = e^{2i\pi/m}$.

18   (Brauer). Let $a, b > 0$ be coprime integers and $c \in \mathbb{Z}_{\geqslant 1}$ such that $a \nmid c$ and $b \nmid c$. Prove that c is of exactly one of the following forms: $c = ax + by$ or $c = ab - ax - by$, with $x, y > 0$.

19   Let $a, b > 0$ be coprime integers and $c \in \mathbb{Z}_{\geqslant 1}$. Prove that[5]

$$c \notin \langle a, b \rangle \iff c = ab - ax - by, \ x, y > 0.$$

20   Let $a, b > 0$ be coprime integers and $c \in \mathbb{Z}_{\geqslant 1}$. Prove that

$$g(a, b) = g(a, b, c) \iff c \in \langle a, b \rangle.$$

21   Let $a, b > 0$ be coprime integers and $c \in \mathbb{Z}_{\geqslant 1}$. Prove that there exist $m, n \in \mathbb{Z}_{\geqslant 0}$ such that

$$g(a, b, c) = g(a, b) - ma - nb.$$

---

[5] See (2.2) for the notation $\langle a, b \rangle$.

22 A safe can only be opened with a code made up of three integers a, b, c satisfying the conditions of Theorem 2.1, such that $5 < a < b < 50 < c < 150$, and the Frobenius number of the set $\{a, b, c\}$ is equal to 234. Can you open the safe?

23 Let a, b be positive integers such that $(a, b) = 1$ and let n be a positive integer. Use Theorem 2.6 with its notation to prove that

$$r = 0 \Longleftrightarrow aa' + bb' \leqslant ab$$

$$r = 1 \Longleftrightarrow aa' + bb' > ab.$$

24 (Beck and Robins). Let a, b be positive integers such that $(a, b) = 1$ and let n be a positive integer. Set $\bar{a}, \bar{b}$ the positive integers such that $a\bar{a} \equiv 1 \pmod{b}$ and $b\bar{b} \equiv 1 \pmod{a}$. In 1953, Popoviciu [19] proved the following very elegant formula

$$\mathcal{D}_2(n) = \frac{n}{ab} + 1 - \left\{ \frac{n\bar{b}}{a} \right\} - \left\{ \frac{n\bar{a}}{b} \right\},$$

where $\{x\}$ is the fractional part of x. The aim of this exercise is to provide an analytic proof of this formula. Recall that $e_a(k) = e^{2\pi i k/a}$ and define the complex-valued function f by

$$f(z) = \frac{1}{z^{n+1}(1 - z^a)(1 - z^b)}.$$

(a) Recall that

$$\sum_{j=0}^{a-1} x^j = \prod_{j=1}^{a-1} (x - e_a(j)). \tag{2.11}$$

Prove that, for positive integers a, k, we have

$$\sum_{j=1}^{a-1} \frac{1}{1 - e_a(j)} = \frac{a-1}{2}$$

$$\prod_{\substack{j=1 \\ j \neq k}}^{a} \frac{1}{e_a(k) - e_a(j)} = \frac{e_a(k)}{a}.$$

(b) Use the generating function $F(z)$ of $\mathcal{D}_2(n)$ to prove that

$$\operatorname*{Res}_{z=0} f(z) = \mathcal{D}_2(n)$$

and compute the residues at all non-zero poles of f.

(c) Prove that

$$\lim_{R \to \infty} \frac{1}{2\pi i} \oint_{|z|=R} f(z)\, dz = 0$$

and deduce the identity (2.4).

(d) Check that if $b = 1$ then we have

$$\mathcal{D}_2(n) = \frac{n}{a} - \left\{\frac{n}{a}\right\} + 1.$$

Deduce that

$$\frac{1}{a} \sum_{k=1}^{a-1} \frac{1}{e_a(kn)(1 - e_a(k))} = \frac{1}{2}\left(1 - \frac{1}{a}\right) - \left\{\frac{n}{a}\right\}.$$

(e) By noticing that

$$\sum_{k=1}^{a-1} \frac{1}{e_a(kn)(1 - e_a(kb))} = \sum_{k=1}^{a-1} \frac{1}{e_a(kn\bar{b})(1 - e_a(k))}$$

finalizes the proof of Popoviciu's identity.

References

1. Agnarsson, G.: On the Sylvester denumerants for general restricted partitions. Congr. Numer. **2002**, 49–60 (2002)
2. Aliev, I.M., Gruber, P.M.: An optimal lower bound for the Frobenius problem. J. Number Theory **123**, 71–79 (2007)
3. Beck, M., Robins, S. A formula related to the Frobenius problem in 2 dimensions. In: Chudnovsky, D., Chudnovsky, G., Nathanson, M. (eds.) Number Theory. New York Seminar, vol. 2003, pp 17–23. Springer, Berlin (2004)
4. Beck, M., Zacks, S.: Refined upper bounds for the linear Diophantine problem of Frobenius. Adv. Appl. Math. **32**, 454–467 (2004)
5. Beck, M., Diaz, R., Robins, S.: The Frobenius problem, rational polytopes, and Fourier–Dedekind sums. J. Number Theory **96**, 1–21 (2002)
6. Bell, E.T.: Interpolated denumerants and Lambert series. Am. J. Math. **65**, 382–386 (1943)
7. Brauer, A, Shockley, J.E.: On a problem of Frobenius. J. Reine Angew. Math. **211**, 215–220 (1962)

8. Curtis, F.: On formulas for the Frobenius problem of a numerical semi-group. Math. Scand. **67**, 190–192 (1990)
9. Davison, J.L.: On the linear Diophantine problem of Frobenius. J. Number Theory **48**, 353–363 (1990)
10. Edwards, A.W.F.: Infinite coprime sequences. Math. Gazette **48**, 416–422 (1964)
11. Ehrhart, E: Sur un problème de géométrie diophantienne linéaire. II. J. Reine Angew. Math. **227**, 25–49 (1967)
12. Erdős, P., Graham, R.L.: On a linear Diophantine problem of Frobenius. Acta Arith. **21**, 399–408 (1972)
13. Fukshansky, L., Robins, S.: Frobenius problem and the covering radius of a lattice. Discret. Comput. Geom. **37**, 471–483 (2007)
14. Gould, H.W.: Combinatorial Identities. A Standardized Set of Tables Listing 500 Binomial Coefficient Summations. Morgantown (1972)
15. Killingbergtrø, H.G.: Betjening av figur i Frobenius' problem. Normat-Nordisk Mat Tidsskrift **48**, 75–82 (2000)
16. Murty, V.N.: Counting the integer solutions of a linear equation with unit coefficients. Math. Mag. **54**, 79–81 (1981)
17. Nathanson, M.B.: Partitions with parts in a finite set. Proc. Am. Math. Soc. **128**, 1269–1273 (2000)
18. Odlyzko, A.M.: Asymptotic Enumeration Methods. Elsevier, Amsterdam (1995)
19. Popoviciu, T.: Asupra unei probleme de patitie a numerelor. Acad Republici Populare Romane, Filiala Cluj, Studii si Cercetari Stiintifice **4**, 7–58 (1953)
20. Rödseth, O.J.: Two remarks on linear forms in non-negative integers. Math. Scand. **51**, 193–198 (1982)
21. Cimpoeaş, M., Nicolae, F.: On the restricted partition function. Ramanujan J. **47**, 565–588 (2018)
22. Selmer, E.S.: On the linear Diophantine problem of Frobenius. J. Reine Angew. Math. **293/294**, 1–17 (1977)
23. Shiu, P.: Moment sums associated with linear binary forms. Am. Math. Mon. **113**, 545–550 (2006)
24. Trimm, J.E.: On Frobenius Numbers in Three Variables. Ph.D. thesis, Auburn University, Auburn (2006)
25. Tripathi, A.: The number of solutions to $ax + by = n$. Fibonacci Q. **38**, 290–293 (2000)
26. Tripathi, A.: Formulae for the Frobenius number in three variables. J. Number Theory **170**, 368–389 (2017)
27. Vitek, Y.: Bounds for a linear Diophantine problem of Frobenius. J. Lond. Math. Soc. **2**, 79–85 (1975)
28. Wilf, H.S. Generatingfunctionology. Academic Press, New York (1990)

Chapter 3
Prime Numbers

The study of prime numbers is one of the main branches of number theory. The literature, very abundant, goes back to Pythagoras and, above all, to Euclid who was the first to show there are infinitely many prime numbers. One can notice now that Euclid's ideas can be used and generalized to prime numbers belonging to certain arithmetic progressions as seen in Sect. 3.6.

This very fruitful idea of prime numbers does not belong solely to the branch of number theory. The reader knowing finite group theory will certainly have noticed the analogy between prime numbers and the so-called *simple groups*.[1] Thus, it is not surprising to notice that these areas of mathematics do really interlace. For example, the decomposition of a positive integer into prime powers also exists in finite group theory, under a slightly different form, and is called the *Jordan–Hölder theorem* which states that given a finite group G, there exists a sequence

$$G = G_0 \supset G_1 \supset \cdots \supset G_{r-1} \supset G_r = \{e_G\}$$

of subgroups G_i of G such that G_{i+1} is a normal subgroup of G_i and the group G_i/G_{i+1} is simple for $i = 0, \ldots, r - 1$. This decomposition is unique, apart from the order of the subgroups. On the other hand, if G is *Abelian* with $|G| = p_1^{\alpha_1} \cdots p_r^{\alpha_r}$, then

$$G \simeq H_1 \oplus \cdots \oplus H_r \qquad (3.1)$$

where the H_i are the p_i-Sylow subgroups of G with $|H_i| = p_i^{\alpha_i}$. This shows how algebra and arithmetic can be related.

---

[1] A non-trivial group G is said to be *simple* if it has no normal subgroup other than $\{e_G\}$ and G itself, where e_G is the identity element of G.

© Springer Nature Switzerland AG 2020
O. Bordellès, *Arithmetic Tales*, Universitext,
https://doi.org/10.1007/978-3-030-54946-6_3

3.1 Primitive Roots

3.1.1 Multiplicative Order

Euler–Fermat's Theorem 2.3 implies the following notion.

Proposition 3.1 (Multiplicative Order) *Let $a \in \mathbb{Z}$ and n be a positive integer such that $(a, n) = 1$. Then the set of positive integers m satisfying $a^m \equiv 1 \pmod{n}$ has a smallest element, denoted by $\mathrm{ord}_n(a)$ and called the* multiplicative order of a *modulo n.*

Proof The set in question is a non-empty subset of $\mathbb{Z}_{\geqslant 0}$ since $\varphi(n)$ belongs to this set by Theorem 2.3. □

The following result gives a characterization of the multiplicative order of a positive integer.

Proposition 3.2 *Let $a \in \mathbb{Z}$ and n be a positive integer such that $(a, n) = 1$. Then*

$$a^m \equiv 1 \pmod{p} \iff \mathrm{ord}_n(a) \mid m.$$

In particular, $\mathrm{ord}_n(a) \mid \varphi(n)$. Furthermore, if $a^t \equiv a^s \pmod{n}$, then $\mathrm{ord}_n(a) \mid t - s$.

Proof Set $\ell := \mathrm{ord}_n(a)$. If $m = k\ell$, then $a^m = (a^\ell)^k \equiv 1 \pmod{n}$. Conversely, define $S = \{m \in \mathbb{Z}_{\geqslant 1} : a^m \equiv 1 \pmod{n}\}$. By Proposition 3.1, ℓ is the smallest element in S. Let (q, r) be the unique pair of positive integers such that $m = q\ell + r$ and $0 \leqslant r < \ell$. We derive $1 \equiv a^m \equiv a^r (a^\ell)^q \equiv a^r \pmod{n}$. Hence, if $r \geqslant 1$, we infer that $r \in S$ and thus $r \geqslant \ell$, contradicting the inequality $r < \ell$. This implies that $r = 0$ and then $\ell \mid m$, and hence $\ell \mid \varphi(n)$ from Theorem 2.3. Finally, suppose for example that $t > s$. Since $a^t - a^s = a^s(a^{t-s} - 1) \equiv 0 \pmod{n}$, we get $a^s \equiv 0 \pmod{n}$ or $a^{t-s} \equiv 1 \pmod{n}$. Since $(a, n) = 1$, the first congruence is never achieved. Hence we obtain $a^{t-s} \equiv 1 \pmod{n}$ and therefore ℓ divides $t - s$ according to the previous statement. □

Lemma 3.1 *Let a, b, n be positive integers and set $\mathrm{ord}_n(a) := k$ and $\mathrm{ord}_n(b) := \ell$. Assume $(k, \ell) = 1$. Then*

$$\mathrm{ord}_n(ab) = \mathrm{ord}_n(a) \times \mathrm{ord}_n(b).$$

Proof Set $h := \mathrm{ord}_n(ab)$. First note that $a^{h\ell} \equiv (ab)^{h\ell} \equiv ((ab)^h)^\ell \equiv 1 \pmod{n}$, so that $k \mid h\ell$ by Proposition 3.2. Since $(k, \ell) = 1$, we get $k \mid h$. Swapping the roles of a and b also yields $\ell \mid h$. Since $(k, \ell) = 1$, we derive $k\ell \mid h$. But $(ab)^{k\ell} = (a^k)^\ell \times (b^\ell)^k \equiv 1 \pmod{n}$, and hence $h \mid k\ell$. Therefore $h = k\ell$ as required. □

Lemma 3.2 *Let p be a prime number. If $d \mid p - 1$, then the congruence $x^d \equiv 1 \pmod{p}$ has exactly d solutions.*

Proof Using Lagrange's theorem we know that this congruence has *at most* d solutions. Let us prove that it has *at least* d solutions. Suppose there are $< d$ incongruent solutions modulo p. Since

$$x^{p-1} - 1 = (x^d - 1)P(x),$$

where $P \in \mathbb{Z}[x]$ has degree $p - 1 - d$, $a \in \mathbb{Z}$ is solution of the equation $x^{p-1} - 1 \equiv 0 \pmod{p}$ if and only if a is solution of $x^d - 1 \equiv 0 \pmod{p}$ or $P(x) \equiv 0 \pmod{p}$. Lagrange's theorem implies that the congruence $P(x) \equiv 0 \pmod{p}$ has at most $p - 1 - d$ solutions. Hence $x^{p-1} - 1 \equiv 0 \pmod{p}$ has $< d + p - 1 - d = p - 1$ solutions, which contradicts Theorem 2.3.   □

3.1.2 Primitive Roots

3.1.2.1 Introduction

Let p be a prime number satisfying $p \neq 2$ and $p \neq 5$. The decimal expansion of $\frac{1}{p}$ is then purely periodic. For example

$$\tfrac{1}{7} = 0.142857142857\ldots \quad \text{and} \quad \tfrac{1}{11} = 0.090909\ldots$$

We shall have a look at the length of the period of the decimal expansion of $\frac{1}{p}$. Since $p \nmid 10$, Proposition 3.1 implies that $\delta = \mathrm{ord}_p(10)$ is well-defined. Hence there exists an integer q such that $10^\delta = 1 + qp$. Therefore

$$\frac{1}{p} = \frac{q}{10^\delta - 1} = q \sum_{k=1}^{\infty} 10^{-\delta k}$$

so that the length of the period of the decimal expansion of $\frac{1}{p}$ is equal to $\delta = \mathrm{ord}_p(10)$. In particular, this length[2] divides $p - 1$ by Proposition 3.2. Therefore one can ask when the period of the decimal expansion of $\frac{1}{p}$ has its maximal length, in other words when $\mathrm{ord}_p(10) = p - 1$. This leads to the following definition.

Definition 3.1 Let n, g be positive integers such that $(g, n) = 1$. The integer g is called *primitive root* modulo n if $\mathrm{ord}_n(g) = \varphi(n)$.

The examples above show that 10 is a primitive root modulo 7, but is not a primitive root modulo 11.

---

[2]This problem has a long history as it goes back at least to J. H. Lambert in 1769 [20, Chapter VI].

3.1.2.2 Basic Facts

Lemma 3.3 *Let* $(g, n) = 1$. *Then* g *is a primitive root modulo* n *if and only if the set* $S := \{g, g^2, \ldots, g^{\varphi(n)}\}$ *is a reduced residue system modulo* n.

Proof First note that every element of S is relatively prime to n, so that S is a reduced residue system modulo n if and only if these powers of g are all distinct modulo n, in which case $\mathrm{ord}_n(g) = \varphi(n)$ since $g^{\varphi(n)} \equiv 1 \pmod{n}$ by Theorem 2.3. On the other hand if S is not a reduced residue system modulo n, say $g^k \equiv g^\ell \pmod{n}$ with $1 \leqslant k < \ell \leqslant \varphi(n)$, then $g^{\ell-k} \equiv 1 \pmod{n}$ with $\ell-k < \varphi(n)$ and so $\mathrm{ord}_n(g) < \varphi(n)$. □

Lemma 3.4 *Let* g *be a primitive root modulo* n *and* $k \in \mathbb{Z}_{\geqslant 1}$. *Then* g^k *is also a primitive root modulo* n *if and only if* $(k, \varphi(n)) = 1$.

Proof Set $\ell := \mathrm{ord}_n(g^k)$ and assume first that $(k, \varphi(n)) = 1$. Since $g^{k\ell} \equiv 1 \pmod{n}$, we get $\varphi(n) \mid k\ell$ by Proposition 3.2, and hence $\varphi(n) \mid \ell$ since $(k, \varphi(n)) = 1$. On the other hand, since $(g^k)^{\varphi(n)} \equiv 1 \pmod{n}$, we similarly derive $\ell \mid \varphi(n)$, and thus $\ell = \varphi(n)$, proving that g^k is a primitive root modulo n. Conversely, assume that $(k, \varphi(n)) := d > 1$. Then $(g^k)^{\varphi(n)/d} = (g^{\varphi(n)})^{k/d} \equiv 1 \pmod{n}$, implying that $\ell \leqslant \varphi(n)/d < \varphi(n)$, so that g^k is not a primitive root modulo n. □

We now are in a position to prove the main result of this section.

Proposition 3.3 *Let* n *be a positive integer and assume that there exists a primitive root modulo* n. *Then the total number of primitive roots modulo* n *is equal to* $\varphi(\varphi(n))$.

Proof Assume that g is a primitive root modulo n. By Lemma 3.3, any other primitive root modulo n is found among the members of the set $S := \{g, g^2, \ldots, g^{\varphi(n)}\}$. Now by Lemma 3.4, the number of powers g^k, $1 \leqslant k \leqslant \varphi(n)$, that have order $\varphi(n)$ is equal to the number of integers k such that $(k, \varphi(n)) = 1$, yielding the asserted result. □

3.1.2.3 Primitive Roots Modulo a Prime

There are essentially three questions about primitive roots. First of all, for which prime number p does there exist a primitive root modulo p? The next result provides a complete answer to this question.

Proposition 3.4 *Let* p *be a prime number. Then there exists a primitive root modulo* p. *According to Proposition 3.3, there are* $\varphi(p-1)$ *primitive roots modulo* p.

Proof Since 1 is a primitive root modulo 2, we may assume $p \geqslant 3$. Let q^α be a prime power dividing $p - 1$. We will first show that there exists an integer m such that $\mathrm{ord}_p(m) = q^\alpha$. By Lemma 3.2, we are led to consider the q^α zeros of the polynomial $x^{q^\alpha} - 1$ modulo p. If m is one of these zeros, then $m^{q^\alpha} \equiv 1 \pmod{p}$, and hence $\operatorname{ord}_p(m) \mid q^\alpha$. Therefore $\operatorname{ord}_p(m) = q^j$ for some $0 \leqslant j \leqslant \alpha$. If $j < \alpha$, then m would be a zero of the polynomial $x^{q^{\alpha-1}} - 1$ modulo p, having only $q^{\alpha-1}$ zeros by Lemma 3.2 again. Hence there are $q^\alpha - q^{\alpha-1} = \varphi(q^\alpha) \geqslant 1$ choices for an integer m such that $\operatorname{ord}_p(a) = q^\alpha$. Now write $p - 1 = q_1^{\alpha_1} \cdots q_r^{\alpha_r}$. By the argument above, for each $q_j^{\alpha_j}$, there exists a number m_j such that $\operatorname{ord}_p(m_j) = q_j^{\alpha_j}$. By Lemma 3.1, the number $g := m_1 \cdots m_r$ has order $q_1^{\alpha_1} \cdots q_r^{\alpha_r} = p - 1$ modulo p, i.e. g is a primitive root modulo p.   □

Another proof will be supplied in Theorem 4.2 as a consequence of the Möbius inversion formula.

3.1.3 Artin's Conjecture

3.1.3.1 Genesis of the Conjecture

In view of the above result, we may ask the following questions.

▷ For a fixed prime p, how many primitive roots are there?
▷ For a fixed integer a, for how many primes will a be a primitive root?

The answer to the first question is given by Proposition 3.4 above. The second question above is far more difficult. Let us first see this with an example, where we ask whether 2 is a primitive root modulo p. With the help of tables of primitive roots, one can see, among primes $p \leqslant 100$, that

▷ 2 is a primitive root modulo 3, 5, 11, 13, 19, 29, 37, 53, 59, 61, 67, 83.
▷ 2 is not a primitive root modulo 7, 17, 23, 31, 41, 43, 47, 71, 73, 79, 89, 97.

Since each list contains 12 elements, one can think that the probability that 2 is a primitive root modulo p is 50%. In fact, extending the list, one should see that this probability could approach a number closer to 37%. More precisely, in 1927, Emil Artin formulated the following conjecture.

Conjecture 3.1 (Artin) Let a be an integer such that $a \neq 0, \pm 1$. If a is not a square, then there are infinitely many prime numbers p such that a is a primitive root modulo p.

This conjecture comes from the following more precise statement also given by Artin.

Conjecture 3.2 (Artin) Let a be a non-zero integer such that $a \neq \pm 1$ and h be the largest integer such that $a = a_0^h$ with $a_0 \in \mathbb{Z}$. If $N_a(x)$ is the number of primes $p \leqslant x$ such that a is a primitive root modulo p, then

$$N_a(x) = \mathcal{A}_h \frac{x}{\log x} + o\left(\frac{x}{\log x}\right),$$

where

$$\mathcal{A}_h = \prod_{p \nmid h} \left(1 - \frac{1}{p(p-1)} \right) \prod_{p \mid h} \left(1 - \frac{1}{p-1} \right). \tag{3.2}$$

Note that the product above is equal to zero when h is even. Hence, the condition that a is not a square is *necessary* for the number of prime numbers p, such that a is a primitive root modulo p, to be infinite. When h is odd, then the product is a positive rational multiple of the so-called *Artin's constant*

$$\mathcal{A} = \mathcal{A}_1 = \sum_{n=1}^{\infty} \frac{\mu(n)}{n \varphi(n)} = \prod_{p} \left(1 - \frac{1}{p(p-1)} \right) \approx 0.373955813619\ldots,$$

where μ is the Möbius function.[3] Artin's conjecture is known to be true as soon as a very difficult hypothesis is proved as seen in Sect. 3.1.3.3 below, but the latter hypothesis is likely to be out of reach for several generations of researchers.

3.1.3.2 Heuristics

Unconditionally, there is no known number a for which the set $S(a)$ of prime numbers p, such that a is a primitive root modulo p, is known to be infinite. Nevertheless, improving on fundamental work by Gupta and Ram Murty [36], Heath-Brown [39] succeeded in showing that there are at most two "bad" prime numbers a such that $S(a)$ is finite, and also at most three "bad" squarefree numbers a such that $S(a)$ is finite. Furthermore, he proved that the set S of integers for which Artin's conjecture does not hold is rather thin by establishing the following estimate:

$$\sum_{\substack{n \leqslant x \\ n \in S}} 1 \ll (\log x)^2.$$

Let us turn our attention for a while to the heuristic arguments that led to Artin's conjecture. A necessary and sufficient condition for a to be a primitive root modulo p is

$$a^{(p-1)/q} \not\equiv 1 \pmod{p} \quad \text{for all primes } q \text{ dividing } p - 1.$$

Indeed, the condition is clearly sufficient by Definition 3.1. It is also necessary by using the following argument. Suppose that $a^{(p-1)/q} \not\equiv 1 \pmod{p}$ for all primes $q \mid p - 1$ and that a is not a primitive root mod p. Then $a^{(p-1)/k} \equiv 1 \pmod{p}$ for some

---

[3] See Chap. 4.

integer $k > 1$ such that $k \mid p - 1$, which in turn implies that $a^{(p-1)/r} \equiv 1 \pmod{p}$ for some prime divisor r of k, resulting in a contradiction.

Hence the heuristic idea is that a is a primitive root mod p if the two following events do not occur simultaneously for any prime q:

$$E_1 : a^{(p-1)/q} \equiv 1 \pmod{p}.$$

$$E_2 : p \equiv 1 \pmod{q}.$$

Now by Dirichlet's Theorem 3.14, the probability of E_2 is equal to $1/(q-1)$, while the probability of E_1 is equal to $\frac{1}{q}$ if $q \nmid h$ and to 1 if $q \mid h$, since, in the first case, we note that $a^{(p-1)/q}$ is a solution of the equation $x^q \equiv 1 \pmod{p}$ and we want a solution congruent to 1 $\pmod{q}$ among the q expected roots, whereas in the second case we always have $a^{(p-1)/q} = a_0^{h(p-1)/q} \equiv 1 \pmod{p}$. Therefore, if we assume that these events are independent, then the probability that both occur is

$$\frac{1}{q(q-1)} \text{ if } q \nmid h \quad \text{and} \quad \frac{1}{q-1} \text{ if } q \mid h$$

leading to the product (3.2).

3.1.3.3 Further Developments

Artin's conjecture is one of the richest problems in number theory, notably on account of the connection between algebraic number theory and analytic number theory. It can be shown that the two conditions $a^{(p-1)/q} \equiv 1 \pmod{p}$ and $p \equiv 1 \pmod{q}$ are equivalent to the fact that p splits completely in Galois extensions $\mathbb{K}_q = \mathbb{Q}(\zeta_q, a^{1/q})$ of $\mathbb{Q}$, where $\zeta_q = e_q(1)$ is a qth primitive root of unity. In the 1930s, great efforts were made to prove this conjecture. In 1935, Erdős tried in vain to combine infinitely many q. Finally, using a special form of Chebotarëv's density theorem[4] with a sharp error term only accessible under the extended Riemann hypothesis for the number fields $\mathbb{K}_q$, Hooley [41] succeeded in proving a somewhat modified conjecture. Indeed, in 1957, Derrick H. Lehmer and his wife Emma numerically tested Artin's conjecture on a computer and found that the naive heuristic approach described above may be wrong in some cases. Artin then introduced a correction factor to explain these discrepancies,[5] and the corrected

---

[4]See Theorem 7.38.

[5]"So I was careless but the machine caught up with me". This is how Artin concluded the fourth letter dated January 28, 1958 to Emma Lehmer, explaining that his conjecture may be false if a is a prime number satisfying $a \equiv 1 \pmod{4}$.

conjectured density under ERH[6] finally turns out to be

$$\sum_{n=1}^{\infty} \frac{\mu(n)}{[\mathbb{K}_n : \mathbb{Q}]},$$

where $[\mathbb{K}_n : \mathbb{Q}]$ is the degree of the number field $\mathbb{K}_n = \mathbb{Q}(\zeta_n, a^{1/n})$. If the discriminant[7] $d_{\mathbb{K}_2}$ of the quadratic field $\mathbb{K}_2 = \mathbb{Q}\left(\sqrt{a}\right)$ is even, then the function $n \mapsto [\mathbb{K}_n : \mathbb{Q}]$ is multiplicative and the expression above reduces to the product $\mathscr{A}_h$ given in (3.2), but when $d_{\mathbb{K}_2}$ is odd, then, for all squarefree n divisible by $d_{\mathbb{K}_2}$, we have

$$[\mathbb{K}_n : \mathbb{Q}] = \tfrac{1}{2} \prod_{p \mid n} [\mathbb{K}_p : \mathbb{Q}]$$

so that the function $n \mapsto [\mathbb{K}_n : \mathbb{Q}]$ is no longer multiplicative. Hooley derived the following expression:

$$\sum_{n=1}^{\infty} \frac{\mu(n)}{[\mathbb{K}_n : \mathbb{Q}]} = \left\{ 1 - \mu\left(|d_{\mathbb{K}_2}|\right) \prod_{p \mid d_{\mathbb{K}_2}} \left([\mathbb{K}_p : \mathbb{Q}] - 1\right)^{-1} \right\} \mathscr{A}_h$$

when $d_{\mathbb{K}_2}$ is odd. The reader interested in Artin's primitive root conjecture should refer to the very fruitful presentation [82] of this subject.

3.1.4 Power Residues

We begin with a first result which has its own interest in itself.

Lemma 3.5 *Let p be a prime number and g a primitive root* mod p. *Then* 1, g, g^2, $\ldots$, g^{p-2} *are incongruent modulo p, and hence are congruent to* 1, 2, $\ldots$, $p - 1$ *in a certain order.*

Proof It suffices to show that, if $0 \leqslant i < j \leqslant p - 2$, then $p \nmid (g^j - g^i)$. Suppose the contrary. Then we have $p \mid g^i \left(g^{j-i} - 1\right)$, and hence $p \mid \left(g^{j-i} - 1\right)$, otherwise p divides g^{p-1} contradicting the equality $g^{p-1} \equiv 1 \pmod{p}$. By using

---

[6]Similarly to the Riemann hypothesis, the *Extended Riemann hypothesis* asserts that the non-trivial zeros of the Dedekind zeta function $\zeta_{\mathbb{K}}$ attached to a number field $\mathbb{K}$ are all on the "critical" line $\sigma = \frac{1}{2}$. It should be mentioned that ERH is sometimes called GRH, for *Generalized Riemann Hypothesis*. However, GRH is referred to in this book as the conjecture asserting that the Dirichlet L-functions have *all* their zeros lying on the line $\sigma = \frac{1}{2}$. See Footnote 45 in Chap. 7.

[7]For the definitions of the usual invariants of a number field, see Chap. 7.

Proposition 3.2, we have $(p - 1) \mid (j - i)$ contradicting the fact that $0 < j - i < p - 1$. □

Definition 3.2 Let p be a prime number and $n \geq 2$ be an integer. An integer a is said to be an *nth power residue* mod p if the congruence $x^n \equiv a \pmod{p}$ has a solution in $\{1, \ldots, p - 1\}$. When $n = 2$, a is a *quadratic residue* mod p.

It is clear that 0 is an *n*th power residue mod p for every prime p and integer $n \geq 2$. On the other hand, if a is an *n*th power residue mod p, then so is every integer congruent to $a \pmod{p}$. The following result gives a useful criterion and also the number of incongruent *n*th power residues $\pmod{p}$.

Proposition 3.5 *Let $n \geq 2$ be an integer, p be a prime number and set $d = (n, p - 1)$.*

▷ *Let a be an integer satisfying $p \nmid a$. Then*

$$a \text{ is an nth power residue modulo } p \iff a^{(p-1)/d} \equiv 1 \pmod{p}.$$

▷ *The number of non-zero integers which are nth power residues mod p is given by $\frac{1}{d}(p - 1)$.*

Proof Let p be a prime and g be a primitive root $\pmod{p}$.

▷ If an integer a is such that $p \nmid a$ and is an *n*th power residue mod p, then there exists a number x such that $p \nmid x$ and $a \equiv x^n \pmod{p}$, so that we have

$$a^{(p-1)/d} \equiv (x^n)^{(p-1)/d} \equiv (x^{p-1})^{n/d} \equiv 1 \pmod{p}$$

by Theorem 2.3.
   Conversely, suppose that $a^{(p-1)/d} \equiv 1 \pmod{p}$. By Lemma 3.5, there exists an integer $k \in \{0, \ldots, p - 2\}$ such that $a \equiv g^k \pmod{p}$, and hence $g^{k(p-1)/d} \equiv 1 \pmod{p}$, which implies that $(p - 1) \mid k(p - 1)/d$ since g is a primitive root mod p. Therefore we have $d \mid k$, so that $k = dh$ for some $h \in \mathbb{Z}_{\geq 0}$. By Bézout's Theorem, there exist two non-negative integers u, v such that $d = nu - v(p - 1)$, so that we get $k = dh = nhu - vh(p - 1)$. By the use of the relation $g^{p-1} \equiv 1 \pmod{p}$, we get

$$a \equiv a\left(g^{p-1}\right)^{vh} \equiv g^k g^{vh(p-1)} \equiv g^{nhu} \equiv \left(g^{hu}\right)^n \pmod{p}$$

and therefore a is an *n*th power residue mod p.
▷ If a is an *n*th power residue mod p, then $a^{(p-1)/d} \equiv 1 \pmod{p}$, hence using Lagrange's theorem, there are at most $(p - 1)/d$ non-zero *n*th power residues mod p.

On the other hand, the numbers g^n, g^{2n}, ..., $g^{n(p-1)/d}$ are $(p-1)/d$ non-zero incongruent nth power residues mod p. Indeed, set $n = dh$ and $p - 1 = dk$ with $(h, k) = 1$, and let $1 \leqslant i < j \leqslant k$ be two integers. Since $(p, g) = 1$, if $g^{in} \equiv g^{jn} \pmod{p}$, then $g^{(j-i)n} \equiv 1 \pmod{p}$, so that $(p - 1) \mid n(j - i)$ and hence $k \mid (j - i)$ by Gauss's theorem, contradicting the fact that $1 \leqslant i < j \leqslant k$. Furthermore, it is clear that the equation $x^n \equiv g^{jn} \pmod{p}$ has the solution $x = g^j$. $\square$

Example 3.1 Let $p = 7$ and $n = 2$, so that $d = 2$. The number $g_7 = 3$ is the smallest primitive root mod 7, so that the three quadratic residues mod 7 are 1, 2 and 4.

Example 3.2 Prove that, if $p \equiv 3 \pmod{4}$, then -1 is not a quadratic residue mod p.

Solution 3.1   Suppose there exists a number x such that $x^2 \equiv -1 \pmod{p}$. We then have

$$x^{p-1} = (x^2)^{(p-1)/2} \equiv (-1)^{(p-1)/2} \equiv -1 \pmod{p}$$

which contradicts Theorem 2.3.

Example 3.3 Use Example 3.2 to show that the equation $y^2 = x^3 + 11$ has no solution in $\mathbb{Z}^2$.

Solution 3.2   Suppose there is a solution $(x, y) \in \mathbb{Z}^2$. Then we have $y^2 \equiv x^3 - 1 \pmod{4}$, and since 0 and 1 are the unique squares modulo 4, we get $x^3 \equiv 1$ or 2 $\pmod{4}$, and then $x \equiv 1 \pmod{4}$ since 2 is not a cube modulo 4. Note that, if $y^2 = x^3 + 11$, then

$$y^2 + 16 = x^3 + 27 = (x + 3)(x^2 - 3x + 9)$$

and since $x \equiv 1 \pmod{4}$, this gives $x^3 - 3x + 9 \equiv 3 \pmod{4}$. Hence there exists a prime number $p \equiv 3 \pmod{4}$ dividing $y^2 + 16$. If a is an integer such that $4a \equiv 1 \pmod{p}$, then the relation $y^2 + 16 \equiv 0 \pmod{p}$ implies that $(ay)^2 \equiv -1 \pmod{p}$. This contradicts the result of Example 3.2. Thus, the Diophantine equation $y^2 = x^3 + 11$ has no solution in $\mathbb{Z}^2$.

3.2   Elementary Prime Numbers Estimates

Between the years 1850 and 1875, a great number of estimates involving some functions of prime numbers were discovered by a few mathematicians including Chebyshev and Mertens. These results are the cornerstones of the whole research on primes which eventually led to the Prime Number Theorem in 1896. Although these estimates were not sufficient to show the PNT, they were, and still are, often used in many problems in number theory.

3.2.1   Chebyshev's Functions of Primes

The following functions are of constant use in analytic number theory.

Definition 3.3

▷ The *von Mangoldt function* Λ is defined by

$$\Lambda(n) = \begin{cases} \log p, & \text{if } n = p^\alpha \text{ for some prime } p \text{ and } \alpha \in \mathbb{Z}_{\geqslant 1} \\ 0, & \text{otherwise.} \end{cases}$$

▷ The *first Chebyshev function* θ is defined for $x \geqslant 2$ by

$$\theta(x) = \sum_{p \leqslant x} \log p.$$

▷ The *second Chebyshev function* Ψ is defined for $x \geqslant 2$ by

$$\Psi(x) = \sum_{n \leqslant x} \Lambda(n).$$

▷ The *prime counting function* π is defined by

$$\pi(x) = \sum_{p \leqslant x} 1.$$

If f is one of the three functions above, it is convenient to set $f(x) = 0$ for $0 < x < 2$.

We will make use of the next lemma.

Lemma 3.6

▷ *For all $x > 0$, we have*

$$\theta(x) \leqslant \Psi(x) \leqslant \theta(x) + \pi\left(\sqrt{x}\right)\log x.$$

▷ *For all $x \geqslant 2$ and all $a > 1$, we have*

$$\frac{\theta(x)}{\log x} \leqslant \pi(x) \leqslant \frac{a\theta(x)}{\log x} + \pi\left(x^{1/a}\right).$$

Proof

▷ One may suppose $x \geqslant 2$. We first have

$$\Psi(x) - \theta(x) = \sum_{p^\alpha \leqslant x} \log p - \sum_{p \leqslant x} \log p = \sum_{p \leqslant \sqrt{x}} \sum_{\alpha=2}^{\lfloor \log x / \log p \rfloor} \log p$$

so that $\Psi(x) \geqslant \theta(x)$. On the other hand

$$\Psi(x) - \theta(x) = \sum_{p \leqslant \sqrt{x}} \sum_{\alpha=2}^{\lfloor \log x / \log p \rfloor} \log p \leqslant \sum_{p \leqslant \sqrt{x}} \log p \left\lfloor \frac{\log x}{\log p} \right\rfloor \leqslant \sum_{p \leqslant \sqrt{x}} \log x = \pi\left(\sqrt{x}\right) \log x.$$

▷ We have

$$\pi(x) = \sum_{p \leqslant x} 1 = \sum_{p \leqslant x} \frac{\log p}{\log p} \geqslant \frac{1}{\log x} \sum_{p \leqslant x} \log p = \frac{\theta(x)}{\log x}.$$

For $2 \leqslant T < x$, we also have

$$\pi(x) = \sum_{p \leqslant T} 1 + \sum_{T < p \leqslant x} 1 = \pi(T) + \sum_{T < p \leqslant x} \frac{\log p}{\log p} \leqslant \pi(T) + \frac{\theta(x)}{\log T}$$

and the choice of $T = x^{1/a}$ implies the asserted estimate.   □

The von Mangoldt function satisfies an important identity which is what we shall call later a *convolution identity* (see Chap. 4).

Lemma 3.7 *For all $x \geqslant 1$, we have* $\sum_{d \leqslant x} \Lambda(d) \left\lfloor \dfrac{x}{d} \right\rfloor = \log\left(\lfloor x \rfloor!\right).$

Proof

▷ We first show that, for all $n \in \mathbb{Z}_{\geqslant 1}$, we have

$$\sum_{d | n} \Lambda(d) = \log n. \tag{3.3}$$

Indeed, the identity is clear for $n = 1$. Suppose $n = 2$ written as

$$n = p_1^{\alpha_1} \cdots p_k^{\alpha_k}$$

with distinct primes p_i and positive integers α_i. By the definition of the von Mangoldt function, it is sufficient to consider in the above sum the divisors d of n written as $d = p_i^{\beta_i}$ with $1 \leqslant \beta_i \leqslant \alpha_i$ for $i = 1, \ldots, k$. Thus we get

$$\sum_{d|n} \Lambda(d) = \sum_{i=1}^{k} \sum_{\beta_i=1}^{\alpha_i} \log p_i = \sum_{i=1}^{k} \log p_i^{\alpha_i} = \log n.$$

▷ One may suppose $x > 1$. The lemma then follows from the previous identity via

$$\log \left(\lfloor x \rfloor! \right) = \sum_{n \leqslant x} \log n = \sum_{n \leqslant x} \sum_{d|n} \Lambda(d) = \sum_{d \leqslant x} \Lambda(d) \sum_{k \leqslant x/d} 1 = \sum_{d \leqslant x} \Lambda(d) \left\lfloor \frac{x}{d} \right\rfloor,$$

where we interchanged the summations as in Proposition 4.3.   □

3.2.2 Chebyshev's Estimates

One of the main goals of Chebyshev was the proof of the Prime Number Theorem which can be stated as

$$\pi(x) \sim \frac{x}{\log x}$$

as $x \to \infty$. Although Chebyshev did not reach this estimate, his work in this direction was crucial and Chebyshev's ideas are still often used.

3.2.2.1 The Functions θ and ψ

One of the well-known results of the elementary theory of the distribution of primes is that

$$\theta(n) < n \log 4 \quad (n \in \mathbb{Z}_{\geqslant 2})$$

for which a very elegant proof was given by Erdős and Kalmar [87, Satz 30]. The idea of the proof rests on the fact that every prime $p \in (m + 1, 2m + 1]$ divides the binomial coefficient $\binom{2m+1}{m}$. Subsequently, many authors improved this bound, such as Hanson who showed in [37] that $n \log 4$ could be replaced by $n \log 3$. As explained below in Proposition 3.16, considerably deeper methods involving the non-trivial zeros of the Riemann zeta-function are used nowadays to derive sharper estimates for the θ-function. The best result to date is the bound [14, Corollary 2.1]

$$\theta(x) \leqslant \left(1 + 1.93378 \times 10^{-8} \right) x \quad (x \geqslant 0).$$

In the spirit of this section dealing with elementary methods in prime number theory, we chose to reproduce here Moser's simple proof [57] with only slight minor changes, somewhat improving Erdős and Kalmar's bound by considering a trinomial coefficient instead of a binomial coefficient.

Lemma 3.8 *For all $x \geqslant 1$, we have $\theta(x) \leqslant x \log\left(432^{1/5}\right)$.*

Proof In view of the relation $\theta(x) = \theta(\lfloor x \rfloor)$, it suffices to show the inequality for all integers $n \geqslant 1$. For any $m \in \mathbb{Z}_{\geqslant 1}$, set

$$C_m := (6m + 1)\binom{6m}{m, 2m, 3m}.$$

Note that $C_m \in \mathbb{Z}$ and $C_1 < 432$. Furthermore, Stirling's formula [72] in the shape

$$n! = \left(ne^{-1}\right)^n \sqrt{2\pi n}\, e^{\lambda_n}, \qquad \tfrac{1}{2n+1} < \lambda_n < \tfrac{1}{2n} \tag{3.4}$$

yields

$$\binom{6m}{m, 2m, 3m} < \frac{432^m}{2\pi m}$$

so that $C_m < 432^m$ for all $m \geqslant 1$. Next observe that, if $p \in (m, 6m + 1]$ is prime, then p divides C_m. Indeed, writing $C_m = \frac{(6m+1)!}{m!(2m)!(3m)!}$,

▷ if $p \in \left(m, \frac{3}{2}m\right]$, then p^4 divides the numerator of C_m while p^3 is the highest power of p which divides the denominator ;
▷ if $p \in \left(\frac{3}{2}m, 2m\right]$, then p^3 divides the numerator while the highest power of p dividing the denominator is p^2 ;
▷ if $p \in (2m, 3m]$, then p^2 divides the numerator while p, but not p^2, divides the denominator ;
▷ and if $p \in (3m, 6m + 1]$, then p divides the numerator of C_m but not the denominator.

Now we are able to prove the lemma by induction on n, the case $n = 2$ and $n = 3$ being trivial. The induction hypothesis will be assumed true up to n. In proving it at $n + 1$, we may assume that $n + 1$ is a prime, otherwise $\theta(n + 1) = \theta(n)$. Since all primes > 3 have the form $6m \pm 1$, we get by the discussion above

$$\theta(6m+1) = \theta(m) + \sum_{m < p \leqslant 6m+1} \log p \leqslant \tfrac{1}{5}m \log 432 + \log C_m < \tfrac{1}{5}(6m+1) \log 432$$

and

$$\theta(6m - 1) = \theta(m) + \sum_{m < p \leqslant 6m+1} \log p - \sum_{6m-1 < p \leqslant 6m+1} \log p$$

$$\leqslant \tfrac{1}{5}m \log 432 + \log C_m - \log(6m + 1) < \tfrac{1}{5}(6m - 1) \log 432$$

completing the proof.   □

Lemma 3.9 *For all $x \geqslant 67$, we have $\Psi(x) > 0.912\,x$ and $\theta(x) > 0.8\,x$.*

Proof Since $\theta(x) = \theta(\lfloor x \rfloor)$ and similarly for Ψ, we first check the inequalities

$$\Psi(n) > 0.912\,(n + 1) \quad \text{and} \quad \theta(n) > 0.8\,(n + 1)$$

for $n \in \{67, \ldots, 2000\}$, so assume $x > 2000$. The proof follows the lines of Chebyshev's work. We first notice that the function f defined by

$$f(x) = \lfloor x \rfloor - \left\lfloor \tfrac{1}{2}x \right\rfloor - \left\lfloor \tfrac{1}{3}x \right\rfloor - \left\lfloor \tfrac{1}{5}x \right\rfloor + \left\lfloor \tfrac{1}{30}x \right\rfloor \tag{3.5}$$

is periodic of period 30, and that $f(30 - x) = 1 - f(x)$ for $x \notin \mathbb{Z}$. An inspection of its values when $x \in [1, 15)$ allows us to infer that $f(x)$ only takes the values 0 or 1 if $x \notin \mathbb{Z}$. Since f is continuous on the right, we also have $f(x) = 0$ or 1 when $x \in \mathbb{Z}$. By periodicity, we infer that $f(x) = 0$ or 1 for all $x \in \mathbb{R}$. Hence we get

$$\Psi(x) \geqslant \sum_{n \leqslant x} \Lambda(n) f(x/n)$$

$$= \sum_{n \leqslant x} \Lambda(n) \left\lfloor \frac{x}{n} \right\rfloor - \sum_{n \leqslant x/2} \Lambda(n) \left\lfloor \frac{x}{2n} \right\rfloor - \sum_{n \leqslant x/3} \Lambda(n) \left\lfloor \frac{x}{3n} \right\rfloor$$

$$- \sum_{n \leqslant x/5} \Lambda(n) \left\lfloor \frac{x}{5n} \right\rfloor + \sum_{n \leqslant x/30} \Lambda(n) \left\lfloor \frac{x}{30n} \right\rfloor$$

where we used the fact that $\lfloor x/kn \rfloor = 0$ as soon as $n > x/k$ for $k \in \{2, 3, 5, 30\}$. Now from Lemma 3.7 and Exercise 11, we deduce that, for all $x \geqslant 30$

$$\Psi(x) \geqslant x \log \left(2^{1/2} 3^{1/3} 5^{1/5} 30^{-1/30} \right) - \tfrac{5}{2} \log x + \log 30 - \tfrac{5}{4}$$

which is $> 0.912\,x$ when $x \geqslant 2000$. Furthermore, Lemma 3.6 and the bound of Exercise 26 yield, for all $x \geqslant 2000 \geqslant 33^2 = 1089$

$$\theta(x) \geqslant \psi(x) - \pi\left(\sqrt{x}\right) \log x$$

$$\geqslant x \log \left(2^{1/2} 3^{1/3} 5^{1/5} 30^{-1/30} \right) - \tfrac{1}{3}x^{1/2} \log x - \tfrac{5}{2} \log x + \log 30 - \tfrac{5}{4}$$

$$> 0.8\,x$$

as claimed.   □

Note that $\log\left(2^{1/2}3^{1/3}5^{1/5}30^{-1/30}\right) \approx 0.92129\ldots$ which is a very good lower bound. It was sufficient to allow Chebyshev to prove Bertrand's famous postulate.

Corollary 3.1 (Bertrand's Postulate)   *Let n be a positive integer. Then the interval $(n, 2n]$ contains a prime number.*

Proof We numerically check the result for $n \in \{1, \ldots, 544\}$ and suppose $n \geqslant 545$. Using Lemmas 3.8 and 3.9, we get

$$\sum_{n < p \leqslant 2n} \log p = \theta(2n) - \theta(n) > n\left(1.6 - \tfrac{1}{5}\log 432\right) > 0$$

which implies the desired result.   □

The method of Lemma 3.9 also provides an upper bound for $\Psi(x)$.

Lemma 3.10 *For $x \geqslant 1$, we have $\Psi(x) < 1.117\,x$.*

Proof We first numerically check the inequality for $x \in [1, 7000]$, and then assume that $x > 7000$. It is not difficult to see that the function f defined in (3.5) satisfies $f(t) = 1$ for all $1 \leqslant t < 6$, and hence

$$\sum_{n \leqslant x} \Lambda(n) f(x/n) \geqslant \sum_{x/6 < n \leqslant x} \Lambda(n) = \Psi(x) - \Psi\left(\tfrac{x}{6}\right)$$

and Lemma 3.7 along with Exercise 11 again yield, for all $x \geqslant 30$

$$\Psi(x) - \Psi\left(\tfrac{x}{6}\right) \leqslant Ax + \tfrac{5}{2}\log x - B,$$

where $A = \log\left(2^{1/2}3^{1/3}5^{1/5}30^{-1/30}\right)$ and $B = \log 30 + \tfrac{5}{8}$. Now replacing x successively by $6^{-1}x$, $6^{-2}x$, $\ldots$, $6^{-K}x$, where $K = \left\lfloor \frac{\log(x/30)}{\log 6} \right\rfloor$, and summing all the resulting inequalities give, for all $x \geqslant 30$

$$\Psi(x) - \Psi\left(\tfrac{x}{6^{K+1}}\right) \leqslant \tfrac{6A}{5}x\left(1 - 6^{-1-K}\right) + (K+1)\left(\tfrac{5}{2}\log x - \tfrac{5}{4}K\log 6 - B\right)$$

and majorizing $\Psi\left(6^{-1-K}x\right) \leqslant \Psi(30) < 28.477$, we derive

$$\Psi(x) \leqslant \tfrac{6}{5}(x - 5)A + \tfrac{5}{4\log 6}(\log x)^2 + 0.253\log x + 26.263 < 1.117\,x$$

as required.   □

As explained in [18], the previous bound may be improved if we use again the function f by taking into account more intervals in which f is identically equal to 1.

Lemma 3.11 *For $x \geqslant 1$, we have $\Psi(x) < 1.095\,x$.*

Proof The inequality is verified for $x \in [1, 2552]$, so assume $x > 2552$. Observe that the function f given in (3.5) satisfies $f(t) = 1$ for all $t \in [1, 6) \cup [7, 10)$, therefore we derive as above

$$\Psi(x) \leqslant Ax + \tfrac{5}{2} \log x - B + \Psi\left(\tfrac{x}{6}\right) - \Psi\left(\tfrac{x}{7}\right) + \Psi\left(\tfrac{x}{10}\right),$$

where A and B are given in the previous proof. Now Lemmas 3.10 and 3.9 yield

$$\Psi(x) < x (A + 0.1676) + \tfrac{5}{2} \log x - B < 1.095 x$$

when $x > 2552$. $\qquad\square$

3.2.2.2 The Functions π and p_n

Theorem 3.1 (Chebyshev's Estimates) *For all $x \geqslant 7$, we have*

$$\frac{4x}{5 \log x} < \pi(x) < \frac{2x}{\log x}.$$

Proof We first check the inequalities

$$\frac{4(n + 1)}{5 \log n} < \pi(n) < \frac{2n}{\log(n + 1)}$$

for all integers $n \in \{7, \ldots, 1089\}$, hence the corollary is true for all $x \in [7, 1089]$, and we may then suppose $x > 1089$. The left-hand side follows directly from the lower bound of the second estimate of Lemma 3.6 and from the lower bound of Lemma 3.9. Using Lemmas 3.6, 3.11, and Exercise 26, we derive for $a > 1$ and $x \geqslant 33^a$

$$\pi(x) \leqslant \frac{1.095 \, ax}{\log x} + \frac{x^{1/a}}{3}$$

and the inequality $\log x < \frac{a}{e(a-1)} x^{1-1/a}$, valid for all $x > 0$, implies that

$$\pi(x) \leqslant \left(1.095 \, a + \frac{a}{3e(a - 1)}\right) \frac{x}{\log x}$$

and the result follows from choosing $a = 1.33$. $\qquad\square$

Remark 3.1 One may wonder whether there exist optimal linear combinations of quantities $\lfloor x/n \rfloor$ to get better positive numbers $c_1 < c_2$ such that

$$c_1 x < \Psi(x) < c_2 x \quad (x \geqslant x_0) \, ?$$

Diamond and Erdős [19] showed that the answer is *yes*, and even provided some constants arbitrarily close to 1. Thus, could Chebyshev really prove the PNT 45 years before Hadamard and de la Vallée Poussin, and without using complex analysis? Unfortunately not, since Diamond and Erdős needed the PNT to provide the constants c_1, c_2. But Chebyshev's ideas played a key part in modern analytic number theory.

The proof of the next result is an easy consequence of Lemma 3.6 and Theorem 3.1.

Corollary 3.2 *For all $x \geqslant 49$, we have $\Psi(x) < \theta(x) + 4\sqrt{x}$.*

Theorem 3.1 enables us to derive fine estimates for the nth prime number p_n.

Corollary 3.3 *For all $n \in \mathbb{Z}_{\geqslant 3}$, we have $\frac{1}{2} n \log n < p_n < 2n \log n$.*

Proof Set $c_0 := \log \left(432^{1/5} \right)$. We first check the inequalities for $3 \leqslant n \leqslant 476$ and assume hereafter $n \geqslant 477$ so that $p_n \geqslant 3389$. The proof rests on the fact that $\pi(p_n) = n$. For the lower bound, we start by using Lemma 3.6 and Theorem 3.1, giving

$$\pi(x) \leqslant \frac{a}{\log x} \left(\theta(x) + 2x^{1/a} \right)$$

for all $a > 1$ and $x \geqslant 7^a$. Now Lemma 3.8, the inequality $x^{1/a} < a^{-7} x$ valid for all $x \geqslant a^{7a/(a-1)}$, and the trivial inequality $p_n > n$ all yield

$$n = \pi(p_n) \leqslant \frac{a p_n}{\log p_n} \left(c_0 + \frac{2}{a^7} \right) < \frac{a p_n}{\log n} \left(c_0 + \frac{2}{a^7} \right)$$

provided that $p_n \geqslant \max \left(7^a, a^{7a/(a-1)} \right)$. The choice of $a = \left(14 c_0^{-1} \right)^{1/8} \approx 1.357$ implies the required lower bound. For the upper bound, we begin by using the inequality

$$\frac{\log x}{x^{3/10}} < \frac{4}{5}$$

valid for all $x \geqslant 3389$. Applied with $x = p_n$ and using the lower bound of Theorem 3.1, we derive

$$\frac{\log p_n}{p_n^{3/10}} < \frac{4}{5} < \frac{\pi(p_n) \log p_n}{p_n} = \frac{n \log p_n}{p_n}$$

which implies that $p_n < n^{10/7}$, and therefore $\log p_n < \frac{10}{7} \log n$. Thus, we get

$$n = \pi(p_n) > \frac{4 p_n}{5 \log p_n} > \frac{14 p_n}{25 \log n} > \frac{p_n}{2 \log n}$$

which implies the desired upper bound. □

3.2.3 An Alternative Approach

In [58], M. Nair had the idea to use $d_n = [1, 2, \ldots, n]$ to get lower bounds for the function π. This estimate rests on the following result.

Lemma 3.12 *For all integers $n \geqslant 7$, we have $d_n \geqslant 2^n$.*

Proof The inequality is clearly true for $n \in \{7, \ldots, 15\}$ so that we suppose $n \geqslant 16$. The formula [31, (1.40)] yields

$$\sum_{k=0}^{n} \binom{n}{k} \frac{(-1)^k}{n+k+1} = \frac{1}{(n+1)\binom{2n+1}{n+1}} = \frac{1}{(2n+1)\binom{2n}{n}}.$$

The left-hand side is a rational number whose denominator is a divisor of d_{2n+1}, so that $(2n+1)\binom{2n}{n} \mid d_{2n+1}$. Now the known lower bound[8] $\binom{2n}{n} \geqslant \frac{4^n}{2\sqrt{n}}$ implies that

$$d_{2n+1} \geqslant n^{1/2} 2^{2n} \quad \text{and} \quad d_{2n+2} \geqslant d_{2n+1} \geqslant n^{1/2} 2^{2n}$$

which yields the asserted inequality. □

Since

$$d_n \leqslant \prod_{p \leqslant n} p^{\log n / \log p} = \prod_{p \leqslant n} n = n^{\pi(n)}$$

---

[8] Which can be proved by induction. See [87, (84) p. 58] for instance.

Lemma 3.12 immediately yields for $n \geqslant 7$

$$\pi(n) \geqslant \frac{\log d_n}{\log n} \geqslant \frac{n \log 2}{\log n}$$

and similarly

$$\Psi(n) = \log d_n \geqslant n \log 2.$$

Remark 3.2 The result of Lemma 3.12 has been generalized in several directions. For instance, in [61, Theorem 1], it is proved that, if $r, u_0 \in \mathbb{Z}_{\geqslant 1}$ are such that $(u_0, r) = 1$, then, for any $n \geqslant r(r+1)$, one has

$$[u_0, u_0 + r, u_0 + 2r, \ldots, u_0 + nr] \geqslant u_0 r^{r+1} (r+1)^n.$$

3.2.4 Mertens's Theorems

Around 1875, Mertens proved two fundamental results on asymptotic formulas for some prime number functions.

Theorem 3.2 (First Mertens's Theorem) *For all $x \geqslant 2$, we have*

$$\sum_{p \leqslant x} \frac{\log p}{p} = \log x + O(1).$$

First Proof Let $n = \lfloor x \rfloor$. By Exercise 35 we have

$$\log(n!) = \sum_{p \leqslant n} \log p \sum_{\alpha=1}^{\infty} \left\lfloor \frac{n}{p^\alpha} \right\rfloor$$

$$= \sum_{p \leqslant n} \log p \left\lfloor \frac{n}{p} \right\rfloor + \sum_{p \leqslant n} \log p \sum_{\alpha=2}^{\infty} \left\lfloor \frac{n}{p^\alpha} \right\rfloor$$

$$= n \sum_{p \leqslant n} \frac{\log p}{p} + O\left(\theta(n)\right) + O\left(n \sum_{p \leqslant n} \log p \sum_{\alpha=2}^{\infty} \frac{1}{p^\alpha}\right)$$

and the use of Exercise 11 and Lemma 3.8 gives

$$\sum_{p \leqslant n} \frac{\log p}{p} = \frac{\log(n!)}{n} + O\left(\frac{\theta(n)}{n}\right) + O\left(\sum_{p \leqslant n} \frac{\log p}{p(p-1)}\right)$$

$$= \frac{1}{n}\left(n \log n + O(n)\right) + O(1) = \log n + O(1)$$

and we conclude the proof with

$$\sum_{p \leqslant x} \frac{\log p}{p} = \sum_{p \leqslant \lfloor x \rfloor} \frac{\log p}{p} = \log\lfloor x \rfloor + O(1) = \log x + O(1)$$

as required.   □

Second Proof We first notice that

$$\sum_{d \leqslant x} \frac{\Lambda(d)}{d} = \sum_{p \leqslant x} \frac{\log p}{p} + \sum_{\substack{p^\alpha \leqslant x \\ \alpha \geqslant 2}} \frac{\log p}{p^\alpha}$$

and the second sum is

$$\leqslant \sum_{p \leqslant \sqrt{x}} \log p \sum_{\alpha = 2}^{\infty} \frac{1}{p^\alpha} = \sum_{p \leqslant \sqrt{x}} \frac{\log p}{p(p-1)} < E - \gamma < 1 \qquad (3.6)$$

where $E \approx 1.332\ldots$ is the constant appearing in Corollary 3.13, so that

$$\sum_{p \leqslant x} \frac{\log p}{p} = \sum_{d \leqslant x} \frac{\Lambda(d)}{d} + O(1).$$

Now appealing to Lemma 3.7 and Corollary 3.3 we get

$$x \log x + O(x) = \sum_{d \leqslant x} \Lambda(d) \left\lfloor \frac{x}{d} \right\rfloor = x \sum_{d \leqslant x} \frac{\Lambda(d)}{d} + O\left(\Psi(x)\right) = x \sum_{d \leqslant x} \frac{\Lambda(d)}{d} + O(x)$$

and hence

$$\sum_{d \leqslant x} \frac{\Lambda(d)}{d} = \log x + O(1)$$

which concludes the proof.   □

By partial summation, it is fairly easy to prove that Theorem 3.2 implies the following result:

$$\sum_{p \leqslant x} \frac{1}{p} = \log \log x + B + O\left(\frac{1}{\log x}\right) \tag{3.7}$$

where $B = 1 - \log \log 2 + \int_2^\infty t^{-1}(\log t)^{-2} R(t)\, dt$ and $|R(t)| \ll 1$. Mertens studied more carefully this sum in order to get a more manageable expression of the constant B, called today the *Mertens constant*.

Corollary 3.4 *For all $x \geqslant e$, we have*

$$\sum_{p \leqslant x} \frac{1}{p} = \log \log x + B + O\left(\frac{1}{\log x}\right),$$

where $B = \gamma + \sum_p \left\{ \log\left(1 - \frac{1}{p}\right) + \frac{1}{p} \right\} \approx 0.2614972128\ldots$ is the Mertens constant.

Proof Let $h > 0$.

The starting point is Euler's formula from Proposition 3.6

$$\prod_p \left(1 - \frac{1}{p^{1+h}}\right)^{-1} = \sum_{n=1}^\infty \frac{1}{n^{1+h}}.$$

Comparing the sum to an integral, we infer

$$\sum_{n=1}^\infty \frac{1}{n^{1+h}} = \int_1^\infty \frac{dt}{t^{1+h}} + O(1) = \frac{1}{h} + O(1)$$

so that

$$\prod_p \left(1 - \frac{1}{p^{1+h}}\right)^{-1} = \frac{1}{h} + O(1)$$

and taking logarithms of both sides we get

$$\sum_p \frac{1}{p^{1+h}} + \sum_p \sum_{\alpha=2}^{\infty} \frac{1}{\alpha p^{\alpha(1+h)}} = \sum_p \log\left(1 - \frac{1}{p^{1+h}}\right)^{-1} = \log\left(\frac{1}{h}\right) + O(h)$$

as $h \to 0$, so that

$$\sum_p \frac{1}{p^{1+h}} = \log\left(\frac{1}{h}\right) - \sum_p \sum_{\alpha=2}^{\infty} \frac{1}{\alpha p^{\alpha(1+h)}} + O(h). \qquad (3.8)$$

Let $X > x \geqslant e$ be large real numbers. By (1.1), Theorem 1.2 and (3.7), we have

$$\sum_{x < p \leqslant X} \frac{1}{p^{1+h}} = \int_x^X \frac{1}{t^h} \, d\left(\sum_{x < p \leqslant t} \frac{1}{p}\right)$$

$$= \frac{1}{X^h} \sum_{x < p \leqslant X} \frac{1}{p} + h \int_x^X \frac{1}{t^{1+h}} \left(\sum_{\log x < n \leqslant \log t} \frac{1}{n} + O\left(\frac{1}{\log x}\right)\right) dt$$

$$= \frac{1}{X^h} \log\left(\frac{\log X}{\log x}\right) + h \int_x^X \left(\sum_{\log x < n \leqslant \log t} \frac{1}{n}\right) \frac{dt}{t^{1+h}} + O\left(\frac{1}{\log x}\right)$$

$$= \frac{1}{X^h} \log\left(\frac{\log X}{\log x}\right) + h \sum_{\log x < n \leqslant \log X} \frac{1}{n} \int_{e^n}^X \frac{dt}{t^{1+h}} + O\left(\frac{1}{\log x}\right)$$

$$= \frac{1}{X^h} \log\left(\frac{\log X}{\log x}\right) + \sum_{\log x < n \leqslant \log X} \frac{e^{-nh} - X^{-h}}{n} + O\left(\frac{1}{\log x}\right)$$

$$= \sum_{\log x < n \leqslant \log X} \frac{e^{-nh}}{n} + O\left(\frac{1}{\log x}\right)$$

so that

$$\sum_{p > x} \frac{1}{p^{1+h}} = \sum_{n > \log x} \frac{e^{-nh}}{n} + O\left(\frac{1}{\log x}\right)$$

$$= -\log(1 - e^{-h}) - \sum_{n \leqslant \log x} \frac{e^{-nh}}{n} + O\left(\frac{1}{\log x}\right)$$

$$= -\log(1 - e^{-h}) - \sum_{n \leqslant \log x} \frac{1}{n} + \sum_{n \leqslant \log x} \frac{1 - e^{-nh}}{n} + O\left(\frac{1}{\log x}\right)$$

$$= -\log(1 - e^{-h}) - \log\log x - \gamma + O(h \log x) + O\left(\frac{1}{\log x}\right).$$

Now combining (3.8) with this estimate, we derive

$$\sum_{p \leqslant x} \frac{1}{p^{1+h}} = \log\left(\frac{1 - e^{-h}}{h}\right) + \log\log x + \gamma - \sum_{p}\sum_{\alpha=2}^{\infty} \frac{1}{\alpha p^{\alpha(1+h)}} + O(h\log x) + O\left(\frac{1}{\log x}\right)$$

and letting $h \to 0$ and noticing that $\displaystyle\sum_{p}\sum_{\alpha=2}^{\infty} \frac{1}{\alpha p^{\alpha}} = -\sum_{p}\left\{\log\left(1 - \frac{1}{p}\right) + \frac{1}{p}\right\}$ we

obtain the asserted result.   □

Mertens discovered the identity

$$\sum_{p}\sum_{\alpha=2}^{\infty} \frac{1}{\alpha p^{\alpha}} = \sum_{n=2}^{\infty} \frac{\mu(n)\log\zeta(n)}{n}$$

which enabled him to compute B very accurately.

From Corollary 3.4 we readily get Mertens's second theorem.

Corollary 3.5 (Mertens's Second Theorem)   *For all $x \geqslant e$, we have*

$$\prod_{p \leqslant x}\left(1 - \frac{1}{p}\right) = \frac{e^{-\gamma}}{\log x}\left\{1 + O\left(\frac{1}{\log x}\right)\right\}.$$

Proof We have

$$\log\prod_{p \leqslant x}\left(1 - \frac{1}{p}\right)^{-1} = \sum_{p \leqslant x}\log\left(1 - \frac{1}{p}\right)^{-1} = \sum_{p \leqslant x}\frac{1}{p} + \sum_{p \leqslant x}\left\{\log\left(1 - \frac{1}{p}\right)^{-1} - \frac{1}{p}\right\}$$

and using the inequalities

$$\log\left(1 - \frac{1}{p}\right)^{-1} - \frac{1}{p} \leqslant \frac{1}{2p(p-1)} \leqslant \frac{1}{p^2}$$

the estimate

$$\sum_{p > x}\frac{1}{p^2} < \frac{3}{x\log x} \quad (x \geqslant e)$$

and Corollary 3.4, we infer that

$$\log \prod_{p \leqslant x} \left(1 - \frac{1}{p}\right)^{-1} = \sum_{p \leqslant x} \frac{1}{p} - \sum_{p} \left\{ \log\left(1 - \frac{1}{p}\right) + \frac{1}{p} \right\} + O\left(\frac{1}{x \log x}\right)$$

$$= \log \log x + \gamma + O\left(\frac{1}{\log x}\right)$$

and the use of $e^h = 1 + O(h)$ for all $h \to 0$ finally gives

$$\prod_{p \leqslant x} \left(1 - \frac{1}{p}\right)^{-1} = e^\gamma \log x \left\{1 + O\left(\frac{1}{\log x}\right)\right\}$$

which is easily seen to be equivalent to the form given in the corollary.   □

3.3 The Riemann Zeta-Function

3.3.1 Euler, Dirichlet and Riemann

Considered as one of the first analytic number theorists, Euler noticed that the function

$$\sigma \mapsto \sum_{n=1}^{\infty} \frac{1}{n^\sigma}$$

well-defined for all $\sigma > 1$, could carry some important arithmetic information, in particular concerning the distribution of primes. Indeed, Euler showed the following fundamental result.

Proposition 3.6 (Euler) *For all $\sigma > 1$, we have*

$$\prod_{p} \left(1 - \frac{1}{p^\sigma}\right)^{-1} = \sum_{n=1}^{\infty} \frac{1}{n^\sigma}.$$

Proof Let N be a positive integer and $\sigma > 1$ be a real number. Expanding the product

$$\prod_{p \leqslant N} \left(1 - \frac{1}{p^\sigma}\right)^{-1} = \prod_{p \leqslant N} \left(1 + \frac{1}{p^\sigma} + \frac{1}{p^{2\sigma}} + \frac{1}{p^{3\sigma}} + \cdots\right)$$

we derive

$$\prod_{p \leqslant N} \left(1 - \frac{1}{p^\sigma}\right)^{-1} = 1 + \frac{1}{n_1^\sigma} + \frac{1}{n_2^\sigma} + \frac{1}{n_3^\sigma} + \cdots,$$

where each integer n_i is such that all its prime factors are $\leqslant N$, so that

$$\prod_{p \leqslant N} \left(1 - \frac{1}{p^\sigma}\right)^{-1} = \sum_{P^+(n) \leqslant N} \frac{1}{n^\sigma}.$$

Since every integer $n \leqslant N$ satisfies this property, we infer that

$$\left| \sum_{n=1}^{\infty} \frac{1}{n^\sigma} - \prod_{p \leqslant N} \left(1 - \frac{1}{p^\sigma}\right)^{-1} \right| = \left| \sum_{n=1}^{\infty} \frac{1}{n^\sigma} - \sum_{P^+(n) \leqslant N} \frac{1}{n^\sigma} \right| \leqslant \sum_{n > N} \frac{1}{n^\sigma}.$$

We conclude the proof by noting that the latter sum tends to 0 as N tends to ∞.   □

Euler used to work with Proposition 3.6 principally as a formal identity and mainly for integer values of σ. Later, Dirichlet based some of his work upon Euler's product formula but used it with $\sigma > 1$ as a real variable, and proved rigorously that this result is true in the interval $(1, \infty)$.

Riemann, who was one of Dirichlet's students, was certainly influenced by his use of Euler's product formula and, as one of the founders of the theory of functions, would naturally consider σ as a complex variable, renamed $s = \sigma + it \in \mathbb{C}$. Since

$$\sum_{p \leqslant x} \frac{1}{|p^s|} = \sum_{p \leqslant x} \frac{1}{p^\sigma}$$

both sides of Proposition 3.6 converge for every complex s such that $\sigma > 1$. This leads to the following definition.

Definition 3.4 (Riemann Zeta-Function)   The Riemann zeta-function $\zeta(s)$ is defined for all complex numbers $s = \sigma + it$ such that $\sigma > 1$ by

$$\zeta(s) = \sum_{n=1}^{\infty} \frac{1}{n^s} = \prod_{p} \left(1 - \frac{1}{p^s}\right)^{-1}.$$

The product representation, called the *Euler product*, enables us to see that $\zeta(s) \neq 0$ in the half-plane $\sigma > 1$. Furthermore, since

$$\left| \sum_{p \leqslant x} \frac{\log p}{p^s - 1} \right| \leqslant \sum_{p \leqslant x} \frac{\log p}{p^{1+\varepsilon} - 1}$$

for all $s = \sigma + it \in \mathbb{C}$ such that $\sigma \geqslant 1 + \varepsilon$, we infer that the series of the left-hand side is uniformly convergent in this half-plane, and this allows us to take the logarithmic derivative of both sides of Definition 3.4 and justifies the change of the order of summations, which gives

$$-\frac{\zeta'(s)}{\zeta(s)} = \sum_p \frac{\log p}{p^s - 1} = \sum_p \sum_{\alpha=1}^\infty \frac{\log p}{p^{\alpha s}} = \sum_{\alpha=1}^\infty \sum_p \frac{\log p}{p^{\alpha s}}$$

and hence we get

$$-\frac{\zeta'(s)}{\zeta(s)} = \sum_{n=1}^\infty \frac{\Lambda(n)}{n^s}. \tag{3.9}$$

The series in (3.9) converges absolutely and uniformly in the half-plane $\sigma \geqslant 1 + \varepsilon$ for all $\varepsilon > 0$, so that (3.9) holds for $\sigma > 1$ by analytic continuation.

3.3.2   The Gamma and Theta Functions

3.3.2.1   Functional Equation

Euler defined the Gamma function for all real numbers $\sigma > 0$ as

$$\Gamma(\sigma) = \int_0^\infty x^{\sigma-1} e^{-x} dx$$

the convergence of the integral being ensured by the estimates $x^{\sigma-1} e^{-x} \sim x^{\sigma-1}$ when $x \to 0$ and $x^{\sigma-1} e^{-x} = o(e^{-x/2})$ when $x \to \infty$. Clearly, we have $\Gamma(\sigma) > 0$ for all $\sigma > 0$, $\Gamma(1) = 1$ and integrating by parts gives the recursion

$$\Gamma(\sigma + 1) = \sigma \Gamma(\sigma)$$

for all $\sigma > 0$. In particular, we get by induction the identity $\Gamma(n) = (n - 1)!$ for all positive integers n.

Let $s = \sigma + it \in \mathbb{C}$. Since $|x^s| = x^\sigma$, the integral above defines an analytic function Γ in the half-plane $\sigma > 0$, which still satisfies the functional equation

$$\Gamma(s + 1) = s\Gamma(s) \quad (\sigma > 0)$$

as can be seen by repeating an integration by parts. For $-1 < \sigma < 0$, we set $\Gamma(s) = s^{-1}\Gamma(s + 1)$ so that the function Γ has a simple pole at $s = 0$. Similarly in the range $-2 < \sigma < -1$, we define $\Gamma(s) = s^{-1}(s + 1)^{-1}\Gamma(s + 2)$ which gives a simple pole at $s = -1$. Continuing in this way, we see that the functional equation extends Γ to a meromorphic function on $\mathbb{C}$, which is analytic except for simple poles at $s = 0, -1, -2, -3, \ldots$ One may check that the residue at the pole $-n$ is given by $(-1)^n/n!$.

Let us mention the following useful formulae [85, 4.41]. For any $s \in \mathbb{C}$ we have[9]

$$\frac{1}{\Gamma(s)\Gamma(1-s)} = \frac{\sin \pi s}{\pi} \tag{3.10}$$

and[10]

$$\Gamma(s) = \pi^{-1/2} 2^{s-1} \Gamma\left(\tfrac{1}{2}s\right) \Gamma\left(\tfrac{1}{2}(s+1)\right). \tag{3.11}$$

3.3.2.2 Complex Stirling's Formula

Stirling's formula may be generalized as follows [85, 4.42].

Theorem 3.3 (Complex Stirling's Formula) *For any $s \in \mathbb{C}$, we have*

$$\log \Gamma(s) = \left(s - \tfrac{1}{2}\right) \log s - s + \log \sqrt{2\pi} + O\left(|s|^{-1}\right)$$

as $|s| \to \infty$ and uniformly for $|\arg s| \leqslant \pi - \delta$ ($\delta > 0$), where the complex logarithm is chosen by taking the principal value of its argument. This implies in particular that, if $\sigma \in [\sigma_1, \sigma_2]$ is fixed, then

$$|\Gamma(\sigma + it)| = |t|^{\sigma - 1/2} e^{-\pi|t|/2} \sqrt{2\pi} \left(1 + O\left(|t|^{-1}\right)\right)$$

when $|t| \geqslant t_0$, the constant implied in the error term depending on σ_1 and σ_2.

We will see later that most of the Dirichlet series arising in analytic number theory satisfy functional equations of the same shape,[11] namely as in Theorem 4.16 where products of Gamma functions appear. Furthermore, this Theorem 3.3, showing that $\Gamma(s)$ tends to zero exponentially fast as $|t| \to \infty$ in vertical strips, is then used to shift the contours of integration. This enables us to get important results which are out of reach by other methods, as in Theorem 7.51.

It should be pointed out that explicit versions of Theorem 3.3 exist in the literature. For instance, Lerch proved the following elegant identity, a proof of which may be found in [30].

---

[9]This identity is called the *reflection formula*.
[10]This identity is called the *duplication formula*.
[11]The set of these functions is now called the *Selberg class*.

Proposition 3.7 *Let $s := \sigma + it \in \mathbb{C}$ such that $0 \leqslant \sigma \leqslant 1$ and $s \neq 0$. Then*

$$|\Gamma(\sigma + it)| = \lambda \frac{\Gamma(1+\sigma)}{\sqrt{\sigma^2 + t^2}} \sqrt{\frac{\pi t}{\sinh(\pi t)}}$$

with $1 < \lambda < \sqrt{1 + t^2}$. Furthermore, when $\sigma = 0$, then $\lambda = 1$ and therefore

$$|\Gamma(it)| = \frac{1}{|t|} \sqrt{\frac{\pi t}{\sinh(\pi t)}} \qquad (t \neq 0).$$

A useful bound is the following estimate [62].

Theorem 3.4 *Let $s := \sigma + it \in \mathbb{C}$ such that $\sigma > 0$. Then*

$$|\Gamma(s)| \leqslant \sqrt{2\pi} \, |s|^{\sigma - \frac{1}{2}} e^{-\frac{1}{2}\pi|t|} \exp\left(\tfrac{1}{6|s|}\right).$$

More precise estimates may be established, such as the following results [83].

Theorem 3.5 *Let $s \in \mathbb{C}$ such that $|\arg s| \leqslant \frac{1}{2}\pi$ and $N \in \mathbb{Z}_{\geqslant 1}$. Then*

$$\log \Gamma(s) = \left(s - \tfrac{1}{2}\right) \log s - s + \log \sqrt{2\pi} + \sum_{k=1}^{N} \frac{B_{2k}}{2k(2k-1)s^{2k-1}} + R_N(s),$$

where

$$|R_N(s)| \leqslant \frac{2^{N+1} |B_{2N+2}|}{(2N+1)(2N+2)|s|^{2N+1}}.$$

For instance with $N = 1$, we derive

$$\log \Gamma(s) = \left(s - \tfrac{1}{2}\right) \log s - s + \log \sqrt{2\pi} + \tfrac{1}{12}s^{-1} + O^\star\left(\tfrac{1}{90}|s|^{-3}\right) \qquad \left(|\arg s| \leqslant \tfrac{1}{2}\pi\right).$$

In [21], the following result is established and may be of interest.

Theorem 3.6 *Let $T \geqslant 1$ and $\sigma \geqslant 0$. Uniformly for all $s = \sigma + it \in \mathbb{C}$ such that $|t| \geqslant T$*

$$|\Gamma(s)| = \sqrt{2\pi} \, |t|^{\sigma - 1/2} e^{-\pi|t|/2} \exp\left\{O^\star\left(\tfrac{\sigma^3}{3T} + \tfrac{\sigma^2}{2T}\left|\sigma - \tfrac{1}{2}\right| + \tfrac{1}{4T}\right)\right\}.$$

For the logarithmic derivative of the Gamma function, we have the following bound [27, Lemme 2.8].

Proposition 3.8 *If $s = \sigma + it \in \mathbb{C} \setminus \mathbb{R}_{\leqslant 0}$ is such that $\sigma \geqslant -1$, then*

$$\left| \frac{\Gamma'}{\Gamma}(s) \right| \leqslant \log|s + 1| + \frac{\pi}{2} + \frac{1}{|s|} + \frac{1}{2|s + 1|} + \frac{1}{12(\sigma + 1)^2} + \frac{1}{12|s + 1|^2}.$$

3.3.2.3 The Theta Function

Replacing x by $\pi n^2 x$ in the integral defining Γ gives

$$\pi^{-s/2} \Gamma\left(\tfrac{1}{2}s\right) n^{-s} = \int_0^\infty x^{\sigma/2 - 1} e^{-n^2 \pi x} \, dx \tag{3.12}$$

for all $\sigma > 0$. The purpose is to sum both sides of this equation. To this end, we define the following two functions. For all $x > 0$, we set

$$\omega(x) = \sum_{n=1}^\infty e^{-n^2 \pi x} \quad \text{and} \quad \theta(x) = 2\omega(x) + 1 = \sum_{n \in \mathbb{Z}} e^{-n^2 \pi x}.$$

The function $g : t \mapsto e^{-t^2 \pi}$ satisfies $\int_{\mathbb{R}} g(t) \, dt = 1$. This implies that its Fourier transform is

$$\widehat{g}(u) = e^{-\pi u^2}.$$

Set $f : t \mapsto e^{-t^2 \pi x}$. Using the transposition formula stating that the Fourier transform of $t \mapsto g(\alpha t)$ is the function $u \mapsto |\alpha|^{-1} \widehat{g}(u/\alpha)$ for all real numbers $\alpha \neq 0$, we obtain with $\alpha = x^{1/2}$

$$\widehat{f}(u) = x^{-1/2} e^{-u^2 \pi / x}.$$

By Lemma 6.8, we derive

$$\theta\left(\tfrac{1}{x}\right) = x^{1/2} \theta(x) \quad (x > 0). \tag{3.13}$$

Now we may return to (3.12). Summing this equation over n and interchanging the sum and integral, we get for all $\sigma > 1$

$$\pi^{-s/2} \Gamma\left(\tfrac{1}{2}s\right) \zeta(s) = \int_0^\infty x^{\sigma/2 - 1} \omega(x) \, dx.$$

The process is justified since the sum and integral converge absolutely in the half-plane $\sigma > 1$. Splitting the integral at $x = 1$ and substituting $\frac{1}{x}$ for x in the first integral yields

$$\pi^{-s/2}\Gamma\left(\tfrac{1}{2}s\right)\zeta(s) = \int_1^\infty x^{\sigma/2-1}\omega(x)\,dx + \int_1^\infty x^{-\sigma/2-1}\omega\left(\frac{1}{x}\right)dx$$

and using (3.13) in the form

$$\omega\left(\frac{1}{x}\right) = x^{1/2}\omega(x) + \tfrac{1}{2}\left(x^{1/2}-1\right)$$

finally gives

$$\pi^{-s/2}\Gamma\left(\tfrac{1}{2}s\right)\zeta(s) = -\frac{1}{s} + \frac{1}{s-1} + \int_1^\infty \omega(x)\left(x^{s/2}+x^{(1-s)/2}\right)\frac{dx}{x} \qquad (3.14)$$

for all $\sigma > 1$.

3.3.3 Functional Equation

The computations above eventually lead to the functional equation of the Riemann zeta-function. Thus, Riemann's idea to consider s as a complex variable is very fruitful. However, it is interesting to note that Riemann did not think of a piece-by-piece extension of the function represented by $\sum_{n=1}^\infty n^{-s}$ in the way that analytic continuation is usually used today, but rather searched a formula which remains valid for all s.[12] The following result, whose real version was conjectured and partially proved by Euler, is fundamental.

Theorem 3.7 (Functional Equation) *It is customary to set*

$$\xi(s) = \pi^{-s/2}\Gamma\left(\tfrac{1}{2}s\right)\zeta(s).$$

Then the function $\xi(s)$ can be analytically continued in the whole complex plane to a meromorphic function having simple poles at $s = 0$ and $s = 1$, and satisfies the functional equation $\xi(s) = \xi(1-s)$.

Thus the Riemann zeta-function can be extended analytically in the whole complex plane to a meromorphic function having a simple pole at $s = 1$ with residue 1. Furthermore, for all $s \in \mathbb{C} \setminus \{1\}$, we have

$$\zeta(s) = 2^s\pi^{s-1}\sin\left(\tfrac{1}{2}\pi s\right)\Gamma(1-s)\zeta(1-s).$$

---

[12] See [25] for instance.

Proof Let $x \geqslant 1$ be a real number and $s = \sigma + it$ with $\sigma > 1$. By (1.3) with $a = 1$, $b = x$ and $f(x) = x^{-s}$, we get

$$\sum_{n \leqslant x} \frac{1}{n^s} = \frac{1}{2} + \frac{1 - x^{1-s}}{s-1} - \frac{\psi(x)}{x^s} - s \int_1^x \frac{\psi(u)}{u^{s+1}} \, du \qquad (3.15)$$

so that making $x \to \infty$ we obtain

$$\zeta(s) = \frac{1}{2} + \frac{1}{s-1} - s \int_1^{\infty} \frac{\psi(u)}{u^{s+1}} \, du. \qquad (3.16)$$

Since $|\psi(x)| \leqslant \frac{1}{2}$, the integral converges for $\sigma > 0$ and is uniformly convergent in any finite region to the right of the line $\sigma = 0$. This implies that it defines an analytic function in the half-plane $\sigma > 0$, and therefore (3.16) extends ζ to a meromorphic function in this half-plane, which is analytic except for a simple pole at $s = 1$ with residue 1. Also notice that (3.16) can be written in the shape

$$\zeta(s) = \frac{s}{s-1} - s \int_1^{\infty} \frac{\{u\}}{u^{s+1}} \, du$$

and since $\lim_{|s| \to \infty} s/(s-1) = 1$, we deduce that

$$|\zeta(s)| \ll |s|$$

as $|s| \to \infty$. By (3.14), we have for $\sigma > 1$

$$\xi(s) = -\frac{1}{s} + \frac{1}{s-1} + \int_1^{\infty} \omega(x) \left(x^{s/2} + x^{(1-s)/2} \right) \frac{dx}{x}.$$

Since $\omega(x) \ll e^{-\pi x}$ as $x \to \infty$, we infer that the integral is absolutely convergent for all $s \in \mathbb{C}$ whereas the left-hand side is a meromorphic function on $\sigma > 0$. This implies that

▷ The identity (3.14) is valid for all $\sigma > 0$.
▷ The function $\xi(s)$ can be defined by this identity as a meromorphic function on $\mathbb{C}$ with simple poles at $s = 0$ and $s = 1$.
▷ Since the right-hand side of (3.14) is invariant under the substitution $s \leftrightarrow 1 - s$, we get $\xi(s) = \xi(1-s)$.
▷ The function $s \longmapsto s(s-1)\xi(s)$ is entire on $\mathbb{C}$. Indeed, if $\sigma > 0$, the factor $s - 1$ counters the pole at $s = 1$, and the result on all $\mathbb{C}$ follows from the functional equation.

It remains to be shown that the functional equation can be written in the form

$$\zeta(s) = 2^s \pi^{s-1} \sin\left(\tfrac{1}{2}\pi s\right) \Gamma(1-s)\zeta(1-s).$$

Since $\xi(s) = \xi(1-s)$, we have

$$\Gamma\left(\tfrac{1}{2}s\right) \zeta(s) = \pi^{s-1/2} \Gamma\left(\tfrac{1}{2}(1-s)\right) \zeta(1-s)$$

and multiplying both sides by $\pi^{-1/2} 2^{s-1} \Gamma\left(\tfrac{1}{2}(1+s)\right)$ and using (3.11) we derive

$$\Gamma(s)\zeta(s) = (2\pi)^{s-1} \Gamma\left(\tfrac{1}{2}(1-s)\right) \Gamma\left(\tfrac{1}{2}(1+s)\right) \zeta(1-s).$$

Now the reflection formula (3.10) implies that

$$\zeta(s) = (2\pi)^{s-1} \left(\frac{\sin \pi s}{\sin(\pi(1+s)/2)}\right) \Gamma(1-s)\zeta(1-s)$$

and the result follows from the identity $\sin \pi s = 2 \sin\left(\tfrac{1}{2}\pi s\right) \sin\left(\tfrac{1}{2}\pi(1+s)\right)$.   □

Remark 3.3 We may deduce the following basic consequences for the Riemann zeta-function.

▷ $\zeta(s)$ has simple zeros at $s = -2, -4, -6, -8, \ldots$ Indeed, since the integral in (3.14) is absolutely convergent for all $s \in \mathbb{C}$ and since $\omega(x) > 0$ for all x, we have

$$\xi(-2n) = \frac{1}{2n} - \frac{1}{2n+1} + \int_1^\infty \omega(x)\left(x^{-n} + x^{n+1/2}\right) \frac{dx}{x} > 0$$

for all positive integers n. The result follows from the fact that $\Gamma(s/2)$ has simple poles at $s = -2n$. These zeros are the only ones lying in the region $\sigma < 0$. They are called *trivial zeros* of the Riemann zeta-function.

▷ For all $0 < \sigma < 1$, we have $\zeta(\sigma) \neq 0$. Indeed, since for all $\sigma > 0$

$$\zeta(s) = \frac{s}{s-1} - s \int_1^\infty \frac{\{x\}}{x^{s+1}} dx$$

we infer that, for all $0 < \sigma < 1$, we get

$$\left|\zeta(\sigma) - \frac{\sigma}{\sigma-1}\right| < \sigma \int_1^\infty \frac{dx}{x^{\sigma+1}} = 1$$

which implies that $\zeta(\sigma) < 1 + \frac{\sigma}{\sigma-1}$ for all $0 < \sigma < 1$. Hence $\zeta(\sigma) < 0$ for all $\tfrac{1}{2} \leqslant \sigma < 1$, and the functional equation implies the asserted result.

3.3.4 Approximate Functional Equations

The functional equation is very important, but may also be insufficient in some applications for it does not express $\zeta(s)$ explicitly. The following tool will be useful to get some estimates of $\zeta(s)$ in the critical strip, especially when σ is close to 1.

Theorem 3.8 (Approximate Functional Equation)   *We have uniformly for $x \geqslant 1$ and $s \in \mathbb{C} \setminus \{1\}$ such that $\sigma > 0$*

$$\zeta(s) = \sum_{n \leqslant x} \frac{1}{n^s} + \frac{x^{1-s}}{s-1} + R_0(s; x)$$

with

$$R_0(s; x) = \frac{\psi(x)}{x^s} - s \int_x^\infty \frac{\psi(u)}{u^{s+1}} \, du$$

and hence

$$|R_0(s; x)| \leqslant \frac{|s|}{\sigma x^\sigma}.$$

Proof Follows by subtracting (3.15) from (3.16).   □

A very easy consequence of this result is the following small first zero-free region for ζ.

Corollary 3.6   *We have $\zeta(s) \neq 0$ when $\sigma > |s - 1|$, i.e. in the region $\sigma > \frac{1}{2}(1 + t^2)$.*

Proof Indeed, using Theorem 3.8 with $x = 1$, we derive

$$\left| \zeta(s) - \frac{s}{s-1} \right| \leqslant \frac{s}{\sigma} \quad (\sigma > 0)$$

implying the asserted result.   □

According to the periodicity of the function ψ, the error term of Theorem 3.8 is expected to be small. Using the techniques of exponential sums from Chap. 6, we may prove the following result bearing out this conjecture.

Theorem 3.9 (Approximate Functional Equation)   *Let $\sigma_0 > 0$. We have uniformly for $x \geqslant 1$, $\sigma \geqslant \sigma_0$ and $|t| \leqslant \pi x$*

$$\zeta(s) = \sum_{n \leqslant x} \frac{1}{n^s} + \frac{x^{1-s}}{s-1} + O(x^{-\sigma}).$$

Proof Let $y > x \geqslant 1$ be real numbers and $s = \sigma + it \in \mathbb{C}$ satisfying the assumptions of the theorem. We have

$$\zeta(s) = \sum_{n \leqslant x} \frac{1}{n^s} + \sum_{x < n \leqslant y} \frac{1}{n^s} + \sum_{n > y} \frac{1}{n^s}.$$

By (1.3), we infer

$$\sum_{n > y} \frac{1}{n^s} = \frac{y^{1-s}}{s-1} - \frac{y^{-s}}{2} - s \int_y^\infty \frac{\psi(u)}{u^{s+1}} \, du = \frac{y^{1-s}}{s-1} - \frac{y^{-s}}{2} + O\left(|s| y^{-\sigma}\right).$$

For the second sum, we use partial summation from Theorem 1.3 which gives

$$\sum_{x < n \leqslant y} \frac{1}{n^s} = \sum_{x < n \leqslant y} n^{-\sigma} n^{-it} = y^{-\sigma} \sum_{x < n \leqslant y} n^{-it} + \sigma \int_x^y \left(\sum_{x < n \leqslant u} n^{-it} \right) \frac{du}{u^{\sigma+1}}.$$

Now Lemma 6.9 is applied with $f(v) = -(2\pi)^{-1} t \log v$ to estimate the sums. Since $|t| \leqslant \pi x$, we have $|f'(v)| < \frac{1}{2}$ for all $x < v \leqslant u \leqslant y$, so that in this case Lemma 6.9 may be written in the following simpler form:

$$\sum_{x < n \leqslant u} n^{-it} = \sum_{x < n \leqslant u} e(f(n)) = \int_x^u e(f(v)) \, dv + O(1) = \frac{u^{1-it} - x^{1-it}}{1 - it} + O(1).$$

Hence we get

$$\sum_{x < n \leqslant y} \frac{1}{n^s} = \frac{y^{1-s} - y^{-\sigma} x^{1-it}}{1 - it} + O\left(y^{-\sigma}\right) + \sigma \int_x^y \left\{ \frac{u^{1-it} - x^{1-it}}{u^{\sigma+1}(1 - it)} + O\left(\frac{1}{u^{\sigma+1}}\right) \right\} du$$

$$= \frac{y^{1-s}}{1 - it} - \frac{x^{1-s}}{1 - it} + \frac{\sigma}{1 - it} \int_x^y u^{-s} \, du + O\left(x^{-\sigma}\right)$$

$$= \frac{y^{1-s}}{1 - s} - \frac{x^{1-s}}{1 - s} + O\left(x^{-\sigma}\right).$$

Therefore we obtain

$$\zeta(s) = \sum_{n \leqslant x} \frac{1}{n^s} - \frac{x^{1-s}}{1 - s} - \frac{y^{-s}}{2} + O\left(|s| y^{-\sigma}\right) + O\left(x^{-\sigma}\right)$$

and letting $y \to \infty$ gives the asserted result.   □

When $0 < \sigma \leqslant 1$ and $|t| \leqslant x$, an explicit version of the above result is proved in [21, Lemma 2.10].

Theorem 3.10 *Let $s = \sigma + it$, and assume that $s \neq 1$, $x \geqslant 1$, $0 < \sigma \leqslant 1$ and $|t| \leqslant x$. Then*

$$\zeta(s) = \sum_{n \leqslant x} \frac{1}{n^s} + \frac{x^{1-s}}{s-1} + O^\star\left(\frac{5}{6x^\sigma}\right).$$

However, it should be noticed that this result has the disadvantage that $\zeta(s)$ is approximated by a sum of length $\gg |t|$, which is difficult to deal with in many applications. Hardy and Littlewood provided another tool that works with some shorter sums.

Theorem 3.11 (Hardy–Littlewood) *Let $x, y, t > c > 0$ such that $2\pi xy = t$ and set $\Theta(s) = 2^s \pi^{s-1} \sin\left(\frac{1}{2}\pi s\right) \Gamma(1-s)$ so that $\zeta(s) = \Theta(s)\zeta(1-s)$ by the functional equation.[13] We have uniformly in $\sigma \in [0, 1]$*

$$\zeta(s) = \sum_{n \leqslant x} \frac{1}{n^s} + \Theta(s) \sum_{n \leqslant y} \frac{1}{n^{1-s}} + O\left(x^{-\sigma} + t^{1/2-\sigma} y^{\sigma-1}\right).$$

3.3.5 Estimates For $|\zeta(s)|$

In this section, we set $\tau = |t| + 3$ for all $t \in \mathbb{R}$.

The next lemma gives inequalities for $|\zeta(s)|$ near the right of the line $\sigma = 1$.

Lemma 3.13 *For all $\sigma > 1$ and $t \in \mathbb{R}$, we have*

$$\frac{1}{\sigma - 1} < \zeta(\sigma) < \frac{\sigma}{\sigma - 1}$$

and

$$\left|-\frac{\zeta'}{\zeta}(\sigma + it)\right| \leqslant -\frac{\zeta'}{\zeta}(\sigma) < \frac{1}{\sigma - 1}.$$

Furthermore, for all $\sigma > 1$ and $t \in \mathbb{R}$, we have

$$|\zeta(\sigma + it)| \leqslant \zeta(\sigma) < \frac{1}{\sigma - 1} + 0.64,$$

---

[13] Some authors use the symbol χ instead of Θ, but we keep this symbol for the Dirichlet characters in this book.

where the last estimate is valid for all $1 < \sigma < 1.12$. Finally, we also have for all $\sigma > 1$ and $t \in \mathbb{R}$

$$\frac{1}{|\zeta(\sigma + it)|} \leqslant \frac{\zeta(\sigma)}{\zeta(2\sigma)}.$$

Proof We start with

$$\int_1^\infty \frac{du}{u^\sigma} < \zeta(\sigma) < \int_1^\infty \frac{du}{u^\sigma} + 1$$

giving the first line of inequalities, and the second line follows from Abel's summation which implies that

$$-\zeta'(\sigma) = \sum_{n=1}^\infty \frac{\log n}{n^\sigma} = \sum_{n=1}^\infty \log\left(1 + \frac{1}{n}\right) \sum_{h=n+1}^\infty \frac{1}{h^\sigma}$$

$$< \sum_{n=1}^\infty \frac{1}{n} \int_n^\infty \frac{du}{u^\sigma} = \sum_{n=1}^\infty \frac{n^{1-\sigma}}{n(\sigma-1)} = \frac{\zeta(\sigma)}{\sigma - 1}.$$

For the next result, we first obviously have

$$|\zeta(\sigma + it)| \leqslant \sum_{n=1}^\infty \frac{1}{n^\sigma} = \zeta(\sigma)$$

and using (3.16) and integrating by parts, we derive

$$\zeta(\sigma) = \frac{1}{\sigma - 1} + \frac{1}{2} - \sigma(\sigma + 1) \int_1^\infty \frac{\psi_2(u)}{u^{\sigma+2}} du$$

and using $0 \leqslant -\psi_2(u) \leqslant \frac{1}{8}$ and $\sigma < 1.12$ we get the asserted estimate. Finally

$$\frac{1}{|\zeta(\sigma + it)|} = \left|\prod_p \left(1 - \frac{1}{p^{\sigma+it}}\right)\right| \leqslant \prod_p \left(1 + \frac{1}{p^\sigma}\right) = \frac{\zeta(\sigma)}{\zeta(2\sigma)}$$

as required.   □

3.3.6   Convexity Bounds

Using Theorem 3.8, we are in a position to estimate $|\zeta(s)|$ near the left of the line $\sigma = 1$.

For $\sigma \geqslant 2$, we have trivially

$$|\zeta(s)| \leqslant \zeta(\sigma) \leqslant \zeta(2) = \tfrac{1}{6}\pi^2.$$

For $1 - c/\log \tau \leqslant \sigma < 2$ for some $0 \leqslant c < \tfrac{1}{3}$, we use Theorem 3.8 with $x = \tau$. Note that $\tau^{1-\sigma} \leqslant e^c$ and, for all $n \leqslant \tau$, we also have $n^{-\sigma} \leqslant e^c n^{-1}$, so that[14]

$$|\zeta(s)| \leqslant \sum_{n \leqslant \tau} \frac{1}{n^\sigma} + \tau^{-\sigma}\left(1 + \frac{|s|}{\sigma}\right)$$

$$\leqslant e^c \sum_{n \leqslant \tau} \frac{1}{n} + \frac{e^c}{\tau}\left(2 + \frac{\tau}{\sigma}\right) \leqslant e^c(\log \tau + 3) \leqslant 4e^c \log \tau.$$

In particular, we have

$$|\zeta(1 + it)| \leqslant 4 \log \tau \qquad (t \in \mathbb{R} \setminus \{0\}). \tag{3.17}$$

For $\sigma = 0$ and $t \neq 0$, we have from the functional equation

$$|\zeta(it)| = \pi^{-1} \sinh\left(\tfrac{1}{2}\pi t\right) |\Gamma(1 - it)||\zeta(1 - it)|$$

and Theorem 3.3 implies that

$$|\Gamma(1 - it)| \ll \tau^{1/2} e^{-\pi\tau/2}$$

so that by above we get

$$\zeta(it) \ll \tau^{1/2} \log \tau \qquad (t \in \mathbb{R}). \tag{3.18}$$

This inequality also holds for $t = 0$ since it can be shown that $\zeta(0) = -\tfrac{1}{2}$.

For $0 < \sigma < 1$ and $t > 0$, we use the *Phragmén–Lindelöf principle*, which may be stated as follows.

Lemma 3.14 *Let $a < b$ be real numbers and suppose that $f : \mathbb{C} \to \mathbb{C}$ is a continuous, bounded function in the strip $a \leqslant \sigma \leqslant b$, and analytic in $a < \sigma < b$. If, for all $t \in \mathbb{R}$, we have*

$$|f(a + it)| \leqslant A \quad \text{and} \quad |f(b + it)| \leqslant B$$

---

[14]The constraint $c < \tfrac{1}{3}$ ensures that $\sigma > \tfrac{1}{2}$ for $\tau \geqslant 3$.

then, for all s such that $a < \sigma < b$, we have

$$|f(s)| \leqslant \left(A^{b-\sigma} B^{\sigma-a} \right)^{1/(b-a)}.$$

Applied here with (3.17) and (3.18), this result immediately yields

$$\zeta(s) \ll \tau^{(1-\sigma)/2} \log \tau \quad (0 < \sigma < 1).$$

By differentiation of the equation of Theorem 3.8, for $s \in \mathbb{C}$ satisfying $1 - c/\log \tau \leqslant \sigma < 2$, we get

$$\zeta'(s) = -\sum_{n \leqslant x} \frac{\log n}{n^s} - \frac{x^{1-s} \log x}{s-1} - \frac{x^{1-s}}{(s-1)^2} + R_1(s; x)$$

with

$$|R_1(s; x)| \leqslant \frac{|s|}{\sigma x^\sigma} \left(\log x + \sigma^{-1} \right) < \frac{|s|}{\sigma x^\sigma} (\log x + 2).$$

Proceeding as before and using the easy inequality

$$\sum_{n \leqslant \tau} \frac{\log n}{n} \leqslant \frac{(\log \tau)^2}{2} + 0.11$$

we obtain $|\zeta'(\sigma + it)| \leqslant 9e^c (\log \tau)^2$.

We may summarize these estimates in the following result.

Theorem 3.12 *Let $\tau = |t| + 3$. Uniformly in σ, we have*

$$\zeta(\sigma + it) \ll \begin{cases} 1, & \text{if } \sigma \geqslant 2 \\ \tau^{(1-\sigma)/2} \log \tau, & \text{if } 0 \leqslant \sigma \leqslant 1. \end{cases}$$

Furthermore, there exists $c \in \left(0, \frac{1}{3} \right)$ such that in the region $1 - \frac{c}{\log \tau} \leqslant \sigma < 2$, we have

$$|\zeta(\sigma + it)| \leqslant 4e^c \log \tau \quad \text{and} \quad |\zeta'(\sigma + it)| \leqslant 9e^c (\log \tau)^2.$$

More elaborate techniques can be used to improve on this result, as will be seen in Chap. 6. It is customary to denote by $\mu(\sigma)$ the lower bound of numbers α such that, for all $|t| \geqslant 1$, we have

$$\zeta(\sigma + it) \ll |t|^{\alpha}.$$

From the theory of Dirichlet series, it may be proved [85, §9.41] that this function is never negative, non-increasing and convex downwards, and hence continuous. By above, we get $\mu(\sigma) = 0$ for all $\sigma > 1$ and using the functional equation, it may be shown that $\mu(\sigma) = \frac{1}{2} - \sigma$ for all $\sigma < 0$. These equalities also hold by continuity for $\sigma = 1$ and $\sigma = 0$, respectively. However, the exact value of $\mu(\sigma)$ in the critical strip is still unknown. The simplest possible hypothesis is that the graph of this function consists of two straight lines $y = \frac{1}{2} - \sigma$ if $\sigma \leqslant \frac{1}{2}$ and $y = 0$ if $\sigma \geqslant \frac{1}{2}$. This is known as the *Lindelöf hypothesis* and is then equivalent to the statement that

$$\zeta\left(\tfrac{1}{2} + it\right) \ll t^{\varepsilon}$$

for all $\varepsilon > 0$ and $t \geqslant 3$. The best value to date is due to Bourgain and Watt [13] who proved that we have[15]

$$\zeta\left(\tfrac{1}{2} + it\right) \ll t^{13/84+\varepsilon}. \tag{3.19}$$

3.3.7   A Zero-Free Region

It can be shown [86, §3.7] that a weak version of the Prime Number Theorem, i.e. $\pi(x) \sim x/\log x$ when $x \to \infty$, follows from the next result, which goes back to Hadamard and de la Vallée Poussin (1896).

Theorem 3.13 *For all $t \in \mathbb{R}$, we have $\zeta(1 + it) \neq 0$. Furthermore, if $\tau = |t| + 3$, we have*

$$|\zeta(1+it)| > \frac{1}{221184(\log \tau)^7}.$$

---

[15] See Exercise 122.

Proof Since 1 is a pole of ζ, we suppose that $1 + it_0$ is a zero or order m of $\zeta(s)$ for some $t_0 \neq 0$, so that there exists $\ell \neq 0$ such that

$$\lim_{\sigma \to 1^+} \frac{\zeta(\sigma + it_0)}{(\sigma - 1)^m} = \ell.$$

Let $\sigma > 1$ and $t \in \mathbb{R}$. Using (3.9), we have

$$\mathrm{Re}\, (3 \log \zeta(\sigma) + 4 \log \zeta(\sigma + it) + \log \zeta(\sigma + 2it))$$

$$= \sum_{n=1}^{\infty} \frac{\Lambda(n)}{n^\sigma \log n} (3 + 4 \cos(t \log n) + \cos(2t \log n))$$

$$= 2 \sum_{n=1}^{\infty} \frac{\Lambda(n)}{n^\sigma \log n} (1 + \cos(t \log n))^2 \geqslant 0$$

which yields

$$\left| \zeta(\sigma)^3 \zeta(\sigma + it)^4 \zeta(\sigma + 2it) \right| \geqslant 1.$$

Therefore we get

$$\left| (\sigma - 1)^3 \zeta(\sigma)^3 (\sigma - 1)^{-4m} \zeta(\sigma + it_0)^4 \zeta(\sigma + 2it_0) \right| \geqslant (\sigma - 1)^{3-4m}.$$

Now making $\sigma \to 1^+$ gives a contradiction, since the left-hand side tends to a finite limit, whereas the right-hand side tends to ∞ if $m \geqslant 1$.

For the lower bound of $|\zeta(1 + it)|$, we proceed as follows. The inequality above and the estimates of Lemma 3.13 and Theorem 3.12 give for $1 < \sigma < 2$

$$\frac{1}{|\zeta(\sigma + it)|} \leqslant \zeta(\sigma)^{3/4} |\zeta(\sigma + 2it)|^{1/4} < (32)^{1/4} (\sigma - 1)^{-3/4} (\log \tau)^{1/4}.$$

so that by Theorem 3.12 we get for $1 < \sigma < 2$

$$|\zeta(1 + it)| \geqslant |\zeta(\sigma + it)| - 9(\sigma - 1)(\log \tau)^2$$

$$> (32 \log \tau)^{-1/4} (\sigma - 1)^{3/4} - 9(\sigma - 1)(\log \tau)^2$$

and the choice of $\sigma = 1 + 663552^{-1} (\log \tau)^{-9}$ gives the stated result. □

The proof given above follows essentially the lines of de la Vallée Poussin's. Hadamard's argument, also exploiting the link between $\zeta(1 + it_0)$ and $\zeta(1 + 2it_0)$, is similar in principle. The trigonometric polynomial used above may be replaced by any trigonometric polynomial

$$P(\theta) = \sum_{m=0}^{M} a_m \cos(m\theta)$$

satisfying $a_m \geqslant 0$ and $P(\theta) \geqslant 0$, which implies that, for $\sigma > 1$, we have

$$\text{Re}\left(\sum_{m=0}^{M} a_m \sum_{n=1}^{\infty} \frac{\Lambda(n)}{n^{\sigma+imt}}\right) \geqslant 0.$$

Since $\zeta(s) \neq 0$ for $\sigma \geqslant 1$, we infer that, apart from the trivial ones, the function $\zeta(s)$ has all its zeros in the so-called *critical strip* $0 < \sigma < 1$. It is customary to call these zeros the *non-trivial zeros* of $\zeta(s)$ and they are denoted by $\rho = \beta + i\gamma$. Since the Riemann zeta-function is real on the real axis, we have $\overline{\zeta(s)} = \zeta(\overline{s})$ by the reflexion principle, so that, if ρ is a zero of $\zeta(s)$, then so is $\overline{\rho}$. By the functional equation, we deduce that $1 - \rho$ and $1 - \overline{\rho}$ are also zeros of $\zeta(s)$.

For a stronger form of the Prime Number Theorem with an estimate of the error term, a larger zero-free region to the left of the line $\sigma = 1$ and some fine estimates of $\zeta(s)$ in this region are needed. This is the purpose of the next result.

Theorem 3.14 *Set $\tau = |t| + 3$. The Riemann zeta-function has no zero in the region*

$$\sigma \geqslant 1 - \frac{1}{5556379(\log \tau)^9} \quad \text{and} \quad t \in \mathbb{R}$$

in which we also have the estimates

$$|\zeta(\sigma + it)| > \frac{1}{442368(\log \tau)^7} \quad \text{and} \quad \left|-\frac{\zeta'}{\zeta}(\sigma + it)\right| < 5556379(\log \tau)^9.$$

Proof Let $0 \leqslant c < \frac{1}{3}$ be the constant appearing in Theorem 3.12 and let s be a complex number such that $1 - c/\log \tau \leqslant \sigma < 2$. From

$$\zeta(1 + it) - \zeta(\sigma + it) = \int_{\sigma}^{1} \zeta'(u + it)\,du$$

and Theorems 3.12 and 3.13, we get

$$|\zeta(\sigma + it)| \geqslant |\zeta(1+it)| - 9e^c(1-\sigma)(\log \tau)^2 > \frac{1}{221\,184(\log \tau)^7} - 9e^c(1-\sigma)(\log \tau)^2$$

so that if $\sigma \geqslant 1 - 3\,981\,312^{-1}e^{-c}(\log \tau)^{-9}$, we infer

$$|\zeta(\sigma + it)| > \frac{1}{442368(\log \tau)^7}$$

and by Theorem 3.12 we also obtain

$$\left| -\frac{\zeta'}{\zeta}(\sigma + it) \right| < 9e^{1/3} \times 442368 \times (\log \tau)^9 < 5556379(\log \tau)^9$$

as required. □

Proposition 3.9 *The function* $-\zeta'(s)/\zeta(s)$ *extends to a meromorphic function that has a simple pole at* $s = 1$ *with residue 1.*

Proof By (3.16), we have $(s - 1)\zeta(s) = \varphi(s)$ for all $\sigma > 0$, where

$$\varphi(s) := s - s(s - 1) \int_1^\infty \frac{\{x\}}{x^{s+1}} \, dx$$

so that $\varphi(s)$ is analytic in the half-plane $\sigma > 0$. Since $\zeta(s) \neq 0$ on $\sigma \geqslant 1$, we derive by logarithmic differentiation

$$-\frac{\zeta'(s)}{\zeta(s)} = \frac{1}{s - 1} - \frac{\varphi'(s)}{\varphi(s)}$$

implying the asserted claim. □

3.3.8 An Improved Zero-Free Region

With more work, one can improve Theorem 3.14 and get the classical zero-free region for the Riemann zeta-function.

Theorem 3.15 *Let* $\tau := |t| + 3$. *There is an absolute constant* $c > 0$ *such that* $\zeta(s) \neq 0$ *in the region* $\sigma \geqslant 1 - c/\log \tau$.

The proof rests on the following estimate, which will be used again later in another problem.

Proposition 3.10 *Let $s = \sigma + it$ with $-1 \leqslant \sigma \leqslant 2$ and s not equal to a zero of the Riemann zeta-function. Set $\tau = |t| + 3$. Then we have*

$$-\frac{\zeta'}{\zeta}(s) = \frac{1}{s-1} - \sum_{\substack{\rho \\ |t-\gamma| \leqslant 1}} \frac{1}{s-\rho} + O(\log \tau).$$

Proof Set $F(s) = s(s-1)\xi(s) = s(s-1)\pi^{-s/2}\Gamma(s/2)\zeta(s)$. $F(s)$ is an entire function of order 1, so that by the *Hadamard factorization theorem*, there exist suitable constants a, b such that

$$F(s) = e^{a+bs} \prod_{\rho} \left(1 - \frac{s}{\rho}\right) e^{s/\rho},$$

where the product runs through all zeros $\rho = \beta + i\gamma$ of $F(s)$ which are exactly the non-trivial zeros of $\zeta(s)$. The logarithmic differentiation provides

$$\frac{F'}{F}(s) = b + \sum_{\rho} \left(\frac{1}{s-\rho} + \frac{1}{\rho}\right) \tag{3.20}$$

where the sum is absolutely convergent. Taking $s = 0$ gives $b = F'(0)/F(0)$, and using Theorem 3.7 gives $F(s) = F(1-s)$, so that

$$b = \frac{F'(0)}{F(0)} = -\frac{F'(1)}{F(1)} = -b - \sum_{\rho} \left(\frac{1}{\rho} + \frac{1}{1-\rho}\right).$$

Now if ρ is a zero of $F(s)$, so are $\bar{\rho}$ and $1 - \rho$. Using this observation in the equation above we get

$$b = -\frac{1}{2} \sum_{\rho} \left(\frac{1}{\rho} + \frac{1}{\bar{\rho}}\right).$$

Therefore (3.20) becomes

$$\frac{F'}{F}(s) = \frac{1}{2} \sum_{\rho} \left(\frac{1}{s-\rho} + \frac{1}{s-\bar{\rho}}\right).$$

Now using the definition of F gives

$$\frac{F'}{F}(s) = \frac{1}{s} + \frac{1}{s-1} - \frac{\log \pi}{2} + \frac{\zeta'}{\zeta}(s) + \frac{1}{2}\frac{\Gamma'}{\Gamma}\left(\frac{s}{2}\right)$$

and by logarithmically differentiating the function equation $\Gamma(s+1) = s\Gamma(s)$ we obtain

$$\frac{\Gamma'}{\Gamma}(s+1) = \frac{1}{s} + \frac{\Gamma'}{\Gamma}(s)$$

so that

$$\frac{F'}{F}(s) = \frac{1}{s-1} - \frac{\log \pi}{2} + \frac{\zeta'}{\zeta}(s) + \frac{1}{2}\frac{\Gamma'}{\Gamma}\left(\frac{s}{2}+1\right)$$

and hence

$$-\frac{\zeta'}{\zeta}(s) = \frac{1}{s-1} - \frac{\log \pi}{2} + \frac{1}{2}\frac{\Gamma'}{\Gamma}\left(\frac{s}{2}+1\right) - \frac{1}{2}\sum_{\rho}\left(\frac{1}{s-\rho} + \frac{1}{s-\bar{\rho}}\right).$$

$$(3.21)$$

By logarithmically differentiating the Hadamard product of $\Gamma(s)$ and denoting temporarily $C \approx 0.577215\ldots$ the Euler–Mascheroni constant, we get using Example 1.1

$$\frac{\Gamma'}{\Gamma}(s) = -\frac{1}{s} - C - \sum_{n=1}^{\infty}\left(\frac{1}{n+s} - \frac{1}{n}\right) = \log N - \frac{1}{s} - \sum_{n=1}^{N}\frac{1}{n+s} + O(|s|N^{-1})$$

for all positive integers $N > |s|$, and by Corollary 1.3 with $f(t) = (t+s)^{-1}$ and letting $N \to \infty$ we infer for all s such that $|s| \geqslant \delta$ and $|\arg s| < \pi - \delta$ (with $\delta > 0$)

$$\frac{\Gamma'}{\Gamma}(s) = \log s - \frac{1}{2s} + O(|s|^{-2})$$

and therefore the Gamma-term in (3.21) is $\ll \log \tau$. Applying this estimate and (3.21) to s and to $2 + it$ and subtracting give

$$-\frac{\zeta'}{\zeta}(s) = \frac{1}{s-1} - \frac{1}{2}\sum_{\rho}\left(\frac{1}{s-\rho} - \frac{1}{2+it-\rho} + \frac{1}{s-\bar{\rho}} - \frac{1}{2+it-\bar{\rho}}\right) + O(\log \tau)$$

$$= \frac{1}{s-1} - \sum_{\rho}\left(\frac{1}{s-\rho} - \frac{1}{2+it-\rho}\right) + O(\log \tau).$$

By (3.51) we have

$$\sum_{\substack{\rho \\ |t-\gamma|\leqslant 1}}\frac{1}{2+it-\rho} \ll \sum_{\substack{\rho \\ |t-\gamma|\leqslant 1}}1 \ll \log \tau.$$

Now let $k \in \mathbb{Z}_{\geq 1}$ and consider the zeros ρ satisfying $k < |\gamma - t| \leq k + 1$. Since

$$\left| \frac{1}{s - \rho} - \frac{1}{2 + it - \rho} \right| = \frac{2 - \sigma}{|(s - \rho)(2 + it - \rho)|} \leq \frac{3}{|\gamma - t|^2} \leq \frac{3}{k^2}$$

we infer that such zeros contribute

$$\ll k^{-2} \left(\sum_{\substack{\rho \\ t+k < \gamma \leq t+k+1}} 1 + \sum_{\substack{\rho \\ t-k-1 \leq \gamma < t-k}} 1 \right) \ll k^{-2} \log(\tau + k)$$

by (3.51). Summing over k gives the asserted result.   □

In fact, such a result is also a consequence of a more general tool due to Borel and Carathéodory which may be stated as follows.[16]

Lemma 3.15 *Let $s_0 \in \mathbb{C}$ and F be an analytic function in the disc $|s - s_0| \leq R$ satisfying in this disc the bound $|F(s)| \leq M$ and such that $|F(s_0)| \geq m$. Then, for all s such that $|s - s_0| \leq \frac{1}{4} R$, we have*

$$\left| \frac{F'(s)}{F(s)} - \sum_{|\rho - s_0| \leq R/2} \frac{1}{s - \rho} \right| \leq \frac{8}{R} \log \left(\frac{M}{m} \right),$$

where the sum runs through the zeros ρ of F such that $|\rho - s_0| \leq \frac{1}{2} R$.

Using this result with $F(s) = \zeta(s)$, $s_0 = 1 + it_0$ for some $t_0 \geq 4$, $R = 2$, a bound $|\zeta(s)| \ll |s|^{3/2}$ for $-1 \leq \sigma \leq 2$ and the lower bound of Theorem 3.14, we get for all s such that $|s - 1 - it_0| \leq \frac{1}{2}$

$$-\frac{\zeta'}{\zeta}(s) = - \sum_{|\rho - 1 - it_0| \leq 1} \frac{1}{s - \rho} + O(\log t_0).$$

Proof of Theorem 3.15 Note first that the Euler product of the Riemann zeta-function enables us to assume $\sigma \leq 1$, and Corollary 3.6 ensures that $\zeta(s) \neq 0$ in any rectangle strictly contained in the parabolic region $\sigma > \frac{1}{2}(1 + t^2)$, for instance in the rectangle $\frac{5}{6} \leq \sigma \leq 1$ and $|t| \leq \frac{4}{5}$. Hence we may assume that $\rho_0 := \beta_0 + i\gamma_0$ is a zero of the zeta function satisfying $\frac{1}{2} \leq \beta_0 \leq 1$ and $|\gamma_0| \geq \frac{4}{5}$. Arguing exactly as in the proof of Theorem 3.13, we first show that, for $\sigma > 1$

$$\operatorname{Re} \left(-3\frac{\zeta'}{\zeta}(\sigma) - 4\frac{\zeta'}{\zeta}(\sigma + it) - \frac{\zeta'}{\zeta}(\sigma + 2it) \right) \geq 0. \tag{3.22}$$

---

[16]This exposition is due to Ramaré [67].

Now let $x \in (0, 1)$ be a parameter at our disposal. By Proposition 3.9, we first have

$$-\frac{\zeta'}{\zeta}(1 + x) = \frac{1}{x} + O(1).$$

Note that, since all zeros ρ of $\zeta(s)$ satisfy $\text{Re}\,\rho \leqslant 1$, it follows that $-\,\text{Re}\left(\frac{1}{s-\rho}\right) < 0$ whenever $\sigma > 1$. Hence Proposition 3.10 with $s = 1 + x + i\gamma_0$ yields

$$-\,\text{Re}\,\frac{\zeta'}{\zeta}(1 + x + i\gamma_0) \leqslant \frac{x}{x^2 + \gamma_0^2} - \frac{1}{1 + x - \beta_0} + c_0 \log(|\gamma_0| + 3)$$

$$\leqslant -\frac{1}{1 + x - \beta_0} + c_1 \log(|\gamma_0| + 3)$$

where we used the fact that $x \in (0, 1)$ and $|\gamma_0| \geqslant \frac{4}{5}$ to derive a bound of the shape $\frac{x}{x^2 + \gamma_0^2} \leqslant \gamma_0^{-2} \leqslant \frac{25}{16} \leqslant c_0' \log(|\gamma_0| + 3)$. Similarly

$$-\,\text{Re}\,\frac{\zeta'}{\zeta}(1 + x + 2i\gamma_0) \leqslant c_2 \log(|2\gamma_0| + 3).$$

Inserting these estimates in (3.22) implies the existence of a constant $c_3 \geqslant 1$ such that

$$\frac{3}{x} - \frac{4}{1 + x - \beta_0} + c_3 \log(|\gamma_0| + 3) \geqslant 0.$$

Choosing $x = (2c_3 \log(|\gamma_0| + 3))^{-1}$ yields

$$7c_3 \log(|\gamma_0| + 3) \geqslant \frac{4}{1 + x - \beta_0}$$

and therefore

$$1 - \beta_0 \geqslant \frac{1}{14c_3 \log(|\gamma_0| + 3)}$$

as required.   □

This zero-free region enables us to improve the bounds of Theorem 3.14.

Theorem 3.16 *Let c be the constant in* Theorem 3.15 *and recall that* $\tau :=$ $|t| + 3$. *If* $\sigma > 1 - c/(2 \log \tau)$, *then*

$$\frac{\zeta'}{\zeta}(s) \ll \log \tau ; \tag{3.23}$$

$$\log |\zeta(s)| \leqslant \log \log \tau + O(1) ; \tag{3.24}$$

$$\zeta(s)^{-1} \ll \log \tau. \tag{3.25}$$

Proof If $|t| \leqslant \frac{4}{5}$, then by Proposition 3.9 we get $-\frac{\zeta'}{\zeta}(s) = \frac{1}{s-1} + O(1)$, so that we may suppose $|t| > \frac{4}{5}$. Furthermore, using Lemma 3.13, we see that the first estimate holds if $\sigma \geqslant 1 + (\log \tau)^{-1}$, hence assume $\sigma < 1 + (\log \tau)^{-1}$ and set $s_0 := 1 + (\log \tau)^{-1} + it$. By Proposition 3.10, we derive

$$\frac{\zeta'}{\zeta}(s) - \frac{\zeta'}{\zeta}(s_0) = \frac{s - s_0}{(s_0 - 1)(s - 1)} + \sum_{\substack{\rho \\ |t-\gamma| \leqslant 1}} \left(\frac{1}{s - \rho} - \frac{1}{s_0 - \rho} \right) + O(\log \tau).$$

Since $|t| > \frac{4}{5}$, the contribution of the first term is $\ll \log \tau$. Also note that, for all zeros in the sum, we have $|s - \rho| \asymp |s_0 - \rho|$, hence, using Proposition 3.10 again

$$\sum_{\substack{\rho \\ |t-\gamma| \leqslant 1}} \left(\frac{1}{s - \rho} - \frac{1}{s_0 - \rho} \right) \ll \sum_{\substack{\rho \\ |t-\gamma| \leqslant 1}} \frac{|s - s_0|}{|s_0 - \rho|^2} \ll \sum_{\substack{\rho \\ |t-\gamma| \leqslant 1}} \mathrm{Re} \left(\frac{1}{s_0 - \rho} \right)$$

$$\ll \frac{\zeta'}{\zeta}(s_0) + \log \tau \ll \log \tau$$

where we used Lemma 3.13 in the last estimate. Inserting into the equality above gives (3.23). To derive (3.24), note that, if $\sigma > 1$

$$\log |\zeta(s)| \leqslant \sum_{n=2}^{\infty} \frac{\Lambda(n)}{n^\sigma \log n} = \log \zeta(\sigma)$$

implying that this estimate is true if $\sigma \geqslant 1 + (\log \tau)^{-1}$. In the case $\sigma < 1 + (\log \tau)^{-1}$, take $s_0 := 1 + (\log \tau)^{-1} + it$ again and

$$\log \zeta(s) - \log \zeta(s_0) = \int_{s_0}^{s} \frac{\zeta'}{\zeta}(z) \, dz,$$

where the path of integration is the line joining the endpoints. It follows that the previous bound yields an error term $O(1)$ for the integral, and hence

$$\log |\zeta(s)| \leqslant \log |\zeta(s_0)| + O(1)$$
$$\leqslant \log \left\{ \zeta \left(1 + (\log \tau)^{-1} \right) \right\} + O(1)$$
$$\leqslant \log \log \tau + O(1)$$

as required. This bound also implies (3.25) via $\log |\zeta(s)|^{-1} = -\operatorname{Re}(\log \zeta(s))$. □

3.3.9 The Resonance Method

One of the most important problems in analytic number theory is the study of the distribution of extreme values of the Riemann zeta-function, which may have crucial repercussions in many branches of arithmetic.

In recent years, the so-called resonance method emerged among the few existing techniques in order to study these large values. The basic idea is to consider the integrals

$$I_1 := \int_{-\infty}^{\infty} |R(t)|^2 \, \Phi\left(\tfrac{t}{T}\right) dt \quad \text{and} \quad I_2 := \int_{-\infty}^{\infty} \zeta(\sigma + it) \, |R(t)|^2 \, \Phi\left(\tfrac{t}{T}\right) dt,$$

where $T \geqslant 1$, Φ is a smooth,[17] [0, 1]-valued function and compactly supported in $[T^{\alpha-1}, 1]$, so that

$$\max_{T^\alpha \leqslant t \leqslant T} |\zeta(\sigma + it)| \geqslant \frac{|I_2|}{I_1}$$

for any given $\alpha \in [0, 1)$, and hence the problem is to find a function $R(t)$ such that the quotient $|I_2|/I_1$ is large. In practice, $R(t)$ is most often a Dirichlet polynomial of the shape

$$R(t) = \sum_{n \in S} \frac{r(n)}{n^{it}}$$

for some set S of integers such that $\max S \leqslant T$, and where $r(n) > 0$. When $|I_2|$ is as large as possible, given that the coefficients $r(n)$ have a fixed square sum and

---

[17]In practice, the only needed properties of Φ are that the Fourier transform $\widehat{\Phi}$ of Φ is positive and that both Φ and $\widehat{\Phi}$ decay fast. A natural choice is the Gaussian $\Phi(t) = e^{-t^2/2}$, for instance.

also subject to whatever a priori restrictions we need to put on the set of integers in S, we say that $r(n)$ "resonates" with ζ.

This method seems to have been first used by Voronin in [91], but it was developed independently and significantly refined by Soundararajan [81] and Hilberdink [40]. The previous tools used to rely on estimates for high moments of the zeta function and on Diophantine approximation arguments. For instance, improving earlier work of Littlewood, Levinson [48] proved that, for arbitrary large t

$$|\zeta(1 + it)| \geqslant e^{\gamma} \log \log t + O(1).$$

The resonance method, with the "long resonator" argument, enables the authors in [1] to sharpen this estimate.

Theorem 3.17 *There exists a constant C such that, for all sufficiently large T, there exists $t \in \left[\sqrt{T}, T\right]$ such that*

$$|\zeta(1 + it)| \geqslant e^{\gamma} \left(\log \log T + \log \log \log T - C \right).$$

This result is in accordance with a conjecture stated by Granville and Soundararajan [33] who surmised that, for $T \to \infty$

$$\max_{t \in [T, 2T]} |\zeta(1 + it)| = e^{\gamma} \left(\log \log T + \log \log \log T + C_1 + o(1) \right)$$

with $C_1 \approx -0.089 \ldots$. These estimates could be compared to the following explicit upper bounds we have at our disposal today [28, 88]. For all $t \geqslant 3$

$$|\zeta(1 + it)| \leqslant \min \left(76.2 (\log t)^{2/3} , \tfrac{3}{4} \log t \right).$$

It is also a "long resonator" which allowed the authors in [9] to derive precise estimates on the line $\sigma = \frac{1}{2}$.

Theorem 3.18 *For $T \to \infty$*

$$\max_{t \in \left[\sqrt{T}, T\right]} \left| \zeta \left(\tfrac{1}{2} + it \right) \right| \geqslant \exp \left(\left(\frac{1}{\sqrt{2}} + o(1) \right) \sqrt{\frac{\log T \log \log \log T}{\log \log T}} \right).$$

This result improves by a factor $\log \log \log T$ the best previous lower estimate given by Soundararajan, who proved in [81] that, for $T \to \infty$

$$\max_{t \in [T, 2T]} \left| \zeta \left(\tfrac{1}{2} + it \right) \right| \geqslant \exp \left((1 + o(1)) \sqrt{\frac{\log T}{\log \log T}} \right).$$

The main novelty is the choice of a suitable resonator whose idea comes from a sharp lower bound of a double gcd-sum as seen in [9, Theorem 2]. Roughly speaking, let $N \in \mathbb{Z}_{\geqslant 1}$ be a large fixed integer and let $\widetilde{P}$ be the set of primes such that

$$e \log N \log_2 N < p \leqslant e^{\sqrt{\log_2 N}} \log N \log_2 N,$$

where $\log_k$ is the kth iterated logarithm function. Now let f be the positive multiplicative function, supported on squarefree numbers, such that

$$f(p) = \begin{cases} \sqrt{\dfrac{\log N \log_2 N}{\log_3 N}} \times \dfrac{1}{\sqrt{p} \left(\log p - \log_2 N - \log_3 N\right)}, & \text{if } p \in \widetilde{P} \\ 0, & \text{otherwise.} \end{cases}$$

Choosing the resonator to be the square-root of the mean square of the function f over a certain restricted set of integers as explained in [9], the authors prove that

$$\frac{|I_2|}{I_1} \gg A_N (\log T)^{-4} + \log T,$$

where

$$A_N := \prod_{p \in \widetilde{P}} \left(1 + \frac{f(p)}{\sqrt{p} \left(1 + f(p)^2\right)}\right)$$

provided that $N \leqslant T^{1/2}$. If $N \geqslant \exp\left(e^{e^{15e}}\right)$, then plainly

$$A_N \geqslant \exp\left(\frac{9}{10} \sum_{p \in \widetilde{P}} \frac{f(p)}{\sqrt{p}}\right) \tag{3.26}$$

with

$$\sum_{p \in \widetilde{P}} \frac{f(p)}{\sqrt{p}} = \sqrt{\frac{\log N \log_2 N}{\log_3 N}} \sum_{p \in \widetilde{P}} \frac{1}{p \left(\log p - \log_2 N - \log_3 N\right)}.$$

Applying Corollary 3.13 in the form $\sum_{a<p\leqslant b} \frac{1}{p} > \log\left(\frac{\log b}{\log a}\right) - \frac{3}{2(\log a)^2}$ with $1 < a < b$, and using a partial summation, we derive

$$\sum_{p\in\widetilde{P}} \frac{1}{p\left(\log p - \log_2 N - \log_3 N\right)} > \frac{\log_3 N}{2(\log_2 N + \log_3 N)} - \frac{1}{(\log_2 N)^{3/2}} - \frac{3}{2(\log_2 N)^2}$$

$$\geqslant \frac{\log_3 N}{4\log_2 N} - \frac{5}{2(\log_2 N)^{3/2}} \geqslant \frac{\log_3 N}{8\log_2 N}$$

since $N \geqslant \exp\left(e^{e^{15e}}\right)$. Inserting this bound in (3.26), we get

$$A_N \geqslant \exp\left(\frac{9}{80}\sqrt{\frac{\log N \log_3 N}{\log_2 N}}\right)$$

and choosing $N = \lfloor T^{1/2}\rfloor$ yields a slightly weaker version of Theorem 3.18.

For the case $\frac{1}{2} < \sigma < 1$, Montgomery [54], using Diophantine approximation arguments, showed that there exists a constant $c(\sigma)$, depending only on σ, such that, for sufficiently large T, there exists $t \in [T^\beta, T]$, $0 < \beta < 1$, such that

$$|\zeta(\sigma + it)| \geqslant \exp\left(c(\sigma)\frac{(\log T)^{1-\sigma}}{(\log\log T)^\sigma}\right).$$

The best constant $c(\sigma)$ previously achieved was of the form $c(\sigma) = \frac{c_0(\sigma)}{1-\sigma}$ for some $c_0(\sigma) < 0.17$.[18] The resonance method also works in this case and improves on Montgomery's result, as it can be shown in [10]. Note that, as pointed out by the authors, the technicalities will differ considerably depending on σ.

Theorem 3.19 *There exists a positive-valued continuous function $\sigma \mapsto \rho(\sigma)$ on $\left(\frac{1}{2}, 1\right)$, bounded below by $(2\sigma - 2)^{-1}$, with the asymptotic behavior*

$$\rho(\sigma) = \begin{cases} (1-\sigma)^{-1} + O\left(|\log(1-\sigma)|\right), & \sigma \nearrow 1; \\ \left(2^{-1/2} + o(1)\right)\sqrt{|\log(2\sigma - 1)|}, & \sigma \searrow \frac{1}{2}; \end{cases}$$

and such that the following holds. If T is sufficiently large, then

▷ *if $\sigma \in \left[\frac{1}{2} + (\log\log T)^{-1}, \frac{3}{4}\right]$, there exists $t \in \left[\sqrt{T}, T\right]$ such that*

$$|\zeta(\sigma + it)| \geqslant \exp\left(\rho(\sigma)\frac{(\log T)^{1-\sigma}}{(\log\log T)^\sigma}\right);$$

---

[18]This result is due to Ramachandra and Sankaranarayanan in 1991.

▷ *if* $\sigma \in \left[\frac{3}{4}, \ 1 - (\log \log T)^{-1} \right]$, *there exists* $t \in \left[\frac{1}{2} T, T \right]$ *such that*

$$|\zeta(\sigma + it)| \geq \exp \left(\rho(\sigma) \frac{(\log T)^{1-\sigma}}{(\log \log T)^{\sigma}} + \log \log \log T + c \right),$$

where c is an absolute constant independent of T.

Let us finally mention that this method may be applied to other Dirichlet *L*-functions, as can be seen in [81] or [2].

3.4 Dirichlet *L*-Functions

Let a and q be two positive integers. The question of the infinity of the prime numbers $p \equiv a \pmod{q}$ arises naturally. First note that we may suppose that $(a, q) = 1$, otherwise the terms of the sequence $(qn + a)$ are all divisible by (a, q). Thus the condition $(a, q) = 1$ is necessary to this problem. Dirichlet's tour de force was precisely to show that this hypothesis is also *sufficient*. The purpose of this section is to provide a proof of Dirichlet's theorem.

> **Theorem 3.20 (Dirichlet)** *Let* a, q *be positive coprime integers. Then there are infinitely many prime numbers* p *such that* $p \equiv a \pmod{q}$.

3.4.1 Euclid vs. Euler

Let us go back in time. Before Euler, there were many attempts to generalize Euclid's method to the problem of primes in some fixed arithmetic progressions. The following lemma shows that the method works for at least two of them.

Lemma 3.16

▷ *There are infinitely many prime numbers* p *such that* $p \equiv 3 \pmod{4}$.
▷ *There are infinitely many prime numbers* p *such that* $p \equiv 1 \pmod{4}$.

Proof

▷ Suppose the contrary and let $\mathcal{P}_{4,3} = \{p_1, \ldots, p_n\}$ the finite set of all prime numbers p such that $p \equiv 3 \pmod{4}$. Set $M = 4p_1 \cdots p_n - 1$. If M is prime, then we get a prime number of the type $M \equiv 3 \pmod{4}$ such that $M > p_n$. Suppose now that M is composite. Then there exists at least an odd prime divisor p of M of the form $p \equiv 3 \pmod{4}$, otherwise all prime factors of M are $\equiv 1 \pmod{4}$ since

M is odd, and we have $M \equiv 1 \pmod{4}$ which is not the case. On the other hand, we have $p \notin \mathcal{P}_{4,3}$, otherwise we have $p \mid M + 1$. Thus we have found a prime number p such that $p \equiv 3 \pmod{4}$ and $p \notin \mathcal{P}_{4,3}$, leading to a contradiction.

▷ Suppose the contrary and let $\mathcal{P}_{4,1} = \{p_1, \ldots, p_n\}$ the finite set of all prime numbers p such that $p \equiv 1 \pmod{4}$. Set $M = 4 (p_1 \cdots p_n)^2 + 1$. If M is prime, then we get a prime number of the type $M \equiv 1 \pmod{4}$ such that $M > p_n$. Suppose that M is composite and let p be a prime factor of M. Since M is odd, we have $p \neq 2$ and also $p \notin \mathcal{P}_{4,1}$, otherwise it divides $M - 1$. Since $p \mid M$, we have $4 (p_1 \cdots p_n)^2 \equiv -1 \pmod{p}$ and hence -1 is a quadratic residue modulo p. By Example 3.2, we infer that $p \equiv 1 \pmod{4}$. Thus we have found a prime number p such that $p \equiv 1 \pmod{4}$ and $p \notin \mathcal{P}_{4,1}$, leading to a contradiction.   □

This argument may be generalized to certain arithmetic progressions.

Proposition 3.11 *Let q be an odd prime number. Then there are infinitely many prime numbers p such that $p \equiv 1 \pmod{q}$.*

Proof Let $a > 1$ be an integer such that $q \mid a$ and set $M = 1 + a + a^2 + \cdots + a^{q-1}$. Let p be a prime divisor of M. Note that $p \nmid a$, otherwise we have $M \equiv 1 \pmod{p}$ which contradicts $p \mid M$. Hence $p \neq q$. Since $p \mid a^q - 1$, we infer that $\mathrm{ord}_p(a)$ divides q and since q is prime, we infer that $\mathrm{ord}_p(a) = 1$ or q. If $\mathrm{ord}_p(a) = 1$, then we get $M \equiv q \pmod{p}$ and hence $q \equiv 0 \pmod{p}$ which is impossible, since p and q are two distinct prime numbers. Therefore $\mathrm{ord}_p(a) = q$ and then $q \mid p - 1$ by Theorem 3.2. In other words, we have proved that

$$p \equiv 1 \pmod{q}.$$

Suppose that there are finitely many primes $p_1, \ldots, p_n$ such that $p_i \equiv 1 \pmod{q}$. Consider the integer $a = q p_1 \cdots p_n$. By above, the integer $M = 1 + a + \cdots + a^{q-1}$ has a prime divisor p such that $p \equiv 1 \pmod{q}$ and also $p \neq p_i$ otherwise we have $p \mid M - 1$, giving a contradiction.   □

The proofs above use a Euclidean argument, in the sense that we have at our disposal a polynomial $P \in \mathbb{Z}[X]$ with degree > 0 and whose integer values have prime divisors, almost all lying in some arithmetic progressions. For instance, Euclid's theorem uses the polynomial $P = X + 1$, Lemma 3.16 uses $P = X - 1$ and $P = X^2 + 1$, respectively, and Proposition 3.11 uses the *cyclotomic polynomial* $\Phi_q = X^{q-1} + X^{q-2} + \cdots + 1$.[19]

   One may wonder whether such arguments may be generalized to *all* arithmetic progressions. The following curious result, due to Schur (1912) for the sufficient condition and Ram Murty (1988) for the necessary condition, shows that such a proof cannot exist for all cases.

---

[19] See Chap. 7, Section 7.1.9.

Proposition 3.12 *Let a, q be positive coprime integers. Then there exists a Euclidean proof for the sequence a (mod q) if and only if $a^2 \equiv 1$ (mod q).*

For instance, Euclid's argument may be used to show the infinity of the set of primes such that $p \equiv 8$ (mod 9) but *cannot be used* to prove that there are infinitely many prime numbers p such that $p \equiv 7$ (mod 9).

Thus, a new idea is needed. Around 1837, Dirichlet succeeded in using a generalization of Euler's proof of Euclid's theorem and some group-theoretic tools. More precisely, Dirichlet proved the divergence of the series

$$\sum_{p \equiv a \,(\mathrm{mod}\, q)} \frac{1}{p}$$

by discovering a clever expression for the characteristic function

$$\mathbf{1}_{q,a}(n) = \begin{cases} 1, & \text{if } n \equiv a \ (\mathrm{mod}\, q) \\ 0, & \text{otherwise} \end{cases} \tag{3.27}$$

as will be seen in Proposition 3.13 and showing

$$\lim_{\sigma \to 1^+} \sum_p \frac{\mathbf{1}_{q,a}(p)}{p^\sigma} = \infty.$$

An alternative way is to estimate partial sums of the above series. Let us examine this in the following example taking $q = 4$ and $a = 1$. Dirichlet used the function $\mathbf{1}_{4,1}$ defined for all *odd* positive integers n by

$$\mathbf{1}_{4,1}(n) = \tfrac{1}{2}\left(1 + \sin\left(\tfrac{1}{2}n\pi\right)\right).$$

One may readily check that, for all odd n, we have

$$\mathbf{1}_{4,1}(n) = \begin{cases} 1, & \text{if } n \equiv 1 \ (\mathrm{mod}\, 4) \\ 0, & \text{if } n \equiv 3 \ (\mathrm{mod}\, 4). \end{cases}$$

Thus, for all $N > 1$, we get

$$\sum_{\substack{p \leqslant N \\ p \equiv 1 \,(\mathrm{mod}\, 4)}} \frac{1}{p} = \sum_{3 \leqslant p \leqslant N} \frac{\mathbf{1}_{4,1}(p)}{p} = \frac{1}{2} \sum_{3 \leqslant p \leqslant N} \frac{1}{p} + \frac{1}{2} \sum_{3 \leqslant p \leqslant N} \frac{\sin(\pi p/2)}{p}.$$

By Corollary 3.4, the first sum tends to ∞ when $N \to \infty$. We will prove in Theorem 3.21 that the series

$$\sum_p \frac{\sin(\pi p/2)}{p}$$

converges, which establishes the divergence of the initial series

$$\sum_{p \equiv 1 \,(\mathrm{mod}\,4)} \frac{1}{p}.$$

3.4.2   Dirichlet Characters

Dirichlet introduced the *characters* which was the main tool in the proof of his theorem. Although these characters may be defined on any finite Abelian group,[20] we may restrict ourselves here to the so-called *Dirichlet characters* as extensions over $\mathbb{Z} \setminus \{0\}$ of homomorphisms of the multiplicative group $(\mathbb{Z}/q\mathbb{Z})^\times$.

Definition 3.5   Let q be a positive integer. A *Dirichlet character* modulo q is a map $\chi : \mathbb{Z} \setminus \{0\} \to \mathbb{C}$ satisfying the following rules for all $a, b \in \mathbb{Z} \setminus \{0\}$.

▷ $\chi(a) = \chi(a \,(\mathrm{mod}\,q))$;
▷ $\chi(ab) = \chi(a)\chi(b)$;
▷ If $(a, q) > 1$, then $\chi(a) = 0$.

In fact, the first two conditions mean that these characters are homomorphisms of the multiplicative group $(\mathbb{Z}/q\mathbb{Z})^\times$ and the third one extends these maps to $\mathbb{Z} \setminus \{0\}$. We shall make frequent use of the q-periodicity of the characters. One can prove that the set of Dirichlet characters modulo q is a group isomorphic to the multiplicative group $(\mathbb{Z}/q\mathbb{Z})^\times$ of the units of the ring $\mathbb{Z}/q\mathbb{Z}$. In particular, there are $\varphi(q)$ Dirichlet characters modulo q. The identity element of this group is called the *principal character* modulo q and is usually denoted by χ_0. Thus, χ_0 is defined for all $a \in \mathbb{Z}$ by

$$\chi_0(a) = \begin{cases} 1, & \text{if } (a, q) = 1 \\ 0, & \text{otherwise.} \end{cases}$$

Let χ be a Dirichlet character modulo q. We define $\overline{\chi}$ by $\overline{\chi}(a) = \overline{\chi(a)}$. Clearly, $\overline{\chi}$ is also a Dirichlet character modulo q called the *conjugate character* of χ. It is also not difficult to see that, if $(a, q) = 1$, then $\chi(a)$ is a $\varphi(q)$th root of unity. Indeed, denoting by a the residue class of the integer a in $(\mathbb{Z}/q\mathbb{Z})^*$, we have

$$(\chi(a))^{\varphi(q)} = \chi\left(a^{\varphi(q)}\right) = \chi(1) = 1.$$

Dirichlet succeeded in proving that a suitable linear combination of these characters provides the desired characteristic functions $\mathbf{1}_{q,a}$.

---

[20] See Definition 7.29 for a more general situation.

Proposition 3.13 *Let a, q be positive coprime integers and define $\mathbf{1}_{q,a}(n)$ as*

$$\mathbf{1}_{q,a}(n) = \begin{cases} 1, & \text{if } n \equiv a \pmod{q} \\ 0, & \text{otherwise.} \end{cases}$$

For all positive integers n we have

$$\mathbf{1}_{q,a}(n) = \frac{1}{\varphi(q)} \sum_{\chi \,(\mathrm{mod}\, q)} \overline{\chi}(a)\chi(n),$$

where the summation is taken over all Dirichlet characters modulo q.

Proof Suppose first that $a = 1$. If $n \equiv 1 \pmod{q}$, then we have by Definition 3.5

$$\sum_{\chi \,(\mathrm{mod}\, q)} \chi(n) = \sum_{\chi \,(\mathrm{mod}\, q)} \chi(1) = \sum_{\chi \,(\mathrm{mod}\, q)} 1 = \varphi(q).$$

Now assume that $n \not\equiv 1 \pmod{q}$ and $(n, q) = 1$. Then there exists a character χ_1 such that $\chi_1(n) \neq 1$. When χ ranges over all Dirichlet characters modulo q, so does the character $\chi_1 \chi$ and hence

$$\sum_{\chi \,(\mathrm{mod}\, q)} \chi(n) = \sum_{\chi \,(\mathrm{mod}\, q)} (\chi_1\chi)(n) = \chi_1(n) \sum_{\chi \,(\mathrm{mod}\, q)} \chi(n)$$

which implies that $\sum_{\chi \,(\mathrm{mod}\, q)} \chi(n) = 0$ as required. Now suppose that $a \neq 1$ and let a^{-1} be the inverse of a in $(\mathbb{Z}/q\mathbb{Z})^*$. From the relations $\chi(a)\overline{\chi}(a) = 1$ and $\chi(a)\chi(a^{-1}) = \chi(1) = 1$, we infer that $\overline{\chi}(a) = \chi(a^{-1})$. Hence

$$\sum_{\chi \,(\mathrm{mod}\, q)} \overline{\chi}(a)\chi(n) = \sum_{\chi \,(\mathrm{mod}\, q)} \chi(a^{-1}n) = \begin{cases} 1, & \text{if } a^{-1}n \equiv 1 \pmod{q} \\ 0, & \text{otherwise} \end{cases}$$

by above, which completes the proof. $\qquad\qquad\qquad\qquad\qquad\qquad\square$

This result thus provides an expression of the partial sums that we wish to estimate.

Corollary 3.7 *Let a, q be positive coprime integers. Then, for any complex-valued arithmetic function f and any $x \geqslant 1$*

$$\sum_{\substack{n \leqslant x \\ n \equiv a \,(\mathrm{mod}\, q)}} f(n) = \frac{1}{\varphi(q)} \sum_{\substack{n \leqslant x \\ (n,q)=1}} f(n) + \frac{1}{\varphi(q)} \sum_{\chi \neq \chi_0} \overline{\chi}(a) \sum_{n \leqslant x} \chi(n)f(n).$$

Proof By Proposition 3.13, the left-hand side is

$$\sum_{n \leqslant x} \mathbb{1}_{q,a}(n) f(n) = \frac{1}{\varphi(q)} \sum_{\chi \,(\mathrm{mod}\, q)} \overline{\chi}(a) \sum_{n \leqslant x} \chi(n) f(n)$$

and we split the first sum according to $\chi = \chi_0$ or $\chi \neq \chi_0$.   □

In particular

$$\sum_{\substack{p \leqslant N \\ p \equiv a \,(\mathrm{mod}\, q)}} \frac{1}{p} = \frac{1}{\varphi(q)} \sum_{\substack{p \leqslant N \\ (p,q)=1}} \frac{1}{p} + \frac{1}{\varphi(q)} \sum_{\chi \neq \chi_0} \overline{\chi}(a) \sum_{p \leqslant N} \frac{\chi(p)}{p}.$$

It is fairly easy to see that the first sum tends to ∞ as $N \to \infty$, since it only differs from the sum $\sum_{p \leqslant N} 1/p$ of a finite number of terms. The crucial point is then to show that the series

$$\sum_{p} \frac{\chi(p)}{p}$$

converges for all $\chi \neq \chi_0$, implying the divergence of the series

$$\sum_{p \equiv a \,(\mathrm{mod}\, q)} \frac{1}{p}.$$

The next result uses the periodicity of the characters to bound partial sums of non-principal Dirichlet characters.

Proposition 3.14 *For all non-principal Dirichlet characters χ modulo q and all integers $M \in \mathbb{Z}_{\geqslant 0}$ and $N \in \mathbb{Z}_{\geqslant 1}$, we have*

$$\left| \sum_{M < n \leqslant M+N} \chi(n) \right| \leqslant \varphi(q).$$

Proof Let $K = q \left\lfloor \frac{1}{q}(N-1) \right\rfloor$. As in Proposition 3.13, one may check that, for all $\chi \neq \chi_0$, we have

$$\sum_{a \,(\mathrm{mod}\, q)} \chi(a) = 0$$

and hence, by periodicity, we get

$$\sum_{n=M+1}^{M+K} \chi(n) = \sum_{j=1}^{K/q} \sum_{n=M+1+(j-1)q}^{M+jq} \chi(n) = \sum_{j=1}^{K/q} \sum_{n=M+1}^{M+q} \chi(n) = 0.$$

The interval $(M+K, M+N]$ contains q integers $n_1, \ldots, n_q$, and denoting by n_i the residue class of the integer n_i in $(\mathbb{Z}/q\mathbb{Z})^\times$, we obtain

$$\left| \sum_{M<n\leqslant M+N} \chi(n) \right| \leqslant \sum_{\substack{i=1 \\ (n_i,q)=1}}^{q} |\chi(n_i)| \leqslant \sum_{\substack{n\leqslant q \\ (n,q)=1}} 1 = \varphi(q)$$

as asserted.   □

Using Abel's summation and the trivial bound $\varphi(q) \leqslant q$, we may readily deduce the following useful consequence.

Corollary 3.8 *Let $F \in C^1[1, +\infty)$ be a decreasing function such that $F > 0$ and $F(x) \to 0$ as $x \to \infty$. For all non-principal Dirichlet characters χ modulo q and all real numbers $x \geqslant 1$, we have*

$$\left| \sum_{n>x} \chi(n) F(n) \right| \leqslant 2q\, F(x).$$

We end this section with the following definitions. A Dirichlet character is called *real* if its values are real, i.e. $\chi(n) \in \{-1, 0, 1\}$. Otherwise a character is *complex*. A character is said to be *quadratic* if it has order 2 in the character group, i.e. $\chi^2 = \chi_0$ and $\chi \neq \chi_0$. Thus a quadratic character is real, and a real character is either principal or quadratic.

3.4.3 Dirichlet L-Functions

It is remarkable that Dirichlet had the idea to introduce generating series of his characters nearly 60 years before the analytic proofs of the Prime Number Theorem and the Prime Number Theorem for Arithmetic Progressions given by Hadamard and de la Vallée Poussin. Not less remarkable is the fact that the infinity of the set of prime numbers in an arithmetic progression is essentially due to the non-vanishing of these generating series at $s = 1$. This very fruitful idea proved to be one of the most crucial points in almost all problems in number theory, whose treatments mimic Dirichlet's work.

Definition 3.6 (*L*-Functions) Let χ be a Dirichlet character modulo $q \geqslant 2$. The *L-function*, or *L-series*, attached to χ is the Dirichlet series of χ, i.e. for all $s = \sigma + it \in \mathbb{C}$ such that $\sigma > 1$, we set

$$L(s, \chi) = \sum_{n=1}^{\infty} \frac{\chi(n)}{n^s}.$$

As in Proposition 3.6, it may be shown that, for all $\sigma > 1$ and all $x \geqslant 2$, we have

$$\prod_{p \leqslant x} \left(1 - \frac{\chi(p)}{p^\sigma}\right)^{-1} = \sum_{P^+(n) \leqslant x} \frac{\chi(n)}{n^\sigma}.$$

Let $s = \sigma + it \in \mathbb{C}$. Since

$$\sum_{p \leqslant x} \left| \frac{\chi(p)}{p^s} \right| = \sum_{p \leqslant x} \frac{1}{p^\sigma}$$

both sides of the above identity converge absolutely for any complex s such that $\sigma > 1$ and therefore, as in Definition 3.4, we have

$$L(s, \chi) = \prod_{p} \left(1 - \frac{\chi(p)}{p^s}\right)^{-1} \tag{3.28}$$

for all $s \in \mathbb{C}$ such that $\sigma > 1$. Note also that

$$L(s, \chi_0) = \sum_{\substack{n=1 \\ (n,q)=1}}^{\infty} \frac{1}{n^s} = \zeta(s) \prod_{p \mid q} \left(1 - \frac{1}{p^s}\right) \tag{3.29}$$

and if $\chi \neq \chi_0$, then the series converges for all $\sigma > 0$ by Corollary 3.8. This result is the best possible since the terms in the series do not tend to 0 when $\sigma = 0$.

3.4.4 The Series $\sum_p \chi(p) p^{-1}$

In this section we intend to prove the main tools which enable us to show Dirichlet's theorem. There are many ways to get the desired result. We choose a rather elementary proof which is due to Shapiro. It has the advantage of relating Dirichlet's theorem to quadratic fields, another area in which Dirichlet showed his talent.

Theorem 3.21 *If $\chi \neq \chi_0$ is a non-principal Dirichlet character modulo q satisfying $L(1, \chi) \neq 0$, then the series*

$$\sum_p \frac{\chi(p)}{p}$$

converges.

Proof Let $N \in \mathbb{Z}_{\geq 2}$. The idea is to estimate the sum $\sum_{n \leq N} n^{-1} \chi(n) \log n$ in two different ways. First, by (3.3) we have

$$\sum_{n \leq N} \frac{\chi(n) \log n}{n} = \sum_{n \leq N} \frac{\chi(n)}{n} \sum_{d \mid n} \Lambda(d).$$

Interchanging the order of summation[21] and using Definition 3.5, we get

$$\sum_{n \leq N} \frac{\chi(n) \log n}{n} = \sum_{d \leq N} \Lambda(d) \sum_{k \leq N/d} \frac{\chi(kd)}{kd}$$

$$= \sum_{d \leq N} \frac{\chi(d) \Lambda(d)}{d} \sum_{k \leq N/d} \frac{\chi(k)}{k}$$

$$= L(1, \chi) \sum_{d \leq N} \frac{\chi(d) \Lambda(d)}{d} - \sum_{d \leq N} \frac{\chi(d) \Lambda(d)}{d} \sum_{k > N/d} \frac{\chi(k)}{k}.$$

Since $L(1, \chi) \neq 0$, we infer that

$$\sum_{d \leq N} \frac{\chi(d) \Lambda(d)}{d} = \frac{1}{L(1, \chi)} \left\{ \sum_{n \leq N} \frac{\chi(n) \log n}{n} + \sum_{d \leq N} \frac{\chi(d) \Lambda(d)}{d} \sum_{k > N/d} \frac{\chi(k)}{k} \right\}.$$

$$(3.30)$$

Using Corollary 3.8 we get

$$\left| \sum_{d \leq N} \frac{\chi(d) \Lambda(d)}{d} \sum_{k > N/d} \frac{\chi(k)}{k} \right| \leq \frac{2q}{N} \sum_{d \leq N} \Lambda(d) = \frac{2q \Psi(N)}{N}.$$

---

[21] See also Chap. 4.

and Lemma 3.11 yields

$$\left| \sum_{d \leqslant N} \frac{\chi(d)\Lambda(d)}{d} \sum_{k > N/d} \frac{\chi(k)}{k} \right| < 4q.$$

Inserting this bound in (3.30) provides the estimate

$$\left| \sum_{d \leqslant N} \frac{\chi(d)\Lambda(d)}{d} \right| < \frac{1}{|L(1,\chi)|} \left(\left| \sum_{n \leqslant N} \frac{\chi(n)\log n}{n} \right| + 4q \right). \tag{3.31}$$

On the other hand, we derive by partial summation

$$\sum_{n \leqslant N} \frac{\chi(n)\log n}{n} = \frac{\chi(2)\log 2}{2} + \sum_{3 \leqslant n \leqslant N} \frac{\chi(n)\log n}{n}$$

$$= \frac{\chi(2)\log 2}{2} + \frac{\log N}{N} \sum_{3 \leqslant n \leqslant N} \chi(n) + \int_3^N \frac{\log t - 1}{t^2} \left(\sum_{3 \leqslant n \leqslant t} \chi(n) \right) dt$$

so that using Proposition 3.14 we obtain

$$\left| \sum_{n \leqslant N} \frac{\chi(n)\log n}{n} \right| \leqslant \frac{\log 2}{2} + q \left(\frac{\log N}{N} + \int_3^N \frac{\log t - 1}{t^2} \, dt \right) = \frac{\log 2}{2} + \frac{q \log 3}{3} < q.$$

Inserting this bound in (3.31) gives

$$\left| \sum_{d \leqslant N} \frac{\chi(d)\Lambda(d)}{d} \right| < \frac{5q}{|L(1,\chi)|}. \tag{3.32}$$

We have

$$\sum_{p \leqslant N} \frac{\chi(p)\log p}{p} = \sum_{d \leqslant N} \frac{\chi(d)\Lambda(d)}{d} - \sum_{p \leqslant N} \log p \sum_{\alpha=2}^{[\log N/\log p]} \frac{\chi(p^\alpha)}{p^\alpha}$$

and the second sum is bounded since

$$\left| \sum_{p \leqslant N} \log p \sum_{\alpha=2}^{[\log N/\log p]} \frac{\chi(p^\alpha)}{p^\alpha} \right| \leqslant \sum_{p \leqslant N} \log p \sum_{\alpha=2}^{[\log N/\log p]} \frac{1}{p^\alpha} \leqslant \sum_p \frac{\log p}{p(p-1)} < 1$$

as already seen in (3.6). Now by partial summation, for any integer $2 \leqslant N < R$, we derive

$$\sum_{N<p\leqslant R} \frac{\chi(p)}{p} = \frac{1}{\log R}\sum_{p\leqslant R}\frac{\chi(p)\log p}{p} - \frac{1}{\log N}\sum_{p\leqslant N}\frac{\chi(p)\log p}{p} + \int_N^R \left(\sum_{p\leqslant t}\frac{\chi(p)\log p}{p}\right)\frac{dt}{t\log^2 t}$$

so that by above we obtain

$$\left|\sum_{N<p\leqslant R}\frac{\chi(p)}{p}\right| \leqslant \frac{1}{\log R}\left(\left|\sum_{d\leqslant R}\frac{\chi(d)\Lambda(d)}{d}\right|+1\right) + \frac{1}{\log N}\left(\left|\sum_{d\leqslant N}\frac{\chi(d)\Lambda(d)}{d}\right|+1\right)$$
$$+ \int_N^R \left(\left|\sum_{d\leqslant t}\frac{\chi(d)\Lambda(d)}{d}\right|+1\right)\frac{dt}{t\log^2 t}$$

and (3.32) provides

$$\left|\sum_{N<p\leqslant R}\frac{\chi(p)}{p}\right| \leqslant \left(\frac{5q}{|L(1,\chi)|}+1\right)\left(\frac{1}{\log R}+\frac{1}{\log N}\right) + \int_N^R\left(\frac{5q}{|L(1,\chi)|}+1\right)\frac{dt}{t\log^2 t}$$
$$= \frac{2}{\log N}\left(\frac{5q}{|L(1,\chi)|}+1\right)$$

and letting $R \to \infty$ gives

$$\left|\sum_{p>N}\frac{\chi(p)}{p}\right| \leqslant \frac{2}{\log N}\left(\frac{5q}{|L(1,\chi)|}+1\right)$$

completing the proof.   □

3.4.5   The Non-vanishing of $L(1, \chi)$

By Theorem 3.21, the non-vanishing of $L(1, \chi)$ for all $\chi \neq \chi_0$ is the main point of the proof of Dirichlet's theorem. Once again, there are many ways to show this.

3.4.5.1   Complex Characters

In this case, one may mimic and adapt the proof of Theorem 3.8 as follows.

Theorem 3.22 *For all* complex *characters* χ *modulo* q *and all* $t \in \mathbb{R}$

$$L(1 + it, \chi) \neq 0.$$

Proof If $\chi \neq \chi_0$, the function $L(s, \chi)$ is analytic for $\sigma > 0$. On the other hand, by (3.29), the function $L(s, \chi_0)$ is analytic in this half-plane except for a simple pole at $s = 1$ with residue $\varphi(q)/q$. Using (3.28), we deduce that in either case a logarithm of $L(s, \chi)$ is given by

$$\sum_{n=2}^{\infty} \frac{\chi(n)\Lambda(n)}{n^s \log n} = \sum_{p} \sum_{\alpha=1}^{\infty} \frac{\chi(p)^\alpha}{\alpha p^{\alpha s}}$$

for $\sigma > 1$. Furthermore, for all characters χ, one may write

$$\chi(n) = \chi_0(n)e^{i\omega(n)},$$

where $\omega : \mathbb{N} \to [0, 2\pi)$. Now let $\sigma > 1$ and $t \in \mathbb{R}$. Setting $\theta(n, t) = \omega(n) - t \log n$, we have

$$\mathrm{Re}\left(3 \log L(\sigma, \chi_0) + 4 \log L(\sigma + it, \chi) + \log L(\sigma + 2it, \chi^2)\right)$$

$$= \sum_{\substack{n=2 \\ (n,q)=1}}^{\infty} \frac{\Lambda(n)}{n^\sigma \log n} \{3 + 4\cos\theta(n, t) + \cos(2\theta(n, t))\}$$

$$= 2 \sum_{\substack{n=2 \\ (n,q)=1}}^{\infty} \frac{\Lambda(n)}{n^\sigma \log n} \{1 + \cos\theta(n, t)\}^2 \geqslant 0$$

which implies that

$$\left| L(\sigma, \chi_0)^3 L(\sigma + it, \chi)^4 L(\sigma + 2it, \chi^2) \right| \geqslant 1.$$

Now suppose that $1 + it_0$ is a zero of order $m \geqslant 1$ of $L(s, \chi)$ for some $t_0 \in \mathbb{R}$, so that there exists $\ell \neq 0$ such that

$$\lim_{\sigma \to 1+} \frac{L(\sigma + it_0, \chi)}{(\sigma - 1)^m} = \ell.$$

By above we infer that

$$\left|(\sigma - 1)^3 L(\sigma, \chi_0)^3 (\sigma - 1)^{-4m} L(\sigma + it_0, \chi)^4 L(\sigma + 2it_0, \chi^2)\right| \geqslant (\sigma - 1)^{3-4m}.$$

Since $\chi^2 \neq \chi_0$, the function $L(s, \chi^2)$ is continuous at all points on the line $\sigma = 1$ and thus does not have a pole at $s = 1$. Therefore letting $\sigma \rightarrow 1^+$ gives a contradiction as in Theorem 3.8. □

It should be mentioned that this proof extends to the case of real characters, proving in this case that $L(1 + it, \chi) \neq 0$ for all $t \in \mathbb{R} \setminus \{0\}$. The difficult point is thus a proof of $L(1, \chi) \neq 0$ for all quadratic Dirichlet characters. Once again, there exist many proofs, almost all related to certain results from algebraic number theory.

3.4.5.2 Primitive Characters

Definition 3.7 Let χ be a Dirichlet character modulo q. The *conductor* of χ is the smallest positive integer $f \mid q$ such that there exists a Dirichlet character $\chi^\star$ modulo f satisfying

$$\chi = \chi_0 \chi^\star. \tag{3.33}$$

A Dirichlet character χ modulo q is said to be *primitive* if the conductor f of χ is such that $f = q$. In (3.33), the character $\chi^\star$ is then primitive and uniquely determined by χ. We will say that $\chi^\star$ *induces* χ or that χ is *induced by* $\chi^\star$.

Example 3.4

▷ It is noteworthy that each Dirichlet character modulo q is induced by a unique primitive Dirichlet character modulo a divisor of q. Furthermore, a character χ is imprimitive if and only if there exists $d \mid q$ with $d < q$ such that, for all positive integers a, b satisfying $(a, q) = (b, q) = 1$ and $a \equiv b \pmod{d}$, we have $\chi(a) = \chi(b)$. This implies that the principal character χ_0 modulo $q > 1$ is imprimitive, by taking $d = 1$.

▷ There is only one primitive character modulo 4 defined for all odd positive integers n by

$$\chi_4(n) = (-1)^{(n-1)/2}.$$

▷ There are two primitive characters[22] modulo 8 defined for all odd positive integers n by

$$\chi_8(n) = (-1)^{(n^2-1)/8} \quad \text{and} \quad \chi_4\chi_8(n) = (-1)^{(n-1)/2+(n^2-1)/8}.$$

---

[22]It is noteworthy that the Dirichlet characters χ_4, χ_8 and $\chi_4\chi_8$ are the characters attached to the quadratic fields $\mathbb{Q}(\sqrt{-1})$, $\mathbb{Q}(\sqrt{2})$ and $\mathbb{Q}(\sqrt{-2})$, respectively. See Chap. 7.

▷ If $q = p^\alpha$ is a prime power, the only real primitive characters of conductor q are $\chi_4, \chi_8, \chi_4\chi_8$ and χ_p. Every real primitive character can be obtained as the product of these characters. This implies that the conductor of a real primitive character is of the form 1, m, $4m$ or $8m$ where m is a positive odd squarefree integer.

▷ Lemma 7.17 gives another useful characterization of the real primitive characters.

▷ Let χ be a Dirichlet character modulo q induced by $\chi^\star$. Then we have for $\sigma > 1$

$$L(s, \chi) = \prod_p \left(1 - \frac{\chi_0\chi^\star(p)}{p^s}\right)^{-1} = \prod_{p \nmid q} \left(1 - \frac{\chi^\star(p)}{p^s}\right)^{-1} = L(s, \chi^\star) \prod_{p \mid q} \left(1 - \frac{\chi^\star(p)}{p^s}\right)$$

and if $\chi \neq \chi_0$, we derive for all $\sigma > 0$

$$L(s, \chi) = L(s, \chi^\star) \prod_{p \mid q} \left(1 - \frac{\chi^\star(p)}{p^s}\right). \tag{3.34}$$

3.4.5.3   Quadratic Characters

We now are in a position to prove the following result.

Theorem 3.23 *For all* quadratic *characters* χ *modulo* q, *we have* $L(1, \chi)$ $\neq 0$.

First Proof By (3.34), we may suppose that χ is primitive since the Euler product of the right-hand side is non-zero. By Lemma 7.17, the real primitive characters modulo q are of the form $(\chi(-1)q/n)$ and (7.27) and the fact that the class number is a positive integer give the asserted result.   □

Second Proof As above, we may suppose that χ is real primitive, and hence is the character attached to a quadratic number field $\mathbb{K}$. Following [53], for all $t \in [0, 1)$, let

$$f(t) = \sum_{n=1}^{\infty} \frac{\chi(n)t^n}{1 - t^n}$$

be the *Lambert series* associated with χ. By [66, Theorem VIII-65] and Proposition 7.33,[23] we have

$$f(t) = \sum_{n=1}^{\infty} \left(\sum_{d \mid n} \chi(d)\right)t^n = \sum_{n=1}^{\infty} r_{\mathbb{K}}(n)t^n$$

---

[23] See also (4.12).

where the function $r_{\mathbb{K}}$ is given in Definition 7.20. This implies that $f(t) \to \infty$ as $t \to 1^-$. Now suppose that $L(1, \chi) = 0$. Then we have

$$-f(t) = \sum_{n=1}^{\infty} \left(\frac{1}{n(1-t)} - \frac{t^n}{1-t^n} \right) \chi(n) = \sum_{n=1}^{\infty} a_n \chi(n).$$

Observe that we have

$$(1-t)(a_n - a_{n+1}) = \frac{1}{n} - \frac{1}{n+1} - \frac{t^n}{1 + \cdots + t^{n-1}} + \frac{t^{n+1}}{1 + \cdots + t^n}$$

$$= \frac{1}{n(n+1)} - \frac{t^n}{(1 + \cdots + t^{n-1})(1 + \cdots + t^n)}$$

and using the arithmetic–geometric mean inequality we get for all $0 \leqslant t < 1$

$$\sum_{j=0}^{n-1} t^j \geqslant n \prod_{j=0}^{n-1} t^{j/n} = nt^{(n-1)/2} \geqslant nt^{n/2}$$

and

$$\sum_{j=0}^{n} t^j \geqslant (n+1)t^{n/2}$$

so that

$$(1-t)(a_n - a_{n+1}) \geqslant \frac{1}{n(n+1)} - \frac{t^n}{n(n+1)t^n} = 0$$

and hence the sequence (a_n) is non-increasing and tends to 0. By Corollary 3.8 we get

$$|-f(t)| \leqslant (2q + 1)a_1 = 2q + 1$$

contradicting the unboundedness of f on $[0, 1)$.   □

Now we may conclude this section.

Proof of Theorem 3.20 Theorems 3.21, 3.22, and 3.23, along with Corollary 3.7, give the complete proof of Dirichlet's theorem.   □

3.4.6   Functional Equation

As for the Riemann zeta-function, the Dirichlet L-function has a functional equation generalizing Theorem 3.7. The proof might mimic the argument of this result, but the following point should be mentioned. Assume that χ is a Dirichlet character modulo q, and define the "twisted" theta function θ_χ by

$$\theta_\chi(x) = \sum_{n \in \mathbb{Z}} \chi(n) e^{-n^2 \pi x}.$$

Then

$$\tfrac{1}{2}\theta_\chi(x) = \sum_{n=1}^{\infty} \tfrac{1}{2}\left(\chi(n) + \chi(-n)\right) e^{-n^2 \pi x} = \begin{cases} \sum_{n=1}^{\infty} \chi(n) e^{-n^2 \pi x}, & \text{if } \chi \text{ is even}; \\ 0, & \text{if } \chi \text{ is odd}; \end{cases}$$

and hence this version of θ_χ cannot be correct for odd χ, since it vanishes identically. The usual proofs then separate the cases when χ is even and odd, a modification of the function θ_χ being used in the latter case to prevent the cancellation of the $\pm n$ summands.[24] This is rather long, and we prefer here to follow the elegant proof given by Berndt in [8].

Theorem 3.24 Let χ be a non-principal, primitive Dirichlet character modulo q and set

$$\delta := \delta_\chi = \begin{cases} 0, & \text{if } \chi \text{ is even}; \\ 1, & \text{if } \chi \text{ is odd}. \end{cases}$$

Then $L(s, \chi)$ can be analytically continued to an entire function, and the completed L-function

$$\Lambda(s, \chi) := \left(\frac{q}{\pi}\right)^{s/2} \pi^{-\delta/2} \Gamma\left(\frac{s + \delta}{2}\right) L(s, \chi)$$

satisfies the functional equation

$$\Lambda(1 - s, \chi) = \frac{i^\delta \sqrt{q}}{\tau(\overline{\chi})} \Lambda(s, \overline{\chi}),$$

where $\tau(\chi)$ is the Gauss sum.

---

[24] More precisely, the Gaussian weight $e^{-\pi x^2}$ is replaced by the following *odd* weight $x e^{-\pi x^2}$.

In order to prove this result, we begin by proceeding to some minor changes. First

$$\Lambda(1-s,\chi) = \frac{i^\delta \sqrt{q}}{\tau(\overline{\chi})} \Lambda(s,\overline{\chi})$$

$$\Longleftrightarrow L(1-s,\chi) = \frac{i^\delta \sqrt{\pi}}{\tau(\overline{\chi})} \left(\frac{q}{\pi}\right)^s \Gamma\left(\frac{s+\delta}{2}\right) \Gamma\left(\frac{1-s+\delta}{2}\right)^{-1} L(s,\overline{\chi}).$$

Then using (3.11) and (3.10) we get

$$\Gamma\left(\frac{s+\delta}{2}\right) \Gamma\left(\frac{1-s+\delta}{2}\right)^{-1} = \frac{\Gamma(s)}{i^\delta 2^s \sqrt{\pi}} \left(e^{i\pi s/2} + \chi(-1)e^{-i\pi s/2}\right)$$

so that

$$\Lambda(1-s,\chi) = \frac{i^\delta \sqrt{q}}{\tau(\overline{\chi})} \Lambda(s,\overline{\chi})$$

$$\Longleftrightarrow L(1-s,\chi) = \frac{1}{\tau(\overline{\chi})} \left(e^{-i\pi s/2} + \chi(-1)e^{i\pi s/2}\right) \left(\frac{q}{2\pi}\right)^s \Gamma(s) L(s,\overline{\chi}).$$

$$(3.35)$$

Therefore it suffices to show that $L(s,\chi)$ can be analytically continued to an entire function which satisfies the functional equation (3.35).

Proof To do this, let $\chi \neq \chi_0$ be any primitive Dirichlet character modulo $q > 1$, and recall the Gauss sum (6.19)

$$\tau(z,\chi) := \sum_{0 \leqslant k < q} \chi(k) e_q(zk).$$

The starting point of the proof is the function F defined by

$$F(z) = \frac{\pi e^{-i\pi z} \tau(z,\overline{\chi})}{\tau(\overline{\chi}) z^s \sin \pi z}$$

which we integrate over the contour C_m given below, for some $m \in \mathbb{Z}_{\geqslant 1}$ and where we first assume that $s > 1$ is *real* (Fig. 3.1).

Here, the contour C_m is the positively oriented, closed contour consisting of a the right half c_m of the circle with centre O and radius $m + \frac{1}{2}$ together with the vertical diameter indented at the origin by a semicircle c_ε of radius $\varepsilon \in (0,1)$ in the right half-plane. On the interior of C_m, F is analytic except for simple poles at $z = 1, \ldots, m$ for which we have

$$\operatorname*{Res}_{z=n}(F(z)) = \frac{\tau(n,\overline{\chi})}{n^s \tau(\overline{\chi})} = \frac{\chi(n)}{n^s} \quad (n = 1, \ldots, m)$$

Fig. 3.1 Contour C_m

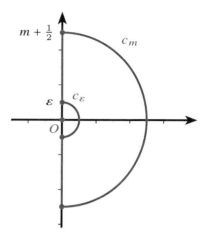

by (6.20). By the Cauchy's residue theorem, we derive

$$\frac{1}{2\pi i} \oint_{C_m} F(z)\, dz = \sum_{n=1}^{m} \frac{\chi(n)}{n^s}. \tag{3.36}$$

The function $z \mapsto \left| e^{-i\pi z} \tau\left(z, \overline{\chi}\right) / \sin \pi z \right|$ has period q and tends to 0 exponentially as $\operatorname{Im} z \to \pm\infty$ so that there exists a positive number M, independent of m, such that for all z on c_m

$$\left| \frac{e^{-i\pi z} \tau\left(z, \overline{\chi}\right)}{\sin \pi z} \right| \leqslant M.$$

Since $s > 1$, the integral of F over c_m tends to 0 as $m \to \infty$, and letting $m \to \infty$ in (3.36) we derive, for all $s > 1$

$$L(s, \chi) = \int_{i\varepsilon}^{i\infty} \frac{\tau\left(z, \overline{\chi}\right) dz}{\tau\left(\overline{\chi}\right) z^s \left(1 - e^{2i\pi z}\right)}$$

$$+ \int_{-i\varepsilon}^{-i\infty} \frac{e^{-2i\pi z} \tau\left(z, \overline{\chi}\right) dz}{\tau\left(\overline{\chi}\right) z^s \left(1 - e^{-2i\pi z}\right)} + \frac{1}{2\pi i} \int_{c_\varepsilon} F(z)\, dz. \tag{3.37}$$

The two first integrals converge uniformly in any compact subset of the complex plane, so that (3.37) shows that $L(s, \chi)$ can be analytically continued to an entire function. In particular, the identity (3.37) is valid for all $s \in \mathbb{C}$. Now assume $s < 0$. Since $\tau(0, \chi) = 0$, the integral over c_ε tends to 0 as $\varepsilon \to 0$. We then get for $s < 0$

$$L(s, \chi) = i e^{-i\pi s/2} \int_0^\infty \frac{\tau\left(iy, \overline{\chi}\right) dy}{\tau\left(\overline{\chi}\right) y^s \left(1 - e^{-2\pi y}\right)} - i e^{i\pi s/2} \int_0^\infty \frac{e^{-2\pi y} \tau\left(-iy, \overline{\chi}\right) dy}{\tau\left(\overline{\chi}\right) y^s \left(1 - e^{-2\pi y}\right)}$$

$$= i e^{-i\pi s/2} \left(\frac{q}{2\pi}\right)^{1-s} \int_0^\infty \frac{\tau\left(\frac{iqx}{2\pi}, \overline{\chi}\right) dx}{\tau(\overline{\chi}) x^s \left(1 - e^{-qx}\right)}$$

$$- i e^{i\pi s/2} \left(\frac{q}{2\pi}\right)^{1-s} \int_0^\infty \frac{e^{-qx} \tau\left(-\frac{iqx}{2\pi}, \overline{\chi}\right) dx}{\tau(\overline{\chi}) x^s \left(1 - e^{-qx}\right)}$$

and replacing k by $q - k$ in the definition of $\tau(z, \chi)$ gives

$$e^{-qx} \tau\left(-\frac{iqx}{2\pi}, \overline{\chi}\right) = \chi(-1) \tau\left(\frac{iqx}{2\pi}, \overline{\chi}\right)$$

so that, for $s < 0$

$$L(s, \chi) = \frac{i}{\tau(\overline{\chi})} \left(e^{-i\pi s/2} - \chi(-1)e^{i\pi s/2}\right) \left(\frac{q}{2\pi}\right)^{1-s} \int_0^\infty \tau\left(\frac{iqx}{2\pi}, \overline{\chi}\right) \frac{x^{-s}}{1 - e^{-qx}} dx. \tag{3.38}$$

Now, if f is any complex-valued arithmetic function such that its Dirichlet series $L(s, f)$ has σ_a for abscissa of absolute convergence, then[25]

$$\Gamma(s) L(s, f) = \int_0^\infty \left(\sum_{n=0}^\infty f(n) e^{-nx}\right) x^{s-1} dx \quad (\sigma > \max(0, \sigma_a))$$

from which we derive

$$\Gamma(s) L(s, \chi) = \int_0^\infty \tau\left(\frac{iqx}{2\pi}, \chi\right) \frac{x^{s-1}}{1 - e^{-qx}} dx \quad (\sigma > 1)$$

and inserting this identity in (3.38) yields, for all $s < 0$

$$L(s, \chi) = \frac{i}{\tau(\overline{\chi})} \left(e^{-i\pi s/2} - \chi(-1)e^{i\pi s/2}\right) \left(\frac{q}{2\pi}\right)^{1-s} \Gamma(1 - s) L(1 - s, \overline{\chi}).$$

Replacing s by $1 - s$, we derive, for all $s > 1$

$$L(1 - s, \chi) = \frac{1}{\tau(\overline{\chi})} \left(e^{-i\pi s/2} + \chi(-1)e^{i\pi s/2}\right) \left(\frac{q}{2\pi}\right)^s \Gamma(s) L(s, \overline{\chi})$$

and using analytic continuation we finally get (3.35).   □

---

[25]This result is a consequence of the "product theorem" of Mellin transforms. See [93, Corollary 8 page 112].

3.5   The Prime Number Theorem

The proof of the Prime Number Theorem was first given independently by
Hadamard and de la Vallée Poussin in 1896. Not only did they provide a proof
of the estimate

$$\pi(x) \sim \frac{x}{\log x}$$

as $x \to \infty$, but they also gave a quite accurate error term which has only been
slightly improved up to now. The strategy follows the lines of estimating some
combinatorial objects, for which one usually computes their generating series giving
in return some information by the use of some extraction theorems. The following
tool is the basic result of the theory.

3.5.1   Perron Summation Formula

> **Theorem 3.25 (Truncated Perron Summation Formula)** *Let $f(n)$ be any
> complex numbers with Dirichlet series $F(s) = \sum_{n=1}^{\infty} f(n)n^{-s}$. Assume that
> $F(s)$ is absolutely convergent in the half-plane $\sigma > \sigma_a$ for some $\sigma_a \in \mathbb{R}$.
> Then for all real numbers $x, T \geqslant 1$ and all $s \in \mathbb{C}$ such that $\sigma \leqslant \sigma_a$, we have*
>
> $$\sum_{n \leqslant x} \frac{f(n)}{n^s} = \frac{1}{2\pi i} \int_{\kappa-iT}^{\kappa+iT} \frac{F(s+u)x^u}{u}\, du + E(x)$$
>
> $$+ O\left\{ x^{-\sigma} \sum_{x/2 < n < 2x} |f(n)| \min\left(1, \frac{x}{T|n-x|}\right) + \frac{x^\kappa}{T} \sum_{n=1}^{\infty} \frac{|f(n)|}{n^{\sigma+\kappa}} \right\},$$
>
> *where $\kappa = \sigma_a - \sigma + \frac{1}{\log x}$ and $E(x) = \frac{1}{2}f(x)$ if $x \in \mathbb{Z}_{\geqslant 1}$, $E(x) = 0$ otherwise.*

Many proofs exist in the literature [56, 84, 86]. In practice, the first sum in the error
term can be handled as follows. Suppose that, for all $n \in \mathbb{Z}_{\geqslant 1}$, we have

$$|f(n)| \leqslant B(n), \tag{3.39}$$

where $B(n)$ is a positive non-decreasing function. Split the sum into three subsums

$$\sum_{x/2 < n < 2x} |f(n)| \min\left(1, \frac{x}{T|n-x|}\right) = \sum_{x/2 < n \leqslant x-1} + \sum_{x-1 < n < x+1} + \sum_{x+1 \leqslant n < 2x}$$

and take 1 for the minimum in the second sum and the other term in the two other sums, so that this error term does not exceed

$$\ll \left(\frac{x}{T} \sum_{x/2 < n \leqslant x-1} \frac{1}{x-n} + \frac{x}{T} \sum_{x+1 \leqslant n < 2x} \frac{1}{n-x} + 1 \right) B(2x)$$

$$\ll \left(\frac{x \log x}{T} + 1 \right) B(2x).$$

If in addition we have for some $\alpha > 0$

$$\sum_{n=1}^{\infty} \frac{|f(n)|}{n^{\sigma}} \ll \frac{1}{(\sigma - \sigma_a)^{\alpha}} \qquad (\sigma > \sigma_a) \qquad (3.40)$$

then, under the hypotheses of Theorem 3.25 with $1 \leqslant T \leqslant x$, we get

$$\sum_{n \leqslant x} \frac{f(n)}{n^s} = \frac{1}{2\pi i} \int_{\kappa-iT}^{\kappa+iT} \frac{F(s+u)x^u}{u} \, du$$

$$+ O \left\{ \frac{x^{\sigma_a - \sigma}}{T} (\log x)^{\alpha} + B(2x) \frac{x^{1-\sigma} \log x}{T} \right\}. \qquad (3.41)$$

It is not always possible to have a bound such as (3.39) at our disposal. For instance, there are some cases where bounds for $|f(n)|$ often require an assumption of the Generalized Ramanujan Conjecture, such as certain automorphic L-functions. The next version, developed by Liu and Ye [51], allows us to sometimes avoid this problem.

Theorem 3.26 *Under the conditions of* Theorem 3.25, *assume* $\kappa > \max(0, \sigma_a)$. *Then, for any* $x, T, H \geqslant 2$

$$\sum_{n \leqslant x} f(n) = \frac{1}{2\pi i} \int_{\kappa-iT}^{\kappa+iT} \frac{F(s)}{s} x^s \, ds$$

$$+ O \left(\sum_{x-x/H < n \leqslant x+x/H} |f(n)| + \frac{Hx^{\kappa}}{T} \sum_{n=1}^{\infty} \frac{|f(n)|}{n^{\kappa}} \right).$$

3.5.2   The Prime Number Theorem

3.5.2.1   The Classical Form

The idea is to apply (3.41) to von Mangoldt's function $\Lambda(n)$. By (3.9), we know that its Dirichlet series is the function $-\zeta'(s)/\zeta(s)$ whose main property is to have an analytic continuation on the line $\sigma = 1$ by Proposition 3.9. We have $\sigma_a = \alpha = 1$, $B(n) = \log n$ and therefore we get with $s = 0, 4 \leqslant T \leqslant x$ and $\kappa = 1 + 1/\log x$

$$\Psi(x) = \frac{1}{2\pi i} \int_{\kappa-iT}^{\kappa+iT} -\frac{\zeta'(s)}{\zeta(s)} \frac{x^s}{s} \, ds + O\left(\frac{x(\log x)^2}{T}\right). \tag{3.42}$$

The strategy is then to apply Cauchy's theorem to treat the integral of (3.42) taken over a rectangle surrounding the point 1. Since the residue at this point is known by Proposition 3.9, it remains to evaluate the integral on the three other sides of the rectangle. This strategy is successful only if we have at our disposal estimates of the function in a region to the left of the line $\sigma = 1$. This is done by Theorem 3.15.

Theorem 3.27 (Prime Number Theorem) *There exists a constant $c_1 > 0$ such that, for $x \to \infty$*

$$\pi(x) = \mathrm{Li}(x) + O\left(x \, e^{-c_1 \sqrt{\log x}}\right).$$

Proof Let c be the constant appearing in Theorem 3.15, and let $0 < c_0 \leqslant c$. Now define $\lambda := 1 - c_0(\log 2T)^{-1}$, and let $\mathcal{R}$ be the rectangle with vertices $\kappa \pm iT$ and $\lambda \pm iT$. By Theorem 3.15 and Proposition 3.9, the function $-\frac{\zeta'}{\zeta}(s)$ has a simple pole with residue 1 at $s = 1$, and is otherwise analytic within $\mathcal{R}$. Integrating over $\mathcal{R}$ in the anticlockwise direction and using Cauchy's residue theorem, we get

$$\frac{1}{2\pi i} \int_{\mathcal{R}} -\frac{\zeta'(s)}{\zeta(s)} \frac{x^s}{s} \, ds = x.$$

If $\mathcal{H}_1$ and $\mathcal{H}_2$ are the two horizontal sides and $\mathcal{V}$ is the other vertical side of $\mathcal{R}$, we deduce that

$$\Psi(x) = x - \frac{1}{2\pi i} \left(\sum_{j=1}^{2} \int_{\mathcal{H}_j} -\frac{\zeta'(s)}{\zeta(s)} \frac{x^s}{s} \, ds + \int_{\mathcal{V}} -\frac{\zeta'(s)}{\zeta(s)} \frac{x^s}{s} \, ds\right) + O\left(\frac{x(\log x)^2}{T}\right).$$

Now using Theorem 3.16, we derive

$$\left| \int_{\mathcal{H}_j} -\frac{\zeta'(s)}{\zeta(s)} \frac{x^s}{s} \, ds \right| \ll \log T \int_\lambda^K \frac{x^\sigma}{|\sigma + iT|} \, d\sigma \ll \frac{\log T}{T} \int_\lambda^K x^\sigma \, d\sigma \ll \frac{x \log T}{T}$$

for $j \in \{1, 2\}$, and

$$\left| \int_{\mathcal{V}} -\frac{\zeta'(s)}{\zeta(s)} \frac{x^s}{s} \, ds \right| \ll x^\lambda \log T \int_{-T}^{T} \frac{dt}{\sqrt{\lambda^2 + t^2}} \ll x^\lambda (\log T)^2$$

so that we get

$$\Psi(x) = x + O\left\{ x(\log 2T)^2 \left(T^{-1} + x^{-c_0/\log(2T)} \right) \right\}.$$

We choose $T = \frac{1}{2} \exp\left(\sqrt{c_0 \log x} \right)$ which implies that the error term is

$$\ll x \log x \, e^{-\sqrt{c_0 \log x}} \ll x \, e^{-c_1 \sqrt{\log x}}$$

for some constant $0 < c_1 \leqslant \frac{1}{2}\sqrt{c_0}$ and x sufficiently large.[26] By Corollary 3.3, we deduce that

$$\theta(x) = x + O\left(x \, e^{-c_1 \sqrt{\log x}} \right)$$

and we complete the proof using Exercise 9.   □

Note that by repeated integration by parts, we derive for any fixed $N \in \mathbb{Z}_{\geqslant 1}$

$$\text{Li}(x) = \frac{x}{\log x} + x \sum_{k=1}^{N-1} \frac{k!}{(\log x)^{k+1}} + O\left(\frac{x}{(\log x)^{N+1}} \right) \tag{3.43}$$

which allows us to have at our disposal several more or less precise versions of the PNT. For instance, for $x \to \infty$, the estimate

$$\pi(x) = \frac{x}{\log x} + \frac{x}{(\log x)^2} + O\left(\frac{x}{(\log x)^3} \right)$$

is useful in many applications.

---

[26]This can be made explicit. Using for instance the inequality $\log \log x \leqslant 4e^{-1}(\log x)^{1/4}$, and since $c_1 \leqslant \frac{1}{2}\sqrt{c_0}$, we see the estimate is valid for any x satisfying $\log x \geqslant 2^{12} e^{-4} c_0^{-2}$.

3.5.2.2   The Korobov–Vinogradov Zero-Free Region

I. M. Vinogradov's method of exponential sums allows us to get estimates of $\zeta(s)$ in the form

$$|\zeta(s)| \leqslant A t^{B(1-\sigma)^{3/2}} (\log t)^{2/3} \tag{3.44}$$

for all $s = \sigma + it \in \mathbb{C}$ such that $\frac{1}{2} \leqslant \sigma \leqslant 1$ and $t \geqslant 3$. This in turn implies the best zero-free region for $\zeta(s)$ up to now which was obtained by Korobov and I. A. Vinogradov, who proved that there exists an absolute constant $c_0 > 0$ such that $\zeta(s)$ has no zero in the region

$$\sigma \geqslant 1 - \frac{c_0}{(\log|t|)^{2/3}(\log\log|t|)^{1/3}} \quad \text{and} \quad |t| \geqslant 3 \tag{3.45}$$

giving the best error term in the PNT to date, namely

$$\pi(x) = \operatorname{Li}(x) + O\left\{ x \exp\left(-c_1 (\log x)^{3/5} (\log\log x)^{-1/5} \right) \right\} \tag{3.46}$$

with $0 < c_1 < 1$ being absolute.

3.5.2.3   Some Applications

The PNT enables us to improve on estimates for some functions of prime numbers by proceeding as follows. Let $x \geqslant 2$ be a large real number and $f \in C^2[2, x]$. Similarly as in Chap. 1, one has

$$\sum_{p \leqslant x} f(p) = \int_{2^-}^{x} f(u)\, d\pi(u)$$

$$= f(x)\pi(x) - \int_{2}^{x} f'(u) \operatorname{Li}(u)\, du - \int_{2}^{x} f'(u)\, (\pi(u) - \operatorname{Li}(u))\, du$$

$$= \int_{2}^{x} \frac{f(u)}{\log u}\, du + f(2) \operatorname{Li}(2) + f(x)\, (\pi(x) - \operatorname{Li}(x))$$

$$\quad - \int_{2}^{x} f'(u)\, (\pi(u) - \operatorname{Li}(u))\, du$$

and if the integral

$$\int_{2}^{\infty} f'(u)\, (\pi(u) - \operatorname{Li}(u))\, du$$

converges, then we get

$$\sum_{p \leqslant x} f(p) = \int_2^x \frac{f(u)}{\log u} \, du + c_f + f(x) \, (\pi(x) - \mathrm{Li}(x)) + \int_x^\infty f'(u) \, (\pi(u) - \mathrm{Li}(u)) \, du$$

with

$$c_f = f(2) \, \mathrm{Li}(2) - \int_2^\infty f'(u) \, (\pi(u) - \mathrm{Li}(u)) \, du.$$

For instance, with the latest version (3.46) of the PNT we get the following estimates.

Corollary 3.9 *There exist absolute constants $c_1, c_2, c_3 > 0$ such that, for all x sufficiently large*

$$\sum_{p \leqslant x} \frac{1}{p} = \log \log x + B + O \left\{ \exp \left(-c_1 (\log x)^{3/5} (\log \log x)^{-1/5} \right) \right\}$$

$$\sum_{p \leqslant x} \frac{\log p}{p} = \log x - E + O \left\{ \exp \left(-c_2 (\log x)^{3/5} (\log \log x)^{-1/5} \right) \right\}$$

$$\prod_{p \leqslant x} \left(1 - \frac{1}{p} \right) = \frac{e^{-\gamma}}{\log x} + O \left\{ \exp \left(-c_3 (\log x)^{3/5} (\log \log x)^{-1/5} \right) \right\}$$

where $B \approx 0.261497212\ldots$ is the Mertens constant and $E \approx 1.332582275\ldots$

The n-th prime may also be estimated asymptotically by the PNT as well.

Corollary 3.10 *For all n sufficiently large*

$$p_n = n \log n + O(n \log \log n).$$

Proof Using Exercise 10 and Theorem 3.27, we infer that there exists an absolute constant $c > 0$ such that, for all n sufficiently large

$$p_n = \mathrm{Li}^{-1}(n) + O \left(n e^{-c\sqrt{\log n}} \right).$$

It remains to show that, for x sufficiently large

$$\mathrm{Li}^{-1}(x) = x \log x + O(x \log \log x). \tag{3.47}$$

From (3.43) with $N = 1$, we derive $\mathrm{Li}(y) = y(\log y)^{-1} \left\{1 + O\left((\log y)^{-1}\right)\right\}$ as $y \to \infty$. Since $\lim_{x \to \infty} \mathrm{Li}^{-1}(x) = +\infty$, we may substitute $y = \mathrm{Li}^{-1}(x)$, giving

$$x = \mathrm{Li}\left(\mathrm{Li}^{-1}(x)\right) = \frac{\mathrm{Li}^{-1}(x)}{\log \mathrm{Li}^{-1}(x)} \left\{1 + O\left(\frac{1}{\log \mathrm{Li}^{-1}(x)}\right)\right\} \tag{3.48}$$

and taking logarithms, we get as $x \to \infty$

$$\log x = \log \mathrm{Li}^{-1}(x) - \log \log \mathrm{Li}^{-1}(x) + O\left(\frac{1}{\log \mathrm{Li}^{-1}(x)}\right) \sim \log \mathrm{Li}^{-1}(x). \tag{3.49}$$

Since $\log x \to \infty$, this implies that $\log \log x \sim \log \log \mathrm{Li}^{-1}(x)$ as $x \to \infty$, and hence, using (3.49), we get

$$\log \mathrm{Li}^{-1}(x) = \log x + O(\log \log x).$$

Now inserting this estimate into (3.48) yields

$$\mathrm{Li}^{-1}(x) = x \log \mathrm{Li}^{-1}(x) \left\{1 + O\left(\frac{1}{\log \mathrm{Li}^{-1}(x)}\right)\right\}$$

$$= x \left(\log x + O(\log \log x)\right) \left\{1 + O\left(\frac{1}{\log x}\right)\right\}$$

implying (3.47), and thus completing the proof.   □

More precise expansions for p_n do exist in the literature. In 1902, Cipolla [17] got the following asymptotic formula:

$$p_n = n \left\{\log n + \log \log n - 1 + \frac{\log \log n - 2}{\log n}\right.$$

$$\left. - \frac{(\log \log n)^2 - 6 \log \log n + 11}{2(\log n)^2} + O\left(\left(\frac{\log \log n}{\log n}\right)^3\right)\right\}.$$

In fact, it is possible to derive asymptotic formulas at any order by using the following general result [75, Théorème 2].

Proposition 3.15 *Let f be a function defined in a neighbourhood of ∞ and having an inverse function in it. Assume that, for $x \to \infty$*

$$f(x) = x(\log x)^{-\alpha} \left\{ P_N\left(\frac{1}{\log x}\right) + o\left(\frac{1}{(\log x)^N}\right) \right\},$$

where $\alpha \neq 0$, $N \in \mathbb{Z}_{\geqslant 1}$, $P_N \in \mathbb{R}[X]$, $\deg P_N \leqslant N$ and $P_N(0) \neq 0$. Then, as $x \to \infty$

$$f^{-1}(x) = \frac{x(\log x)^{\alpha}}{P_N(0)} \left\{ 1 + \sum_{n=1}^{N} \alpha^n \frac{Q_n(\log \log x)}{(\log x)^n} + o\left(\frac{1}{(\log x)^N}\right) \right\},$$

where $Q_n \in \mathbb{R}[X]$, $Q_0 = 1$ and, for $0 \leqslant n \leqslant N - 1$, $Q_{n+1}' = Q_n' - (n - \alpha)Q_n$.

By (3.43), we may use this result with $P_N = 1 + X + 2X^2 + \cdots + N!X^N$ and $\alpha = 1$, implying that, for $x \to \infty$ and any $N \in \mathbb{Z}_{\geqslant 1}$

$$\mathrm{Li}^{-1}(x) = x \log x \left\{ 1 + \sum_{n=1}^{N} \frac{Q_n(\log \log x)}{(\log x)^n} + o\left(\frac{1}{(\log x)^N}\right) \right\},$$

where the polynomials Q_n satisfy $Q_{n+1}' = Q_n' - (n - 1)Q_n$ with $Q_0 = 1$.

3.5.2.4 Landau's Explicit Formula

Let us return to (3.42). Instead of integrating over a rectangle containing only 1, suppose we integrate over a contour that proceeds by straight lines from $\kappa - iT$ to $\kappa + iT$ to $-(2K + 1) + iT$ to $-(2K + 1) - iT$ with $K \geqslant 1$ integer. In the interior of this contour, the integrand has poles at $s = 1$, at zeros ρ of $\zeta(s)$ and at the trivial zeros $s = -2k$. Since x^s decays quickly as $\sigma \to -\infty$, one may expect that we can pull the contour to the left and thus get a totally explicit formula for $\psi(x)$.

We start by using Proposition 3.10 to derive a result which allows us to get a finer estimate of $-\zeta'(s)/\zeta(s)$ in a larger region than that of Theorem 3.14.

Corollary 3.11 *For every real number $T \geqslant 2$, there exists $T' \in [T, T + 1]$ such that, uniformly for $-1 \leqslant \sigma \leqslant 2$, we have*

$$-\frac{\zeta'}{\zeta}(\sigma + iT') = O(\log^2 T).$$

Proof Indeed, by (3.51), the number of zeros ρ such that $\gamma \in [T, T+1]$ is $\ll \log T$. Subdividing the interval into $\ll \log T$ equal parts of length $c/\log T$ for some $c > 0$ chosen so that the number of parts exceeds the number of zeros, we deduce that there is a part that contains no zeros by the Dirichlet pigeon-hole principle. Hence for T' lying in this part, we must have $|T' - \gamma| \gg 1/\log T$. We infer that each summand in Proposition 3.10 is $\ll \log T$ and since there are $\ll \log T$ summands by (3.51), we get the stated result.   □

It should also be noticed that, using an asymmetric form of the functional equation of $\zeta(s)$, it can be proved that for $\sigma \leqslant -1$, we have

$$-\frac{\zeta'}{\zeta}(s) = O\left(\log(|s| + 1)\right) \tag{3.50}$$

provided that circles of radii $\frac{1}{4}$ around the trivial zeros $s = -2k$ are excluded (see [56, Lemma 12.4]).

We now are in a position to prove Landau's explicit formula for $\Psi(x)$.

Theorem 3.28 (Landau) *Let $T_0 \geqslant 2$ be a real number. Uniformly for $T \geqslant T_0$, we have*

$$\Psi(x) = x - \sum_{|\gamma| \leqslant T} \frac{x^\rho}{\rho} - \log 2\pi - \frac{1}{2}\log\left(1 - \frac{1}{x^2}\right) + O\left(\frac{x}{T}(\log xT)^2 + \log x\right).$$

Proof We may suppose that $x \notin \mathbb{Z}$. Let T' be the number supplied by Corollary 3.11, K be a large positive integer and call $\mathfrak{R}$ the rectangle with vertices $\kappa - iT'$, $\kappa + iT'$, $-(2K+1) + iT'$ and $-(2K+1) - iT'$. By (3.21), we see that $-\zeta'(s)/\zeta(s)$ has a simple pole at $s = -2k$ with residue -1. Since the residue at $s = 1$ is equal to 1, we get by Cauchy's residue theorem

$$\frac{1}{2\pi i}\int_\mathfrak{R} -\frac{\zeta'(s)}{\zeta(s)}\frac{x^s}{s}\,ds = x - \sum_{|\gamma| \leqslant T'}\frac{x^\rho}{\rho} - \sum_{1 \leqslant k < K+1/2}\frac{x^{-2k}}{-2k} - \frac{\zeta'}{\zeta}(0).$$

It can be shown that $\zeta'(0)/\zeta(0) = \log 2\pi$. By (3.42), we get

$$\Psi(x) = x - \sum_{|\gamma| \leqslant T'}\frac{x^\rho}{\rho} + \sum_{1 \leqslant k < K+1/2}\frac{x^{-2k}}{2k} - \log 2\pi - \sum_{j=1}^{2}I_{\mathcal{H}_j} - I_{\mathcal{V}} + O\left(\frac{x(\log x)^2}{T'} + \log x\right),$$

where $I_{\mathcal{H}_j}$ denotes the integrals taken over the two horizontal sides and $I_{\mathcal{V}}$ is the integral taken over the vertical side. Using Corollary 3.11, (3.50) and the easy estimate $T' \asymp T$, we obtain

$$I_{\mathcal{H}_j} \ll \int_{-2K-1}^{K} \left| -\frac{\zeta'}{\zeta}(\sigma + iT') \right| \frac{x^\sigma}{|\sigma + iT'|} \, d\sigma$$

$$\ll \int_{-2K-1}^{-1} x^\sigma \frac{\log|\sigma + iT'|}{|\sigma + iT'|} \, d\sigma + (\log T)^2 \int_{-1}^{K} \frac{x^\sigma}{|\sigma + iT'|} \, d\sigma$$

$$\ll \frac{(\log T)^2}{T} \int_{-2K-1}^{K} x^\sigma \, d\sigma \ll \frac{x(\log T)^2}{T}$$

for $j \in \{1, 2\}$ and

$$I_{\mathcal{V}} \ll \int_{-T'}^{T'} \left| -\frac{\zeta'}{\zeta}(-2K - 1 + it) \right| \frac{x^{-2K-1}}{|-2K - 1 + it|} \, dt \ll \frac{x^{-2K-1}T}{2K + 1} \log(KT)$$

and letting $K \to \infty$ completes the proof. □

3.5.3 Counting the Non-trivial Zeros

3.5.3.1 The Riemann–von Mangoldt Formula

We have seen that the non-trivial zeros $\rho = \beta + i\gamma$ of the Riemann zeta-function are of crucial importance in the distribution of prime numbers. Let $\sigma \in [0, 1]$ and $T \geqslant 1$ be real numbers. It is customary to define

$$N(\sigma, T) = |\{\rho = \beta + i\gamma : \beta \geqslant \sigma \text{ and } |\gamma| \leqslant T\}| = \sum_{\substack{\beta \geqslant \sigma \\ |\gamma| \leqslant T}} 1$$

and

$$\mathcal{N}(T) = |\{\rho = \beta + i\gamma : \beta \in {]}0, 1{[} \text{ and } 0 < \gamma \leqslant T\}|.$$

If $T = \gamma$, then we set $\mathcal{N}(T) = \frac{1}{2}\left(\mathcal{N}(T^+) + \mathcal{N}(T^-)\right)$.

The first important result was conjectured by Riemann and proved by von Mangoldt. For a proof, see [42, 56, 84, 86].

Theorem 3.29 (Riemann–von Mangoldt Formula) *We have*

$$N(T) = \frac{T}{2\pi} \log \frac{T}{2\pi} - \frac{T}{2\pi} + \frac{7}{8} + S(T) + O\left(\frac{1}{T}\right)$$

with

$$S(t) = \frac{1}{\pi} \arg \zeta \left(\tfrac{1}{2} + it\right),$$

where the argument is defined by continuous variation of s in $\zeta(s)$ starting at $s = 2$, then vertically to $s = 2 + it$ and then horizontally to $s = \tfrac{1}{2} + it$. Furthermore, we have $S(t) = O(\log t)$.

This implies in particular that

$$N(T + 1) - N(T) = O(\log T) \tag{3.51}$$

and using partial summation, we also get

$$\sum_{\substack{\rho \\ |\gamma| \leqslant T}} \frac{1}{|\rho|} \ll \int_1^T \frac{\log u}{u} \, du + \log T \ll (\log T)^2. \tag{3.52}$$

Similarly, since $N(\gamma_n - 1) < n \leqslant N(\gamma_n + 1)$, we infer that, for $n \to \infty$, we have

$$\gamma_n \sim \frac{2\pi n}{\log n}.$$

By adapting the method to the L-functions attached to primitive Dirichlet characters, one may also show that

$$N(T, \chi) = \frac{T}{2\pi} \log \frac{qT}{2\pi} - \frac{T}{2\pi} + O(\log qT), \tag{3.53}$$

where χ is a primitive Dirichlet character modulo $q > 1$ and $N(T, \chi)$ is the number of zeros $\rho = \beta + i\gamma$ of $L(s, \chi)$ in the rectangle $0 < \beta < 1$ and $0 < \gamma \leqslant T$.

3.5.3.2   The Density Hypothesis

Let us now have a look at estimates of the form

$$N(\sigma, T) \ll T^{A(\sigma)(1-\sigma)}(\log T)^D \tag{3.54}$$

for some $A(\sigma) \geqslant 0$ and $D \geqslant 0$. In view of Theorem 3.29, we must have $A(\sigma) \geqslant 2$. In many applications, such as the gaps between consecutive primes, the results that we may get using the Lindelöf or the Riemann hypothesis can be obtained via a weaker conjecture, namely the *density hypothesis* stating that, for all $\frac{1}{2} \leqslant \sigma \leqslant 1$ and $T \geqslant 3$, we have $A(\sigma) \leqslant 2$ so that

$$\mathcal{N}(\sigma, T) \ll T^{2(1-\sigma)} \log T.$$

A great deal of effort has been made to establish estimates of the form (3.54). We may summarize the main results in the following table.

$(A(\sigma), D)$	$\left(\frac{3}{2-\sigma}, 5\right)$	$\left(\frac{3}{3\sigma-1}, 44\right)$	$\left(\frac{12}{5}, 9\right)$
Range of validity	$\frac{1}{2} \leqslant \sigma \leqslant \frac{3}{4}$	$\frac{3}{4} \leqslant \sigma \leqslant 1$	$\frac{1}{2} \leqslant \sigma \leqslant 1$
Author	Ingham (1940)	Huxley (1972)	Huxley–Ingham–Ivić

3.5.4 The Siegel–Walfisz Theorem

3.5.4.1 Main Result

Let $a, q \geqslant 1$ be integers such that $(a, q) = 1$. It is customary to define the function

$$\pi(x; q, a) := \sum_{\substack{p \leqslant x \\ p \equiv a \,(\mathrm{mod}\, q)}} 1.$$

By Theorem 3.20, we have $\lim_{x \to \infty} \pi(x; q, a) = \infty$ and hence the question of its order of magnitude arises naturally. We expect the prime numbers to be well distributed in the $\varphi(q)$ reduced residue classes modulo q. Therefore, applying the method of the former section to the function

$$\Psi(x, \chi) = \sum_{n \leqslant x} \Lambda(n) \chi(n), \tag{3.55}$$

where χ is a non-principal Dirichlet character modulo q and to its Dirichlet series $-\dfrac{L'}{L}(s, \chi)$, we get

$$\pi(x; q, a) = \frac{\text{Li}(x)}{\varphi(q)} + O_q\left(xe^{-c(q)\sqrt{\log x}}\right)$$

for some constant $0 < c(q) < 1$ depending on q, the constants implied in the error term depending also on q. This dependence makes this result useless in practice. A great deal of effort has been made to prove some efficient estimates where the constants do not depend on the modulus. One of the most important results in the theory is called the *Siegel–Walfisz–Page theorem* or *Siegel–Walfisz theorem*. For a proof, see [26, Theorem 55].

Theorem 3.30 (Siegel–Walfisz) *Let $a, q \geqslant 1$ be coprime integers. Then there exists an absolute constant $c_0 > 0$ such that, for any $A > 0$, we have*

$$\pi(x; q, a) = \frac{\text{Li}(x)}{\varphi(q)} + O_A\left(xe^{-c_0\sqrt{\log x}}\right)$$

provided that $q \leqslant (\log x)^A$ and $x \geqslant 3$.

Obviously, if we define the Chebyshev type functions

$$\theta(x; q, a) := \sum_{\substack{p \leqslant x \\ p \equiv a \,(\text{mod}\, q)}} \log p \quad \text{and} \quad \Psi(x; q, a) := \sum_{\substack{n \leqslant x \\ n \equiv a \,(\text{mod}\, q)}} \Lambda(n)$$

similar estimates hold for these functions, namely, if $q \leqslant (\log x)^A$, then

$$\left.\begin{array}{c}\theta(x; q, a) \\ \Psi(x; q, a)\end{array}\right\} = \frac{x}{\varphi(q)} + O_A\left(xe^{-c_0\sqrt{\log x}}\right). \tag{3.56}$$

The proof of Theorem 3.30 rests on an explicit formula for $\Psi(x, \chi)$ similar to that of Theorem 3.28, namely

$$\Psi(x, \chi) = E_0(\chi)x - \sum_{|\gamma| \leqslant T} \frac{x^\rho}{\rho} + O\left(\frac{x}{T}(\log qx)^2 + x^{1/4}\log x\right),$$

where

$$E_0(\chi) = \begin{cases} 1, & \text{if } \chi = \chi_0 \\ 0, & \text{otherwise} \end{cases} \tag{3.57}$$

which implies using Proposition 3.13

$$\Psi(x;q,a) = \frac{x}{\varphi(q)} + \sum_{\chi \,(\mathrm{mod}\,q)} \sum_{|\gamma| \leqslant T} \frac{x^{\rho}}{\rho} + O\left(\frac{x}{T}(\log qx)^2 + x^{1/4}\log x\right).$$

3.5.4.2 A Zero-Free Region for L-Functions

The other important tool is the knowledge of a zero-free region for the function $L(s,\chi)$. The arguments generalize those of the function $\zeta(s)$, except that there is an unforeseen difficulty in connection with the possible existence, still unproven, of an exceptional zero $\beta_1 \in \mathbb{R}$ near the point 1 of a function $L(s,\chi)$ attached to a quadratic Dirichlet character. More precisely, we have the following result.

Theorem 3.31 (Zero-Free Region for L-Functions) *Let $q \in \mathbb{N}$ and $\tau = |t| + 3$. There exists an absolute constant $c_0 > 0$ such that if χ is a Dirichlet character modulo q, then the function $L(s,\chi)$ has no zero in the region*

$$\sigma \geqslant 1 - \frac{c_0}{\log(q\tau)}$$

unless χ is a quadratic character, in which case $L(s,\chi)$ has at most one, necessarily real, zero $\beta_1 < 1$ in this region. This zero is called exceptional.

At the present time, we do not know much more about this exceptional zero. Nevertheless, Landau, Page and Siegel provided some very important results, showing in particular that such a zero occurs at most rarely. We summarize their main discoveries in the next theorem.

Theorem 3.32 *Let q be a positive integer and $\tau = |t| + 3$.*

▷ *(Landau). There exists an absolute constant $c_1 > 0$ such that the function*
$$\prod_{\chi \,(\mathrm{mod}\,q)} L(s,\chi) \text{ has at most one zero in the region}$$

$$\sigma \geqslant 1 - \frac{c_1}{\log(q\tau)}.$$

If such a zero β_1 exists, then it is necessarily real and associated with a quadratic character χ_1. This character is called the exceptional character.

(continued)

Theorem 3.32 (continued)

▷ (Page). *If χ is a Dirichlet character modulo q, then $L(\sigma, \chi) \neq 0$ in the region*

$$\sigma \geqslant 1 - \frac{c_2}{q^{1/2}(\log(q+1))^2},$$

where $c_2 > 0$ is an effectively computable absolute constant.

▷ (Siegel). *Let χ be a quadratic Dirichlet character modulo q. For all $\varepsilon > 0$, there exists a* non-effectively computable *constant $c_\varepsilon > 0$ such that*

$$L(1, \chi) > \frac{c_\varepsilon}{q^\varepsilon}.$$

This implies that, if χ is a quadratic character modulo q, then $L(\sigma, \chi) \neq 0$ in the region

$$\sigma \geqslant 1 - \frac{c_\varepsilon}{q^\varepsilon}.$$

The proof relies on the following lemma due to Estermann providing the lower bound of certain Dirichlet series at $s = 1$, which may have its own interest [56, 92].

Lemma 3.17 (Estermann) *Let $f(s)$ be an analytic function in the disc $|s - 2| \leqslant \frac{4}{3}$ satisfying the bound $|f(s)| \leqslant M$ in this disc. Assume that, for all $\sigma > 1$, we have*

$$f(s)\zeta(s) = \sum_{n=1}^{\infty} \frac{a_n}{n^s}$$

with $a_1 \geqslant 1$ and $a_n \geqslant 0$ for all n. Finally, suppose that there exists $\alpha \in \left[\frac{26}{27}, 1\right[$ such that $f(\alpha) \geqslant 0$. Then we have

$$f(1) \geqslant \frac{1 - \alpha}{4M^{4(1-\alpha)}}.$$

In the proof of Theorem 3.32, one usually considers a primitive quadratic character χ_1 modulo q_1 such that the attached Dirichlet L-function has a real zero $\beta_1 \geqslant 1 - \frac{1}{4}\varepsilon$, and one applies Lemma 3.17 to $f(s) = L(s, \chi)L(s, \chi_1)L(s, \chi\chi_1)$ where $\chi \neq \chi_1$ is a primitive quadratic character. This implies that $f(1) \geqslant c_1(\varepsilon)q^{-\varepsilon}$ and since $f(1) \ll L(1, \chi)(\log qq_1)^2$, the result of Theorem 3.32 follows for primitive characters. It may be extended to imprimitive Dirichlet characters by using (3.34).

It should be noticed that we have no way of estimating the size of the smallest possible modulus q_1, so that the constant c_ε of Siegel's theorem is ineffective when

$\varepsilon < \frac{1}{2}$. All attempts at providing a value to c_ε for a sufficiently small $\varepsilon > 0$ have been unsuccessful.

3.5.5 Explicit Estimates

3.5.5.1 L-Functions

In the last decade, several explicit bounds for (3.44), (3.45), and (3.46) were discovered. The best results until now were obtained by Ford [29, 28] which proved, respectively, for (3.44), (3.45), and (3.46) the following estimates.

▷ For all $s = \sigma + it \in \mathbb{C}$ such that $\frac{1}{2} \leqslant \sigma \leqslant 1$ and $t \geqslant 3$, we have

$$|\zeta(s)| \leqslant 76.2\, t^{4.45(1-\sigma)^{3/2}} (\log t)^{2/3}. \tag{3.58}$$

▷ $\zeta(s)$ has no zero in the region

$$\sigma \geqslant 1 - \frac{1}{57.54\, (\log |t|)^{2/3} (\log \log |t|)^{1/3}} \quad \text{and} \quad |t| \geqslant 3. \tag{3.59}$$

▷ The PNT can be written as follows:

$$\pi(x) = \mathrm{Li}(x) + O\left\{ x \exp\left(-0.2098\, (\log x)^{3/5} (\log \log x)^{-1/5} \right) \right\}.$$

In another direction, one may consider zero-free regions of the type

$$\sigma \geqslant 1 - \frac{1}{R \log(|t|/d)} \quad \text{and} \quad |t| \geqslant t_0 \tag{3.60}$$

trying to get the lowest positive real number R in order to increase the region. Such zero-free regions are generally determined with the help of the real part of the function $-\zeta'(s)/\zeta(s)$. There are two classical ways to deal with this real part.

▷ The *global* one. Used by de la Vallée Poussin in 1899, the formula has the form

$$\mathrm{Re}\left(-\frac{\zeta'}{\zeta}(s)\right) = \frac{1}{2}\,\mathrm{Re}\left(\frac{\Gamma'}{\Gamma}(s-1)\right) - \frac{\log \pi}{2} + \mathrm{Re}\left(\frac{1}{s-1}\right) - \sum_\rho \mathrm{Re}\left(\frac{1}{s-\rho}\right),$$

where the sum runs through all non-trivial zeros ρ of the Riemann zeta-function.

▷ The *local* one. Used by Landau, the formula takes the form

$$\mathrm{Re}\left(-\frac{\zeta'}{\zeta}(s)\right) = - \sum_{\substack{\rho \\ |s-\rho| \leqslant c/\log |t|}} \mathrm{Re}\left(\frac{1}{s-\rho}\right) + O\left(\log |t|\right).$$

In [44], an intermediate way is considered with circles of radius $\asymp 1$. The main tools are then the knowledge of more and more zeros of $\zeta(s)$ lying on the critical line $\sigma = \frac{1}{2}$ and Weil type explicit formulae as follows.

Theorem 3.33 *Let $r > 0$ and $f \in C^2[0, r]$ with compact support in $[0, r)$ and satisfying $f(r) = f'(0) = f'(r) = f''(r) = 0$. We denote $F(s)$ the Laplace transform of f, i.e.*

$$F(s) = \int_0^r e^{-st} f(t) \, dt$$

and set F_2 the Laplace transform of f''. Then, for all complex numbers s, we have

$$\operatorname{Re}\left(\sum_{n=1}^{\infty} \frac{\Lambda(n)}{n} f(\log n) \right) = f(0) \left\{ \operatorname{Re}\left(\frac{1}{2} \frac{\Gamma'}{\Gamma}\left(\frac{s}{2}+1\right) \right) - \frac{\log \pi}{2} \right\}$$

$$+ \operatorname{Re} F(s-1) - \sum_{\rho} \operatorname{Re} F(s-\rho) + \operatorname{Re}\left(\frac{F_2(s)}{s^2} \right)$$

$$+ \frac{1}{2\pi} \int_{-\infty}^{\infty} \operatorname{Re}\left(\frac{\Gamma'}{\Gamma}\left(\frac{1}{4}+\frac{it}{2}\right) \right) \operatorname{Re}\left(\frac{F_2(s-1/2-it)}{(s-1/2-it)^2} \right) dt.$$

With an appropriate choice of f, it is proved in [44] that $\zeta(s)$ has no zero in the region

$$\sigma \geqslant 1 - \frac{1}{5.7 \log |t|} \quad \text{and} \quad |t| \geqslant 3.$$

This region supersedes the above region given by Ford as long as $|t| \leqslant e^{9400}$ and the method may be generalized to zero-free regions of L-functions, as can be seen in [43] where it is proved that the functions $L(s, \chi)$ never vanish in the region

$$\sigma \geqslant 1 - \frac{1}{6.4355 \log(\max(q, q\tau))}$$

except for at most one of them which should be real and vanishes at most once in this region.

3.5.5.2 Functions of Primes

In [77, Lemma 8], the authors established the following effective version of Theorem 3.28.

Proposition 3.16 *Let $x > 1$, $0 < \delta < 1 - x^{-1}$ be real numbers and R, d be real numbers satisfying (3.60). Set*

$$E(x) = \Psi(x) - \left\{ x - \log 2\pi - \tfrac{1}{2} \log \left(1 - x^{-2} \right) \right\}.$$

Then we have

$$|E(x)| \leqslant (2 + 2\delta^{-1} + \delta) \left(x \sum_{\gamma > A} F(\gamma) + 0.0463\sqrt{x} \right) + \tfrac{1}{2}x\delta$$

with

$$F(u) = \frac{1}{u^2} \exp \left(-\frac{\log x}{R \log(u/d)} \right)$$

and A is the unique solution of the equation

$$N(A) = \tfrac{A}{2\pi} \log \left(\tfrac{A}{2\pi} \right) - \tfrac{A}{2\pi} + \tfrac{7}{8},$$

where $N(A)$ is the number of zeros $\rho = \beta + i\gamma$ of $\zeta(s)$ such that $0 < \gamma \leqslant A$.

Rosser and Schœnfeld showed that one can take $R \approx 9.6459\ldots$, $d = 17$ and with the knowledge of non-trivial zeros at that time, they were able to take $A \approx 1894438.512\ldots$. The sum is treated by partial summation, which enables them to estimate integrals of the form

$$\int_A^\infty F(u) \log \left(\frac{u}{2\pi} \right) du$$

which in turn may be written as combinations of Bessel functions of the second kind. After using fine estimates of these quantities and choosing δ appropriately, we get the following explicit result.

Theorem 3.34 (Rosser and Schœnfeld) *Define*

$$\epsilon(x) = 0.110123 \left(1 + \frac{3.0015}{\sqrt{\log x}} \right) (\log x)^{3/8} \exp \left(-\sqrt{\frac{\log x}{R}} \right)$$

with $R \approx 9.6459\ldots$ Then, for all $x > 0$, we have

$$\theta(x) - x \leqslant \Psi(x) - x \leqslant x\epsilon(x)$$

(continued)

Theorem 3.34 (continued)
and for all x ⩾ 39.4, we have

$$\Psi(x) - x \geqslant \theta(x) - x \geqslant -x\epsilon(x).$$

For instance, the next result summarizes some of the estimates the authors obtained with Theorem 3.34.

Corollary 3.12 *The following estimates hold in the specified range of validity.*

Inequality	Validity
$\theta(x) < 1.000028\,x$	$x > 0$
$\theta(x) > 0.75\,x$	$x \geqslant 36$
$\dfrac{x}{\log x} \leqslant \pi(x) \leqslant \dfrac{1.25506\,x}{\log x}$	lower bound:$x \geqslant 17$ upper bound:$x > 1$
$\|\theta(x) - x\| < \dfrac{8.686\,x}{(\log x)^2}$	$x > 1$
$\|\psi(x) - x\| < \dfrac{8.686\,x}{(\log x)^2}$	$x > 1$

The usual functions of prime numbers may be handled as in the previous section [76].

Corollary 3.13 *The following estimates hold in the specified range of validity. p_n is the n-th prime number, $\gamma \approx 0.5772\ldots$ is the Euler–Mascheroni constant, $B \approx 0.261\,497\,212\ldots$ is the Mertens constant and $E \approx 1.332\,582\,275\ldots$.*

Inequality	**Validity**
$n(\log n + \log \log n - 1) \leqslant p_n \leqslant n(\log n + \log \log n)$	lower bound: $n \geqslant 2$ upper bound: $n \geqslant 6$
$\log x - E - \dfrac{1}{2 \log x} < \displaystyle\sum_{p \leqslant x} \dfrac{\log p}{p} < \log x - E + \dfrac{1}{\log x}$	$x \geqslant 32$
$\log \log x + B - \dfrac{1}{2(\log x)^2} < \displaystyle\sum_{p \leqslant x} \dfrac{1}{p} < \log \log x + B + \dfrac{1}{(\log x)^2}$	$x > 1$
$e^\gamma \log x \left(1 - \dfrac{1}{2(\log x)^2}\right) < \displaystyle\prod_{p \leqslant x} \left(1 - \dfrac{1}{p}\right)^{-1} < e^\gamma \log x \left(1 + \dfrac{1}{(\log x)^2}\right)$	$x > 1$

The lower bound for p_n is due to Dusart [22].

More precise estimates may be obtained in some restricted range of the variable. For instance, from [76, Theorems 21 and 22], we derive for $0 < x < 10^8$

$$\log x - E < \sum_{p \leqslant x} \frac{\log p}{p} < \log x - E + \frac{2}{\sqrt{x}}, \tag{3.61}$$

the upper bound being valid if one also assume that $x \notin [113, 113.8)$.

Explicit estimates for the functions $\pi(x; q, a)$, $\theta(x; q, a)$ and $\Psi(x; q, a)$ were established by McCurley [52], Ramaré, and Rumely [68] and Dusart [23]. In the second paper, the authors showed an analogue of Proposition 3.16 for $\Psi(x; q, a)$. This result is refined in the third paper, where the following estimates are proved.

Theorem 3.35 (Dusart) *Let $q \in \mathbb{Z}_{\geqslant 1}$ and set*

$$\epsilon(x) = \left(\frac{q^2 \log x}{R \varphi(q)^2}\right)^{1/4} \exp\left(-\sqrt{\frac{\log x}{R}}\right),$$

where $R \approx 9.6459\ldots$ is as in Theorem 3.34. Then, for all $x > x_0(q)$ where $x_0(q)$ is an effectively computable constant, we have

$$\left| \Psi(x; q, a) - \frac{x}{\varphi(q)} \right| < x\epsilon(x).$$

The same inequality holds with $\Psi(x; q, a)$ replaced by $\theta(x; q, a)$.

In fact, Dusart's result is slightly more accurate and makes use of a constant $C_1(q)$ in the function $\epsilon(x)$ which is quite complicated to define. We use here the fact that, under the conditions of the theorem, we always have $C_1(q) \geqslant 9.14$. Dusart then gave some applications of Theorem 3.35. For instance, if $a \in \{1, 2\}$, then we have

$$\pi(x; 3, a) < \frac{0.55\, x}{\log x} \quad (x \geqslant 229869)$$

and

$$\pi(x; 3, a) > \frac{x}{2 \log x} \quad (x \geqslant 151).$$

Also

$$\left| \theta(x; 3, a) - \frac{x}{2} \right| < \frac{0.262\, x}{\log x} \quad (x \geqslant 1531).$$

3.6 The Riemann Hypothesis

3.6.1 The Genesis of the Conjecture

The following quotation is attributed to Hilbert.

> If I were to awaken after having slept for a thousand years, my first question would be: Has the Riemann hypothesis been proven?

This shows the crucial importance of what has proved to be one of the most difficult problems in mathematics. In 1859, in his benchmarking Memoir, Bernhard Riemann formulated his conjecture, called today *the Riemann hypothesis*, which makes a very precise connection between two seemingly unrelated objects. There exist many great old unsolved problems in mathematics, but none of them has the stature of the Riemann hypothesis. This is probably due to the large number of ways in which this conjecture may be formulated. This is also certainly due to the personality of Riemann, a true genius ahead of his time and one of the most extraordinary mathematical talents. Finally, the Riemann hypothesis was highlighted at the 1900 International Congress of Mathematicians, in which Hilbert raised 23 problems that he thought would shape the next centuries. In 2000, the Clay Mathematics Institute listed seven hard open problems and promised a one million-dollar prize. Curiously, any disproof of the Riemann hypothesis does not earn the prize.

Riemann's formulation of his conjecture does not make arithmetic statements appear directly. We have seen that the non-trivial zeros of $\zeta(s)$ are all in the strip $0 < \sigma < 1$, and by the functional equation, if there is a zero in $0 < \sigma \leqslant \frac{1}{2}$, there is also a zero in the region $\frac{1}{2} \leqslant \sigma < 1$. Therefore, the following conjecture represents the "best of all possible worlds" for the zeros of the Riemann zeta-function.

> **Conjecture 3.3 (Riemann, 1859)**   All non-trivial zeros of $\zeta(s)$ are on the "critical line"
>
> $$\sigma = \tfrac{1}{2}.$$

Numerical computations have been made since 1859. It was Riemann himself who calculated the first zero of $\zeta(s)$ on the critical line, whose imaginary part is ≈ 14.13 [86]. The usual way is the use of the *Hardy function* $Z(t)$, also sometimes called the *Riemann–Siegel function*, defined in the following way. First set for $t \in \mathbb{R}$

$$\vartheta(t) = \arg\left(\pi^{-it/2}\Gamma\left(\tfrac{1}{4} + \tfrac{it}{2}\right)\right),$$

where the argument is defined by continuous variation of t starting with the value 0 at $t = 0$, and let

$$Z(t) = e^{i\vartheta(t)}\zeta\left(\tfrac{1}{2} + it\right).$$

The functional equation of $\zeta(s)$ implies that $\overline{Z(t)} = Z(t)$ so that $Z(t)$ is a real-valued function for $t \in \mathbb{R}$ and we have

$$|Z(t)| = \left|\zeta\left(\tfrac{1}{2} + it\right)\right|.$$

Therefore, if $Z(t_1)$ and $Z(t_2)$ have opposite signs, $\zeta(s)$ has a zero on the critical line between $\tfrac{1}{2} + it_1$ and $\tfrac{1}{2} + it_2$. Using Theorem 3.11 at $s = \tfrac{1}{2} + it$ with $x = y = \sqrt{|t|/(2\pi)}$ and multiplying out by $e^{i\vartheta(t)}$, we deduce that for $N = \lfloor\sqrt{|t|/(2\pi)}\rfloor$ we have

$$Z(t) = e^{i\vartheta(t)}\sum_{n=1}^{N}\frac{1}{n^{1/2+it}} + e^{-i\vartheta(t)}\sum_{n=1}^{N}\frac{1}{n^{1/2-it}} + O(t^{-1/4})$$

$$= 2\sum_{n=1}^{N}\frac{\cos(\vartheta(t) - t\log n)}{\sqrt{n}} + O(t^{-1/4})$$

which is a concise form of the *Riemann–Siegel formula*. Siegel discovered this identity among Riemann's private papers in 1932. It enables us to get an improvement over Euler–MacLaurin's summation formula in approximating values of $\zeta(s)$. More precise formulae exist (see [42, 86]) that aid computations and add to the empirical evidence for the Riemann hypothesis. The best result up to now is due to Trudgian and Platt [64] who found out that all zeros $\beta + i\gamma$ of the Riemann zeta-function with $0 < \gamma \leqslant 3 \times 10^{12}$ have $\beta = \tfrac{1}{2}$.

3.6.2   Hardy's Theorem

The minimal necessary condition for the Riemann hypothesis was proved by Hardy in 1914 [38].

Theorem 3.36 (Hardy) *There are infinitely many zeros of $\zeta(s)$ on the critical line.*

There exist several proofs of this result (see [86]), all of which rely on the consideration of moments of the form $\int t^n f(t)\,dt$. A useful tool is then the following lemma due to Fejér.

Lemma 3.18 *Let $a > 0$ and $n \in \mathbb{Z}_{\geqslant 1}$. The number of sign changes in the interval $]0, a[$ of a continuous function f is at least the number of sign changes of the sequence*

$$f(0), \quad \int_0^a f(t)\,dt, \quad \int_0^a t f(t)\,dt, \ldots, \int_0^a t^n f(t)\,dt.$$

Proof of Theorem 3.36   Recall the function

$$\omega(x) = \sum_{n=1}^{\infty} e^{-n^2 \pi x}$$

of the functional equation (3.14). One can show [86] that

$$\int_0^\infty \xi\left(\tfrac{1}{2} + it\right)\cos(xt)\,dt = \pi\left(2e^{-x/2}\omega(e^{-2x}) - e^{x/2}\right).$$

Putting $x = -iy$ gives

$$\int_0^\infty \xi\left(\tfrac{1}{2} + it\right)\cosh(yt)\,dt = 2\pi\left\{e^{iy/2}\left(\omega(e^{2iy}) + \tfrac{1}{2}\right) - \cos\tfrac{y}{2}\right\}.$$

By Theorem 3.12, we have $\zeta\left(\tfrac{1}{2} + it\right) \ll |t|^{1/4}$ so that $\xi\left(\tfrac{1}{2} + it\right) \ll |t|^{1/4} e^{-\pi|t|/4}$ and hence the above integral may be differentiated with respect to y any number of times provided that $y < \pi/4$. We then get

$$\int_0^\infty \xi\left(\tfrac{1}{2} + it\right) t^{2n}\cosh(yt)\,dt = 2\pi\left\{\frac{d^{2n}}{dy^{2n}}\left(e^{iy/2}\left(\omega(e^{2iy}) + \tfrac{1}{2}\right)\right) + (-1)^{n+1}2^{-2n}\cos\tfrac{y}{2}\right\}.$$

Now using (3.13) one can prove that the first term on the right-hand side tends to 0 as y tends to $\frac{1}{4}\pi$ for a fixed integer n, and thus

$$\lim_{y \to \pi/4} \int_0^\infty \xi\left(\tfrac{1}{2} + it\right) t^{2n} \cosh(yt)\, dt = (-1)^{n+1} 2^{1-2n} \pi \cos\left(\tfrac{1}{8}\pi\right).$$

Let m be a large positive integer. From above we infer that, if $a_m > 0$ is large enough and y_m is close enough to $\frac{1}{4}\pi$, the integral

$$\int_0^{a_m} \xi\left(\tfrac{1}{2} + it\right) t^{2n} \cosh(y_m t)\, dt$$

has the same sign as $(-1)^{n+1}$ for $n = 0, 1, \ldots, m$. By Lemma 3.18, we infer that $\xi\left(\frac{1}{2} + it\right)$ has at least m changes of sign in $]0, a_m[$, as required. $\qquad\square$

It is customary to define $\mathcal{N}_0(T)$ to be the number of zeros $\rho = \frac{1}{2} + i\gamma$ with $0 < \gamma < T$, so that the Riemann hypothesis may be written as

$$\mathcal{N}(T) = \mathcal{N}_0(T)$$

for all $T > 0$. Hardy's theorem states that $\mathcal{N}_0(T) \to \infty$ as $T \to \infty$ and his method yields

$$\mathcal{N}_0(T) > CT$$

for some constant $C > 0$.

The second important result was given by Selberg who proved that a positive proportion of zeros lies on the critical line and more precisely that

$$\mathcal{N}_0(T) > CT \log T$$

for T sufficiently large. Selberg's method could be used to yield an effective estimate of C, but this value was not made specific until 1974 when Levinson found an explicit estimate of this constant by proving that at least $1/3$ of the zeros lie on the critical line.[27] The best result to date is due to Bui et al. [16] who proved that more than 41% of the zeros lie on the critical line, slightly improving the previous record due to Conrey who showed in 1989 that at least $2/5$ of the zeros of ζ are on the critical line.

---

[27] See [42, Chapter 10].

3.6.3   Some Consequences of the Riemann Hypothesis

3.6.3.1   Sharpening the Prime Number Theorem

The first corollary of the Riemann hypothesis is to provide the best error term in the Prime Number Theorem. Indeed, suppose that all the zeros of $\zeta(s)$ have real parts equal to $\frac{1}{2}$. By (3.52), we then have for all $T \geqslant 2$

$$\sum_{|\gamma| \leqslant T} \frac{x^\rho}{\rho} \ll x^{1/2}(\log T)^2$$

so that by Theorem 3.28 we get

$$\Psi(x) = x + O\left(x^{1/2}(\log T)^2 + \frac{x}{T}(\log xT)^2 + \log x\right)$$

and choosing $T = x^{1/2}$ we obtain

$$\Psi(x) = x + O\left(x^{1/2}(\log x)^2\right).$$

This implies that

$$\theta(x) = x + O\left(x^{1/2}(\log x)^2\right)$$

and

$$\pi(x) = \mathrm{Li}(x) + O\left(x^{1/2}\log x\right)$$

by Exercise 9, a result proved by von Koch in 1901 [45]. In fact, one can prove that the Riemann hypothesis is *equivalent* to the estimate

$$\Psi(x) = x + O\left(x^{1/2+\varepsilon}\right)$$

for all $\varepsilon > 0$.[28] Also using [73, Théorème 3], we derive the following important result.

---

[28] See [84].

Theorem 3.37 *The Riemann hypothesis is equivalent to the estimate*

$$\pi(x) = \mathrm{Li}(x) + O\left(x^{1/2}\log x\right)$$

or equivalently

$$p_n = \mathrm{Li}^{-1}(n) + O\left(n^{1/2}(\log n)^{5/2}\right).$$

There exist explicit versions of this criterion in the literature. For instance, based upon [80, Theorem 10], the following result is proved in [15, Theorem 4.9].

Theorem 3.38 *The Riemann hypothesis is equivalent to the inequality*

$$|\Psi(x) - x| \leqslant \tfrac{1}{8\pi} x^{1/2}(\log x)^2 \quad (x \geqslant 74).$$

Unsurprisingly, a similar bound holds for the function π [80, Corollary 1].

Theorem 3.39 *If the Riemann hypothesis is true, then*

$$|\pi(x) - \mathrm{Li}(x)| \leqslant \tfrac{1}{8\pi} x^{1/2}\log x \quad (x \geqslant 2657).$$

This implies an analog estimate for the n-th prime [70, Theorem 6.1].

Theorem 3.40 *The Riemann hypothesis is equivalent to the inequality*

$$\left|p_n - \mathrm{Li}^{-1}(n)\right| \leqslant \tfrac{1}{\pi} n^{1/2}(\log n)^{5/2} \quad (n \geqslant 11).$$

These two estimates were recently refined by Dusart [24].

Theorem 3.41 *If the Riemann hypothesis is true, then*

$$|\pi(x) - \mathrm{Li}(x)| \leqslant \tfrac{1}{8\pi} x^{1/2}\log\left(\frac{x}{\log x}\right) \quad (x \geqslant 5639)$$

and

$$\left|p_n - \mathrm{Li}^{-1}(n)\right| \leqslant \tfrac{1}{8\pi} n^{1/2}(\log(n\log n))^{5/2} \quad (n \geqslant 230).$$

In the case of primes in arithmetic progressions, we have a similar result for the *generalized Riemann hypothesis*, i.e.

all zeros of every Dirichlet L-function $L(s, \chi)$ such that $0 < \sigma < 1$ are on the line $\sigma = \tfrac{1}{2}$.

Theorem 3.42 (Titchmarsh)   *The generalized Riemann hypothesis is equiv-*
alent to the statement that, for all $0 < q \leqslant x$ and positive integers a with
$(a, q) = 1$, we have

$$\Psi(x; q, a) = \frac{x}{\varphi(q)} + O\left(x^{1/2}(\log x)^2\right),$$

where the implied constant is absolute.

3.6.3.2   Mertens' Conjecture

Perhaps one of the most arithmetic formulations of the Riemann hypothesis can
be made with the Möbius function. This function $\mu(n)$ will be defined in Chap. 4
where we shall see that the Dirichlet series attached to this function is $\zeta(s)^{-1}$. We
need here the summatory function $M(x)$ of $\mu(n)$, called the *Mertens function*, and
hence defined by

$$M(x) = \sum_{n \leqslant x} \mu(n).$$

We will prove unconditionally in Theorem 4.25 the estimate

$$M(x) = O\left(xe^{-c_0(\log x)^{3/5}(\log\log x)^{-1/5}}\right)$$

for some constant $c_0 \in (0, 1)$. The Riemann hypothesis considerably improves this
bound.

Theorem 3.43 (Littlewood)   *The Riemann hypothesis is equivalent to the*
estimate

$$M(x) = O_\varepsilon\left(x^{1/2+\varepsilon}\right)$$

for all $\varepsilon > 0$.

Proof Indeed, this estimate implies by partial summation the convergence of the
series $\sum_{n=1}^{\infty} \mu(n)n^{-s} = \zeta(s)^{-1}$ in the half-plane $\sigma > \frac{1}{2}$, and hence $\zeta(s)$ has no
zero in this half-plane, which is the Riemann hypothesis. Now let $x, T \geqslant 2$, with
x large and $T \leqslant x^2$. If the Riemann Hypothesis is true, then by Perron's formula

Theorem 3.25, we get

$$\sum_{n \leqslant x} \mu(n) = \frac{1}{2\pi i} \int_{2-iT}^{2+iT} \frac{1}{\zeta(s)} \frac{x^s}{s} \, ds + O_\varepsilon \left(\frac{x^{2+\varepsilon}}{T} \right).$$

We shift the line $\sigma = 2$ to the line $\sigma = \frac{1}{2} + \varepsilon$. In the rectangle with vertices $2 \pm iT$, $\frac{1}{2} + \varepsilon \pm iT$, the Riemann Hypothesis implies that $\zeta(s)^{-1} \ll |t|^\varepsilon$, so that

$$\sum_{n \leqslant x} \mu(n) \ll_\varepsilon x^\varepsilon \left(x^2 T^{-1} + x^{1/2} \right)$$

and the choice of $T = x^2$ gives the asserted estimate.   □

Several authors improved the necessary condition. The best result is given in [7].

Theorem 3.44 *If the Riemann hypothesis is true, then, for all $\varepsilon > 0$*

$$M(x) \ll x^{1/2} \exp \left\{ (\log x)^{1/2} (\log \log x)^{5/2+\varepsilon} \right\}.$$

It has long been believed that the *Mertens conjecture* stating that, for all $n \in \mathbb{Z}_{\geqslant 1}$, the inequality

$$|M(n)| < n^{1/2}$$

would be true. In 1885, Stieltjes announced he had found a proof of the weaker bound $M(n) \ll n^{1/2}$ but he died without publishing his result, and no proof was found in his papers posthumously. We know now that Mertens' conjecture is false since Odlyzko and te Riele [60] showed that

$$\limsup_{n \to \infty} \frac{M(n)}{\sqrt{n}} > 1.06 \quad \text{and} \quad \liminf_{n \to \infty} \frac{M(n)}{\sqrt{n}} < -1.009.$$

These bounds were subsequently improved to 1.218 and -1.229, respectively [46], and it is also known that the smallest number for which the Mertens conjecture is false is $\exp \left(1.59 \times 10^{40} \right)$, improving on a previous result of Pintz [63].

A similar estimate holds for the *Liouville's function* defined by $\lambda(n) := (-1)^{\Omega(n)}$. If $L(x)$ is the summatory function of the λ-function, then it is an exercise in number theory to verify that

$$L(x) = \sum_{n \leqslant \sqrt{x}} M \left(\frac{x}{n^2} \right) \quad \text{and} \quad M(x) = \sum_{n \leqslant \sqrt{x}} \mu(n) L \left(\frac{x}{n^2} \right).$$

Now taking Theorem 3.43 into account, we clearly derive the following result.

Theorem 3.45 *The Riemann hypothesis is equivalent to the estimate*

$$L(x) = O_\varepsilon \left(x^{1/2+\varepsilon} \right).$$

3.6.3.3 Arithmetic Functions

The Riemann hypothesis also relies on arithmetic functions. The next two results deal with $\sigma(n)$, defined to be the sum of the positive divisors of n. The first criterion is due to Robin [74].

Theorem 3.46 *The Riemann hypothesis is equivalent to the inequality*

$$\sigma(n) < e^\gamma n \log \log n \quad (n \geqslant 5041).$$

Lagarias [47] derived a subtle reformulation of Theorem 3.46.

Theorem 3.47 *Let $H_n = \sum_{k=1}^n k^{-1}$ be the nth harmonic number. The Riemann hypothesis is equivalent to the inequality*

$$\sigma(n) \leqslant e^{H_n} \log H_n + H_n \quad (n \geqslant 1).$$

In [59], Nicolas established a criterion dealing with the Euler totient function.

Theorem 3.48 *For $k \in \mathbb{Z}_{\geqslant 1}$, denote p_k the kth prime and $N_k = 2 \times 3 \times \cdots \times p_k$ the primorial number of order k. The Riemann hypothesis is equivalent to the inequality*

$$\frac{N_k}{\varphi(N_k)} > e^\gamma \log \log N_k \quad (k \geqslant 1).$$

A less usual arithmetic function is investigated in [11]. Let p be a prime satisfying $p \equiv \pm 5 \pmod{24}$ and set

$$\lambda_p(n) := \left(\frac{\tau(n)}{p} \right),$$

where τ is the usual divisor counting function and $(\cdot/p)$ is the Legendre symbol. Finally, define

$$f_p(n) := \sum_{d \mid n} \lambda_p(d).$$

The function f_p is easily seen to be multiplicative, and since $|\lambda_p| \leqslant 1$, we get $|f_p(n)| \leqslant \tau(n)$. The Dirichlet series of f_p is studied in [11, Proposition 3], from which we derive the next criterion.

Theorem 3.49 *Let p be a fixed prime such that $p \equiv \pm 5 \pmod{24}$. The Riemann hypothesis is equivalent to the estimate*

$$\sum_{n \leqslant x} f_p(n) \ll x^{1/4+\varepsilon}$$

for all $\varepsilon > 0$.

The distribution of k-free numbers is also impacted by the Riemann hypothesis. In Theorem 4.27, we get an error term which seems to be hard to improve unconditionally without new ideas. On the other hand, if one assumes that the Riemann hypothesis is true, Montgomery and Vaughan [55] proved that, for any $T > 0$

$$\sum_{n \leqslant x} \mu_k(n) = \frac{x}{\zeta(k)} - \sum_{d \leqslant T} \mu(d)\psi\left(\frac{x}{d^k}\right) + O\left(x^{1/2+\varepsilon}T^{1/2-k/2} + T^{1/2+\varepsilon}\right)$$

for any $\varepsilon > 0$. Estimating the sum with tools we shall see in Chap. 6 and choosing T optimally lead to non-trivial improvements of Theorem 4.27. The trivial estimate of the sum already provides the asymptotic formula

$$\sum_{n \leqslant x} \mu_k(n) = \frac{x}{\zeta(k)} + O\left(x^{\frac{1}{k+1}+\varepsilon}\right)$$

obtained by Montgomery and Vaughan [55, Corollary]. Subsequently, several authors improved this estimate by treating the above sum non-trivially, essentially with new results concerning sums of type I and type II as seen in Chap. 6. The next theorem summarizes the latest results to date.

Theorem 3.50 *If the Riemann hypothesis is true, then*

$$\sum_{n \leqslant x} \mu_k(n) = \frac{x}{\zeta(k)} + O\left(x^{\theta_k+\varepsilon}\right),$$

where

k	2	3	4	5	6	$\geqslant 7$
θ_k	$\frac{11}{35}$	$\frac{17}{74}$	$\frac{17}{94}$	$\frac{23}{154}$	$\frac{4}{31}$	$\frac{11}{11k+18}$

The case $k = 2$ is shown in [50], the cases $k = 3, 4$ are from [6] and the cases $k \geqslant 5$ are proved in [49]. Note that for sufficiently large values of k, Vinogradov–Korobov's method gives better results [32, Theorem 2] of the shape

$$\sum_{n \leqslant x} \mu_k(n) = \frac{x}{\zeta(k)} + O\left(x^{\frac{1}{k+ck^{1/3}}}\right)$$

for some constant c.

3.6.3.4   Redheffer's Matrix

In 1977, Redheffer [69] introduced the matrix $R_n = (r_{ij}) \in \mathcal{M}_n(\{0, 1\})$ defined by

$$r_{ij} = \begin{cases} 1, & \text{if } i \mid j \text{ or } j = 1 \\ 0, & \text{otherwise} \end{cases}$$

and has shown that

$$\det R_n = M(n),$$

where $M(n)$ is the Mertens function.[29] Hence by Theorem 3.43 the Riemann hypothesis is equivalent to the estimate

$$|\det R_n| = O\left(n^{1/2+\varepsilon}\right)$$

for all $\varepsilon > 0$. This bound remains unproven, but Vaughan [89] showed that 1 is an eigenvalue of R_n with (algebraic) multiplicity $n - \lfloor \log n / \log 2 \rfloor - 1$, that R_n has two "dominant" eigenvalues $\lambda_{\pm}$ such that $|\lambda_{\pm}| \asymp n^{1/2}$, and that the other eigenvalues satisfy $\lambda \ll (\log n)^{2/5}$. However, it seems to be very difficult to extract more information from the eigenvalues of R_n by using tools from matrix analysis. For instance, Hadamard's inequality, which states that

$$|\det M|^2 \leqslant \prod_{i=1}^{n} \|L_i\|_2^2$$

---

[29] See Exercise 94.

for all matrices $M \in \mathcal{M}_n(\mathbb{C})$, where L_i is the ith row of M and $\|\ldots\|_2$ is the Euclidean norm on $\mathbb{C}^n$, gives

$$(M(n))^2 \leqslant n \prod_{i=2}^{n} \left(1 + \left\lfloor \frac{n}{i} \right\rfloor \right) = 2^{n - \lfloor n/2 \rfloor} n \prod_{i=2}^{\lfloor n/2 \rfloor} \left(1 + \left\lfloor \frac{n}{i} \right\rfloor \right) \leqslant 2^{n - \lfloor n/2 \rfloor} \binom{n + \lfloor n/2 \rfloor}{n}$$

which is very far from the trivial bound $|M(n)| \leqslant n$.

In another direction, the authors in [12] investigated the following integer upper triangular matrix. For all integers $i, j \geqslant 1$, set $\mathrm{mod}(j, i)$ to be the remainder in the Euclidean division of j by i. Let $T_n = (t_{ij}) \in \mathcal{M}_n(\mathbb{Z})$ be the upper triangular matrix of size n such that

$$t_{ij} = \begin{cases} \mathrm{mod}(j, 2) - 1, & \text{if } i = 1 \text{ and } 2 \leqslant j \leqslant n \\ \mathrm{mod}(j, i + 1) - \mathrm{mod}(j, i), & \text{if } 2 \leqslant i \leqslant n - 1 \text{ and } 1 \leqslant j \leqslant n \\ 1, & \text{if } (i, j) \in \{(1, 1), (n, n)\} \\ 0, & \text{otherwise.} \end{cases}$$

For instance with $n = 8$, we have

$$T_8 = \begin{pmatrix} 1 & -1 & 0 & -1 & 0 & -1 & 0 & -1 \\ 0 & 2 & -1 & 1 & 1 & 0 & 0 & 2 \\ 0 & 0 & 3 & -1 & -1 & 2 & 2 & -2 \\ 0 & 0 & 0 & 4 & -1 & -1 & -1 & 3 \\ 0 & 0 & 0 & 0 & 5 & -1 & -1 & -1 \\ 0 & 0 & 0 & 0 & 0 & 6 & -1 & -1 \\ 0 & 0 & 0 & 0 & 0 & 0 & 7 & -1 \\ 0 & 0 & 0 & 0 & 0 & 0 & 0 & 1 \end{pmatrix}.$$

One can prove the following result [12, Corollary 2.6].

Theorem 3.51 *Let σ_n be the smallest singular value of T_n. If the estimate*

$$\sigma_n \gg n^{-1/2 - \varepsilon}$$

holds for all $\varepsilon > 0$, then the Riemann hypothesis is true.

This criterion is probably at least as tricky to show as the Riemann hypothesis itself. However, it is interesting to note that this result is in accordance with the heuristic proofs concerning the smallest singular value[30] of a random matrix of size n. For instance, it is shown in [78, 79] that this singular value has a high probability of

---

[30]The singular values $\sigma_1 \geqslant \cdots \geqslant \sigma_n \geqslant 0$ of a matrix $A \in \mathcal{M}_n(\mathbb{C})$ are the square-roots of the eigenvalues of the Hermitian positive semidefinite matrix AA^*, where A^* is the hermitian adjoint of A.

satisfying the estimate $\sigma_n \asymp n^{-1/2}$. More precisely, let $A \in M_n(\mathbb{R})$ whose entries are i.i.d centred random variables with unit variance and fourth moment bounded by B. Let σ_n be the smallest singular value of A. The authors show that, for all $\varepsilon > 0$, there exist $K_1, K_2 > 0$ and positive integers n_1, n_2, depending polynomially only on B and ε, such that

$$\mathbb{P}\left(\sigma_n > K_1 n^{-1/2}\right) \leqslant \varepsilon \quad (n \geqslant n_1)$$

and

$$\mathbb{P}\left(\sigma_n < K_2 \, n^{-1/2}\right) \leqslant \varepsilon \quad (n \geqslant n_2)$$

where $\mathbb{P}(E)$ is the probability of the event E. But there are loads of examples in number theory showing that, from a heuristic argument to a rigorous proof, the way is very long and hard.

3.6.3.5   Riesz's Criterion

In 1916, Riesz [71] introduced the function $R(x)$ defined by

$$R(x) := \sum_{n=0}^{\infty} \frac{(-1)^n x^{n+1}}{n! \, \zeta(2n+2)}.$$

Theorem 3.52 (Riesz) *The Riemann hypothesis is equivalent to the bound*

$$R(x) \ll x^{1/4+\varepsilon}$$

for all $\varepsilon > 0$.

A discrete version of this result was discovered by Báez-Duarte [5].

Theorem 3.53 (Báez-Duarte) *The Riemann hypothesis is equivalent to the bound*

$$\sum_{k=0}^{n} (-1)^k \binom{n}{k} \frac{1}{\zeta(2k+2)} \ll n^{-3/4+\varepsilon}$$

for all $\varepsilon > 0$. Furthermore, if

$$\sum_{k=0}^{n} (-1)^k \binom{n}{k} \frac{1}{\zeta(2k+2)} \ll n^{-3/4}$$

then the zeros of the Riemann zeta-function are simple.

3.6.3.6 Báez-Duarte Convolution

Let f be a locally Lebesgue integrable function over $(0, \infty)$. The *Mellin transform* $\mathcal{M}_f$ of f is defined for $s \in \mathbb{C}$ by

$$\mathcal{M}_f(s) := \int_0^\infty f(x) x^{s-1} \, dx.$$

This tool is of constant use in analytic number theory, as seen in Theorem 3.25. In [4], unifying and generalizing the series of Riesz and Hardy–Littlewood, Báez-Duarte defines a convolution based upon the Mellin transform. Set

$$m(x) := \sum_{n \leqslant x} \frac{\mu(n)}{n}.$$

For any continuous positive-valued function f, define

$$(m \circledast f)(x) := \int_0^\infty m(xt) f(1/t) \frac{dt}{t}.$$

Theorem 3.54 (Báez-Duarte) *Let f be a continuous positive-valued function such that, for all $-\frac{1}{2} < \sigma \leqslant 0$, $\mathcal{M}_{|f|}(\sigma) < \infty$. Assume also that, for $-\frac{1}{2} < \sigma < 0$, we have $\mathcal{M}_f(-s) \neq 0$. Then, the Riemann hypothesis is equivalent to the bound*

$$(m \circledast f)(x) \ll x^{-1/2 + \varepsilon}$$

for all $\varepsilon > 0$.

3.6.3.7 The Cyclotomic Polynomials

The cyclotomic polynomial Φ_n is defined in Sect. 7.2.7.2 in Chap. 7. For all $N \in \mathbb{Z}_{\geqslant 1}$, define A_N to be the polynomial

$$A_N = \Phi_1 \times \cdots \times \Phi_N.$$

For any polynomial P, define the following measure, introduced by Mahler in 1964, by

$$\widetilde{h}(P) := \frac{1}{2\pi} \int_0^{2\pi} \log^+ \left| P\left(e^{ix}\right) \right| \, dx,$$

where $\log^+ x := \max(\log x, 0)$. In [3], Amoroso proved the following criterion.

Theorem 3.55 *Let* $\lambda \in \left(\frac{1}{2}, 1\right)$. *For all* $\varepsilon > 0$, *the bound*

$$\widetilde{h}(A_N) \ll N^{\lambda + \varepsilon}$$

holds if and only if the Riemann zeta-function has no zero in the half-plane $\sigma > \lambda$.

3.6.3.8   The Jensen Polynomials

Let $f : \mathbb{Z}_{\geqslant 1} \to \mathbb{R}$ be any real-valued arithmetic function and $n, d \in \mathbb{Z}_{\geqslant 0}$. The *degree* d *and shift* n *Jensen polynomial for* f is given by

$$J_f^{d,n}(X) := \sum_{j=0}^{d} f(n+j)\binom{d}{j}X^j.$$

A polynomial with real coefficients is said to be *hyperbolic* if all of its roots are real. Now define the coefficients $\gamma(k)$ to be the sequence of Taylor coefficients given by

$$\left(-1 + 4z^2\right)\xi\left(\frac{1}{2} + z\right) = \sum_{k=0}^{\infty} \frac{\gamma(k)}{k!}z^{2k},$$

where the Riemann ξ-function is defined in Theorem 3.7. In [65], Pólya proved the following criterion.

Theorem 3.56 (Pólya) *The Riemann hypothesis is equivalent to the hyperbolicity of the Jensen polynomials* $J_\gamma^{d,n}(X)$ *for all degrees* $d \geqslant 1$ *and all shifts* $n \geqslant 0$.

Subsequently, many researchers focus on the hyperbolicity of the Jensen polynomials. In [34], the authors proved that $J_\gamma^{d,n}(X)$ are hyperbolic for all d and all *sufficiently large* n. Furthermore, their work extends to a large class of arithmetic functions, including the partition function $p(n)$, and answering positively to a conjecture of Chen, Jia, and Wang stating that, for all $d \in \mathbb{Z}_{\geqslant 1}$, there exists an integer $N(d)$ such that $J_p^{d,n}(X)$ is hyperbolic for all $n \geqslant N(d)$.[31] The proof rests on the fact that, if f satisfies a certain regularity property, then the Jensen polynomials $J_f^{d,n}$ are related to the modified Hermite polynomials $H_d(X)$ defined as the orthogonal polynomials for the measure $e^{-X^2/4}$, i.e. $H_0(X) = 1$, $H_1(X) = X$, $H_2(X) = X^2 - 2$, $H_3(X) = X^3 - 6X$, $H_4(X) = X^4 - 12X^2 + 12$ and so on. The result follows by using the fact that each polynomial $H_d(X)$ is hyperbolic with d

---

[31]It is known that $N(1) = 1$, $N(2) = 25$, and Chen, Jia, and Wang proved that $N(3) = 94$. This result was also discovered independently by Larson and Wagner, who also calculated that $N(4) = 206$ and $N(5) = 381$ and proved that $N(d) \leqslant (3d)^{24d}(50d)^{3d^2}$.

distinct roots. This result is made effective in [35] where it is proved that, if $d \geqslant 1$ and $n \gg e^{8d/9}$, then the Jensen polynomials $J_\gamma^{d,n}(X)$ are hyperbolic.

3.6.3.9 An Integral

In [90], the author proves this curious criterion.

Theorem 3.57 *The Riemann hypothesis is equivalent to the evaluation*

$$\int_0^\infty \int_{1/2}^\infty \frac{1 - 12y^2}{(1 + 4y^2)^3} \log |\zeta(x + iy)| \, dx \, dy = \frac{\pi(3 - \gamma)}{32}.$$

Exercises

25 Let p be an odd prime number. Prove that every prime divisor of $2^p - 1$ is of the type $1 + 2kp$ for some positive integer k.

26 Show that, for each integer $n \geqslant 33$, we have $\pi(n) \leqslant \frac{1}{3}n$.

27 (H. Laurent, 1898). Prove that, for $n \geqslant 3$

$$\pi(n) = 2 + \sum_{k=5}^n \frac{e^{2i\pi(k-1)!/k} - 1}{e^{-2i\pi/k} - 1}.$$

28 Let π_k be the number of prime numbers in the sequence $6^2 + 2, 7^2 + 2, \ldots, (6k)^2 + 2$, with $k \in \mathbb{Z}_{\geqslant 2}$. Prove that $\pi_k < k$.

29 Show that the series $\displaystyle\sum_p \frac{1}{p \log p}$ is convergent.

30 Using Exercise 7 and Corollary 3.12, prove that[32] $\displaystyle\sum_p \frac{\log p}{p^2} < \frac{1}{2}$.

31 For any $n \in \mathbb{Z}_{\geqslant 1}$, set $H_n := 1 + \frac{1}{2} + \cdots + \frac{1}{n}$ be the n-th harmonic number. Prove that, if $p > 3$ is prime and if $H_p = \frac{a}{b}$ with $(a, b) = 1$, then $ap \equiv b \pmod{p^4}$.

---

[32]One may also use the inequality

$$\sum_{p \leqslant 100} \frac{\log p}{p^2} < 0.484$$

evaluated using the PARI/GP system for instance.

32   Let $n \in \mathbb{Z}_{\geqslant 1}$. Prove that, if the smallest prime divisor $p(n)$ of n satisfies $p(n) > n^{1/3}$, then $n/p(n)$ is prime or equal to 1.

33

(a) Let $n \in \mathbb{Z}_{\geqslant 1}$ and $p \geqslant 5$ be a prime number such that $p \mid (n^2 + n + 1)$. Show that $\text{ord}_p(n) = 3$ and deduce that $p \equiv 1 \pmod 6$.
(b) Show that there are infinitely many prime numbers of the form $p \equiv 1 \pmod 6$.

34   Let $a > b \geqslant 1$ and $n \geqslant 1$ be integers. Show that $n \mid \varphi(a^n - b^n)$.

35   (Legendre). Let p be a prime number and n be a positive integer. Prove that the p-adic valuation of $n!$ is given by

$$v_p(n!) = \sum_{k=1}^{\infty} \left\lfloor \frac{n}{p^k} \right\rfloor.$$

36   Let $n \in \mathbb{Z}_{\geqslant 1}$. Show that, for all primes $p \leqslant n$, we have

$$v_p(n!) = \frac{n}{p-1} + O(\log n).$$

37   Let p be a prime number and $r \in \mathbb{Z}_{\geqslant 1}$ such that p^r divides $\binom{2n}{n}$. Prove that

$$p^r \leqslant 2n.$$

38   Show that $n \in \mathbb{Z}_{\geqslant 1}$ is a difference of two squares if and only if $n \not\equiv 2 \pmod 4$. Deduce the values of n for which $n!$ is a difference of two squares.

39   For any $n \in \mathbb{Z}_{\geqslant 1}$, set $d_n := p_{n+1} - p_n$, where p_n is the nth prime number. Prove

$$\limsup_{n \to \infty} \frac{d_n}{\log n} \geqslant 1.$$

40   Let p be a prime number satisfying $p \equiv 3 \pmod 4$ and $(m, n) \in \left(\mathbb{Z}_{\geqslant 1}\right)^2$ such that $\sqrt{p} - \frac{m}{n} > 0$. Show that

$$\sqrt{p} - \frac{m}{n} \geqslant \frac{2}{3mn}.$$

41 Let $P \in \mathbb{Z}[X]$ be a polynomial of degree $n \geqslant 1$. Show that if there exist at least $2n + 1$ distinct integers m such that $|P(m)|$ is prime, then P is irreducible[33] over $\mathbb{Z}$. Are the polynomials $P_1 = X^4 - X^3 + 2X - 1$ and $P_2 = X^4 - 4X^2 - X + 1$ irreducible?

42 Using Dirichlet's theorem[34] and the sequence $7 + 15n$, show that there exist infinitely many prime numbers p not lying in a pair of twin prime numbers (q, r).

43 Let $n \geqslant 3$ be an integer and $1 = d_1 < d_2 < \cdots < d_k = n$ be all the divisors of n in increasing order. Show that

$$\sum_{j=2}^{k} d_{j-1} d_j < n^2.$$

44 With the help of Proposition 4.21, prove that every integer $n \geqslant 2$ can be written as the sum of two squarefree integers.

45 The aim of this exercise is to prove that the set $F(\mathbb{P})$ of all quotients of prime numbers is dense in $\mathbb{R}_{\geqslant 0}$.

(a) Let $\alpha > 1$. Show that there exists $m_\alpha \in \mathbb{Z}_{\geqslant 1}$ such that

$$n > m_\alpha \implies \left[\alpha^n, \alpha^{n+1}\right] \cap \mathbb{P} \neq \emptyset.$$

(b) Deduce that the set $F(\mathbb{P})$ is dense in $\mathbb{R}_{\geqslant 0}$.

46 For all $(x, y) \in (\mathbb{Z}_{\geqslant 1})^2$, set $T(x, y) := x(y + 1) - (y! + 1)$ and define a function $f : (\mathbb{Z}_{\geqslant 1})^2 \to \mathbb{Z}_{\geqslant 0}$ by

$$f(x, y) := \begin{cases} 2, & \text{if } |T(x, y)| \geqslant 1 \\ y + 1, & \text{otherwise.} \end{cases}$$

Show that all values of f are prime numbers, and that every odd prime number has a unique pre-image.

47 (Suryanarayana). With the help of (3.4), prove that

$$\sum_{k=2}^{\infty} \frac{(-1)^k \zeta(k)}{k + 1} = 1 + \tfrac{1}{2}(\gamma - \log 2\pi).$$

---

[33] See Definition 7.1 and Proposition 7.56.
[34] See Theorem 3.20.

48 Let $\sigma > 1$ be a real number and N be a positive integer. Show that

$$\sum_{n=N}^{\infty} \frac{1}{n^{\sigma}} \leqslant \frac{\zeta(\sigma)}{N^{\sigma-1}}$$

and if $N \geqslant 3$, then

$$\sum_{n=N}^{\infty} \frac{\log n}{n^{\sigma}} \leqslant \frac{\zeta(\sigma) \log N}{N^{\sigma-1}}.$$

49 Prove that, for any prime $p \geqslant 3$, the Fermat equation $x^p + y^p = z^p$ has no non-trivial solution $(x, y, z) \in \left(\mathbb{Z}_{\geqslant 1}\right)^3$ such that $z \geqslant \min\left(x^2, y^2\right)$.

50 Assume that there exists $\alpha \in \left[0, \frac{1}{2}\right]$ such that, for all $\varepsilon > 0$ and $t \geqslant e$, we have

$$\zeta\left(\frac{1}{2} + it\right) \ll_{\varepsilon} t^{\alpha+\varepsilon}.$$

Using Lemma 3.14, prove that, for all $\varepsilon > 0$, $\sigma \in [0, 1]$ and $t \geqslant e$

$$\zeta(\sigma + it) \ll_{\varepsilon} \begin{cases} t^{2\alpha(1-\sigma)+\varepsilon}, & \text{if } \frac{1}{2} \leqslant \sigma \leqslant 1 \\[2mm] t^{\sigma(2\alpha-1)+\frac{1}{2}+\varepsilon}, & \text{if } 0 \leqslant \sigma \leqslant \frac{1}{2}. \end{cases}$$

References

1. Aistleitner, C., Mahatab, K., Munsch, M.: Extreme values of the Riemann zeta function on the 1-line. Int. Math. Res. Not. **20**, 6924–6932 (2019)
2. Aistleitner, C., Mahatab, K., Munsch, M., Peyrot, A.: On large values of $L(\sigma, \chi)$. Q. J. Math. **70**, 831–848 (2019)
3. Amoroso, F.: On the heights of the product of cyclotomic polynomials. Rend. Sem. Mat. Univ. Politec. Torino **53**, 183–191 (1995)
4. Báez-Duarte, L.: Möbius convolutions and the Riemann hypothesis. Int. J. Math. Math. Sci. **22**, 3599–3608 (2005)
5. Báez-Duarte, L.: A sequential Riesz-like criterion for the Riemann hypothesis. Int. J. Math. Math. Sci. **21**, 3527–3537 (2005)
6. Baker, R.C., Powell, K.: The distribution of k-free numbers. Acta Math. Hungar. **126**, 181–197 (2010)
7. Balazard, M., de Roton, A.: Notes de lecture de l'article 'Partial sums of the Möbius function' de Kannan Soundararajan, p. 25 (2008). Preprint. https://arxiv.org/abs/0810.3587
8. Berndt, B.C.: A new proof of the functional equation of Dirichlet L-functions. Proc. Am. Math. Soc. **37**, 355–357 (1973)
9. Bondarenko, A., Seip, K.: Large greatest common divisor sums and extreme values of the Riemann zeta function. Duke Math. J. **166**, 1685–1701 (2017)

10. Bondarenko, A., Seip, K.: Note on the resonance method for the Riemann zeta function. In: 50 years with Hardy spaces. A tribute to Victor Havin, baranov, anton, kisliakov, sergei, nikolski, nikolai edn, no. 261 in Operator Theory: Advances and Applications, pp. 121–139. Birkhäuser, Basel (2018)
11. Bordellès, O.: Long and short sums of a twisted divisor function. J. Integer Seq. **20**, Article 17.7.1 (2017)
12. Bordellès, O., Cloitre, B.: A matrix inequality for Möbius functions. J. Inequal Pure Appl. Math. **10**, 62 (2009)
13. Bourgain, J., Watt, N.: Mean square of zeta function, circle problem and divisor problem revisited, p. 3. (2017). Preprint. https://arxiv.org/abs/1709.04340
14. Broadbent, S., Kadiri, H., Lumley, A., Ng, N., Wilk, K.: Sharper bounds for the Chebyshev function $\theta(x)$, p. 34 (2020). Preprint. https://arxiv.org/abs/2002.11068
15. Broughan, K.: Equivalents to the Riemann Hypothesis. Volume 1: Arithmetic Equivalents. No. 164. Encyclopedia of Mathematics and Its Applications. Cambridge University Press, Cambridge (2017)
16. Bui, H.M., Conrey, J.B., Young, M.P.: More than 41% of the zeros of the zeta function are on the critical line. Acta Arith. **150**, 35–64 (2011)
17. Cipolla, M.: La determinazione assintotica dell'n^{imo} numero primo. Mat. Napoli **3**, 132–166 (1902)
18. Deshouillers, J.M.: Retour sur la méthode de Čebyšev. Bull. Soc. Math. France **49–50**, 41–45 (1977)
19. Diamond, H., Erdős, P.: On sharp elementary prime number estimates. L'ens Math. **26**, 313–321 (1980)
20. Dickson, L.E.: History of the Theory of Numbers. Volume 1: Divisibility and Primality. Dover, New York (2005)
21. Dona, D, Helfgott, H.A., Alterman, S.Z.: Explicit L^2 bounds for the Riemann zeta function, p. 43 (2019). Preprint. https://arxiv.org/abs/1906.01097
22. Dusart, P.: Autour de la fonction qui compte le nombre de nombres premiers. Ph.D. thesis, Univ. Limoges (1998)
23. Dusart, P.: Estimates for $\theta(x; k, l)$ for large values of x. Math. Comp. **239**, 1137–1168 (2001)
24. Dusart, P.: Estimates of the kth prime under the Riemann hypothesis. Ramanujan J. **47**, 141–154 (2018)
25. Edwards, H.M.: Riemann's Zeta Function, 2nd edn, vol. 2001. Academic Press, Dover, Mineola, New York (1974)
26. Estermann, T.: Introduction to Modern Prime Number Theory. No. 41. Cambridge Tracts in Mathematics and Mathematical Physics. Cambridge University Press, Cambridge (1952)
27. Euvrard, C.: Majoration explicite sur le nombre de coefficients suffisants pour déterminer une fonction L. J. Théor. Nombres Bordeaux **29**, 51–83 (2017)
28. Ford, K.: Vinogradov's integral and bounds for the Riemann zeta function. Proc. Lond. Math. Soc. **85**, 565–633 (2002)
29. Ford, K.: Zero-free regions for the Riemann zeta function. In: Bennett, M.A., et al. (eds.) Number Theory for the Millennium, pp. 25–56. A. K. Peters, Boston (2002)
30. Godefroy, M.: La fonction Gamma; Théorie, Histoire, Bibliographie. Gauthier-Villars, Paris (1901)
31. Gould, H.W.: Combinatorial Identities. A Standardized Set of Tables Listing 500 Binomial Coefficient Summations. Morgantown (1972)
32. Graham, S.W., Pintz, J.: The distribution of r-free numbers. Acta Math . Hungar. **53**, 213–236 (1989)
33. Granville, A., Soundararajan, K.: Extreme values of $|\zeta(1 + it)|$. In: The Riemann Zeta Function and Related Themes: Papers in Honour of Professor K. Ramachandra, vol 2. Ramanujan Math. Soc. Lect. Notes Ser. Edn. Ramanujan Mathematical Society, Mysore (2006)
34. Griffin, M., Ono, K., Rolen, L., Zagier, D.: Jensen polynomials for the Riemann Zeta function and other sequences. Proc. Natl. Acad. Sci. U.S.A. **113**, 11103–11110 (2019)

35. Griffin, M.J., Ono, K., Rolen, L., Thorner, J., Tripp, Z., Wagner, I.: Jensen polynomials for the Riemann Xi function, p. 12 (2019). Preprint. https://arxiv.org/pdf/1910.01227.pdf
36. Gupta, R., Murty, M.R.: A remark on Artin's conjecture. Invent. Math. **78**, 127–130 (1984)
37. Hanson, D.: On the product of primes. Can. Math. Bull. **15**, 33–37 (1972)
38. Hardy, G.H.: Sur les zéros de la fonction $\zeta(s)$ de Riemann. C.R. Acad. Sci. Paris **158**, 1012–1014 (1914)
39. Heath-Brown, D.R.: Artin's conjecture for primitive roots. Q. J. Math. Oxf. **37**, 27–38 (1986)
40. Hilberdink, T.: An arithmetical mapping and applications to Ω-results for the Riemann zeta function. Acta Arith. **139**, 341–367 (2009)
41. Hooley, C.: On Artin's conjecture. J. Reine Angew. Math. **226**, 207–220 (1967)
42. Ivić, A.: The Riemann Zeta-Function. Theory and Applications, 2nd edn, vol. 2003. Wiley, New York; Dover, New York (1985)
43. Kadiri, H.: Une région explicite sans zéro pour les fonctions L de Dirichlet. Ph.D. thesis, Univ. Lille 1 (2002)
44. Kadiri, H.: An explicit region without zeros for the Riemann ζ function. Acta Arith. **117**, 303–339 (2005)
45. von Koch, H.: Sur la distribution des nombres premiers. Acta Math. **24**, 159–182 (1901)
46. Kotnik, T., te Riele, H.J.J.: The Mertens conjecture revisited. In: Algorithmic Number Theory, Seventh International Symposium, ANTS-VII, Berlin, 2006, pp 156–167. Lecture Notes in Computer Science, vol. 4076. Springer, Berlin (2006)
47. Lagarias, J.C.: An elementary problem equivalent to the Riemann hypothesis. Am. Math. Mon. **109**, 534–543 (2002)
48. Levinson, N.: Ω-theorems for the Riemann zeta-function. Acta Arith. **20**, 317–330 (1972)
49. Liu, H.Q.: On the distribution of k-free integers. Acta Math. Hungar. **144**, 269–284 (2014)
50. Liu, H.Q.: On the distribution of squarefree numbers. J. Number Theory **159**, 202–222 (2016)
51. Liu, J., Ye, Y.: Perron's formula and the prime number theorem for automorphic L-functions. Pure Appl. Math. Q. **3**, 481–497 (2007)
52. McCurley, K.S.: Explicit estimates for the error term in the prime number theorem for arithmetic progressions. Math. Comp. **423**, 265–285 (1984)
53. Monsky, P.: Simplifying the proof of Dirichlet's theorem. Am. Math. Mon. **100**, 861–862 (1993)
54. Montgomery, H.L.: Extreme values of the Riemann zeta function. Comment Math. Helv. **52**, 511–518 (1977)
55. Montgomery, H.L., Vaughan, R.C.: The distribution of squarefree numbers. In: Halberstam, H., Hooley, C. (eds.) Recent Progress in Analytic Number Theory, vol. 1, pp. 247–256. Academic Press, New York (1981)
56. Montgomery, H.L., Vaughan, R.C.: Multiplicative Number Theory Vol. I. Classical Theory, vol. 97. Cambridge Studies in Advanced Mathematics (2007)
57. Moser, L.: On the product of the primes not exceeding n. Can. Math. Bull. **2**, 119–121 (1959)
58. Nair, M.: A new method in elementary prime number theory. J. Lond. Math. Soc. **25**, 385–391 (1982)
59. Nicolas, J.L.: Petites valeurs de la fonction d'Euler. J. Number Theory **17**, 375–388 (1983)
60. Odlyzko, A., te Riele, H.J.J.: Disproof of the Mertens conjecture. J. Reine Angew. Math. **357**, 138–160 (1985)
61. Oon, S.M.: Note on the lower bound of least common multiple. Abstr. Appl. Anal. **2013**, 218125 (2013)
62. Paris, R.B., Kaminski, D.: Asymptotics and Mellin-Barnes Integrals. Encyclopedia of Mathematics and Its Applications. Cambridge University Press, Cambridge (2001)
63. Pintz, J.: An effective disproof of the Mertens conjecture. Astérisque **147/148**, 325–333 (1987)
64. Platt, D., Trudgian, T.: The Riemann hypothesis is true up to $3 \cdot 10^{12}$, p. 8 (2020). Preprint. https://arxiv.org/pdf/2004.09765.pdf
65. Pólya, G.: Über die algebraisch-funcktionentheoretischen Untersuchungen von J. L. W. V. Jensen. Kgl Danske Vid Sel Math-Fys Medd **7**, 3–33 (1927)

66. Pólya, G., Szegő, G.: Problems and Theorems in Analysis II. Springer, Berlin (1998)
67. Ramaré, O.: La méthode de Balasubramanian pour une région sans zéro. In: Groupe de Travail d'Analyse Harmonique et de Théorie Analytique des Nombres de Lille (2010/2011)
68. Ramaré, O., Rumely, R.: Primes in arithmetic progressions. Math. Comp. **65**, 397–425 (1996)
69. Redheffer, R.M.: Eine explizit lösbare Optimierungsaufgabe. Internat Schiftenreihe Numer. Math. **36**, 203–216 (1977)
70. de Reyna, J.A., Toulisse, J.: The n-th prime asymptotically. J. Théor. Nombres Bordeaux **25**(3), 521–555 (2013)
71. Riesz, M.: Sur l'hypothèse de Riemann. Acta Math. **40**:185–190 (1916)
72. Robbins, H.: A remark on Stirling's formula. Am. Math. Mon. **62**, 26–29 (1955)
73. Robin, G.: Estimation de la fonction de Tchebychef θ sur le k-ième nombre premier et grandes valeurs de la fonction $\omega(n)$ nombre de diviseurs premiers de n. Acta Arith. **42**(4), 367–389 (1983)
74. Robin, G.: Grandes valeurs de la fonction somme des diviseurs et hypothèse de Riemann. J. Math. Pures Appl. **63**, 187–213 (1983)
75. Robin, G.: Permanence de relations de récurrence dans certains développements asymptotiques. Pub. Inst. Math. **43**, 17–25 (1988)
76. Rosser, J.B., Schoenfeld, S.L.: Approximate formulas for some functions of prime numbers. Ill. J. Math. **6**, 64–94 (1962)
77. Rosser, J.B., Schoenfeld, S.L.: Sharper bounds for the Chebyshev functions $\theta(x)$ and $\psi(x)$. Math. Comp. **29**, 243–269 (1975)
78. Rudelson, M., Vershynin, R.: The least singular value of a random square matrice is $o(n^{-1/2})$. C.R. Math. Acad. Sci. Paris **346**, 893–896 (2008)
79. Rudelson, M., Vershynin, R.: The Littlewood–Offord Problem and invertibility of random matrices. Adv. Math. **218**, 600–633 (2008)
80. Schoenfeld, L.: Sharper bounds for the Chebyshev functions $\theta(x)$ and $\psi(x)$ II. Math. Comp. **309**, 337–360 (1976)
81. Soundararajan, K.: Extreme values of zeta and L-functions. Math. Ann. **342**, 467–486 (2008)
82. Stevenhagen, P.: The correction factor in Artin's primitive root conjecture. J. Théor. Nombres Bordeaux **15**, 383–391 (2003)
83. Stieltjes, T.J.: Sur le développement de $\log \gamma(a)$. J. Math. Pures Appl. **5**, 425–444 (1889)
84. Tenenbaum, G.: Introduction à la Théorie Analytique et Probabiliste des Nombres. Belin (2007)
85. Titchmarsh, E.C.: The Theory of Functions, 2nd edn, vol. 1979. Oxford University Press, Oxford (1939)
86. Titchmarsh, E.C.: The Theory of the Riemann Zeta-Function, 2nd edn, vol. 1986. Oxford University Press, Oxford (1951). Revised by D. R. Heath-Brown
87. Trost, E.: Primzahlen. Elemente der Mathematik vom höheren Standpunkt aus. II. Verlag Birkhäuser, Basel/Stuttgart (1953)
88. Trudgian, T.: A new upper bound for $|\zeta(1 + it)|$. Bull. Aust. Math. Soc. **89**, 259–264 (2014)
89. Vaughan, R.C.: On the eigenvalues of Redheffer's matrix I. In: Number Theory with an Emphasis on the Markoff Spectrum, Dekker, New York, Provo, Utah, 1991, vol. 147. Lecture Notes in Pure and Applied Mathematics (1993)
90. Volchkov, V.V.: On an equality equivalent to the Riemann hypothesis. Ukrain. Math. Zh. **47**, 422–423 (1995). English Trans. in Ukrain. Math. Zh. **47**, 491–493 (1995)
91. Voronin, S.M.: Lower bounds in Riemann zeta-function theory. Izv. Akad. Nauk. SSSR Ser. Mat. **52**, 882–892, 896 (1988)
92. Washington, L.C.: Introduction to Cyclotomic Fields. GTM 83, 2nd edn, vol. 1997. Springer, Berlin (1982)
93. Widder, D.V.: An Introduction to Transform Theory. Academic Press, New York and London (1971)

Chapter 4
Arithmetic Functions

4.1 The Basic Theory

4.1.1 The Ring of Arithmetic Functions

4.1.1.1 Definition

Definition 4.1 An *arithmetic function* is a map $f : \mathbb{Z}_{\geqslant 1} \to \mathbb{C}$, i.e. a sequence of complex numbers, although this viewpoint is not very useful.

In the subsections below, we list the main arithmetic functions the reader may encounter in his studies in analytic number theory. In what follows, $k \geqslant 1$ is a fixed integer.

4.1.1.2 Fundamental Arithmetic Functions

▷ The function e_1 is defined by[1] $e_1(n) = \begin{cases} 1, & \text{if } n = 1 \\ 0, & \text{otherwise.} \end{cases}$

▷ The constant function $\mathbf{1}$ is defined by $\mathbf{1}(n) = 1$.
▷ The k-th powers $\mathrm{id}_k(n) = n^k$. We also set $\mathrm{id}_1 = \mathrm{id}$.
▷ The function ω is defined by $\omega(1) = 0$ and, for all $n \geqslant 2$, $\omega(n)$ counts the number of distinct prime factors of n.
▷ The function Ω is defined by $\Omega(1) = 0$ and, for all $n \geqslant 2$, $\Omega(n)$ counts the number of prime factors of n, including multiplicities.

---

[1]Some authors also use the notation δ or i.

© Springer Nature Switzerland AG 2020
O. Bordellès, *Arithmetic Tales*, Universitext,
https://doi.org/10.1007/978-3-030-54946-6_4

▷ The *Möbius function* μ is defined by $\mu(1) = 1$ and, for all $n \geqslant 2$, by

$$\mu(n) = \begin{cases} (-1)^{\omega(n)}, & \text{if } n \text{ is squarefree} \\ 0, & \text{otherwise.} \end{cases}$$

This is one of the most important arithmetic functions of the theory.

▷ The *Liouville function* λ is defined by $\lambda(n) = (-1)^{\Omega(n)}$.

▷ The function μ_k is the *characteristic function of the set of k-free numbers* $(k \geqslant 2)$. Note that $\mu_2 = \mu^2$.

▷ The function s_k is the *characteristic function of the set of k-full numbers*[2] $(k \geqslant 2)$.

▷ The *generalized von Mangoldt functions* Λ_k are defined by

$$\Lambda_k(n) = \sum_{d|n} \mu(d) \left(\log \frac{n}{d} \right)^k.$$

Note that, by (3.3) and Möbius inversion formula,[3] we have $\Lambda_1 = \Lambda$.

▷ The *Dirichlet characters* $\chi(n)$ modulo q.

4.1.1.3   Divisor Functions

▷ The *Dirichlet–Piltz divisor function* τ_k is defined by[4] $\tau_1 = \mathbf{1}$, and for $k \geqslant 2$ and $n \geqslant 1$, by

$$\tau_k(n) = \sum_{d|n} \tau_{k-1}(d).$$

It is customary to denote τ_2 by τ.

▷ The *Hooley divisor function* Δ_k is defined by $\Delta_1 = \mathbf{1}$, and for $k \geqslant 2$ and $n \geqslant 1$, by

$$\Delta_k(n) = \max_{u_1,\dots,u_{k-1} \in \mathbb{R}} \sum_{\substack{d_1 d_2 \cdots d_{k-1}|n \\ e^{u_i} < d_i \leqslant e^{u_i+1}}} 1.$$

It is customary to denote Δ_2 by Δ.

---

[2]There is no official notation for this function in the literature. For instance, Ivić [51] uses the notation f_k.

[3]See Theorem 4.2 below.

[4]Some authors also use the notation d_k.

▷ Let $1 \leqslant a_1 \leqslant \cdots \leqslant a_k$ be fixed integers. The *generalized divisor function* is given by

$$\tau(a_1, \ldots, a_k; n) := \sum_{n=n_1^{a_1} \cdots n_k^{a_k}} 1.$$

In particular, we often set

$$\tau^{(k)}(n) = \sum_{d^k \mid n} 1.$$

▷ The function $a(n)$ is *the number of non-isomorphic abelian groups* of order n.
▷ The function $\tau_{(k)}(n)$ is the number of k-free divisors of n ($k \geqslant 2$). Note that $\tau_{(2)} = 2^\omega$.
▷ The function $\gamma_k(n)$ is the greatest k-free divisor of n ($k \geqslant 2$). The function γ_2 is sometimes called the *core*, or *squarefree kernel*, of n.
▷ The *divisor functions* σ_k are defined by

$$\sigma_k(n) = \sum_{d \mid n} d^k.$$

It is customary to denote σ_1 by σ. Also note that $\sigma_0 = \tau$.
▷ The functions $\tau^{(e)}(n)$ and $\sigma^{(e)}(n)$ are, respectively, *the number and the sum of exponential divisors*.[5]
▷ The function β is defined by $\beta(1) = 1$ and, for all $n \geqslant 2$, $\beta(n)$ is *the number of square-full divisors* of n.
▷ Let $\mathbb{K}/\mathbb{Q}$ be an algebraic number field. The function $r_{\mathbb{K}}$ is defined in Chap. 7 by $r_{\mathbb{K}}(1) = 1$ and, for all $n \geqslant 2$, $r_{\mathbb{K}}(n)$ is *the number of non-zero integral ideals of $\mathcal{O}_{\mathbb{K}}$ of norm equal to n*.

4.1.1.4   Totients

▷ The *Euler totient function* φ is defined by

$$\varphi(n) = \sum_{\substack{m \leqslant n \\ (m,n)=1}} 1.$$

---

[5]If $n = p_1^{\alpha_1} \cdots p_r^{\alpha_r}$ and $d = p_1^{\beta_1} \cdots p_r^{\beta_r}$ is a divisor of n, then d is called an *exponential divisor* if $\beta_i \mid \alpha_i$ for all $i \in \{1, \ldots, r\}$.

▷ The *Jordan totient function* J_k is defined by

$$J_k(n) = \sum_{\substack{(m_1,\dots,m_k)\in\mathbb{Z}_+^k,\, m_i\leqslant n \\ (m_1,\dots,m_k,n)=1}} 1.$$

Note that $J_1 = \varphi$.

▷ The *Dedekind totient function* Ψ[6] is defined by

$$\Psi(n) = n \prod_{p|n}\left(1 + \frac{1}{p}\right).$$

▷ The *generalized Dedekind function* Ψ_k is defined by

$$\Psi_k(n) = n^k \prod_{p|n}\left(1 + \frac{1}{p^k}\right).$$

Note that $\Psi_1 = \Psi$.

Remark 4.1 Clearly, sums and products of arithmetic functions are still arithmetic functions. It should also be mentioned that the behaviour of such functions can be erratic, as we can see in the following table showing values taken by τ at certain consecutive integers.

n	1007	1008	1009	1010	1011	1012	1013	1014
$\tau(n)$	4	30	2	8	4	12	2	12

Hence the first idea is to study these functions *on average*, i.e. to get estimates for the sums

$$\sum_{n\leqslant x} f(n).$$

4.1.2   Additive and Multiplicative Functions

4.1.2.1   Definitions

Definition 4.2 Let f be an arithmetic function.

---

[6]This notation should not be confused with the second Chebyshev function given in Definition 3.3.

▷ f is *multiplicative* if $f(1) \neq 0$ and if, for all positive integers m, n such that $(m, n) = 1$, we have

$$f(mn) = f(m)f(n). \tag{4.1}$$

▷ f is *completely multiplicative* if $f(1) \neq 0$ and if the condition $f(mn) = f(m)f(n)$ holds for *all* positive integers m, n.
▷ f is *strongly multiplicative* if f is multiplicative and if $f(p^\alpha) = f(p)$ for all prime powers p^α.

The condition $f(1) \neq 0$ is a convention to exclude the zero-function of the set of multiplicative functions. Furthermore, it is easily seen that, if f and g are multiplicative, then so are fg and f/g (with $g \neq 0$ for the quotient).
The dual notion of *additivity* is similar.

Definition 4.3 Let f be an arithmetic function.

▷ f is *additive* if, for all positive integers m, n such that $(m, n) = 1$, we have

$$f(mn) = f(m) + f(n). \tag{4.2}$$

▷ f is *completely additive* if the condition $f(mn) = f(m) + f(n)$ holds for *all* positive integers m, n.
▷ f is *strongly additive* if f is additive and if $f(p^\alpha) = f(p)$ for all prime powers p^α.

Similarly, f is said to be *sub-multiplicative* if, for all positive coprime integers m, n, we have

$$f(mn) \leqslant f(m)f(n)$$

and f is said to be *super-multiplicative* if, for all positive coprime integers m, n, we have

$$f(mn) \geqslant f(m)f(n).$$

The notions of sub-additivity and super-additivity are defined in a similar way, as well as the notions of *completely* sub-multiplicative, sub-additive, super-multiplicative and super-additive.

4.1.2.2 A Useful Criterion

In order to show that a given function is multiplicative or additive, one can use the relations (4.1) or (4.2). However, one often needs to know that a function is multiplicative or additive precisely *in order to* use (4.1) or (4.2). The following result is then a useful criterion.

Lemma 4.1 *Let f be an arithmetic function.*

▷ *f is multiplicative if and only if $f(1) = 1$ and for all $n = p_1^{\alpha_1} \cdots p_r^{\alpha_r}$ where the p_i are distinct primes, we have*

$$f(n) = \prod_{k=1}^{r} f\left(p_k^{\alpha_k}\right). \tag{4.3}$$

▷ *f is additive if and only if $f(1) = 0$ and for all $n = p_1^{\alpha_1} \cdots p_r^{\alpha_r}$ where the p_i are distinct primes, we have*

$$f(n) = \sum_{k=1}^{r} f\left(p_k^{\alpha_k}\right). \tag{4.4}$$

Proof The two proofs are similar, so that we only show the first one. Assume first that f satisfies $f(1) = 1$ and (4.3). Let $n = p_1^{\alpha_1} \cdots p_r^{\alpha_r}$ and $m = q_1^{\beta_1} \cdots q_r^{\beta_r}$ be two positive coprime integers. Using (4.3) and the fact that $p_i \neq q_j$ we get

$$f(nm) = f\left(p_1^{\alpha_1} \cdots p_r^{\alpha_r} q_1^{\beta_1} \cdots q_r^{\beta_r}\right) = \prod_{k=1}^{r} f\left(p_k^{\alpha_k}\right) \prod_{k=1}^{r} f\left(q_k^{\beta_k}\right) = f(n)f(m)$$

and hence f satisfies (4.1). Since $f(1) = 1 \neq 0$, we infer that f is multiplicative.

Conversely, let f be a multiplicative function. Using (4.1) with $m = n = 1$ gives $f(1) = f(1)f(1)$ so that $f(1) = 1$ since $f(1) \neq 0$. Now let $n_1, \ldots, n_k$ be pairwise coprime integers. By induction using (4.1), we get

$$f(n_1 \cdots n_k) = f(n_1) \cdots f(n_k).$$

Thus, if $n = p_1^{\alpha_1} \cdots p_r^{\alpha_r}$ where the p_i are distinct primes, we infer that f satisfies (4.3) as required.   □

4.1.2.3   Fundamental Examples

▷ The functions ω and Ω are additive, the first one strongly, the second one completely. All the other functions, except the Hooley divisor function Δ_k and the generalized von Mangoldt function Λ_k, are multiplicative.
▷ The functions e_1, **1**, id_k, λ and the Dirichlet characters are completely multiplicative.
▷ It is easily seen that, for all positive integers m, n, we have

$$\omega(mn) = \omega(m) + \omega(n) - \omega((m, n))$$

since in the sum $\omega(m) + \omega(n)$, the prime factors of (m, n) have been counted twice. This implies easily the additivity of ω.

▷ The functions μ_k and s_k are clearly multiplicative.

▷ Now let us have a look at the Möbius function. Using its definition, we have $\mu(1) = 1$ and, for all prime powers p^α, we also have

$$\mu\left(p^\alpha\right) = \begin{cases} -1, & \text{if } \alpha = 1 \\ 0, & \text{otherwise} \end{cases}$$

so that, if $n = p_1^{\alpha_1} \cdots p_r^{\alpha_r}$ where the p_i are distinct primes, we have

$$\mu\left(p_1^{\alpha_1}\right) \cdots \mu\left(p_r^{\alpha_r}\right) = \begin{cases} (-1)^r, & \text{if } \alpha_1 = \cdots = \alpha_r = 1 \\ 0, & \text{otherwise} \end{cases}$$

and hence $\mu\left(p_1^{\alpha_1}\right) \cdots \mu\left(p_r^{\alpha_r}\right) = \mu(n)$. Therefore, the μ-function is multiplicative by Lemma 4.1.

▷ The multiplicativity of the function $r_\mathbb{K}$ will be established in Proposition 7.29.

▷ The function τ is multiplicative via

$$\tau(n) = \prod_{p^\alpha \| n} (\alpha + 1) = \prod_{p^\alpha \| n} \tau\left(p^\alpha\right).$$

▷ In order to prove the multiplicativity of the function $\tau_{(k)}$, let us write uniquely n in the form $n = ab$ where $(a, b) = 1$ and $\mu_k(a) = s_k(b) = 1$ with $a = p_1^{\alpha_1} \cdots p_r^{\alpha_r}$ and $b = p_{r+1}^{\alpha_{r+1}} \cdots p_s^{\alpha_s}$ where the p_i are distinct primes, and consider the product

$$\prod_{i=1}^r \left(1 + p_i + \cdots + p_i^{\alpha_i}\right) \times \prod_{i=r+1}^s \left(1 + p_i + \cdots + p_i^{k-1}\right).$$

Each term in the expansion of this product is a k-free divisor of n and, conversely, each k-free divisor of n is a term in the expansion of this product. This implies

$$\tau_{(k)}(n) = \tau(a)k^{\omega(b)}$$

and therefore $\tau_{(k)}$ is multiplicative.

▷ Let n be uniquely written in the form $n = ab$ where $(a, b) = 1$ and $\mu_2(a) = s_2(b) = 1$ with $a = p_1^{\alpha_1} \cdots p_r^{\alpha_r}$ and $b = p_{r+1}^{\alpha_{r+1}} \cdots p_s^{\alpha_s}$ where the p_i are distinct primes and $\alpha_i \in \{0, 1\}$ for $i = 1, \ldots, r$ and $\alpha_i \geqslant 2$ for $i = r+1, \ldots, s$. A divisor $d = p_1^{\beta_1} \cdots p_s^{\beta_s}$ of n is square-full if and only if $d = 1$ or $d \mid b$. This is equivalent to $\beta_i = 0$ or $2 \leqslant \beta_i \leqslant \alpha_i$ for $i = r+1, \ldots, s$. Therefore we get

$$\beta(n) = \prod_{i=r+1}^s \alpha_i$$

and hence β is multiplicative.

▷ Let n be uniquely written in the form $n = p_1^{\alpha_1} \cdots p_s^{\alpha_s}$. A divisor $d = p_1^{\beta_1} \cdots p_s^{\beta_s}$ is an exponential divisor of n if and only if $\beta_i \mid \alpha_i$. We readily deduce that

$$\tau^{(e)}(n) = \prod_{i=1}^{s} \tau(\alpha_i)$$

and hence $\tau^{(e)}$ is multiplicative.

▷ Let G be an abelian group of order $n = p_1^{\alpha_1} \cdots p_s^{\alpha_s}$. From (3.1), we have the decomposition

$$G \simeq H_1 \oplus \cdots \oplus H_s,$$

where H_i is the p_i-Sylow subgroup of G of order $p_i^{\alpha_i}$, which implies that

$$a(n) = \prod_{i=1}^{s} a\left(p_i^{\alpha_i}\right)$$

so that the function a is multiplicative by Lemma 4.1. Furthermore, let G be an abelian group of order p^{α}. Then G may be factorized in the form

$$G \simeq G_1 \oplus \cdots \oplus G_r$$

for some positive integer r, where the $G_i \neq \{e_G\}$ are cyclic subgroups of G of order p^{β_i} for some $\beta_i \mid \alpha$ by Lagrange's theorem.[7] We infer that the number of abelian groups of order p^{α} is equal to the number of decompositions of the form

$$\alpha = \beta_1 + \cdots + \beta_r,$$

i.e. the number of unrestricted partitions of α. Denoting by $P(\alpha)$ this number, we finally get

$$a(n) = \prod_{i=1}^{s} P(\alpha_i).$$

▷ The multiplicativity of the other functions is not obvious. We shall see another tool which will prove to be very useful in this problem.

---

[7]Theorem 7.1.

4.1.3 The Dirichlet Convolution Product

4.1.3.1 Definition

Definition 4.4 Let f and g be two arithmetic functions. The Dirichlet convolution product of f and g is the arithmetic function $f \star g$ defined by

$$(f \star g)(n) = \sum_{d \mid n} f(d) g\left(\frac{n}{d}\right) = \sum_{d \mid n} f\left(\frac{n}{d}\right) g(d).$$

It should be noticed that the second equality above follows from the fact that the map $d \mapsto d'$ such that $dd' = n$ is one-to-one.

Example 4.1 We have $\Lambda_k = \mu \star \log^k$, $\tau = \mathbf{1} \star \mathbf{1}$ and by induction

$$\tau_k = \underbrace{\mathbf{1} \star \cdots \star \mathbf{1}}_{k \text{ times}}.$$

Also $\sigma_k = \mathrm{id}_k \star \mathbf{1}$. We shall see later that other arithmetic functions may be written as a convolution product of two simpler arithmetic functions. This will be a powerful tool to get estimates for averages of these functions.

The next result shows that this operation behaves well.

Lemma 4.2 *The Dirichlet convolution product is commutative, associative and has an identity element which is the function e_1. Furthermore, if $f(1) \neq 0$, then f is invertible.*

Proof The commutativity follows at once from Definition 4.4. Now let f, g and h be three arithmetic functions and n be a positive integer. We have

$$((f \star g) \star h)(n) = \sum_{d \mid n} (f \star g)(d) h\left(\frac{n}{d}\right) = \sum_{d \mid n} \sum_{\delta \mid d} f(\delta) g\left(\frac{d}{\delta}\right) h\left(\frac{n}{d}\right)$$

and

$$(f \star (g \star h))(n) = \sum_{d \mid n} f(d)(g \star h)\left(\frac{n}{d}\right) = \sum_{d \mid n} f(d) \sum_{\delta \mid (n/d)} g(\delta) h\left(\frac{n}{d\delta}\right).$$

Setting $d' = d\delta$ in the last inner sum gives

$$(f \star (g \star h))(n) = \sum_{d' \mid n} \sum_{d \mid d'} f(d) g\left(\frac{d'}{d}\right) h\left(\frac{n}{d'}\right) = ((f \star g) \star h)(n).$$

establishing the associativity. We also have obviously

$$(e_1 \star f)(n) = \sum_{d|n} e_1(d) f\left(\frac{n}{d}\right) = f(n).$$

Finally, we prove the invertibility by constructing by induction the inverse g of an arithmetic function f satisfying $f(1) \neq 0$. The function g is the inverse of f if and only if $(f \star g)(1) = 1$ and $(f \star g)(n) = 0$ for all $n > 1$. This is equivalent to

$$f(1)g(1) = 1 \quad \text{and} \quad \sum_{d|n} g(d) f\left(\frac{n}{d}\right) = 0 \ (n \geqslant 2).$$

Since $f(1) \neq 0$, we have $g(1) = f(1)^{-1}$ by the first equation. Now let $n > 1$ and assume that we have proved that there exist unique values $g(1), \ldots, g(n-1)$ satisfying the above equations. Since $f(1) \neq 0$, the second equation above is equivalent to

$$g(n) = -\frac{1}{f(1)} \sum_{\substack{d|n \\ d<n}} g(d) f\left(\frac{n}{d}\right)$$

which determines $g(n)$ in a unique way by the induction hypothesis, and this definition of $g(n)$ shows that the equations above are satisfied, which completes the proof.   □

Therefore the condition $f(1) \neq 0$ is necessary and sufficient to the invertibility of f. By Lemma 4.1, we infer that every multiplicative function is invertible. Also note that there is a self-contained formula for the Dirichlet inverse [45, Theorem 2.2], namely

$$f^{-1}(n) = \sum_{k=1}^{\Omega(n)} \frac{(-1)^k}{f(1)^{k+1}} \sum_{\substack{d_1 \cdots d_k = n \\ d_1, \ldots, d_k \geqslant 2}} f(d_1) \cdots f(d_k) \quad (n \geqslant 2).$$

4.1.3.2   The Algebraic Point of View

Lemma 4.3 *The set $(\mathcal{A}, +, \star)$ of arithmetic functions is a unitary commutative ring with identity element e_1, and it is also an integral domain.*

Proof The first part of the Lemma follows immediately from Lemma 4.2. In order to prove the integrality, we may proceed as follows. Define a map $N : \mathcal{A} \to \mathbb{Z}_{\geqslant 0}$ by setting $N(0) = 0$ and, if $f \neq 0$, then $N(f)$ is the smallest non-negative integer n such that $f(n) \neq 0$. Observe that, if $f, g \in \mathcal{A}$, then $N(f \star g) = N(f)N(g)$, and thus, if $f \neq 0$ and $g \neq 0$ are such that $N(f) = a$ and $N(g) = b$, then we have

$(f \star g)(n) = 0$ for all $n < ab$, for if $n < ab$ and $d \mid n$, then either $d < a$ or $n/d < b$. Furthermore, we also have $(f \star g)(ab) = f(a)g(b) \neq 0$, so that $\mathcal{A}$ is an integral domain. One can also prove that this ring is a UFD.   □

One may lose the integrity property with a slight change in the convolution product. For instance, define the *unitary convolution product* by

$$(f \circledast g)(n) = \sum_{\substack{d \mid n \\ (d,n/d)=1}} f(d)g\left(\frac{n}{d}\right).$$

Then it can be shown [99] that the unitary ring $(\mathcal{A}, +, \circledast)$ *is not* an integral domain.

One may wonder whether the ring $(\mathcal{A}, +, \star)$ is nœtherian. Let $p_1 < p_2 < \dots$ be the increasing sequence of the prime numbers and, for all positive integers k, define the subsets S_k of $\mathcal{A}$ as follows:

$$S_k = \{ f \in \mathcal{A} : f(n) = 0 \text{ for all } n \text{ such that } (n, p_1 \cdots p_k) = 1 \}.$$

Then the sets S_k are pairwise distinct ideals of $\mathcal{A}$ since $(f - g)(n) = f(n) - g(n) = 0$ for all n such that $(n, p_1 \cdots p_k) = 1$ and, for all $f, g \in S_k$ and all $h \in \mathcal{A}$, we have

$$(f \star h)(n) = \sum_{d \mid n} f(d)h\left(\frac{n}{d}\right) = 0$$

since if $d \mid n$ and $(n, p_1 \cdots p_k) = 1$, then $(d, p_1 \cdots p_k) = 1$. Now one may check that

$$\{0\} = S_0 \subseteq S_1 \subseteq S_2 \subseteq \cdots \subseteq S_k \subseteq S_{k+1} \subseteq \cdots$$

and

$$\bigcup_{k \in \mathbb{Z}_{\geqslant 1}} S_k = \mathcal{A}$$

and hence the ring $(\mathcal{A}, +, \star)$ is not nœtherian. One may prove that this ring is also not *artinian* by considering the sequence of pairwise distinct ideals $\mathcal{T}_k$ of $\mathcal{A}$ defined by

$$\mathcal{T}_k = \{ f \in \mathcal{A} : f(n) = 0 \text{ for all } n \text{ suchthat } \Omega(n) < k \}.$$

4.1.3.3   Convolution Product and Multiplicativity

The next result is of crucial importance.

Theorem 4.1 *If f and g are multiplicative, then so is $f \star g$.*

Proof Let f and g be two multiplicative functions and let m, n be positive integers. Each divisor d of mn can be written uniquely in the form $d = ab$ with $a \mid m, b \mid n$ and $(a, b) = 1$ so that

$$(f \star g)(mn) = \sum_{d \mid mn} f(d) g\left(\frac{mn}{d}\right) = \sum_{a \mid m} \sum_{b \mid n} f(ab) g\left(\frac{mn}{ab}\right)$$

and since f and g are multiplicative and $(a, b) = (m/a, n/b) = 1$, we infer that

$$(f \star g)(mn) = \sum_{a \mid m} \sum_{b \mid n} f(a) f(b) g\left(\frac{m}{a}\right) g\left(\frac{n}{b}\right) = (f \star g)(m)(f \star g)(n)$$

as required.   □

This result enables us:

▷ to show that a given arithmetic function is multiplicative by writing it as a product of at least two multiplicative functions;
▷ to show by multiplicativity several arithmetic identities.

4.1.3.4   Fundamental Examples

▷ Since $\tau = \mathbf{1} \star \mathbf{1}$ and $\mathbf{1}$ is multiplicative, we deduce that τ is multiplicative. The same conclusion obviously holds for τ_k.
▷ We intend to prove the following identity

$$\sum_{d \mid n} \mu(d) = \begin{cases} 1, & \text{if } n = 1 \\ 0, & \text{if } n > 1, \end{cases} \tag{4.5}$$

i.e.

$$\mu \star \mathbf{1} = e_1. \tag{4.6}$$

Now since μ and $\mathbf{1}$ are multiplicative, so is the function $\mu \star \mathbf{1}$ by Theorem 4.1 and hence (4.6) is true for $n = 1$. It is sufficient to prove (4.6) for prime powers by Lemma 4.5, which is easy to check. Indeed, for all prime powers p^α, we have

$$(\mu \star \mathbf{1})\left(p^\alpha\right) = \sum_{j=0}^{\alpha} \mu\left(p^j\right) = \mu(1) + \mu(p) = 1 - 1 = 0$$

as asserted.

▷ Let us have a look at Euler's totient function. By (4.5), we have

$$\sum_{d|(m,n)} \mu(d) = \begin{cases} 1, & \text{if } (m,n) = 1 \\ 0, & \text{otherwise} \end{cases}$$

so that

$$\varphi(n) = \sum_{\substack{m \leqslant n \\ (m,n)=1}} 1 = \sum_{m \leqslant n} \sum_{d|(m,n)} \mu(d) = \sum_{d|n} \mu(d) \sum_{\substack{m \leqslant n \\ d|m}} 1 = \sum_{d|n} \mu(d) \left\lfloor \frac{n}{d} \right\rfloor = \sum_{d|n} \mu(d) \frac{n}{d}$$

and therefore

$$\varphi = \mu \star \text{id} . \tag{4.7}$$

We first deduce that φ is multiplicative by Theorem 4.1. Furthermore, if p^α is a prime power, then $\varphi(p^\alpha)$ counts the number of integers $m \leqslant p^\alpha$ such that $p \nmid m$, and hence $\varphi(p^\alpha)$ is equal to p^α minus the number of multiples of p less than p^α, so that

$$\varphi(p^\alpha) = p^\alpha - \left\lfloor \frac{p^\alpha}{p} \right\rfloor = p^\alpha - p^{\alpha-1} = p^\alpha \left(1 - \frac{1}{p} \right)$$

which gives using Lemma 4.1

$$\varphi(n) = n \prod_{p|n} \left(1 - \frac{1}{p} \right) . \tag{4.8}$$

▷ Let $k \geqslant 2$ be an integer and let us prove the following identity

$$\sum_{d^k|n} \mu(d) = \mu_k(n), \tag{4.9}$$

i.e.

$$\mu_k = f_k \star \mathbf{1}, \tag{4.10}$$

where

$$f_k(n) = \begin{cases} \mu(d), & \text{if } n = d^k \\ 0, & \text{otherwise.} \end{cases}$$

Since f_k is clearly multiplicative, we deduce that $f_k \star \mathbf{1}$ is also multiplicative. We easily check (4.10) for $n = 1$. For all prime powers p^α, we have

$$(f_k \star \mathbf{1})(p^\alpha) = 1 + \sum_{j=1}^{\alpha} f_k(p^j) = 1 + \begin{cases} 0, & \text{if } \alpha < k \\ \mu(p), & \text{if } \alpha \geqslant k \end{cases} = \begin{cases} 1, & \text{if } \alpha < k \\ 0, & \text{otherwise} \end{cases} = \mu_k(p^\alpha)$$

and therefore (4.10) holds by Lemma 4.1.

▷ By Example 4.1, we have $\sigma = \text{id} \star \mathbf{1}$ and hence σ is multiplicative. Now let us prove the following identity

$$\sigma = \tau \star \varphi. \tag{4.11}$$

(4.11) is true for $n = 1$ and if p^α is a prime power, then we have

$$(\tau \star \varphi)(p^\alpha) = \sum_{j=0}^{\alpha} \varphi(p^j) \tau(p^{\alpha-j})$$

$$= \alpha + 1 + \left(1 - \frac{1}{p}\right) \sum_{j=1}^{\alpha} p^j (\alpha - j + 1)$$

$$= \frac{p^{\alpha+1} - 1}{p - 1} = \sigma(p^\alpha)$$

establishing (4.11).

▷ Let $d \in \mathbb{Z} \setminus \{0, 1\}$ be squarefree and $\mathbb{K} = \mathbb{Q}(\sqrt{d})$ be a quadratic field with discriminant $d_\mathbb{K}$. Let χ be the quadratic character associated with $\mathbb{K}$. Now Proposition 7.33 implies

$$r_\mathbb{K} = \chi \star \mathbf{1}. \tag{4.12}$$

▷ Let us prove the convolution identity

$$\tau_{(k)} = \mathbf{1} \star \mu_k. \tag{4.13}$$

(4.13) is easy for $n = 1$. Since the two functions are multiplicative, it suffices to check this identity for prime powers. Now

$$(\mathbf{1} \star \mu_k)(p^\alpha) = \sum_{j=0}^{\alpha} \mu_k(p^j) = \begin{cases} \alpha + 1, & \text{if } \alpha < k \\ k, & \text{if } \alpha \geqslant k. \end{cases} = \tau_{(k)}(p^\alpha)$$

as required.

▷ For all prime powers p^α, we have

$$(f \star \mu)\left(p^\alpha\right) = f\left(p^\alpha\right) - f\left(p^{\alpha-1}\right). \qquad (4.14)$$

4.1.4 The Möbius Inversion Formula

The identity (4.6) may be seen as an arithmetic form of the inclusion-exclusion principle, and then is the starting point of Brun's sieve (4.50). On the other hand, this identity means algebraically that the Möbius function is the inverse of the function **1** for the Dirichlet convolution product. One can exploit this information in the following way. Suppose that f and g are two arithmetic functions such that

$$g = f \star \mathbf{1}.$$

By Lemma 4.2, we infer that

$$g = f \star \mathbf{1} \Longleftrightarrow g \star \mu = f \star (\mathbf{1} \star \mu) = f.$$

This relation is called the *Möbius inversion formula*, and is a key part in the estimates of average orders of certain multiplicative functions. Let us summarize this result in the following theorem.

> **Theorem 4.2 (Möbius Inversion Formula)** *Let f and g be two arithmetic functions. Then we have*
>
> $$g = f \star \mathbf{1} \Longleftrightarrow f = g \star \mu,$$
>
> *i.e. for all positive integers n*
>
> $$g(n) = \sum_{d \mid n} f(d) \Longleftrightarrow f(n) = \sum_{d \mid n} g(d)\mu\left(\frac{n}{d}\right).$$

Our first application is a second proof of Proposition 3.4 stating that, for each prime number p, there exists a primitive root modulo p.

Proof (Proof of Proposition 3.4) One can suppose that $p \geqslant 3$. If $d \mid (p-1)$, let $N(d)$ be the number of elements of a reduced residue system modulo p having an order equal to d. For each divisor δ of d, a solution of the congruence $x^\delta \equiv$

1 (mod p) is also a solution of $x^d \equiv 1$ (mod p). By Lemma 3.2, this congruence has exactly d solutions, so that

$$\sum_{\delta \mid d} N(\delta) = d$$

or equivalently

$$N \star \mathbf{1} = \mathrm{id}\,.$$

By the Möbius inversion formula and (4.7), we derive $N = \mu \star \mathrm{id} = \varphi$ and therefore $N(p-1) = \varphi(p-1)$, so that there are $\varphi(p-1) \geqslant 1$ primitive roots modulo p. □

Our second application deals with the number of monic irreducible polynomials of degree n in $\mathbb{F}_p[X]$.

Proposition 4.1 *Let p be a prime number, n be a positive integer and $N_{n,p}$ be the number of monic irreducible polynomials of degree n in $\mathbb{F}_p[X]$. Then we have*

$$N_{n,p} = \frac{1}{n} \sum_{d \mid n} \mu(d) p^{n/d}\,.$$

Proof Let $P \in \mathbb{F}_p[X]$ be a monic polynomial. Since P may be factorized uniquely as a product of monic irreducible polynomials in $\mathbb{F}_p[X]$, we have

$$\sum_{\substack{P \in \mathbb{F}_p[X] \\ P \text{ monic}}} T^{\deg P} = \prod_{\substack{Q \in \mathbb{F}_p[X] \\ Q \text{ monic irreducible}}} \left(1 + T^{\deg Q} + T^{2 \deg Q} + \cdots \right)$$

$$= \prod_{\substack{Q \in \mathbb{F}_p[X] \\ Q \text{ monic irreducible}}} \left(1 - T^{\deg Q} \right)^{-1}$$

$$= \prod_{d=1}^{\infty} (1 - T^d)^{-N_{d,p}}\,.$$

Since the number of monic polynomials of degree n in $\mathbb{F}_p[X]$ is equal to p^n, we deduce that

$$\sum_{n=1}^{\infty} p^n T^n = \prod_{d=1}^{\infty} (1 - T^d)^{-N_{d,p}}$$

and taking logarithms of both sides gives

$$\log(1 - pT) = \sum_{d=1}^{\infty} N_{d,p} \log(1 - T^d) = \sum_{d=1}^{\infty} N_{d,p} \sum_{\delta=1}^{\infty} \delta^{-1} T^{d\delta}$$

and hence

$$\sum_{n=1}^{\infty} n^{-1} p^n T^n = \sum_{d=1}^{\infty} N_{d,p} \sum_{\delta=1}^{\infty} \delta^{-1} T^{d\delta} = \sum_{n=1}^{\infty} n^{-1} T^n \sum_{d|n} d N_{d,p}$$

and comparing the coefficients of T^n implies that

$$p^n = \sum_{d|n} d N_{d,p}.$$

Now Theorem 4.2 gives the asserted result. □

The Möbius inversion formula may also be used to count the number of *primitive* Dirichlet characters modulo q. We shall actually prove the following slightly more general result.

Proposition 4.2 *Let $q \geq 2$ be an integer and $k \in \mathbb{Z}_{\geq 1}$ such that $(k, q) = 1$. Then*

$$\sideset{}{^{\star}}\sum_{\chi \,(\mathrm{mod}\, q)} \chi(k) = \sum_{d|(q,k-1)} \varphi(d)\mu(q/d),$$

where the star indicates a summation over primitive Dirichlet characters. In particular, the number $\varphi^{\star}(q)$ of primitive Dirichlet characters modulo q is given by

$$\varphi^{\star}(q) = (\varphi \star \mu)(q).$$

Hence $\varphi^{\star}$ is multiplicative and there is no primitive Dirichlet character modulo $2m$ where m is an odd positive integer.

Proof Let $k \in \mathbb{Z}_{\geq 1}$ such that $(k, q) = 1$. Since each Dirichlet character modulo q is induced by a unique primitive Dirichlet character modulo a divisor of q, we have

$$\sum_{d|q} \sideset{}{^{\star}}\sum_{\chi \,(\mathrm{mod}\, d)} \chi(k) = \sum_{\chi \,(\mathrm{mod}\, q)} \chi(k) = \mathbf{1}_{q,1}(k)\varphi(q) = \begin{cases} \varphi(q), & \text{if } k \equiv 1 \ (\mathrm{mod}\, q) \\ 0, & \text{otherwise,} \end{cases}$$

where we used Proposition 3.13 in the second equality. Now by Theorem 4.2 we get

$$\sideset{}{^\star}\sum_{\chi \,(\mathrm{mod}\, q)} \chi(k) = \sum_{d|q} \mathbf{1}_{d,1}(k)\varphi(d)\mu(q/d) = \sum_{d|(q,k-1)} \varphi(d)\mu(q/d)$$

as required. Taking $k = 1$ gives the formula for $\varphi^\star(q)$ and hence $\varphi^\star$ is multiplicative by Theorem 4.1. Furthermore, $\varphi^\star(2) = \varphi(2) - \varphi(1) = 1 - 1 = 0$ and therefore by multiplicativity, for all odd positive integers m, we get $\varphi^\star(2m) = \varphi^\star(2)\varphi^\star(m) = 0$ as asserted.   □

Example 4.2

▷ By (3.3), one has $\log = \Lambda \star \mathbf{1}$ and hence $\Lambda = \mu \star \log$ by Theorem 4.2.
▷ Let us prove that

$$\sum_{d|n} \varphi(d) = n$$

and

$$\sum_{d|n} \frac{\mu(d)}{d} = \prod_{p|n} \left(1 - \frac{1}{p}\right).$$

Indeed, from (4.7) and Theorem 4.2, we get $\mathrm{id} = \varphi \star \mathbf{1}$ which is the first identity. For the second one, we have using (4.7)

$$\sum_{d|n} \frac{\mu(d)}{d} = \frac{1}{n} \sum_{d|n} \mu(d)\frac{n}{d} = \frac{(\mu \star \mathrm{id})(n)}{n} = \frac{\varphi(n)}{n}$$

which implies the asserted result.

4.1.5   The Dirichlet Hyperbola Principle

In this section, we intend to provide some tools to estimate sums of the form

$$S(x) = \sum_{n \leqslant x} (f \star g)(n) = \sum_{n \leqslant x} \sum_{d|n} f(d)g(n/d)$$

with $x \geqslant 1$. The first idea which comes to mind is to interchange the order of summation, which amounts to rearranging the terms of the sum favouring the factorization by the terms $f(1), \ f(2), \ldots$ We thus get

$$S(x) = \sum_{d \leqslant x} f(d) \sum_{\substack{n \leqslant x \\ d \mid n}} g(n/d).$$

Making the change of variable $n = kd$ in the inner sum and noticing that $n \leqslant x$ and $d \mid n$ is equivalent to $k \leqslant x/d$, we finally obtain

$$S(x) = \sum_{d \leqslant x} f(d) \sum_{k \leqslant x/d} g(k).$$

We may thus state the following result.

Proposition 4.3 *Let* $x \geqslant 1$ *be a real number and* f *and* g *be two arithmetic functions. Then*

$$\sum_{n \leqslant x} (f \star g)(n) = \sum_{d \leqslant x} f(d) \sum_{k \leqslant x/d} g(k).$$

Example 4.3

▷ Since $\mu \star \mathbf{1} = e_1$, we get using Proposition 4.3

$$1 = \sum_{n \leqslant x} (\mu \star \mathbf{1})(n) = \sum_{d \leqslant x} \mu(d) \sum_{k \leqslant x/d} 1 = \sum_{d \leqslant x} \mu(d) \left\lfloor \frac{x}{d} \right\rfloor$$

so that

$$\sum_{n \leqslant x} \mu(n) \left\lfloor \frac{x}{n} \right\rfloor = 1 \tag{4.15}$$

which is a very important identity. A variant of this identity may be obtained as follows:

$$\sum_{n \leqslant x+y} \mu(n) \left(\left\lfloor \frac{x+y}{n} \right\rfloor - \left\lfloor \frac{x}{n} \right\rfloor \right) = \sum_{n \leqslant x+y} \mu(n) \left\lfloor \frac{x+y}{n} \right\rfloor - \sum_{n \leqslant x+y} \mu(n) \left\lfloor \frac{x}{n} \right\rfloor$$

$$= 1 - \sum_{n \leqslant x} \mu(n) \left\lfloor \frac{x}{n} \right\rfloor - \sum_{x < n \leqslant x+y} \mu(n) \left\lfloor \frac{x}{n} \right\rfloor$$

$$= 1 - 1 = 0$$

since in the last sum we have $\lfloor x/n \rfloor = 0$. This implies that

$$\sum_{x < n \leqslant x+y} \mu(n) = - \sum_{n \leqslant x} \mu(n) \left(\left\lfloor \frac{x+y}{n} \right\rfloor - \left\lfloor \frac{x}{n} \right\rfloor \right).$$

Set $N = \lfloor x \rfloor$. Using (4.15) we get

$$N \sum_{n \leqslant N} \frac{\mu(n)}{n} = \sum_{n \leqslant N} \mu(n) \left\lfloor \frac{N}{n} \right\rfloor + \sum_{n \leqslant N} \mu(n) \left\{ \frac{N}{n} \right\} = 1 + \sum_{n \leqslant N-1} \mu(n) \left\{ \frac{N}{n} \right\}$$

so that

$$\left| \sum_{n \leqslant x} \frac{\mu(n)}{n} \right| \leqslant \frac{1}{N} \left(1 + \sum_{n \leqslant N-1} \mu(n) \left\{ \frac{N}{n} \right\} \right) \leqslant \frac{1 + N - 1}{N} = 1. \qquad (4.16)$$

▷ Using Proposition 4.7, we shall see that the Dirichlet series of the Möbius function is $\zeta(s)^{-1}$ which is absolutely convergent in the half-plane $\sigma > 1$. This implies in particular that, for all integers $k \geqslant 2$, we have

$$\sum_{n=1}^{\infty} \frac{\mu(n)}{n^k} = \frac{1}{\zeta(k)}. \qquad (4.17)$$

Using Proposition 4.3, one may provide another proof of (4.17). It is first easily seen that the series of the left-hand side converges, since $|\mu(n)|/n^k \leqslant 1/n^k$. Now applying Proposition 4.3 with $f(n) = \mu(n)/n^k$ and $g(n) = 1/n^k$, we derive for all $N \in \mathbb{Z}_{\geqslant 1}$

$$1 = \sum_{n \leqslant N} (f \star g)(n)$$

$$= \sum_{d \leqslant N} \frac{\mu(d)}{d^k} \sum_{m \leqslant N/d} \frac{1}{m^k}$$

$$= \sum_{d \leqslant N} \frac{\mu(d)}{d^k} \sum_{m=1}^{\infty} \frac{1}{m^k} - \sum_{d \leqslant N} \frac{\mu(d)}{d^k} \sum_{m > N/d} \frac{1}{m^k}$$

$$= \zeta(k) \sum_{d \leqslant N} \frac{\mu(d)}{d^k} - \frac{1}{(k-1)N^{k-1}} \sum_{d \leqslant N} \frac{\mu(d)}{d} - \sum_{d \leqslant N} \frac{\mu(d) r_{k,N}(d)}{d^k}$$

with $|r_{k,N}(d)| \leqslant (d/N)^k$, and where we used Corollary 1.2 to estimate the last sum. Now (4.16) yields

$$\left| 1 - \zeta(k) \sum_{d \leqslant N} \frac{\mu(d)}{d^k} \right| \leqslant \frac{1}{(k-1)N^{k-1}} + \frac{1}{N^k} \sum_{d \leqslant N} |\mu(d)| \leqslant \frac{2}{N^{k-1}}$$

giving (4.17) by letting $N \to \infty$.

Remark 4.2 Tao [102] generalized the inequality (4.16) in the following sense.

Theorem 4.3 (Tao) *Let $\mathcal{P}$ be an arbitrary set of primes, either finite or infinite, and let $\langle \mathcal{P} \rangle \subset \mathbb{Z}_{\geqslant 1}$ be the multiplicative semigroup generated by $\mathcal{P}$, i.e. the set of natural numbers whose prime factors all lie in $\mathcal{P}$. For all $x \geqslant 1$, we have*

$$\left| \sum_{\substack{n \leqslant x \\ n \in \langle \mathcal{P} \rangle}} \frac{\mu(n)}{n} \right| \leqslant 1.$$

As a corollary, we derive the following bounds.

Corollary 4.1 *For all $q, m \in \mathbb{Z}_{\geqslant 1}$ and all $x \in \mathbb{R}_{\geqslant 1}$*

$$\left| \sum_{\substack{n \leqslant x \\ (n,q)=1}} \frac{\mu(n)}{n} \right| \leqslant 1, \qquad \left| \sum_{\substack{n \leqslant x \\ n|q}} \frac{\mu(n)}{n} \right| \leqslant 1 \quad and \quad \left| \sum_{n \leqslant x} \frac{\mu(mn)}{n} \right| \leqslant 1.$$

One may improve on Proposition 4.3 by inserting a parameter which may be optimized. This is indeed a very fruitful idea leading to the next result, called the *Dirichlet hyperbola principle*.

Theorem 4.4 (Dirichlet Hyperbola Principle) *Let f, g be complex-valued arithmetic functions and h be any complex-valued map. For all $1 \leqslant U \leqslant x$*

$$\sum_{n \leqslant x} (f \star g)(n)h(n) = \sum_{n \leqslant U} f(n) \sum_{m \leqslant x/n} g(m)h(mn)$$

$$+ \sum_{n \leqslant x/U} g(n) \sum_{m \leqslant x/n} f(m)h(mn) - \sum_{n \leqslant U} \sum_{m \leqslant x/U} f(n)g(m)h(mn).$$

Proof Splitting the sum of the right-hand side gives

$$\sum_{n\leqslant x}(f\star g)(n)h(n) = \sum_{n\leqslant x}\left(\sum_{d\mid n}f(d)g(n/d)\right)h(n) = \sum_{d\leqslant x}f(d)\sum_{k\leqslant x/d}g(k)h(kd)$$

$$= \left(\sum_{d\leqslant U} + \sum_{U<d\leqslant x}\right)f(d)\sum_{k\leqslant x/d}g(k)h(kd)$$

with

$$\sum_{U<d\leqslant x}f(d)\sum_{k\leqslant x/d}g(k)h(kd) = \sum_{k\leqslant x/U}g(k)\sum_{U<d\leqslant x/k}f(d)h(kd)$$

$$= \sum_{k\leqslant x/U}g(k)\left(\sum_{d\leqslant x/k}-\sum_{d\leqslant U}\right)f(d)h(kd)$$

implying the announced result.   □

Historically, it was Dirichlet who discovered this principle when he succeeded in improving the error term in the sum

$$\sum_{n\leqslant x}\tau(n).$$

The name comes from the following observation. Since, by above,

$$\sum_{n\leqslant x}\tau(n) = \sum_{d\leqslant x}\left\lfloor\frac{x}{d}\right\rfloor$$

we are led to count the number of integer points (m, n) with $1 \leqslant m \leqslant x$ lying under the hyperbola $mn = x$.

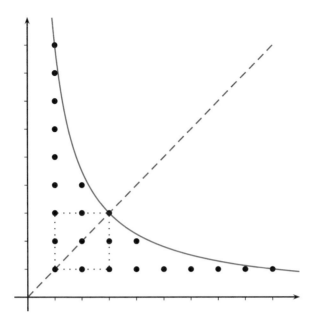

Dirichlet used the symmetry of the hyperbola to deduce that the number of integer points is equal to that of the interior of the square $[1, \sqrt{x}]^2$ plus twice the number of integer points (m, n) such that $\sqrt{x} < m \leqslant x$. See Theorem 4.28 for an direct application of Theorem 4.4.

4.2   Dirichlet Series

4.2.1   The Formal Viewpoint

We have seen in Chap. 2 that the concept of generating function may be fruitful to capture the information of a sequence. In view of the multiplicative properties of certain arithmetic functions, we use *Dirichlet series* rather than power series in number theory.

Definition 4.5 Let f be an arithmetic function. The *formal Dirichlet series* of a variable s associated with f is defined by

$$L(s, f) = \sum_{n=1}^{\infty} \frac{f(n)}{n^s}.$$

As always for formal mathematical objects, we ignore here convergence problems, and $L(s, f)$ is the complex number equal to the sum when it converges.

Remark 4.3 The notation $L(s, f)$ may be sometimes abbreviated in $F(s)$, for instance, when the coefficients $f(n)$ are not explicitly mentioned.

The following proposition reveals the importance of the Dirichlet convolution product.

Proposition 4.4 *Let f, g and h be three arithmetic functions. Then*

$$h = f \star g \iff L(s, h) = L(s, f)L(s, g).$$

Proof We have

$$L(s, f)L(s, g) = \sum_{k,d=1}^{\infty} \frac{f(k)g(d)}{(kd)^s} = \sum_{n=1}^{\infty} \frac{1}{n^s} \sum_{d|n} f\left(\frac{n}{d}\right) g(d) = \sum_{n=1}^{\infty} \frac{(f \star g)(n)}{n^s}$$

as required.   □

Proposition 4.5 *An arithmetic function f is multiplicative if and only if*

$$L(s, f) = \prod_p \left(1 + \sum_{\alpha=1}^{\infty} \frac{f(p^\alpha)}{p^{s\alpha}} \right).$$

The above product is called the Euler product *of $L(s, f)$.*

Proof Expanding the product shows that it is equivalent to the conditions $f(1) = 1$ and $f(n) = f\left(p_1^{\alpha_1}\right) \cdots f\left(p_r^{\alpha_r}\right)$ for all $n = p_1^{\alpha_1} \cdots p_r^{\alpha_r}$. Use Lemma 4.1 to complete the proof.   □

4.2.2   Absolute Convergence

If we wish to deal with convergence problems, we need to have at our disposal some tools to determine precisely the region of convergence of a Dirichlet series. Recall that for power series $\sum_{n \geq 0} f(n)z^n$, the domain of convergence is a disc on the boundary of which the behaviour of the sum is a priori undetermined. For Dirichlet series, there is an analogous result, except that the domain of absolute convergence is a half-plane.

Proposition 4.6 *For each Dirichlet series $F(s)$, there exists $\sigma_a \in \mathbb{R} \cup \{\pm\infty\}$, called the* abscissa of absolute convergence, *such that $F(s)$ converges absolutely in the half-plane $\sigma > \sigma_a$ and does not converge absolutely in the half-plane $\sigma < \sigma_a$.*

Proof Let S be the set of complex numbers s at which $F(s)$ converges absolutely. If $S = \emptyset$, then put $\sigma_a = +\infty$. Otherwise define

$$\sigma_a = \inf \{\sigma : s = \sigma + it \in S\}.$$

By the definition of σ_a, $F(s)$ does not converge absolutely if $\sigma < \sigma_a$. On the other hand, suppose that $F(s)$ is absolutely convergent for some $s_0 = \sigma_0 + it_0 \in \mathbb{C}$ and let $s = \sigma + it$ such that $\sigma \geqslant \sigma_0$. Since

$$\left|\frac{f(n)}{n^s}\right| = \left|\frac{f(n)}{n^{s_0}}\right| \times \frac{1}{n^{\sigma-\sigma_0}} \leqslant \left|\frac{f(n)}{n^{s_0}}\right|$$

we infer that $F(s)$ converges absolutely at any point s such that $\sigma \geqslant \sigma_0$. Now by the definition of σ_a, there exist points arbitrarily close to σ_a at which $F(s)$ converges absolutely, and therefore by above $F(s)$ converges absolutely at *each* point s such that $\sigma > \sigma_a$. □

Remark 4.4 By abuse of notation, this function is still denoted by $F(s)$.

The next result is an immediate consequence of Proposition 4.6.

Corollary 4.2

▷ *The Dirichlet series $L(s, f)$ defines an analytic function in the half-plane $\sigma > \sigma_a$.*
▷ *If $|f(n)| \leqslant \log n$, then the series $F(s)$ is absolutely convergent in the half-plane $\sigma > 1$, and hence $\sigma_a \leqslant 1$.*

At $\sigma = \sigma_a$, the series may or may not converge absolutely. For instance, $\zeta(s)$ converges absolutely in the half-plane $\sigma > \sigma_a = 1$ but does not converge on the line $\sigma = 1$. On the other hand, the Dirichlet series associated with the function $f(n) = (1 + \log n)^{-2}$ has also $\sigma_a = 1$ for the abscissa of absolute convergence, but converges absolutely at $\sigma = 1$.

The partial sums $\sum_{x<n\leqslant y} f(n)$ and the Dirichlet series $F(s)$ are strongly related to each other, as shown in the next result below.

Theorem 4.5 *Let $F(s) = \sum_{n=1}^{\infty} f(n)n^{-s}$ be a Dirichlet series. Assume that for all $0 < x < y$, we have*

$$\left|\sum_{x<n\leqslant y} f(n)\right| \leqslant My^{\alpha}$$

for some $\alpha \geqslant 0$ and $M > 0$ independent of x and y. Then $F(s)$ converges absolutely in the half-plane $\sigma > \alpha$. Furthermore, we have in this half-plane

$$\sum_{n=1}^{\infty} \frac{f(n)}{n^s} = s \int_{1}^{\infty} \left(\sum_{n\leqslant t} f(n)\right) \frac{dt}{t^{s+1}}$$

(continued)

Theorem 4.5 (continued)
and then

$$|F(s)| \leqslant \frac{M|s|}{\sigma - \alpha} \quad and \quad \left| \sum_{x < n \leqslant y} \frac{f(n)}{n^s} \right| \leqslant \frac{M}{x^{\sigma - \alpha}} \left(\frac{|s|}{\sigma - \alpha} + 1 \right).$$

Proof Set $A(x) = \sum_{n \leqslant x} f(n)$ and $S(x, y) = A(y) - A(x)$. A partial summation yields

$$\sum_{x < n \leqslant y} \frac{f(n)}{n^s} = \frac{S(x, y)}{y^s} + s \int_x^y \frac{S(x, u)}{u^{s+1}} \, du$$

and by hypothesis we have $|S(x, y)/y^s| \leqslant M y^{\alpha - \sigma}$, so that $S(x, y)/y^s$ tends to 0 as $y \to \infty$ in the half-plane $\sigma > \alpha$. Therefore if one of

$$\sum_{n \geqslant 1} \frac{f(n)}{n^s} \quad \text{or} \quad s \int_1^\infty \frac{A(u)}{u^{s+1}} \, du$$

converges absolutely, then so does the other, and the two quantities converge to the same limit. But since

$$\left| \frac{A(u)}{u^{s+1}} \right| \leqslant \frac{M}{u^{\sigma - \alpha + 1}}$$

we infer that the integral converges absolutely for $\sigma > \alpha$, and hence $F(s)$ is absolutely convergent in this half-plane. Therefore for all $\sigma > \alpha$, we get

$$\sum_{n=1}^\infty \frac{f(n)}{n^s} = s \int_1^\infty \frac{A(u)}{u^{s+1}} \, du$$

and hence

$$|F(s)| \leqslant M|s| \int_1^\infty \frac{du}{u^{\sigma - \alpha + 1}} = \frac{M|s|}{\sigma - \alpha}$$

and similarly

$$\left| \sum_{x < n \leqslant y} \frac{f(n)}{n^s} \right| \leqslant \frac{M}{y^{\sigma - \alpha}} + M|s| \int_x^\infty \frac{du}{u^{\sigma - \alpha + 1}} \leqslant \frac{M}{x^{\sigma - \alpha}} \left(\frac{|s|}{\sigma - \alpha} + 1 \right)$$

as required.   □

We have seen in Proposition 4.4 that the product of two Dirichlet series is of great importance in the theory of arithmetic functions. The next result gives the domain of absolute convergence of this product.

Proposition 4.7 *Let f and g be two arithmetic functions with associated Dirichlet series $F(s)$ and $G(s)$. Set $h = f \star g$ and let $H(s)$ be the associated Dirichlet series. If $F(s)$ and $G(s)$ are absolutely convergent at a point s_0, then $H(s)$ converges absolutely at s_0 and we have $H(s_0) = F(s_0)G(s_0)$.*

Proof From Proposition 4.4, we have

$$F(s_0)G(s_0) = \sum_{n=1}^{\infty} \frac{h(n)}{n^{s_0}} = H(s_0),$$

where the rearrangement of the terms in the double sums is justified by the absolute convergence of the two series $F(s)$ and $G(s)$ at $s = s_0$. Furthermore, we have

$$\sum_{n=1}^{\infty} \left| \frac{h(n)}{n^{s_0}} \right| \leqslant \left(\sum_{n=1}^{\infty} \left| \frac{f(n)}{n^{s_0}} \right| \right) \left(\sum_{n=1}^{\infty} \left| \frac{g(n)}{n^{s_0}} \right| \right)$$

proving the absolute convergence of $H(s_0)$.   □

In 1885/1887, Stieltjes proved the following stronger results.

Proposition 4.8 (Stieltjes) *Let $F(s)$ and $G(s)$ be two Dirichlet series.*

▷ *If F and G converge at the point $\sigma_0 \in \mathbb{R}$ and if they converge absolutely at $\sigma_0 + c$ where $c \geqslant 0$, then the series $F(s)G(s)$ converges at $\sigma_0 + \frac{c}{2}$.*
▷ *If F and G converge at the point $\sigma_0 \in \mathbb{R}$, then the series $F(s)G(s)$ converges at $\sigma_0 + \frac{1}{2}$.*

Using Proposition 4.7 with $h = e_1$ gives the following result.

Proposition 4.9 *Let f be an arithmetic function such that $f(1) \neq 0$ and let $F(s)$ be its associated Dirichlet series. Let f^{-1} be the convolution inverse of f, i.e. the arithmetic function such that $f \star f^{-1} = e_1$ and let $G(s)$ be the associated Dirichlet series of f^{-1}. Then*

$$G(s) = \frac{1}{F(s)}$$

at every point s where $F(s)$ and $G(s)$ converge absolutely.

For instance, this proposition enables us to get the Dirichlet series of the Möbius function. Indeed, by (4.6), we have $\mu^{-1} = \mathbf{1}$ and hence, for all $\sigma > 1$, we have

$$\sum_{n=1}^{\infty} \frac{\mu(n)}{n^s} = \frac{1}{\zeta(s)}$$

as stated in Example 4.3. We end this section with two examples of computation of abscissa of the absolute convergence.

Proposition 4.10 *Let f be an arithmetic function and $L(s, f)$ be its associated Dirichlet series with abscissa of absolute convergence σ_a.*

▷ *If $|f(n)| \leqslant Mn^\alpha$ for some real numbers $M > 0$ and $\alpha \geqslant 0$, then $\sigma_a \leqslant \alpha + 1$.*
▷ *We have*

$$\sigma_a \leqslant \limsup_{n \to \infty} \left(1 + \frac{\log|f(n)|}{\log n}\right).$$

Proof

▷ The first result follows from Theorem 4.5 since we have

$$\left|\sum_{n \leqslant x} f(n)\right| \leqslant M \sum_{n \leqslant x} n^\alpha \leqslant Mx^{\alpha+1}.$$

▷ Set

$$L = \limsup_{n \to \infty} \left(1 + \frac{\log|f(n)|}{\log n}\right)$$

and we may suppose that $L < \infty$. Fix a small real number $\varepsilon > 0$ and let $\sigma > L$. Hence $\sigma \geqslant L + \varepsilon$ and therefore there exists a large positive integer $n_0 = n_0(\varepsilon)$ such that, for all $n \geqslant n_0$, we have

$$1 + \frac{\log|f(n)|}{\log n} < L \leqslant \sigma - \varepsilon$$

and hence, for all $n \geqslant n_0$, we get

$$\left|\frac{f(n)}{n^s}\right| < \frac{n^{\sigma-1-\varepsilon}}{n^\sigma} = \frac{1}{n^{1+\varepsilon}}$$

so that $F(s)$ is absolutely convergent in the half-plane $\sigma > L$.   □

4.2.3   Conditional Convergence

This is one the most important differences between power series and Dirichlet series. For power series, the regions of convergence and absolute convergence are identical, except possibly for the boundaries. For Dirichlet series, there may be a non-trivial region, in the form of a vertical strip, in which the series converges but does not converge absolutely. However, we shall see that the width of this strip does not exceed 1.

Proposition 4.11 *For each Dirichlet series $F(s)$, there exists $\sigma_c \in \mathbb{R} \cup \{\pm\infty\}$, called the* abscissa of convergence, *such that $F(s)$ converges in the half-plane $\sigma > \sigma_c$ and does not converge in the half-plane $\sigma < \sigma_c$. Furthermore, we have*

$$\sigma_c \leqslant \sigma_a \leqslant \sigma_c + 1.$$

Proof

▷ Suppose first that $F(s)$ converges at a point $s_0 = \sigma_0 + it_0$ and fix a small real number $\varepsilon > 0$. By Cauchy's theorem, there exists $x_0 = x_0(\varepsilon) \geqslant 1$ such that, for all $y > x \geqslant x_0$, we have

$$\left| \sum_{x < n \leqslant y} \frac{f(n)}{n^{s_0}} \right| \leqslant \varepsilon.$$

Let $s = \sigma + it \in \mathbb{C}$ such that $\sigma > \sigma_0$. Using Theorem 4.5 with s replaced by $s - s_0$ and $\alpha = 0$, we get

$$\left| \sum_{x < n \leqslant y} \frac{f(n)}{n^s} \right| \leqslant \varepsilon \left(\frac{|s - s_0|}{\sigma - \sigma_0} + 1 \right)$$

so that $F(s)$ converges by Cauchy's theorem.

▷ Now we may proceed as in Proposition 4.6. Let S be the set of complex numbers s at which $F(s)$ converges. If $S = \varnothing$, then put $\sigma_c = +\infty$. Otherwise define

$$\sigma_c = \inf \{\sigma : s = \sigma + it \in S\}.$$

By the definition of σ_c, $F(s)$ does not converge if $\sigma < \sigma_c$. On the other hand, there exist points s_0 with σ_0 being arbitrarily close to σ_c at which $F(s)$ converges. By above, $F(s)$ converges at any point s such that $\sigma > \sigma_0$. Since σ_0 may be chosen as close to σ_c as we want, it follows that $F(s)$ converges at any point s such that $\sigma > \sigma_c$.

▷ The inequality $\sigma_c \leqslant \sigma_a \leqslant \sigma_c + 1$ remains to be shown. The left-hand side is obvious. For the right-hand side, it suffices to show that if $F(s_0)$ converges for some s_0, then it converges absolutely for all s such that $\sigma > \sigma_0 + 1$. Now if $F(s)$ converges at some point s_0, then $f(n)/n^{s_0}$ tends to 0 as $n \to \infty$. Thus there exists a positive integer n_0 such that, for all $n \geqslant n_0$, we have $|f(n)/n^{s_0}| \leqslant 1$. We deduce that, for any s, we have $|f(n)/n^s| \leqslant n^{\sigma_0 - \sigma}$ so that $F(s)$ is absolutely convergent in the half-plane $\sigma > \sigma_0 + 1$ as required.   □

As in the case of absolute convergence, the behaviour of the series on the line $\sigma = \sigma_c$ is a priori undetermined. Furthermore, although there exist series such that

$\sigma_c = +\infty$, such as $f(n) = n!$, for instance, or $\sigma_c = -\infty$, such as $f(n) = 1/n!$, these extreme cases do not appear in almost all number-theoretic applications in which we shall have $\sigma_c \in \mathbb{R}$ and $\sigma_a \in \mathbb{R}$. Hence we must deal with half-planes of absolute convergence, of convergence and with a strip in which the series converges conditionally but not absolutely.

The inequality $\sigma_c \leqslant \sigma_a \leqslant \sigma_c + 1$ is sharp. Indeed,

▷ If $f(n) \geqslant 0$, then the absolute convergence and the conditional convergence of $F(s)$ are equivalent at any point $s \in \mathbb{R}$. But since the regions of convergence and absolute convergence are half-planes, we infer that in the case of non-negative coefficients, we have $\sigma_c = \sigma_a$.
▷ On the other hand, if $f(n) = \chi(n) \neq \chi_0(n)$ is a non-principal Dirichlet character, then we have seen in Chap. 3 that $\sigma_c = 0$, and we clearly have $\sigma_a = 1$.

The next result is an easy consequence of partial summation techniques.

Theorem 4.6 *Let $\sigma_0 > 0$. If the series $\sum_{n=1}^{\infty} f(n)n^{-\sigma_0}$ converges, then*

$$\sum_{n=1}^{N} f(n) = o\left(N^{\sigma_0}\right) \quad (N \to \infty).$$

The proof rests on the following lemma due to Kronecker.

Lemma 4.4 *Let (α_n) be a sequence of complex numbers such that $\sum_{n \geqslant 1} \alpha_n$ converges, and let (β_n) be an increasing sequence of positive real numbers such that $\beta_n \to \infty$. Then*

$$\sum_{n=1}^{N} \alpha_n \beta_n = o\left(\beta_N\right) \quad (N \to \infty).$$

Proof Let $R_n = \sum_{k > n} \alpha_k$. By assumption, we have $R_n = o(1)$ as $n \to \infty$, and by Proposition 1.4 we derive

$$\sum_{n=1}^{N} \alpha_n \beta_n = \sum_{n=1}^{N} \beta_n \left(R_n - R_{n+1}\right)$$

$$= \beta_N \sum_{n=1}^{N} \left(R_n - R_{n+1}\right) - \sum_{n=1}^{N-1} \left(\beta_{n+1} - \beta_n\right) \sum_{h=1}^{n} \left(R_h - R_{h+1}\right)$$

$$= R_1 \beta_1 - R_{N+1} \beta_N + \sum_{n=1}^{N-1} R_{n+1} \left(\beta_{n+1} - \beta_n\right).$$

Since (β_n) increases to infinity with n, we get $\beta_{n+1} - \beta_n > 0$ and $\sum_{n=1}^{\infty} (\beta_{n+1} - \beta_n) = \infty$, so that as $N \to \infty$

$$\sum_{n=1}^{N-1} |R_{n+1}| (\beta_{n+1} - \beta_n) = o \left(\sum_{n=1}^{N-1} (\beta_{n+1} - \beta_n) \right) = o(\beta_N)$$

as required. Now Theorem 4.6 follows by applying Lemma 4.4 with $\alpha_n = f(n)n^{-\sigma_0}$ and $\beta_n = n^{\sigma_0}$. □

Combining Theorems 4.5 and 4.6, we derive the next useful tool for Dirichlet series with non-negative coefficients.

Corollary 4.3 Let $f(n) \geqslant 0$ and $\sigma_0 > 0$ be a fixed real number. Then

$$L(s, f) \text{ converges for } \sigma > \sigma_0 \iff \sum_{n \leqslant x} f(n) \ll x^{\sigma_0 + \varepsilon}$$

for all $\varepsilon > 0$, where the implied constant may depend on ε.

Proof The necessary condition is a consequence of Theorem 4.5. Conversely, if $L(s, f)$ converges in the half-plane $\sigma > \sigma_0$, then it converges at $\sigma_0 + \varepsilon$ for all $\varepsilon > 0$ and the result follows by applying Theorem 4.6. □

Remark 4.5 Note that the ε cannot be removed. For instance, when $f = \tau$ the counting divisor function, then $L(s, \tau) = \zeta(s)^2$ converges in the half-plane $\sigma > 1$, but

$$\sum_{n \leqslant x} \tau(n) = \sum_{d \leqslant x} \left\lfloor \frac{x}{d} \right\rfloor \geqslant \frac{x}{2} \sum_{d \leqslant x} \frac{1}{d} \geqslant \frac{x \log x}{2}.$$

4.2.4 Analytic Properties

It has been seen above that a Dirichlet series $F(s)$ defines an analytic function in its half-plane of absolute convergence. The next result proves much better: $F(s)$ defines in fact an analytic function in its half-plane of *conditional* convergence, which enables us to apply the powerful tools from complex analysis to arithmetic functions. The proof makes use of Theorem 2.

Theorem 4.7 *A Dirichlet series $F(s) = \sum_{n=1}^{\infty} f(n)n^{-s}$ defines an analytic function of s in the half-plane $\sigma > \sigma_c$, in which $F(s)$ can be differentiated termwise so that, for all positive integers k, we have*

$$F^{(k)}(s) = \sum_{n=1}^{\infty} \frac{(-1)^k (\log n)^k f(n)}{n^s}$$

for all $s = \sigma + it$ such that $\sigma > \sigma_c$.

Proof Let $s_1 = \sigma_1 + it_1$ such that $\sigma_1 > \sigma_c$ and we consider the disc $\mathcal{D}$ with s_1 as centre and of radius r contained entirely in the half-plane $\sigma > \sigma_c$. We will show that $F(s)$ converges uniformly in $\mathcal{D}$. Fix $\varepsilon > 0$. We must prove that there exists $n_0 = n_0(\varepsilon) \in \mathbb{Z}_{\geq 1}$, independent of $s \in \mathcal{D}$, such that for all $N \geq n_0$, we have

$$\left| \sum_{n>N} \frac{f(n)}{n^s} \right| \leq \varepsilon$$

for all $s \in \mathcal{D}$. To this end, we choose $s_0 = \sigma_0 + it_0$ such that $t_0 = t_1$ and $\sigma_c < \sigma_0 < \sigma_1 - r$.

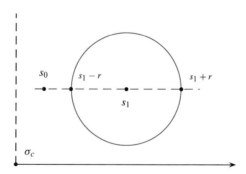

Since $F(s)$ converges at s_0, the partial sums are bounded at this point in absolute value by $M > 0$. Using Theorem 4.5 with s replaced by $s - s_0$ and $\alpha = 0$, we get for all $s \in \mathcal{D}$ and $N_1 > N$

$$\left| \sum_{N < n \leq N_1} \frac{f(n)}{n^s} \right| \leq \frac{2M}{N^{\sigma-\sigma_0}} \left(\frac{|s - s_0|}{\sigma - \sigma_0} + 1 \right) \leq \frac{2M}{N^{\sigma_1-\sigma_0-r}} \left(\frac{\sigma_1 - \sigma_0 + r}{\sigma_1 - \sigma_0 - r} + 1 \right).$$

The right-hand side is independent of s and tends to 0 as $N \to \infty$, which ensures the existence of n_0. By Theorem 2, we deduce that $F(s)$ is analytic in $\mathcal{D}$, and term-by-term differentiation is allowed there. Since s_1 is arbitrary, the asserted result follows.

$\square$

The next result implies that, if $F(s) = \sum_{n=1}^{\infty} f(n)n^{-s}$ and $G(s) = \sum_{n=1}^{\infty} g(n)n^{-s}$ having the same abscissa of convergence σ_c satisfy $F(s) = G(s)$, then $f(n) = g(n)$ for all $n \in \mathbb{Z}_{\geq 1}$.

Proposition 4.12 *Let $F(s) = \sum_{n=1}^{\infty} f(n)n^{-s}$ be a Dirichlet series with abscissa of convergence σ_c. If $F(s) = 0$ for all s such that $\sigma > \sigma_c$, then $f(n) = 0$ for all $n \in \mathbb{Z}_{\geq 1}$.*

Proof Suppose the contrary and let k be the smallest positive integer such that $f(k) \neq 0$. Let

$$G(s) = k^s \sum_{n=k}^{\infty} \frac{f(n)}{n^s}$$

defined for $\sigma > \sigma_c$. By assumption we have $G(s) = 0$ in this half-plane. On the other hand, if $\sigma > \sigma_c$, we have

$$G(s) = f(k) + \sum_{n=k+1}^{\infty} f(n) \left(\frac{k}{n} \right)^s$$

so that allowing s to become infinite along the real axis, we obtain

$$\lim_{\sigma \to \infty} G(\sigma) = f(k)$$

and thus we get $f(k) = 0$, giving a contradiction.

$\square$

This result enables us to show arithmetic identities of the form $f(n) = g(n)$ by considering their Dirichlet series $F(s)$ and $G(s)$ and by showing that, for σ sufficiently large, we have $F(s) = G(s)$. For instance, using Proposition 4.5, one may check that the Dirichlet series of the Euler totient function φ is given by

$$\frac{\zeta(s-1)}{\zeta(s)} = \zeta(s-1) \times \frac{1}{\zeta(s)}.$$

Now $\zeta(s-1)$ is the Dirichlet series of id and $\zeta(s)^{-1}$ is that of μ, and hence $\zeta(s-1)/\zeta(s)$ is the Dirichlet series of $\mu \star \mathrm{id}$. Since both series converge for $\sigma > 2$, we get $\varphi = \mu \star \mathrm{id}$ by Proposition 4.12 as already proved in (4.7).

Another important problem is the investigation of the behaviour of a Dirichlet series $F(s)$ on vertical lines. The following result shows that $F(\sigma + it)$ cannot increase more rapidly than $|t|$ as $|t| \to \infty$.

Proposition 4.13 *Let $F(s) = \sum_{n=1}^{\infty} f(n)n^{-s}$ be a Dirichlet series with abscissa of convergence σ_c and let $\sigma_1 > \sigma_c$. Then, uniformly in σ such that $\sigma \geqslant \sigma_1$ and as $|t| \to \infty$, we have*

$$F(\sigma + it) = o(|t|).$$

Furthermore, for any $\varepsilon > 0$ and $\sigma_c < \sigma \leqslant \sigma_c + 1$ and as $|t| \to \infty$, we have

$$F(\sigma + it) = O\left(|t|^{1-(\sigma-\sigma_c)+\varepsilon}\right).$$

Proof Let $\sigma_0 \in \mathbb{R}$ such that $\sigma_c < \sigma_0 < \sigma_1$. Using Theorem 4.5 with s replaced by $s - \sigma_0$ and $\alpha = 0$, we get for all $x \geqslant 1$ and $\sigma > \sigma_0$

$$\left| \sum_{n>x} \frac{f(n)}{n^s} \right| \leqslant \frac{M}{x^{\sigma-\sigma_0}} \left(\frac{|\sigma - \sigma_0 + it|}{\sigma - \sigma_0} + 1 \right) \leqslant \frac{M}{x^{\sigma-\sigma_0}} \left(\frac{|t|}{\sigma - \sigma_0} + 2 \right)$$

for some $M > 0$, and hence for all $\sigma \geqslant \sigma_1$, we get

$$\left| \frac{F(\sigma + it)}{t} \right| \leqslant \frac{1}{|t|} \left\{ \sum_{n \leqslant x} \frac{|f(n)|}{n^\sigma} + \frac{M}{x^{\sigma-\sigma_0}} \left(\frac{|t|}{\sigma - \sigma_0} + 2 \right) \right\}$$

$$\leqslant \frac{1}{|t|} \sum_{n \leqslant x} \frac{|f(n)|}{n^{\sigma_1}} + \frac{M}{x^{\sigma_1-\sigma_0}} \left(\frac{2}{|t|} + \frac{1}{\sigma_1 - \sigma_0} \right),$$

where the right-hand side is independent of $\sigma \in [\sigma_1, +\infty[$. This inequality implies that

$$\limsup_{|t| \to \infty} \left| \frac{F(\sigma + it)}{t} \right| \leqslant \frac{M}{x^{\sigma_1-\sigma_0}(\sigma_1 - \sigma_0)}$$

and since the left-hand side is independent of x, letting $x \to \infty$ gives

$$\limsup_{|t| \to \infty} \left| \frac{F(\sigma + it)}{t} \right| = 0$$

uniformly in $\sigma \geqslant \sigma_1$, implying the first asserted result. For the second one, since $\sigma_c + 1 + \varepsilon > \sigma_a$ by Proposition 4.11, we infer that

$$F(\sigma_c + 1 + \varepsilon) = O(1).$$

By above, we also have $F(\sigma_c + \varepsilon + it) = O(|t|)$ so that the asserted estimate follows from the Phragmén–Lindelöf principle seen in Lemma 3.14. □

We know that a function defined by a power series has a singularity on the circle of convergence. The situation is rather different for Dirichlet series which need have no singularity on the axis of convergence. For instance, it may be shown that the function $\eta(s) = \sum_{n=1}^{\infty}(-1)^{n+1}n^{-s}$ is entire although we easily see that $\sigma_c = 0$. However, if the coefficients of $F(s)$ are real non-negative, then the next result, due to Landau, shows that the series does have a singularity on the axis of convergence.

Proposition 4.14 (Landau) *Let $F(s) = \sum_{n=1}^{\infty} f(n)n^{-s}$ be a Dirichlet series with abscissa of convergence $\sigma_c \in \mathbb{R}$ and suppose that $f(n) \geqslant 0$ for all n. Then $F(s)$ has a singularity at $s = \sigma_c$, i.e. $F(s)$ cannot be continued analytically to a neighbourhood of σ_c.*

Proof Without loss of generality, we may assume that $\sigma_c = 0$ and suppose that the origin is not a singularity of $F(s)$. Thus the Taylor expansion of $F(s)$ about $a > 0$

$$F(s) = \sum_{k=0}^{\infty} \frac{(s-a)^k}{k!} F^{(k)}(a) = \sum_{k=0}^{\infty} \frac{(s-a)^k}{k!} \sum_{n=1}^{\infty} \frac{(-1)^k (\log n)^k f(n)}{n^a}$$

must converge at some point $s = b < 0$ so that the double series

$$\sum_{k=0}^{\infty} \sum_{n=1}^{\infty} \frac{((a-b)\log n)^k f(n)}{n^a k!}$$

must converge, and then may be summed in any order since each term is non-negative. But interchanging the summation implies that this sum is equal to

$$\sum_{n=1}^{\infty} \frac{f(n)}{n^a} \sum_{k=0}^{\infty} \frac{((a-b)\log n)^k}{k!} = \sum_{n=1}^{\infty} \frac{f(n)}{n^b}$$

which do not converge by Proposition 4.11 since $b < 0 = \sigma_c$, giving a contradiction.
□

For instance, the Riemann zeta-function $\zeta(s)$ has a singularity at $\sigma_c = 1$, as we already proved in Theorem 3.7. In fact, this result proves much more, telling us that this singularity is a *pole*.

Landau's theorem is often used in its contrapositive form.

Corollary 4.4 *Let $F(s) = \sum_{n=1}^{\infty} f(n)n^{-s}$ be a Dirichlet series with abscissa of convergence $\sigma_c < \infty$ and suppose that $f(n) \geqslant 0$ for all n. If $F(s)$ can be analytically continued to the half-plane $\sigma > \sigma_0$, then $F(s)$ converges for $\sigma > \sigma_0$.*

By the previous results, we infer that a Dirichlet series converges *uniformly* on every compact subset of the half-plane of convergence. The next result, due to Cahen and Jensen, shows that it also converges uniformly in certain regions which extend to infinity (Fig. 4.1).

Fig. 4.1 Stolz domain

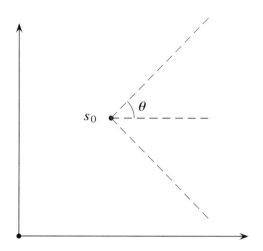

Proposition 4.15 (Cahen–Jensen) *If the Dirichlet series* $F(s) = \sum_{n=1}^{\infty} f(n) n^{-s}$ *converges at* s_0, *then it converges uniformly on every domain* $C(s_0, \theta)$, *sometimes called the* Stolz domain, *defined by*

$$C(s_0, \theta) = \left\{ s \in \mathbb{C} : |\arg(s - s_0)| \leqslant \theta < \tfrac{1}{2}\pi \right\}.$$

Proof The map $s \mapsto s + s_0$ transforms $F(s)$ into a Dirichlet series of the same shape where the coefficient a_n is replaced by the coefficient $a_n n^{-s_0}$, so that we may assume that $s_0 = 0$ without loss of generality. Let $\varepsilon > 0$. By Cauchy's theorem, there exists $N_0 = N_0(\varepsilon) \in \mathbb{N}$ such that, for all $N_1 > N > N_0$, we have

$$\left| \sum_{N < n \leqslant N_1} f(n) \right| \leqslant \frac{\varepsilon \cos \theta}{2}$$

since the series converges at 0. Hence using Theorem 4.5 with $\alpha = 0$ and $M = \frac{\varepsilon}{2} \cos \theta$, we get if $\sigma > 0$

$$\left| \sum_{N < n \leqslant N_1} \frac{f(n)}{n^s} \right| \leqslant \frac{\varepsilon \cos \theta}{2N^\sigma} \left(\frac{|s|}{\sigma} + 1 \right) \leqslant \frac{\varepsilon |s| \cos \theta}{\sigma N^\sigma}$$

and the result follows by noticing that, if $s \in C(0, \theta)$, then $\sigma > 0$ and $|s| \cos \theta \leqslant \sigma$.
$\square$

Among other things, this result implies that, if $F(s)$ converges at s_0, then

$$\lim_{s \to s_0} F(s) = F(s_0)$$

provided that s remains inside $C\left(s_0, \theta\right)$ as it approaches its limit. When s_0 is off the axis of convergence, this result is clearly a consequence of Theorem 4.7, but when s_0 is on that axis, then the limit above is the analogue of Abel's theorem for power series.

Finally, it should be mentioned that explicit formulæ do exist for σ_c and σ_a, but they are not much used in number theory. For instance, Cahen proved that, if the series $\sum_{n \geqslant 1} f(n)$ diverges, then

$$
\sigma_c = \limsup_{N \to \infty} \frac{\log\left|\sum_{n=1}^{N} f(n)\right|}{\log N} = \inf\left\{\sigma_0 : \sum_{n=1}^{N} f(n) = O\left(N^{\sigma_0}\right)\right\}.
$$

If the series $\sum_{n \geqslant 1} f(n)$ converges, then[8]

$$
\sigma_c = \limsup_{N \to \infty} \frac{\log\left|\sum_{n > N} f(n)\right|}{\log N}.
$$

This should be compared to *Hadamard's formula* for the radius R of convergence of the power series $\sum_{n=1}^{\infty} a_n z^n$ given by

$$
R^{-1} = \limsup_{n \to \infty} |a_n|^{1/n}.
$$

4.2.5 Multiplicative Aspects

In this section, we will now focus on the Dirichlet series $L(s, f)$ with multiplicative coefficients $f(n)$. The first result is of crucial importance.

Theorem 4.8 *Let f be a multiplicative function satisfying*

$$
\sum_{p} \sum_{\alpha=1}^{\infty} \left|f\left(p^\alpha\right)\right| < \infty. \tag{4.18}
$$

Then the series $\sum_{n \geqslant 1} f(n)$ is absolutely convergent and we have

$$
\sum_{n=1}^{\infty} f(n) = \prod_{p} \left(1 + \sum_{\alpha=1}^{\infty} f\left(p^\alpha\right)\right).
$$

---

[8]This result is due to Pincherle.

Proof Let us first notice that (4.18) implies the convergence of the product

$$\prod_{p}\left(1 + \sum_{\alpha=1}^{\infty}|f\left(p^{\alpha}\right)|\right).$$

Now let $x \geqslant 2$ be a real number and set

$$P(x) = \prod_{p \leqslant x}\left(1 + \sum_{\alpha=1}^{\infty}|f\left(p^{\alpha}\right)|\right).$$

The convergence of the series $\sum_{\alpha=1}^{\infty}|f\left(p^{\alpha}\right)|$ enables us to rearrange the terms when we expand $P(x)$ and since $|f|$ is multiplicative, we deduce that

$$P(x) = \sum_{P^{+}(n) \leqslant x}|f(n)|.$$

Since each integer $n \leqslant x$ satisfies the condition $P^{+}(n) \leqslant x$, we infer that

$$\sum_{n \leqslant x}|f(n)| \leqslant P(x).$$

Since $P(x)$ has a finite limit as $x \to \infty$, the above inequality implies that the left-hand side is bounded as $x \to \infty$, proving the absolute convergence of the series $\sum_{n \geqslant 1} f(n)$. The second part of the theorem follows from the inequality

$$\left|\sum_{n=1}^{\infty}f(n) - \prod_{p \leqslant x}\left(1 + \sum_{\alpha=1}^{\infty}f\left(p^{\alpha}\right)\right)\right| \leqslant \sum_{n>x}|f(n)|$$

and the fact that the right-hand side tends to 0 as $x \to \infty$.   □

Applying this theorem to absolutely convergent Dirichlet series readily implies the next result which is an analytic version of Proposition 4.5.

Corollary 4.5 *Let f be a multiplicative function with associated Dirichlet series $L(s, f)$ and $s_0 \in \mathbb{C}$. The three following assertions are equivalent.*

▷

$$\sum_{p}\sum_{\alpha=1}^{\infty}\left|\frac{f\left(p^{\alpha}\right)}{p^{s_0\alpha}}\right| < \infty. \tag{4.19}$$

(continued)

Corollary 4.5 (continued)
▷ *The series $L(s, f)$ is absolutely convergent in the half-plane $\sigma > \sigma_0$.*
▷ *The product*

$$\prod_p \left(1 + \sum_{\alpha=1}^{\infty} \frac{f(p^\alpha)}{p^{s\alpha}} \right)$$

is absolutely convergent in the half-plane $\sigma > \sigma_0$.

If one of these conditions holds, then we have for all $\sigma > \sigma_0$

$$L(s, f) = \prod_p \left(1 + \sum_{\alpha=1}^{\infty} \frac{f(p^\alpha)}{p^{s\alpha}} \right). \tag{4.20}$$

In particular, if σ_a is the abscissa of absolute convergence of $L(s, f)$, then (4.20) holds for all $\sigma > \sigma_a$.

We list in the following tables several arithmetic functions f with their Dirichlet series $L(s, f)$ and some convolution identities often used. The number Θ is the infimum of the real parts of the non-trivial zeros $\rho = \beta + i\gamma$ of ζ such that $\frac{1}{2} \leqslant \beta < 1$. The Riemann hypothesis is $\Theta = \frac{1}{2}$.

In these tables, we have

▷ $f_k(n) = \begin{cases} \mu(m), & \text{if } n = m^k \\ 0, & \text{otherwise.} \end{cases}$

▷ $g_k(n) = \begin{cases} \mu(a), & \text{if } n = ab, \ (a, b) = 1, \ \mu_2(a) = s_k(b) = 1 \\ 0, & \text{otherwise.} \end{cases}$

▷ $G_2(s) = 1$ and, for $k \geqslant 3$, $G_k(s)$ is a Dirichlet series absolutely convergent in the half-plane $\sigma > (2k + 3)^{-1}$. For all $k \geqslant 2$, $H_k(s)$ is a Dirichlet series absolutely convergent in the half-plane $\sigma > \frac{1}{3}$.

▷ h is the multiplicative function defined by $h(p) = 0$ and $h(p^\alpha) = P(\alpha) - P(\alpha - 1)$.

Table 4.1 Usual multiplicative functions

f	Convolution	$L(s, f)$	σ_c	σ_a
e_1	Identity element	1	$-\infty$	$-\infty$
$\mathbf{1}$	$\mathbf{1} \star \mu = e_1$	$\zeta(s)$	1	1
μ	$\mu \star \mathbf{1} = e_1$	$\zeta(s)^{-1}$	Θ	1
λ	$\lambda \star \mu_2 = e_1$	$\dfrac{\zeta(2s)}{\zeta(s)}$	Θ	1
μ_k	$\mu_k \star \mu = f_k$	$\dfrac{\zeta(s)}{\zeta(ks)}$	1	1
s_k	$s_k \star \mu = g_k$	$\displaystyle\prod_{j=k}^{2k-1} \zeta(js) \times \dfrac{G_k(s)}{\zeta((2k+2)s)}$	$\dfrac{1}{k}$	$\dfrac{1}{k}$
β	$\beta \star \mu = s_2$	$\dfrac{\zeta(s)\zeta(2s)\zeta(3s)}{\zeta(6s)}$	1	1
a	$a \star \mu = h$	$\displaystyle\prod_{j=1}^{\infty} \zeta(js)$	1	1
$\tau_{(k)}$	$\tau_{(k)} = \mathbf{1} \star \mu_k$	$\dfrac{\zeta(s)^2}{\zeta(ks)}$	1	1
k^{ω}	$k^{\omega} = \mu_2(k-1)^{\omega} \star \mathbf{1}$	$\zeta(s)^k \zeta(2s)^{-\frac{1}{2}k(k-1)} H_k(s)$	1	1
τ_k	$\tau_k = \tau_{k-1} \star \mathbf{1}$ $\mathrm{id} \times \tau = \sigma \star \varphi$	$\zeta(s)^k$	1	1
φ	$\varphi = \mu \star \mathrm{id}$ $\sigma = \varphi \star \tau$	$\dfrac{\zeta(s-1)}{\zeta(s)}$	2	2
id_k	$\mathrm{id}_k = \sigma_k \star \mu$ $\mathrm{id} = \varphi \star \mathbf{1}$	$\zeta(s-k)$	$k+1$	$k+1$
σ_k	$\sigma_k = \mathrm{id}_k \star \mathbf{1}$ $\sigma = \tau \star \varphi$	$\zeta(s)\zeta(s-k)$	$k+1$	$k+1$
Λ_k	$\Lambda_k = \mu \star \log^k$	$(-1)^k \dfrac{\zeta^{(k)}(s)}{\zeta(s)}$	1	1
$\chi \neq \chi_0$	$\chi \times \log \times (\mu \star \mathbf{1}) = 0$	$L(s, \chi)$	0	1
$r_{\mathbb{K}}$	$r_{\mathbb{K}} = \chi \star \mathbf{1}$ if $\mathbb{K}$ is a quadratic field	$\zeta_{\mathbb{K}}(s)$	1	1

4.3 General Mean Value Results

4.3.1 A Useful Upper Bound

4.3.1.1 The Main Result

Definition 4.6 Let f be an arithmetic function. We will look at asymptotic formulæ of the form

$$\sum_{n \leqslant x} f(n) = g(x) + O(R(x)),$$

where the main term $g(x)$ lies in the set of usual functions and the error term $R(x)$ satisfies $R(x) = o(g(x))$ for x sufficiently large. Under these conditions, we shall say that the function $x \mapsto x^{-1}g(x)$ is an *average order* of f on $[1, x]$ for x sufficiently large.

For instance, as will be seen in Theorem 4.28, the function $x \mapsto \log x + 2\gamma - 1$ is an average order of the divisor function τ on $[1, x]$ for x sufficiently large.

One of the most important problems in number theory is to find the smallest error term admissible in the above asymptotic estimate. This problem is sometimes open, as, for instance, in the PNT, where the best error term to date (3.46) is presumably far from the conjectured remainder term given by the Riemann hypothesis.

In this section, we consider a certain class of arithmetic functions satisfying the following hypotheses. We shall say that $f \in \mathcal{M}$ if f is a non-negative multiplicative function such that there exist $a, b > 0$ such that, for all $x \geqslant 1$

$$\frac{1}{x} \sum_{p \leqslant x} f(p) \log p \leqslant a \tag{4.21}$$

$$\sum_{p \leqslant x} \sum_{\alpha=2}^{\infty} \frac{f(p^{\alpha}) \log p^{\alpha}}{p^{\alpha}} \leqslant b. \tag{4.22}$$

The purpose of this section is to provide a proof of the following important result.

Theorem 4.9 *Let f be a non-negative multiplicative function satisfying* (4.21) *and* (4.22). *Then, for all $x \geqslant 1$, we have*

$$\sum_{n \leqslant x} f(n) \leqslant e^{b}(a + b + 1)\frac{x}{\log ex} \exp\left(\sum_{p \leqslant x} \frac{f(p)}{p}\right).$$

In other words, only knowing the values of $f \in \mathcal{M}$ at prime numbers is sufficient to determine an upper bound which in many cases proves to be of the right order of magnitude.

It should be mentioned that $\mathcal{M}$ is not empty. Indeed, the function $\mathbf{1}$ is positive, completely multiplicative and using Lemma 3.8 or Corollary 3.12 and (2.7) with $k = 2$ and $x = \frac{1}{p}$ we get

$$\frac{1}{x} \sum_{p \leqslant x} \mathbf{1}(p) \log p = \frac{\theta(x)}{x} \leqslant \log 4$$

and

$$\sum_{p \leqslant x} \sum_{\alpha=2}^{\infty} \frac{\mathbf{1}(p^{\alpha}) \log p^{\alpha}}{p^{\alpha}} = \sum_{p \leqslant x} \frac{(2p-1) \log p}{p(p-1)^2} \leqslant 6 \sum_{p \leqslant x} \frac{\log p}{p^2} < 3,$$

where we used the inequality

$$\sum_{p \leqslant x} \frac{\log p}{p^2} < \frac{1}{2} \tag{4.23}$$

from Exercise 30, so that one may take $(a, b) = (\log 4, 3)$. The reader may also check that the functions 2^{ω} and τ lie in $\mathcal{M}$ since one can take $(a, b) = (\log 16, 6)$ for the first one and $(a, b) = (\log 16, 14)$ for the second one.

4.3.1.2   The Wirsing Conditions

The next result provides a useful sufficient condition for a function f to lie in $\mathcal{M}$.

Lemma 4.5 *Let f be a multiplicative function satisfying the* Wirsing conditions, *i.e.*

$$0 \leqslant f\left(p^{\alpha}\right) \leqslant \lambda_1 \lambda_2^{\alpha-1} \tag{4.24}$$

for all prime powers p^{α} and some real numbers $\lambda_1 > 0$ and $0 \leqslant \lambda_2 < 2$. Then $f \in \mathcal{M}$ with

$$(a, b) = \left(\lambda_1 \log 4, \ \frac{\lambda_1 \lambda_2 (4 - \lambda_2)}{(2 - \lambda_2)^2}\right).$$

Proof We will use the inequality

$$\frac{1}{p - \lambda_2} \leqslant \frac{2}{p(2 - \lambda_2)} \tag{4.25}$$

which readily comes from the fact that $\lambda_2 < 2 \leqslant p$. Note first that f is non-negative by multiplicativity. Furthermore, we have

$$\frac{1}{x} \sum_{p \leqslant x} f(p) \log p \leqslant \frac{\lambda_1 \theta(x)}{x} \leqslant \lambda_1 \log 4$$

by Lemma 3.8. Next we have, using (2.7), (4.25), and (4.23),

$$\sum_{p \leqslant x} \sum_{\alpha=2}^{\infty} \frac{f(p^\alpha) \log p^\alpha}{p^\alpha} \leqslant \lambda_1 \sum_{p \leqslant x} \log p \sum_{\alpha=2}^{\infty} \frac{\alpha}{p} \left(\frac{\lambda_2}{p}\right)^{\alpha-1} = \lambda_1 \lambda_2 \sum_{p \leqslant x} \frac{(2p - \lambda_2) \log p}{p(p - \lambda_2)^2}$$

$$\leqslant \frac{\lambda_1 \lambda_2 (4 - \lambda_2)}{2 - \lambda_2} \sum_{p \leqslant x} \frac{\log p}{p(p - \lambda_2)} \leqslant \frac{\lambda_1 \lambda_2 (4 - \lambda_2)}{(2 - \lambda_2)^2}$$

as asserted.   □

Lemma 4.5 enables us to increase the number of arithmetic functions lying in $\mathcal{M}$.

Lemma 4.6 *The following arithmetic functions lie in* $\mathcal{M}$.

$$e_1, \ \mathbf{1}, \ \mu_k, \ s_k, \ \beta, \ a, \ k^\omega, \ \tau_{(k)}, \ \tau^{(e)} \ and \ \tau_k.$$

Proof All these functions are non-negative multiplicative as seen in Sect. 4.1.2.3. Furthermore, if f is such that $0 \leqslant f(n) \leqslant 1$, then the Wirsing conditions (4.24) are obviously satisfied, and this is the case for the first four arithmetic functions. We have

$$\beta(p^\alpha) = \begin{cases} 1, & \text{if } \alpha \in \{0, 1\} \\ \alpha, & \text{if } \alpha \geqslant 2 \end{cases}$$

and hence $\beta(p^\alpha) \leqslant \max(1, \alpha)$, so that the Wirsing conditions are readily satisfied. By Krätzel [63], we have $P(\alpha) \leqslant 5^{\alpha/4}$ so that

$$a(p^\alpha) \leqslant 5^{1/4} \left(5^{1/4}\right)^{\alpha-1}$$

and therefore the function a satisfies (4.24) with $\lambda_1 = \lambda_2 = 5^{1/4}$. Since

$$k^{\omega(p^\alpha)} = k$$

k^ω satisfies (4.24) with $\lambda_1 = k$ and $\lambda_2 = 1$. Similarly, since $\tau^{(e)}(p^\alpha) = \tau(\alpha)$ and

$$\tau_{(k)}(p^\alpha) = \begin{cases} k, & \text{if } \alpha \geqslant k \\ \alpha + 1, & \text{if } \alpha < k \end{cases}$$

we easily see that these two functions satisfy (4.24). Finally, $\tau_k(p^\alpha)$ is the number of solutions of the equation $x_1 \cdots x_k = p^\alpha$, and setting $x_i = p^{\beta_i}$ for some $\beta_i \in \mathbb{Z}_{\geqslant 0}$, we see that we have to count the number of solutions in $\left(\mathbb{Z}_{\geqslant 0}\right)^k$ of the Diophantine equation

$$\sum_{i=1}^{k} \beta_i = \alpha$$

whose number of solutions is equal to $\mathcal{D}_{(1,\ldots,1)}(\alpha) = \binom{k+\alpha-1}{\alpha}$ by Theorem 2.7, where $\mathcal{D}_{(1,\ldots,1)}$ is the denumerant defined in Chap. 2. Therefore[9]

$$\tau_k(p^\alpha) = \binom{k+\alpha-1}{\alpha} \tag{4.26}$$

and hence

$$\frac{1}{x} \sum_{p \leqslant x} \tau_k(p) \log p = \frac{k\theta(x)}{x} \leqslant k \log 4.$$

Using (2.7) we get

$$\sum_{\alpha=2}^{\infty} \frac{\alpha \tau_k(p^\alpha)}{p^\alpha} = \frac{k}{p} \left\{ \left(1 - \frac{1}{p}\right)^{-k-1} - 1 \right\}$$

and the inequality

$$(1-x)^{-k} \leqslant 1 + kx + k(k+1)2^{k+1}x^2$$

valid for all $0 \leqslant x \leqslant \frac{1}{2}$, implies that

$$\sum_{\alpha=2}^{\infty} \frac{\alpha \tau_k(p^\alpha)}{p^\alpha} \leqslant \frac{k(k+1)^2 2^{k+2}}{p^2}$$

so that

$$\sum_{p \leqslant x} \log p \sum_{\alpha=2}^{\infty} \frac{\alpha \tau_k(p^\alpha)}{p^\alpha} \leqslant k(k+1)^2 2^{k+2}$$

and therefore $\tau_k \in \mathcal{M}$. $\qquad\qquad\qquad\qquad\qquad\qquad\qquad\qquad\qquad\qquad\qquad\qquad$ □

---

[9]It is noteworthy that $\tau_k(p^\alpha) = \mathcal{D}_{(1,\ldots,1)}(\alpha)$ where the vector $(1,\ldots,1)$ has k components. See also Proposition 7.29.

4.3.1.3   Auxiliary Lemmas

Lemma 4.7 *Let $n \in \mathbb{Z}_{\geqslant 1}$. For all $x \geqslant n$, we have*

$$\log(ex) \leqslant \log n + \tfrac{x}{n}.$$

Proof Indeed, the function $x \mapsto \log n + x/n - \log(ex)$ is non-decreasing on $[n, +\infty)$ and vanishes at $x = n$. □

Lemma 4.8 *Let $x \geqslant 1$ be a real number and f be a positive multiplicative function satisfying* (4.22). *Then*

$$\sum_{n \leqslant x} \frac{f(n)}{n} \leqslant e^b \exp\left(\sum_{p \leqslant x} \frac{f(p)}{p}\right).$$

Proof Expanding the product

$$\prod_{p \leqslant x} \left(1 + \sum_{\alpha=1}^{\infty} \frac{f(p^{\alpha})}{p^{\alpha}}\right)$$

we infer that this product is equal to

$$\sum_{P^+(n) \leqslant x} \frac{f(n)}{n}.$$

Since each positive integer $n \leqslant x$ satisfies the condition $P^+(n) \leqslant x$ and since $f \geqslant 0$, we get

$$\sum_{n \leqslant x} \frac{f(n)}{n} \leqslant \prod_{p \leqslant x}\left(1 + \sum_{\alpha=1}^{\infty} \frac{f(p^{\alpha})}{p^{\alpha}}\right) \leqslant \exp\left(\sum_{p \leqslant x}\sum_{\alpha=1}^{\infty} \frac{f(p^{\alpha})}{p^{\alpha}}\right)$$

$$= \exp\left(\sum_{p \leqslant x} \frac{f(p)}{p} + \sum_{p \leqslant x}\sum_{\alpha=2}^{\infty} \frac{f(p^{\alpha})}{p^{\alpha}}\right) \leqslant e^b \exp\left(\sum_{p \leqslant x} \frac{f(p)}{p}\right),$$

where we used (4.22). □

4.3.1.4   Proof of Theorem 4.9

We are now in a position to prove Theorem 4.9, following essentially the ideas from
[39, 41]. Instead of dealing with the sum of the theorem, we shall estimate the sum

$$\sum_{n \leqslant x} f(n) g(n)$$

with a suitable choice of the weight $g(n)$ which makes the treatment of the new
sum easier. The function g is often chosen among the functions $\log n$ or $(N/n)^\beta$ for
some real $\beta > 0$. This last choice is called *Rankin's trick* and proves to be a very
fruitful idea.

We will choose here the function $g(n) = \log n$ on account of its complete
additivity. Hence, if $n = p_1^{\alpha_1} \cdots p_r^{\alpha_r}$, then

$$\log n = \sum_{i=1}^{r} \log p_i^{\alpha_i}$$

and therefore

$$\sum_{n \leqslant x} f(n) \log n = \sum_{p^\alpha \leqslant x} \sum_{\substack{k \leqslant x/p^\alpha \\ p \nmid k}} f\left(kp^\alpha\right) \log p^\alpha$$

$$= \sum_{p^\alpha \leqslant x} \sum_{\substack{k \leqslant x/p^\alpha \\ p \nmid k}} f(k) f\left(p^\alpha\right) \log p^\alpha.$$

Interchanging the summations and neglecting the condition $p \nmid k$ gives

$$\sum_{n \leqslant x} f(n) \log n \leqslant \sum_{k \leqslant x} f(k) \sum_{p^\alpha \leqslant x/k} f\left(p^\alpha\right) \log p^\alpha. \tag{4.27}$$

We split the inner sum into two subsums according to either $\alpha = 1$ or $\alpha \geqslant 2$ and,
in the second subsum, we use the fact that $p^\alpha \leqslant x/k$ is equivalent to $1 \leqslant x/(kp^\alpha)$
which gives

$$\sum_{p^\alpha \leqslant x/k} f\left(p^\alpha\right) \log p^\alpha = \sum_{p \leqslant x/k} f\left(p\right) \log p + \sum_{\substack{p^\alpha \leqslant x/k \\ \alpha \geqslant 2}} f\left(p^\alpha\right) \log p^\alpha$$

$$\leqslant \frac{ax}{k} + \frac{x}{k} \sum_{\substack{p^\alpha \leqslant x/k \\ \alpha \geqslant 2}} \frac{f\left(p^\alpha\right) \log p^\alpha}{p^\alpha} \leqslant \frac{(a+b)x}{k},$$

where we used (4.21) and (4.22). Reporting this estimate in (4.27) we get

$$\sum_{n \leqslant x} f(n) \log n \leqslant (a+b)x \sum_{k \leqslant x} \frac{f(k)}{k}. \tag{4.28}$$

By Lemma 4.7, we infer

$$\sum_{n \leqslant x} f(n) \leqslant \frac{1}{\log ex} \sum_{n \leqslant x} f(n) \log n + \frac{x}{\log ex} \sum_{k \leqslant x} \frac{f(k)}{k}$$

and combining this inequality with (4.28) we get

$$\sum_{n \leqslant x} f(n) \leqslant (a+b+1)\frac{x}{\log ex} \sum_{k \leqslant x} \frac{f(k)}{k}$$

and we conclude the proof by using Lemma 4.8.   □

Example 4.4 Theorem 4.9 enables us to get several bounds for average orders of usual arithmetic functions. The sole tool we need is Corollary 3.4 used under the weaker form $\sum_{p \leqslant x} 1/p \ll \log \log x$. We leave the details to the reader.

▷ Let $k \geqslant 2$. We have

$$\sum_{n \leqslant x} s_k(n) \ll \frac{x}{\log x}.$$

▷ The estimate

$$\sum_{n \leqslant x} f(n) \ll x$$

holds with $f = \beta$, $f = a$, $f = \tau^{(e)}$ and $f = \mu_k$ ($k \geqslant 2$).
▷ Let $k \geqslant 2$. We have

$$\sum_{n \leqslant x} \tau_{(k)}(n) \ll x \log x.$$

▷ Let $k \geqslant 1$. The estimate

$$\sum_{n \leqslant x} f(n) \ll x (\log x)^{k-1}$$

holds with $f = k^\omega$ and $f = \tau_k$.

It should be noticed that almost all these bounds are of the right order of magnitude. Indeed, we shall see in Theorem 4.27 that

$$\sum_{n \leqslant x} \mu_k(n) = \frac{x}{\zeta(k)} + o\left(x^{1/k}\right).$$

It can also be shown that [90, 125, 101]

$$\sum_{n \leqslant x} a(n) = A_1 x + A_2 x^{1/2} + A_3 x^{1/3} + O\left(x^{1/4+\varepsilon}\right) \tag{4.29}$$

$$\sum_{n \leqslant x} \tau^{(e)}(n) = B_1 x + B_2 x^{1/2} + O\left(x^{1057/4785+\varepsilon}\right)$$

$$\sum_{n \leqslant x} \tau_{(k)}(n) = \frac{x}{\zeta(k)} \left(\log x + 2\gamma - 1 - k\frac{\zeta'}{\zeta}(k)\right) + O\left(x^{\max\left(\frac{517}{1648}, \frac{1}{k}\right)}\right),$$

where $A_i = \prod_{\substack{j=1 \\ j \neq i}}^{\infty} \zeta\left(\frac{j}{i}\right)$, $B_1 = \prod_p \left(1 + \sum_{\alpha=2}^{\infty} \frac{\tau(\alpha) - \tau(\alpha-1)}{p^\alpha}\right)$ and

$$B_2 = \prod_p \left(1 + \sum_{\alpha=5}^{\infty} \frac{\widetilde{\tau}(\alpha)}{p^{\alpha/2}}\right)$$

with

$$\widetilde{\tau}(\alpha) = \tau(\alpha) - \tau(\alpha-1) - \tau(\alpha-2) + \tau(\alpha-3) \tag{4.30}$$

for $\alpha \geqslant 5$.[10] Furthermore we shall see in Exercise 70 that

$$\sum_{n \leqslant x} \tau_k(n) = \frac{x(\log x)^{k-1}}{(k-1)!} + O\left(x(\log x)^{k-2}\right)$$

and using Theorem 4.10 below yields

$$\sum_{n \leqslant x} k^{\omega(n)} = \frac{x(\log x)^{k-1}}{(k-1)!} \prod_p \left(1 - \frac{1}{p}\right)^k \left(1 + \frac{k}{p-1}\right) + O\left(x(\log x)^{k-2}\right).$$

---

[10]For the average order of $\tau_{(k)}$, see also Theorem 4.10 below.

Only the function s_k is overestimated, since we shall prove in Exercise 85 that

$$\sum_{n \leqslant x} s_k(n) \ll x^{1/2}.$$

4.3.2 A Simple Asymptotic Formula

4.3.2.1 Main Result

We consider here a class of arithmetic functions which are "not too far", in a certain sense, from the Piltz–Dirichlet divisor functions τ_k. As will be seen in Chap. 6, for each $k \in \mathbb{Z}_{\geqslant 1}$, there exists $\theta_k \in [0, 1)$ such that

$$\sum_{n \leqslant x} \tau_k(n) = x P_{k-1}(\log x) + O\left(x^{\theta_k + \varepsilon}\right), \tag{4.31}$$

where P_{k-1} is a polynomial of degree $k - 1$ and leading coefficient $\frac{1}{(k-1)!}$. Also, for any polynomial P, let $\|P\|$ be the maximum of the absolute values of the coefficients of P. The next result generalizes (4.31).

Theorem 4.10 *Let f be an arithmetic function such that there exist $k \in \mathbb{Z}_{\geqslant 1}$ and a number $\delta \in [0, 1)$ such that $L(s, f) = \zeta(s)^k G(s)$, where $G(s)$ is a Dirichlet series absolutely convergent in the half-plane $\sigma > \delta$. Then, for all $0 < \varepsilon \leqslant \frac{1}{2}(1 - \delta)$*

$$\sum_{n \leqslant x} f(n) = x Q_{k-1}(\log x) + O_{\varepsilon, \delta, k}\left(x^{\max(\theta_k, \delta) + \varepsilon}\right),$$

where θ_k is given in (4.31) and Q_{k-1} is a polynomial of degree $k - 1$ and leading coefficient $\frac{1}{(k-1)!} \times G(1)$.

Remark 4.6 For some values of θ_k, see Theorem 4.62.

4.3.2.2 Lemmas

In this section, set $G(s) = \sum_{n=1}^{\infty} \dfrac{g(n)}{n^s}$, and assume that all the conditions of Theorem 4.10 are fulfilled.

Lemma 4.9 *For all $z \geqslant 3$, $\ell \in \mathbb{Z}_{\geqslant 0}$, and $\beta \in (\delta, 1]$, we have*

$$\sum_{n>z} \frac{|g(n)|(\log n)^{\ell}}{n^{\beta}} \ll_{\varepsilon,\delta,\beta,\ell} z^{\delta-\beta+2\varepsilon}.$$

Proof Since $G(s)$ is absolutely convergent in the half-plane $\sigma > \delta$, we infer from Corollary 4.3 that

$$\sum_{n \leqslant z} |g(n)| \ll_{\varepsilon} z^{\delta+\varepsilon}.$$

Now by partial summation we derive

$$\sum_{n>z} \frac{|g(n)|(\log n)^{\ell}}{n^{\beta}} = -\frac{(\log z)^{\ell}}{z^{\beta}} \sum_{n \leqslant z} |g(n)|$$

$$+ \int_{z}^{\infty} \left(\frac{\beta(\log t)^{\ell}}{t^{\beta+1}} - \frac{\ell(\log t)^{\ell-1}}{t^{\beta+1}} \right) \left(\sum_{n \leqslant t} |g(n)| \right) dt$$

$$\ll_{\varepsilon,\delta,\beta,\ell} z^{\delta-\beta+2\varepsilon} + \int_{z}^{\infty} t^{-1-\beta+\delta+\varepsilon} dt \ll_{\varepsilon,\delta,\beta,\ell} z^{\delta-\beta+2\varepsilon}$$

as required.   □

Lemma 4.10 *Let P_{k-1} be the polynomial given in (4.31). Then*

$$\sum_{n \leqslant x} \frac{g(n)}{n} P_{k-1}\left(\log \frac{x}{n}\right) = Q_{k-1}(\log x) + O_{\varepsilon,\delta,k}\left(x^{\delta-1+3\varepsilon} \|P_{k-1}\|\right),$$

where Q_{k-1} is a polynomial of degree $k-1$ and leading coefficient $\frac{1}{(k-1)!} \times G(1)$.

Proof Write $P_{k-1} := \sum_{j=0}^{k-1} a_j X^j$ with $a_{k-1} = \frac{1}{(k-1)!}$. Since

$$(\log x - \log n)^j = \sum_{h=0}^{j} (-1)^{j-h} \binom{j}{h} (\log x)^h (\log n)^{j-h}$$

we get

$$
\sum_{n \leqslant x} \frac{g(n)}{n} P_{k-1}\left(\log \frac{x}{n}\right)
$$

$$
= \sum_{j=0}^{k-1} a_j \sum_{h=0}^{j} (-1)^{j-h} \binom{j}{h} (\log x)^h \sum_{n \leqslant x} \frac{g(n)(\log n)^{j-h}}{n}
$$

$$
= \sum_{j=0}^{k-1} a_j \sum_{h=0}^{j} (-1)^{j-h} \binom{j}{h} (\log x)^h \left\{ \sum_{n=1}^{\infty} \frac{g(n)(\log n)^{j-h}}{n} - \sum_{n>x} \frac{g(n)(\log n)^{j-h}}{n} \right\}
$$

$$
= \sum_{j=0}^{k-1} a_j \sum_{h=0}^{j} (-1)^{j-h} \binom{j}{h} (\log x)^h \left\{ (-1)^{j-h} G^{(j-h)}(1) + O_{\varepsilon,\delta,j,h}\left(x^{\delta-1+2\varepsilon} \right) \right\},
$$

where we used Theorem 4.7 and Lemma 4.9 with $\beta = 1$. Set

$$
Q_{k-1} := \sum_{j=0}^{k-1} a_j \sum_{h=0}^{j} \binom{j}{h} G^{(j-h)}(1)\, X^h = \sum_{h=0}^{k-1} \left(\sum_{j=h}^{k-1} a_j \binom{j}{h} G^{(j-h)}(1) \right) X^h.
$$

Then obviously Q_{k-1} is a polynomial of degree $k-1$ and leading coefficient $\frac{1}{(k-1)!} \times G(1)$, and we have

$$
\sum_{n \leqslant x} \frac{g(n)}{n} P_{k-1}\left(\log \frac{x}{n}\right) = Q_{k-1}(\log x) + O_{\varepsilon,\delta,k}\left(x^{\delta-1+3\varepsilon} \sum_{j=0}^{k} |a_j| \right)
$$

as asserted.   □

4.3.2.3   Proof of Theorem 4.10

Since $f = \tau_k \star g$, we derive using Proposition 4.3 and (4.31)

$$
\sum_{n \leqslant x} f(n) = \sum_{d \leqslant x} g(d) \sum_{h \leqslant x/d} \tau_k(h)
$$

$$
= \sum_{d \leqslant x} g(d) \left\{ \frac{x}{d} P_{k-1}\left(\log \frac{x}{d}\right) + O_{\varepsilon,k}\left(\left(\frac{x}{d}\right)^{\theta_k+\varepsilon} \right) \right\}
$$

$$
= x \sum_{d \leqslant x} \frac{g(d)}{d} P_{k-1}\left(\log \frac{x}{d}\right) + O_{\varepsilon,k}\left(x^{\theta_k+\varepsilon} \sum_{d \leqslant x} \frac{|g(d)|}{d^{\theta_k+\varepsilon}} \right).
$$

By partial summation

$$\sum_{d \leqslant x} \frac{|g(d)|}{d^{\theta_k+\varepsilon}} = x^{-\theta_k-\varepsilon} \sum_{d \leqslant x} |g(d)| + (\theta_k + \varepsilon) \int_1^x \frac{1}{t^{1+\theta_k+\varepsilon}} \left(\sum_{d \leqslant t} |g(d)| \right) dt$$

$$\ll_{\varepsilon,\delta,k} x^{\delta-\theta_k+\varepsilon} + \int_1^x \frac{dt}{t^{1-\delta+\theta_k-\varepsilon}} \ll_{\varepsilon,\delta,k} x^{\max(0,\delta-\theta_k)+\varepsilon}$$

completing the proof.   □

4.3.2.4   Some Applications

▷ If $f = \tau^2$, then $L(s, \tau^2) = \zeta(s)^4 \zeta(2s)^{-1}$ and hence, by Theorem 4.10 applied
with $k = 4$ and $\delta = \frac{1}{2}$

$$\sum_{n \leqslant x} \tau^2(n) = \frac{x(\log x)^3}{6\zeta(2)} + Ax(\log x)^2 + Bx \log x + Cx + O_\varepsilon\left(x^{1/2+\varepsilon}\right)$$

for some $A, B, C \in \mathbb{R}$.
▷ If $f = \tau_{(k)}$, then Theorem 4.10 can be applied with $G(s) = \zeta(ks)^{-1}$, so that
$\delta = \frac{1}{k}$ and then

$$\sum_{n \leqslant x} \tau_{(k)}(n) = \frac{x \log x}{\zeta(k)} + B_k x + O_{\varepsilon,k}\left(x^{\max\left(\frac{517}{1648}, \frac{1}{k}\right)+\varepsilon}\right)$$

for some $B_k \in \mathbb{R}$.
▷ If $f = \tau^{(k)}$, then $L(s, \tau^{(k)}) = \zeta(s)\zeta(ks)$, so that, by Theorem 4.10

$$\sum_{n \leqslant x} \tau^{(k)}(n) = x\zeta(k) + O_{\varepsilon,k}\left(x^{1/k+\varepsilon}\right).$$

▷ If $f = k^\omega$, then $G(s) = \prod_p \left(1 - \frac{1}{p^s}\right)^k \left(1 + \frac{k}{p^s - 1}\right)$, so that Theorem 4.10
can be applied with $\delta = \frac{1}{2}$, and then

$$\sum_{n \leqslant x} k^{\omega(n)} = x Q_{k-1}(\log x) + O_{\varepsilon,k}\left(x^{\max(\theta_k, 1/2)+\varepsilon}\right),$$

where Q_{k-1} is a polynomial of degree $k - 1$ and leading coefficient

$$\frac{1}{(k-1)!} \prod_p \left(1 - \frac{1}{p}\right)^k \left(1 + \frac{k}{p-1}\right).$$

▷ If $f = \tau(1, 2, 3; n)$, then $L(s, \tau(\cdot; 1, 2, 3)) = \zeta(s)\zeta(2s)\zeta(3s)$, and therefore, applying Theorem 4.10 with $k = 1$, $G(s) = \zeta(2s)\zeta(3s)$ and $\delta = \frac{1}{2}$

$$\sum_{n \leqslant x} \tau(1, 2, 3; n) = x\zeta(2)\zeta(3) + O_\varepsilon\left(x^{1/2+\varepsilon}\right).$$

▷ If $f = a$, then Theorem 4.10 with $k = 1$, $G(s) = \zeta(2s)\zeta(3s)\cdots$ and $\delta = \frac{1}{2}$ yields

$$\sum_{n \leqslant x} a(n) = x \prod_{j=2}^{\infty} \zeta(j) + O_\varepsilon\left(x^{1/2+\varepsilon}\right).$$

▷ Let $S(n)$ be the number of finite semisimple rings of order n. From [11], we know that $L(s, S) = \prod_{q=1}^{\infty}\prod_{m=1}^{\infty} \zeta\left(qm^2 s\right)$, so that, from Theorem 4.10

$$\sum_{n \leqslant x} S(n) = x \prod_{q=2}^{\infty} \zeta(q) \prod_{q=1}^{\infty}\prod_{m=2}^{\infty} \zeta\left(qm^2\right) + O_\varepsilon\left(x^{1/2+\varepsilon}\right).$$

For more precise results concerning $\tau(1, 2, 3; \cdot)$, a and S, see Sect. 4.7.7.4.

4.3.3   Vinogradov's Lemma

In this section, we derive a useful tool dealing with sum running over integers coprime to a fixed natural number.

Proposition 4.16 *Let $(G, +)$ be an abelian group, A be a finite set, $f : A \to G$ and $g : A \to \mathbb{Z}_{\geqslant 0}$ any maps. Then*

$$\sum_{\substack{a \in A \\ g(a)=1}} f(a) = \sum_{m=1}^{\infty} \mu(m) \sum_{\substack{a \in A \\ m \mid g(a)}} f(a).$$

Proof Since the sum $\sum_{m|g(a)} \mu(m)$ is equal to 1 if $g(a) = 1$ and is 0 otherwise, we get

$$\sum_{\substack{a \in A \\ g(a)=1}} f(a) = \sum_{a \in A} f(a) \sum_{m|g(a)} \mu(m) = \sum_{m=1}^{\infty} \mu(m) \sum_{\substack{a \in A \\ m|g(a)}} f(a)$$

as asserted.   □

In practice, the following corollary provides a useful formulation of this principle.

Corollary 4.6 *Let* $k \in \mathbb{Z}_{\geq 1}$ *and* $f : \mathbb{Z}_{\geq 0} \to \mathbb{C}$ *be an arithmetic function. Then*

$$\sum_{\substack{n \leq x \\ (n,k)=1}} f(n) = \sum_{d|k} \mu(d) \sum_{m \leq x/d} f(md).$$

Proof Since

$$\sum_{d|n, d|k} \mu(d) = \begin{cases} 1, & \text{if } (n,k) = 1 \\ 0, & \text{otherwise} \end{cases}$$

we infer

$$\sum_{\substack{n \leq x \\ (n,k)=1}} f(n) = \sum_{n \leq x} f(n) \sum_{d|n, d|k} \mu(d) = \sum_{\substack{md \leq x \\ d|k}} \mu(d) f(md) = \sum_{d|k} \mu(d) \sum_{m \leq x/d} f(md)$$

as required.   □

Example 4.5 Let $k \in \mathbb{Z}_{\geq 1}$ and set $E(k) := \sum_{p|k} \frac{\log p}{p-1}$. Let us prove that

$$\sum_{\substack{n \leq x \\ (n,k)=1}} \frac{1}{n} = \frac{\varphi(k)}{k} (\log x + \gamma + E(k)) + O\left(\frac{2^{\omega(k)}}{x}\right). \tag{4.32}$$

Indeed, using Corollary 4.6, we derive

$$
\sum_{\substack{n \leqslant x \\ (n,k)=1}} \frac{1}{n} = \sum_{d \mid k} \frac{\mu(d)}{d} \sum_{h \leqslant x/d} \frac{1}{h} = \sum_{d \mid k} \frac{\mu(d)}{d} \left(\log \frac{x}{d} + \gamma + O\left(\frac{d}{x}\right) \right)
$$

$$
= (\log x + \gamma) \sum_{d \mid k} \frac{\mu(d)}{d} - \sum_{d \mid k} \frac{\mu(d) \log d}{d} + O\left(\frac{1}{x} \sum_{d \mid k} |\mu(d)| \right)
$$

$$
= \frac{\varphi(k)}{k} (\log x + \gamma + E(k)) + O\left(\frac{2^{\omega(k)}}{x} \right),
$$

where we used Exercise 65 or 66.

4.3.4   Wirsing and Halász Results

In view of the erratic behaviour of most multiplicative functions, the question of the existence of a mean value of such functions arises naturally. Since higher prime powers p^α with $\alpha \in \mathbb{Z}_{\geqslant 2}$ are quite rare, one may wonder how the values of f at the prime numbers influence the behaviour of f in general. The interest in this problem increased as Erdős and Wintner [26] formulated the following conjecture:

Conjecture 4.1 (Erdős–Wintner) Any multiplicative function assuming only the values -1 and $+1$ has a mean value.

On the other hand, one may wonder whether the quantity

$$
\prod_{p \leqslant x} \left(1 - \frac{1}{p} \right) \left(1 + \sum_{\alpha=1}^{\infty} \frac{f(p^\alpha)}{p^\alpha} \right)
$$

of Theorem 4.10 would be close to the right mean value of multiplicative functions f provided that $f(p)$ is sufficiently close to 1. This product is sometimes called the *heuristic value* for the mean $x^{-1} \sum_{n \leqslant x} f(n)$.

The search for accurate asymptotic formulae for average orders really began in the early 1960s with the following satisfactory result due to Wirsing [124].

Theorem 4.11 (Wirsing) *Let f be a positive multiplicative function satisfying the Wirsing conditions* (4.24) *and*

$$\sum_{p \leqslant x} f(p) \log p = \kappa x + o(x)$$

for some $\kappa > 0$. Then we have as $x \to \infty$

$$\sum_{n \leqslant x} f(n) = (1 + o(1)) \frac{e^{-\gamma \kappa}}{\Gamma(\kappa)} \frac{x}{\log x} \prod_{p \leqslant x} \left(1 + \sum_{\alpha=1}^{\infty} \frac{f(p^{\alpha})}{p^{\alpha}} \right).$$

Wirsing's ideas are similar to those of Theorem 4.9, using equalities instead of inequalities and making also use of the Hardy–Littlewood–Karamata tauberian theorem, although he used elementary arguments in his original paper of 1961. In the same paper, Wirsing also deduced theorems for complex-valued multiplicative functions, but these results neither contained the PNT nor settled the Erdős–Wintner conjecture.

Six years later, Wirsing was able to settle this conjecture, proving in an elementary but tricky way that if f is a real-valued multiplicative function such that $|f(n)| \leqslant 1$, then f has a mean value. The following year, Halász [37] extended Wirsing's results by establishing the *Wirsing conjecture*.

Theorem 4.12 (Halász) *Let f be a complex-valued multiplicative function such that $|f(n)| \leqslant 1$, then there exist $\kappa = \kappa_f \in \mathbb{C}$, $a \in \mathbb{R}$ and a slowly varying function[11] L with $|L(u)| = 1$, so that*

$$\sum_{n \leqslant x} f(n) = \kappa \, L(\log x) \, x^{1+ia} + o(x) \quad (x \to \infty).$$

The proof involves contour integration using an asymptotic formula for the Dirichlet series $\sum_{n=1}^{\infty} f^*(n) n^{-s}$, where f^* is a completely multiplicative function associated with f. Various authors [85, 100, 116] subsequently improved on Halász's theorem by partly removing the unpleasant constraint $|f(n)| \leqslant 1$.

---

[11]A *slowly varying function* is a non-zero Lebesgue measurable function $L : [x_0, +\infty) \to \mathbb{C}$ for some $x_0 > 0$ for which

$$\lim_{x \to \infty} \frac{L(cx)}{L(x)} = 1$$

for any $c > 0$.

4.3.5   The Selberg–Delange Method

The exposition of this method follows the lines of [1].

Let $z \in \mathbb{C}$. First note that Theorem 3.15 allows us to define the function $\log \zeta(s)$, and then $\zeta(s)^z = \exp(z \log \zeta(s))$, in the region given by

$$\sigma \geqslant 1 - \frac{c}{\log(|t| + 3)} \quad \text{and} \quad \sigma \notin (1 - c, 1].$$

Now the classic arithmetical function τ_k, with $k \in \mathbb{Z}_{\geqslant 1}$, can be generalized as follows. Let $z \in \mathbb{C}$ and define $\tau_z(n)$ to be the nth coefficient of the Dirichlet series $\zeta(s)^z$. Therefore τ_z is multiplicative and, for all prime powers p^α

$$\tau_z(p^\alpha) = \binom{z + \alpha - 1}{\alpha} := \frac{z(z+1)\cdots(z+\alpha-1)}{\alpha!}.$$

The asymptotic formula of Exercise 70 was first extended by Kienast [59] when z is a positive real number, and then by Selberg [97] who proved the next result.

Theorem 4.13 *Let $A > 0$. Uniformly for $x \geqslant 2$ and $z \in \mathbb{C}$ such that $|z| \leqslant A$*

$$\sum_{n \leqslant x} \tau_z(n) = \frac{x(\log x)^{z-1}}{\Gamma(z)} + O_A\left(x(\log x)^{\operatorname{Re} z - 2}\right).$$

Note that this theorem contains the PNT, since applying it with $z = -1$ yields

$$\sum_{n \leqslant x} \mu(n) \ll x(\log x)^{-3}.$$

Now working as in Theorem 4.10, one may derive an asymptotic formula for the summatory function of an arithmetic function whose Dirichlet series is "close" to a *complex* power of the Riemann zeta function.

Theorem 4.14 *Let $A > 0$, E be a set and $\phi : E \to \mathbb{C}$ be any map. Define $L_f(s; z)$ to be a Dirichlet series with coefficients $f_z(n)$ convergent for $\sigma > 1$ and $z \in E$. Assume also that $L_f(s; z) = \zeta(s)^{\phi(z)} G(s; z)$ where $G(s; z)$ is a Dirichlet series with coefficients $g_z(n)$ such that there exists a constant $B > 0$ such that*

$$\sum_{n=1}^{\infty} \frac{|g_z(n)|}{n} (\log 3n)^{A+2} \leqslant B \quad (z \in E, \ |\phi(z)| \leqslant A).$$

Then, uniformly for $x \geqslant 2$ and $z \in E$ such that $|\phi(z)| \leqslant A$

$$\sum_{n \leqslant x} f_z(n) = \frac{G(1; z)}{\Gamma(\phi(z))} x (\log x)^{\phi(z)-1} + O_A \left(Bx (\log x)^{\mathrm{Re}\,\phi(z)-2} \right).$$

For instance, applying this result with $f_z(n) := z^{\omega(n)}$ and $\phi(z) = z$ yields

$$\sum_{n \leqslant x} z^{\omega(n)} = z H(z) x (\log x)^{z-1} + O_A \left(x (\log x)^{\mathrm{Re}\,z-2} \right) \tag{4.33}$$

uniformly for $x \geqslant 2$ and $|z| \leqslant A$, where

$$H(z) := \frac{1}{\Gamma(z+1)} \prod_p \left(1 - \frac{1}{p} \right)^z \left(1 + \frac{z}{p-1} \right). \tag{4.34}$$

The initial motivation of Selberg was to derive a quick proof of some results obtained by Sathe giving asymptotic formulas for local laws of the usual functions of primes. In the very long paper [94], Sathe proved by induction on k that

$$\pi_k(x) := \sum_{\substack{n \leqslant x \\ \omega(n)=k}} 1 = H \left(\frac{k}{\log \log x} \right) \frac{x}{\log x} \frac{(\log \log x)^{k-1}}{(k-1)!} \left\{ 1 + O_A \left(\frac{1}{\log \log x} \right) \right\}$$

uniformly for $x \geqslant 3$ and $1 \leqslant k \leqslant A \log \log x$ for $A < 2$, where the function H is given in (4.34). Selberg's idea rests on the use of the generating function

$$\sum_{k=0}^{\infty} \pi_k(x) z^k = \sum_{n \leqslant x} z^{\omega(n)}.$$

Now (4.33) along with a lemma[12] dealing with extractions of the coefficients of a power series allow Selberg to recover Sathe's result.

As pointed out in [1], Theorem 4.14 has other numerous applications in number theory. Let us mention the following one, dealing with a result first proved by Pillai, stating that

$$\sum_{\substack{n \leqslant x \\ \omega(n) \equiv a \,(\mathrm{mod}\, q)}} 1 \sim \frac{x}{q} \quad (q \geqslant 2 \text{ fixed}, \ 1 \leqslant a \leqslant q, \ x \to \infty).$$

Delange noticed that this estimate follows immediately from (4.33) since the sum of the left-hand side is equal to

$$\sum_{n \leqslant x} \frac{1}{q} \sum_{k=1}^{q} e_q\left(k\left(\omega(n) - a\right)\right) = \frac{1}{q} \sum_{k=1}^{q} e_q\left(-ka\right) \sum_{n \leqslant x} e_q\left(k\omega(n)\right)$$

$$= \frac{x}{q} + \frac{1}{q} \sum_{k=1}^{q-1} e_q\left(-ka\right) \sum_{n \leqslant x} \left(e_q(k)\right)^{\omega(n)}$$

$$= \frac{x}{q} + o(x).$$

Selberg's theorems were extended by Delange in [22], showing among other things some asymptotic formulæ for sums running over arithmetic progressions. For a complete exposition of the Selberg–Delange Method, see also [106].

⋅

4.3.6 Logarithmic Mean Values

Martin [74] proved the following precise estimates using some ideas developed by Iwaniec.

Theorem 4.15 (Martin) *Let f be a complex-valued positive multiplicative function satisfying*

$$\sum_{p \leqslant x} \frac{f(p) \log p}{p} = \kappa \log x + O(1),$$

(continued)

---

[12] See [1, Lemme].

Theorem 4.15 (continued)
where $\kappa = \sigma_\kappa + it_\kappa \in \mathbb{C}$ is such that $t_\kappa^2 < 2\sigma_\kappa + 1$. Assume also that

$$\sum_p \frac{|f(p)|\log p}{p} \sum_{\alpha=1}^{\infty} \frac{|f(p^\alpha)|}{p^\alpha} + \sum_p \sum_{\alpha=2}^{\infty} \frac{|f(p^\alpha)|\log p^\alpha}{p^\alpha} < \infty$$

and

$$\prod_{p \leqslant x} \left(1 + \frac{|f(p)|}{p}\right) \ll (\log x)^\lambda$$

for all $x \geqslant 2$ and some real number $0 \leqslant \lambda < \sigma_\kappa + 1$. Then we have

$$\sum_{n \leqslant x} \frac{f(n)}{n} = \frac{(\log x)^\kappa}{\Gamma(\kappa + 1)} \prod_p \left(1 - \frac{1}{p}\right)^\kappa \left(1 + \sum_{\alpha=1}^{\infty} \frac{f(p^\alpha)}{p^\alpha}\right) + O\left((\log x)^{\lambda-1}\right)$$

and if $q \in \mathbb{N}$, we also have

$$\sum_{\substack{n \leqslant x \\ (n,q)=1}} \frac{f(n)}{n} = \frac{(\log x)^\kappa}{\Gamma(\kappa + 1)} \left(\frac{\varphi(q)}{q}\right)^\kappa \prod_p \left(1 - \frac{1}{p}\right)^\kappa \left(1 + \sum_{\alpha=1}^{\infty} \frac{f(p^\alpha)}{p^\alpha}\right)$$

$$+ O\left\{\left(1 + \sum_{p|q} \frac{|f(p)|\log p}{p}\right)(\log x)^{\lambda-1}\right\}.$$

In the case of positive multiplicative functions, the statement is simpler.

Corollary 4.7 *Let f be a positive multiplicative function such that $f(n) \ll n^\lambda$ for some $\lambda < \frac{1}{2}$ and satisfying*

$$\sum_{p \leqslant x} \frac{f(p)\log p}{p} = \kappa \log x + O(1),$$

where $\kappa > 0$. Then uniformly in $x \geqslant 2$, we have

$$\sum_{n \leqslant x} \frac{f(n)}{n} = \frac{(\log x)^\kappa}{\Gamma(\kappa + 1)} \prod_p \left(1 - \frac{1}{p}\right)^\kappa \left(1 + \sum_{\alpha=1}^{\infty} \frac{f(p^\alpha)}{p^\alpha}\right) + O\left((\log x)^{\kappa-1}\right).$$

Example 4.6 Let us apply this result to the function $f(n) = n/\mathcal{P}(n)$ where

$$\mathcal{P}(n) := \sum_{i=1}^{n} (i, n)$$

is the so-called *Pillai function*. Using Exercise 73, we have $\mathcal{P} = \varphi \star \mathrm{id}$, so that $\mathcal{P}$ is multiplicative and since $(i, n) \geqslant 1$, we clearly get $0 \leqslant f(n) \leqslant 1$ for all $n \geqslant 1$. Furthermore, we have, using Theorem 3.2

$$\sum_{p \leqslant x} \frac{f(p) \log p}{p} = \sum_{p \leqslant x} \frac{\log p}{p} \left(\frac{1}{2} + \frac{1}{4p - 2} \right) = \frac{1}{2} \sum_{p \leqslant x} \frac{\log p}{p} + O(1) = \frac{\log x}{2} + O(1)$$

so that the conditions of Corollary 4.7 are satisfied with $\kappa = \frac{1}{2}$, and using $\Gamma \left(\frac{3}{2} \right) = \frac{\sqrt{\pi}}{2}$, we derive the asymptotic formula

$$\sum_{n \leqslant x} \frac{1}{\mathcal{P}(n)} = \frac{2}{\sqrt{\pi}} \prod_{p} \left(1 - \frac{1}{p} \right)^{1/2} \left(1 + \sum_{\alpha=1}^{\infty} \frac{1}{\mathcal{P}(p^\alpha)} \right) \sqrt{\log x} + O \left((\log x)^{-1/2} \right)$$

for all $x \geqslant 2$. Note that, since $\varphi = \mu \star \mathrm{id}$, we get $\mathcal{P} = \mu \star \mathrm{id} \star \mathrm{id} = \mu \star (\mathrm{id} \times \tau)$, so that

$$\mathcal{P} \left(p^\alpha \right) = (\alpha + 1) p^\alpha - \alpha p^{\alpha - 1}.$$

4.3.7   *Using the Functional Equation*

In [31], the authors have another approach to this problem. They consider a complex-valued multiplicative function f verifying the Ramanujan condition $f(n) \ll n^\varepsilon$ and whose Dirichlet series $F(s)$ satisfies the following properties.

▷ $F(s)$ is holomorphic in the half-plane $\sigma > 1$, and has analytic continuation to the whole complex plane where it is holomorphic except possibly for a pole at $s = 1$, not necessarily simple.
▷ $F(s)$ satisfies the functional equation in the half-plane $\sigma > 1$

$$F(1 - s) = w\gamma(s)G(s),$$

where w is the so-called *root-number* such that $|w| = 1$, $G(s) = \sum_{n=1}^{\infty} g(n)n^{-s}$ with $g(n) \ll n^{\varepsilon}$ and $\gamma(s)$ is holomorphic in the half-plane $\sigma > \frac{1}{2}$ having the shape

$$\gamma(s) = \left(\pi^{-m}D\right)^{s-1/2} \prod_{j=1}^{m} \Gamma\left(\frac{s+\kappa_j}{2}\right) \Gamma\left(\frac{1-s+\kappa_j}{2}\right)^{-1},$$

where $D \geqslant 1$ is an integer called the *conductor* of $F(s)$, $m \in \mathbb{N}$ is the *degree* of $F(s)$, $\kappa_1, \ldots, \kappa_m$ are the *spectral parameters* of $F(s)$, i.e. complex numbers such that $\operatorname{Re} \kappa_j \geqslant 0$.

We may now state the main result of [31].

Theorem 4.16 (Friedlander–Iwaniec) *With the above hypotheses and for all $\varepsilon > 0$ and $x \geqslant \sqrt{D}$, we have*

$$\sum_{n \leqslant x} f(n) = \operatorname*{Res}_{s=1}\left(F(s)x^s s^{-1}\right) + O_{\varepsilon,\kappa_j}\left(D^{\frac{1}{m+1}} x^{\frac{m-1}{m+1}+\varepsilon}\right),$$

where the implied constant depends only on ε and the spectral parameters $\kappa_1, \ldots, \kappa_m$.

The strength of this result lies in the fact that the error term depends only on ε and the spectral parameters $\kappa_1, \ldots, \kappa_m$. We may apply it to the multiplicative function $r_{\mathbb{K}}$ studied in Chap. 7, where $\mathbb{K}$ is an algebraic number field of degree $d \geqslant 1$. It is shown in Corollary 7.8 that $r_{\mathbb{K}}(n) \leqslant \tau_d(n) \ll n^{\varepsilon}$. Furthermore, the Dirichlet series of $v_{\mathbb{K}}$ is the Dedekind zeta-function $\zeta_{\mathbb{K}}(s)$,[13] and using Theorem 7.14 and the duplication formula (3.11), we infer that $\zeta_{\mathbb{K}}(s)$ satisfies the following functional equation

$$\zeta_{\mathbb{K}}(1-s) = \left(\pi^{-d}|d_{\mathbb{K}}|\right)^{s-1/2} \left(\Gamma\left(\tfrac{1}{2}s\right)\Gamma\left(\tfrac{1}{2}(1-s)\right)^{-1}\right)^{r_1+r_2}$$

$$\times \left(\Gamma\left(\tfrac{1}{2}(s+1)\right)\Gamma\left(\tfrac{1}{2}(2-s)\right)^{-1}\right)^{r_2} \zeta_{\mathbb{K}}(s), \qquad (4.35)$$

---

[13] See Definition 7.21.

where (r_1, r_2) is the signature and $d_{\mathbb{K}}$ is the discriminant of $\mathbb{K}$. Hence the function $v_{\mathbb{K}}$ satisfies the above conditions with $w = 1$, $G(s) = \zeta_{\mathbb{K}}(s)$, $D = |d_{\mathbb{K}}|$, $m = d$ and

$$
\kappa_j = \begin{cases} 0, & \text{if } 1 \leqslant j \leqslant r_1 + r_2 \\ 1, & \text{if } 1 \leqslant j \leqslant r_2. \end{cases}
$$

As an example, we derive a weaker version of the Ideal Theorem.[14]

Corollary 4.8 (Ideal Theorem)   *Let $\mathbb{K}$ be an algebraic number field of degree d and discriminant $d_{\mathbb{K}}$. For all $x \geqslant |d_{\mathbb{K}}|^{1/2}$ and all $\varepsilon > 0$, we have*

$$
\sum_{n \leqslant x} r_{\mathbb{K}}(n) = \kappa_{\mathbb{K}} x + O_\varepsilon \left(|d_{\mathbb{K}}|^{\frac{1}{d+1}} x^{1 - \frac{2}{d+1} + \varepsilon} \right),
$$

where $\kappa_{\mathbb{K}}$ is given by (7.19) and the implied constant depends only on ε.

This result was first proved by Landau in [68, Satz 210], but with an unspecified constant in the error term. Both proofs start using the same tools, namely Theorem 3.25 and contour integration methods, but Friedlander and Iwaniec next used the stationary phase[15] to estimate some integrals more precisely. For a more precise version of the Ideal Theorem, see Theorem 7.28.

4.3.8   Lower Bounds

Lower bounds for sums $\sum_{n \leqslant x} f(n)$ are not surprisingly harder to obtain than upper bounds. The next result provides a lower bound for the logarithmic mean value on squarefree numbers.

Proposition 4.17   *Let f be a positive multiplicative function satisfying $f(p) \leqslant \lambda < 1$ for all primes p. Then*

$$
\sum_{n \leqslant x} \frac{\mu^2(n) f(n)}{n} \geqslant (1 - \lambda) \prod_{p \leqslant x} \left(1 + \frac{f(p)}{p} \right).
$$

Note that we have

$$
\sum_{n \leqslant x} \frac{\mu^2(n) f(n)}{n} \leqslant \sum_{P^+(n) \leqslant x} \frac{\mu^2(n) f(n)}{n} = \prod_{p \leqslant x} \left(1 + \frac{f(p)}{p} \right)
$$

---

[14]See Theorem 7.28 for more precise estimates.

[15]See Lemma 6.10.

so that the result above is of the right order of magnitude.

Proof We have

$$0 \leqslant \prod_{p \leqslant x} \left(1 + \frac{f(p)}{p} \right) - \sum_{n \leqslant x} \frac{\mu^2(n) f(n)}{n} = \sum_{\substack{n > x \\ P^+(n) \leqslant x}} \frac{\mu^2(n) f(n)}{n}$$

and using $\log n > \log x$ in the right-hand side implies that

$$\prod_{p \leqslant x} \left(1 + \frac{f(p)}{p} \right) - \sum_{n \leqslant x} \frac{\mu^2(n) f(n)}{n}$$

$$< \sum_{\substack{n > x \\ P^+(n) \leqslant x}} \frac{\mu^2(n) f(n)}{n} \frac{\log n}{\log x} = \frac{1}{\log x} \sum_{\substack{n > x \\ P^+(n) \leqslant x}} \frac{\mu^2(n) f(n)}{n} \sum_{p \mid n} \log p$$

$$\leqslant \frac{1}{\log x} \sum_{p \leqslant x} \frac{f(p) \log p}{p} \sum_{\substack{k \geqslant x/p \\ P^+(k) \leqslant x}} \frac{\mu^2(k) f(k)}{k}$$

$$\leqslant \frac{\lambda}{\log x} \sum_{p \leqslant x} \frac{\log p}{p} \sum_{P^+(k) \leqslant x} \frac{\mu^2(k) f(k)}{k}$$

$$= \frac{\lambda}{\log x} \sum_{p \leqslant x} \frac{\log p}{p} \prod_{p \leqslant x} \left(1 + \frac{f(p)}{p} \right)$$

and Corollary 3.13 in the form

$$\sum_{p \leqslant x} \frac{\log p}{p} < \log x$$

completes the proof.   □

Let us state without proof the next result, which is a consequence of a theorem due to Barban.[16]

Proposition 4.18 *Let f be a positive multiplicative function such that $f(p) \geqslant \lambda_1 > 0$ for all primes $p \geqslant p_0$ and $0 \leqslant f(p^\alpha) \leqslant \lambda_2$ for all prime powers p^α, where $\lambda_1 > 0$ and $\lambda_2 > 0$. Then we have*

$$\sum_{n \leqslant x} f(n) \gg x \exp \left(\sum_{p \leqslant x} \frac{f(p) - 1}{p} \right).$$

---

[16]See [95].

Lemma 4.13 below provides another lower bound using elementary means of a certain class of multiplicative functions.

4.3.9 Short Sums

4.3.9.1 Upper Bounds

By *short sums*, we mean sums having the shape

$$\sum_{x < n \leqslant x+y} f(n),$$

where $x \geqslant 1$, $y > 0$ are real numbers such that $y = o(x)$ as $x \to \infty$. Compared to the case of long sums, there are fewer results for such sums in the literature. In particular, there is no known asymptotic formula for the case of positive multiplicative functions satisfying the Wirsing conditions (4.24). The first very important theorem is due to Shiu [98] and provides only an upper bound.

Theorem 4.17 (Shiu) *Let f be a positive multiplicative function, $\delta > 0$, $0 < \varepsilon, \theta < \frac{1}{2}$ and $0 < a < q$ be two positive coprime integers. Assume that*

▷ *there exists $\lambda_1 = \lambda_1(\delta) > 0$ such that $f(n) \leqslant \lambda_1 n^\delta$ for all $n \geqslant 1$*
▷ *there exists $\lambda_2 > 0$ such that $f(p^\alpha) \leqslant \lambda_2^\alpha$ for all prime powers p^α.*

Then, for $x \geqslant 1$ sufficiently large and uniformly in a, q and y such that $q < y^{1-\theta}$ and $x^\varepsilon \leqslant y \leqslant x$, we have

$$\sum_{\substack{x < n \leqslant x+y \\ n \equiv a \,(\mathrm{mod}\, q)}} f(n) \ll \frac{y}{\varphi(q) \log x} \exp\left(\sum_{\substack{p \leqslant x \\ p \nmid q}} \frac{f(p)}{p} \right).$$

The implied constant depends on λ_1, λ_2 and ε.

This result has turned out to be very useful. One may easily check that all the functions of Lemma 4.6 satisfy the conditions of Shiu's theorem. In order to apply this result to them, the following lemma will be useful.

Lemma 4.11 *Let $q \geqslant 2$ be an integer. For all $x \geqslant \max(e, q)$, we have*

$$\sum_{\substack{p \leqslant x \\ p \nmid q}} \frac{1}{p} < \log\left(2e^{\gamma}\frac{\varphi(q)}{q}\log x\right).$$

Proof On the one hand, we have

$$\prod_{\substack{p \leqslant x \\ p \nmid q}}\left(1 - \frac{1}{p}\right)^{-1} = \exp\left(\sum_{\substack{p \leqslant x \\ p \nmid q}}\frac{1}{p} + \sum_{\substack{p \leqslant x \\ p \nmid q}}\sum_{\alpha=2}^{\infty}\frac{1}{\alpha p^{\alpha}}\right) \geqslant \exp\left(\sum_{\substack{p \leqslant x \\ p \nmid q}}\frac{1}{p}\right)$$

and on the other hand, since $q \leqslant x$, we also have

$$\prod_{\substack{p \leqslant x \\ p \nmid q}}\left(1 - \frac{1}{p}\right)^{-1} = \prod_{p \leqslant x}\left(1 - \frac{1}{p}\right)^{-1}\prod_{\substack{p \leqslant x \\ p \mid q}}\left(1 - \frac{1}{p}\right) = \frac{\varphi(q)}{q}\prod_{p \leqslant x}\left(1 - \frac{1}{p}\right)^{-1}$$

and we use Corollary 3.13 to conclude the proof.   □

With this lemma at our disposal, Shiu's theorem implies at once the following estimates.

Corollary 4.9 *Let $x^{\varepsilon} \leqslant y \leqslant x$, $(a, q) = 1$ and $q < y^{1-\theta}$ as in* Theorem 4.17. *Then we have for x sufficiently large the following estimates.*

▷ *Let $k \geqslant 2$. We have*

$$\sum_{\substack{x < n \leqslant x+y \\ n \equiv a \,(\mathrm{mod}\, q)}} s_k(n) \ll \frac{y}{\varphi(q)\log x}.$$

▷ *The estimate*

$$\sum_{\substack{x < n \leqslant x+y \\ n \equiv a \,(\mathrm{mod}\, q)}} f(n) \ll \frac{y}{q}$$

holds with $f = \beta$, $f = a$, $f = \tau^{(e)}$ and $f = \mu_k$ ($k \geqslant 2$).
▷ *Let $k \geqslant 2$. We have*

$$\sum_{\substack{x < n \leqslant x+y \\ n \equiv a \,(\mathrm{mod}\, q)}} \tau_{(k)}(n) \ll \frac{\varphi(q)}{q^2}\, y \log x.$$

▷ *Let* $k \geqslant 1$. *If* $f = k^{\omega}$ *and* $f = \tau_k$, *then*

$$\sum_{\substack{x < n \leqslant x+y \\ n \equiv a \,(\mathrm{mod}\,q)}} f(n) \ll \frac{y}{q} \left(\frac{\varphi(q)}{q} \log x \right)^{k-1} .$$

Once again, only the first bound is overestimated. We shall see in Chap. 5 how to get some improvements on this problem.

The idea of the proof of Shiu's theorem is a combination of sieve methods,[17] estimates of particular parts of sums as in Theorem 4.9 and a method used earlier by Wolke, whose ideas go back to Erdős, starting with a decomposition of each integer $n \in (x, x + y]$ as follows:

$$n = ab = \left(p_1^{\alpha_1} \cdots p_r^{\alpha_r} \right) \left(p_{r+1}^{\alpha_{r+1}} \cdots p_s^{\alpha_s} \right)$$

with $a \leqslant y^{\theta/10} < ap_{r+1}^{\alpha_{r+1}}$. The original sum is then split into four classes according to

▷ $P^-(b) > y^{\theta/20}$
▷ $P^-(b) \leqslant y^{\theta/20}$ and $a \leqslant y^{\theta/20}$
▷ $P^-(b) \leqslant \log x \log \log x$ and $a > y^{\theta/20}$
▷ $\log x \log \log x < P^-(b) \leqslant y^{\theta/20}$ and $a > y^{\theta/20}$.

Shiu's theorem was later generalized in two directions. First, Nair proved an analogous estimate for sums of the shape

$$\sum_{x < n \leqslant x+y} f\left(|P(n)| \right) ,$$

where $P \in \mathbb{Z}[X]$ is an integer polynomial having non-zero discriminant and no fixed prime divisors, and f is a multiplicative function satisfying the same conditions as in Shiu's theorem. Then, Nair and Tenenbaum [82] weakened the property of multiplicativity by showing the next result.

Theorem 4.18 (Nair–Tenenbaum) *Let* f *be a positive arithmetic function,* $0 < \delta \leqslant 1$, $d \in \mathbb{N}$, $0 < \varepsilon < \frac{1}{8d^2}$, $c_0 > 0$ *and* $P = a_d X^d + \cdots + a_0 \in \mathbb{Z}[X]$ *be irreducible. Let* $\rho(n)$ *be the number of solutions of the congruence* $P(x) \equiv 0 \,(\mathrm{mod}\,n)$. *Assume that* $\rho(p) < p$ *for all primes* p *and that there exist* $\lambda_1, \lambda_2 \geqslant 1$ *such that, for all positive coprime integers* m, n, *we have*

$$f(mn) \leqslant \min \left(\lambda_1 m^{\varepsilon \delta/3} , \lambda_2^{\Omega(m)} \right) f(n).$$

(continued)

---

[17]See the Selberg sieve below.

Theorem 4.18 (continued)

Then, for $x \geqslant c_0 \max \{|a_0|, \ldots, |a_d|\}^\delta$ and uniformly in $x^{4d^2\varepsilon} \leqslant y \leqslant x$, we have

$$\sum_{x < n \leqslant x+y} f\left(|P(n)|\right) \ll y \prod_{p \leqslant x} \left(1 - \frac{\rho(p)}{p}\right) \sum_{n \leqslant x} \frac{\rho(n) f(n)}{n}.$$

In particular, if $0 < \varepsilon < \frac{1}{2}$ and $x \geqslant c_0$, we have uniformly in $x^\varepsilon \leqslant y \leqslant x$

$$\sum_{x < n \leqslant x+y} f(n) \ll y \prod_{p \leqslant x} \left(1 - \frac{1}{p}\right) \sum_{n \leqslant x} \frac{f(n)}{n}.$$

This result enables us to treat short sums of non-necessarily multiplicative functions, such as the Hooley divisor Δ-function. See Theorem 4.35.

4.3.9.2 Asymptotic Formulæ

The problem of getting asymptotic formulæ for the usual arithmetic functions in short intervals is far more difficult. Essentially, we can have very different results according to the size of the arithmetic function with respect to the divisor function τ_k. More precisely, when $f(p)$ is close to 1, the next result has been proved in [8].

Theorem 4.19 *Let $k \geqslant 2$ be an integer, $A > 0$, $\varepsilon > 0$, and consider the class $\mathcal{S}(k; A, \varepsilon)$ of complex-valued multiplicative functions f satisfying*

$$\left| f\left(p^\alpha\right) - f\left(p^{\alpha-1}\right) \right| \leqslant A p^{-\alpha} \quad \text{for } 1 \leqslant \alpha < k\,;$$

$$|(f \star \mu)(n)| \leqslant c(\varepsilon, k)\, n^\varepsilon \quad \text{if } n \text{ is } k\text{-full}\,;$$

for some $c(\varepsilon, k) > 0$ depending only on ε and k. For $f \in \mathcal{S}(k; A, \varepsilon)$, define

$$\mathcal{M}_f := \prod_p \left(1 - \frac{1}{p}\right) \left(1 + \sum_{\alpha=1}^{\infty} \frac{f\left(p^\alpha\right)}{p^\alpha}\right).$$

(continued)

> **Theorem 4.19** (continued)
> *Then, uniformly for $4^k < y \leqslant x$*
>
> $$\sum_{x < n \leqslant x+y} f(n) = y M_f + O\left(x^{\frac{1}{2k+1}+\varepsilon} + yx^{-\frac{1}{6(4k-1)(2k-1)}+\varepsilon} + y^{1-\frac{2(k-1)}{k(3k-1)}} x^{\varepsilon}\right).$$

The proof uses in a crucial way Theorem 5.11 which will be seen in Chap. 5. This result may be applied to the functions μ_k, a, $\tau_{(k)}$, $\tau^{(e)}$, $\mu^{(e)}$ and so on.

For other arithmetic functions, the method using Perron summation formula may be quite well adapted. Using the so-called Huxley-Hooley contour, several authors [57] proved the next result, for which we need the following extension of the definition of $N(\sigma, T)$ seen in (3.53) and (3.54). Let χ be a Dirichlet character modulo $q \geqslant 1$, $\sigma \in [0, 1]$, $T \geqslant 3$ be real numbers, and set

$$N(\sigma, T, \chi) = |\{\rho = \beta + i\gamma : L(\rho, \chi) = 0, \ \beta \geqslant \sigma \text{ and } |\gamma| \leqslant T\}|.$$

Now define $B > 0$ and $D > 0$ such that

$$N(\sigma, T, \chi) \ll T^{B(1-\sigma)} (\log T)^D. \tag{4.36}$$

Theorem 4.20 *Let f a complex-valued arithmetic function such that*

$$L(s, f) = P_1(s) P_2(s) P_3(s) G(s)$$

and satisfying the following conditions.

▷ $G(s) := \sum_{n=1}^{\infty} g(n) n^{-s}$ *such that $g(n) \ll_{\varepsilon} n^{\varepsilon}$ and*

$$\sum_{n=1}^{\infty} \frac{|g(n)|}{n^{\sigma}} \ll 1 \quad \left(\sigma > \tfrac{1}{2}\right).$$

▷ *The functions P_i are finite products of powers of functions of the shape*

◇ $i = 1 : L(s, \chi)$, *the exponents being complex numbers or not ;*
◇ $i = 2 : L'(s, \chi)$, *the exponents being non-negative integers ;*
◇ $i = 3 : \log L(s, \chi)$, *the exponents being non-negative integers.*[18]

---

[18] $\log L(s, \chi)$ is defined by analytic continuation in any simply connected domain containing the half-plane $\sigma > 1$, and containing no zero or pole of $L(s, \chi)$.

Let $0 < \varepsilon < \frac{1}{12}$ and $0 < r < \frac{1}{8} - \frac{3}{2}\varepsilon$, and assume $L(s, f)$ can be written as

$$L(s, f) = U(s)(s - 1)^{-z},$$

where $z \in \mathbb{C}$ and U is analytic in $|s - 1| \leqslant r$ in which

$$U(s) = A_0 + A_1(s - 1) + \cdots + A_N(s - 1)^N + O\left(|s - 1|^{N+1}\right)$$

with $N \in \mathbb{Z}_{\geqslant 0}$. Then, for $x^{1-1/B+\varepsilon} \leqslant y \leqslant x^{2(1-r)/3}$ and $\operatorname{Re} z \leqslant N + 1$

$$\sum_{x < n \leqslant x+y} f(n) = y \sum_{\ell=0}^{N} \frac{(-1)^{\ell+1} A_\ell \sin \pi z}{\pi} \frac{\Gamma(\ell - z)}{(\log x)^{\ell - z + 1}} + O\left(\frac{y}{(\log x)^{N - \operatorname{Re} z + 2}}\right),$$

where B is given in (4.36).

Note that Huxley's value $B = \frac{12}{5}$ proved for the Riemann zeta function is also valid for any Dirichlet L-function.

As an example, Katai and Subbarao [57] showed, among other things, the following estimate. Assume $x^{7/12+\varepsilon} \leqslant y \leqslant x^{2(1-r)/3}$ where $r > 0$ is as in the above theorem. Then, as $x \to \infty$

$$\sum_{x < n\tau(n) \leqslant x+y} 1 = (1 + o_x(1)) \frac{y}{\sqrt{\log x}}.$$

4.3.10 Sub-multiplicative Functions

Halberstam and Richert [39] proved the next result, which is a generalization of Theorem 4.9 to sub-multiplicative functions and provides an improvement of the constants appearing in this result.

Theorem 4.21 *Let f be a non-negative sub-multiplicative function such that $f(1) = 1$ and satisfying the Wirsing conditions (4.24). Assume that, for all $x \geqslant 2$, we have*

$$\sum_{p \leqslant x} f(p) \log p \leqslant \kappa x + O\left(\frac{x}{(\log x)^2}\right)$$

(continued)

Theorem 4.21 (continued)
for some constant $\kappa > 0$. Then

$$\sum_{n \leqslant x} f(n) \leqslant \frac{\kappa x}{\log x}\left(1 + O\left(\frac{1}{\log x}\right)\right)\sum_{n \leqslant x} \frac{f(n)}{n}.$$

The sum of the right-hand side may be estimated by Theorem 4.15 or Corollary 4.7, or by using the usual inequality

$$\sum_{n \leqslant x} \frac{f(n)}{n} \leqslant \prod_{p \leqslant x}\left(1 + \sum_{\alpha=1}^{\infty} \frac{f(p^\alpha)}{p^\alpha}\right).$$

For instance, applying Theorem 4.21 to the function f_q defined by

$$f_q(n) = \begin{cases} 1, & \text{if } (n, q) = 1 \\ 0, & \text{if } (n, q) > 1 \end{cases},$$

where q is a fixed positive integer such that $P^+(q) \leqslant x$, we get using Corollary 3.5

$$\sum_{\substack{n \leqslant x \\ (n,q)=1}} 1 \leqslant e^\gamma \frac{\varphi(q)}{q} x\left(1 + O\left(\frac{1}{\log x}\right)\right).$$

4.3.11   Additive Functions

The average orders of additive functions are in general easier to estimate than those of multiplicative functions.

Proposition 4.19   *Let f be an additive function and $x \geqslant 2$ be a real number. Then*

$$\sum_{n \leqslant x} f(n) = x \sum_{p \leqslant x} \frac{f(p)}{p}$$

$$+ x \sum_{\substack{p^\alpha \leqslant x \\ \alpha \geqslant 2}} \frac{f(p^\alpha) - f(p^{\alpha-1})}{p^\alpha} + O\left(\sum_{\substack{p^\alpha \leqslant x \\ \alpha \geqslant 1}} \left|f(p^\alpha) - f(p^{\alpha-1})\right|\right).$$

Proof Let $g = f \star \mu$. By Exercise 59, g is supported on prime powers. Furthermore, since f is additive, we have $g(p) = f(p)$. Hence, by Proposition 4.3, we have

$$\sum_{n \leqslant x} f(n) = \sum_{n \leqslant x} (g \star 1)(n) = \sum_{d \leqslant x} g(d) \left\lfloor \frac{x}{d} \right\rfloor$$

$$= \sum_{p \leqslant x} g(p) \left\lfloor \frac{x}{p} \right\rfloor + \sum_{\substack{p^\alpha \leqslant x \\ \alpha \geqslant 2}} g\left(p^\alpha\right) \left\lfloor \frac{x}{p^\alpha} \right\rfloor$$

$$= x \sum_{p \leqslant x} \frac{f(p)}{p} + x \sum_{\substack{p^\alpha \leqslant x \\ \alpha \geqslant 2}} \frac{g\left(p^\alpha\right)}{p^\alpha} + O\left(\sum_{\substack{p^\alpha \leqslant x \\ \alpha \geqslant 1}} \left| g\left(p^\alpha\right) \right| \right)$$

and we conclude by using $g\left(p^\alpha\right) = f\left(p^\alpha\right) - f\left(p^{\alpha-1}\right)$. $\square$

For instance, with $f = \omega$, using a weak form of the PNT, we get

$$\sum_{n \leqslant x} \omega(n) = x \log \log x + Bx + O\left(\frac{x}{\log x} \right),$$

where $B \approx 0.261497 \ldots$ is the Mertens constant from Corollary 3.4.

As for the functions ω and Ω, some more precise estimates appeared in the literature. In [93], Saffari used the Dirichlet hyperbola principle to show the next estimate.

Theorem 4.22 *For any $N \in \mathbb{Z}_{\geqslant 1}$*

$$\sum_{n \leqslant x} \omega(n) = x \log \log x + Bx + x \sum_{k=1}^{N} \frac{a_k}{(\log x)^k} + O\left(\frac{x}{(\log x)^{N+1}} \right),$$

where

$$a_k := -\int_1^\infty \frac{\{t\}}{t^2} (\log t)^{k-1} \, \mathrm{d}t \quad \left(k \in \mathbb{Z}_{\geqslant 1} \right).$$

Note that $a_1 = \gamma - 1$ and $a_2 = \gamma + \gamma_1 - 1$ where $\gamma_1 \approx -0.072816 \ldots$ is the first Riemann–Stieltjes constant. Furthermore, for any $s \in \mathbb{C}$ satisfying $\sigma > 1$

$$L(s, \omega) = \zeta(s) \sum_p \frac{1}{p^s} = \zeta(s) \left(\log \zeta(s) - G(s) \right),$$

where $G(s) = \sum_p \sum_{\alpha=2}^{\infty} \frac{1}{\alpha p^{\alpha s}}$ is a Dirichlet series absolutely convergent in the half-plane $\sigma > \frac{1}{2}$. Also

$$L(s, \Omega) = \zeta(s) \sum_p \frac{1}{p^s - 1}.$$

More recently, Hassani [43] investigated the average order of the functions Ω_ℓ defined by

$$\Omega_\ell(n) = \sum_{p^\alpha \| n} \alpha^\ell \quad (\ell \in \mathbb{Z}_{\geqslant 0}).$$

Note that $\Omega_0 = \omega$ and $\Omega_1 = \Omega$. By studying the difference

$$\sum_{n \leqslant x} (\Omega_\ell(n) - \omega(n))$$

and applying Theorem 4.22, the author proved the following result.

Theorem 4.23   *For any $N \in \mathbb{Z}_{\geqslant 1}$*

$$\sum_{n \leqslant x} \Omega_\ell(n) = x \log \log x + (B + B_\ell)x + x \sum_{k=1}^{N} \frac{a_k}{(\log x)^k} + O\left(\frac{x}{(\log x)^{N+1}}\right),$$

where $B_\ell := \sum_p \sum_{\alpha=2}^{\infty} \frac{\alpha^\ell - (\alpha-1)^\ell}{p^\alpha}$ and the a_ks are given in Theorem 4.22.

As for an upper bound for ω, the best result is due to Robin [92].

Proposition 4.20   *For any $n \in \mathbb{Z}_{\geqslant 3}$*

$$\omega(n) \leqslant \frac{1.38402 \log n}{\log \log n}.$$

4.3.12   Refinements

There are several asymptotic formulæ in the literature, as one may see in [69, 107, 111, 117], for instance. On the other hand, the Selberg–Delange method seen in Sect. 4.3.5 deals with complex-valued multiplicative functions satisfying the following hypotheses.

▷ $|f|$ does not grow too fast ;
▷ There exist constants $\alpha \in \mathbb{C}$ and $c > 0$ such that $(s-1)^{\alpha} L(s,f)$ is analytic in a region of the shape $\sigma > 1 - \frac{c}{\log(|t|+2)}$.

In practice, the second condition is not always easily verified. With this in mind, Granville and Koukoulopoulos [36] proved a similar result with a more flexible condition.

Theorem 4.24 *Let f be a complex-valued multiplicative function such that there exist constants $\kappa \in \mathbb{C}$ and $A, k > 0$ such that $|f| \leqslant \tau_k$ and, for any $x \geqslant 2$*

$$\sum_{p \leqslant x} f(p) \log p = \kappa x + O\left(x(\log x)^{-A}\right).$$

Set $N = \lfloor A \rfloor$ if $A \notin \mathbb{Z}$ and $N = A - 1$ otherwise. Then

$$\sum_{n \leqslant x} f(n) = \frac{x(\log x)^{\kappa-1}}{\Gamma(\kappa)} \prod_p \left(1 - \frac{1}{p}\right)^{\kappa} \left(1 + \sum_{\alpha=1}^{\infty} \frac{f(p^{\alpha})}{p^{\alpha}}\right)$$

$$+ x \sum_{j=1}^{N} \frac{a_j(\log x)^{\kappa-j-1}}{\Gamma(\kappa-j)} + O\left(x(\log x)^{k-1-A}(\log\log x)^{\delta}\right),$$

where $\delta = 1$ if $N = A - 1$, $\delta = 0$ otherwise. The coefficient $a_1, \dots, a_N$ are the first N Taylor coefficients about 1 of the function $s \mapsto s^{-1}(s-1)^k L(s,f)$ given by

$$a_j = \frac{1}{j!}\left[\frac{d^j}{ds^j}\left(\frac{(s-1)^{\kappa} L(s,f)}{s}\right)\right]_{[s=1]} \quad (1 \leqslant j \leqslant N).$$

Example 4.7 As an example, let us recover an old well-known result of Ramanujan [87], proved by Wilson [123]. Take $f = \tau^{-1}$, hence $\kappa = k = \frac{1}{2}$ and

$$\frac{1}{\Gamma(\kappa)} \prod_p \left(1 - \frac{1}{p}\right)^{\kappa} \left(1 + \sum_{\alpha=1}^{\infty} \frac{f(p^{\alpha})}{p^{\alpha}}\right)$$

$$= \frac{1}{\sqrt{\pi}} \prod_p (p(p-1))^{1/2} \log\left(\frac{p}{p-1}\right) := C \approx 0.54685\dots$$

so that we derive, for any $A \in \mathbb{Z}_{\geqslant 1}$

$$\sum_{n \leqslant x} \frac{1}{\tau(n)} = \frac{Cx}{\sqrt{\log x}} + \frac{x}{\sqrt{\log x}} \sum_{j=1}^{A-1} \frac{a_j (\log x)^{-j}}{\Gamma\left(\frac{1}{2} - j\right)} + O\left(\frac{x \log \log x}{(\log x)^{1/2+A}}\right).$$

4.4 Usual Multiplicative Functions

4.4.1 The Möbius Function

Theorem 4.25 *There exists an absolute constant $c_0 > 0$ such that*

$$\sum_{n \leqslant x} \mu(n) \ll x \delta_{c_0}(x),$$

where $\delta_{c_0}(x) := e^{-c_0 (\log x)^{3/5} (\log \log x)^{-1/5}}$.

Proof Set $c := 57.54^{-1}$, $\kappa := 1 + \frac{1}{\log x}$,

$$T := e^{c(\log x)^{3/5}(\log \log x)^{-1/5}} \quad \text{and} \quad \alpha := \alpha(T) = c(\log T)^{-2/3}(\log \log T)^{-1/3}.$$

We use (3.41) with $B(x) = 1$ yielding

$$\sum_{n \leqslant x} \mu(n) = \frac{1}{2\pi i} \int_{\kappa - iT}^{\kappa + iT} \frac{1}{\zeta(s)} \frac{x^s}{s} \, ds + O\left(\frac{x \log x}{T}\right).$$

By (3.59), $\zeta(s)$ has no zero in the region $\sigma \geqslant 1 - c(\log |t|)^{-2/3}(\log \log |t|)^{-1/3}$ and $|t| \geqslant 3$, so that we may shift the line of integration to the left and apply Cauchy's theorem in the rectangle with vertices $\kappa \pm iT$, $1 - \alpha \pm iT$. In this region

$$\zeta(s)^{-1} \ll (\log(|t| + 2))^{2/3} (\log \log(|t| + 2))^{1/3}.$$

Therefore, the contribution of the horizontal sides does not exceed

$$\ll x T^{-1} (\log T)^{2/3} (\log \log T)^{1/3}$$

and the contribution of the vertical side is bounded by

$$\ll x^{1-\alpha} (\log T)^{5/3} (\log \log T)^{1/3}.$$

Since the path of integration does not surround the origin nor the poles of the integrand, Cauchy's theorem and the choice of T give the asserted estimate for any $c_0 \leqslant \frac{1}{4}c$ and any real number x satisfying $x \geqslant e^{e^{11}}$, say. $\square$

This result may be used to derive estimates on more general situations. As an example, let us show how to handle sums of the form

$$\sum_{n \leqslant x} \mu(n) f(n),$$

where f is a multiplicative function taking small values at prime numbers. We will use Theorem 4.25 under the following weaker form

$$\sum_{n \leqslant x} \mu(n) \ll \frac{x}{(\log e^2 x)^A} \tag{4.37}$$

for all $A > 0$. Note that there exist explicit bounds of this result. For instance, it is shown in [73] that, for all $x > 1$, we have

$$\left| \sum_{n \leqslant x} \mu(n) \right| \leqslant \frac{362.7\,x}{(\log x)^2}$$

and for the logarithmic mean value it is proved [10] that

$$\left| \sum_{n \leqslant x} \frac{\mu(n)}{n} \right| < \frac{546}{(\log x)^2}.$$

When $f = \mu \times \mathrm{id}^{-1}$, we have the following inequalities

$$\frac{\log x}{\zeta(2)} + 0.83211 < \sum_{n \leqslant x} \frac{\mu^2(n)}{n} < \frac{\log x}{\zeta(2)} + 1.165471. \tag{4.38}$$

We intend to show the following result.

Theorem 4.26 *Let f be a real-valued multiplicative function such that $0 \leqslant f(p) \leqslant 1$ for all prime numbers p. Then we have for all $x \geqslant e$*

$$\sum_{n \leqslant x} \mu(n) f(n) \ll \frac{x}{\log x} \exp\left(\sum_{p \leqslant x} \frac{1 - f(p)}{p} \right).$$

Proof Let $g = \mu f \star \mathbf{1}$. By Theorem 4.1, g is multiplicative and, for all prime powers p^α, we have

$$g\left(p^\alpha\right) = 1 + \sum_{j=1}^{\alpha} \mu\left(p^j\right) f\left(p^j\right) = 1 - f(p).$$

Hence by assumption on f we infer that $0 \leqslant g\left(p^\alpha\right) \leqslant 1$ and, using multiplicativity, we get $0 \leqslant g(n) \leqslant 1$. Using the Möbius inversion formula, Proposition 4.3, the bound (4.37) with $A = 2$ and partial summation, we get

$$\sum_{n \leqslant x} \mu(n) f(n) = \sum_{n \leqslant x} (g \star \mu)(n) = \sum_{d \leqslant x} g(d) \sum_{k \leqslant x/d} \mu(k)$$

$$\ll x \sum_{d \leqslant x} \frac{g(d)}{d \log^2(e^2 x/d)}$$

$$\ll \sum_{d \leqslant x} g(d) + x \int_1^x \frac{\log(x/t)}{t^2 (\log e^2 x/t)^3} \left(\sum_{d \leqslant t} g(d)\right) dt$$

$$\ll \sum_{d \leqslant x} g(d) + x \left(\int_1^{\sqrt{x}} + \int_{\sqrt{x}}^x\right) \frac{\log(x/t)}{t^2 (\log e^2 x/t)^3} \left(\sum_{d \leqslant t} g(d)\right) dt.$$

In the first integral, we use the trivial inequality $\sum_{d \leqslant t} g(d) \leqslant t$, while in the second integral and the first sum, we use Theorem 4.9 which gives

$$\sum_{d \leqslant t} g(d) \ll \frac{t}{\log(et)} \exp\left(\sum_{p \leqslant t} \frac{g(p)}{p}\right) \ll \frac{t}{\log(et)} \exp\left(\sum_{p \leqslant x} \frac{1 - f(p)}{p}\right)$$

for any $1 \leqslant t \leqslant x$, and therefore

$$\sum_{n \leqslant x} \mu(n) f(n) \ll \frac{x}{\log x} \exp\left(\sum_{p \leqslant x} \frac{1 - f(p)}{p}\right) + \frac{x}{(\log x)^2}$$

$$\ll \frac{x}{\log x} \exp\left(\sum_{p \leqslant x} \frac{1 - f(p)}{p}\right)$$

as required. $\qquad\square$

4.4.2   Distribution of k-free Numbers

> **Theorem 4.27** *Let* $k \in \mathbb{Z}_{\geqslant 2}$ *be fixed. There exists an absolute constant* $c_1 > 0$ *such that*
>
> $$\sum_{n \leqslant x} \mu_k(n) = \frac{x}{\zeta(k)} + O\left(x^{1/k}\delta_{c_1,k}(x)\right),$$
>
> *where* $\delta_{c_1,k}(x) := e^{-c_1 k^{-8/5}(\log x)^{3/5}(\log\log x)^{-1/5}}$.

Proof Using (4.10) and Theorem 4.4 with $f(n) = \mu(m)$ if $n = m^k$ and 0 otherwise, and $g = h = \mathbf{1}$, we derive for any $5^k \leqslant U \leqslant x$

$$\sum_{n \leqslant x} \mu_k(n) = \sum_{m \leqslant U^{1/k}} \mu(m) \left\lfloor \frac{x}{m^k} \right\rfloor + \sum_{n \leqslant x/U} M\left(\left(\frac{x}{n}\right)^{1/k}\right) - \left\lfloor \frac{x}{U} \right\rfloor M\left(U^{1/k}\right)$$

$$:= S_1 + S_2 - S_3$$

say, where $M(x) := \sum_{n \leqslant x} \mu(n)$ is the Mertens function. Now using partial summation as in Exercise 3 along with Theorem 4.25, we get for all $z \geqslant 3$ and all $\kappa > 1$

$$\sum_{m > z} \frac{\mu(m)}{m^\kappa} = -z^{-\kappa}M(z) + \kappa \int_z^\infty t^{-\kappa-1}M(t)\,\mathrm{d}t$$

$$\ll z^{1-\kappa}\delta_{c_0}(z) + \int_z^\infty t^{-\kappa}\delta_{c_0}(t)\,\mathrm{d}t \ll z^{1-\kappa}\delta_{c_0}(z), \tag{4.39}$$

where $c_0 \in (0, 1)$ is the constant given in Theorem 4.25, so that

$$S_1 = x \sum_{m=1}^\infty \frac{\mu(m)}{m^k} - x \sum_{m > U^{1/k}} \frac{\mu(m)}{m^k} + O\left(U^{1/k}\right)$$

$$= \frac{x}{\zeta(k)} + O\left(xU^{-1+1/k}\delta_{c_0}\left(U^{1/k}\right) + U^{1/k}\right)$$

and

$$S_2 \ll x^{1/k} \sum_{n \leqslant x/U} \frac{1}{n^{1/k}} \delta_{c_0} \left(\left(\frac{x}{n} \right)^{1/k} \right)$$

$$\ll x^{1/k} \delta_{c_0} \left(U^{1/k} \right) \sum_{n \leqslant x/U} \frac{1}{n^{1/k}} \ll x U^{-1+1/k} \delta_{c_0} \left(U^{1/k} \right)$$

and

$$S_3 \ll x U^{-1+1/k} \delta_{c_0} \left(U^{1/k} \right).$$

Therefore

$$\sum_{n \leqslant x} \mu_k(n) = \frac{x}{\zeta(k)} + O \left(x U^{-1+1/k} \delta_{c_0} \left(U^{1/k} \right) + U^{1/k} \right).$$

Choose $U = x e^{-\frac{1}{4} c_0 k^{-3/5} (\log x)^{3/5} (\log \log x)^{-1/5}}$ and assume $x \geqslant e^{e^{5/2}}$. Then

$$\log U = \left(1 - \frac{c_0}{4 k^{3/5} (\log x)^{2/5} (\log \log x)^{1/5}} \right) \log x \geqslant \left(1 - \frac{c_0}{4 e k^{3/5}} \right) \log x \geqslant \tfrac{1}{2} \log x$$

and $\log \log U^{1/k} \leqslant \log \log x$, so that

$$x U^{-1+1/k} \delta_{c_0} \left(U^{1/k} \right)$$

$$= x^{1/k} e^{\frac{1}{4} c_0 k^{-3/5} \left(1 - \frac{1}{k} \right) (\log x)^{3/5} (\log \log x)^{-1/5} - c_0 k^{-3/5} (\log U)^{3/5} (\log \log U^{1/k})^{-1/5}}$$

$$\leqslant x^{1/k} e^{-c_0 k^{-3/5} (\log x)^{3/5} (\log \log x)^{-1/5} \left\{ 2^{-3/5} - \frac{1}{4} \left(1 - \frac{1}{k} \right) \right\}}$$

$$\leqslant x^{1/k} e^{-\frac{2}{5} c_0 k^{-3/5} (\log x)^{3/5} (\log \log x)^{-1/5}}$$

and hence

$$x U^{-1+1/k} \delta_{c_0} \left(U^{1/k} \right) + T^{1/k} \ll x^{1/k} e^{-\frac{1}{4} c_0 k^{-8/5} (\log x)^{3/5} (\log \log x)^{-1/5}}$$

as required.   □

When $k = 2$, explicit versions of this result do exist in the literature. Let us state two of them in the following proposition. For a proof, see [80, 19].

Proposition 4.21

▷ *For $x \geqslant 8$*

$$\sum_{n \leqslant x} \mu(n)^2 = \frac{x}{\zeta(2)} + O^\star\left(0.5\sqrt{x}\right).$$

▷ *For $x \geqslant 1664$*

$$\sum_{n \leqslant x} \mu(n)^2 = \frac{x}{\zeta(2)} + O^\star\left(0.1333\sqrt{x}\right).$$

4.4.3   The Number of Divisors

Theorem 4.28 (Dirichlet) *For all $x \geqslant 1$, we have*

$$\sum_{n \leqslant x} \tau(n) = x(\log x + 2\gamma - 1) + O\left(\sqrt{x}\right).$$

Proof Using Theorem 4.4 with $f = g = h = 1$ and $U = \sqrt{x}$, we derive

$$\sum_{n \leqslant x} \tau(n) = 2\sum_{m \leqslant \sqrt{x}} \left\lfloor \frac{x}{m} \right\rfloor - \left\lfloor \sqrt{x} \right\rfloor^2 = 2x \sum_{m \leqslant \sqrt{x}} \frac{1}{m} + O\left(\sqrt{x}\right) - \left(\sqrt{x} + O(1)\right)^2$$

$$= 2x\left(\log\sqrt{x} + \gamma + O\left(x^{-1/2}\right)\right) - x + O\left(\sqrt{x}\right)$$

$$= x\log x + x(2\gamma - 1) + O\left(\sqrt{x}\right)$$

as asserted.   □

For an improvement of this result, see Theorem 6.6. For the best result to date in this problem, see Corollary 6.12. For the logarithmic mean value of the function τ, the same method leads to the next estimate.

Theorem 4.29 *For all $x \geqslant e$, we have*

$$\sum_{n \leqslant x} \frac{\tau(n)}{n} = \tfrac{1}{2}(\log x)^2 + 2\gamma \log x + \gamma^2 - 2\gamma_1 + O\left(x^{-1/2}(\log x)^2\right),$$

where $\gamma_1 \approx -0.0728$ is the first Riemann–Stieltjes constant.

Proof Using Theorem 4.4 with $f = g = \mathbf{1} \times \mathrm{id}^{-1}$, $h = \mathbf{1}$ and $U = \sqrt{x}$, we get

$$\sum_{n \leqslant x} \frac{\tau(n)}{n} = 2 \sum_{n \leqslant \sqrt{x}} \frac{1}{n} \sum_{m \leqslant x/n} \frac{1}{m} - \left(\sum_{n \leqslant \sqrt{x}} \frac{1}{n} \right)^2$$

and using the estimate of Example 1.1 yields

$$\sum_{n \leqslant x} \frac{\tau(n)}{n} = 2 \sum_{n \leqslant \sqrt{x}} \frac{1}{n} \left\{ \log \frac{x}{n} + \gamma + O\left(\frac{n}{x} \right) \right\} - \left(\frac{\log x}{2} + \gamma + O\left(x^{-1/2} \right) \right)^2$$

$$= 2 \sum_{n \leqslant \sqrt{x}} \frac{1}{n} \log \frac{x}{n} + 2\gamma \left(\frac{\log x}{2} + \gamma + O\left(x^{-1/2} \right) \right)$$

$$- \frac{\log^2 x}{4} - \gamma \log x - \gamma^2 + O\left(x^{-1/2} \log x \right)$$

$$= 2 \sum_{n \leqslant \sqrt{x}} \frac{1}{n} \log \frac{x}{n} - \frac{\log^2 x}{4} + \gamma^2 + O\left(x^{-1/2} \log x \right)$$

with

$$\sum_{n \leqslant \sqrt{x}} \frac{1}{n} \log \frac{x}{n} = \log x \sum_{n \leqslant \sqrt{x}} \frac{1}{n} - \sum_{n \leqslant \sqrt{x}} \frac{\log n}{n}$$

$$= \log x \left(\frac{\log x}{2} + \gamma + O\left(x^{-1/2} \right) \right) - \frac{(\log \sqrt{x})^2}{2} - \gamma_1$$

$$+ O\left(x^{-1/2} (\log x)^2 \right)$$

$$= \tfrac{3}{8} \log^2 x + \gamma \log x - \gamma_1 + O\left(x^{-1/2} (\log x)^2 \right).$$

Inserting this estimate above yields the desired result.   □

As for a pointwise upper bound of $\tau(n)$ and $\tau_3(n)$, Robin [91, 83] derived the following very precise estimate with the help of Ramanujan's highly composite numbers, providing an explicit form of Exercise 54. The second bound below improves Exercise 55.

Theorem 4.30 (Robin) *For all* $n \in \mathbb{Z}_{\geqslant 3}$

$$\tau(n) \leqslant n^{\frac{1.5379 \log 2}{\log \log n}} \quad and \quad \tau_3(n) \leqslant n^{\frac{1.59141 \log 3}{\log \log n}}.$$

4.4.4   Totients

4.4.4.1   Euler's and Jordan's Totients

Theorem 4.31 *For all* $x \in \mathbb{R}_{\geqslant e}$ *and* $\alpha > 0,\ \beta \geqslant 0$

$$\sum_{n \leqslant x} \frac{J_\alpha(n)}{n^\beta} = \begin{cases} \dfrac{1}{\zeta(1+\alpha)} \dfrac{x^{1+\alpha-\beta}}{1+\alpha-\beta} + O_{\alpha,\beta}\left(x^\theta \log x\right), & \textit{if } \beta - \alpha < 1; \\[4mm] \dfrac{\log x + \gamma}{\zeta(1+\alpha)} - \dfrac{\zeta'}{\zeta^2}(1+\alpha) + O_\alpha\left(\dfrac{\log x}{x^{\min(1,\alpha)}}\right), & \textit{if } \beta - \alpha = 1; \\[4mm] \dfrac{\zeta(\beta-\alpha)}{\zeta(\beta)} + O_{\alpha,\beta}\left(x^{\alpha-\beta+1}\right), & \textit{if } \beta - \alpha > 1, \end{cases}$$

where $\theta := \max\left(0, \max(1,\alpha) - \beta\right)$.

Proof Since $J_\alpha(n) = n^\alpha \sum_{d|n} \dfrac{\mu(d)}{d^\alpha}$, we get

$$\sum_{n \leqslant x} \frac{J_\alpha(n)}{n^\beta} = \sum_{n \leqslant x} \left(n^{\alpha-\beta} \sum_{d|n} \frac{\mu(d)}{d^\alpha} \right) = \sum_{d \leqslant x} \frac{\mu(d)}{d^\beta} \sum_{k \leqslant x/d} k^{\alpha-\beta}.$$

▷ If $\beta - \alpha = 1$, then

$$\sum_{n \leqslant x} \frac{J_\alpha(n)}{n^\beta} = \sum_{d \leqslant x} \frac{\mu(d)}{d^\beta} \left(\log \frac{x}{d} + \gamma + O\left(\frac{d}{x}\right) \right)$$

$$= (\log x + \gamma) \sum_{d \leqslant x} \frac{\mu(d)}{d^\beta} - \sum_{d \leqslant x} \frac{\mu(d) \log d}{d^\beta} + O\left(\frac{1}{x} \sum_{d \leqslant x} \frac{\mu(d)^2}{d^{\beta-1}} \right)$$

and since $\beta - 1 = \alpha > 0$, we derive

$$\sum_{n \leqslant x} \frac{J_\alpha(n)}{n^\beta} = (\log x + \gamma) \sum_{d=1}^{\infty} \frac{\mu(d)}{d^\beta} - \sum_{d=1}^{\infty} \frac{\mu(d) \log d}{d^\beta} + O\left(\frac{\log x}{x^{\beta-1}} + \frac{\log x}{x}\right)$$

$$= \frac{\log x + \gamma}{\zeta(1+\alpha)} - \frac{\zeta'}{\zeta^2}(1+\alpha) + O\left(\frac{\log x}{x^{\min(1,\alpha)}}\right).$$

▷ If $\beta - \alpha > 1$, then

$$\sum_{n \leqslant x} \frac{J_\alpha(n)}{n^\beta} = \sum_{d \leqslant x} \frac{\mu(d)}{d^\beta}\left(\zeta(\beta - \alpha) + O\left(\left(\frac{x}{d}\right)^{\alpha-\beta+1}\right)\right)$$

$$= \zeta(\beta - \alpha) \sum_{d \leqslant x} \frac{\mu(d)}{d^\beta} + O\left(x^{\alpha-\beta+1} \sum_{d \leqslant x} \frac{1}{d^{\alpha+1}}\right)$$

and since $\beta > \alpha + 1 > 1$, we get

$$\sum_{n \leqslant x} \frac{J_\alpha(n)}{n^\beta} = \frac{\zeta(\beta - \alpha)}{\zeta(\beta)} + O\left(x^{1-\beta} + x^{\alpha-\beta+1}\right)$$

and we easily check that the second term dominates.

▷ If $\beta - \alpha < 1$, then

$$\sum_{n \leqslant x} \frac{J_\alpha(n)}{n^\beta} = \sum_{d \leqslant x} \frac{\mu(d)}{d^\beta}\left(\frac{(x/d)^{\alpha-\beta+1}}{\alpha - \beta + 1} + O\left(\left(\frac{x}{d}\right)^{\alpha-\beta} + 1\right)\right)$$

$$= \frac{x^{\alpha-\beta+1}}{\alpha - \beta + 1} \sum_{d \leqslant x} \frac{\mu(d)}{d^{\alpha+1}} + O\left(x^{\alpha-\beta} \sum_{d \leqslant x} \frac{1}{d^\alpha} + \sum_{d \leqslant x} \frac{1}{d^\beta}\right).$$

Since $\alpha + 1 > 1$, the main term is equal to

$$\frac{1}{\zeta(1+\alpha)} \frac{x^{1+\alpha-\beta}}{1 + \alpha - \beta} + O\left(x^{1-\beta}\right)$$

while the error term does not exceed

$$\ll x^{\alpha-\beta} \begin{cases} x^{1-\alpha}, & \text{if } 0 < \alpha < 1 \\ \log x, & \text{if } \alpha = 1 \\ 1, & \text{if } \alpha > 1 \end{cases} + \begin{cases} x^{1-\beta}, & \text{if } 0 < \beta < 1 \\ \log x, & \text{if } \beta = 1 \\ 1, & \text{if } \beta > 1 \end{cases}$$

$$\ll \begin{cases} x^{1-\beta}, & \text{if } 0 < \alpha < 1 \\ x^{1-\beta} \log x, & \text{if } \alpha = 1 \\ x^{\alpha-\beta}, & \text{if } \alpha > 1 \end{cases} + \begin{cases} x^{1-\beta}, & \text{if } 0 < \beta < 1 \\ \log x, & \text{si } \beta = 1 \\ 1, & \text{if } \beta > 1 \end{cases}$$

completing the proof.   □

Since $\varphi = J_1$, we derive the following estimates for the Euler's totient.

Corollary 4.10 *For all* $x \in \mathbb{R}_{\geq e}$ *and* $\beta \geq 0$

$$\sum_{n \leq x} \frac{\varphi(n)}{n^\beta} = \begin{cases} \dfrac{x^{2-\beta}}{(2-\beta)\zeta(2)} + O_\beta \left(x^{\max(0,1-\beta)} \log x \right), & \text{if } \beta < 2 ; \\[3ex] \dfrac{\log x + \gamma}{\zeta(2)} - \dfrac{\zeta'}{\zeta^2}(2) + O \left(\dfrac{\log x}{x} \right), & \text{if } \beta = 2 ; \\[3ex] \dfrac{\zeta(\beta-1)}{\zeta(\beta)} + O_\beta \left(x^{2-\beta} \right), & \text{if } \beta > 2. \end{cases}$$

4.4.4.2 Dedekind's Totient

Theorem 4.32 *For all* $x \in \mathbb{R}_{\geq e}$ *and* $\alpha > 0, \beta \geq 0$

$$\sum_{n \leq x} \frac{\Psi_\alpha(n)}{n^\beta} = \begin{cases} \dfrac{\zeta(1+\alpha)}{\zeta(2+2\alpha)} \dfrac{x^{1+\alpha-\beta}}{1+\alpha-\beta} + O_{\alpha,\beta} \left(x^\theta \log x \right), & \text{if } \beta - \alpha < 1 ; \\[3ex] \dfrac{\zeta(1+\alpha)\,(\log x + \gamma)}{\zeta(2+2\alpha)} + C_\alpha + O_\alpha \left(\dfrac{\log x}{x^{\min(1,\alpha)}} \right), & \text{if } \beta - \alpha = 1 ; \\[3ex] \dfrac{\zeta(\beta-\alpha)\zeta(\beta)}{\zeta(2\beta)} + O_{\alpha,\beta} \left(x^{\alpha+1-\beta} \right), & \text{if } \beta - \alpha > 1, \end{cases}$$

where $\theta := \max \left(0, \max(1, \alpha) - \beta \right)$ *and*

$$C_\alpha := \zeta(2+2\alpha)^{-2} \left(\zeta'(1+\alpha)\zeta(2+2\alpha) - 2\zeta(1+\alpha)\zeta'(2+2\alpha) \right).$$

Proof Similar to that of Theorem 4.31, except that $\Psi_\alpha(n) = n^\alpha \sum_{d|n} \frac{\mu(d)^2}{d^\alpha}$, so that

$$\sum_{n\leqslant x} \frac{\Psi_\alpha(n)}{n^\beta} = \sum_{d\leqslant x} \frac{\mu(d)^2}{d^\beta} \sum_{k\leqslant x/d} k^{\alpha-\beta}$$

and hence every μ has to be replaced by μ^2. □

4.4.5 The Sum of Divisors

Theorem 4.33 *For all $x \in \mathbb{R}_{\geqslant e}$ and $\alpha > 0$, $\beta \geqslant 0$*

$$\sum_{n\leqslant x} \frac{\sigma_\alpha(n)}{n^\beta} = \begin{cases} \dfrac{\zeta(1+\alpha)}{1+\alpha-\beta} x^{1+\alpha-\beta} + O_{\alpha,\beta}\left(x^\theta \log x\right), & \text{if } \beta - \alpha < 1; \\[2ex] \zeta(1+\alpha)\left(\log x + \gamma\right) + C_\alpha + O_\alpha\left(\dfrac{\log x}{x^{\min(1,\alpha)}}\right), & \text{if } \beta - \alpha = 1; \\[2ex] \zeta(\beta-\alpha)\zeta(\beta) + O_{\alpha,\beta}\left(x^{\alpha-\beta+1}\right), & \text{if } \beta - \alpha > 1, \end{cases}$$

where $\theta := \max\left(0, \max(1,\alpha) - \beta\right)$ and $C_\alpha := \zeta'(1+\alpha)$.

Proof Similar to that of Theorem 4.31, except that $\sigma_\alpha(n) = n^\alpha \sum_{d|n} \frac{1}{d^\alpha}$, so that

$$\sum_{n\leqslant x} \frac{\sigma_\alpha(n)}{n^\beta} = \sum_{d\leqslant x} \frac{1}{d^\beta} \sum_{k\leqslant x/d} k^{\alpha-\beta}$$

and hence every μ has to be replaced by 1. □

For the sum of divisors function $\sigma = \sigma_1$, we infer the following result.

Corollary 4.11 *For all $x \in \mathbb{R}_{\geq e}$ and $\beta \geq 0$*

$$\sum_{n \leq x} \frac{\sigma(n)}{n^{\beta}} = \begin{cases} \dfrac{\zeta(2)}{2-\beta} x^{2-\beta} + O_{\beta}\left(x^{\max(0,1-\beta)} \log x\right), & \text{if } \beta < 2; \\[3ex] \zeta(2) (\log x + \gamma) + \zeta'(2) + O\left(\dfrac{\log x}{x}\right), & \text{if } \beta = 2; \\[3ex] \zeta(\beta - 1)\zeta(\beta) + O_{\beta}\left(x^{2-\beta}\right), & \text{if } \beta > 2. \end{cases}$$

4.4.6 The Hooley Divisor Function

4.4.6.1 Background

Recall that the *Hooley divisor function* Δ_r is defined by $\Delta_1 = 1$, and for $r \geq 2$ and $n \geq 1$, by

$$\Delta_r(n) := \max_{u_1, \ldots, u_{r-1} \in \mathbb{R}} \sum_{\substack{d_1 d_2 \cdots d_{r-1} \mid n \\ e^{u_i} < d_i \leq e^{u_i + 1}}} 1.$$

We also set $\Delta_2 := \Delta$. Hooley introduced this function in [48] and showed that it has numerous applications in number theory. Among other things, he proved the following results.

▷ If $\nu(x)$ is the number of integers $\leq x$ that are expressible as the sum of a square and two biquadrates, then, for all $x > x_0(\varepsilon)$

$$\nu(x) > x(\log x)^{1-4/\pi-\varepsilon}$$

improving the previous known lower bound $\nu(x) \gg x^{1-\varepsilon}$.

▷ Let $\gamma \in \mathbb{R}$ be arbitrary. Generalizing Kronecker's theorem regarding the approximation, mod 1, to γ by numbers of the shape $n^2\theta$ with θ irrational, Hooley showed that the inequality

$$\|n^2\theta - \gamma\| < n^{-1/2}(\log n)^{-1/2+2/\pi+\varepsilon}$$

holds for infinitely many numbers n.

▷ Let $r_8(n)$ be the number of representations of the integer n as the sum of eight positive cubes. Then, for $n > n_0(\varepsilon)$

$$r_8(n) < n^{5/3}(\log n)^{\sqrt{3}-1+\varepsilon}$$

improving the previous bound $r_8(n) \ll n^{5/3+\varepsilon}$.

Trivially, $\Delta_r(n) \leqslant \tau_r(n)$ and hence

$$\sum_{n \leqslant x} \Delta_r(n) \ll x (\log x)^{r-1}.$$

Thus, the question of an improvement of this estimate arises naturally. Since $\Delta_r(n) \geqslant 1$, only a power of the logarithm can be saved. In what follow, we survey the different techniques used to derive such an improvement.

4.4.6.2 Hooley's Kernel

The main idea of Hooley, developed here in the case $r = 2$ for the sake of clearness, is to pick up from Fourier analysis the kernel

$$\kappa(x) := \left(\frac{\sin \frac{1}{2} x}{\frac{1}{2} x} \right)^2 = \int_{-1}^{1} (1 - |v|) \, e^{ixv} \, dv \quad (|x| \leqslant 1)$$

satisfying

$$\kappa(x) \geqslant \kappa(1) > \tfrac{9}{10} \quad (|x| \leqslant 1). \tag{4.40}$$

Now set for all $u \in \mathbb{R}$

$$\Delta(n; u) := \sum_{\substack{d \mid n \\ e^u < d \leqslant e^{u+1}}} 1.$$

Using (4.40), we derive

$$\Delta(n; u) \leqslant \tfrac{10}{9} \sum_{d \mid n} \kappa (\log d - u) \ll \int_{-1}^{1} |\tau(n; v)| \, dv,$$

where $\tau(n; v) := \sum_{d \mid n} d^{iv}$ for all $v \in [-1, 1]$, and hence

$$\Delta(n) \ll \int_{-1}^{1} |\tau(n; v)| \, dv. \tag{4.41}$$

This idea is very fruitful, since the function $n \mapsto |\tau(n; v)|$ is multiplicative, and since $|\tau(n; v)| \leqslant \tau(n)$, we may apply Theorem 4.9 yielding

$$\sum_{n \leqslant x} \Delta(n) \ll \frac{x}{\log x} \int_{-1}^{1} \exp\left(\sum_{p \leqslant x} \frac{|\tau(p; v)|}{p}\right) dv.$$

Now by [41, Lemma 30.2], if $(\log x)^{-1} < |v| \leqslant 1$, we have

$$\sum_{p \leqslant x} \frac{|\tau(p; v)|}{p} = \frac{4}{\pi} \log\log x + \left(2 - \frac{4}{\pi}\right) \log \frac{1}{|v|} + O(1)$$

and trivially, if $|v| \leqslant (\log x)^{-1}$

$$\sum_{p \leqslant x} \frac{|\tau(p; v)|}{p} \leqslant 2 \sum_{p \leqslant x} \frac{1}{p} = 2 \log\log x + O(1).$$

Therefore

$$\sum_{n \leqslant x} \Delta(n) \ll \frac{x}{\log x} \left\{ \int_{0}^{1/\log x} e^{2\log\log x}\, dv + \int_{1/\log x}^{1} \left((\log x)^{4/\pi} + v^{-2+4/\pi}\right) dv \right\}$$

$$\ll x(\log x)^{-1+4/\pi} \approx x(\log x)^{0.273239}.$$

Theorem 4.34 (Hooley) *For $x \geqslant 3$*

$$\sum_{n \leqslant x} \Delta(n) \ll x(\log x)^{-1+4/\pi}.$$

Note that by partial summation, we derive

$$\sum_{n \leqslant x} \frac{\Delta(n)}{n} \ll (\log x)^{4/\pi}$$

and using Theorem 4.18 we get

$$\sum_{x < n \leqslant x+y} \Delta(n) \ll y(\log x)^{-1+4/\pi}.$$

whenever $x^{\varepsilon} \leqslant y \leqslant x$. In the general case, Hooley proved that, for $r \geqslant 3$

$$\sum_{n \leqslant x} \Delta_r(n) \ll x (\log x)^{\sqrt{r}-1}. \qquad (4.42)$$

4.4.6.3   The Method of the Differential Inequality

The weakness of Hooley's method lies in (4.41) with the introduction of the modulus sign, but, as pointed out by the authors in [41], Hooley's idea is very useful for the weighted sum $\sum_{n \leqslant x} \Delta(n) z^{\omega(n)}$ when z is not too small. For instance, [41, Theorem 66] asserts that, if $z > \frac{1}{2} + \frac{1}{\pi-2} \approx 1.3759$, then

$$\sum_{n \leqslant x} \Delta(n) z^{\omega(n)} \ll x (\log x)^{2z-2}.$$

Hall and Tenenbaum [41] came with a new idea, essentially resting on the following result [41, Lemma 70.2], which may have its own interest.

Lemma 4.12 *Let $L(\sigma)$ and $X(\sigma)$ be continuously differentiable for $\sigma \in (1, \sigma_0]$ satisfying, respectively,*

$$-L'(\sigma) \leqslant \varphi(\sigma, L(\sigma)) \quad and \quad -X'(\sigma) = \varphi(\sigma, X(\sigma)),$$

where $x \mapsto \varphi(\sigma, x)$ is a non-decreasing function for each fixed σ. Then $L(\sigma) < X(\sigma)$ throughout the range $1 < \sigma \leqslant \sigma_0$.

Proof Assume on the contrary that there exists $\sigma_1 \in (1, \sigma_0)$ such that $L(\sigma_1) = X(\sigma_1)$ and $L(\sigma) < X(\sigma)$ for $\sigma \in (\sigma_1, \sigma_0]$. Then

$$L(\sigma_1) = L(\sigma_0) - \int_{\sigma_1}^{\sigma_0} L'(\sigma)\, d\sigma \leqslant L(\sigma_0) + \int_{\sigma_1}^{\sigma_0} \varphi(\sigma, L(\sigma))\,(\sigma)\, d\sigma$$

$$< X(\sigma_0) + \int_{\sigma_1}^{\sigma_0} \varphi(\sigma, X(\sigma))\,(\sigma)\, d\sigma = X(\sigma_1)$$

leading to a contradiction.   □

This result is applied to the function

$$L(\sigma) := \sum_{n=1}^{\infty} M_q(n)^{1/q} \frac{\mu(n)^2 z^{\omega(n)}}{n^{\sigma}} \qquad (\sigma > 1),$$

where $M_q(n)$ is the moment

$$M_q(n) := \int_{-\infty}^{\infty} \cdots \int_{-\infty}^{\infty} \Delta(n; u_1, \ldots, u_{r-1})^q \, du_1 \cdots du_{r-1}$$

and for which a fundamental inequality [41, Theorem 73] is proved. This enables the authors to reduce the exponent $\sqrt{r} - 1$ in (4.42) to any positive real number.

Theorem 4.35 (Hall–Tenenbaum) *Let* $r \in \mathbb{Z}_{\geq 2}$ *fixed. Define the k-fold iterate logarithm* $\log_k x := \log\left(\log_{k-1} x\right)$ *with* $\log_1 x = \log x$. *For* $x > e^e$, *set*

$$\epsilon_r(x) := \left(r - 1 + \frac{30}{\log_3 x}\right) \sqrt{\frac{r \log_3 x}{\log_2 x}}.$$

Then

$$\sum_{n \leq x} \Delta_r(n) \ll_r x (\log x)^{\epsilon_r(x)}.$$

A partial summation gives at once

$$\sum_{n \leq x} \frac{\Delta_r(n)}{n} \ll (\log x)^{1 + \epsilon_r(x)}$$

so that using Theorem 4.18 and the second Mertens theorem,[19] we derive

$$\sum_{x < n \leq x+y} \Delta_r(n) \ll y (\log x)^{\epsilon_r(x)}$$

for $0 < \varepsilon < \frac{1}{2}$, $x \geq c_0$ and uniformly in $x^\varepsilon \leq y \leq x$. With $r = 2$ and noticing that, for all $N \in \mathbb{Z}_{\geq 1}$, we have

$$\sum_{N < n \leq 2N} \left(\left\lfloor \frac{x+y}{n} \right\rfloor - \left\lfloor \frac{x}{n} \right\rfloor\right) = \sum_{N < d \leq 2N} \sum_{x < dk \leq x+y} 1$$

$$= \sum_{x < n \leq x+y} \sum_{\substack{d \mid n \\ N < d \leq 2D}} 1 \leq \sum_{x < n \leq x+y} \Delta(n)$$

---

[19] See Corollary 3.5.

we infer the bound

$$\sum_{N < n \leqslant 2N} \left(\left\lfloor \frac{x+y}{n} \right\rfloor - \left\lfloor \frac{x}{n} \right\rfloor \right) \ll y (\log x)^{\epsilon_2(x)} = y (\log x)^{o(1)} \tag{4.43}$$

for $0 < \varepsilon < \frac{1}{2}$, $x \geqslant c_0$ and uniformly in $x^\varepsilon \leqslant y \leqslant x$ and $N \in \mathbb{Z}_{\geqslant 1}$. This is the best result to date for the left-hand side.[20]

4.4.6.4 The Generalization of Daniel and Brüdern

During a seminar in Stuttgart, on November 24, 2000, Stephan Daniel proposed a far-reaching generalisation of Hooley's Delta function. He investigated the functions

$$\Delta_g(n) := \max_{u \in \mathbb{R}} \max_{0 < v \leqslant 1} \left| \sum_{\substack{d \mid n \\ e^u < d \leqslant e^{u+v}}} g(d) \right| \quad \text{and} \quad M_g(n) := \max_{u \in \mathbb{R}} \left| \sum_{\substack{d \mid n \\ d \leqslant e^u}} g(d) \right|,$$

where g is any complex-valued arithmetic function satisfying $|g(n)| \leqslant 1$, and showed how further savings can be made when g has mean value 0 in some suitable sense.

Several months after the event, Daniel left the academic world, and his work was never published. In 2010, Brüdern rediscovered a set of notes that he took during Daniel's seminar, and finally published them in [14], adding on this occasion a number of generalizations of Daniel's initial theorems.

In order to tackle the next results, some specific notation is needed. For all $a, b > 0$ and $c \geqslant 1$, define $\mathcal{F}_{a,b,c}$ to be the set of multiplicative functions $\rho : \mathbb{Z}_{\geqslant 1} \to \mathbb{R}_+$ such that $\rho(p) \leqslant c$, $\rho(p^\alpha) \leqslant c\, p^{(1-b)\alpha - 1}$ for all $\alpha \geqslant 2$, and

$$\sum_p \frac{\rho(p) \log p}{p^\sigma} \leqslant \frac{a}{\sigma - 1} + c \quad (\sigma \in (1, 2]).$$

For any multiplicative function g such that $|g(n)| \leqslant 1$, denote $\mathcal{F}_g$ to be the set of $\rho \in \mathcal{F}_{a,b,c}$ satisfying

$$\left| \sum_{n \leqslant x} \mu(n)^2 \rho(n) g(n) \right| \leqslant c\, x (\log x)^{-a}.$$

---

[20]See also Exercise 112.

As a useful example, one may take $\rho(n) = 1$, or $\rho(n) = \rho_P(n)$ the number of solutions of $P(x) \equiv 0 \pmod{n}$ where $P \in \mathbb{Z}[X]$ is an irreducible polynomial over $\mathbb{Z}$. In both cases, $a = 1$.

Finally, for all $a, t > 0$, set

$$\kappa := \max\left(a,\, a + \tfrac{1}{2}t(a-1),\, 2^t a - \tfrac{3}{2}t\right) \quad \text{and} \quad \nu := \left(\tfrac{1}{2}at\right)^{1/2}$$

and, for $z > 30$

$$L_1(z) := (\log z)^{\sqrt{\log z}} \quad \text{and} \quad L_2(z) := e^{\sqrt{\log z \log_2 z}}$$

and if $z \in [0, 30]$, then set $L_1(z) = L_2(z) = 1$.

> **Theorem 4.36** *Let* $t \in [1, 2]$. *There exists* $d \in \mathbb{R}$ *such that, for any multiplicative function* g *such that* $|g(n)| \leqslant 1$ *and any* $\rho \in \mathcal{F}_g$, *we have*
>
> $$\sum_{n \leqslant x} \rho(n)\Delta_g(n)^t \ll x(\log x)^{\kappa-1} L_1(\log x)^{\nu} L_2(\log x)^d.$$

As a corollary, we derive the next estimate.

Corollary 4.12 *Let* $t \in [1, 2]$. *There exists* $d \in \mathbb{R}$ *such that, for any multiplicative function* g *satisfying* $|g(n)| \leqslant 1$ *and*

$$\sum_{n \leqslant x} \mu(n)^2 g(n) \ll x(\log x)^{-1}$$

we have

$$\sum_{n \leqslant x} \Delta_g(n)^t \ll x(\log x)^{\epsilon_{d,t}(x)},$$

where $\epsilon_{d,t}(x) := \dfrac{\sqrt{t}\,\log_3 x + d\sqrt{2\log_3 x}}{\sqrt{2\log_2 x}} = o(1)$ *if* $x > 30$, $\epsilon_{d,t}(x) = 0$ *if* $x \in [0, 30]$.

For the function M_g, more additional conditions must be taken into account, as can be seen with $g = \chi_4$ the Dirichlet character modulo 4. Let $\mathcal{F}_g^{\star}$ be the set of $\rho \in \mathcal{F}_g$ such that

$\triangleright$ for all $\sigma \in (1, 2]$, $\displaystyle\sum_{n=1}^{\infty} \mu(n)^2 \rho(n) g(n) n^{-\sigma} \ll \sqrt{\sigma - 1}$;

▷ for all $\varepsilon > 0$, $\sum_{n \leqslant x} \mu(n)^2 \rho(n) g(n) \ll x (\log x)^{-1/\varepsilon}$;

▷ there exists $\delta > 0$ such that, for each prime p, we have $|g(p)| \rho(p) \leqslant (1 - \delta) p$.

Once again, one may take $\rho(n) = 1$ or $\rho(n) = \rho_P(n)$ as above.

Theorem 4.37 *Let $t \in [1, 2]$. There exists $d \in \mathbb{R}$ such that, for any multiplicative function g satisfying $|g(n)| \leqslant 1$ and any $\rho \in \mathcal{F}_g^\star$, we have*

$$\sum_{n \leqslant x} \rho(n) M_g(n)^t \ll x (\log x)^{\kappa - 1} L_1 (\log x)^\nu L_2 (\log x)^d.$$

As above, this result implies the next estimate.

Corollary 4.13 *Let $t \in [1, 2]$. There exists $d \in \mathbb{R}$ such that, for any multiplicative function g satisfying $|g(n)| \leqslant 1$ and*

▷ *for all $\sigma \in (1, 2]$, $\sum_{n=1}^{\infty} \mu(n)^2 g(n) n^{-\sigma} \ll \sqrt{\sigma - 1}$;*

▷ *for all $\varepsilon > 0$, $\sum_{n \leqslant x} \mu(n)^2 g(n) \ll x (\log x)^{-1/\varepsilon}$;*

we have

$$\sum_{n \leqslant x} M_g(n)^t \ll x (\log x)^{\epsilon_{d,t}(x)},$$

where $\epsilon_{d,t}(x) := \dfrac{\sqrt{t} \log_3 x + d \sqrt{2 \log_3 x}}{\sqrt{2 \log_2 x}} = o(1)$ if $x > 30$, $\epsilon_{d,t}(x) = 0$ if $x \in [0, 30]$.

4.4.6.5 The Generalization of de la Bretèche and Tenenbaum

The conditions on the function ρ are quite restrictive in the previous results. In [21], de la Bretèche and Tenenbaum enhanced these conditions to finally derive a large number of estimates when $g = \mu$ or $g = \chi$ a Dirichlet character.

In what follows, let $A > 0$, $c > 0$, $\eta \in (0, 1)$ and let $\mathcal{M}_A(c, \eta)$ be the class of multiplicative functions $g : \mathbb{Z}_{\geqslant 1} \to \mathbb{R}_{\geqslant 0}$ satisfying $g(p^\alpha) \leqslant A^\alpha$, $g(n) \ll_\varepsilon n^\varepsilon$ and

$$\sum_{p \leqslant x} g(p) = \kappa \operatorname{Li}(x) + O \left(x e^{-2c(\log x)^\eta} \right)$$

with $\kappa := \kappa_g \geqslant 0$. For any real Dirichlet character $\chi \neq \chi_0$, let $\mathcal{M}_A(\chi; c, \eta)$ be the subclass of $\mathcal{M}_A(c, \eta)$ such that

$$\sum_{p \leqslant x} \chi(p) g(p) = \theta \operatorname{Li}(x) + O\left(x e^{-c(\log x)^\eta}\right)$$

with $\theta := \theta_g \in \mathbb{R}$. Finally, set

$$\nu_t := \frac{t}{2^{2t-1} - 1} \quad (t \geqslant 1)$$

$$\lambda_t := \frac{1}{2\pi} \int_{-\pi}^{\pi} \left|1 + e^{iu}\right|^{2t} \, du = \frac{4^t \Gamma\left(t + \frac{1}{2}\right)}{\sqrt{\pi}\, \Gamma(t+1)} \quad (t > 0).$$

Theorem 4.38 *Let $t \geqslant 1$.*

▷ *Let $g \in \mathcal{M}_A(\chi; c, \eta)$, $\kappa > 0$ and $\theta \leqslant 2^{1-2t}$. Then there exists a number $\alpha = \alpha(g, \chi; c, t, \eta) > 0$ such that, for all $x \geqslant 16$,*

$$\sum_{n \leqslant x} g(n) \Delta_\chi(n)^{2t} \ll x (\log x)^{\kappa - 1 + \max(0,\, 2^{2t-1}\kappa - \kappa - 1)} e^{\alpha \sqrt{\log_2 x \log_3 x}}$$

the $\log_3 x$ being omitted if $\kappa > \nu_t$. Furthermore, if $t \in \mathbb{Z}_{\geqslant 2}$, $\theta \leqslant 0$ and $\kappa > \max\left(2^{1-t}, 2\nu_t\right)$, then for $x \geqslant 16$,

$$\sum_{n \leqslant x} g(n) \Delta_\chi(n)^{2t} \ll x (\log x)^{2^{2t-1}\kappa - 2t - 1} e^{\alpha \sqrt{\log_2 x}}.$$

▷ *Let $g \in \mathcal{M}_A(c, \eta)$. Then there exists $\alpha = \alpha(g; c, t, \eta) > 0$ such that, for all $x \geqslant 16$*

$$\sum_{n \leqslant x} g(n) \Delta_\mu(n)^{2t} \ll x (\log x)^{\kappa - 1 + \max(0,\, \lambda_t\kappa - \kappa - t)} e^{\alpha \sqrt{\log_2 x \log_3 x}}$$

the term $\log_3 x$ being omitted if $\kappa > \dfrac{t}{\lambda_t - 1}$.

Specifying $t = 1$ and $g = \mu^2 \times z^\omega$, we derive the following more precise estimates.

Theorem 4.39 *If $z > 0$, then*

$$\sum_{n \leqslant x} \mu(n)^2 z^{\omega(n)} \Delta_\mu(n)^2 = x(\log x)^{z-1+\max(0,\,z-1)+o(1)}.$$

Theorem 4.40 *Let $\gamma_0 := \log 2 \left| \log \left(1 - \frac{1}{\log 3} \right) \right|^{-1} \approx 0.28754$. Then*

$$\sum_{n \leqslant x} \mu(n)^2 \Delta_\mu(n)^2 > x(\log \log x)^{2\gamma_0+o(1)}.$$

4.5   Arithmetic Functions of Several Variables

The ideas of this section are essentially picked up from [112]. Let $r \in \mathbb{Z}_{\geqslant 1}$ and $\mathcal{A}_r$ be the set of arithmetic functions of r variables, i.e. the functions $f : (\mathbb{Z}_{\geqslant 1})^r \to \mathbb{C}$.

4.5.1   Definitions

Definition 4.7 Let $r \in \mathbb{Z}_{\geqslant 1}$ et $f \in \mathcal{A}_r$.

▷ f is *multiplicative* if $f \neq 0$ and if

$$f(m_1 n_1, \ldots, m_r n_r) = f(m_1, \ldots, m_r)\, f(n_1, \ldots, n_r)$$

for all r-uples $(m_1, \ldots, m_r), (n_1, \ldots, n_r) \in (\mathbb{Z}_{\geqslant 1})^r$ satisfying $(m_1 \cdots m_r, n_1 \cdots n_r) = 1$. In this case, $f(1, \ldots, 1) = 1$ and

$$f(n_1, \ldots, n_r) = \prod_p f\left(p^{v_p(n_1)}, \ldots, p^{v_p(n_r)} \right).$$

The set of multiplicative functions of r variables is denoted by $\mathcal{M}_r$.

▷ f is *firmly multiplicative* if $f \neq 0$ and if

$$f(m_1 n_1, \ldots, m_r n_r) = f(m_1, \ldots, m_r)\, f(n_1, \ldots, n_r)$$

for all r-uples $(m_1, \ldots, m_r), (n_1, \ldots, n_r) \in (\mathbb{Z}_{\geqslant 1})^r$ satisfying $(m_1, n_1) = \cdots = (m_r, n_r) = 1$. In this case, $f(1, \ldots, 1) = 1$ and

$$f(n_1, \ldots, n_r) = \prod_p \left(f\left(p^{v_p(n_1)}, 1, \ldots, 1 \right) \cdots f\left(1, 1, \ldots, p^{v_p(n_r)} \right) \right).$$

The set of firmly multiplicative functions of r variables is denoted by $\mathcal{F}_r$.

▷ f is *completely multiplicative* if $f \neq 0$ and if

$$f(m_1 n_1, \ldots, m_r n_r) = f(m_1, \ldots, m_r) f(n_1, \ldots, n_r)$$

for all r-uples $(m_1, \ldots, m_r), (n_1, \ldots, n_r) \in (\mathbb{Z}_{\geqslant 1})^r$. In this case, $f(1, \ldots, 1) = 1$ and

$$f(n_1, \ldots, n_r) = \prod_p \left(f(p, 1, \ldots, 1)^{v_p(n_1)} \cdots f(1, 1, \ldots, p)^{v_p(n_r)} \right).$$

The set of completely multiplicative functions of r variables is denoted by C_r.

Example 4.8 Let $r \in \mathbb{Z}_{\geqslant 1}$.

▷ The functions $(n_1, \ldots, n_r) \mapsto (n_1, \ldots, n_r)$ and $(n_1, \ldots, n_r) \mapsto [n_1, \ldots, n_r]$ are multiplicative, but not firmly multiplicative.
▷ The functions $(n_1, \ldots, n_r) \mapsto \tau(n_1) \cdots \tau(n_r)$ and $(n_1, n_2) \mapsto \tau(n_1)\sigma(n_2)$ are firmly multiplicative, but not completely multiplicative.
▷ The functions $(n_1, \ldots, n_r) \mapsto n_1 \cdots n_r$ and $(n_1, n_2) \mapsto n_1 \lambda(n_2)$ are completely multiplicative.
▷ The product and the quotient of two non-vanishing multiplicative functions are multiplicative functions.
▷ If $h \in \mathcal{M}_1$, then the functions $(n_1, \ldots, n_r) \mapsto h((n_1, \ldots, n_r))$ and $(n_1, \ldots, n_r) \mapsto h([n_1, \ldots, n_r])$ are multiplicative.
▷ If $f \in \mathcal{M}_r$ and if we fix $s \geqslant 1$ variables, then the resulting function *is not necessary* multiplicative.
▷ If $f \in \mathcal{M}_r$, then the function $\overline{f} \in \mathcal{A}_1$ defined by $\overline{f}(n) = f(n, \ldots, n)$ is multiplicative.
▷ The *Ramanujan sum*[21]

$$(k, n) \mapsto c_n(k) := n \sum_{d \mid (k,n)} \frac{\mu(d)}{d}$$

is a multiplicative function of two variables.

Proposition 4.22 *Let $r \in \mathbb{Z}_{\geqslant 1}$ and $f \in \mathcal{A}_r$.*

▷ *$f \in \mathcal{F}_r$ if and only if there exist $f_1, \ldots, f_r \in \mathcal{M}_1$ such that*

$$f(n_1, \ldots, n_r) = f_1(n_1) \cdots f_r(n_r).$$

In this case, $f_j(n) = f(1, \ldots, n, \ldots, 1)$, n being placed in jth position.

---

[21] See Definition 6.3.

▷ $f \in C_r$ if and only if there exist $f_1, \ldots, f_r \in C_1$ such that

$$f (n_1, \ldots, n_r) = f_1 (n_1) \cdots f_r (n_r).$$

In this case, $f_j (n) = f (1, \ldots, n, \ldots, 1)$, n being placed in jth position.

Proof Voir [112, Propositions 1 and 2].   □

Proposition 4.23 *The set* $(\mathcal{A}_r, +, \cdot)$ *is a* $\mathbb{C}$-*linear space.*

Proof See [112, p. 490].   □

4.5.2   Dirichlet Convolution

Definition 4.8 Let $r \in \mathbb{Z}_{\geqslant 1}$ and $f, g \in \mathcal{A}_r$. The *Dirichlet convolution product* of f and g is defined by

$$(f \star g) (n_1, \ldots, n_r) = \sum_{d_1 | n_1, \ldots, d_r | n_r} f (d_1, \ldots, d_r) g (n_1/d_1, \ldots, n_r/d_r).$$

Proposition 4.24 *The set* $(\mathcal{A}_r, +, \cdot, \star)$ *forms a unital commutative* $\mathbb{C}$-*algebra, the unity being the function*

$$\delta_r : (n_1, \ldots, n_r) \mapsto \delta_r (n_1, \ldots, n_r) := \begin{cases} 1, & \text{if } n_1 = \cdots = n_r = 1 \\ 0, & \text{if } n_1 \cdots n_r > 1. \end{cases}$$

Furthermore, the ring $(\mathcal{A}_r, +, \star)$ *is an integral domain and a unique factorization domain. Set* $\mathcal{A}_r^\times$ *group of invertible functions of* $\mathcal{A}_r$. *Finally, one has the following subgroup relations.*

$$(\mathcal{F}_r, \star) \subseteq (\mathcal{M}_r, \star) \subseteq \left(\mathcal{A}_r^\times, \star \right).$$

In particular, the Dirichlet convolution product preserves the multiplicativity of functions.

Proof See [112, p. 490 and Proposition 3].   □

Remark 4.7 The set $(C_r, \star)$ does not form a group under the Dirichlet convolution. If $f \in C_r$

$$f^{-1} = \mu_r f,$$

where μ_r is the generalized Möbius function defined in Theorem 4.41 below. Also note that, for all $f \in C_r$

$$(f \star f)(n_1, \ldots, n_r) = f(n_1, \ldots, n_r)\,\tau(n_1) \cdots \tau(n_r).$$

Theorem 4.41 (Möbius Inversion) *Let* $r \in \mathbb{Z}_{\geqslant 1}$. *If we define* $\mathbf{1}_r$ *and* μ_r *as* $\mathbf{1}_r(n_1, \ldots, n_r) := 1$ *and* $\mu_r(n_1, \ldots, n_r) := \mu(n_1) \cdots \mu(n_r)$, *then*

$$\mathbf{1}_r \star \mu_r = \delta_r.$$

Proof See [112, p. 491]. □

4.5.3 Dirichlet Convolute

Let $r \in \mathbb{Z}_{\geqslant 2}$ and $f \in \mathcal{A}_r$. It is fairly easy to associate to f a single-variable function, for instance, by taking $n_1 = \cdots = n_r = n$. A less trivial way to retrieve from f a function of a single variable is to consider

$$C_f : n \mapsto \sum_{d_1 \cdots d_r = n} f(d_1, \ldots, d_r). \tag{4.44}$$

Definition 4.9 Let $r \in \mathbb{Z}_{\geqslant 2}$ and $f \in \mathcal{A}_r$. The function C_f defined in (4.44) is called the *Dirichlet convolute* of f.

The next result shows that the Dirichlet convolute of a multiplicative function of r variables is multiplicative.

Proposition 4.25 *Let* $r \in \mathbb{Z}_{\geqslant 2}$ *and* $f \in M_r$. *Then* C_f *is multiplicative.*

Proof We follow the lines of [112, Proposition 12]. Let $m, n \in \mathbb{Z}_{\geqslant 1}$ such that $(m, n) = 1$. If $d_1 \cdots d_r = mn$, then there exist unique integers $a_1, b_1, \ldots, a_r, b_r$ such that $a_1, \ldots, a_r \mid n$, $b_1, \ldots, b_r \mid m$ and $d_k = a_k b_k$ with $(a_1 \cdots a_r, b_1 \cdots b_r) = 1$. Since $f \in M_r$, we then derive

$$C_f(mn) = \sum_{\substack{a_1 \cdots a_r = n \\ b_1 \cdots b_r = m}} f(a_1 b_1, \ldots, a_r b_r)$$

$$= \sum_{a_1 \cdots a_r = n} f(a_1, \ldots, a_r) \sum_{b_1 \cdots b_r = m} f(b_1, \ldots, b_r)$$

$$= C_f(n) C_f(m).$$

as required.   □

   The following examples are due to Tóth [113].

Example 4.9 Let $r \in \mathbb{Z}_{\geqslant 2}$.

▷ Take $f(n_1, \ldots, n_r) = \mu(n_1 \cdots n_r)$. Then

$$\sum_{n_1 \cdots n_r \mid n} \mu(n_1 \cdots n_r) = (1 - r)^{\omega(n)}.$$

Indeed, by Proposition 4.25, it suffices to verify the equality for prime powers p^α.
But, writing only the non-zero terms, the left-hand side is equal to

$$\mu(1 \cdots 1) + \mu(p \cdot 1 \cdots 1) + \mu(1 \cdot p \cdots 1) + \cdots + \mu(1 \cdots 1 \cdot p)$$
$$= 1 + (-1)r = (1 - r)^{\omega(p^\alpha)}$$

as required.

▷ Take $f(n_1, \ldots, n_r) = \mu^2(n_1 \cdots n_r)$. Then, similarly as above, we derive

$$\sum_{n_1 \cdots n_r \mid n} \mu^2(n_1 \cdots n_r) = (1 + r)^{\omega(n)}.$$

Proposition 4.26 *Let $r \in \mathbb{Z}_{\geqslant 2}$ and $f, g \in \mathcal{A}_r$. Then $C_{f \star g} = C_f \star C_g$. Furthermore,
the map*

$$C_f : (\mathcal{A}_r, +, \cdot, \star) \to (\mathcal{A}_1, +, \cdot, \star)$$

is a surjective algebra homomorphism.

Proof See [112, Propositions 15 and 16].   □

Corollary 4.14 *Let $r \in \mathbb{Z}_{\geqslant 2}$. The maps $C_f : (\mathcal{M}_r, \star) \to (\mathcal{M}_1, \star)$ and $C_f :
(\mathcal{F}_r, \star) \to (\mathcal{M}_1, \star)$ are surjective group homomorphisms.*

Proof See [112, Corollary 2].   □

4.5.4   Dirichlet Series

Definition 4.10 Let $r \in \mathbb{Z}_{\geqslant 1}$ and $f \in \mathcal{A}_r$. The *multiple Dirichlet series* of f is
defined by

$$L(s_1, \ldots, s_r; f) := \sum_{n_1 = 1, \ldots, n_r = 1}^{\infty} \frac{f(n_1, \ldots, n_r)}{n_1^{s_1} \cdots n_r^{s_r}}.$$

Proposition 4.27 *Let* $r \in \mathbb{Z}_{\geqslant 1}$ *et* $f, g \in \mathcal{A}_r$. *If* $L(s_1, \ldots, s_r; f)$ *and* $L(s_1, \ldots, s_r; g)$ *are absolutely convergent, then* $L(s_1, \ldots, s_r; f \star g)$ *is also absolutely convergent and*

$$L(s_1, \ldots, s_r; f \star g) = L(s_1, \ldots, s_r; f) \times L(s_1, \ldots, s_r; g).$$

Similarly, if $f \in \mathcal{A}_r^{\times}$, *then*

$$L\left(s_1, \ldots, s_r; f^{-1}\right) = \frac{1}{L(s_1, \ldots, s_r; f)}.$$

Proof See [112, Proposition 10]. □

Proposition 4.28 (Euler Product) *Let* $r \in \mathbb{Z}_{\geqslant 1}$ *and* $f \in \mathcal{M}_r$. *For all* $(s_1, \ldots, s_r) \in \mathbb{C}^r$, *the Dirichlet series* $L(s_1, \ldots, s_r; f)$ *is absolutely convergent if and only if*

$$\prod_p \sum_{\substack{\alpha_1, \ldots, \alpha_r = 0 \\ \alpha_1 + \cdots + \alpha_r \geqslant 1}}^{\infty} \frac{|f(p^{\alpha_1}, \ldots, p^{\alpha_r})|}{p^{\alpha_1 \sigma_1 + \cdots + \alpha_r \sigma_r}} < \infty.$$

In this case

$$L(s_1, \ldots, s_r; f) = \prod_p \sum_{\alpha_1, \ldots, \alpha_r = 0}^{\infty} \frac{f(p^{\alpha_1}, \ldots, p^{\alpha_r})}{p^{\alpha_1 s_1 + \cdots + \alpha_r s_r}}.$$

Proof See [112, Proposition 11]. □

4.5.5   Mean Values

Definition 4.11 Let $r \in \mathbb{Z}_{\geqslant 1}$ and $f \in \mathcal{A}_r$. If the limit exists, the *mean value* of f is the number

$$M(f) := \lim_{x_1, \ldots, x_r \to \infty} \frac{1}{x_1 \cdots x_r} \sum_{n_1 \leqslant x_1, \ldots, n_r \leqslant x_r} f(n_1, \ldots, n_r).$$

Proposition 4.29 *Let* $r \in \mathbb{Z}_{\geqslant 1}$.

▷ *If* $f \in \mathcal{A}_r$ *is such that*

$$\sum_{n_1, \ldots, n_r = 1}^{\infty} \frac{|(\mu_r \star f)(n_1, \ldots, n_r)|}{n_1 \cdots n_r} < \infty,$$

then the mean value $M(f)$ exists and is equal to

$$M(f) = \sum_{n_1,\ldots,n_r=1}^{\infty} \frac{(\mu_r \star f)(n_1,\ldots,n_r)}{n_1\cdots n_r}.$$

▷ *If $f \in \mathcal{M}_r$ is such that*

$$\sum_{p} \sum_{\substack{\alpha_1,\ldots,\alpha_r=0 \\ \alpha_1+\cdots+\alpha_r \geqslant 1}}^{\infty} \frac{|(\mu_r \star f)(p^{\alpha_1},\ldots,p^{\alpha_r})|}{p^{\alpha_1+\cdots+\alpha_r}} < \infty,$$

then the mean value $M(f)$ exists and is equal to

$$M(f) = \prod_{p}\left(1-\frac{1}{p}\right)^{r} \sum_{\alpha_1,\ldots,\alpha_r=0}^{\infty} \frac{f(p^{\alpha_1},\ldots,p^{\alpha_r})}{p^{\alpha_1+\cdots+\alpha_r}}.$$

Proof See [112, Propositions 18 and 19].   □

The next result is useful to establish asymptotic formulæ of a certain class of arithmetic functions of several variables. For a proof, see [114, Theorem 3.3].

Theorem 4.42 *Let $r \in \mathbb{Z}_{\geqslant 2}$, $f,g \in \mathcal{A}_r$ and $h_j : \mathbb{Z}_{\geqslant 1} \longrightarrow \mathbb{C}$, $j = 1,\ldots,r$ be arithmetic functions such that, for all $n_1,\ldots,n_r \in \mathbb{Z}_{\geqslant 1}$, we have*

$$f(n_1,\ldots,n_r) = \sum_{d_1|n_1,\ldots,d_r|n_r} g(d_1,\ldots,d_r)\,h_1(n_1/d_1)\cdots h_r(n_r/d_r).$$

Assume that

▷ *there exist constants $0 < b_j < a_j$, $j = 1,\ldots,r$, such that*

$$\sum_{n\leqslant x} h_j(n) = x^{a_j} P_j(\log x) + O\left(x^{b_j}\right) \quad (1 \leqslant j \leqslant r),$$

where P_j is a polynomial of degree δ_j and leading coefficient α_j ;

(continued)

Theorem 4.42 (continued)
▷ *the Dirichlet series*

$$G(s_1, \ldots, s_r) := \sum_{n_1=1,\ldots,n_r=1}^{\infty} \frac{g(n_1, \ldots, n_r)}{n_1^{s_1} \cdots n_r^{s_r}}$$

is absolutely convergent for

$$(s_1, \ldots, s_r) = \left(a_1 - \varepsilon, \ldots, a_{j-1} - \varepsilon, b_j - \varepsilon, a_{j+1} - \varepsilon, \ldots, a_r - \varepsilon\right)$$

for all $j \in \{1, \ldots, r\}$ and all $\varepsilon \in \left]0, \frac{1}{2}\right[$.

Set $A := a_1 + \cdots + a_r$, $m := \min(a_1 - b_1, \ldots, a_r - b_r)$ and $D := \delta_1 + \cdots + \delta_r$. Then

$$\sum_{n_1 \leqslant x, \ldots, n_r \leqslant x} f(n_1, \ldots, n_r) = x^A Q_D(\log x) + O\left(x^{A-m}(\log x)^D\right),$$

where Q_D is a polynomial of degree D and leading coefficient $\alpha_1 \cdots \alpha_r G(a_1, \ldots, a_r)$.

In certain particular cases, the result is simpler.

Proposition 4.30 *Let $r \in \mathbb{Z}_{\geqslant 1}$ and $f : \mathbb{Z}_{\geqslant 1} \longrightarrow \mathbb{C}$ be a completely multiplicative function such that, for any prime number p, $|f(p)| \leqslant 1$. Set $\theta := \max_{p \text{ prime}} |1 - f(p)|$ and assume that $\theta r \geqslant 1$. Then*

$$\sum_{n \leqslant x} \sum_{i_1=1}^{n} \cdots \sum_{i_r=1}^{n} f\left((i_1, n) \cdots (i_r, n)\right) = \kappa_r(f) x^{r+1} + O\left(\left(x \log^\theta x\right)^r\right),$$

where

$$\kappa_r(f) := \frac{1}{r+1} \prod_p \left(1 - \frac{1}{p}\right) \left(1 + \sum_{\alpha=1}^{\infty} \frac{1}{p^\alpha} \left(1 - \left(1 - \left(\frac{f(p)}{p}\right)^\alpha\right) \frac{1 - f(p)}{p - f(p)}\right)^r\right).$$

Proof See [9, Theorem 3]. □

4.6 Sieves

4.6.1 Combinatorial Sieve

Let $n \geqslant 2$ be a fixed integer. The well-known sieve of Eratosthenes asserts that an integer $m \in (\sqrt{n}, n]$ which is not divisible by any prime number $p \leqslant \sqrt{n}$ is prime. Let P_n be the set of prime numbers $p \leqslant n$ and S_n be the set of positive integers $m \leqslant n$ which are not divisible by all prime numbers $\leqslant \sqrt{n}$. We then have

$$P_n \subseteq S_n \cup \{1, \ldots, \sqrt{n}\}$$

and hence

$$\pi(n) \leqslant |S_n| + \lfloor \sqrt{n} \rfloor.$$

More generally, let $r \geqslant 2$ be an integer. We define $\pi(n, r)$ to be the number of positive integers $m \leqslant n$ which are not divisible by prime numbers $\leqslant r$, so that $|S_n| = \pi(n, \lfloor \sqrt{n} \rfloor)$. Similar arguments as above give

$$\pi(n) \leqslant \pi(n, r) + r. \qquad (4.45)$$

One may bound $\pi(n, r)$ by appealing to the *inclusion-exclusion principle* which generalizes the well-known formula

$$|A \cup B| = |A| + |B| - |A \cap B|.$$

There exist many statements of this result, but in number theory we often use the following one.

Proposition 4.31 *Consider N objects and r properties denoted by $p_1, \ldots, p_r$. Suppose that N_1 objects satisfy the property p_1, N_2 objects satisfy the property p_2, ..., N_{12} objects satisfy the properties p_1 and p_2, ..., N_{123} objects satisfy the properties p_1, p_2 and p_3, and so on. Then, the number of objects which satisfy none of those properties is equal to*

$$N - N_1 - N_2 - \cdots - N_r + N_{12} + N_{13} + \cdots + N_{r-1,r} - N_{123} - N_{124} - \cdots$$

For instance, the identity $\max(a, b) = a + b - \min(a, b)$ can be generalized into the following one

$$\max(a_1, \ldots, a_r) = a_1 + \cdots + a_r - \min(a_1, a_2) - \cdots - \min(a_{r-1}, a_r)$$
$$+ \cdots \pm \min(a_1, \ldots, a_r).$$

Applied to $\pi(n, r)$, we derive

$$\pi(n, r) = n - \sum_{p \leqslant r} \left\lfloor \frac{n}{p} \right\rfloor + \sum_{p_1 < p_2 \leqslant r} \left\lfloor \frac{n}{p_1 p_2} \right\rfloor - \sum_{p_1 < p_2 < p_3 \leqslant r} \left\lfloor \frac{n}{p_1 p_2 p_3} \right\rfloor + \cdots$$

$$(4.46)$$

Since $x - 1 < \lfloor x \rfloor \leqslant x$, we obtain

$$\pi(n, r) < n - \sum_{p \leqslant r} \frac{n}{p} + \sum_{p_1 < p_2 \leqslant r} \frac{n}{p_1 p_2} + \cdots + \sum_{p \leqslant r} 1 + \sum_{p_1 < p_2 \leqslant r} 1 + \cdots$$

$$= n - \sum_{p \leqslant r} \frac{n}{p} + \sum_{p_1 < p_2 \leqslant r} \frac{n}{p_1 p_2} + \cdots + \binom{\pi(r)}{1} + \binom{\pi(r)}{2} + \cdots$$

$$= n \prod_{p \leqslant r} \left(1 - \frac{1}{p}\right) + 2^{\pi(r)} - 1.$$

Now inserting this bound in (4.45) implies that

$$\pi(n) < n \prod_{p \leqslant r} \left(1 - \frac{1}{p}\right) + 2^{\pi(r)} + r - 1.$$

Using the inequalities $\log(1 - x) \leqslant -x$, we get

$$\prod_{p \leqslant r} \left(1 - \frac{1}{p}\right) \leqslant \exp\left(-\sum_{p \leqslant r} \frac{1}{p}\right) < \frac{e^{1/2}}{\log r}$$

so that

$$\pi(n) < \frac{n e^{1/2}}{\log r} + 2^r + r - 1.$$

Choosing $r = 1 + \lfloor \log n \rfloor$ with $n \geqslant 7$ implies that

$$\pi(n) < \frac{3n}{\log \log n}.$$

This is a weaker result than Theorem 3.1, but the ideas developed above are very fruitful and eventually gave birth to an efficient new branch in number theory called *sieve methods*. Nevertheless, it is interesting to note that this inequality is sufficient to assert that the prime numbers *rarefy*, i.e. $\pi(n) = o(n)$ as $n \to \infty$.

Let us adopt a more arithmetical point of view and let $\mathcal{A}$ be a finite set of integers, and define

$$S(\mathcal{A}, \mathcal{P}; z) = \sum_{\substack{n \in \mathcal{A} \\ (n, P_z) = 1}} 1,$$

where $P_z = \prod_{p \in \mathcal{P}, \, p \leqslant z} p$ and $\mathcal{P}$ is a set of primes. By Vinogradov's Lemma[22] we derive

$$S(\mathcal{A}, \mathcal{P}; z) = \sum_{d | P_z} \mu(d) A_d, \tag{4.47}$$

where

$$A_d = \sum_{\substack{n \in \mathcal{A} \\ d | n}} 1. \tag{4.48}$$

Assume that, for all positive integers d, we have

$$A_d = \frac{X \rho(d)}{d} + r_d, \tag{4.49}$$

where $X > 0$, $\rho(d) \geqslant 0$ is multiplicative and the remainder term satisfies $|r_d| \leqslant \rho(d)$. Inserting this estimate into (4.47) implies that

$$S(\mathcal{A}, \mathcal{P}; z) = X \sum_{d | P_z} \frac{\mu(d) \rho(d)}{d} + \sum_{d | P_z} \mu(d) r_d = X \prod_{p \leqslant z} \left(1 - \frac{\rho(p)}{p}\right) + \sum_{d | P_z} \mu(d) r_d.$$

Suppose we want to count the number of prime numbers in an interval $(x, x + y]$ with $x, y \in \mathbb{Z}_{\geqslant 0}$. We take $\mathcal{A} = (x, x + y] \cap \mathbb{Z}$, $\mathcal{P}$ the set of all primes, and since

$$A_d = \sum_{\substack{x < n \leqslant x + y \\ d | n}} 1 = \frac{y}{d} + \left\{\frac{x}{d}\right\} - \left\{\frac{x + y}{d}\right\}$$

we have $X = y$, $\rho(d) = 1$ which gives with $z = \sqrt{x}$ and setting $P = P_{\sqrt{x}}$

$$\pi(x + y) - \pi(x) = y \prod_{p \leqslant x^{1/2}} \left(1 - \frac{1}{p}\right) + \sum_{d | P} \mu(d) \left(\left\{\frac{x}{d}\right\} - \left\{\frac{x + y}{d}\right\}\right)$$

---

[22] See Corollary 4.6.

and using Corollary 3.5 gives

$$
\pi(x+y) - \pi(x) = (1 + o(1)) \frac{2ye^{-\gamma}}{\log x} + \sum_{d\mid P} \mu(d) \left(\left\{ \frac{x}{d} \right\} - \left\{ \frac{x+y}{d} \right\} \right).
$$

A crude estimate of the remainder term then shows that it is larger than the main term, partly because of the fact that as z increases, the factors of P_z become very large and so does their number. It seems to be very difficult to take account of some cancellations of the summands. This is the limitation of the sieve of Eratosthenes.

In 1915, Brun [15] came up with a simple, but very efficient idea. Suppose that we have at our disposal a function g such that

$$
\sum_{d\mid n} \mu(d) \leqslant \sum_{d\mid n} g(d).
$$

Then by repeating the computations above, we get

$$
S(\mathcal{A}, \mathcal{P}; z) \leqslant X \prod_{p\mid P_z} \left(1 - \frac{\rho(p)}{p} \right) + O\left(\sum_{d\mid P_z} |g(d)r_d| \right).
$$

The problem is then to find a suitable function g which is easier to handle than the Möbius function, and to minimize the right-hand side of this inequality. Moving from the classic world of exactness to the world of inequalities, Brun proved that

$$
\sum_{\substack{d\mid(n,P_z)\\ \omega(d)\leqslant 2k+1}} \mu(d) \leqslant \sum_{d\mid(n,P_z)} \mu(d) \leqslant \sum_{\substack{d\mid(n,P_z)\\ \omega(d)\leqslant 2k}} \mu(d)
$$

for all integers $n, k \geqslant 0$ and all $z \geqslant 2$. This implies that

$$
\sum_{\substack{d\mid P_z\\ \omega(d)\leqslant 2k+1}} \mu(d)A_d \leqslant S(\mathcal{A}, \mathcal{P}; z) \leqslant \sum_{\substack{d\mid P_z\\ \omega(d)\leqslant 2k}} \mu(d)A_d \tag{4.50}
$$

for all integers $k \geqslant 0$ and all $z \geqslant 2$. These inequalities, named *Brun's pure sieve* , are related to the so-called *Bonferroni's inequalities* stating that, if $A_1, \ldots, A_n$ are subsets of a finite set S, then we have

$$
\sum_{j=0}^{2k+1} (-1)^j S_j \leqslant \left| \overline{A_1} \cap \cdots \cap \overline{A_n} \right| \leqslant \sum_{j=0}^{2k} (-1)^j S_j \quad \left(k \in \mathbb{Z}_{\geqslant 0} \right),
$$

where $\overline{A_j}$ is the complement of A_j in S and $S_h := \sum_{\{i_1,\ldots,i_h\}\in\{1,\ldots,n\}} \left| A_{i_1} \cap \cdots \cap A_{i_h} \right|$.

Let us try Brun's ideas on the example of *twin primes*,[23] and let $\pi_2(x)$ be the number of twin primes $\leqslant x$. In this problem, we work with $\mathcal{P}$ the set of all primes, and

$$\mathcal{A} = \{n(n+2) : 1 \leqslant n < x - 2\}$$

so that A_d is the number of solutions of the congruence $n(n+2) \equiv 0 \pmod{d}$ with d squarefree. Writing $d = 2^e d'$ with d' odd and $e \in \{0, 1\}$ and applying the Chinese remainder theorem, we deduce that (4.49) holds with $X = x$ and the strongly multiplicative function ρ defined by $\rho(2) = 1$ and $\rho(p) = 2$ for all odd prime numbers p, and we have $r_d \ll \rho(d)$. Now if p is a twin prime less than x, then either $p \leqslant z$ or the number $n(n+2)$ has no prime factor $\leqslant z$, and hence

$$\pi_2(x) \leqslant S(\mathcal{A}, \mathcal{P}; z) + z.$$

Using (4.50) we get

$$\pi_2(x) \leqslant \sum_{\substack{d \mid P_z \\ \omega(d) \leqslant 2k}} \mu(d) A_d + z$$

$$= x \sum_{\substack{d \mid P_z \\ \omega(d) \leqslant 2k}} \frac{\mu(d)\rho(d)}{d} + z + O\left(\sum_{\substack{d \mid P_z \\ \omega(d) \leqslant 2k}} \mu^2(d)\rho(d) \right)$$

$$= x \sum_{d \mid P_z} \frac{\mu(d)\rho(d)}{d} + z + O\left(\sum_{\substack{d \mid P_z \\ \omega(d) \leqslant 2k}} \mu^2(d)\rho(d) + x \sum_{\substack{d \mid P_z \\ \omega(d) > 2k}} \frac{\mu^2(d)\rho(d)}{d} \right)$$

$$= \frac{x}{2} \prod_{3 \leqslant p \leqslant z} \left(1 - \frac{2}{p}\right) + z + O\left(\sum_{\substack{d \mid P_z \\ \omega(d) \leqslant 2k}} \mu^2(d)\rho(d) + x \sum_{\substack{d \mid P_z \\ \omega(d) > 2k}} \frac{\mu^2(d)\rho(d)}{d} \right).$$

Using $\mu^2(d)\rho(d) \leqslant \mu^2(d) 2^{\omega(d)}$ gives

$$\sum_{\substack{d \mid P_z \\ \omega(d) \leqslant 2k}} \mu^2(d)\rho(d) \leqslant \sum_{d \leqslant z^{2k}} \mu^2(d) 2^{\omega(d)} < 3k\, z^{2k} \log z$$

---

[23] The twin primes are the pairs $(p, p+2)$ for which both p and $p+2$ are primes.

and

$$\sum_{\substack{d \mid P_z \\ \omega(d) > 2k}} \frac{\mu^2(d)\rho(d)}{d} \leqslant \frac{1}{4^k} \sum_{d \mid P_z} \frac{\mu^2(d)\rho(d)2^{\omega(d)}}{d}$$

$$\leqslant \frac{1}{4^k} \sum_{d \leqslant z^{2k}} \frac{\mu^2(d)4^{\omega(d)}}{d} < 82(\log z)^4$$

for all $z \geqslant 2$, where we used Exercises 80 and 81.[24] By Corollary 3.5 we have

$$\frac{x}{2} \prod_{3 \leqslant p \leqslant z} \left(1 - \frac{2}{p}\right) \leqslant 2x \prod_{p \leqslant z} \left(1 - \frac{1}{p}\right)^2 \ll \frac{x}{(\log z)^2}$$

and choosing $k = \lfloor \log x / \log(z^3) \rfloor$ and $z = \exp\left(\dfrac{\log x}{20 \log \log x}\right)$ implies that

$$\pi_2(x) \ll x \left(\frac{\log \log x}{\log x}\right)^2.$$

By partial summation, we obtain Brun's theorem [16].

Theorem 4.43 (Brun)   *The sum of reciprocals of twin primes converges.*

4.6.2   The Selberg's Sieve

The aim of this section is to introduce another powerful tool developed by Atle Selberg in the late 1940s. We first recall the specific notation in sieve methods.

Let $z \geqslant 2$ be a real number, $\mathcal{P}$ be a set of primes, $P_z = \displaystyle\prod_{p \in \mathcal{P},\, p \leqslant z} p$ and $\mathcal{A}$ be a finite set of integers such that, for all $d \mid P_z$, we have

$$\sum_{\substack{n \in \mathcal{A} \\ d \mid n}} 1 = \frac{X\rho(d)}{d} + r_d \tag{4.51}$$

---

[24] Implying that $\sum_{d \leqslant x} 2^{\omega(d)} < \frac{3}{2} x \log x$ and $\sum_{d \leqslant x} \mu^2(d)4^{\omega(d)}d^{-1} < 4(\log x)^4$ whenever $x \geqslant 4$.

with $X > 0$, ρ being a non-negative multiplicative function and r_d being a remainder term satisfying $|r_d| \leqslant \rho(d)$. We set

$$S(\mathcal{A}, \mathcal{P}; z) = \sum_{\substack{n \in \mathcal{A} \\ (n, P_z) = 1}} 1. \tag{4.52}$$

Instead of using Bonferroni-like inequalities (4.50), Selberg observed that if $\lambda :$ $\mathbb{Z}_{\geqslant 1} \to \mathbb{R}$ is any function such that $\lambda(1) = 1$, then, for all n, we have

$$\sum_{d \mid n} \mu(d) \leqslant \left(\sum_{d \mid n} \lambda(d) \right)^2 \tag{4.53}$$

and squaring out the inner sum implies that

$$S(\mathcal{A}, \mathcal{P}; z) \leqslant \sum_{n \in \mathcal{A}} \left(\sum_{d \mid (n, P_z)} \lambda(d) \right)^2 = \sum_{d \mid P_z} A_d \sum_{[d_1, d_2] = d} \lambda(d_1) \lambda(d_2).$$

Using (4.49), we get

$$S(\mathcal{A}, \mathcal{P}; z) \leqslant X \sum_{d \mid P_z} \frac{\rho(d)}{d} \sum_{[d_1, d_2] = d} \lambda(d_1) \lambda(d_2) + O\left(\sum_{d \mid P_z} |r_d| \sum_{[d_1, d_2] = d} |\lambda(d_1) \lambda(d_2)| \right).$$

Assume $1 \leqslant \rho(d) < d$ for $d \mid P_z$. Since ρ is multiplicative, we have

$$\rho(d_1) \rho(d_2) = \rho((d_1, d_2)) \rho([d_1, d_2])$$

and setting $\phi(d) = d / \rho(d)$ and $f = \mu \star \phi$ we get

$$\sum_{d \mid P_z} \frac{\rho(d)}{d} \sum_{[d_1, d_2] = d} \lambda(d_1) \lambda(d_2) = \sum_{d_1, d_2 \mid P_z} \frac{\lambda(d_1) \lambda(d_2)}{[d_1, d_2]} \rho([d_1, d_2])$$

$$= \sum_{d_1, d_2 \mid P_z} \left(\frac{\lambda(d_1) \lambda(d_2) (d_1, d_2)}{d_1 d_2} \times \frac{\rho(d_1) \rho(d_2)}{\rho((d_1, d_2))} \right)$$

$$= \sum_{d_1, d_2 \mid P_z} \frac{\lambda(d_1) \lambda(d_2) \phi((d_1, d_2))}{\phi(d_1) \phi(d_2)}.$$

Now by the Möbius inversion formula, we have $\phi = f \star \mathbf{1}$ and hence

$$
\sum_{d|P_z} \frac{\rho(d)}{d} \sum_{[d_1,d_2]=d} \lambda(d_1)\lambda(d_2) = \sum_{d_1,d_2|P_z} \sum_{\delta|(d_1,d_2)} \frac{f(\delta)\lambda(d_1)\lambda(d_2)}{\phi(d_1)\phi(d_2)}
$$

$$
= \sum_{\delta|P_z} f(\delta) \left(\sum_{\substack{d|P_z \\ \delta|d}} \frac{\lambda(d)}{\phi(d)} \right)^2 .
$$

We have then proved the estimate

$$
\mathcal{S}(\mathcal{A}, \mathcal{P}; z) \leqslant X \sum_{d|P_z} f(d) \left(\sum_{\substack{\delta|P_z \\ d|\delta}} \frac{\lambda(\delta)}{\phi(\delta)} \right)^2 + O\left(\sum_{d_1,d_2|P_z} |\lambda(d_1)\lambda(d_2)r_{[d_1,d_2]}| \right)
$$

(4.54)

for all $z \geqslant 2$, if (4.49) holds with $1 \leqslant \rho(d) < d$ for $d \mid P_z$, and where $\lambda : \mathbb{Z}_{\geqslant 1} \to \mathbb{R}$ satisfies $\lambda(1) = 1$. This leads to an optimization problem of a quadratic form which was solved by Selberg. Note that by (4.53) we have the freedom to choose any function λ we like, subject only to the constraint $\lambda(1) = 1$. We begin by setting

$$
y_d = \sum_{\substack{\delta|P_z \\ d|\delta}} \frac{\lambda(\delta)}{\phi(\delta)} .
$$

(4.55)

For all positive integers $d \mid P_z$, we have by the Möbius inversion formula

$$
\sum_{\substack{n|P_z \\ d|n}} \mu\left(\frac{n}{d}\right) y_n = \frac{\lambda(d)}{\phi(d)}
$$

so that

$$
\lambda(d) = \phi(d) \sum_{\substack{n|P_z \\ d|n}} \mu\left(\frac{n}{d}\right) y_n
$$

(4.56)

which implies in particular that the linear transformation (4.55) is invertible. From (4.55) and (4.56), we have

$$
\lambda(d) = 0 \text{ for } d > z \iff y_d = 0 \text{ for } d > z
$$

which we may suppose for convenience. Since $\phi(d) = d/\rho(d)$, we get $\phi(1) = 1$. By (4.56), the condition $\lambda(1) = 1$ may be written as

$$\sum_{\substack{n|P_z \\ n \leqslant z}} \mu(n) y_n = 1. \tag{4.57}$$

Hence we set about minimizing the quadratic form

$$\sum_{\substack{d|P_z \\ d \leqslant z}} f(d) y_d^2 \tag{4.58}$$

subject to (4.57). Let

$$M_z = \sum_{\substack{d|P_z \\ d \leqslant z}} \frac{1}{f(d)}.$$

Using (4.57) and the fact that $d \mid P_z \Rightarrow \mu^2(d) = 1$, we have

$$\sum_{\substack{d|P_z \\ d \leqslant z}} \frac{1}{f(d)} \left(f(d) y_d - \frac{\mu(d)}{M_z} \right)^2 = \sum_{\substack{d|P_z \\ d \leqslant z}} f(d) y_d^2 - \frac{2}{M_z} \sum_{\substack{d|P_z \\ d \leqslant z}} \mu(d) y_d + \frac{1}{M_z^2} \sum_{\substack{d|P_z \\ d \leqslant z}} \frac{\mu^2(d)}{f(d)}$$

$$= \sum_{\substack{d|P_z \\ d \leqslant z}} f(d) y_d^2 - \frac{1}{M_z}$$

which implies that (4.58) is minimized when $y_d = M_z^{-1} \mu(d)/f(d)$, the minimum being equal to M_z^{-1}. Note that (4.57) is satisfied with this choice of y_d. By (4.56), we infer that the choice of $\lambda(d)$ ensuring the minimum of (4.58) is given by

$$\lambda(d) = \frac{\phi(d)}{M_z} \sum_{\substack{n|P_z \\ d|n \\ n \leqslant z}} \left\{ \mu\left(\frac{n}{d}\right) \times \frac{\mu(n)}{f(n)} \right\} = \frac{\phi(d)}{M_z} \sum_{\substack{m|(P_z/d) \\ m \leqslant z/d}} \frac{\mu(m)\mu(md)}{f(md)}$$

$$= \frac{\phi(d)}{M_z} \sum_{\substack{m|P_z \\ (m,d)=1 \\ m \leqslant z/d}} \frac{\mu(m)\mu(md)}{f(md)} = \frac{\phi(d)\mu(d)}{f(d)M_z} \sum_{\substack{m|P_z \\ (m,d)=1 \\ m \leqslant z/d}} \frac{\mu^2(m)}{f(m)},$$

where we used the equivalence

$$m \mid (P_z/d) \iff \begin{cases} m \mid P_z \\ (m, d) = 1. \end{cases}$$

The above sum may be estimated with the following useful tool.

Lemma 4.13 *Let f be a positive multiplicative function, $x \geqslant 2$ be a real number and k be a positive integer. Then*

$$\prod_{p \mid k} (1 + f(p))^{-1} \sum_{n \leqslant x} \mu^2(n) f(n) \leqslant \sum_{\substack{n \leqslant x \\ (n,k)=1}} \mu^2(n) f(n)$$

$$\leqslant \prod_{p \mid k} (1 + f(p))^{-1} \sum_{n \leqslant kx} \mu^2(n) f(n).$$

Proof Each positive integer n may be uniquely written as $n = d_1 d_2$ with $d_1 \mid k$ and $(d_2, k) = 1$, and hence $(d_1, d_2) = 1$. Since f is multiplicative, we get for all $y \geqslant 1$

$$\sum_{n \leqslant y} \mu^2(n) f(n) = \sum_{\substack{d_1 \leqslant y \\ d_1 \mid k}} \mu^2(d_1) f(d_1) \sum_{\substack{d_2 \leqslant y/d_1 \\ (d_2,k)=1}} \mu^2(d_2) f(d_2).$$

Now since f is positive, using the identity above with $y = kx$ gives

$$\sum_{n \leqslant kx} \mu^2(n) f(n) = \sum_{d_1 \mid k} \mu^2(d_1) f(d_1) \sum_{\substack{d_2 \leqslant kx/d_1 \\ (d_2,k)=1}} \mu^2(d_2) f(d_2)$$

$$\geqslant \sum_{d_1 \mid k} \mu^2(d_1) f(d_1) \sum_{\substack{d_2 \leqslant x \\ (d_2,k)=1}} \mu^2(d_2) f(d_2)$$

and with $y = x$ we get

$$\sum_{n \leqslant x} \mu^2(n) f(n) \leqslant \sum_{d_1 \mid k} \mu^2(d_1) f(d_1) \sum_{\substack{d_2 \leqslant x \\ (d_2,k)=1}} \mu^2(d_2) f(d_2)$$

and the identity

$$\sum_{d_1 \mid k} \mu^2(d_1) f(d_1) = \prod_{p \mid k} (1 + f(p))$$

completes the proof.   □

Thus, for all $d \mid P_z$ and $d \leqslant z$, we get

$$\sum_{\substack{m|P_z \\ (m,d)=1 \\ m \leqslant z/d}} \frac{\mu^2(m)}{f(m)} \leqslant \prod_{p|d} \left(1 + \frac{1}{f(p)}\right)^{-1} \sum_{\substack{m|P_z \\ m \leqslant z}} \frac{\mu^2(m)}{f(m)} = \frac{f(d)}{\phi(d)} \times M_z$$

and hence $|\lambda(d)| \leqslant 1$ for all $d \mid P_z$ such that $d \leqslant z$. Therefore, using $|r_d| \leqslant \rho(d)$ and $\rho(d) \geqslant 1$ for $d \mid P_z$, we infer that the error term of (4.54) is bounded by

$$\sum_{d_1,d_2|P_z} |\lambda(d_1)\lambda(d_2)| \frac{\rho(d_1)\rho(d_2)}{\rho((d_1,d_2))} \leqslant \sum_{d_1,d_2|P_z} \rho(d_1)\rho(d_2) \leqslant z^2 \left(\sum_{\substack{d|P_z \\ d \leqslant z}} \frac{\rho(d)}{d}\right)^2$$

$$\leqslant z^2 \prod_{p|P_z} \left(1 + \frac{\rho(p)}{p}\right)^2 \leqslant z^2 \prod_{p \leqslant z} \left(1 - \frac{\rho(p)}{p}\right)^{-2}.$$

Finally, using $1 \leqslant \rho(d) < d$, we have

$$\sum_{\substack{d|P_z \\ d \leqslant z}} \frac{\mu^2(d)\rho(d)}{\varphi(d)} \leqslant \sum_{\substack{d|P_z \\ d \leqslant z}} \frac{\mu^2(d)\rho(d)}{d} \prod_{p|d} \left(1 - \frac{\rho(p)}{p}\right)^{-1}$$

$$= \sum_{\substack{d|P_z \\ d \leqslant z}} \frac{\mu^2(d)\rho(d)}{df(d)} < \sum_{\substack{d|P_z \\ d \leqslant z}} \frac{\mu^2(d)}{f(d)} = M_z.$$

Putting all together we have proved Selberg's upper bound sieve.

Theorem 4.44 (Selberg) *Let $z \geqslant 2$, P_z be a set of primes $p \leqslant z$, $P_z = \prod_{p \in P_z} p$ and $\mathcal{A}$ be a finite set of integers assuming (4.51) for all $d \mid P_z$, with $X > 0$, ρ is a multiplicative function such that $1 \leqslant \rho(d) < d$ and $|r_d| \leqslant \rho(d)$. Setting $S(\mathcal{A}, \mathcal{P}; z)$ as in (4.52), we have*

$$S(\mathcal{A}, \mathcal{P}; z) \leqslant X \left(\sum_{\substack{d|P_z \\ d \leqslant z}} \frac{\mu^2(d)\rho(d)}{\varphi(d)}\right)^{-1} + z^2 \prod_{p \leqslant z} \left(1 - \frac{\rho(p)}{p}\right)^{-2}.$$

In order to have a bound for the sum of the main term, the following lemma will be useful.

Lemma 4.14 *Let $z \geqslant 2$ be a real number and $k \in \mathbb{N}$. Then*

$$\sum_{\substack{n \leqslant z \\ (n,k)=1}} \frac{\mu^2(n)}{\varphi(n)} > \frac{\varphi(k)}{k} \log z.$$

Proof The lower bound of Lemma 4.13 first gives

$$\sum_{\substack{n \leqslant z \\ (n,k)=1}} \frac{\mu^2(n)}{\varphi(n)} \geqslant \frac{\varphi(k)}{k} \sum_{n \leqslant z} \frac{\mu^2(n)}{\varphi(n)}.$$

Now each positive integer n can be uniquely written as $n = qd$ with q squarefree and $d \mid q^\infty$. Note also that

$$\sum_{d \mid q^\infty} \frac{1}{d} = \prod_{p \mid q} \left(1 + \frac{1}{p} + \frac{1}{p^2} + \cdots \right) = \prod_{p \mid q} \left(1 - \frac{1}{p} \right)^{-1} = \frac{q}{\varphi(q)}.$$

Therefore we get

$$\log z < \sum_{n \leqslant z} \frac{1}{n} \leqslant \sum_{q \leqslant z} \frac{\mu^2(q)}{q} \sum_{d \mid q^\infty} \frac{1}{d} = \sum_{q \leqslant z} \frac{\mu^2(q)}{\varphi(q)}$$

which concludes the proof.   □

When $\rho(d) = 1$, the above results yield a useful estimate.

Corollary 4.15 *With the notation of* Theorem 4.44 *assuming $\rho(d) = 1$ and $|r_d| \leqslant 1$ for all $d \mid P_z$, we have*

$$S(\mathcal{A}, \mathcal{P}; z) < \frac{X}{\log z} \prod_{\substack{p \leqslant z \\ p \notin \mathcal{P}}} \left(1 - \frac{1}{p} \right)^{-1} + z^2.$$

Proof Set

$$Q_z = \prod_{\substack{p \leqslant z \\ p \notin \mathcal{P}}} p.$$

When $\rho(d) = 1$, then $\phi(d) = d$ and hence $f(d) = \varphi(d)$, so that we have by Lemma 4.14

$$M_z = \sum_{\substack{d \leqslant z \\ (Q_z, d)=1}} \frac{\mu^2(d)}{\varphi(d)} > \frac{\varphi(Q_z)}{Q_z} \log z = \prod_{\substack{p \leqslant z \\ p \notin \mathcal{P}}} \left(1 - \frac{1}{p}\right) \log z.$$

Furthermore, since $|\lambda(d)| \leqslant 1$ and $|r_d| \leqslant 1$ for all $d \mid P_z$, we infer that the error term in (4.54) is bounded by z^2, concluding the proof. $\qquad\square$

One of the most famous applications of Selberg's sieve is Brun–Titchmarsh's theorem.

> **Theorem 4.45 (Brun–Titchmarsh)** *Let a, q be positive coprime integers and set*
>
> $$\pi(x; q, a) = \sum_{\substack{p \leqslant x \\ p \equiv a \,(\mathrm{mod}\, q)}} 1.$$
>
> *Assume that $x > q$. Then*
>
> $$\pi(x; q, a) \leqslant \frac{6x}{\varphi(q) \log(x/q)}.$$

Proof If $1 < x/q \leqslant 20$, then the trivial estimate implies

$$\pi(x; q, a) \leqslant \frac{2x}{q} \leqslant \frac{6x}{\varphi(q) \log(20)} \leqslant \frac{6x}{\varphi(q) \log(x/q)}$$

so we may suppose that $x/q > 20$. Let $z \geqslant 2$ be a real number and we take $\mathcal{P}$ the set of primes $p \nmid q$ and $\mathcal{A}$ the set of numbers $n \leqslant x$ such that $n \equiv a \pmod{q}$. We then have

$$\pi(x; q, a) \leqslant S(\mathcal{A}, \mathcal{P}; z) + z.$$

Now if $(d, q) = 1$, then

$$\sum_{\substack{n \in \mathcal{A} \\ d \mid n}} 1 = \frac{x}{qd} + r_d$$

with $|r_d| \leqslant 1$, so that Selberg's sieve may be used with $X = x/q$ and $\rho(d) = 1$ if $(d, q) = 1$. Therefore by Corollary 4.15 we get

$$\pi(x; q, a) \leqslant \frac{x}{\varphi(q) \log z} + z^2 + z$$

and choosing $z = (x/q)^{1/2} \log^{-1/2}(x/q)$ gives the asserted result, using $z > (x/q)^{1/4}$ in the first term.   □

For an extensive account on sieve methods, the reader is referred to [4, 40].

4.6.3   The Large Sieve

4.6.3.1   Introduction

The large sieve was invented by Ju. V. Linnik in 1941 [70] while investigating the distribution of quadratic non-residues.[25] In a series of papers beginning in the late 1940s, Renyi was the first to systematically study the large sieve and to prove that each large even number $2k$ may be expressed in the form $2k = p + P_k$ where $\omega(P_k) \leqslant k$. Subsequently, values of k were given by several authors until Chen proved that $k = 2$ is admissible. The large sieve always played a key part. The large sieve remained in the area of a few specialists until 1965 when major contributions were made by Roth and then by Bombieri. They paved the way for the recognition that these results rely on an underlying analytic inequality dealing with trigonometric polynomials. The analytic principle of the large sieve was formulated explicitly 1 year later by Davenport and Halberstam. For an account of the theory of the large sieve, we may refer the reader to [4, 32, 77]. Nowadays, the large sieve is one of the most powerful tools in multiplicative number theory. Roughly speaking, it may be regarded as Fourier analysis of arithmetic progressions, both from the additive and multiplicative points of view. Hence the usual tools from Fourier analysis[26] are the main ingredients of the large sieve.

4.6.3.2   Additive Large Sieve

The analytic additive form of the large sieve can be described as follows. Let $M \in \mathbb{Z}$, N be a positive integer, (a_n) be a sequence of arbitrary complex numbers supported on the interval $(M, M + N]$ and let

$$S(\alpha) = \sum_{M < n \leqslant M+N} a_n e(n\alpha) \tag{4.59}$$

---

[25] see Example 4.10.

[26] Quasi-orthogonal systems, Bessel's and Hilbert's inequalities, etc.

be a trigonometric polynomial.

Definition 4.12 Let $\delta > 0$ and assume that $\alpha_1, \ldots, \alpha_R$ are δ-*well spaced*, i.e. for all $r \neq s \in \{1, \ldots, R\}$, we have

$$\|\alpha_r - \alpha_s\| \geqslant \delta. \tag{4.60}$$

The large sieve is an inequality of the shape

$$\sum_{r=1}^{R} |S(\alpha_r)|^2 \leqslant \Delta(N, \delta) \sum_{n=M+1}^{M+N} |a_n|^2. \tag{4.61}$$

Since $|S|$ is 1-periodic, one may assume that $0 < \alpha_1 < \alpha_2 < \cdots < \alpha_R \leqslant 1$ with $\delta \leqslant R^{-1}$, and since

$$\left| \sum_{n=M+K+1}^{M+K+N} a_{n-K} e(n\alpha) \right| = |e(K\alpha)S(\alpha)| = |S(\alpha)|$$

we see that the parameter M is irrelevant, and hence we are interested in determining how $\Delta(N, \delta)$ depends on N and δ. The next lemma is a version of Parseval's well-known identity.

Lemma 4.15 (Parseval's Identity) *We have*

$$\int_0^1 |S(t)|^2 \, dt = \sum_{M < n \leqslant M+N} |a_n|^2.$$

Proof Squaring out we get

$$|S(t)|^2 = \sum_{M < n \leqslant M+N} |a_n|^2 + \sum_{n_1 \neq n_2} a_{n_1} \overline{a_{n_2}} \, e((n_1 - n_2)t)$$

and the result follows from the identity

$$\int_0^1 e(kt) \, dt = \begin{cases} 1, & \text{if } k = 0 \\ 0, & \text{otherwise} \end{cases} \tag{4.62}$$

which concludes the proof. $\square$

Observe that $\Delta(N, \delta)$ cannot be too small. If $a_n = e(-n\alpha_1)$, then

$$|S(\alpha_1)|^2 = N^2 = N \sum_{M < n \leqslant M+N} |a_n|^2$$

and hence $\Delta(N, \delta) \geqslant N$. On the other hand, assume that the α_r are equally spaced so that $\delta = R^{-1}$. By periodicity and Lemma 4.15, we have

$$\int_0^1 \sum_{r=1}^R |S(\alpha_r + t)|^2 \, dt = R \int_0^1 |S(t)|^2 \, dt = R \sum_{M < n \leqslant M+N} |a_n|^2$$

so that for some values of t we have

$$\sum_{r=1}^R |S(\alpha_r + t)|^2 \geqslant R \sum_{M < n \leqslant M+N} |a_n|^2$$

and therefore $\Delta(N, \delta) \geqslant \lfloor \delta^{-1} \rfloor > \delta^{-1} - 1$. The following optimal result, attributed to Selberg, shows that we may take $\Delta(N, \delta)$ to be barely larger than required by the above observation.

Theorem 4.46 (Large Sieve) *Under the hypothesis* (4.60), *the inequality* (4.61) *holds with*

$$\Delta(N, \delta) = N + \delta^{-1} - 1.$$

A possible proof uses generalizations of Bessel's and Hilbert's inequalities. We shall see that one can prove a weaker version of this result, although sufficiently strong to be used in many applications, with the following quicker and elegant ideas due to Gallagher. We start with a useful lemma.

Lemma 4.16 *Let $f \in C^1[a, b]$ with $a < b$. Then*

$$\left| f\left(\frac{a+b}{2} \right) \right| \leqslant \frac{1}{b-a} \int_a^b |f(t)| \, dt + \frac{1}{2} \int_a^b |f'(t)| \, dt.$$

Proof An integration by parts provides for all $x \in [a, b]$

$$\int_x^b \frac{t-b}{b-a} f'(t) \, dt + \int_a^x \frac{t-a}{b-a} f'(t) \, dt = f(x) - \frac{1}{b-a} \int_a^b f(t) \, dt$$

and the result follows by noticing that, if $x = \frac{a+b}{2}$, then $\left| \frac{t-b}{b-a} \right| \leqslant \frac{1}{2}$ in the first integral and $\left| \frac{t-a}{b-a} \right| \leqslant \frac{1}{2}$ in the second one. □

We are now in a position to prove Gallagher's version of the large sieve [33].

Theorem 4.47 (Gallagher) *Under the hypothesis* (4.60), *the inequality* (4.61) *holds with*

$$\Delta(N, \delta) = \pi N + \delta^{-1}.$$

Proof Lemma 4.16 with $a = \alpha_r - \frac{1}{2}\delta$ and $b = \alpha_r + \frac{1}{2}\delta$ implies that

$$|f(\alpha_r)| \leqslant \frac{1}{\delta} \int_{\alpha_r-\delta/2}^{\alpha_r+\delta/2} |f(t)|\, dt + \frac{1}{2} \int_{\alpha_r-\delta/2}^{\alpha_r+\delta/2} |f'(t)|\, dt$$

and taking $f(\alpha) = S(\alpha)^2$ we get

$$|S(\alpha_r)|^2 \leqslant \frac{1}{\delta} \int_{\alpha_r-\delta/2}^{\alpha_r+\delta/2} |S(t)|^2\, dt + \int_{\alpha_r-\delta/2}^{\alpha_r+\delta/2} |S(t)S'(t)|\, dt.$$

By (4.60), the intervals $\mathcal{I}_r = \left(\alpha_r - \frac{1}{2}\delta, \alpha_r + \frac{1}{2}\delta\right)$ do not overlap modulo 1, meaning that if $r \neq s$, then no point of $\mathcal{I}_r$ differs by an integer from another point of $\mathcal{I}_s$. Hence summing over r and using Lemma 4.15, we derive

$$\sum_{r=1}^{R} |S(\alpha_r)|^2 \leqslant \frac{1}{\delta} \int_0^1 |S(t)|^2\, dt + \int_0^1 |S(t)S'(t)|\, dt$$

$$= \frac{1}{\delta} \sum_{M < n \leqslant M+N} |a_n|^2 + \int_0^1 |S(t)S'(t)|\, dt.$$

Now Cauchy–Schwarz's inequality[27] and Parseval's identity again yield

$$\int_0^1 |S(t)S'(t)|\, dt \leqslant \left(\int_0^1 |S(t)|^2\, dt\right)^{1/2} \left(\int_0^1 |S'(t)|^2\, dt\right)^{1/2}$$

$$= \left(\sum_{M < n \leqslant M+N} |a_n|^2\right)^{1/2} \left(\sum_{M < n \leqslant M+N} |2\pi n a_n|^2\right)^{1/2}$$

$$\leqslant 2\pi \left(\max_{M < n \leqslant M+N} |n|\right) \sum_{M < n \leqslant M+N} |a_n|^2.$$

---

[27] Integral analogue of Lemma 6.1.

This gives

$$\sum_{r=1}^{R} |S(\alpha_r)|^2 \leqslant \left\{ \delta^{-1} + 2\pi \left(\max_{M < n \leqslant M+N} |n| \right) \right\} \sum_{M < n \leqslant M+N} |a_n|^2$$

and we conclude the proof by noticing that, since $\Delta(N, \delta)$ is independent of M, we may suppose that $M = -\left\lfloor \frac{1}{2}(N+1) \right\rfloor$ so that $\max\limits_{M < n \leqslant M+N} |n| \leqslant \frac{1}{2} N$. □

4.6.3.3 Arithmetic Large Sieve

The arithmetic version of the large sieve often takes the following form. Consider a finite set $\mathcal{P}$ of prime numbers and, for all $p \in \mathcal{P}$, let Ω_p be a subset of $\mathbb{F}_p$ of residue classes to sieve out. The main purpose of sieve methods is to provide an estimate of

$$\sum_{n \in \mathcal{S}(N, \mathcal{P}, \Omega_p)} a_n,$$

where (a_n) is any sequence of complex numbers and $\mathcal{S}(N, \mathcal{P}, \Omega_p)$ is the so-called *sifted set*

$$\mathcal{S}(N, \mathcal{P}, \Omega_p) = \left\{ n \in (M, M+N] \cap \mathbb{Z} : n \ (\mathrm{mod}\ p) \notin \Omega_p \text{ for all } p \in \mathcal{P} \right\}.$$

We apply Theorems 4.46 or 4.47 with $\alpha_r = a/q$ for some integers a, q such that $1 \leqslant a \leqslant q \leqslant Q$ and $(a, q) = 1$. (4.60) is then satisfied with $\delta = Q^{-2}$ since, for $r \neq s$, we have

$$\|\alpha_r - \alpha_s\| = \left\| \frac{a}{q} - \frac{a'}{q'} \right\| = \left\| \frac{aq' - a'q}{qq'} \right\| \geqslant \frac{1}{qq'} \geqslant \frac{1}{Q^2},$$

where the inequality $\|\frac{n}{m}\| \geqslant \frac{1}{m}$ with $(m, n) = 1$ comes from Exercise 104. Using Theorem 4.46 we get

$$\sum_{q \leqslant Q} \sum_{\substack{a=1 \\ (a,q)=1}}^{q} \left| S\left(\frac{a}{q} \right) \right|^2 \leqslant (N - 1 + Q^2) \sum_{M < n \leqslant M+N} |a_n|^2. \tag{4.63}$$

Now [56, Lemma 7.15] states that, for all positive integers q, we have

$$\sum_{\substack{a=1 \\ (a,q)=1}}^{q} \left| S\left(\frac{a}{q} \right) \right|^2 \geqslant \mu(q)^2 \prod_{p|q} \left(1 - \frac{\rho(p)}{p} \right)^{-1} \left| \sum_{n \in \mathcal{S}(N, \mathcal{P}, \Omega_p)} a_n \right|^2,$$

where, for all primes $p \in \mathcal{P}$, we set

$$\rho(p) = |\Omega_p| \tag{4.64}$$

and hence we obtain the following useful result.

Corollary 4.16 (Arithmetic Large Sieve) *Let (a_n) be any complex-valued sequence supported on $(M, M + N]$. For all positive integers Q, set*

$$L = L(Q) = \sum_{q \leqslant Q} \mu(q)^2 \prod_{p|q} \left(1 - \frac{\rho(p)}{p}\right)^{-1}, \tag{4.65}$$

where $\rho(p)$ is defined in (4.64). If $\rho(p) < p$ for all primes $p \in \mathcal{P}$, then for all $Q \geqslant 1$ we have

$$\left| \sum_{n \in \mathcal{S}(N, \mathcal{P}, \Omega_p)} a_n \right|^2 \leqslant \frac{N - 1 + Q^2}{L} \sum_{M < n \leqslant M + N} |a_n|^2.$$

In particular, we have

$$\sum_{n \in \mathcal{S}(N, \mathcal{P}, \Omega_p)} 1 \leqslant \frac{N - 1 + Q^2}{L}.$$

It is important to keep in mind the following examples given in [4] (with $M = 0$).

Example 4.10

▷ *Sieve of Eratosthenes. $\mathcal{P}$ is the set of primes $p \leqslant \sqrt{N}$ and $\Omega_p = \{\bar{0}\}$.*
▷ *Twin primes. $\mathcal{P}$ is the set of primes $p \leqslant \sqrt{N}$ and $\Omega_p = \{\bar{0}, \bar{2}\}$.*
▷ *Linnik's example on the least quadratic non-residue.*

$$\mathcal{P} = \left\{ p \leqslant \sqrt{N} : \left(\frac{n}{p}\right) = 1 \text{ for all } n \leqslant N^\varepsilon \right\}$$

and $\Omega_p = \{\bar{h} : (h/p) = -1\}$.

A more elaborate large sieve inequality enables the authors in [78] to prove the most elegant version of the Brun–Titchmarsh theorem.

Theorem 4.48 (Montgomery and Vaughan)   *Let* $x, y \geqslant 2$ *be real numbers and* a, q *be positive integers such that* $(a, q) = 1$ *and* $y > q$. *Then*

$$\pi(x + y; q, a) - \pi(x; q, a) < \frac{2y}{\varphi(q) \log(y/q)}.$$

This is a powerful version of Brun–Titchmarsh's inequality which may replace Siegel–Walfisz–Page's theorem[28] in many cases. The factor 2 is of great importance in number theory. More precisely, if the Siegel zero β_1 exists, then it is very close to 1 since we may have

$$1 - \frac{c_1}{\log q} \leqslant \beta_1 < 1.$$

On the other hand, it is proved in [81] that if the estimate

$$\pi(x; q, a) \leqslant \frac{(2 - \varepsilon) x}{\varphi(q) \log(x/q)} \tag{4.66}$$

holds for $x \geqslant q^{c_2}$, where $\varepsilon > 0$ is an absolute constant, then

$$\beta_1 \leqslant 1 - \frac{c_3 \varepsilon}{\log q}$$

and hence (4.66) enables us to disprove the existence of this exceptional zero (4.66) also implies that

$$\pi(x; q, a) = \left\{ 1 + O\left(e^{-c_4 \varepsilon \log x / \log q} \right) \right\} \frac{x}{\varphi(q) \log x}$$

as long as $x \geqslant q^{c_2}$, and thus the constant $2 - \varepsilon$ will be automatically reduced to $1 + \varepsilon$ for all x larger than a sufficiently high power of q.

Vaughan [118] obtained the following useful lower bound for L, in which only a lower estimate of the average values of $\rho(p)$ is sufficient. For all $Q \in \mathbb{Z}_{\geqslant 1}$, we have

$$L \geqslant \max_{m \in \mathbb{N}} \exp\left(m \log \left(\frac{1}{m} \sum_{p \leqslant Q^{1/m}} \frac{\rho(p)}{p} \right) \right). \tag{4.67}$$

This allows us to get the following effective version of Corollary 4.16.

---

[28] See Theorem 3.30.

Corollary 4.17 *Along with the notation of* Corollary 4.16, *let* $\beta > 0$ *and assume that there exists a constant* $c_0 > 0$ *such that, for all sufficiently large* R

$$\sum_{p \leqslant R} \frac{\rho(p)}{p} > c_0 (\log \log R)^\beta.$$

Then there exists a constant $c_1 > 0$ *such that for all sufficiently large* N, *we have*

$$\sum_{n \in \mathcal{S}(N, \mathcal{P}, \Omega_p)} 1 < \frac{N}{\exp \left(c_1 (\log \log N)^\beta \right)}.$$

4.6.3.4 Multiplicative Large Sieve

A multiplicative version of the inequality (4.63) is the following result, also due to Gallagher, which is an important application to estimating averages of character sums.

Theorem 4.49 *For* $M \in \mathbb{Z}$, $Q, N \in \mathbb{Z}_{\geqslant 1}$ *and any sequence* (a_n) *of complex numbers, we have*

$$\sum_{q \leqslant Q} \frac{q}{\varphi(q)} \sideset{}{^\star}\sum_{\chi \,(\mathrm{mod}\, q)} \left| \sum_{M < n \leqslant M+N} a_n \chi(n) \right|^2 \leqslant \left(N - 1 + Q^2 \right) \sum_{M < n \leqslant M+N} |a_n|^2,$$

where the star indicates that the sum is restricted to primitive characters.

Proof From Proposition 6.7, we derive

$$\sum_{M < n \leqslant M+N} a_n \chi(n) = \frac{1}{\tau(\overline{\chi})} \sum_{a \,(\mathrm{mod}\, q)} \overline{\chi}(a) S\left(\frac{a}{q} \right),$$

where $S(\alpha)$ is given in (4.59). Since χ is primitive, we have $\tau(\overline{\chi}) = q^{1/2}$, and hence

$$\sideset{}{^\star}\sum_{\chi \,(\mathrm{mod}\, q)} \left| \sum_{M < n \leqslant M+N} a_n \chi(n) \right|^2 = \frac{1}{q} \sideset{}{^\star}\sum_{\chi \,(\mathrm{mod}\, q)} \left| \sum_{a \,(\mathrm{mod}\, q)} \overline{\chi}(a) S\left(\frac{a}{q} \right) \right|^2$$

$$\leqslant \frac{1}{q} \sum_{\chi \,(\mathrm{mod}\, q)} \left| \sum_{a \,(\mathrm{mod}\, q)} \overline{\chi}(a) S\left(\frac{a}{q} \right) \right|^2$$

$$= \frac{1}{q} \sum_{\chi \,(\mathrm{mod}\, q)} \sum_{a \,(\mathrm{mod}\, q)} \sum_{b \,(\mathrm{mod}\, q)} \overline{\chi}(a) \chi(b) S\left(\frac{a}{q} \right) \overline{S\left(\frac{b}{q} \right)}$$

$$= \frac{1}{q} \sum_{a \,(\mathrm{mod}\, q)} \sum_{b \,(\mathrm{mod}\, q)} S\left(\frac{a}{q}\right) \overline{S\left(\frac{b}{q}\right)} \sum_{\chi \,(\mathrm{mod}\, q)} \overline{\chi}(a) \chi(b)$$

$$= \frac{\varphi(q)}{q} \sum_{\substack{a \,(\mathrm{mod}\, q) \\ (a,q)=1}} \left| S\left(\frac{a}{q}\right) \right|^2 ,$$

where we used Proposition 3.13 in the last line. Therefore

$$\sum_{q \leqslant Q} \frac{q}{\varphi(q)} \sideset{}{^\star}\sum_{\chi \,(\mathrm{mod}\, q)} \left| \sum_{M < n \leqslant M+N} a_n \chi(n) \right|^2 \leqslant \sum_{q \leqslant Q} \sum_{\substack{a \,(\mathrm{mod}\, q) \\ (a,q)=1}} \left| S\left(\frac{a}{q}\right) \right|^2$$

and the result follows with (4.63).   □

Using the Cauchy–Schwarz's inequality and a result dealing with completion of incomplete sums, we derive from Theorem 4.49 the next technical corollary [20, p. 164].

Corollary 4.18 *For* $x \geqslant 2$, $Q, M, N \in \mathbb{Z}_{\geqslant 1}$ *and all sequences* (a_m), (b_n) *of complex numbers, we have*

$$\sum_{q \leqslant Q} \frac{q}{\varphi(q)} \sideset{}{^\star}\sum_{\chi \,(\mathrm{mod}\, q)} \max_{y \leqslant x} \left| \sum_{m \leqslant M} \sum_{\substack{n \leqslant N \\ mn \leqslant y}} a_m b_n \chi(mn) \right|$$

$$\leqslant \left(M + Q^2\right)^{1/2} \left(N + Q^2\right)^{1/2} \left(\sum_{m \leqslant M} |a_m|^2\right)^{1/2} \left(\sum_{n \leqslant N} |b_n|^2\right)^{1/2} \log(xMN),$$

where the star indicates that the sum is restricted to primitive characters.

4.7   Selected Problems in Multiplicative Number Theory

4.7.1   Squarefree Values of $n^2 + 1$

The aim of this section is to prove the next result, ingeniously discovered by Estermann in [27].

Proposition 4.32 *For any* $x \geqslant e$

$$\sum_{n \leqslant x} \mu^2 \left(n^2 + 1\right) = Cx + O\left(x^{2/3} \log x\right),$$

where

$$C := \prod_{p \equiv 1 \ (\mathrm{mod}\, 4)} \left(1 - \frac{2}{p^2}\right) \approx 0.89484. \qquad (4.68)$$

In what follows, set $D_0 := \sqrt{x^2 + 1}$ and let T be any parameter satisfying $\sqrt{2x} \leqslant T \leqslant D_0$, and, for any positive integer d, define

$$N(x; d) := \sum_{\substack{n \leqslant x \\ d^2 | n^2 + 1}} 1 \quad \text{and} \quad \rho(d) := \sum_{\substack{n=1 \\ d^2 | n^2 + 1}}^{d^2} 1.$$

The proof of Proposition 4.32 rests on the following lemma [27, Hilfssatz 2], and which can also be found in [47, (3) page 4].

Lemma 4.17 *Let* $h \in \mathbb{Z}$ *and* $k, a, b \in \mathbb{Z}_{\geqslant 1}$ *such that* ab *is not a square. Then*

$$\sum_{\substack{m,n \\ an^2 - bm^2 = h \\ an^2 \leqslant k}} 1 \leqslant 2\,\tau\,(|h|)\,(\log\,(k + |h|) + 1).$$

Note that if $h \geqslant 0$, then $\log(k + h)$ may be replaced by $\log k$.

Proof (Proof of Proposition 4.32) Using (4.9) with $k = 2$ we first derive

$$\sum_{n \leqslant x} \mu^2 \left(n^2 + 1\right) = \sum_{d \leqslant D_0} \mu(d) N(x; d)$$

$$= \left(\sum_{d \leqslant T} + \sum_{T < d \leqslant D_0}\right) \mu(d) N(x; d) := \Sigma_1 + \Sigma_2,$$

say. We first treat Σ_1. To this end, let $0 \leqslant n_1 < \cdots < n_{\rho(d)} < d^2$ be a set of distinct solutions of $n^2 + 1 \equiv 0 \pmod{d^2}$. We have

$$N(x; d) = \sum_{j=1}^{\rho(d)} \sum_{\substack{n \leqslant x \\ n \equiv n_j \pmod{d^2}}} 1 = \sum_{j=1}^{\rho(d)} \left(\left\lfloor \frac{x - n_j}{d^2} \right\rfloor - \left\lfloor \frac{-n_j}{d^2} \right\rfloor \right)$$

$$= \sum_{j=1}^{\rho(d)} \left(\frac{x}{d^2} + O(1) \right) = \frac{x\rho(d)}{d^2} + O\left(\rho(d)\right)$$

so that

$$\Sigma_1 = x \sum_{d \leqslant T} \frac{\mu(d)\rho(d)}{d^2} + O\left(\sum_{d \leqslant T} \rho(d) \right).$$

The bound $\sum_{d \leqslant T} \rho(d) \ll T$ can easily be derived from Theorem 7.28, and partial summation as in Exercise 3 yields

$$\sum_{d > T} \frac{\rho(d)}{d^2} = -\frac{1}{T^2} \sum_{d \leqslant T} \rho(d) + 2 \int_T^{\infty} \frac{1}{t^3} \left(\sum_{d \leqslant t} \rho(d) \right) dt \ll T^{-1}$$

and therefore

$$\Sigma_1 = x \sum_{d=1}^{\infty} \frac{\mu(d)\rho(d)}{d^2} + O\left(T + xT^{-1} \right).$$

Note that, expanding the series in Euler product, we get

$$\sum_{d=1}^{\infty} \frac{\mu(d)\rho(d)}{d^2} = \prod_p \left(1 - \frac{\rho(p)}{p^2} \right) = C.$$

Next, by a splitting argument, we derive

$$\Sigma_2 \ll \sum_{T < d \leqslant D_0} \sum_{\substack{b \\ d^2 b \asymp x^2}} \sum_{\substack{n \leqslant x \\ d^2 b - n^2 = 1}} 1 \ll \max_{T < D \leqslant D_0} \max_{D^2 B \asymp x^2} \left(\sum_{B < b \leqslant 2B} \sum_{\substack{n \leqslant x \\ D < d \leqslant 2D \\ d^2 b - n^2 = 1}} 1 \right) (\log x)^2$$

and using Lemma 4.17 with $a = 1$, $h = 1$ and $k = \lfloor x^2 \rfloor$ we get

$$\Sigma_2 \ll \max_{T < D \leqslant D_0} \left(\max_{D^2 B \asymp x^2} B \right) (\log x)^3 \ll x^2 T^{-2} (\log x)^3.$$

Putting altogether we finally derive

$$\sum_{n \leqslant x} \mu^2 \left(n^2 + 1 \right) = Cx + O \left(T + xT^{-1} + x^2 T^{-2} (\log x)^3 \right)$$

and choosing $T = x^{2/3} \log x$ gives the asserted result. $\qquad\square$

Estermann's result remained unbeaten for more than 80 years, until Heath-Brown [47] estimated Σ_2 using his so-called determinant method to count the number of integer points on the algebraic variety defined by $d^2 b - n^2 = 1$. This enabled the author to prove that

$$\sum_{n \leqslant x} \mu^2 \left(n^2 + 1 \right) = Cx + O \left(x^{\frac{1}{81} \left(\sqrt{433} + 26 \right) + \varepsilon} \right),$$

where C is given in (4.68). Note that $\frac{1}{81} \left(\sqrt{433} + 26 \right) \approx 0.57788$. The process has then been generalized by Reuss [89] allowing the author to show related estimates such as

$$\sum_{n \leqslant x} \mu_k(n) \mu_k(n + h) = C_{h,k} x + O_{\varepsilon, h, k} \left(x^{\alpha_k + \varepsilon} \right),$$

where $k \geqslant 2$ and $h \neq 0$ are fixed integers, $C_{h,k}$ is a constant depending only on h and k, and

$$\alpha_k := \begin{cases} \frac{1}{81} \left(\sqrt{433} + 26 \right), & \text{if } k = 2 ; \\ \frac{169}{144k}, & \text{if } k \geqslant 3. \end{cases}$$

4.7.2   The Bombieri–Vinogradov Theorem

4.7.2.1   Introduction

For positive integers a, q such that $(a, q) = 1$, set

$$\Delta(x; q, a) := \Psi(x; a, q) - \frac{x}{\varphi(q)} \; ;$$

$$\Delta(x; q) := \max_{a:(a,q)=1} |\Delta(x; q, a)| \; ;$$

$$\Delta^\star(x; q) := \sup_{y \leqslant x} \Delta(y; q).$$

Theorem 3.30 asserts that there exists $c > 0$ such that

$$\Delta(x; q, a) \ll x e^{-c\sqrt{\log x}}$$

whenever $q \leqslant (\log x)^A$ and the generalized Riemann hypothesis profoundly improve the error term to

$$\Delta(x; q, a) \ll x^{1/2} (\log x)^2$$

whenever $q \leqslant x$, as seen in Theorem 3.42. As the authors point out in [56, p. 419], the GRH may be out of reach for a long time. On the other hand, the Bombieri–Vinogradov Theorem, a mainstay of prime number theory, allows one to replace the GRH in many cases, and with somewhat the same strength. It can be formulated as follows [3, 121].

Theorem 4.50 (Bombieri, Vinogradov) *For any constant $A > 0$*

$$\sum_{q \leqslant Q} \sup_{y \leqslant x} \max_{a:(a,q)=1} \left| \Psi(x; a, q) - \frac{x}{\varphi(q)} \right| \ll x(\log x)^{-A} + x^{1/2} Q(\log x Q)^5$$

the implied constant depending on A.

Bombieri and Vinogradov independently discovered this result, deriving it from new density theorems for zeros of L-functions. Subsequently, the argument was simplified by taking new theorems on estimates of bilinear forms over arithmetic progressions into account. In particular, Vaughan [120] used his identity as seen in Proposition 6.15 to derive a simplified proof of Theorem 4.50.

Note that the bound of Theorem 4.50 is quite as good as having GRH for all χ to all moduli $q \leqslant x^{1/2}(\log x)^{5-A}$, so that it is not surprising that this result has many applications.

It is also important to note that the bound of Theorem 4.50 is trivial for Q large. Indeed, for $y \geqslant 1$ and $q \in \mathbb{Z}_{\geqslant 1}$, there are at most $\frac{y}{q} + 1$ positive integers n such that $n \leqslant y$ and $n \equiv a \pmod q$. Thus

$$\Psi(y; q, a) \leqslant \log y \sum_{\substack{n \leqslant y \\ n \equiv a \ (\mathrm{mod}\, q)}} 1 \ll q^{-1}x \log x + \log x$$

for $y \leqslant x$. In addition, using Exercise 63

$$\frac{y}{\varphi(q)} \ll \frac{x \log q}{q}$$

for $y \leqslant x$, and hence $\Delta^\star(x; q) \ll q^{-1}x \log x + \log x$, so that

$$\sum_{q \leqslant Q} \Delta^\star(x; q) \ll x(\log x Q)^2 + Q \log x$$

which is better than Theorem 4.50 when $Q > x^{1/2}$. Hence we may assume that $Q \leqslant x^{1/2}$.

The three subsections below provide a proof of Theorem 4.50, essentially following [119, 120]. The next two subsections deal with some possible extensions of this important result.

4.7.2.2   Reduction to Primitive Characters

The starting point of the proof of Theorem 4.50 is the use of Corollary 3.7 with $f = \Lambda$ which yields

$$\Psi(y; q, a) = \frac{1}{\varphi(q)} \sum_{\chi \ (\mathrm{mod}\, q)} \overline{\chi}(a)\Psi(y; \chi),$$

where the function $\Psi(y; \chi)$ is defined in (3.55), and noting that

$$\frac{y}{\varphi(q)} = \frac{1}{\varphi(q)} \sum_{\chi \ (\mathrm{mod}\, q)} \overline{\chi}(a)E_0(\chi)y,$$

where $E_0(\chi)$ is given in (3.57), we derive

$$\Psi(y; q, a) - \frac{y}{\varphi(q)} = \frac{1}{\varphi(q)} \sum_{\chi \,(\mathrm{mod}\, q)} \overline{\chi}(a)\, (\Psi(y; \chi) - E_0(\chi)y)$$

and hence

$$\Delta(y; q) \leqslant \frac{1}{\varphi(q)} \sum_{\chi \,(\mathrm{mod}\, q)} \big|\Psi(y; \chi) - E_0(\chi)y\big|.$$

Now let $\chi^\star$ denote the primitive character inducing χ and let $q^\star$ be the conductor of $\chi^\star$. Recall that $q^\star \mid q$. We have

$$\psi(y; \chi^\star) - \psi(y; \chi) = \sum_{\substack{p \mid q \\ p \nmid q^\star}} \sum_{\alpha \leqslant \lfloor \log y / \log p \rfloor} \chi^\star\big(p^\alpha\big) \log p$$

$$\ll \log y \sum_{p \mid q} \log p \ll \log y \times \log q \ll (\log yq)^2$$

so that

$$\Delta(y; q) \ll \frac{1}{\varphi(q)} \sum_{\chi \,(\mathrm{mod}\, q)} \big|\Psi(y; \chi^\star) - E_0(\chi)y\big| + (\log yq)^2$$

from which we derive

$$\sum_{q \leqslant Q} \Delta^\star(x; q) \ll \sum_{q \leqslant Q} \frac{1}{\varphi(q)} \sum_{\chi \,(\mathrm{mod}\, q)} \sup_{y \leqslant x} \big|\Psi(y; \chi^\star) - E_0(\chi)y\big| + Q(\log xQ)^2.$$

We can count the Dirichlet characters modulo $q \leqslant Q$ by considering all characters induced by primitive characters of conductor $q^\star \leqslant Q$, which gives

$$\sum_{q \leqslant Q} \Delta^\star(x; q) \ll \sum_{q \leqslant Q} \frac{1}{\varphi(q)} \sum_{q^\star \mid q} \sum_{\chi^\star \,(\mathrm{mod}\, q^\star)} \sup_{y \leqslant x} \big|\Psi(y; \chi^\star) - E_0(\chi^\star)y\big|$$

$$+ Q(\log xQ)^2$$

$$\ll \sum_{q^\star \leqslant Q} \sum_{\chi^\star \,(\mathrm{mod}\, q^\star)} \sup_{y \leqslant x} \big|\Psi(y; \chi^\star) - E_0(\chi^\star)y\big|$$

$$\times \left(\sum_{m \leqslant Q/q^\star} \frac{1}{\varphi(mq^\star)} \right) + Q(\log xQ)^2$$

$$\ll \sum_{q^\star \leqslant Q} \frac{1}{\varphi(q^\star)} \sum_{\chi^\star \,(\mathrm{mod}\, q^\star)} \sup_{y \leqslant x} \left| \Psi(y; \chi^\star) - E_0(\chi^\star) y \right|$$

$$\times \left(\sum_{m \leqslant Q/q^\star} \frac{1}{\varphi(m)} \right) + Q(\log x Q)^2$$

$$\ll \log Q \sum_{q^\star \leqslant Q} \frac{1}{\varphi(q^\star)} \sum_{\chi^\star \,(\mathrm{mod}\, q^\star)} \sup_{y \leqslant x} \left| \Psi(y; \chi^\star) - E_0(\chi^\star) y \right|$$

$$+ Q(\log x Q)^2,$$

where we used the fact that, for all $m, n \geqslant 1$, $\varphi(mn) \geqslant \varphi(m)\varphi(n)$[29] and Exercise 72. Since $Q \leqslant x^{1/2}$, we finally derive

$$\sum_{q \leqslant Q} \Delta^\star(x; q) \ll \log x \sum_{q \leqslant Q} \frac{1}{\varphi(q)} \sideset{}{^\star}\sum_{\chi \,(\mathrm{mod}\, q)} \sup_{y \leqslant x} |\Psi(y; \chi) - E_0(\chi) y| + Q(\log x Q)^2,$$

$$(4.69)$$

where the star indicates that the sum is restricted to primitive characters.

4.7.2.3 Vaughan's Mean Value Theorem

The key ingredient to estimate (4.69) is the following inequality due to Vaughan [120, Theorem 1].

Theorem 4.51 (Vaughan) *Let $Q \geqslant 1$, $x \geqslant 2$. Then*

$$\sum_{q \leqslant Q} \frac{q}{\varphi(q)} \sideset{}{^\star}\sum_{\chi \,(\mathrm{mod}\, q)} \sup_{y \leqslant x} |\Psi(y; \chi)| \ll \left(x + x^{5/6} Q + x^{1/2} Q^2 \right) (\log x Q)^4.$$

Proof If $Q^2 > x$, then Corollary 4.18 with $M = 1$, $a_1 = 1$, $b_n = \Lambda(n)$ and $N = \lfloor x \rfloor$ implies that the left-hand side does not exceed

$$\ll Q\left(x^{1/2} + Q \right) \left(\sum_{n \leqslant x} \Lambda(n)^2 \right)^{1/2} \log x \ll Q^2 x^{1/2} (\log x)^{3/2},$$

---

[29]This property is easily proved by the definition of φ. We say that the function φ is *completely super-multiplicative*.

where we used $Q > x^{1/2}$ and

$$\sum_{n \leqslant x} \Lambda(n)^2 \leqslant \Psi(x) \log x \ll x \log x$$

so that we may assume $Q^2 \leqslant x$. Using (6.39) with $f(n) = \chi(n)$, we get

$$\sum_{q \leqslant Q} \frac{q}{\varphi(q)} \sum_{\chi \,(\mathrm{mod}\, q)}^{\star} \sup_{y \leqslant x} |\Psi(y; \chi)| \leqslant \sum_{i=1}^{4} S_i,$$

where

$$S_i := \sum_{q \leqslant Q} \frac{q}{\varphi(q)} \sum_{\chi \,(\mathrm{mod}\, q)}^{\star} \sup_{y \leqslant x} \left| \sum_{n \leqslant y} \Lambda(n) a_i(n) \right|$$

and where the functions $a_i(n)$ are given in (6.39). Recall that these functions depend on a parameter $U \leqslant x$ at our disposal. The sums S_i have been estimated by many authors. For instance, from [96, Lemmas 2.1 and 2.2], we derive

$$S_1 \ll U Q^2$$

$$S_2 \ll \left(x + U Q^{5/2} \right) (\log x U)^2$$

$$S_3 \ll \left(x + U Q^{5/2} + x^{1/2} Q U + x Q U^{-1/2} + x^{1/2} Q^2 \right) (\log x U)^2$$

$$S_4 \ll \left(x + x Q U^{-1/2} + x^{1/2} Q^2 \right) (\log x)^{5/2}.$$

▷ If $x^{1/3} < Q \leqslant x^{1/2}$, choose $U = x^{2/3} Q^{-1}$, so that the sum of the theorem does not exceed

$$\ll \left(x + x^{2/3} Q^{3/2} + x^{7/6} + x^{1/2} Q^2 \right) (\log x)^{5/2}$$

and note that the condition $Q > x^{1/3}$ implies $x^{2/3} Q^{3/2} < x^{1/2} Q^2$ and $x^{7/6} < x^{5/6} Q$.

▷ If $Q \leqslant x^{1/3}$, choose $U = x^{1/3}$, so that the sum is bounded by

$$\ll \left(x + x^{1/3} Q^{5/2} + x^{5/6} Q + x^{1/2} Q^2 \right) (\log x)^{5/2}$$

and the condition $Q \leqslant x^{1/3}$ implies $x^{1/3} Q^{5/2} \leqslant x^{1/2} Q^2$, as required. $\square$

4.7.2.4 Completion of the Proof

Assume $Q > (\log x)^{6+A}$ and split the sum (4.69) in two subsums as

$$= \left(\sum_{q \leqslant (\log x)^{6+A}} + \sum_{(\log x)^{6+A} < q \leqslant Q} \right) \frac{\log x}{\varphi(q)} \sideset{}{^*}\sum_{\chi \pmod q} \sup_{y \leqslant x} |\Psi(y; \chi) - E_0(\chi)y| := S_1 + S_2$$

say.

▷ S_1 is treated with a version of the Siegel–Walfisz theorem suitable for the function $\Psi(y; \chi)$. As in Theorem 3.30, one can prove [79, Corollary 11.18] that there exists an absolute constant $c_0 > 0$ such that, for any $A > 0$, there is an $x_0(A) > 0$, depending only on A, such that

$$\Psi(y; \chi) = yE_0(\chi) + O_A\left(ye^{-c_0\sqrt{\log y}} \right) \tag{4.70}$$

provided that $q \leqslant (\log x)^A$ and $x \geqslant x_0(A)$. Now inserting this estimate in S_1 yields

$$S_1 \ll x(\log x)^{7+A}e^{-c_0\sqrt{\log x}} \ll xe^{-c_1\sqrt{\log x}}$$

for some $0 < c_1 < c_0$ and x sufficiently large.[30]

▷ S_2 is treated with Theorem 4.51 in the following way. First note that in this case we have $q > 1$ and any primitive character modulo q is then non-principal, so that $E_0(\chi) = 0$. Now set $L := \log x$ and

$$F(q) := \frac{q}{\varphi(q)} \sideset{}{^*}\sum_{\chi \pmod q} \sup_{y \leqslant x} |\Psi(y; \chi)|.$$

By partial summation and Theorem 4.51, we derive

$$L^{-1}S_2 = Q^{-1} \sum_{L^{6+A} < q \leqslant Q} F(q) + \int_{L^{6+A}}^{Q} t^{-2} \left(\sum_{L^{6+A} < q \leqslant t} F(q) \right) dt$$

$$\ll L^4 \left(xQ^{-1} + x^{5/6} + x^{1/2}Q + \int_{L^{6+A}}^{Q} \left(xt^{-2} + x^{5/6}t^{-1} + x^{1/2} \right) dt \right)$$

$$\ll xQ^{-1}L^4 + xL^{-A-2} + x^{1/2}QL^4 \ll xL^{-A-2} + x^{1/2}QL^4$$

since $Q > (\log x)^{6+A}$, completing the proof. □

---

[30]If $Q \leqslant (\log x)^{6+A}$, then $S_2 = 0$ and the sum (4.69) is $\ll xe^{-c_1\sqrt{\log x}}$ as above.

4.7.2.5 Other Formulations

The following formulation is quite more suitable for applications.

Theorem 4.52 *For any $A > 0$, there is a constant $B = B(A) > 0$ such that, if $Q \leqslant x^{1/2}(\log x)^{-B}$, we have*

$$\sum_{q \leqslant Q} \sup_{y \leqslant x} \max_{a:(a,q)=1} \left| \Psi(x; a, q) - \frac{x}{\varphi(q)} \right| \ll x(\log x)^{-A}$$

the implied constant depending on A.

Define

$$\widetilde{\pi}(y) := \sum_{2 \leqslant n \leqslant y} \frac{\Lambda(n)}{\log n}$$

and similarly for $\widetilde{\pi}(y; q, a)$. Note that

$$\widetilde{\pi}(y) = \pi(y) + \sum_{2 \leqslant \alpha \leqslant \log y / \log 2} \sum_{p^{\alpha} \leqslant y} \frac{1}{\alpha} = \pi(y) + O\left(y^{1/2}\right)$$

and also $\widetilde{\pi}(y; q, a) = \pi(y; q, a) + O\left(y^{1/2}\right)$. Furthermore, by partial summation

$$\widetilde{\pi}(y; q, a) - \frac{\widetilde{\pi}(y)}{\varphi(q)} = \frac{\Psi(y; q, a) - \Psi(y)}{\varphi(q) \log y} + \int_2^y \frac{1}{t(\log t)^2} \left(\Psi(t; q, a) - \frac{\Psi(t)}{\varphi(q)} \right) dt$$

and hence

$$\left| \widetilde{\pi}(y; q, a) - \frac{\widetilde{\pi}(y)}{\varphi(q)} \right| \ll \sup_{2 \leqslant t \leqslant y} \left| \Psi(t; q, a) - \frac{\Psi(t)}{\varphi(q)} \right|$$

then

$$\left| \pi(y; q, a) - \frac{\pi(y)}{\varphi(q)} \right| \ll \sup_{2 \leqslant t \leqslant y} \left| \Psi(t; q, a) - \frac{\Psi(t)}{\varphi(q)} \right| + y^{1/2}.$$

Now using the PNT in the shape

$$\left. \begin{matrix} \pi(y) \\ \Psi(y) \end{matrix} \right\} = \left. \begin{matrix} \mathrm{Li}(y) \\ y \end{matrix} \right\} + O\left(y e^{-c\sqrt{\log y}}\right)$$

with c absolute and y sufficiently large, yields

$$\left| \pi(y; q, a) - \frac{\mathrm{Li}(y)}{\varphi(q)} \right| \ll \sup_{2 \leqslant t \leqslant y} \left| \Psi(t; q, a) - \frac{t}{\varphi(q)} \right| + y\varphi(q)^{-1} e^{-c\sqrt{\log y}} + y^{1/2}.$$

Therefore Theorem 4.52 implies the following version of the Bombieri–Vinogradov theorem.

Theorem 4.53 *For any $A > 0$, there is a constant $B = B(A) > 0$ such that, if $Q \leqslant x^{1/2}(\log x)^{-B}$, we have*

$$\sum_{q \leqslant Q} \sup_{y \leqslant x} \max_{a:(a,q)=1} \left| \pi(y; q, a) - \frac{\mathrm{Li}(y)}{\varphi(q)} \right| \ll x(\log x)^{-A}$$

the implied constant depending on A.

Note that there also exist completely explicit versions of the Bombieri–Vinogradov theorem, all resting on explicit versions of the Vaughan inequality in Theorem 4.51. See [96], for instance.

4.7.2.6 Extensions

The Bombieri–Vinogradov theorem shows that the prime numbers up to x are "well-distributed" in all arithmetic progressions modulo q, for almost all integers $q \leqslant x^{1/2}(\log x)^{-B}$. The problem of proving results of the shape

$$\sum_{q \leqslant Q} \max_{a:(a,q)=1} \left| \pi(x; q, a) - \frac{\mathrm{Li}(x)}{\varphi(q)} \right| \ll x(\log x)^{-A},$$

where $Q = x^\theta$ with $\theta > \frac{1}{2}$ is both crucial to certain results, as seen in Theorem 4.55 below, and a staggeringly difficult one. This leads to the following conjecture [25].

Conjecture 4.2 (Elliott–Halberstam) The Bombieri–Vinogradov theorem holds for all $Q \leqslant x(\log x)^{-C}$.

This conjecture is probably very far from reach, but one possible way to get past the $x^{1/2}$-barrier is to relax the condition on a, since, in practice, one often only needs the bound of Theorem 4.52 for a fixed value of a.

A first answer towards Conjecture 4.2 was given by Bombieri, Friedlander and Iwaniec in [6] with the next result.

Theorem 4.54 *Let $a \neq 0$ and $x \geqslant y \geqslant 3$ and $Q \leqslant \sqrt{xy}$. Then*

$$\sum_{\substack{Q \leqslant q < 2Q \\ (q,a)=1}} \left| \Psi(x; a, q) - \frac{x}{\varphi(q)} \right| \ll_a x \left(\frac{\log y}{\log x} \right)^2 (\log \log x)^B,$$

where B is an absolute constant.

This theorem means that, in most applications, one can go up to $Q = x^{1/2+(\log\log x)^{-B}}$. Two years later, the same authors [7] refined their result by proving that one can take $Q = x^{1/2+o(1)}$, whatever the nature of the $o(1)$ is.

Another way to get some improvements is to replace the absolute values by some suitably chosen weights. This is done independently in [30] and [5]. In the latter paper, the authors define the concept of "well-factorable" arithmetic functions, and prove in this case that, for any $a \neq 0$, $\varepsilon > 0$, well-factorable function $\lambda(q)$ of level $Q = x^{4/7-\varepsilon}$ and any $A > 0$, we have

$$\sum_{(a,q)=1} \lambda(q) \left(\Psi(x; a, q) - \frac{x}{\varphi(q)} \right) \ll_{\varepsilon,a,A} x (\log x)^{-A}.$$

In [30], it is shown that, if $a \in \mathbb{Z}$, $a \neq 0$, $r \in \mathbb{Z}_{\geqslant 1}$ are such that $(a, r) = 1$, then, for any $x \geqslant 2$, $A > 0$ and $Q \leqslant x (\log x)^{-200(A+1)}$, we have

$$\sum_{\substack{q \leqslant Q \\ (a,q)=1}} \left(\pi(x; qr, a) - \frac{\mathrm{Li}(x)}{\varphi(qr)} \right) \ll x (\log x)^{-A} \tag{4.71}$$

uniformly for $1 \leqslant |a| \leqslant (\log x)^A$. Note the impressively large range for Q.

Finally, Fiorilli [29] proved that these results yield a Bombieri–Vinogradov type theorem for the Titchmarsh divisor problem in arithmetic progressions, up to level $Q = x^{1/10-\varepsilon}$.

The Bombieri–Vinogradov theorem has also been extended to certain arithmetic functions. For all positive integers a, q such that $(a, q) = 1$, define in an obvious way

$$\Delta(f; x; q, a) := \sum_{\substack{n \leqslant x \\ n \equiv a \,(\mathrm{mod}\, q)}} f(n) - \frac{1}{\varphi(q)} \sum_{\substack{n \leqslant x \\ (n,q)=1}} f(n). \tag{4.72}$$

A Bombieri–Vinogradov type result is a bound of the form

$$\sum_{q \leqslant Q} \max_{a:(a,q)=1} |\Delta(f; x; q, a)| \ll x (\log x)^{-A} \tag{4.73}$$

for any $A > 0$ and $Q \leqslant x^{1/2}(\log x)^{-B}$ for some $B = B(A) > 0$, and Theorem 4.50 shows that this bound holds when f is the characteristic function for the primes. It is also shown that such a bound holds when $f = \mu$ [120], or f is the characteristic function for the y-smooth numbers. Recall that a positive integer n is said to be y-smooth if $P^+(n) \leqslant y$. Several examples have also been proved when f is a complex-valued arithmetic function such that $|f(n)| \leqslant 1$, but even in this case one needs to be careful: indeed, if f strongly correlates with some Dirichlet character χ of small modulus, i.e. if

$$\sum_{n \leqslant x} f(n)\overline{\chi}(n) \gg x(\log x)^{-C},$$

then it is proved that (4.73) cannot hold for any $A > 0$.

4.7.3 Bounded Gaps Between Primes

In this section, we define the following function close to the von Mangoldt Λ-function.

$$\mathcal{L}(n) := \begin{cases} \log n, & \text{if } n \text{ is prime}; \\ 0, & \text{otherwise.} \end{cases} \tag{4.74}$$

4.7.3.1 de Polignac's Conjecture

When looking carefully at a list of primes, some patterns begin to emerge. For instance, one sees that they often come in pairs, such as $(3, 5)$, $(5, 7)$, $(11, 13)$, $\ldots$, $(101999, 102001)$, etc. This naturally leads to the famous *twin prime conjecture*, which asserts that there are infinitely primes p, such that $p + 2$ is also a prime. More generally, the de Polignac conjecture is the following one.

Conjecture 4.3 (de Polignac, 1849) For any positive even number h, there are infinitely many primes p such that $p + h$ is also a prime.

4.7.3.2 Admissible Tuples

An obvious question is to ask ourselves whether there is a way of predicting which patterns can occur and which do not. For instance, looking at the set $\{n, n+2, n+4\}$, the only prime triple is only given when $n = 3$, i.e. $\{3, 5, 7\}$, because one of the three numbers is always a multiple of 3. This becomes apparent when looking at the triple $\mathcal{H} = \{0, 2, 4\}$ modulo 3, giving the set $\{0, 2, 1\}$ which is a complete residue system for the prime 3. In other words, the prime 3 is an obstruction to the fact that there may be infinitely many primes in the set $\{n, n+2, n+4\}$. This leads to the following definition.

Definition 4.13 Let $n \in \mathbb{Z}_{\geq 0}$, p be a prime, $k \in \mathbb{Z}_{\geq 1}$, $\mathcal{H} = \{h_1, \ldots, h_k\}$ where $h_i \in \mathbb{Z}_{\geq 0}$, and consider the k-tuple $(n + h_1, \ldots, n + h_k)$.

▷ If every component of the tuple is a prime, we call this a *prime tuple*.
▷ We say that p is an *obstruction* if $p \mid P_{\mathcal{H}}(n) := (n + h_1) \cdots (n + h_k)$ for all n.
▷ We say that $\mathcal{H}$ is *admissible* if no p is an obstruction, i.e. it does not include a complete set of residues modulo p for any prime $p \leqslant k$. In this case the associated tuple $(n + h_1, \ldots, n + h_k)$ is called an *admissible tuple*.

Example 4.11

▷ Let $\mathcal{H} = \{5, 7, 11, 13, 17\}$. Reduction modulo 2 yields $\{1, 1, 1, 1, 1\}$, modulo 3 we get $\{2, 1, 2, 1, 2\}$ and modulo 5 we obtain $\{0, 2, 1, 3, 2\}$, so that $\mathcal{H}$ is admissible.
▷ Let $\mathcal{H} = \{0, 2, 8, 10, 12, 16\}$. Then modulo 3 we get $\{0, 2, 2, 1, 0, 1\}$, and hence $\mathcal{H}$ is not admissible.

From this definition, one surmises the following statement, called the *prime k-tuple conjecture*.

Conjecture 4.4 Admissible tuples will infinitely often be prime tuples, i.e. if a set $\{x + h_1, \ldots, x + h_k\}$ is an admissible tuple, then there are infinitely many integers n such that $n + h_1, \ldots, n + h_k$ are all prime numbers.

This conjecture is a long-standing problem. A first step towards it is the study of the smallest gap between primes that occur in any interval, that is, the behaviour of $\inf_{N < n \leqslant 2N} (p_{n+1} - p_n)$ when $N \to \infty$. Surprisingly, this requires the following definition.

Definition 4.14 Let $\theta > 0$. We say that the primes have *level of distribution* θ if Theorem 4.52 holds for any $A > 0$ and any $\varepsilon > 0$ with $Q = x^{\theta - \varepsilon}$.

4.7.3.3   The GPY Method

Many authors studied the difference $p_{n+1} - p_n$, but a first breakthrough came from [34] who showed that

$$\liminf_{n \to \infty} \frac{p_{n+1} - p_n}{\log p_n} = 0,$$

i.e. there are infinitely many pairs of primes $p < q$ such that $q - p < c \log p$ for any given $c > 0$. They also proved the following result.

Theorem 4.55 (GPY) *Suppose the primes have level of distribution $\theta > \frac{1}{2}$. Then there exists an explicitly calculable constant $C(\theta)$, depending only on θ, such that any admissible k-tuple with $k \geqslant C(\theta)$ contains at least two primes infinitely often.*

The proof shows in particular that, if $\theta \geqslant 0.971$, then the theorem is true for $k \geqslant 6$. Since the 6-tuple $\{n, n+4, n+6, n+10, n+12, n+16\}$ is admissible, Theorem 4.55 and Conjecture 4.2 imply that $\liminf_{n\to\infty} (p_{n+1} - p_n) \leqslant 16$.

Let us have a brief glance at the GPY strategy. Let $\mathcal{H} = \{h_1, \ldots, h_k\}$ be an admissible k-tuple, $N \in \mathbb{Z}_{\geqslant 1}$ such that $N > \max(h_j)$, and let $w(n)$ be an arithmetic function satisfying $w(n) \geqslant 0$ for all n, and set

$$S_1 := \sum_{N < n \leqslant 2N} w(n) \quad \text{and} \quad S_2 := \sum_{N < n \leqslant 2N} \left(\sum_{j=1}^{k} \mathcal{L}(n + h_j) \right) w(n),$$

where $\mathcal{L}(n)$ is defined in (4.74). The main aim is to select a good function $w(n)$ which maximises

$$\rho := \frac{1}{\log 3N} \frac{S_2}{S_1}$$

for the purpose of obtaining a good lower bound for ρ. If $\rho > m$ for some $N > \max(h_j)$ and some positive integer m, then

$$\sum_{N < n \leqslant 2N} w(n) \left(\sum_{j=1}^{k} \mathcal{L}(n + h_j) - m \log 3N \right) > 0$$

and hence there must exist an integer $n_0 \in (N, 2N]$ such that

$$w(n_0) \left(\sum_{j=1}^{k} \mathcal{L}(n_0 + h_j) - m \log 3N \right) > 0.$$

We infer that $w(n_0) \neq 0$, and then $w(n_0) > 0$, so that

$$\sum_{j=1}^{k} \mathcal{L}(n_0 + h_j) > m \log 3N.$$

But for each $j = 1, \ldots, k$, one has $n_0 + h_j < 3N$, and hence $\mathcal{L}(n_0 + h_j) < \log 3N$, from which we derive that at least $m + 1$ of $\mathcal{L}(n_0 + h_j)$ are non-zero, i.e. at least $m + 1$ of $n_0 + h_1, \ldots, n_0 + h_k$ are prime numbers.

The main difficulty in this method is to find a suitable function $w(n)$. In [34], they choose

$$w(n) = \left(\frac{1}{(k+\ell)!} \sum_{\substack{d \mid P_{\mathcal{H}}(n) \\ d < D}} \mu(d) \left(\log \frac{D}{d} \right)^{k+\ell} \right)^2, \tag{4.75}$$

where $P_{\mathcal{H}}(n)$ is given in Definition 4.13, $\ell \in \mathbb{Z}_{\geqslant 1}$ and $D = D(N)$ is a parameter at our disposal.

4.7.3.4   Zhang's Breakthrough

In 2014, a landmark article [128] was then published, proving for the first time that there are infinitely many pairs of primes $p < q$ such that $q - p$ is bounded. Let us sift through Zhang's argument. Firstly, the evaluations of the sums S_1 and S_2 above, with $m = 1$, lead to a relation of the shape

$$S_2 - (\log 3N)S_1 = M + O\left(N(\log N)^{k+2\ell} + E \right)$$

for $D < N^{1/2-\varepsilon}$, where $M = M(N)$ is the "main term" and

$$E := \sum_{j=1}^{k} \sum_{d < D^2} \mu(d)^2 \tau_{3k-3}(d) \sum_{\substack{c \leqslant d \\ (c,d)=1 \\ d \mid P_{\mathcal{H}}(c-h_j)}} |\Delta\left(\mathfrak{L}; N; d, c\right)|,$$

where $\Delta\left(\mathfrak{L}; N; d, c\right)$ is given in (4.72). Now let $\varpi > 0$ be a fixed small real number.

▷ If $D = N^{1/4+\varpi}$ and k is sufficiently large in terms of ϖ, then it can be proved that $M \gg N(\log N)^{k+2\ell+1}$ and E can be efficiently bounded if the primes have level of distribution $\theta > \frac{1}{2} + 2\varpi$.

▷ If $D = N^{1/4-\varepsilon}$, then Theorem 4.52 is good enough to majorize E, but in this case the lower bound $M \gg N(\log N)^{k+2\ell+1}$ is not valid, even if a more general form of $w(n)$ is considered.

With these two problems in mind, Zhang had the idea to combine the GPY method with Bombieri–Friedlander–Iwaniec's work leading to Theorem 4.54 seen in the previous section.

Let $\varpi = \frac{1}{1168}$, $D = N^{1/4+\varpi}$ and

$$\mathcal{P} := \prod_{p < N^\varpi} p.$$

Zhang made two crucial observations. First, the contributions in M from the terms with d having a large prime divisor are relatively small. This allows him to add the condition $d \mid \mathcal{P}$ in (4.75), the bound $M \gg N(\log N)^{k+2\ell+1}$ still being valid. Secondly, with this choice of D and the condition $d \mid \mathcal{P}$, the error term can be efficiently bounded, thanks to his following new extension of Theorem 4.52, which lies at the heart of his proof.

Theorem 4.56 Let $\varpi = \frac{1}{1168}$, $D = N^{1/4+\varpi}$. For $1 \leqslant j \leqslant k$ and any $A > 0$

$$\sum_{d < D^2} \sum_{\substack{c \leqslant d \\ (c,d)=1 \\ d \mid \mathcal{P}_{\mathcal{H}}(c-h_j)}} |\Delta(\mathcal{L}; N; d, c)| \ll N(\log N)^{-A}.$$

All these efforts enable Zhang to show that

$$\liminf_{n \to \infty} (p_{n+1} - p_n) < 7 \times 10^7.$$

This bound was subsequently improved by Maynard [76] and by the Polymath8 project [86], respectively, showing that the constant 7×10^7 can be replaced by 600 and 246. But, as Heath-Brown pointed out in his review of Zhang's paper, this article is, and will ever stay, "an outstanding landmark in the history of prime number theory".

4.7.4 The Titchmarsh Divisor Problem

A famous application of the Bombieri–Vinogradov theorem is the Titchmarsh divisor problem dealing with an asymptotic formula of the sum

$$\sum_{p \leqslant x} \tau(p - 1).$$

Theorem 4.57 *For all x sufficiently large*

$$\sum_{p\leqslant x}\tau(p-1) = x\frac{\zeta(2)\zeta(3)}{\zeta(6)} + O\left(\frac{x\log\log x}{\log x}\right).$$

Proof By (7.48), we have

$$\sum_{p\leqslant x}\tau(p-1) = 2\sum_{\substack{p\leqslant x \ d\mid p-1 \\ d<\sqrt{p-1}}}\sum 1 + \sum_{p\leqslant x}\mathbf{1}_{\square}(p-1)$$

$$= 2\sum_{\substack{d<\sqrt{x-1} \ d^2+1<p\leqslant x \\ p\equiv 1\,(\mathrm{mod}\,d)}}\sum 1 + O\left(x^{1/2}\right)$$

$$= 2\sum_{d\leqslant\sqrt{x-1}}\left(\pi(x;d,1) - \pi(d^2+1;d,1)\right) + O\left(x^{1/2}\right).$$

Let $A > 0$ and $B = B(A)$ be the constant in Theorem 4.53. If $R := R(x) = x(\log x)^{-B}$, write the sum above as

$$\sum_{p\leqslant x}\tau(p-1) = 2\sum_{d\leqslant R}\frac{\mathrm{Li}(x)}{\varphi(d)} + 2\sum_{d\leqslant R}\left(\pi(x;d,1) - \frac{\mathrm{Li}(x)}{\varphi(d)}\right)$$

$$+ 2\sum_{R<d\leqslant\sqrt{x-1}}\pi(x;d,1) - 2\sum_{d\leqslant\sqrt{x-1}}\pi(d^2+1;d,1) + O\left(x^{1/2}\right)$$

$$:= 2\,\mathrm{Li}(x)\sum_{d\leqslant R}\frac{1}{\varphi(d)} + 2\,(S_1 + S_2 - S_3) + O\left(x^{1/2}\right)$$

say. Now use Exercise 72 for the first sum, Theorem 4.53 for S_1 and Theorem 4.45 for S_2 and S_3, which yield, with $C := \frac{\zeta(2)\zeta(3)}{\zeta(6)}$

$$\sum_{p\leqslant x}\tau(p-1) = 2\,\mathrm{Li}(x)\,(C\log R + O(1))$$

$$+ O\left(x(\log x)^{-A} + \sum_{R<d\leqslant\sqrt{x}}\frac{x}{\varphi(d)\log(x/d)} + \sum_{d\leqslant\sqrt{x}}\frac{d^2+1}{\varphi(d)\log(ed)}\right)$$

$$= Cx + O\left(\frac{x\log\log x}{\log x}\right) + O\left(\frac{x}{\log x}\sum_{R<d\leqslant\sqrt{x}}\frac{1}{\varphi(d)} + \sum_{d\leqslant\sqrt{x}}\frac{d^2}{\varphi(d)\log(ed)}\right)$$

and using Exercise 72 again yields

$$\sum_{R<d\leqslant\sqrt{x}}\frac{1}{\varphi(d)}\ll\log\log x$$

and with the help of Exercise 63 we derive

$$\sum_{d\leqslant\sqrt{x}}\frac{d^2}{\varphi(d)\log(ed)}=\left(\sum_{d\leqslant x^{1/4}}+\sum_{x^{1/4}<d\leqslant\sqrt{x}}\right)\frac{d^2}{\varphi(d)\log(ed)}$$

$$\ll\sum_{d\leqslant x^{1/4}}\frac{d^2}{\varphi(d)}+\frac{1}{\log x}\sum_{x^{1/4}<d\leqslant\sqrt{x}}\frac{d^2}{\varphi(d)}$$

$$\ll\sum_{d\leqslant x^{1/4}}d\log\log d+\frac{1}{\log x}\sum_{d\leqslant\sqrt{x}}d\log\log d$$

$$\ll\frac{x\log\log x}{\log x}$$

as required.   □

This result has been improved and extended by several authors. Titchmarsh [108] was the first to prove this asymptotic formula, but under the generalized Riemann hypothesis for L-functions. Unconditionally, Linnik [71] showed this result as an application of his "dispersion method".[31] The simplified proof above is due to Halberstam [38]. Independently in [30] and [5], the authors used an extended version of the Bombieri–Vinogradov theorem in the shape (4.71) to derive the following asymptotic formula.

Theorem 4.58 *Let $a\in\mathbb{Z}$, $a\neq 0$. Then, for any $A>0$*

$$\sum_{p\leqslant x}\tau(p+a)=C_1 x+2C_2\operatorname{Li}(x)+O\left(x(\log x)^{-A}\right),$$

where $C_1:=\dfrac{\zeta(2)\zeta(3)}{\zeta(6)}\displaystyle\prod_{p\mid a}\left(1-\frac{p}{p(p-1)+1}\right)$ and

$$C_2:=C_1\left(\gamma-\sum_p\frac{\log p}{p(p-1)+1}+\sum_{p\mid a}\frac{p^2\log p}{(p-1)(p^2-p+1)}\right).$$

---

[31] Therefore this problem should be named "the Titchmarsh-Linnik divisor problem".

Recently, some generalizations of the Titchmarsh divisor problem have been investigated. For instance, Felix [28] proved that, for any integers $a \neq 0$ and $k \geqslant 1$, we have

$$\sum_{\substack{p \leqslant x \\ p \equiv a \,(\mathrm{mod}\, k)}} \tau\left(\frac{p-a}{k}\right) = C_1 \frac{x}{k} \prod_{p|k}\left(1 + \frac{p-1}{p(p-1)+1}\right) + O\left(\frac{x}{\log x}\right),$$

where C_1 is given in Theorem 4.58, and a more precise formula has been proved by Fiorilli in [29].

The Titchmarsh divisor problem belongs to a vast class of problems of correlation of arithmetic functions. Sums of the shape

$$x^{-1} \sum_{n \leqslant x} f_1\,(n+h_1) \cdots f_k\,(n+h_k),$$

where $(h_1, \ldots, h_k) \in \mathbb{Z}^k$ are often called *cross-correlations* of the arbitrary arithmetic functions $f_1, \ldots, f_k$. Some of the most famous theorems and problems in number theory are about these statistical properties for certain specific arithmetic functions. For instance, the study of prime pairs amounts to the study of correlation

$$x^{-1} \sum_{n \leqslant x} \Lambda(n)\Lambda(n+2)$$

of the von Mangoldt function. Similarly, the Titchmarsh divisor problem is the study of the correlation

$$x^{-1} \sum_{n \leqslant x} \Lambda(n)\tau(n+a).$$

Another famous example is the following conjecture surmised by Chowla [17].

Conjecture 4.5 (Chowla) If $h_1, \ldots, h_k$ are any k fixed distinct integers, then

$$\sum_{n \leqslant x} \lambda\,(n+h_1) \cdots \lambda\,(n+h_k) = o(x) \quad (x \to \infty),$$

where $\lambda = (-1)^{\Omega}$ is the Liouville function.

This conjecture is believed to be at least as deep as the twin prime conjecture, and is to date out of reach. In a series of several papers [104, 103, 75], the authors investigated these correlations further, and proved among other things the following difficult results.

▷ *Logarithmic Chowla 2-point correlation conjecture* [103]. Let a_1, a_2 be positive integers, $b_1, b_2 \in \mathbb{Z}$ such that $a_1 b_2 \neq a_2 b_1$, $1 \leqslant w(x) \leqslant x$ be a quantity depending on x that goes to infinity as $x \to \infty$. Then

$$\sum_{x/w(x)<n\leqslant x} \frac{\lambda\,(a_1 n + b_1)\,\lambda\,(a_2 n + b_2)}{n} = o\,(\log w(x)) \quad (x \to \infty)\,.$$

▷ *Erdős discrepancy problem* [104]. Every arithmetic function f taking his values in $\{-1, 1\}$ satisfies

$$\sup_{n,d\in\mathbb{Z}_{\geqslant 1}} \left| \sum_{k=1}^{n} f(kd) \right| = \infty.$$

▷ *Higher order Titchmarsh divisor problem* [75]. Let $A > 0$, $\varepsilon \in \left(0, \frac{1}{2}\right)$, $k \in \mathbb{Z}_{\geqslant 2}$, $x \geqslant 2$, $1 \leqslant h_0 \leqslant x^{1-\varepsilon}$ and H satisfying $x^{8/33+\varepsilon} \leqslant H \leqslant x^{1-\varepsilon}$. Then the asymptotic formula

$$\sum_{x<n\leqslant 2x} \Lambda(n)\tau_k(n+h) = x P_{k,h}(\log x) + O_{A,\varepsilon,k}\left(x(\log x)^{-A}\right),$$

where $Q_{k,h}$ is a polynomial of degree $k - 1$, holds for all but $O\left(H(\log x)^{-A}\right)$ values of h such that $|h - h_0| \leqslant H$.

4.7.5   Power Means of the Riemann Zeta Function

4.7.5.1   The Mean Square

A possible approach for bounding $\zeta\left(\frac{1}{2} + it\right)$ rests on the study of the integral

$$I_1(T) := \int_0^T \left| \zeta\left(\tfrac{1}{2} + it\right) \right|^2 dt.$$

Indeed, in view of the following inequalities

$$\left| \zeta\left(\tfrac{1}{2} + it\right) \right|^2 \ll \left\{ \int_{t-(\log t)^2}^{t+(\log t)^2} \left| \zeta\left(\tfrac{1}{2} + iu\right) \right|^2 du + 1 \right\} \log t$$

$$\ll \left\{ I_1(t + \log^2 t) - I_1(t - \log^2 t) + 1 \right\} \log t$$

$$\ll \left\{ E(t + \log^2 t) - E(t - \log^2 t) + (\log t)^3 \right\} \log t,$$

where $E(t)$ is the error term in the asymptotic formula

$$I_1(T) = T\left(\log \tfrac{T}{2\pi} + 2\gamma - 1\right) + E(T) \qquad (4.76)$$

proved in 1918 by Hardy and Littlewood, we see that any estimate of the form $E(t) \ll t^\alpha$ implies the bound

$$\zeta\left(\tfrac{1}{2} + it\right) \ll t^{\alpha/2}(\log t)^{1/2}.$$

Numerous bounds have been provided by many authors for $E(t)$. For instance, Ingham proved in 1926 that $E(t) \ll t^{1/2}\log t$ whereas Balasubramanian obtained in 1978 the bound $E(t) \ll t^{346/1067+\varepsilon} \approx t^{0.324273+\varepsilon}$. The best result till now is due to Bourgain and Watt [13] who showed that

$$E(t) \ll t^{1515/4816+\varepsilon} \approx t^{0.314576+\varepsilon}.$$

More generally, set

$$I_k(T) = \int_0^T \left|\zeta\left(\tfrac{1}{2} + it\right)\right|^{2k} dt.$$

The long-standing conjecture is that $I_k(T) \sim T P_{k^2}(\log T)$ where P_{k^2} is a polynomial of degree k^2. So far asymptotic formulae have been only proved for the cases $k = 1$ and $k = 2$. The ambition of this section is quite modest, since we intend to prove the following weak version of (4.76).

Proposition 4.33 *Let $T > 0$. Then*

$$\int_0^T \left|\zeta\left(\tfrac{1}{2} + it\right)\right|^2 dt = T \log T + O(T).$$

The proof, for which we follow the lines of Heath-Brown in [109, p. 175], rests on the next useful lemma dealing with the evaluation of some Dirichlet polynomial. See [56, Theorem 9.1], for instance.

Lemma 4.18 *Let $T > 0$, $\sigma \in \mathbb{R}$, $N \in \mathbb{Z}_{\geqslant 1}$ and $a_1, \ldots, a_N$ be arbitrary complex numbers. Then*

$$\int_0^T \left|\sum_{n=1}^N \frac{a_n}{n^{\sigma+it}}\right|^2 dt = \sum_{n=1}^N \frac{|a_n|^2}{n^{2\sigma}} \left(T + O(n)\right).$$

Proof (Proof of Proposition 4.33) Let $T > 0$ and assume $T \leqslant t \leqslant 2T$. By Theorem 3.9 applied with $\sigma = \frac{1}{2} + it$ and $x = 2T$, we get

$$\zeta\left(\tfrac{1}{2} \pm it\right) = \sum_{n \leqslant 2T} \frac{1}{n^{1/2 \pm it}} + R$$

with $R \ll T^{-1/2}$, so that

$$\int_T^{2T} \left| \zeta\left(\tfrac{1}{2} + it\right) \right|^2 dt = \int_T^{2T} \zeta\left(\tfrac{1}{2} + it\right) \zeta\left(\tfrac{1}{2} - it\right) dt$$

$$= \int_T^{2T} \left\{ \left| \sum_{n \leqslant T} \frac{1}{n^{1/2+it}} \right|^2 + 2\,\mathrm{Re}\left(R \sum_{n \leqslant T} \frac{1}{n^{1/2+it}} \right) + O\left(T^{-1}\right) \right\} dt$$

and using Lemma 4.18 we derive

$$\int_T^{2T} \left| \sum_{n \leqslant T} \frac{1}{n^{1/2+it}} \right|^2 dt = \sum_{n \leqslant 2T} n^{-1}\left(T + O(n)\right) = T \log T + O(T)$$

and

$$\int_T^{2T} 2\,\mathrm{Re}\left(R \sum_{n \leqslant T} \frac{1}{n^{1/2+it}} \right) dt \ll T^{-1/2} \int_T^{2T} T^{1/2}\, dt \ll T.$$

Therefore

$$\int_T^{2T} \left| \zeta\left(\tfrac{1}{2} + it\right) \right|^2 dt = T \log T + O(T)$$

and Proposition 4.33 follows on summing over $\frac{1}{2}T, \frac{1}{4}T, \ldots$, producing an error term of $O(T)$. $\square$

4.7.5.2 Higher Power Means

We intend to provide some other results without proof, being beyond the scope of this book. For the next result, see [52, 54].

Theorem 4.59 *There exists an effectively computable constant $\beta > 0$ such that, for $T \to \infty$*

$$\int_0^T \left| \zeta \left(\tfrac{1}{2} + it \right) \right|^4 dt = T \sum_{k=0}^4 c_k (\log T)^{4-k} + O\left(T^{2/3} (\log T)^\beta \right),$$

where $c_0 = (2\pi^2)^{-1}$ et $c_1, \ldots, c_4$ are computable. Furthermore, for $\frac{1}{2} < \sigma \leqslant 1$

$$\int_1^T |\zeta(\sigma + it)|^4 dt = \frac{\zeta(2\sigma)^2}{\zeta(4\sigma)} T + O\left(T^{2-2\sigma} \log^3 T \right).$$

Note that using the Cauchy–Schwarz inequality enables us to derive upper bounds for some other exponents. For instance, with the help of the previous results, we easily get

$$\int_0^T \left| \zeta \left(\tfrac{1}{2} + it \right) \right|^3 dt = \int_0^T \left| \zeta \left(\tfrac{1}{2} + it \right) \right| \left| \zeta \left(\tfrac{1}{2} + it \right) \right|^2 dt$$

$$\leqslant \left(\int_0^T \left| \zeta \left(\tfrac{1}{2} + it \right) \right|^2 dt \right)^{1/2} \left(\int_0^T \left| \zeta \left(\tfrac{1}{2} + it \right) \right|^4 dt \right)^{1/2}$$

$$\ll (T \log T)^{1/2} \left(T \log^4 T \right)^{1/2} \ll T (\log T)^{5/2}.^{32}$$

Some other upper bounds have been derived in the literature. For the proof, see, respectively, [53, 46].

Theorem 4.60 *Let $T > 0$, $n \in \mathbb{Z}_{\geqslant 2}$, $2 \leqslant k \leqslant 6$ and $\varepsilon > 0$. Then*

$$\int_0^T \left| \zeta \left(\tfrac{1}{2} + it \right) \right|^{2k} dt \ll_\varepsilon T^{\frac{k+2}{4} + \varepsilon}.$$

$$\int_0^T \left| \zeta \left(\tfrac{1}{2} + it \right) \right|^{2/n} dt \ll T (\log T)^{1/n^2}.$$

Explicit bounds have recently been proved in [23], in which among other things one finds the following estimates.

Theorem 4.61 *If $T \geqslant 4$*

$$\int_1^T \left| \zeta \left(\tfrac{1}{2} + it \right) \right|^2 dt \leqslant T \log T + 7.9548\, T + 20.10576\, T^{1/2} \log T$$

---

[32]In [2], the authors show that the exponent $\frac{5}{2}$ may be replaced by $\frac{9}{4}$.

and

$$\int_1^T \left| \zeta \left(\tfrac{1}{2} + it \right) \right|^2 dt \geqslant T \log T - 8.80037\, T - 5.2576\, T^{1/2} \log T.$$

4.7.6 The Dirichlet–Piltz Divisor Problem

4.7.6.1 Introduction

Let $r \in \mathbb{Z}_{\geqslant 2}$. Generalizing Theorem 4.28, the *Dirichlet–Piltz divisor problem* is about the smallest exponent $\theta_r \in [0, 1)$ for which the following estimate

$$\sum_{n \leqslant x} \tau_r(n) = x P_{r-1} (\log x) + O \left(x^{\theta_r} (\log x)^{\beta_r} \right)$$

holds for some $\beta_r \geqslant 0$, where P_{r-1} is a polynomial of degree $r - 1$ and whose coefficients depend on r. By Dirichlet's hyperbola principle, we get

$$\sum_{n \leqslant x} \tau_r(n) = x \sum_{d \leqslant x^{1-1/r}} \frac{\tau_{r-1}(d)}{d} + \sum_{m \leqslant x^{1/r}} \sum_{d \leqslant x/m} \tau_{r-1}(d)$$

$$- x^{1/r} \sum_{d \leqslant x^{1-1/r}} \tau_{r-1}(d) + O \left(\sum_{x^{1-1/r}} \tau_{r-1}(d) \right)$$

so that by induction one can show that the pair

$$(\theta_r, \beta_r) = \left(1 - \tfrac{1}{r}, 0 \right) \tag{4.77}$$

is admissible.

4.7.6.2 Estimates Using Contour Integration

Improvements in this problem arise when using results for power mean values of the Riemann zeta function in the critical strip, such as seen in the previous section. For instance, using Theorem 4.59, Hardy and Littlewood showed that $\theta_r = 1 - \frac{3}{r+2}$ is admissible[33] as soon as $r \geqslant 4$. As for the function τ, one can show that $\theta_r \geqslant \frac{1}{2} - \frac{1}{2r}$ with some $\beta_r > 0$, and one conjectures this is the right value in this problem.[34]

We list in the next theorem the more recent results obtained for θ_r. For a proof of each of these estimates, see [12, 60] and [53, Chapter 8].

Theorem 4.62 *When $2 \leqslant r \leqslant 10$, we have the following results.*

r	2	3	$\in [4, 8]$	9	10
$\theta_r \leqslant \ldots$	$\frac{517}{1648} \approx 0.3137$	$\frac{43}{96} \approx 0.4479$	$\frac{3}{4} - \frac{1}{r}$	$\frac{35}{54} \approx 0.6481$	$\frac{27}{40} \approx 0.675$

To illustrate the method, let us prove the following Hardy–Littlewood like result. The reader should notice how this problem and estimates of the Riemann zeta function on the critical line are related.

Theorem 4.63 *Let $r \in \mathbb{Z}_{\geqslant 4}$ and assume that there exists $\alpha \in \left(0, \frac{1}{4}\right]$ such that, for all $\varepsilon > 0$ and $t \geqslant e$, we have*

$$\zeta\left(\tfrac{1}{2} + it\right) \ll_\varepsilon t^{\alpha + \varepsilon}.$$

Then, for all $\varepsilon > 0$

$$\sum_{n \leqslant x} \tau_r(n) = x P_{r-1}(\log x) + O_{\varepsilon, r}\left(x^{1 - \frac{1}{2(\alpha(r-4)+1)} + \varepsilon}\right),$$

where P_{r-1} is a polynomial of degree $r - 1$ and leading coefficient $\frac{1}{(r-1)!}$.

Proof Let $1 \leqslant T \leqslant \left(e^{-1}x\right)^{\frac{1}{2\alpha r}}$. Using (3.41), we derive

$$\sum_{n \leqslant x} \tau_r(n) = \frac{1}{2\pi i} \int_{\kappa - iT}^{\kappa + iT} \zeta(s)^r \frac{x^s}{s} \, ds + O_{\varepsilon, r}\left(\frac{x^{1+\varepsilon}}{T}\right)$$

---

[33] See Corollary 4.19 below.

[34] The Dirichlet–Piltz divisor problem can be generalized to number fields where it is usually called the *Number fields divisor problem*. It was originally investigated by Hasse and Suetuna [44], and the asymptotics are of the same order of magnitude as in the rational case.

with $\kappa := 1 + \frac{1}{\log x}$. We shift the line of integration to the line $\sigma = \frac{1}{2}$ and apply Cauchy's theorem in the rectangle with vertices $\kappa \pm iT$, $\frac{1}{2} \pm iT$. The residue is given by

$$\operatorname*{Res}_{s=1} \left(\zeta(s)^r x^s s^{-1} \right) = x P_{r-1} (\log x)$$

and using Exercise 50, the contribution of the horizontal segments does not exceed

$$\ll \frac{1}{T} \int_{1/2}^{1+\kappa} x^\sigma \, |\zeta(\sigma + iT)|^r \, d\sigma$$

$$\ll \frac{1}{T} \left\{ \int_{1/2}^{1} x^\sigma T^{2r\alpha(1-\sigma)+\varepsilon} \, d\sigma + x (\log T)^r \right\}$$

$$\ll x^{1/2} T^{r\alpha - 1 + \varepsilon} + x^{1+\varepsilon} T^{-1}.$$

Since $r \geqslant 4$, the contribution of the vertical segment is bounded by

$$\ll x^{1/2} \left\{ 1 + \int_e^T \left| \zeta \left(\tfrac{1}{2} + it \right) \right|^r \frac{dt}{t} \right\}$$

$$\ll x^{1/2} \left\{ 1 + \max_{e \leqslant t \leqslant T} \left| \zeta \left(\tfrac{1}{2} + it \right) \right|^{r-4} \int_e^T \left| \zeta \left(\tfrac{1}{2} + it \right) \right|^4 \frac{dt}{t} \right\}$$

$$\ll x^{1/2} \left\{ 1 + T^{\alpha(r-4)+\varepsilon} \int_e^T \left| \zeta \left(\tfrac{1}{2} + it \right) \right|^4 \frac{dt}{t} \right\}$$

and setting $F(T) := \int_e^T \left| \zeta \left(\tfrac{1}{2} + it \right) \right|^4 dt$, we get integrating by parts and using Theorem 4.59

$$\int_e^T \left| \zeta \left(\tfrac{1}{2} + it \right) \right|^4 \frac{dt}{t} = \frac{F(t)}{t} \bigg|_e^T + \int_e^T \frac{F(t)}{t^2} \, dt \ll (\log T)^5.$$

Hence

$$\sum_{n \leqslant x} \tau_r(n) = x P_{r-1} (\log x) + O_{\varepsilon, r} \left(x^{1/2} T^{\alpha(r-4)+\varepsilon} + x^{1/2} T^{r\alpha - 1 + \varepsilon} + x^{1+\varepsilon} T^{-1} \right)$$

and note that the second term is absorbed by the first one since $0 < \alpha \leqslant \frac{1}{4}$. Choosing $T = x^{\frac{1}{2(\alpha(r-4)+1)}}$ yields the asserted result. $\square$

Taking (3.19) into account, we derive the next corollary.

Corollary 4.19 *Let $r \in \mathbb{Z}_{\geqslant 4}$. Then, for all $\varepsilon > 0$*

$$\sum_{n \leqslant x} \tau_r(n) = x P_{r-1}(\log x) + O_{\varepsilon, r}\left(x^{1 - \frac{42}{13r+32} + \varepsilon}\right),$$

where P_{r-1} is a polynomial of degree $r - 1$ and leading coefficient $\frac{1}{(k-1)!}$.

For larger values of r, it is preferable to borrow Vinogradov–Korobov's techniques. The *Karatsuba's constant* is the number $c_0 > 0$ such that

$$\theta_r \leqslant 1 - c_0 (Dr)^{-2/3},$$

where $D \geqslant 0$ satisfies the bound

$$|\zeta(\sigma + it)| \ll |t|^{D(1-\sigma)^{3/2}} (\log |t|)^{2/3} \quad (0.9 < \sigma \leqslant 1, \ |t| \geqslant 1).$$

From (3.58), one may take $D = 4.45$. Therefore, we have the following results [55, 61].

r	$\in [79, 119]$	$\geqslant 120$	$\geqslant 160$
$\theta_r \leqslant \ldots$	$1 - \frac{165}{28r}$	$1 - \frac{1}{3} \times 2^{2/3} (Dr)^{-2/3}$	$1 - \left(\frac{2}{3D(r-159.9)}\right)^{2/3}$

4.7.6.3   Estimates Using Voronoï's Formula

The Dirichlet's hyperbola principle can be used with $h(n)$ being of the form $e(F(n))$.

Proposition 4.34 *Let f, g be complex-valued arithmetic functions and $F : \mathbb{R}_{\geqslant 1} \to \mathbb{R}_{\geqslant 0}$ be any map. For all $N < N_1 \in \mathbb{Z}_{\geqslant 1}$ and $1 \leqslant U \leqslant N$*

$$\sum_{N < n \leqslant N_1} (f \star g)(n) e(F(n))$$

$$= \sum_{n \leqslant U} f(n) \sum_{\frac{N}{n} < m \leqslant \frac{N_1}{n}} g(m) e(F(mn)) + \sum_{n \leqslant \frac{N}{U}} g(n) \sum_{\frac{N}{n} < m \leqslant \frac{N_1}{n}} f(m) e(F(mn))$$

$$+ \sum_{\frac{N}{U} < n \leqslant \frac{N_1}{U}} g(n) \sum_{m \leqslant \frac{N_1}{n}} f(m) e(F(mn)) - \sum_{n \leqslant U} f(n) \sum_{\frac{N}{U} < m \leqslant \frac{N_1}{U}} g(m) e(F(mn)).$$

Proof From Proposition 4.4 with $h(n) = e(F(n))$, we derive

$$\sum_{N < n \leqslant N_1} (f \star g)(n) e(F(n))$$

$$= \sum_{n \leqslant U} f(n) \sum_{m \leqslant \frac{N_1}{n}} g(m) e(F(mn)) + \sum_{n \leqslant \frac{N_1}{U}} g(n) \sum_{m \leqslant \frac{N_1}{n}} f(m) e(F(mn))$$

$$- \sum_{n \leqslant U} f(n) \sum_{m \leqslant \frac{N_1}{U}} g(m) e(F(mn)) - \sum_{n \leqslant U} f(n) \sum_{m \leqslant \frac{N}{n}} g(m) e(F(mn))$$

$$- \sum_{n \leqslant \frac{N}{U}} g(n) \sum_{m \leqslant \frac{N}{n}} f(m) e(F(mn)) + \sum_{n \leqslant U} f(n) \sum_{m \leqslant \frac{N}{U}} g(m) e(F(mn))$$

giving the asserted result.   □

Applying Proposition 4.34 to the function τ_r, which has the simple induction relation $\tau_r = \tau_{r-1} \star 1$, we derive the following estimate.[35]

Proposition 4.35 *Let $N, N_1 \in \mathbb{Z}_{\geqslant 1}$ such that $N < N_1 \leqslant 2N$, and (X_r) be a sequence of positive real numbers. Let (k, ℓ) be an exponent pair. Then, for any $r \in \mathbb{Z}_{\geqslant 1}$ and any sequence of functions (f_r) satisfying*

▷ *$f_r \in C^\infty([N, N_1])$ and $f_r > 0$;*
▷ *for any $j \in \mathbb{Z}_{\geqslant 0}$ and any $x \in [N, N_1]$, $\left| f_r^{(j)}(x) \right| \asymp X_r N^{-j}$*

we have

$$\sum_{N < n \leqslant N_1} \tau_r(n) e(f_r(n)) \ll X_r^k N^{\frac{\ell-k}{r}+1-\frac{1}{r}} (\log N)^r + N X_r^{-1} (\log N)^{r+1}.$$

Proof We can assume that $(k, \ell) \neq (0, 1)$, otherwise the result is trivial. By induction on r, the case $r = 1$ being precisely Definition 6.2. Assume the estimate for $r \geqslant 1$ and let f_{r+1} be a function satisfying the hypotheses of the proposition with

---

[35] For the definition of *exponent pairs*, see Sect. 6.5.2 of Chap. 6.

X_r replaced by X_{r+1}. Using Proposition 4.34 with $f = \tau_r$, $g = 1$ and $U = N^{\frac{r}{r+1}}$, we get

$$\sum_{N < n \leqslant N_1} \tau_{r+1}(n) e\left(f_{r+1}(n)\right)$$

$$\ll \sum_{n \leqslant N^{\frac{r}{r+1}}} \tau_r(n) \left| \sum_{\frac{N}{n} < m \leqslant \frac{N_1}{n}} e\left(f_{r+1}(mn)\right) \right| + \sum_{n \leqslant N^{\frac{1}{r+1}}} \left| \sum_{\frac{N}{n} < m \leqslant \frac{N_1}{n}} \tau_r(m) e\left(f_{r+1}(mn)\right) \right|$$

$$+ \sum_{N^{\frac{1}{r+1}} < n \leqslant N_2} \left| \sum_{m \leqslant \frac{N_1}{n}} \tau_r(m) e\left(f_{r+1}(mn)\right) \right| + \sum_{n \leqslant N^{\frac{r}{r+1}}} \tau_r(n) \left| \sum_{N^{\frac{1}{r+1}} < m \leqslant N_2} e\left(f_{r+1}(mn)\right) \right|,$$

where $N_2 := N_1 N^{\frac{-r}{r+1}} \leqslant 2N^{\frac{1}{r+1}}$, and the induction hypothesis implies that

$$\sum_{N < n \leqslant N_1} \tau_{r+1}(n) e\left(f_{r+1}(n)\right)$$

$$\ll \sum_{n \leqslant N^{\frac{r}{r+1}}} \tau_r(n) \left(X_{r+1}^k N^{\ell-k} n^{k-\ell} + \frac{N}{n X_{r+1}} \right)$$

$$+ \sum_{n \leqslant N^{\frac{1}{r+1}}} \left(X_{r+1}^k N^{\frac{\ell-k}{r}+1-\frac{1}{r}} n^{-\frac{\ell-k}{r}-1+\frac{1}{r}} (\log N)^r + \frac{N}{n X_{r+1}} (\log N)^{r+1} \right)$$

$$+ \sum_{N^{\frac{1}{r+1}} < n \leqslant 2N^{\frac{1}{r+1}}} \left(X_{r+1}^k N^{\frac{\ell-k}{r}+1-\frac{1}{r}} n^{-\frac{\ell-k}{r}-1+\frac{1}{r}} (\log N)^{r+1} + \frac{N}{n X_{r+1}} (\log N)^{r+2} \right)$$

$$+ \sum_{n \leqslant N^{\frac{r}{r+1}}} \tau_r(n) \left(X_{r+1}^k N^{\frac{\ell-k}{r+1}} + \frac{N^{\frac{1}{r+1}}}{X_{r+1}} \right).$$

The bound

$$\sum_{n \leqslant x} \frac{\tau_r(n)}{n^\alpha} \ll x^{1-\alpha} (\log x)^r \quad (0 \leqslant \alpha < 1)$$

allows us to get

$$\sum_{N < n \leqslant N_1} \tau_{r+1}(n) e\left(f(n)\right) \ll X_{r+1}^k N^{\frac{\ell-k}{r+1}+1-\frac{1}{r+1}} (\log N)^{r+1} + N X_{r+1}^{-1} (\log N)^{r+2}$$

completing the proof. $\square$

Corollary 4.20 *Let $r \in \mathbb{Z}_{\geqslant 2}$ and (k, ℓ) be an exponent pair. Then, for all $\varepsilon > 0$*

$$\sum_{n \leqslant x} \tau_r(n) = x P_{r-1}(\log x) + O_{\varepsilon, r}\left(x^{1 - \frac{1}{r} - \frac{r-2k-1}{r(2\ell+r-1)} + \varepsilon}\right),$$

where P_{r-1} is a polynomial of degree $r - 1$ and leading coefficient $\frac{1}{(k-1)!}$.

Proof Let $r \in \mathbb{Z}_{\geqslant 2}$. It is customary to set[36]

$$\Delta_r(x) := \sum_{n \leqslant x} \tau_r(n) - \operatorname*{Res}_{s=1}\left(\zeta(s)^r x^s s^{-1}\right).$$

The *Voronoï's truncated formula*[53] implies the bound

$$\Delta_r(x) \ll x^{\frac{r-1}{2r}} \left| \sum_{n \leqslant y} \tau_r(n) n^{-\frac{r+1}{2r}} e\left(r(nx)^{1/r}\right) \right| + x^\varepsilon \left(x^{1-1/r} y^{-1/r} + 1\right)$$

for any integer parameter y satisfying $1 \ll y \leqslant x^A$, with $A > 0$. Now using Proposition 4.35 and partial summation, the sum does not exceed

$$\ll \max_{N \leqslant y} N^{-\frac{r+1}{2r}} \max_{N \leqslant N_1 \leqslant 2N} \left| \sum_{N < n \leqslant N_1} \tau_r(n) e\left(r(nx)^{1/r}\right) \right| \log y$$

$$\ll \max_{N \leqslant y} \left\{ x^{k/r} N^{\frac{2\ell+r-3}{2r}} (\log N)^r + x^{-1/r} N^{\frac{r-3}{2r}} (\log N)^{r+1} \right\} \log y$$

$$\ll x^{k/r} y^{\frac{2\ell+r-3}{2r}} (\log y)^{r+1} + \begin{cases} x^{-1/r} y^{\frac{r-3}{2r}} (\log y)^{r+2}, & \text{if } r \geqslant 3; \\ x^{-1/4} (\log y)^4, & \text{if } r = 2; \end{cases}$$

$$\ll x^{k/r} y^{\frac{2\ell+r-3}{2r}} (\log y)^{r+2}$$

so that

$$\Delta_r(x) \ll x^{\frac{r+2k-1}{2r}} y^{\frac{2\ell+r-3}{2r}} (\log y)^{r+2} + x^\varepsilon \left(x^{1-1/r} y^{-1/r} + 1\right)$$

---

[36] A slightly different definition is sometimes given. For all $x > 1$

$$\Delta_r(x) := \sum_{n \leqslant x}^{*} \tau_r(n) - \operatorname*{Res}_{s=1}\left(\zeta(s)^r x^s s^{-1}\right) - (-1)^r 2^{-r},$$

where the star means that the last term in the sum has to be halved if $x \in \mathbb{Z}_{\geqslant 2}$.

and the result follows from choosing $y = \left\lfloor x^{\frac{r-2k-1}{r+2\ell-1}} \right\rfloor$.   □

With Bourgain's exponent pair $(k, \ell) = \left(\frac{13}{84} + \varepsilon, \frac{55}{84} + \varepsilon\right)$, we derive the following estimate.

Corollary 4.21 *Let $r \in \mathbb{Z}_{\geqslant 2}$ and (k, ℓ) be an exponent pair. Then, for all $\varepsilon > 0$*

$$\sum_{n \leqslant x} \tau_r(n) = x P_{r-1}(\log x) + O_{\varepsilon, r}\left(x^{1 - \frac{42(2r-1)}{r(42r+13)} + \varepsilon}\right),$$

where P_{r-1} is a polynomial of degree $r - 1$ and leading coefficient $\frac{1}{(k-1)!}$.

Note that this result is slightly better than Landau's exponent $1 - \frac{2}{r+1}$ [110, Theorem 12.2], but weaker than Corollary 4.19 for $r \geqslant 4$, and significantly worse than Ivić's results [53, Chapter 13].

On the other hand, this result slightly improves contour integration techniques in the case $r = 3$ for which calculations as in Theorem 4.63 yield $\Delta_3(x) \ll_\varepsilon x^{1/2+\varepsilon}$. Taking

$$(k, \ell) = A^2 \left(B A^2\right)^2 B(0, 1) = \left(\frac{13}{238}, \frac{97}{119}\right)$$

or

$$(k, \ell) = A\left(\frac{13}{84} + \varepsilon, \frac{55}{84} + \varepsilon\right) = \left(\frac{13}{194} + \varepsilon, \frac{76}{97} + \varepsilon\right)$$

in Corollary 4.21, respectively, yields

$$\Delta_3(x) \ll_\varepsilon x^{71/144+\varepsilon} \approx x^{0.49305...+\varepsilon}$$

and

$$\Delta_3(x) \ll_\varepsilon x^{511/1038+\varepsilon} \approx x^{0.49229...+\varepsilon}$$

which is also slightly better than Rankin's estimate who obtained $\Delta_3(x) \ll x^{0.4931466...+\varepsilon}$ in [88], but is still far from Kolesnik's result seen in Theorem 4.62.

4.7.7 Multidimensional Divisor Problem

4.7.7.1 Introduction

Let $r \in \mathbb{Z}_{\geqslant 2}$. The Dirichlet–Piltz divisor problem seen in the previous section may be generalized as follows. Let $\mathfrak{a} = \{a_1, \ldots, a_r\}$ be a set of fixed positive integers such that $1 \leqslant a_1 \leqslant \cdots \leqslant a_r$, and denote $\tau(\mathfrak{a}; n)$ to be the number of representations of an integer $n \geqslant 1$ in the form

$$n = n_1^{a_1} \cdots n_r^{a_r}$$

with natural numbers $n_1, \ldots, n_r$.[37] It is customary to set $A_j := a_1 + \cdots + a_j$ for $j = 1, \ldots, r$. Finally, define

$$D(\mathfrak{a}; x) := \sum_{n \leqslant x} \tau(\mathfrak{a}; n).$$

Note that

$$L(s, \tau(\mathfrak{a}; \cdot)) = \prod_{j=1}^{r} \zeta(a_j s) \quad \left(\sigma > \tfrac{1}{a_1}\right)$$

and hence one may write

$$D(\mathfrak{a}; x) = H(\mathfrak{a}; x) + \Delta(\mathfrak{a}; x), \tag{4.78}$$

where

$$H(\mathfrak{a}; x) = \sum_{j=1}^{r} \operatorname*{Res}_{s=1/a_j} \left(L(s, \tau(\mathfrak{a}; \cdot)) x^s s^{-1} \right)$$

is the main term and $\Delta(\mathfrak{a}; x)$ is the error term. If all the a_is are distinct, then the poles $a_1^{-1}, \ldots, a_r^{-1}$ of $L(s, \tau(\mathfrak{a}; \cdot))$ are simple, and it is an easy exercise to verify that, in this case

$$H(\mathfrak{a}; x) = \sum_{j=1}^{r} \left(\prod_{\substack{i=1 \\ i \neq j}}^{r} \zeta\left(\frac{a_i}{a_j}\right) \right) x^{1/a_j}.$$

---

[37] In fact, we could take $a_1, \ldots, a_r$ to be *real* numbers, but the function $d(\mathfrak{a}; \cdot)$ is not a *divisor* function anymore.

In case of some equalities in the a_is, we may take the appropriate limit values in the sum.

Definition 4.15 The study of the remainder $\Delta\,(\mathfrak{a};x)$ given in (4.78) is usually called the *asymmetric many dimensional divisor problem*.

4.7.7.2 A ψ-Expression of $\Delta\,(\mathfrak{a};x)$

As in the case of the classical Dirichlet–Piltz divisor problem, the remainder $\Delta\,(\mathfrak{a};x)$ may be written with the use of the ψ-function. For a proof, see [122, 65].

Proposition 4.36 *For $r \geqslant 2$ such that $(r-2)a_r < A_r$*

$$
\Delta\,(\mathfrak{a};x) = -\sum_{\sigma\in\pi(\{1,\dots,r\})}\;{\sum_{}}^{*}\;\psi\left(\left(\frac{x}{n_1^{a_{\sigma(1)}}\cdots n_{r-1}^{a_{\sigma(r-1)}}}\right)^{1/a_{\sigma(r)}}\right) + O\left(x^{\frac{r-2}{A_r}}\right),
$$

where the outer sum runs over all permutations σ of the set $\{1,\dots,r\}$, and the condition summation of the starred inner sum is given by

$$
n_1^{a_{\sigma(1)}}\cdots n_{r-2}^{a_{\sigma(r-2)}} n_{r-1}^{a_{\sigma(r-1)}+a_{\sigma(r)}} \leqslant x \quad \text{and} \quad n_1(\leqslant)n_2(\leqslant)\cdots(\leqslant)n_{r-1},
$$

where the notation $n_j(\leqslant)n_{j+1}$ means $n_j \leqslant n_{j+1}$ for $\sigma(j) < \sigma(j+1)$ and $n_j < n_{j+1}$ for $\sigma(j) > \sigma(j+1)$.

Example 4.12 Choose $r = 2$ and rename the set $\mathfrak{a}$ in $\mathfrak{a} = \{a,b\}$ with $1 \leqslant a < b$. Then

$$
D(a,b;x) = \zeta\left(\tfrac{b}{a}\right)x^{1/a} + \zeta\left(\tfrac{a}{b}\right)x^{1/b} + \Delta(a,b;x)
$$

with

$$
\Delta(a,b;x) = -\sum_{n^{a+b}\leqslant x}\left\{\psi\left(\left(\frac{x}{n^b}\right)^{1/a}\right)+\psi\left(\left(\frac{x}{n^a}\right)^{1/b}\right)\right\} + O(1).
$$

Example 4.13 Choose $r = 3$ and rename the set $\mathfrak{a}$ in $\mathfrak{a} = \{a,b,c\}$ with $1 \leqslant a < b < c$. Then

$$
D(a,b,c;x) = \zeta\left(\tfrac{b}{a}\right)\zeta\left(\tfrac{c}{a}\right)x^{1/a}+\zeta\left(\tfrac{a}{b}\right)\zeta\left(\tfrac{c}{b}\right)x^{1/b}+\zeta\left(\tfrac{a}{c}\right)\zeta\left(\tfrac{b}{c}\right)x^{1/c}+\Delta(a,b,c;x)
$$

with

$$
\Delta(a,b,c;x) = -\sum_{(u,v,w)\in\pi(\{a,b,c\})}\;\sum_{\substack{m^u n^{v+w}\leqslant x\\ m\leqslant n}}\psi\left(\left(\frac{x}{m^u n^v}\right)^{1/w}\right)+O\left(x^{\frac{1}{a+b+c}}\right),
$$

where (u, v, w) runs over all permutations of $\{a, b, c\}$.

4.7.7.3 A Survey of Estimates for $\Delta(\mathfrak{a}; x)$

Several techniques may provide non-trivial bounds for $\Delta(\mathfrak{a}; x)$, the effectiveness of which depends essentially on the size of r. When r is small, the theory of exponential sums that will be seen in Chap. 6 would be preferable. When r is large, then Perron summation formula and the theory of the Riemann zeta function can yield better results. Recall that $A_r = a_1 + \cdots + a_r$.

The Trivial Estimate Using $\psi(t) \ll 1$ in Proposition 4.36 gives a proper result.

Lemma 4.19 *If $r \geqslant 2$ and $(r-1)a_r < A_r$, then*

$$\Delta(\mathfrak{a}; x) \ll x^{\frac{r-1}{A_r}}.$$

Proof Let $\left(a_{\sigma(1)}, \ldots, a_{\sigma(r)}\right) = (u_1, \ldots, u_r)$ be a fixed permutation of $\mathfrak{a}$. It suffices to show that, for all $r \geqslant 2$

$$\sum_{\substack{n_1^{u_1} \cdots n_{r-2}^{u_{r-2}} n_{r-1}^{u_{r-1}+u_r} \leqslant x \\ n_1 \leqslant \cdots \leqslant n_{r-1}}} 1 \ll x^{\frac{r-1}{u_1 + \cdots + u_r}}.$$

We use induction on r, the case $r = 2$ being obvious. Assume that the inequality is true for $r - 1$. Then the sum of the left-hand side does not exceed, by induction hypothesis

$$\ll \sum_{n_1 \leqslant x^{\frac{1}{u_1 + \cdots + u_r}}} \left(\frac{x}{n_1^{u_1}}\right)^{\frac{r-2}{u_2 + \cdots + u_r}}$$

and the hypothesis $(r-1)a_r < A_r$ implies that, for any permutation $(u_1, \ldots, u_r)$ of $\mathfrak{a}$, we have $(r-2)u_1 < u_2 + \cdots + u_r$, and therefore the sum is

$$\ll x^{\frac{r-2}{u_2 + \cdots + u_r}} x^{\frac{1}{u_1 + \cdots + u_r}\left(1 - \frac{u_1(r-2)}{u_2 + \cdots + u_r}\right)} = x^{\frac{r-1}{u_1 + \cdots + u_r}}$$

as required. □

Note that the condition $(r-1)a_r < A_r$ is somewhat irrelevant, since it only ensures that the error term is smaller than the main term.

A Result by Landau Using [67, (13)] with $\beta = a_1^{-1}$ and $\eta = \frac{1}{2}r$, we immediately derive for $r \geqslant 2$ and all $\varepsilon > 0$

$$\Delta(\mathfrak{a}; x) \ll x^{\frac{1}{a_1}\frac{r-1}{r+1}+\varepsilon}.$$

This bound has the disadvantage that only the smallest number a_1 is preferred, and Lemma 4.19 is better as soon as $A_r > (r+1)a_1$.   □

Using the Theory of the Riemann Zeta Function Krätzel [64] proves the next estimate.

Theorem 4.64 *If $r \geqslant 4$ and $(r-2)a_r < A_r \leqslant 2(r-2)a_1$, then for all $\varepsilon > 0$*

$$\Delta(\mathfrak{a}; x) \ll x^{\frac{r-2}{A_r}+\varepsilon}.$$

This bound is very good as can be seen in Proposition 4.36, but the condition $a_r < 2a_1$ is very hard.   □

Using Exponential Sums Results The following bounds are proved in [65] using similar techniques as will be seen in Chap. 6.

Theorem 4.65 *If $r \geqslant 2$ and $(r-1)a_r < a_1 + A_r$, then*

$$\Delta(\mathfrak{a}; x) \ll x^{\frac{r-1}{a_1+A_r}} (\log x)^r.$$

More precise estimates can be obtained with the help of exponent pairs[38]

Theorem 4.66 *Let $r \geqslant 3$ and (k, ℓ) be an exponent pair. Assume furthermore that*

▷ $(k + \ell + r - 2)a_r < (k+\ell)a_1 + A_r$;
▷ $2(k + \ell + 1)a_1 \leqslant (2k+\ell)(a_2 + a_3)$;
▷ *Assume also one of the following two assumptions.*

  ◇ $\ell a_1 \leqslant k a_2$ *and* $(k + \ell + 1)a_1 \geqslant k(a_2 + a_3)$;
  ◇ $\ell a_1 \geqslant k a_2$ *and* $(\ell - k)(2k + 1)a_3 \leqslant (2\ell - 2k - 1)(k + \ell + 1)a_1 + (2k(k - \ell + 1) + 1)a_2$.

Then

$$\Delta(\mathfrak{a}; x) \ll x^{\frac{k+\ell+r-2}{(k+\ell)a_1+A_r}} (\log x)^r.$$

---

[38]See Definition 6.1.

Example 4.14 Setting $r = 3$ and choosing the exponent pair $(k, \ell) = \left(\frac{1}{6}, \frac{2}{3}\right)$, we derive

$$\Delta(a, b, c; x) \ll x^{\frac{11}{11a+6(b+c)}} (\log x)^3$$

with $1 \leqslant a \leqslant b \leqslant c$ satisfying $4a \geqslant b$, $7b \geqslant 2c$, $5c < 11a+6b$ and $11a \geqslant 3(b+c)$.

4.7.7.4   Application to Some Number-Theoretic Problems

A numerous number of arithmetic problems are related to some divisor functions estimates.

Distribution of Square-Full Numbers   Since $L(s, s_2) = \zeta(2s)\zeta(3s)\zeta(6s)^{-1}$, we derive

$$s_2(n) = \sum_{d^6 | n} \mu(d)\tau\left(2, 3; n/d^6\right)$$

and hence

$$\sum_{n \leqslant x} s_2(n) = \sum_{d \leqslant x^{1/6}} \mu(d) \sum_{k \leqslant x/d^6} \tau(2, 3; k)$$

and by Example 4.12 and Theorem 4.65 we get

$$\sum_{n \leqslant x} s_2(n) = \sum_{d \leqslant x^{1/6}} \mu(d) \left\{ \zeta\left(\frac{3}{2}\right)\left(\frac{x}{d^6}\right)^{1/2} + \zeta\left(\frac{2}{3}\right)\left(\frac{x}{d^6}\right)^{1/3} + \Delta\left(2, 3; \frac{x}{d^6}\right) \right\}$$

$$= \frac{\zeta(3/2)}{\zeta(3)} x^{1/2} + \frac{\zeta(2/3)}{\zeta(2)} x^{1/3} + \sum_{d \leqslant x^{1/6}} \mu(d)\Delta\left(2, 3; \frac{x}{d^6}\right) + O\left(x^{1/6}\right)$$

$$= \frac{\zeta(3/2)}{\zeta(3)} x^{1/2} + \frac{\zeta(2/3)}{\zeta(2)} x^{1/3} + O\left(x^{1/6}(\log x)^2\right).$$

$\square$

Abelian Groups of Given Order   Since $L(s, a) = \zeta(s)\zeta(2s)\zeta(3s)\zeta(4s)\cdots$, we infer

$$a(n) = \sum_{d | n} g(n)\tau(1, 2, 3; n/d),$$

where

$$L(s, g) = \prod_{j=4}^{\infty} \zeta(sj) \quad \left(\sigma > \tfrac{1}{4}\right)$$

so that $G(s)$ is absolutely convergent in the half-plane $\sigma > \tfrac{1}{4}$ and hence

$$\sum_{n \leqslant x} |g(n)| \ll x^{1/4+\varepsilon}.$$

By partial summation, we derive, for $\alpha > \tfrac{1}{4}$

$$\sum_{n > x} \frac{|g(n)|}{n^\alpha} = -x^{-\alpha} \sum_{n \leqslant x} |g(n)| + \alpha \int_x^{\infty} t^{-\alpha-1} \left(\sum_{n \leqslant t} |g(n)|\right) dt \ll x^{1/4-\alpha+\varepsilon}.$$

$$(4.79)$$

Now Examples 4.13 and 4.14, along with (4.79), yield

$$\sum_{n \leqslant x} a(n) = \sum_{d \leqslant x} g(d) \sum_{k \leqslant x/d} \tau(1, 2, 3; k)$$

$$= A_1 x + A_2 x^{1/2} + A_3 x^{1/3} + O\left(x^{1/4+\varepsilon}\right) + \sum_{d \leqslant x} g(d) \Delta(1, 2, 3; x/d)$$

$$= A_1 x + A_2 x^{1/2} + A_3 x^{1/3} + O\left(x^{11/41} (\log x)^3\right),$$

where $A_i = \prod_{j \neq i} \zeta\left(\tfrac{j}{i}\right)$ for $i = 1, 2, 3$. Note that $\tfrac{11}{41} \approx 0.26829$, which is weaker than Robert and Sargos's result [90] seen in (4.29).

Number of Finite Semisimple Rings It has previously been seen in Sect. 4.3.2.4 that $L(s, S) = \prod_{q=1}^{\infty} \prod_{m=1}^{\infty} \zeta\left(qm^2 s\right)$, so that we may write

$$S(n) = \sum_{d \mid n} g(d) a(n/d),$$

where

$$L(s, g) = \prod_{q=1}^{\infty} \prod_{m=2}^{\infty} \zeta\left(qm^2 s\right)$$

so that $G(s)$ is absolutely convergent in the half-plane $\sigma > \frac{1}{4}$ and hence, as above

$$\sum_{n>x} \frac{|g(n)|}{n^\alpha} \ll x^{1/4-\alpha+\varepsilon} \qquad \left(\alpha > \tfrac{1}{4}\right).$$

Taking the previous estimate into account, we get

$$\sum_{n\leqslant x} S(n) = \sum_{d\leqslant x} g(d) \sum_{k\leqslant x/d} a(k)$$

$$= \sum_{d\leqslant x} g(d) \left\{ A_1(x/d) + A_2(x/d)^{1/2} + A_3(x/d)^{1/3} + O\left((x/d)^{11/41}(\log x)^3\right)\right\}$$

$$= B_1 x + B_2 x^{1/2} + B_3 x^{1/3} + O\left(x^{1/4+\varepsilon}\right) + O\left(x^{11/41}(\log x)^3 \sum_{d\leqslant x} \frac{|g(d)|}{d^{11/41}}\right)$$

$$= B_1 x + B_2 x^{1/2} + B_3 x^{1/3} + O\left(x^{11/41}(\log x)^3\right),$$

where $B_i := A_i L\left(\frac{1}{i}, g\right) = \prod_{j\neq i} \zeta\left(\frac{j}{i}\right) \prod_{q=1}^\infty \prod_{m=2}^\infty \zeta\left(\frac{qm^2}{i}\right)$ for $i \in \{1, 2, 3\}$.

Remark 4.8 More generally, following the previous argument, one can see that, if we have at our disposal an asymptotic formula of the shape

$$\sum_{n\leqslant x} a(n) = A_1 x + A_2 x^{1/2} + A_3 x^{1/3} + O\left(x^\alpha (\log x)^\beta\right)$$

for some $\frac{1}{4} < \alpha < \frac{1}{3}$ and $\beta > 0$, then we also have

$$\sum_{n\leqslant x} S(n) = B_1 x + B_2 x^{1/2} + B_3 x^{1/3} + O\left(x^\alpha (\log x)^\beta\right)$$

with B_i as above. Hence from Robert and Sargos's estimate we derive

$$\sum_{n\leqslant x} S(n) = B_1 x + B_2 x^{1/2} + B_3 x^{1/3} + O\left(x^{1/4+\varepsilon}\right).$$

Direct Factors of a Finite Abelian Group Let G be a finite abelian group, and denote $N(G)$ to be the number of direct factors of G, i.e. the total number of decompositions $G = A \times B$ where $A \times B$ is the direct product of the subgroups A and B of G, and set

$$T(x) := \sum_{|G|\leqslant x} N(G).$$

It can be shown [18] that

$$T(x) := \sum_{n \leqslant x} t(n),$$

where $t(n)$ is a multiplicative function whose Dirichlet series is given by

$$L(s, t) = L(s, a)^2 = \prod_{j=1}^{\infty} \zeta(js)^2 \quad (\sigma > 1).$$

As above, we can thus write

$$t(n) = \sum_{d \mid n} g(d) \tau(1, 1, 2, 2; n/d),$$

where $L(s, g) = \zeta(3s)^2 \zeta(4s)^2 \cdots$ is a Dirichlet series absolutely convergent in the half-plane $\sigma > \frac{1}{3}$. As in (4.79), partial summation yields for $\alpha > \frac{1}{3}$

$$\sum_{n>x} \frac{|g(n)|}{n^{\alpha}} + \sum_{n>x} \frac{|g(n)| \log n}{n^{\alpha}} \ll x^{1/3 - \alpha + \varepsilon}. \tag{4.80}$$

Therefore

$$T(x) = \sum_{d \leqslant x} g(d) \sum_{k \leqslant x/d} \tau(1, 1, 2, 2; k)$$

$$= x \zeta(2)^2 \sum_{d \leqslant x} \frac{g(d)}{d} \left(\log \frac{x}{d} + 2\gamma - 1 \right)$$

$$+ x^{1/2} \zeta \left(\tfrac{1}{2} \right)^2 \sum_{d \leqslant x} \frac{g(d)}{d^{1/2}} \left(\frac{1}{2} \log \frac{x}{d} + 2\gamma - 1 \right)$$

$$+ \sum_{d \leqslant x} g(d) \Delta(1, 1, 2, 2; x/d)$$

$$= x \zeta(2)^2 L(1, g) \left(\log x + 2\gamma - 1 + \frac{L'}{L}(1, g) \right)$$

$$+ \frac{1}{2} x^{1/2} \zeta \left(\tfrac{1}{2} \right)^2 L \left(\tfrac{1}{2}, g \right) \left(\log x + 4\gamma - 2 + \frac{L'}{L} \left(\tfrac{1}{2}, g \right) \right)$$

$$+ \sum_{d \leqslant x} g(d) \Delta(1, 1, 2, 2; x/d) + O \left(x^{1/3 + \varepsilon} \right),$$

where we used (4.80) to estimate the residual sums. Note that

$$\zeta(2)^2 L(1, g) = \prod_{j=2}^{\infty} \zeta(j)^2 \approx 5.266 \quad \text{and} \quad \zeta\left(\tfrac{1}{2}\right)^2 L\left(\tfrac{1}{2}, g\right)$$

$$= \prod_{j \neq 2} \zeta\left(\tfrac{j}{2}\right)^2 \approx 214.551$$

and

$$\frac{L'}{L}(1, g) = 2 \sum_{j=3}^{\infty} j \frac{\zeta'}{\zeta}(j) \approx -2.111 \quad \text{and}$$

$$\frac{L'}{L}\left(\tfrac{1}{2}, g\right) = 2 \sum_{j=3}^{\infty} j \frac{\zeta'}{\zeta}\left(\tfrac{j}{2}\right) \approx -23.763.$$

Now Theorem 4.65 with $r = 4$ yields

$$\Delta(1, 1, 2, 2; y) \ll y^{3/7} (\log y)^4$$

and inserting this estimate above, we finally derive

$$T(x) = x \prod_{j=2}^{\infty} \zeta(j)^2 (\log x + 2\gamma - 1 + c_1)$$

$$+ \tfrac{1}{2} x^{1/2} \prod_{j \neq 2} \zeta\left(\tfrac{j}{2}\right)^2 (\log x + 4\gamma - 2 + c_2) + O\left(x^{3/7}(\log x)^4\right)$$

with $c_1 \approx -2.111$ and $c_2 \approx -23.763$.

Remark 4.9 The best result to date in this problem is due to Liu [72] who shows that the error term is $\ll x^{13/36+\varepsilon}$. The proof rests on the following asymmetric expression of $\Delta(1, 1, 2, 2; x)$ proved in [127]

$$\Delta(1, 1, 2, 2; x) = \sum_{n \leqslant x^{1/3}} \tau(n) \Delta\left(\frac{x}{n^2}\right) + \sum_{n \leqslant x^{1/3}} \tau(n) \Delta\left(\sqrt{\frac{x}{n}}\right) + O\left(x^{1/3} \log x\right),$$

where $\Delta(x)$ is the remainder in the classical Dirichlet divisor problem. It can also be proved [127, Proposition 2] that the second sum does not exceed $\ll x^{1/3+\varepsilon}$.

4.7.8 The Hardy–Ramanujan Inequality

4.7.8.1 Background

Dealing with the normal number of prime factors of an integer, Hardy and Ramanujan [42] derived the following fundamental inequality.

Theorem 4.67 *For all $x \geqslant 3$ and $k \in \mathbb{Z}_{\geqslant 1}$, set*

$$\pi_k(x) := \sum_{\substack{n \leqslant x \\ \omega(n)=k}} 1.$$

Then there exist two absolute constants $c_1, c_2 > 0$ such that, uniformly for $x \geqslant 3$ and $k \in \mathbb{Z}_{\geqslant 1}$

$$\pi_k(x) \leqslant \frac{c_1 \, x}{\log x} \frac{(\log \log x + c_2)^{k-1}}{(k-1)!}.$$

The proof uses induction and the fact that

$$\pi_{k+1}(x) \leqslant \frac{1}{k+1} \sum_{p^\alpha \leqslant x} \pi_k\left(\frac{x}{p^\alpha}\right).$$

The constants have been calculated by some authors, showing that $(c_1, c_2) = (2, 5)$ is admissible.

Tudesq [115] proved the following short interval version of Theorem 4.67.

Theorem 4.68 *Uniformly for $2 \leqslant y \leqslant x$ and any $k \in \mathbb{Z}_{\geqslant 1}$*

$$\sum_{\substack{x < n \leqslant x+y \\ \omega(n)=k}} 1 \leqslant \frac{23y}{\log y} \frac{(\log \log y + 6)^{k-1}}{(k-1)!} \exp\left(\frac{(k-1)(k-2)}{2 \log \log y + 12}\right).$$

Note that the right-hand side is a $o(y)$ whenever $1 \leqslant k \leqslant 0.524 \log \log y$.

4.7.8.2 Generalization

It could be useful to have at our disposal a result dealing with a certain class of multiplicative function. The next theorem, proved by Tenenbaum [105], is similar to Theorem 4.9.

Theorem 4.69 *Let $a, b > 0$ and f be a non-negative multiplicative function satisfying*

$$x^{-1} \sum_{p^\alpha \leqslant x} f\left(p^\alpha\right) \log p^\alpha \leqslant a \quad and \quad \sum_p \sum_{\alpha=2}^{\infty} \frac{f\left(p^\alpha\right)}{p^\alpha} \leqslant b.$$

Then, uniformly for $k \in \mathbb{Z}_{\geqslant 1}$ and $x \geqslant 2$

$$\sum_{\substack{n \leqslant x \\ \omega(n)=k}} f(n) \leqslant \frac{2ax}{(k-1)! \log x} \left(\sum_{p \leqslant x} \frac{f(p)}{p} + b \right)^{k-1}.$$

Proof We follow the lines of [105, Lemma 1]. Set $S_k(x)$ the sum of the left-hand side and define

$$T_k(x) := \sum_{\substack{n \leqslant x \\ \omega(n)=k}} f(n) \log n.$$

Proceeding as in (4.27), we first have

$$T_k(x) \leqslant \sum_{\substack{n \leqslant x \\ \omega(n)=k-1}} f(n) \sum_{p^\alpha \leqslant x/n} f\left(p^\alpha\right) \log p^\alpha$$

$$\leqslant ax \sum_{\substack{n \leqslant x \\ \omega(n)=k-1}} \frac{f(n)}{n}$$

$$\leqslant \frac{ax}{(k-1)!} \left(\sum_{p^\alpha \leqslant x} \frac{f\left(p^\alpha\right)}{p^\alpha} \right)^{k-1}$$

$$\leqslant \frac{ax}{(k-1)!} \left(E_f(x) + b\right)^{k-1},$$

where $E_f(x) := \sum_{p \leqslant x} \dfrac{f(p)}{p}$. Now by partial summation

$$
\begin{aligned}
S_k(x) &= \frac{T_k(x)}{\log x} + \int_2^x \frac{T_k(t)}{t(\log t)^2}\, dt \\[2mm]
&\leqslant \frac{a x \left(E_f(x) + b\right)^{k-1}}{(k-1)!\log x} + \frac{1}{(k-1)!}\int_2^x \frac{a \left(E_f(t) + b\right)^{k-1}}{(\log t)^2}\, dt \\[2mm]
&\leqslant \frac{a \left(E_f(x) + b\right)^{k-1}}{(k-1)!}\left(\frac{x}{\log x} + \int_2^x \frac{dt}{(\log t)^2}\right)
\end{aligned}
$$

and using (7.42) it is not difficult to show that the integral does not exceed $\leqslant \dfrac{x}{\log x}$.

$\square$

4.7.9   Prime-Independent Multiplicative Functions

4.7.9.1   Definition

Definition 4.16 A prime-independent multiplicative function is a multiplicative arithmetic function f satisfying $f(1) = 1$ and such that there exists a map $g : \mathbb{Z}_{\geqslant 0} \to \mathbb{R}$ such that $g(0) = 1$ and, for any prime powers p^α

$$
f\left(p^\alpha\right) = g(\alpha).
$$

In this section, we only consider *integer-valued* prime-independent multiplicative functions.

Example 4.15 The functions τ_k and a obviously are prime-independent, as well as the functions β and $\tau^{(e)}$, $\tau_{(k)}$ and $\tau^{(k)}$. A related function to the function a is the multiplicative function $n \mapsto S(n)$, counting the number of non-isomorphic finite semi-simple rings with n elements. Indeed, we have

$$
S\left(p^\alpha\right) = P^\star(\alpha),
$$

where $P^\star(\alpha)$ is the number of partitions of α into parts which are square.

4.7.9.2   Local Density

One of the long-standing problems in number theory concerning these prime-independent multiplicative functions is the study of the distribution of their values. To this end, we define the concept of *local density*.

Definition 4.17 Let f be an integer-valued prime-independent multiplicative function. Let $k \in \mathbb{Z}_{\geq 1}$ and

$$\Sigma_{f,k}(x) := \sum_{\substack{n \leqslant x \\ f(n)=k}} 1.$$

We say that f possesses a *local density* $d_{f,k}$ if

$$d_{f,k} := \lim_{x \to \infty} \left(x^{-1} \Sigma_{f,k}(x) \right)$$

exists.

When $f = a$, the existence of the local density $d_{a,k}$ was first established in [58] and later Ivić [49] showed that

$$\Sigma_{a,k}(x) = d_{a,k} x + O\left(x^{1/2} \log x \right).$$

Many authors improved on this estimate, such as [51, 66, 84] in which the best error term to date was established. The general case was introduced by Ivić in [51] and improved in [84] for a certain class of arithmetic functions.

4.7.9.3   Long Sums

The aim of this section is the proof of the following two results. The first one comes from [51].

Theorem 4.70 *Let* $k \in \mathbb{Z}_{\geq 1}$ *and* f *be an integer-valued prime-independent multiplicative function such that* $g(1) > 1$. *Then* $d_k = 0$.

Proof By assumption, $g(1)$ has a prime divisor p. Write $k = p^\alpha k'$ with $p \nmid k'$ and $\alpha \in \mathbb{Z}_{\geq 1}$. If $f(n) = k$, then $n = ab$ with $a \geqslant 1$ squarefree, $b \geqslant 1$ square-full, $(a,b) = 1$ and $a = p_1 \cdots p_h$ with $h \leqslant \alpha$. Hence

$$\Sigma_{f,k}(x) \ll \sum_{h=0}^{\alpha} \sum_{\substack{ab \leqslant x \\ \omega(a)=h}} \mu_2(a) s_2(b) \ll \sum_{h=0}^{\alpha} \sum_{b \leqslant x} s_2(b) \sum_{\substack{a \leqslant x/b \\ \omega(a)=h}} \mu_2(a)$$

and using Theorem 4.69 we derive

$$\Sigma_{f,k}(x) \ll x \sum_{h=0}^{\alpha} \sum_{b\leqslant x} \frac{s_2(b)}{b} \frac{(\log\log(ex/b))^{h-1}}{\log(ex/b)}$$

$$\ll \frac{x}{\log x} \sum_{h=0}^{\alpha} (\log\log x)^{h-1} \sum_{b\leqslant x} \frac{s_2(b)}{b}$$

$$\ll \frac{x}{\log x} \sum_{h=0}^{\alpha} (\log\log x)^{h-1} \ll \frac{x(\log\log x)^{\alpha-1}}{\log x}$$

for x sufficiently large, implying the asserted result since α is fixed.   □

Theorem 4.71 *Let $k \in \mathbb{Z}_{\geqslant 1}$ and f be an integer-valued prime-independent multiplicative function such that there exists $r \in \mathbb{Z}_{\geqslant 2}$ such that*

$$g(1) = \cdots = g(r-1) = 1 \quad \text{and} \quad \alpha \geqslant r \Rightarrow g(\alpha) > 1. \tag{4.81}$$

Then

$$\sum_{\substack{n\leqslant x \\ f(n)=k}} 1 = d_{f,k}x + O_r\left(x^{1/r}(\log x)^2\right).$$

Proof By multiplicativity, (4.81) implies that $f(a) = 1$ for all r-free integers a. Writing uniquely $n = ab$, with a r-free, b r-full and $(a,b) = 1$, yields $k = f(n) = f(a)f(b) = f(b)$, and hence

$$\Sigma_{f,k}(x) = \sum_{\substack{b\leqslant x \\ f(b)=k}} s_r(b) \sum_{\substack{a\leqslant x/b \\ (a,b)=1}} \mu_r(a). \tag{4.82}$$

Now define the *Legendre totient function* $\varphi(x,n)$ by[39]

$$\varphi(x,n) := \sum_{\substack{k\leqslant x \\ (k,n)=1}} 1 = \sum_{d|n} \mu(d) \left\lfloor \frac{x}{d} \right\rfloor.$$

---
[39] See also Exercise 74.

Note that

$$\varphi(x, n) = x \sum_{d|n} \frac{\mu(d)}{d} + O\left(\sum_{d|n} |\mu(d)|\right) = \frac{x\varphi(n)}{n} + O\left(2^{\omega(n)}\right).$$

Using (4.9), we derive

$$\sum_{\substack{a \leqslant x/b \\ (a,b)=1}} \mu_r(a) = \sum_{\substack{d \leqslant (x/b)^{1/r} \\ (d,b)=1}} \mu(d) \sum_{\substack{k \leqslant x/(bd^r) \\ (k,b)=1}} 1$$

$$= \sum_{\substack{d \leqslant (x/b)^{1/r} \\ (d,b)=1}} \mu(d) \varphi\left(\frac{x}{bd^r}, b\right)$$

$$= \sum_{\substack{d \leqslant (x/b)^{1/r} \\ (d,b)=1}} \mu(d) \left\{\frac{x\varphi(b)}{b^2 d^r} + O\left(2^{\omega(b)}\right)\right\}$$

$$= \frac{x\varphi(b)}{b^2} \sum_{\substack{d \leqslant (x/b)^{1/r} \\ (d,b)=1}} \frac{\mu(d)}{d^r} + O\left(x^{1/r} b^{-1/r} 2^{\omega(b)}\right)$$

$$= \frac{x\varphi(b)}{b^2} \sum_{\substack{d=1 \\ (d,b)=1}}^{\infty} \frac{\mu(d)}{d^r} + O\left(x^{1/r} b^{-1/r} 2^{\omega(b)} + x^{1/r} b^{-1-1/r} \varphi(b)\right)$$

$$= \frac{x\varphi(b)}{\zeta(r) b^2} \prod_{p|b} \left(1 - \frac{1}{p^r}\right)^{-1} + O\left(x^{1/r} b^{-1/r} 2^{\omega(b)}\right)$$

$$= \frac{x}{b\zeta(r)} \prod_{p|b} \left(1 + \frac{1}{p} + \cdots + \frac{1}{p^{r-1}}\right)^{-1} + O\left(x^{1/r} b^{-1/r} 2^{\omega(b)}\right)$$

$$= \frac{x}{\zeta(r) \Psi_r(b)} + O\left(x^{1/r} b^{-1/r} 2^{\omega(b)}\right).$$

inserting this estimate in (4.82), we get

$$\Sigma_{f,k}(x) = \frac{x}{\zeta(r)} \sum_{\substack{b \leqslant x \\ f(b)=k}} \frac{s_r(b)}{\Psi_r(b)} + O\left(x^{1/r} \sum_{b \leqslant x} \frac{s_r(b) 2^{\omega(b)}}{b^{1/r}}\right). \tag{4.83}$$

Note that every r-full integer n may be uniquely written as $n = a_1^r a_2^{r+1} \cdots a_r^{2r-1}$ with $a_2 \cdots a_r$ squarefree and $(a_i, a_j) = 1$ for $2 \leqslant i < j \leqslant r$. Since the function 2^ω is completely sub-multiplicative, we infer that

$$\sum_{b \leqslant x} s_r(b) 2^{\omega(b)} \ll \sum_{a_r \leqslant x^{\frac{1}{2r-1}}} 2^{\omega\left(a_r^{2r-1}\right)}$$

$$\times \sum_{a_{r-1} \leqslant \left(\frac{x}{a_r^{2r-1}}\right)^{\frac{1}{2r-2}}} 2^{\omega\left(a_{r-1}^{2r-2}\right)} \cdots \sum_{a_1 \leqslant \left(\frac{x}{a_2^{r+1} \cdots a_r^{2r-1}}\right)^{1/r}} 2^{\omega(a_1^r)}$$

and the easy bound from Theorem 4.9

$$\sum_{n \leqslant z} 2^{\omega(n^r)} \ll z \log z$$

applied to the last inner sum yields

$$\sum_{b \leqslant x} s_r(b) 2^{\omega(b)} \ll x^{1/r} \log x$$

and hence

$$x^{1/r} \sum_{b \leqslant x} \frac{s_r(b) 2^{\omega(b)}}{b^{1/r}} \ll x^{1/r} (\log x)^2$$

by partial summation. Also note that

$$\sum_{\substack{b \leqslant x \\ f(b)=k}} \frac{b s_r(b)}{\Psi_r(b)} \leqslant \sum_{b \leqslant x} s_r(b) \ll x^{1/r}$$

so that the Dirichlet series of the multiplicative function $b \mapsto b s_r(b) \mathbf{1}_{f(b)=k} \Psi_r(b)^{-1}$ is absolutely convergent in the half-plane $\sigma > \frac{1}{r}$ by Theorem 4.5. Hence the series

$$\sum_{\substack{b \geqslant 1 \\ f(b)=k}} \frac{s_r(b)}{\Psi_r(b)}$$

converges absolutely. Inserting these estimates in (4.83) implies that the limit of

$$\frac{1}{x} \sum_{\substack{n \leqslant x \\ f(n)=k}} 1$$

exists as $x \to \infty$ and is equal to

$$d_{f,k} = \frac{1}{\zeta(r)} \sum_{\substack{b=1 \\ f(b)=k}}^{\infty} \frac{s_r(b)}{\Psi_r(b)}$$

as required.   □

4.7.9.4   Short Sums

The next step was the study of the distribution of values of f in short intervals. As usual, by "short intervals" we mean the study of sums of the shape

$$\Sigma_{f,k}(x+y) - \Sigma_{f,k}(x) = \sum_{\substack{x < n \leqslant x+y \\ f(n)=k}} 1,$$

where $y = o(x)$ as $x \to \infty$. In the case of $f = a$, Ivić [50] first showed that

$$\sum_{\substack{x < n \leqslant x+y \\ a(n)=k}} 1 = d_{a,k} y + o(y)$$

holds for $y \geqslant x^{581/1744} \log x \approx x^{0.33314} \log x$. This value was successfully improved and generalized by many authors. For instance, Zhai [126, Theorem 2.5] states without detailed proof that

$$\sum_{\substack{x < n \leqslant x+y \\ f(n)=k}} 1 = d_{f,k} y + o(y)$$

holds for $y \geqslant x^{\frac{1}{2r+1}+\varepsilon}$ where r is given in (4.81). An effective version of Zhai's result has been given in [11] who showed the following estimate.

Theorem 4.72 *Let* $k \in \mathbb{Z}_{\geqslant 1}$ *fixed and* f *be an integer-valued prime-independent multiplicative function such that* $f(p) = 1$ *for any prime* p *and let* $r \in \mathbb{Z}_{\geqslant 2}$ *as in (4.81). Let* $x^{\frac{1}{2r+1}+\varepsilon} \leqslant y \leqslant 4^{-2r^2} x$ *be real numbers. Then*

$$\sum_{\substack{x < n \leqslant x+y \\ f(n)=k}} 1 = d_{f,k} y$$

$$+ O_{r,\varepsilon} \left\{ \left(x^{r-1} y^{r+1} \right)^{\frac{1}{2r^2}} x^\varepsilon + yx^{-\frac{1}{6(4r-1)(2r-1)}+\varepsilon} + y^{1-\frac{2(r-1)}{r(3r-1)}} x^\varepsilon \right\}.$$

4.7.10   Smooth Numbers

Owing to their crucial importance in many problems, the smooth numbers are increasingly studied in number theory. *Recall that* $P(n)$ *is the greatest prime divisor of* n, *with the convention* $P(1) = 1$.

4.7.10.1   Definition

Definition 4.18 Let $y \geqslant 1$. An integer $n \geqslant 2$ is said to be *y-smooth* if $P(n) \leqslant y$. If $1 \leqslant y \leqslant x$, the set of y-smooth numbers $n \leqslant x$ is denoted by $S(x, y)$ and we customary define

$$\Psi(x, y) := \sum_{n \in S(x,y)} 1 = \sum_{\substack{n \leqslant x \\ P(n) \leqslant y}} 1.$$

When $y > 1$, it is customary to set $u := \frac{\log x}{\log y}$.

4.7.10.2   Bounds

A rather simple, but efficient device due to Rankin yields an almost sharp result. We give here an explicit version due to Granville and Soundararajan.

Theorem 4.73 *If* $3 \leqslant y \leqslant x$ *and* $\frac{1}{\log y} \leqslant a \leqslant 1$*, then*

$$\Psi(x, y) < \frac{1.5 \, y^{1-a} x^a}{\log x} \, \zeta(a, y),$$

where

$$\zeta(s, y) := \prod_{p \leqslant y} \left(1 - \frac{1}{p^s}\right)^{-1} \quad (\sigma > 0).$$

Proof Using the convolution $\log = \mathbf{1} \star \Lambda$, we get

$$\sum_{\substack{n \leqslant x \\ P(n) \leqslant y}} \log n = \sum_{\substack{d \leqslant x \\ P(d) \leqslant y}} \sum_{\substack{k \leqslant x/y \\ P(d) \leqslant y}} \Lambda(k)$$

$$= \sum_{\substack{d \leqslant x \\ P(d) \leqslant y}} \sum_{p \leqslant \min(y, x/d)} \log p \left\lfloor \frac{\log(x/d)}{\log p} \right\rfloor$$

$$\leqslant \sum_{\substack{d \leqslant x \\ P(d) \leqslant y}} \pi(\min(y, x/d)) \log(x/d)$$

so that

$$\Psi(x, y) \log x = \sum_{\substack{d \leqslant x \\ P(d) \leqslant y}} (\log d + \log(x/d))$$

$$\leqslant \sum_{\substack{d \leqslant x \\ P(d) \leqslant y}} \{1 + \pi(\min(y, x/d))\} \log(x/d).$$

Now using Corollary 3.12, we derive

$$1 + \pi(t) < \frac{1.5 \, t}{\log t} \quad (t > 1)$$

and hence

$$\Psi(x, y) \log x < 1.5 \sum_{\substack{d \leqslant x/y \\ P(d) \leqslant y}} \frac{y \log(x/d)}{\log y} + 1.5 \sum_{\substack{x/y < d \leqslant x \\ P(d) \leqslant y}} \frac{x}{d}.$$

Now notice that, if $\frac{1}{\log y} \leqslant a \leqslant 1$, then

$$
y^{1-a} \left(\frac{x}{d}\right)^a \geqslant
\begin{cases}
\frac{y \log(x/d)}{\log y}, & \text{if } d \leqslant x/y \\
x/d, & \text{if } x/y < d \leqslant x
\end{cases}
$$

and therefore

$$
\Psi(x, y) \log x < 1.5 \sum_{\substack{d \leqslant x \\ P(d) \leqslant y}} y^{1-a} \left(\frac{x}{d}\right)^a < 1.5\, y^{1-a} x^a\, \zeta(a, y)
$$

as required.   □

Corollary 4.22 *For all* $7.75 \leqslant y \leqslant x$

$$
\Psi(x, y) < 62\, x^{1 - \frac{1}{2\log y}}.
$$

Proof First note that $\frac{1}{\log y} < \frac{1}{2}$. Let $a \in \left[\frac{3}{4}, 1\right[$ and we have

$$
\zeta(a, y) = \exp\left(-\sum_{p \leqslant y} \log\left(1 - \frac{1}{p^a}\right)\right) \leqslant \exp\left(\sum_{p \leqslant y} \frac{1}{p^a} + \sum_{p \leqslant y} \frac{1}{p^{2a}}\right).
$$

Now

$$
\sum_{p \leqslant y} \frac{1}{p^{2a}} \leqslant \log\left(\zeta(2a)\right) < \log 3
$$

and

$$
\sum_{p \leqslant y} \frac{1}{p^a} = \sum_{p \leqslant y} \frac{1}{p} + \sum_{p \leqslant y} \frac{p - p^a}{p^{a+1}} \leqslant \sum_{p \leqslant y} \frac{1}{p} + (1 - a) \sum_{p \leqslant y} \frac{\log p}{p^a}.
$$

By partial summation, and writing $\theta(t) = t + R(t)$, we derive

$$\sum_{p \leqslant y} \frac{\log p}{p^a} = y^{-a}\theta(y) + a \int_2^y \frac{\theta(t)}{t^{a+1}} \, dt$$

$$= y^{1-a} + y^{-a} R(y) + a \int_2^y \left(\frac{1}{t^a} + \frac{R(t)}{t^{a+1}} \right) dt$$

$$= \frac{y^{1-a}}{1-a} - \frac{a 2^{1-a}}{1-a} + \frac{R(y)}{y^a} + a \int_2^y \frac{R(t)}{t^{a+1}} \, dt.$$

From [24], we have $|R(t)| \leqslant \dfrac{1.2323 t}{\log t}$ as soon as $t \geqslant 2$, so that

$$\int_2^y \frac{|R(t)|}{t^{a+1}} \, dt \leqslant 1.2323 \int_2^y \frac{dt}{t^a \log t}$$

$$= 1.2323 \left(\int_2^{\sqrt{y}} + \int_{\sqrt{y}}^y \right) \frac{dt}{t^a \log t}$$

$$\leqslant 1.2323 \left(\frac{1}{\log 2} \int_2^{\sqrt{y}} \frac{dt}{t^a} + \frac{2}{\log y} \int_{\sqrt{y}}^y \frac{dt}{t^a} \right)$$

$$= \frac{1.2323}{1-a} \left(\frac{y^{(1-a)/2}}{\log 2} - \frac{2^{1-a}}{\log 2} + \frac{2y^{1-a}}{\log y} - \frac{2y^{(1-a)/2}}{\log y} \right).$$

Noticing that $\left| \frac{y^{1-a}}{1-a} - \frac{a2^{1-a}}{1-a} \right| = \frac{y^{1-a}}{1-a} - \frac{a2^{1-a}}{1-a}$, we get

$$\sum_{p \leqslant y} \frac{1}{p^a} \leqslant \sum_{p \leqslant y} \frac{1}{p} + y^{1-a} - a2^{1-a}$$

$$+ 1.2323 \left(\frac{(3-a)y^{1-a}}{\log y} + \frac{y^{(1-a)/2}}{\log 2} - \frac{2^{1-a}}{\log 2} - \frac{2y^{(1-a)/2}}{\log y} \right)$$

and the assumption $y \geqslant 7.75$ implies that the first sum does not exceed $\log \log y + \frac{1}{2}$ by Corollary 3.13. Now choose $a = 1 - \frac{1}{2\log y}$, which yields

$$\zeta\left(1 - \tfrac{1}{2\log y}, y\right) < 25 \log y$$

and the asserted estimate follows from Theorem 4.73.   □

As a lower bound, we have the next estimate [62].

Theorem 4.74 *For all $x \geqslant 4$ and $2 \leqslant y \leqslant x$*

$$\Psi(x, y) > x^{1 - \frac{\log \log x}{\log y}}.$$

Exercises

51 Prove that

(a) For all $n \in \mathbb{Z}_{\geqslant 1}$, $\tau(n) \leqslant 2\sqrt{n}$.
(b) For all *composite* integers $n \in \mathbb{Z}_{\geqslant 1}$, we have $\varphi(n) \leqslant n - \sqrt{n}$.

52 Let $n \in \mathbb{Z}_{\geqslant 1}$ and, for all $t \in [1, n]$, define

$$\tau^+(n; t) := \sum_{\substack{d \mid n \\ d \geqslant t}} 1.$$

Prove that, for all $n \in \mathbb{Z}_{\geqslant 1}$

$$\tau_3(n) = 2 \sum_{\substack{d \mid n \\ d \leqslant \sqrt{n}}} \tau^+(n/d; d) - \sum_{d^2 \mid n} 1.$$

53 Prove that, for any $x \in \mathbb{R}_{\geqslant 1}$

$$\sum_{n \leqslant x} 2^n \tau(n)^{-n} < \pi(x) + \tfrac{61}{27}.$$

54 Let f be a multiplicative function such that $\lim\limits_{p^\alpha \to \infty} f\left(p^\alpha\right) = 0$. Show that $\lim\limits_{n \to \infty} f(n) = 0$, and deduce that, for all $\varepsilon > 0$ and $n \in \mathbb{Z}_{\geqslant 1}$, we have $\tau(n) \ll n^\varepsilon$.

55 Let $k \in \mathbb{Z}_{\geqslant 2}$. Show that, for all $n \in \mathbb{Z}_{\geqslant 3}$

$$\tau_k(n) \leqslant n^{\frac{(k-1)1.5379 \log 2}{\log \log n}}.$$

56 Prove the following identities.

(a) $\displaystyle\sum_{d|n} \tau^3(d) = \left(\sum_{d|n} \tau(d)\right)^2$ (b) $\displaystyle\sum_{d|n} \beta(d)\mu\left(\frac{n}{d}\right) = s_2(n)$

(c) $\displaystyle\sum_{d|n} \sigma_k(d)\mu\left(\frac{n}{d}\right) = n^k$ (d) $\displaystyle\sum_{d|n} \mu_2(d)k^{\omega(d)} = (k+1)^{\omega(n)}$

57 Let k be a positive integer.

(a) Prove that for all $n \in \mathbb{Z}_{\geqslant 1}$, we have

$$\Lambda_k(n) = \Lambda_{k-1}(n)\log n + (\Lambda_{k-1} \star \Lambda)(n).$$

(b) Prove that for all $m, n \in \mathbb{Z}_{\geqslant 1}$ such that $(m, n) = 1$, we have

$$\Lambda_k(mn) = \sum_{j=0}^{k} \binom{k}{j} \Lambda_j(m)\Lambda_{k-j}(n).$$

58 With the help of the identity [35, (3.93)]

$$\sum_{j=0}^{n} \frac{\binom{2j}{j}\binom{2n-2j}{n-j}}{(2j-1)(2n-2j-1)} = \begin{cases} 1, & \text{if } n = 0; \\ -4, & \text{if } n = 1; \\ 0, & \text{if } n \geqslant 2; \end{cases}$$

determine the multiplicative function f such that $\mu = f \star f$.

59 Let f be an additive function and g be a multiplicative function. Prove that, for all $m, n \in \mathbb{Z}_{\geqslant 1}$ such that $(m, n) = 1$, we have

$$(f \star g)(mn) = (f \star g)(m)(g \star \mathbf{1})(n) + (f \star g)(n)(g \star \mathbf{1})(m).$$

Deduce that, if $n \neq p^\alpha$, then $(f \star \mu)(n) = 0$.

60 Prove that for all integers m, n not necessarily coprime, we have

$$\Delta_k(mn) \leqslant \tau_k(m)\Delta_k(n).$$

61 Let $n \in \mathbb{Z}_{\geqslant 3}$. Prove

$$\sum_{p|n} \frac{1}{p} < \max\left(1, \log\log\log n + 1\right).$$

62 Let $n \in \mathbb{Z}_{\geqslant 3}$. Prove

$$\sum_{p|n} \frac{\log p}{p-1} < \log \log n + 1.$$

63 Let $n \in \mathbb{Z}_{\geqslant 16}$. Prove

$$\varphi(n) > \frac{n}{e\zeta(2) \log \log n} \quad \text{and} \quad \sigma(n) < e\zeta(2) n \log \log n.$$

64 Let n be a positive integer.

(a) Prove that

$$\frac{n^{n+1}}{\zeta(n+1)} \leqslant \varphi(n) \sigma\left(n^n\right) \leqslant n^{n+1}.$$

(b) Deduce the nature of the series $\sum_{n \geqslant 1} f(n)$ where

$$f(n) = \frac{n}{\varphi(n)} - \frac{\sigma\left(n^n\right)}{n^n}.$$

65 Let $n \in \mathbb{Z}_{\geqslant 1}$. Prove

$$\sum_{d|n} \frac{\mu(d) \log d}{d} = -\frac{\varphi(n)}{n} \sum_{p|n} \frac{\log p}{p-1}.$$

66 Let $n \in \mathbb{Z}_{\geqslant 1}$, $e \in \{1, 2\}$, f be a multiplicative function satisfying $f(p) \neq (-1)^{e+1}$ and g be an additive function. Prove

$$\sum_{d|n} \mu(d)^e f(d) g(d) = \prod_{p|n} \left(1 + (-1)^e f(p)\right) \sum_{p|n} \frac{f(p) g(p)}{f(p) + (-1)^e}.$$

67 Let f be a positive multiplicative function, $n \in \mathbb{Z}_{\geqslant 2}$ and $t \in [1, n)$. Prove

$$\sum_{\substack{d|n \\ d \leqslant t}} f(d) \leqslant \frac{(f \star \mathbf{1})(n)}{\log(n/t)} \times \Gamma_n,$$

where

$$\Gamma_n := \log n + \sum_{p^\alpha \| n} \left[\frac{d}{ds} \log \left(\sum_{j=0}^{\alpha} \frac{f(p^j)}{p^{js}} \right) \right]_{[s=0]}.$$

68 Let $k \in \mathbb{Z}_{\geqslant 1}$. Describe the coefficients $f_k(n)$ and $g_k(n)$ such that

$$L(s, f_k) = \zeta(ks) \quad \text{and} \quad L(s, g_k) = \zeta(ks)^{-1}.$$

69 Let f be a multiplicative function satisfying the Wirsing conditions (4.24) and

$$f(p^\alpha) \geqslant f(p^{\alpha-1}) \tag{4.84}$$

for all primes p and $\alpha \in \mathbb{N}$. Prove that

$$\sum_{n \leqslant x} f(n) \leqslant x \prod_{p \leqslant x} \left(1 - \frac{1}{p}\right) \left(1 + \sum_{\alpha=1}^{\infty} \frac{f(p^\alpha)}{p^\alpha}\right).$$

What kind of upper bounds can we get when $f = \tau_k$? $f = k^\omega$?

70 Prove that for all $k \geqslant 2$

$$\sum_{n \leqslant x} \tau_k(n) = \frac{x(\log x)^{k-1}}{(k-1)!} + O\left(x(\log x)^{k-2}\right).$$

71 Let χ be a non-principal Dirichlet character modulo q. Use Theorem 4.4 to show that, for each real number $x \geqslant q^{1/2}$ sufficiently large, we have

$$\sum_{n \leqslant x} (\chi \star \mathrm{id})(n) = \frac{x^2 L(2, \chi)}{2} - x \sum_{n \leqslant x^{2/3}q^{1/6}} \frac{\chi(n)}{n} \psi\left(\frac{x}{n}\right) + O\left(x^{2/3}q^{1/6} \log q\right).$$

72 (E. Landau, 1900).

(a) Prove that $\displaystyle\sum_{n \leqslant x} \frac{1}{\varphi(n)} \ll \log x$.

(b) Let $g := \dfrac{\mathrm{id}}{\varphi} \star \mu$. Show that $L(s, g)$ is absolutely convergent in the half-plane $\sigma > 0$.

(c) Deduce that

$$\sum_{n\leqslant x}\frac{1}{\varphi(n)} = \frac{\zeta(2)\zeta(3)}{\zeta(6)}\left(\log x + \gamma - \sum_{p}\frac{\log p}{p(p-1)+1}\right) + O\left(\frac{(\log x)^2}{x}\right).$$

(d) Let $q \in \mathbb{Z}_{\geqslant 1}$. Use Example 4.32 to deduce that

$$\sum_{\substack{n\leqslant x \\ (n,q)=1}}\frac{1}{\varphi(n)} = \frac{\varphi(q)}{q}\,(C_1\log x + C_2) + O\left(\frac{\log x}{x}\left(\log x + 2^{\omega(q)}\right)\right),$$

where

$$C_1 = \frac{\zeta(2)\zeta(3)}{\zeta(6)}\prod_{p\mid q}\left(1 - \frac{1}{p(p-1)+1}\right);$$

$$C_2 = C_1\left(\gamma - \sum_{p}\frac{\log p}{p(p-1)+1} + \sum_{p\mid q}\frac{p^2\log p}{(p-1)\left(p^2-p+1\right)}\right).$$

73 (Cesáro, 1885). Let f be an arithmetic function. Show that

$$\sum_{k=1}^{n} f\,((k,n)) = (f \star \varphi)(n).$$

74 Define the *Legendre totient function* $\varphi(x,n)$ by

$$\varphi(x,n) := \sum_{\substack{k\leqslant x \\ (k,n)=1}} 1 = \sum_{d\mid n}\mu(d)\left\lfloor\frac{x}{d}\right\rfloor.$$

Prove that

$$\sum_{k=1}^{n}(k,\lfloor n/k\rfloor) = \sum_{d\leqslant n}d\sum_{k\leqslant n/d}\left\{\varphi\left(\frac{n}{kd^2},k\right) - \varphi\left(\frac{n}{d(kd+1)},k\right)\right\}.$$

75 (Chern). Let $1 \leqslant T \leqslant x$ and f be a positive-valued arithmetic function.
Prove

$$\sum_{T<n\leqslant x} f\,(\lfloor x/n\rfloor) = \sum_{d\leqslant x/T} f(d)\left(\left\lfloor\frac{x}{d}\right\rfloor - \left\lfloor\frac{x}{d+1}\right\rfloor\right) + O^{\star}\left\{f\,(\lfloor x/T\rfloor)\left(1 + T^2 x^{-1}\right)\right\}.$$

76 The *Pillai function* $\mathcal{P}(n)$ is defined by

$$\mathcal{P}(n) = \sum_{k=1}^{n}(k, n).$$

Let θ be the exponent in the Dirichlet divisor problem,[40] i.e. the smallest positive real number θ such that the asymptotic estimate

$$\sum_{n \leqslant x} \tau(n) = x(\log x + 2\gamma - 1) + O\left(x^{\theta + \varepsilon}\right) \tag{4.85}$$

holds for all $\varepsilon > 0$ and $x \geqslant 1$. The aim of this exercise is to prove the following result.

$$\sum_{n \leqslant x} \mathcal{P}(n) = \frac{x^2}{2\zeta(2)}\left(\log x + 2\gamma - \frac{1}{2} - \frac{\zeta'(2)}{\zeta(2)}\right) + O\left(x^{1+\theta+\varepsilon}\right). \tag{4.86}$$

(a) Prove that, for all $\varepsilon > 0$ and $z \geqslant 1$, we have

$$\sum_{n \leqslant z} n\tau(n) = \frac{z^2 \log z}{2} + z^2\left(\gamma - \frac{1}{4}\right) + O\left(z^{\theta+1+\varepsilon}\right).$$

(b) Using Exercise 73 and (4.7) check that $\mathcal{P} = \mu \star (\tau \times \mathrm{id})$.
(c) Prove (4.86).

77 (A. Selberg).

(a) Let $k \in \mathbb{Z}_{\geqslant 1}$. Prove that, for all $x \geqslant 1$

$$\sum_{\substack{n \leqslant x \\ (n,k)=1}} \tau(n) = \sum_{d|k}\sum_{e|k} \mu(d)\mu(e) \sum_{n \leqslant x/(ed)} \tau(n).$$

(b) Using (4.85), deduce that

$$\sum_{\substack{n \leqslant x \\ (n,k)=1}} \tau(n) = \left(\frac{\varphi(k)}{k}\right)^2 x\left(\log x + 2\gamma - 1 + 2\sum_{p|k}\frac{\log p}{p-1}\right)$$

$$+ O\left(x^{\theta+\varepsilon}\prod_{p|k}\left(1 + \frac{1}{p^\theta}\right)^2\right),$$

---

[40] The best inequalities to date for θ are $\frac{1}{4} \leqslant \theta \leqslant \frac{517}{1648}$. See Chap. 6.

where θ is the exponent in the Dirichlet divisor problem given in (4.85).

78 Let k be a fixed positive integer. For all $x \geqslant 1$, define

$$S_k(x) = \sum_{n \leqslant x} \tau_k(n).$$

(a) Prove that

$$x \int_1^x \frac{S_k(t)}{t^2}\, dt \; < \; S_{k+1}(x) \; \leqslant \; x \int_1^x \frac{S_k(t)}{t^2}\, dt + S_k(x).$$

(b) Using induction, deduce that

$$x \sum_{j=0}^{k-1} (-1)^{k+j+1} \frac{(\log x)^j}{j!} + (-1)^k \; < \; \sum_{n \leqslant x} \tau_k(n) \; \leqslant \; x \sum_{j=0}^{k-1} \binom{k-1}{j} \frac{(\log x)^j}{j!}.$$

(c) Deduce that

$$\sum_{n \leqslant x} \tau_k(n) \; \leqslant \; \frac{x(\log x + k - 1)^{k-1}}{(k-1)!}.$$

79 For all $k, n \in \mathbb{Z}_{\geqslant 1}$, let $\tau_k^\star(n)$ be the kth *strict divisor function* of n, i.e. the number of choices of $n_1, \ldots, n_k$ satisfying $n = n_1 \cdots n_k$ with $n_j \geqslant 2$. Prove that, for any $x \geqslant 1$, we have

$$\sum_{n \leqslant x} \tau_k^\star(n) \; \leqslant \; \frac{x(\log x)^{k-1}}{(k-1)!}.$$

80 Let $k \in \mathbb{Z}_{\geqslant 1}$. Prove that, for all $x \geqslant 1$

$$\sum_{n \leqslant x} \frac{\mu(n)^2 k^{\omega(n)}}{n} \; \leqslant \; \frac{1}{\zeta(2)^k} \left(\frac{(\log x)^k}{k!} + 2^k \sum_{j=0}^{k-1} \frac{1}{j!} \binom{k}{j} \left(\frac{\log x}{2} \right)^j \right).$$

81 Use Exercise 80 to show that, for any $k \in \mathbb{Z}_{\geqslant 1}$ and any $x > 1$

$$\sum_{n \leqslant x} k^{\omega(n)} \; \leqslant \; \frac{x(\log x)^{k-1}}{(k-1)!\, \zeta(2)^{k-1}} \, (1 + R_k(x)).$$

with $R_k(x) := \dfrac{2(k-1)^2}{\log x} \left(1 + \dfrac{2k-4}{\log x}\right)^{k-2}$.

82 Let $k \in \mathbb{Z}_{\geqslant 1}$. Show that there exists an absolute constant $c_0 > 0$ and a constant $c_k \geqslant 1$, depending on k, such that, for any $x \geqslant c_k$ sufficiently large

$$\sum_{n \leqslant x} (-k)^{\omega(n)} \ll_k x e^{-c_0 (\log x)^{3/5} (\log \log x)^{-1/5}}.$$

83

(a) Let (f, F) be a couple of arithmetic functions satisfying $F = f \star \mathbf{1}$, and let G be the summatory function of F. Prove that, for any $n \in \mathbb{Z}_{\geqslant 1}$ and any $a \in \mathbb{Z}_{\geqslant 2}$

$$\sum_{j=1}^{a-1} j \sum_{\substack{k \\ j/a \leqslant \{n/k\} < (j+1)/a}} f(k) = G(an) - a G(n).$$

(b) Using (4.85), deduce that, for any $n \in \mathbb{Z}_{\geqslant 1}$, any $a \in \mathbb{Z}_{\geqslant 2}$ and $\varepsilon \in (0, 1)$ fixed

$$\sum_{j=1}^{a-1} j \sum_{\substack{k \\ j/a \leqslant \{n/k\} < (j+1)/a}} 1 = an \log a + O\left((an)^{517/1648+\varepsilon}\right).$$

84 Prove that, for all $x \geqslant 1$

$$\left| \sum_{n > x} \frac{\mu(n)}{n^2} \right| < \frac{2}{25x} + \frac{1}{x^{3/2}}.$$

85 Prove that, for all $x \geqslant 1$, we have

$$\sum_{n \leqslant x} s_2(n) < 3\sqrt{x}$$

and deduce by partial summation that

$$\sum_{n > x} \frac{s_2(n)}{n} < \frac{6}{\sqrt{x}}.$$

86   Let f be a multiplicative function such that $|f(p) - 1| \leqslant p^{-1}$ for all primes p and $|(f \star \mu)(p^\alpha)| \leqslant 1$ for all prime powers p^α with $\alpha \geqslant 2$. Prove that

$$\sum_{n \leqslant x} f(n) = x \prod_p \left(1 - \frac{1}{p}\right) \left(1 + \sum_{\alpha=1}^\infty \frac{f(p^\alpha)}{p^\alpha}\right) + O\left(x^{1/2}\right).$$

87   Prove that

$$\sum_{n \leqslant x} \beta(n) = x \frac{\zeta(2)\zeta(3)}{\zeta(6)} + O\left(x^{1/2}\right).$$

88   Prove that

$$\sum_{n \leqslant x} \frac{\varphi(n)\gamma_2(n)}{n^2} = x \prod_p \left(1 - \frac{1}{p}\right)\left(1 + \frac{1}{p+1}\right) + O\left(x^{1/2}\right).$$

89   Let $k \in \mathbb{Z}_{\geqslant 2}$ be fixed and define the multiplicative function D_k by $D_k := \tau_k k^{-\omega}$. The aim of this exercise is to show the asymptotic formula

$$\sum_{n \leqslant x} D_k(n) = x \prod_p \left(1 - \frac{1}{p}\right)\left(1 - \frac{1}{k} + \frac{1}{k}\left(1 - \frac{1}{p}\right)^{-k}\right) + O_k\left(x^{1/2}(\log x)^{k-2}\right).$$

$$\text{(4.87)}$$

(a) Check that, for all $m, n \in \mathbb{Z}_{\geqslant 1}$, $D_k(mn) \leqslant D_k(m)D_k(n)k^{\omega(\min(m,n))}$.
(b) Check that $D_k \star \mu = \frac{k-1}{k}(s_2 \times D_{k-1}) := g_k$.
(c) Prove that

$$\sum_{n \leqslant x} D_k(n) = x \sum_{n=1}^\infty \frac{g_k(n)}{n} + O\left(\sum_{n \leqslant x} g_k(n) + x \int_x^\infty \left(\sum_{n \leqslant t} g_k(n)\right)\frac{dt}{t^2}\right).$$

(d) Deduce (4.87).

90   Let $f : n \mapsto \dfrac{\sigma(n)}{\tau(n)}$.

(a) Show that

$$\sum_{n \leqslant x} f(n) \ll \frac{x^2}{\sqrt{\log x}}.$$

(b) Prove that, if n is an odd squarefree number, then $f(n) \in \mathbb{Z}$.

(c) Prove that, if $\alpha \geqslant 1$ and p is prime verifying $P^-(\alpha+1) \nmid p-1$, then $f(p^\alpha) \notin \mathbb{Z}$.

91 (S. Selberg). For all $x \geqslant 1$, define

$$S(x) = \sum_{n \leqslant x} \frac{\mu(n)}{n\tau(n)}.$$

(a) Show that

$$S(x) = \frac{1}{x} \left(\sum_{n \leqslant x} 2^{-\omega(n)} + \sum_{n \leqslant x} \frac{\mu(n)}{\tau(n)} \left\{ \frac{x}{n} \right\} \right).$$

(b) Using Theorem 4.9, deduce that there exist $c_0 > 0$ and $x_0 \geqslant e$ such that, for all $x \geqslant x_0$, we have

$$0 < S(x) < c_0 (\log x)^{-1/2}.$$

92 (Hall and Tenenbaum). For all $x \geqslant 2$ and $k \in \mathbb{Z}_{\geqslant 1}$, define

$$N_k(x) = \sum_{\substack{n \leqslant x \\ \Omega(n)=k}} 1.$$

The aim of this exercise is to prove the following bound.

$$N_k(x) \ll k2^{-k} x \log x. \tag{4.88}$$

(a) Let $t \in [1, 2)$. Using $f = t^\Omega \star \mu$ and Proposition 4.3, prove that

$$\sum_{n \leqslant x} t^{\Omega(n)} \ll \frac{x \log x}{2 - t}.$$

(b) Prove (4.88).

93 (Alladi, Erdős and Vaaler). Let $n, a \geqslant 2$ be integers with n squarefree and $\lambda \geqslant 0$ be a real number. Let f be a multiplicative function such that, for all primes p, we have

$$0 \leqslant f(p) \leqslant \lambda < \frac{1}{a - 1}. \tag{4.89}$$

The aim of this exercise is to show the following inequality.

$$\sum_{d|n} f(d) \leqslant \left(\frac{1+\lambda}{1+\lambda-a\lambda}\right) \sum_{\substack{d|n \\ d \leqslant n^{1/a}}} f(d). \tag{4.90}$$

1. (a) Let p be a prime factor of n. Show that

$$\sum_{d|n} f(d) = \sum_{d|(n/p)} f(d) + \sum_{k|(n/p)} f(kp)$$

and then

$$\sum_{k|(n/p)} f(k) = \left(\frac{1}{1+f(p)}\right) \sum_{d|n} f(d). \tag{4.91}$$

(b) Deduce that

$$\sum_{d|n} f(d) \log d = \left(\sum_{d|n} f(d)\right) \left(\sum_{p|n} \frac{f(p) \log p}{1+f(p)}\right)$$

and then

$$\sum_{d|n} f(d) \log d \leqslant \frac{\lambda \log n}{1+\lambda} \left(\sum_{d|n} f(d)\right). \tag{4.92}$$

2. Prove that

$$\sum_{d|n} f(d) \frac{\log(n^{1/a}/d)}{\log(n^{1/a})} \leqslant \sum_{\substack{d|n \\ d \leqslant n^{1/a}}} f(d)$$

and conclude with (4.92).

94   (Redheffer). Let $R_n = (r_{ij}) \in M_n(\{0, 1\})$ be the matrix defined by

$$r_{ij} = \begin{cases} 1, & \text{if } i \mid j \text{ or } j = 1 \\ 0, & \text{otherwise.} \end{cases}$$

Show that

$$\det R_n = M(n)$$

where $M(n) = \sum_{k=1}^{n} \mu(k)$ is the Mertens function.

95   Let $t > 0$.

(a) Prove that, for all $n \in \mathbb{Z}_{\geqslant 1}$, we have

$$\sum_{d \mid n} t^{\omega(d)} \leqslant (1 + t)^{\Omega(n)}.$$

(b) Let $k \in \mathbb{Z}_{\geqslant 1}$. Show that

$$\sum_{\substack{d \mid n \\ \omega(d) \leqslant k}} t^{\omega(d)} \leqslant \sum_{j=0}^{k} \binom{\Omega(n)}{j} t^{j}.$$

96   Let $n \in \mathbb{Z}_{\geqslant 16}$. Prove that

$$\left| \sum_{k=1}^{n} \frac{1}{k} \mu\left(\frac{k}{(n, k)} \right) \right| < e\zeta(2) \log \log n.$$

97   Let f be the arithmetic function defined by $f(1) = 1$ and, if $n = p_1^{a_1} \cdots p_r^{a_r}$ where the p_is are distinct primes and $a_i \in \mathbb{Z}_{\geqslant 1}$, then $f(n) = a_1^{p_1} \cdots a_r^{p_r}$.

(a) Prove that $f(f(n)) \leqslant n$.
(b) Show that, if n is square-full, then there exists m square-full such that $f(m) = n$.

98   Show that

$$\zeta(3) = \frac{2}{5} \sum_{m,n=1}^{\infty} \frac{1}{mn[m, n]}.$$

99

(a) Let $f : (\mathbb{Z}_{\geqslant 1})^2 \to \mathbb{R}$ be an arithmetic function of two variables. Prove

$$\sum_{mn \leqslant x} f(m, n) = \sum_{n \leqslant x} \sum_{d \mid n} f(d, n/d).$$

(b) Using Corollary 4.19, deduce that, for all $\varepsilon > 0$

$$\sum_{mn \leqslant x} \tau(m)\tau(n) = x P_3(\log x) + O_\varepsilon\left(x^{1/2+\varepsilon} \right),$$

where P_3 is a polynomial of degree 3 and leading coefficient $\frac{1}{6}$.

100   Let $r \in \mathbb{Z}_{\geqslant 2}$.

(a) Prove that, for any $x \geqslant 1$

$$\sum_{n_1 \leqslant x, \ldots, n_r \leqslant x} \mu(n_1 \cdots n_r) \left\lfloor \frac{x}{n_1 \cdots n_r} \right\rfloor = \sum_{n \leqslant x} (1 - r)^{\omega(n)}.$$

(b) Prove that, for any $x \geqslant 1$

$$\sum_{1 \leqslant n_1 < \cdots < n_r \leqslant x} \mu(n_1 \cdots n_r) \left\lfloor \frac{x}{n_1 \cdots n_r} \right\rfloor = \frac{1}{r!} \sum_{n_1 \leqslant x, \ldots, n_r \leqslant x} \mu(n_1 \cdots n_r) \left\lfloor \frac{x}{n_1 \cdots n_r} \right\rfloor$$

$$- \frac{1}{r!} \sum_{j=2}^{r-2} (-1)^j (j-1) \binom{r}{j} \sum_{n \leqslant x} (1 - r + j)^{\omega(n)} - \frac{(-1)^r \lfloor x \rfloor}{r(r-2)!} + \frac{(-1)^r (r-2)}{(r-1)!}$$

and deduce with the help of Exercise 82 that there exists an absolute constant $c_0 > 0$ and a constant $c_r \geqslant 1$, depending on r, such that, for any $x \geqslant c_r$ sufficiently large

$$\sum_{1 \leqslant n_1 < \cdots < n_r \leqslant x} \mu(n_1 \cdots n_r) \left\lfloor \frac{x}{n_1 \cdots n_r} \right\rfloor = \frac{(-1)^{r-1} x}{r(r-2)!}$$

$$+ O_r \left(x e^{-c_0 (\log x)^{3/5} (\log \log x)^{1/5}} \right).$$

(c) Prove that, for any $x \geqslant 1$

$$\sum_{n_1 \leqslant x, \ldots, n_r \leqslant x} \mu(n_1 \cdots n_r)^2 \left\lfloor \frac{x}{n_1 \cdots n_r} \right\rfloor = \sum_{n \leqslant x} (1 + r)^{\omega(n)}.$$

101   Let $r \in \mathbb{Z}_{\geqslant 2}$, f be a complex-valued arithmetic function and set

$$S_r(x) = \sum_{n_1 \leqslant x, \ldots, n_r \leqslant x} (f(n_1) + \cdots + f(n_r)) \left\lfloor \frac{x}{n_1 \cdots n_r} \right\rfloor \qquad (x \geqslant 1).$$

Let $(T_r(x))$ be the sequence recursively defined by

$$T_1(x) = \sum_{n \leqslant x} \left\lfloor \frac{x}{n} \right\rfloor (f \star \mathbf{1})(n) \quad \text{and} \quad T_r(x) = \sum_{n \leqslant x} T_{r-1}\left(\frac{x}{n}\right) \quad (r \in \mathbb{Z}_{\geqslant 2}).$$

Show that, for any $r \in \mathbb{Z}_{\geqslant 2}$ and $x \in \mathbb{R}_{\geqslant 1}$

$$S_r(x) = r T_{r-1}(x).$$

Apply this identity with $f = \mu$.

References

1. Balazard, M.: Les formules sommatoires de Selberg et applications. In: Colloque de Théorie des Nombres, pp. 5–16 (1989)
2. Bettin, S., Chandee, V., Radziwiłł, M.: The mean square of the Riemann zeta-function with Dirichlet polynomials. J. Reine. Angew. Math. **729**, 51–79 (2015)
3. Bombieri, E.: On the large sieve. Mathematika **12**, 201–225 (1965)
4. Bombieri, E.: Le Grand Crible dans la Théorie Analytique des Nombres, vol. 18. Société Mathématique de France, France (1974)
5. Bombieri, E., Friedlander, J., Iwaniec, H.: Primes in arithmetic progressions to large moduli. Acta Math. **156**, 203–251 (1986)
6. Bombieri, E., Friedlander, J., Iwaniec, H.: Primes in arithmetic progressions to large moduli, II. Math. Ann. **277**, 361–393 (1987)
7. Bombieri, E., Friedlander, J., Iwaniec, H.: Primes in arithmetic progressions to large moduli, III. J. Am. Math. Soc. **2**, 215–224 (1989)
8. Bordellès, O.: Multiplicative functions over short segments. Acta Arith. **157**, 1–10 (2013)
9. Bordellès, O.: A multidimensional Cesáro type identity and applications. J. Integer Seq. **18**, Art. 15.3.7 (2015a)
10. Bordellès, O.: Some explicit estimates for the Möbius function. J. Integer Seq. **18**, Art. 15.11.1 (2015b)
11. Bordellès, O.: Short interval results for certain prime-independent multiplicative functions. J. Number Theory **177**, 100–111 (2017)
12. Bourgain, J., Watt, N.: Mean square of zeta function, circle problem and divisor problem revisited, Preprint p. 23 (2017). https://arxiv.org/abs/1709.04340.
13. Bourgain, J., Watt, N.: Decoupling for perturbed cones and the mean square of $\left| \zeta \left(\frac{1}{2} + it \right) \right|$. Int. Math. Res. Not. **17**, 5219–5296 (2018)
14. Brüdern, J.: Daniel's twists of Hooley's Delta function. In: Blomer, V., et al. (eds.) Contributions in Analytic and Algebraic Number Theory. Springer Proceedings in Mathematics, University of Göttingen, Germany, July 2009, vol. 9, pp 31–82. Springer, Berlin (2012)
15. Brun, V.: Über das Goldbachsche Gesetz und die Anzahl der Primzahlpaare. Arch. Mat. Natur. **34**(8), pp 19 (1915)
16. Brun, V.: La série $1/5 + 1/7 + 1/11 + 1/13 + 1/17 + 1/19 + 1/29 + 1/31 + 1/41 + 1/43 + 1/59 + 1/61 + \cdots$ où les dénominateurs sont nombres premiers jumeaux EST convergente ou finie. Bull. Sci. Math. **43**, 124–128 (1919)
17. Chowla, S.: The Riemann Hypothesis and Hilbert's tenth problem. In: Mathematics and its Applications, vol. 4. Gordon and Breach Science Publishers, New York (1965)
18. Cohen, E.: On the average number of direct factors of a finite abelian group. Acta Arith. **6**, 159–173 (1960)
19. Cohen, H., Dress, F.: Estimations numériques du reste de la fonction sommatoire relative aux entiers sans facteur carré. In: Publication Mathematical Orsay: Colloque de Théorie Analytique des Nombres, Marseille, 1985, pp. 73–76 (1988)
20. Davenport, H.: Multiplicative number theory. In: Montgomery, H.L. (ed.) GTM, vol. 74, 2nd edn. Springer, Berlin (1980)

21. de la Bretèche, R., Tenenbaum, G.: Oscillations localisées sur les diviseurs. J. Lond. Math. Soc. **85**, 669–693 (2012)
22. Delange, H.: Sur des formules de Atle Selberg. Acta Arith. **19**, 105–146 (1971)
23. Dona, D., Helfgott, H.A., Alterman, S.Z.: Explicit L^2 bounds for the Riemann zeta function, p. 43 (2019). https://arxiv.org/abs/1906.01097
24. Dusart, P.: Explicit estimates of some functions over primes. Ramanujan J. **45**, 227–251 (2018)
25. Elliott, P.D.T.A., Halberstam, H.: A conjecture in prime number theory. Symposia Math. **4**, 59–72 (1970)
26. Erdős, P.: Some unsolved problems. Michigan Math. J. **4**, 291–300 (1957)
27. Estermann, T.: Einige sätze über quadratfreie zahlen. Math. Ann. **105**, 653–662 (1931)
28. Felix, A.T.: Generalizing the Titchmarsh divisor problem. Int. J. Number Theory **8**, 613–629 (2012)
29. Fiorilli, D.: On a theorem of Bombieri, Friedlander, and Iwaniec. Can. J. Math. **64**, 1019–1035 (2012)
30. Fouvry, E.: Sur le problème des diviseurs de Titchmarsh. J. Reine Angew. Math. **357**, 51–76 (1985)
31. Friedlander, J., Iwaniec, H.: Summation formulæ for coefficients of L-functions. Canad. J. Math. **57**, 494–505 (2005)
32. Friedlander, J., Iwaniec, H.: Opera de Cribro. In: Colloquium Publications, vol. 57. American Mathematical Society, New York (2010)
33. Gallagher, P.X.: The large sieve. Mathematika **14**, 14–20 (1967)
34. Goldston, D., Pintz, J., Yildirim, C.: Primes in tuples I. Ann. Math. **170**, 819–862 (2009)
35. Gould, H.W.: Combinatorial Identities. A standardized set of tables listing 500 binomial coefficient summations. Morgantown, West Virginia (1972)
36. Granville, A., Koukoulopoulos, D.: Beyond the LSD method for the partial sums of multiplicative functions, p. 26 (2018)
37. Halász, G.: Über die Mittelwerte multiplikativer zahlentheoretischer Funktionen. Acta Math. Acad. Sci. Hung. **19**, 365–403 (1968)
38. Halberstam, H.: Footnote to the Titchmarsh-Linnik divisor problem. Proc. Am. Math. Soc. **18**, 187–188 (1967)
39. Halberstam, H., Richert, H.E.: On a result of R. R. Hall. J. Number Theory **11**, 76–89 (1979)
40. Halberstam, H., Richert, H.E.: Sieve Methods. Dover Publications, New York (2011)
41. Hall, R.R., Tenenbaum, G.: Divisors. Cambridge University, Cambridge (1988)
42. Hardy, G.H., Ramanujan, S.: The normal number of prime factors of a number n. Q. J. Math. **48**, 76–92 (1917)
43. Hassani, M.: Asymptotic expansions for the average of the generalized omega function. Integers **18**, Art. A.23 (2018)
44. Hasse, H., Suetuna, Z.: Ein allgemeines Teilerproblem der Idealtheorie. J. Fac. Sci. Uni. Tokyo **2**, 133–154 (1931)
45. Haukkanen, P.: Expressions for the Dirichlet inverse of arithmetical functions. Notes on Number Theory and Discrete Math. **6**, 118–124 (2000)
46. Heath-Brown, D.R.: Fractional moments of the Riemann zeta function. J. Lond. Math. Soc. **24**, 65–78 (1981)
47. Heath-Brown, D.R.: Square-free values of $n^2 + 1$. Acta Arith. **155**, 1–13 (2012)
48. Hooley, C.: On a new technique and its applications to the theory of numbers. Proc. Lond. Math. Soc. **38**, 115–151 (1979)
49. Ivić, A.: The distribution of values of the enumerating function of non-isomorphic abelian groups of finite order. Arch. Math. (Basel) **30**, 374–379 (1978)
50. Ivić, A.: On the number of finite non-isomorphic abelian groups in short intervals. Math. Nachr. **101**, 257–271 (1981)
51. Ivić, A.: On the number of abelian groups of a given order and on certain related multiplicative functions. J. Number. Theory **16**, 119–137 (1983)

52. Ivić, A.: Some problems on mean values of the Riemann zeta-function. J. Théorie des Nombres de Bordeaux **8**, 101–123 (1996)
53. Ivić, A.: The Riemann Zeta-Function, Theory and Applications. Dover Publication, Unabridged republication of the work first published in 1985. The Riemann Zeta-Function: The Theory of the Riemann Zeta-Function with Applications. Wiley, New-York (2003)
    Ivit'c, A.: Hybrid moments of the Riemann zeta-function. J. Numbers 14, Article ID 892324 (2015)
54. Ivić, A., Motohashi, Y.: On the fourth power moment of the Riemann zeta-function. J. Number Theory **51**, 16–45 (1995)
55. Ivić, A., Ouellet, M.: Some new estimates in the Dirichlet divisor problem. Acta Arith. **52**, 241–253 (1989)
56. Iwaniec, H., Kowalski, E.: Analytic Number Theory. In: Colloquium Publications, vol. 53. American Mathematical Society, New York (2004)
57. Kátai, I., Subbarao, M.V.: Some remarks on a paper of Ramachandra. Lith. Math. J. **43**, 410–418 (2003)
58. Kendall, D.G., Rankin, R.A.: On the number of abelian groups of a given order. Quart. J. Math. **18**, 197–208 (1947)
59. Kienast, A,: Über die asymptotische Darstellung der summatorischen Funktion von Dirichletreihen mit positiven Koeffizienten. Math. Z. **45**, 115–126 (1939)
60. Kolesnik, G.: On the estimation of multiple exponential sums. In: Halberstam, H., Hooley, C.: (eds.) Recent Progress in Analytic Number Theory, vol. 1. Academic Press, New York (1981)
61. Kolpakova, O.V.: New estimates of the remainder in an asymptotic formula in the multidimensional Dirichlet divisor problem. Math. Notes **89**, 504–518 (2011)
62. Konyagin, S., Pomerance, C.: Primes recognizable in polynomial deterministic time. In: Graham, R.L., Nesetril, J. (eds.) The mathematics of Paul Erdős. Springer, Berlin, pp. 176–198 (1997)
63. Krätzel, E.: Die maximale Ordnung der Anzahl der wesentlich verschiedenen Abelschen Gruppen n-ter Ordnung. Q. J. Math. (2) Oxford Ser. **21**, 273–275 (1970)
64. Krätzel, E.: Divisor problems and powerful numbers. Math. Nachr. **114**, 97–104 (1983)
65. Krätzel, E.: Estimates in the general divisor problem. Abh. Math. Sem. Univ. Hamburg **82**, 191–206 (1992)
66. Krätzel, E., Wolke, D.: Über die anzahl del abelschen gruppen gegebener ordnung. Analysis **14**, 257–266 (1994)
67. Landau, E.: Über die anzahl der gitterpunkte in gewissen bereichen (zweite abhandlung). Nachr. K. Ges Wiss Göttingen, pp. 209–243, 137–150 (Klasse, 1915)
68. Landau, E.: Einführung in die elementare und analytische Theorie der algebraischen Zahlen und der Ideale. In: Leipzig und Berlin, 2nd edn. Chelsea Publishing Company, Philadelphia (1927/1949)
69. Levin, B.V., Fainleb, A.S.: Application of some integral equations to problems of number theory. Russ. Math. Surv. **22**, 119–204 (1967)
70. Linnik, J.V.: The large sieve. Dokl. Akad. Nauk. SSSR **30**, 292–294 (1941)
71. Linnik, J.V.: The dispersion method in binary additive problems. In: American Mathematical Society, Providence, Rhode Island, Translate Mathematical monographs, vol. 4 (1963)
72. Liu, H.Q. On multiple exponential sums and their applications. Functiones et Approximatio **45**(2), 155–163 (2011)
73. Marraki, M.E.: Fonction sommatoire de la fonction de Möbius, 3. Majorations effectives fortes. J. Théor Nombres Bordeaux **7**, 407–433 (1995)
74. Martin, G.: An asymptotic formula for the number of smooth values of a polynomial. J. Number. Theory **93**, 108–182 (2002)
75. Matomäki, K., Radziwiłł, M., Tao, T.: Correlations of the von Mangoldt and higher divisor functions I. Long shift ranges. Proc. Lond. Math. Soc. **118**, 284–350 (2018)
76. Maynard, J.: Small gaps between primes. Ann. Math. **181**, 383–413 (2015)
77. Montgomery, H.L.: The analytic principle of the large sieve. Bull. Am. Math. Soc. **84**, 547–567 (1978)

78. Montgomery, H.L., Vaughan, R.C.: The large sieve. Mathematika **20**, 119–134 (1973)
79. Montgomery, H.L., Vaughan, R.C.: Multiplicative Number Theory I. Classical Theory. In: Cambridge studies in advanced mathematics, vol. 97. Cambridge University, Cambridge (2007)
80. Moser, L., MacLeod, R.A.: The error term for the squarefree integers. Canad. Math. Bull. **9**, 303–306 (1966)
81. Motohashi, Y.: A note on Siegel's zeros. Proc. Jpn Acad. **55**, 190–192 (1979)
82. Nair, M., Tenenbaum, G.: Short sums of certain arithmetic functions. Acta Math. **180**, 119–144 (1998)
83. Nicolas, J.L., Robin, G.: Majorations explicites pour le nombre de diviseurs de N. Canad. Math. Bull. **26**, 485–492 (1983)
84. Nowak, G.: On the value distribution of a class of arithmetic functions. Comment Math. Univ. Carolin **37**, 117–134 (1996)
85. Parson, A., Tull, J.: Asymptotic behavior of multiplicative functions. J. Number Theory **10**, 395–420 (1978)
86. Polymath, D.H.J.: Variants of the Selberg sieve, and bounded intervals containing many primes. Res. Math. Sci. **1**(12), 83 (2014)
87. Ramanujan, S.: Some formulæ in the analytic theory of numbers. Mess. Math. **45**, 81–84 (1916)
88. Rankin, R.A.: Van der Corput's method and the theory of exponent pairs. Quart. J. Math. **6**, 147–153 (1955)
89. Reuss, T.: Pairs of k-free numbers, consecutive square-full numbers, p. 28 (2018). https://arxiv.org/abs/1212.3150
90. Robert, O., Sargos, P.: Three-dimensional exponential sums with monomials. J. Reine Angew. Math. **591**, 1–20 (2006)
91. Robin, G.: Majorations explicites du nombre de diviseurs d'un entier. Publ. Dept. Math. Limoges **2**, 6 (1980)
92. Robin, G.: Estimation de la fonction de Tchebychef θ sur le k-ième nombre premier et grandes valeurs de la fonction $\omega(n)$ nombre de diviseurs de n. Acta Arith. **42**, 367–389 (1983)
93. Saffari, B.: Sur quelques applications de la "méthode de l'hyperbole" de Dirichlet à la théorie des nombres premiers. Enseignement Math. **14**, 205–224 (1970)
94. Sathe, L.G.: On a problem of Hardy on the distribution of integer with a given number of prime factors. J. Indian Math. Soc. **17**, 63–141 (1953)
95. Schwarz, L., Spilker, J.: Arithmetical Functions. Cambridge University, Cambridge (1994)
96. Sedunova, A.: A partial Bombieri-Vinogradov theorem with explicit constants. Pub. Math. Besançon, 101–110 (2018)
97. Selberg, A.: A note on a paper by L. G. Sathe. J. Indian Math. Soc. **18**, 83–87 (1954)
98. Shiu, P.: A Brun–Titchmarsh theorem for multiplicative functions. J. Reine Angew Math. **313**, 161–170 (1980)
99. Sivaramakrishnan, R.: Classical theory of arithmetical functions. In: Pure and Applied Mathematics, vol. 126. Marcel Dekker, New York (1989)
100. Skrabutenas, R.: Asymptotic expansion of sums of multiplicative functions. Liet. Mat. Rinkinys **14**, 115–126 (1974)
101. Suryanarayana, D.: The number of k-free divisors of an integer. Acta Arith. **47**, 345–354 (1971)
102. Tao, T.: A remark on partial sums involving the Möbius function. Bull. Aust. Math. Soc. **81**, 343–349 (2010)
103. Tao, T.: The logarithmically averaged Chowla and Elliott conjectures for two-point correlations. Forum. Math. Pi. **4**(Article ID e8), 36 (2016a)
104. Tao, T.: The Erdős discrepancy problem. Discrete Anal. **1**, 29 (2016b)
105. Tenenbaum, G.: A rate estimate in Billingsley's theorem for the size distribution of large prime factors. Quart. J. Math. **51**, 385–403 (2000)
106. Tenenbaum, G.: Introduction à la Théorie Analytique et Probabiliste des Nombres, Berlin (2007)

107. Tenenbaum, G.: Moyennes effectives de fonctions multiplicatives complexes. Ramanujan J. **44**, 641–701 (2017)
108. Titchmarsh, E.C.: A divisor problem. Rend di Palermo **54**, 414–429 (1931)
109. Titchmarsh, E.C.: The Theory of Functions. Oxford University Press (1939); 2nd edition (1979)
110. Titchmarsh, E.C.: The Theory of the Riemann Zeta-function, 2nd edn. Oxford Science Publication, Oxford (1986) Revised by D. R. Heath-Brown
111. Tóth, L.: On certain arithmetical functions involving exponential divisors II. Annales Univ. Sci. Budapest Sect. Comp. **27**, 155–166 (2007)
112. Tóth, L.: Multiplicative arithmetic functions of several variables: a survey. In: Rassias, T.M., Pardalos, P. (eds.) Mathematics Without Boundaries, Surveys in Pure Mathematics. Springer, New-York, pp. 483–514 (2014)
113. Tóth, L.: Simple proofs of identities. Personal communication (2018)
114. Tóth, L., Zhai, W.: On multivariable averages of divisor functions. J. Number Theory **192**, 251–269 (2018)
115. Tudesq, C.: Étude de la loi locale de $\omega(n)$ dans de petits intervalles. Ramanujan J. **4**, 277–290 (2000)
116. Tuljaganova, M.I.: A generalization of a theorem of Halász. Izv. Akad. Nauk. USSR **4**, 35–40 (1978)
117. Tulyaganov, S.T.: On the summation of multiplicative arithmetic functions. Number theory Vol I Elementary and analytic. Proceeding of the Conference Budapest/Hung 1987. Coll. Math. Soc. János Bolyai **51**, 539–573 (1990)
118. Vaughan, R.C.: Some applications of Montgomery's sieve. J. Number Theory **5**, 641–679 (1973)
119. Vaughan, R.C.: An elementary method in prime number theory. Acta Arith. **37**, 111–115 (1980)
120. Vaughan, R.C.: An elementary method in prime number theory. In: Halberstam, H., Hooley, C. (eds.) Recent Progress in Analytic Number Theory, vol. 1. Academic Press, New York, pp. 341–347 (1981)
121. Vinogradov, A.I.: The density hypothesis for Dirichlet L-series. Izv. Akad. Nauk. SSSR Ser. Mat. **29**, 903–934 (1965)
122. Vogts. M.: Many-dimensional generalized divisor problems. Math. Nachr. **124**, 103–121 (1985)
123. Wilson, B.M.: Proofs of some formulæ enunciated by Ramanujan. Proc. Lond. Math. Soc. **2**, 235–255 (1922)
124. Wirsing, E.: Das asymptotische Verhalten vun Summen über multiplikative Funktionen. Math. Ann. **143**, 75–102 (1961)
125. Wu, J.: Problèmes de diviseurs exponentiels et entiers exponentiellement sans facteur carré. J. Théor Nombres Bordeaux **7**, 133–141 (1995)
126. Zhai, W.: On prime-independent multiplicative functions. In: Diophantine Problems and Analytic Number Theory. Proceeding of a symposium held at the Research Institute for Mathematical Sciences. Kyoto University, Kyoto (2002)
127. Zhai, W., Cao, X.: On the average number of direct factors of finite Abelian groups. Acta Arith. **82**, 45–55 (1997)
128. Zhang, Y.: Bounded gaps between primes. Ann. Math. **179**, 1121–1174 (2014)

Chapter 5
Lattice Points

5.1 Introduction

5.1.1 Multiplicative Functions over Short Segments

Let x, y be two real numbers satisfying $2 \leqslant y \leqslant x$ with $y = o(x)$ as $x \to \infty$. As seen in Chap. 4, Sect. 4.3.9, the problem of deriving an asymptotic formula for the sum

$$\sum_{x < n \leqslant x+y} f(n)$$

is quite intricate and heavily depends on the arithmetic nature of f. Mimicking the long-sum case, one of the first ideas we have to deal with is the use of the *Eratosthenes transform* g of f given by $g = f \star \mu$. Standard computations as seen in Chap. 4 yield

$$\sum_{x < n \leqslant x+y} f(n) = \sum_{x < n \leqslant x+y} (g \star 1)(n) = \sum_{d \leqslant x+y} g(d) \left(\left\lfloor \frac{x+y}{d} \right\rfloor - \left\lfloor \frac{x}{d} \right\rfloor \right).$$

Now assume that the Dirichlet series of g has an abscissa of absolute convergence $\sigma_a(g) = \delta \in (0, 1)$. Then, by Corollary 4.3 and partial summation, we derive for all $z \geqslant 1$ and $\varepsilon > 0$

$$\sum_{d \leqslant z} |g(d)| \ll z^{\delta + \varepsilon} \quad \text{and} \quad \sum_{d > z} \frac{|g(d)|}{d} \ll z^{-1 + \delta + \varepsilon},$$

© Springer Nature Switzerland AG 2020
O. Bordellès, *Arithmetic Tales*, Universitext,
https://doi.org/10.1007/978-3-030-54946-6_5

and hence

$$\sum_{x<n\leqslant x+y} f(n) = \left(\sum_{d\leqslant 2y} + \sum_{2y<d\leqslant x} + \sum_{x<d\leqslant x+y} \right) g(d) \left(\left\lfloor \frac{x+y}{d} \right\rfloor - \left\lfloor \frac{x}{d} \right\rfloor \right)$$

$$= \sum_{d\leqslant 2y} g(d) \left(\frac{y}{d} + O(1) \right) + O\left(\max_{2y<D\leqslant x} (S_D(g)) \log x + \sum_{x<d\leqslant x+y} |g(d)| \right)$$

$$= y \sum_{d=1}^{\infty} \frac{g(d)}{d} + O\left(y \sum_{d>2y} \frac{|g(d)|}{d} + \sum_{d\leqslant 2y} |g(d)| + \max_{2y<D\leqslant x} (S_D(g)) \log x + \sum_{x<d\leqslant x+y} |g(d)| \right)$$

$$= y \sum_{d=1}^{\infty} \frac{g(d)}{d} + O\left(y^{\delta+\varepsilon} + \max_{2y<D\leqslant x} (S_D(g)) \log x + \sum_{x<d\leqslant x+y} |g(d)| \right),$$

where

$$S_D(g) := \sum_{D<d\leqslant 2D} |g(d)| \left(\left\lfloor \frac{x+y}{d} \right\rfloor - \left\lfloor \frac{x}{d} \right\rfloor \right). \tag{5.1}$$

The second sum in the error term is usually easily estimated, for example, by Shiu's Theorem 4.17. Also note that

$$\sum_{d=1}^{\infty} \frac{g(d)}{d} = \prod_{p} \left(1 - \frac{1}{p} \right) \sum_{\alpha=0}^{\infty} \frac{f(p^{\alpha})}{p^{\alpha}},$$

which is believed to be the mean value of f in the interval $(x, x+y]$. For instance, when $f = \mu^2 = \mu_2$, we have $L(s, g) = L(s, f_2^{-1}) = \zeta(2s)^{-1}$ so that $\delta = \frac{1}{2}$, and hence we get

$$\sum_{x<n\leqslant x+y} \mu_2(n) = \frac{y}{\zeta(2)} + O\left(\max_{2y<D\leqslant x} \left(S_D(f_2^{-1}) \right) \log x + y^{1/2+\varepsilon} \right),$$

where f_k^{-1} is given in Table 4.1 in Chap. 4 and in Exercise 68. Hence, the main difficulty is to derive a non-trivial bound for (5.1). Observe that, since $2y < d$, the difference

$$\left\lfloor \frac{x+y}{d} \right\rfloor - \left\lfloor \frac{x}{d} \right\rfloor$$

is equal to 1 or 0 depending on the fact that there is either an integer between x/d and $(x+y)/d$ or not. Hence, to have a chance to do better than the trivial estimate, we must take these cancellations into account. This leads to estimate the number of *lattice points*, also called *integer points*, lying near the curve of some smooth functions. This chapter is devoted to providing results counting integer points close to sufficiently regular plane curves, especially focusing our results on the functions $f = \mu_2$ and $f = s_2$.

Remark 5.1 The above argument is not possible if $\sigma_c(g) \geqslant 1$, for the series $\sum_{d=1}^{\infty} g(d)d^{-1}$ does not converge. Furthermore, the expected main term is $y \sum_{d \leqslant x} g(d)d^{-1}$ rather than $y \sum_{d \leqslant y} g(d)d^{-1}$. A possible way to overcome this problem is to resort to the ψ-function from Chap. 6 in the following manner:

$$\sum_{x < n \leqslant x+y} f(n) = y \sum_{d \leqslant x} \frac{g(d)}{d} - \sum_{y < d \leqslant x} g(d) \left(\psi \left(\frac{x+y}{d} \right) - \psi \left(\frac{x}{d} \right) \right)$$

$$+ \sum_{x < d \leqslant x+y} g(d) + O \left(\sum_{d \leqslant y} |g(d)| \right),$$

and then to make use of the results of Chap. 6.

Remark 5.2 The argument of the previous remark is not suitable when $f = \mu$ or $f = \lambda$, since in these cases the error term $\sum_{d \leqslant y} |g(d)|$ is $\asymp y \log y$, and hence is greater than the expected main term. For these functions, belonging in a more general class of multiplicative functions, K. Matomäki and J. Teräväinen [14] developed a new tool that requires the next definition.

Definition 5.1 A multiplicative function $f : \mathbb{Z}_{\geqslant 1} \to \mathbb{C}$ is said to be *eventually periodic on the primes* if there exist $n_0, m_0 \in \mathbb{Z}_{\geqslant 1}$ such that, for all primes $p, q \geqslant n_0$ satisfying $p \equiv q \pmod{m_0}$, we have $f(p) = f(q)$.

Note that the class of multiplicative functions under consideration includes the generalized divisor functions τ_z for any complex z, as well as the indicator of those integers that can be represented as the norm of an element in a given abelian extension $\mathbb{K}/\mathbb{Q}$. The main result in [14] may be stated as follows.

Theorem 5.1 *Let* $f : \mathbb{Z}_{\geqslant 1} \to \mathbb{C}$ *be a multiplicative function eventually periodic on the primes such that there exists* $r \in \mathbb{Z}_{\geqslant 1}$ *such that, for all* $n \in \mathbb{Z}_{\geqslant 1}$, *we have* $|f(n)| \leqslant \tau_r(n)$. *Then, for all* $\varepsilon > 0$ *and all* $2 \leqslant y \leqslant x$ *satisfying* $y \geqslant x^{0.55+\varepsilon}$, *we have*

$$\sum_{x < n \leqslant x+y} f(n) = \frac{y}{x} \sum_{x < n \leqslant 2x} f(n) + O \left\{ \frac{y}{\log x} \prod_{\substack{p \leqslant x \\ p \notin [P, Q]}} \left(1 + \frac{|f(p)|}{p} \right) \right\},$$

where $P := e^{(\log x)^{2/3+\varepsilon/2}}$ *et* $Q := x^{1/(\log \log x)^2}$.

As a corollary, the authors show that, if $\varepsilon > 0$ and $\theta > 0.55$ are fixed, then, for x large and $y \geqslant x^\theta$

$$\sum_{x < n \leqslant x+y} \mu(n) \ll y(\log x)^{-1/3+\varepsilon}.$$

5.1.2   The Number $\mathcal{R}(f, N, \delta)$

In what follows, $N \geqslant 4$ is a large integer and δ and c_0 are small positive real numbers. We will always suppose that

$$0 < \delta < \tfrac{1}{4}. \tag{5.2}$$

Note that the constant c_0 may take different values according to the section in which it appears.

x, y are large real numbers satisfying $2 \leqslant y \leqslant x$, except in the examples where we impose the following more restricted range:

$$16 \leqslant y < \tfrac{1}{4}\sqrt{x}. \tag{5.3}$$

Finally, we shall need the following integer.

Definition 5.2 Let $f : [N, 2N] \to \mathbb{R}$ be any map and δ be a real number satisfying (5.2). We define

$$\mathcal{S}(f, N, \delta) = \{n \in [N, 2N] \cap \mathbb{Z} : \|f(n)\| < \delta\}$$

and $\mathcal{R}(f, N, \delta) = |\mathcal{S}(f, N, \delta)|$.

The following result deals with small differences of integral parts. The proof is easy.

Lemma 5.1 *Let* $0 \leqslant \delta < \tfrac{1}{2}$ *be any small real number. Then*

$$\begin{cases} \lfloor x + \delta \rfloor - \lfloor x \rfloor = 1 \iff \{x\} \geqslant 1 - \delta \\[2mm] \lfloor x \rfloor - \lfloor x - \delta \rfloor = 1 \iff \{x\} < \delta. \end{cases}$$

In particular

$$\lfloor x + \delta \rfloor - \lfloor x - \delta \rfloor = \begin{cases} 1, & \text{if } \|x\| < \delta \\ 0, & \text{if } \|x\| > \delta \\ 0 \text{ or } 1, & \text{if } \|x\| = \delta. \end{cases}$$

We shall always make use of the following lemma.

Lemma 5.2 *Let $f : [N, 2N] \to \mathbb{R}$ be any map, δ be a real number satisfying (5.2) and (δ_n) be a sequence of real numbers supported on $[N, 2N]$ such that $0 \leqslant \delta_n < \delta$. Then*

$$\sum_{N \leqslant n \leqslant 2N} (\lfloor f(n) + \delta_n \rfloor - \lfloor f(n) - \delta_n \rfloor) \leqslant \mathcal{R}(f, N, \delta) \leqslant \sum_{N \leqslant n \leqslant 2N} (\lfloor f(n) + \delta \rfloor - \lfloor f(n) - \delta \rfloor).$$

Proof From Lemma 5.1, we derive

$$\sum_{N \leqslant n \leqslant 2N} (\lfloor f(n) + \delta_n \rfloor - \lfloor f(n) - \delta_n \rfloor)$$

$$= \sum_{\substack{N \leqslant n \leqslant 2N \\ \|f(n)\| < \delta_n}} 1 + \sum_{\substack{N \leqslant n \leqslant 2N \\ \|f(n)\| = \delta_n}} (\lfloor f(n) + \delta_n \rfloor - \lfloor f(n) - \delta_n \rfloor)$$

$$\leqslant \sum_{\substack{N \leqslant n \leqslant 2N \\ \|f(n)\| \leqslant \delta_n}} 1 \leqslant \mathcal{R}(f, N, \delta)$$

$$= \sum_{N \leqslant n \leqslant 2N} (\lfloor f(n) + \delta \rfloor - \lfloor f(n) - \delta \rfloor) - \sum_{\substack{N \leqslant n \leqslant 2N \\ \|f(n)\| = \delta}} (\lfloor f(n) + \delta \rfloor - \lfloor f(n) - \delta \rfloor)$$

$$\leqslant \sum_{N \leqslant n \leqslant 2N} (\lfloor f(n) + \delta \rfloor - \lfloor f(n) - \delta \rfloor)$$

as asserted. □

It should be noticed that the trivial estimate yields

$$\mathcal{R}(f, N, \delta) \leqslant N + 1. \tag{5.4}$$

This inequality must be kept in mind whenever we obtain a new estimate for $\mathcal{R}(f, N, \delta)$.

5.1.3 Basic Lemmas

5.1.3.1 The Squarefree Number Problem

With the definitions and notation of the previous section, we may state the basic result we shall use to estimate the number of squarefree integers in short intervals.

Lemma 5.3 *Let* x, y *satisfy (5.3) and* $2\sqrt{y} \leqslant A < B \leqslant 2\sqrt{x}$. *Then*

$$\sum_{x<n\leqslant x+y} \mu_2(n) = \frac{y}{\zeta(2)} + O\left((R_1 + R_2)\log x + A\right),$$

where $R_1 = R_1(x, y; A, B)$ *and* $R_2 = R_2(x, y; B)$ *are given by*

$$R_1 = \max_{A<N\leqslant B} \mathcal{R}\left(\frac{x}{n^2}, N, \frac{y}{N^2}\right) \quad and \quad R_2 = \max_{N\leqslant 2x/B^2} \mathcal{R}\left(\sqrt{\frac{x}{n}}, N, \frac{y}{\sqrt{Nx}}\right).$$

Proof By the computations made in the previous sections, we have

$$\sum_{x<n\leqslant x+y} \mu_2(n) = \frac{y}{\zeta(2)} + S_2 + O\left(\sqrt{y}\right)$$

with

$$S_2 = \sum_{2\sqrt{y}<d\leqslant\sqrt{x}} \mu(d)\left(\left\lfloor \frac{x+y}{d^2} \right\rfloor - \left\lfloor \frac{x}{d^2} \right\rfloor\right).$$

Inserting the parameters A and B and estimating the first sum trivially, we get

$$|S_2| \leqslant \left(\sum_{2\sqrt{y}<d\leqslant A} + \sum_{A<d\leqslant B} + \sum_{B<d\leqslant\sqrt{x}}\right)\left(\left\lfloor \frac{x+y}{d^2} \right\rfloor - \left\lfloor \frac{x}{d^2} \right\rfloor\right)$$

$$\leqslant A + \sum_{A<d\leqslant B}\left(\left\lfloor \frac{x+y}{d^2} \right\rfloor - \left\lfloor \frac{x}{d^2} \right\rfloor\right) + \sum_{B<d\leqslant\sqrt{x}}\sum_{x/d^2<k\leqslant(x+y)/d^2} 1$$

$$= A + \sum_{A<d\leqslant B}\left(\left\lfloor \frac{x+y}{d^2} \right\rfloor - \left\lfloor \frac{x}{d^2} \right\rfloor\right) + \sum_{B<d\leqslant\sqrt{x}}\sum_{\substack{k \\ x<kd^2\leqslant x+y}} 1,$$

and interchanging the summations gives

$$|S_2| \leqslant A + \sum_{A<d\leqslant B}\left(\left\lfloor \frac{x+y}{d^2} \right\rfloor - \left\lfloor \frac{x}{d^2} \right\rfloor\right) + \sum_{k\leqslant(x+y)/B^2}\sum_{\sqrt{x/k}<d\leqslant\sqrt{(x+y)/k}} 1$$

$$\leqslant A + \sum_{A<d\leqslant B}\left(\left\lfloor \frac{x+y}{d^2} \right\rfloor - \left\lfloor \frac{x}{d^2} \right\rfloor\right) + \sum_{k\leqslant 2x/B^2}\left(\left\lfloor \sqrt{\frac{x+y}{k}} \right\rfloor - \left\lfloor \sqrt{\frac{x}{k}} \right\rfloor\right).$$

Now for $k \in [N, 2N]$ and using (5.3), we have

$$\sqrt{\frac{x+y}{k}} - \sqrt{\frac{x}{k}} \leqslant \frac{y}{2\sqrt{Nx}} < \frac{y}{\sqrt{Nx}} < \frac{1}{4},$$

and splitting the sums into $O(\log x)$ subintervals $(N, 2N]$ gives the asserted result.

□

Remark 5.3 If we choose $B = x^{1/3}$, then the two sums have a range of the same order of magnitude. Also note that the functions $u \mapsto x/u^2$ and $u \mapsto \sqrt{x/u}$ are inverse. Hence, one may expect that $R_1(A, x^{1/3})$ and $R_2(x^{1/3})$ have the same order of magnitude. Indeed they have by Exercise 105, so that Lemma 5.3 enables us to reduce the range of summation. Choosing $A = 4\sqrt{y}$ and using (5.4) to estimate R_1 and R_2 trivially, we get at once

$$\sum_{x < n \leqslant x+y} \mu_2(n) = \frac{y}{\zeta(2)} + O\left(x^{1/3} \log x\right)$$

if x, y satisfy (5.3). Hence, there exists $c_0 > 0$ such that, if $c_0 x^{1/3} \log x \leqslant y < \frac{1}{4} x^{1/2}$, the interval $(x, x+y]$ contains a squarefree number.

5.1.3.2 The r-Free Number Problem

Lemma 5.3 may easily be generalized to the problem of r-free numbers in short intervals, with $r \geqslant 2$. The proof is exactly the same as in Lemma 5.3, so we leave the details to the reader.

Lemma 5.4 *Let $r \geqslant 2$ be a fixed integer. For all $(4y)^{1/r} \leqslant A < B \leqslant x^{1/r}$, we have*

$$\sum_{x < n \leqslant x+y} \mu_r(n) = \frac{y}{\zeta(r)} + O\left((R_1 + R_2) \log x + A\right),$$

where

$$R_1 = \max_{A < N \leqslant B} \mathcal{R}\left(\frac{x}{n^r}, N, \frac{y}{N^r}\right) \quad \text{and} \quad R_2 = \max_{N \leqslant 2x/B^r} \mathcal{R}\left(\left(\frac{x}{n}\right)^{1/r}, N, \frac{y}{N^{1/r} x^{1-1/r}}\right).$$

5.1.3.3 The Square-Full Number Problem

The problem of square-full numbers in short intervals is quite similar in nature to that of the squarefree number problem, except with the following major difference. In [15], the author proved that there exist infinitely many positive integers n such that there is no square-full number between n^2 and $(n+1)^2$. It follows that, if $y < \sqrt{x}$, the interval $(x, x + y]$ may contain no square-full number at all. On the other hand, since there is a square in the interval $(x, x + 2\sqrt{x} + 1]$ for all $x \geqslant 0$, it follows that there exists a constant $c_0 > 0$ such that, for all $x \geqslant 1$, the interval $(x, x + c_0\sqrt{x}]$ contains a square-full integer. Thus, the maximum size of gaps between square-full numbers is known, and one may ask for the distribution of square-full numbers in intervals $(x, x + y]$ with $y > \sqrt{x}$ and the ratio $yx^{-1/2}$ being as small as possible. In Exercise 109, the following basic lemma similar to Lemma 5.3 is established.

Lemma 5.5 *Let* $L = L(x, y) = y(x \log x)^{-1/2}$, *and assume*

$$16 x^{1/2}(\log x)^3 \leqslant y \leqslant 4^{-3}x(\log x)^{-1}.$$

Then

$$\sum_{x < n \leqslant x+y} s_2(n) = \frac{\zeta(3/2)}{2\zeta(3)} \frac{y}{\sqrt{x}} + O\left\{(R_1 + R_2) \log x + \frac{y}{\sqrt{x} \log x}\right\},$$

where

$$R_1 = \max_{L < N \leqslant (2x)^{1/5}} \mathcal{R}\left(\sqrt{\frac{x}{n^3}}, N, \frac{y}{\sqrt{x}N^3}\right) \quad and$$

$$R_2 = \max_{L < N \leqslant (2x)^{1/5}} \mathcal{R}\left(\left(\frac{x}{n^2}\right)^{1/3}, N, \frac{y}{(Nx)^{2/3}}\right).$$

5.1.4   Srinivasan's Optimization Lemma

We end this section with a useful optimization lemma due to Srinivasan [16, Lemma 4], which generalizes the following well-known situation. Suppose we have an estimate of the shape

$$E(H) \ll AH^a + BH^{-b},$$

where $A, B, a, b > 0$ and $H > 0$ is a parameter at our disposal. Choosing H to equalize both terms, we get

$$E(H) \ll \left(A^b B^a\right)^{\frac{1}{a+b}},$$

and this is best possible apart from the value of the implied constant.

Lemma 5.6 (Srinivasan) *Let*

$$E(H) = \sum_{i=1}^{m} A_i H^{a_i} + \sum_{j=1}^{n} B_j H^{-b_j},$$

where $m, n \in \mathbb{N}$ and A_i, B_j, a_i and b_j are positive real numbers. Suppose that $0 \leqslant H_1 \leqslant H_2$. Then

$$\min_{H_1 \leqslant H \leqslant H_2} E(H) \leqslant (m+n) \left\{ \sum_{i=1}^{m} \sum_{j=1}^{n} \left(A_i^{b_j} B_j^{a_i} \right)^{\frac{1}{a_i + b_j}} + \sum_{i=1}^{m} A_i H_1^{a_i} + \sum_{j=1}^{n} B_j H_2^{-b_j} \right\}.$$

This inequality corresponds to the best possible choice of H in the interval $[H_1, H_2]$. Srinivasan pointed out that the case $H_1 = 0$ and $H_2 = \infty$ was already shown by van der Corput in 1922.

5.1.5 Divided Differences

The concept of divided differences will play a key part in this chapter. The next result is a consequence of Rolle's theorem applied to the function $x \mapsto f(x) - \mathcal{P}(x)$.

Theorem 5.2 *Let k be a positive integer, $x_0 < x_1 < \cdots < x_k$ be real numbers, and $f \in C^k[x_0, x_k]$. Set $\mathcal{P}(x) = b_k x^k + \cdots + b_0$ the unique polynomial of degree $\leqslant k$ such that $\mathcal{P}(x_i) = f(x_i)$ for $i = 0, \ldots, k$. Then there exists a real number $t \in (x_0, x_k)$ such that*

$$b_k = \frac{f^{(k)}(t)}{k!}.$$

Remark 5.4 The polynomial $\mathcal{P}$ is the *Lagrange polynomial*, and b_k is the *divided difference* of f at the points $x_0, x_1, \ldots, x_k$ and usually denoted by $f[x_0, x_1, \ldots, x_k]$. One can show that

$$b_k = \sum_{j=0}^{k} \frac{f(x_j)}{\prod_{\substack{0 \leqslant i \leqslant k \\ i \neq j}} (x_j - x_i)} = \frac{A}{\prod_{0 \leqslant i < j \leqslant k} (x_j - x_i)}, \qquad (5.5)$$

and it is important to note that $A \in \mathbb{Z}$ as soon as $x_j \in \mathbb{Z}$ and $f(x_j) \in \mathbb{Z}$.

Remark 5.5 Theorem 5.2 shows that there exists a real number $t \in (x_0, x_k)$ such that

$$\sum_{\substack{j=0}}^{k} \frac{f(x_j)}{\prod_{\substack{0 \leqslant i \leqslant k \\ i \neq j}} (x_j - x_i)} = \frac{f^{(k)}(t)}{k!}. \tag{5.6}$$

This generalizes the mean value theorem.

5.2 Criteria for Integer Points

5.2.1 The First Derivative Test

The first result we will establish follows from the classical mean value theorem. It is only useful when λ_1 is very small and hence is rather restrictive. However, it will turn out to be the starting point of more elaborate estimates.

Theorem 5.3 (First Derivative Test) *Let* $f \in C^1[N, 2N]$ *such that there exist* $\lambda_1 > 0$ *and* $c_1 \geqslant 1$ *such that, for all* $x \in [N, 2N]$, *we have*

$$\lambda_1 \leqslant |f'(x)| \leqslant c_1 \lambda_1. \tag{5.7}$$

Then

$$\mathcal{R}(f, N, \delta) \leqslant 2c_1 N \lambda_1 + 4c_1 N \delta + \frac{2\delta}{\lambda_1} + 1.$$

In practice, the implied constant c_1 is useless, so using Titchmarsh–Vinogradov's notation, this result may be rewritten as follows. If $|f'(x)| \asymp \lambda_1$, then we have

$$\mathcal{R}(f, N, \delta) \ll N\lambda_1 + N\delta + \frac{\delta}{\lambda_1} + 1.$$

Proof If $4c_1\delta \geqslant 1$, then $4c_1 N\delta + 1 \geqslant N + 1 \geqslant \mathcal{R}(f, N, \delta)$ by (5.4). Similarly, if $2c_1\lambda_1 \geqslant 1$, then $2c_1 N\lambda_1 + 1 \geqslant N + 1 \geqslant \mathcal{R}(f, N, \delta)$. Therefore, we may suppose that $\max(4c_1\delta, 2c_1\lambda_1) < 1$. Let n and $n + a$ be any integers in $\mathcal{S}(f, N, \delta)$ with $a \in \mathbb{Z}_{\geqslant 1}$. Using the mean value theorem, we will prove that either

$$a > \frac{1}{2c_1\lambda_1} = a_1 \tag{5.8}$$

or

$$a < \frac{2\delta}{\lambda_1} = a_2. \tag{5.9}$$

We postpone the proof of these inequalities and assume that either (5.8) or (5.9) holds. Note that the condition $\max(4c_1\delta, 2c_1\lambda_1) < 1$ implies $a_1 > \max(1, a_2)$. Subdividing the interval $[N, 2N]$ into $s = \lfloor N/a_1 \rfloor + 1$ subintervals $I_1, \ldots, I_s$ of lengths $\leqslant a_1$, two elements of $\mathcal{S}(f, N, \delta) \cap I_j$ have a distance $\leqslant a_2$ and hence lie in an interval of length $\leqslant a_2$. We infer that

$$\left| \mathcal{S}(f, N, \delta) \cap I_j \right| \leqslant a_2 + 1$$

and thus

$$\mathcal{R}(f, N, \delta) \leqslant \left(\frac{N}{a_1} + 1 \right)(a_2 + 1) = 2c_1 N \lambda_1 + 4c_1 N \delta + \frac{2\delta}{\lambda_1} + 1.$$

The rest of the text is devoted to the proof of the inequalities (5.8) and (5.9). Since $n, n + a \in \mathcal{S}(f, N, \delta)$, there exist two integers m_1 and m_2 and two real numbers δ_1 and δ_2 such that

$$f(n) = m_1 + \delta_1 \quad \text{and} \quad f(n + a) = m_2 + \delta_2,$$

with $|\delta_i| < \delta$ for $i \in \{1, 2\}$. Thus there exist $m_3 \in \mathbb{Z}$ and $\delta_3 \in \mathbb{R}$ such that

$$f(n + a) - f(n) = m_3 + \delta_3,$$

with $|\delta_3| < 2\delta$. By the mean value theorem, there exists $t \in (n, n + a)$ such that

$$f(n + a) - f(n) = a f'(t).$$

Since $n + a \in \mathcal{S}(f, N, \delta)$, we have $n + a \leqslant 2N$ and thus $t \in [N, 2N]$. Hence, there exist $t \in [N, 2N]$, $m \in \mathbb{Z}$ and δ_3 such that $|\delta_3| < 2\delta < \frac{1}{2}$ satisfying

$$a f'(t) = m + \delta_3.$$

Now two cases may occur.

▷ $m \neq 0$. Since $m \in \mathbb{Z}$, we have $|m| \geqslant 1$ and then using (5.7)

$$a c_1 \lambda_1 \geqslant a |f'(t)| \geqslant |m| - |\delta_3| > 1 - \tfrac{1}{2} = \tfrac{1}{2},$$

which gives (5.8).

▷ $m = 0$. Then we have using (5.7) again

$$a\lambda_1 \leqslant a|f'(t)| = |\delta_3| < 2\delta,$$

which gives (5.9) as required.   □

We use Lemma 5.3 with $A = x^{1/3}$ and $B = 2x^{1/2}$. By (5.3), we have $A > 2y^{1/2}$ and hence

$$\sum_{x<n\leqslant x+y} \mu_2(n) = \frac{y}{\zeta(2)} + O\left(R_1\left(x^{1/3}, 2x^{1/2}\right)\log x + x^{1/3}\right),$$

and the use of Theorem 5.3 and (5.3) implies that

$$R_1\left(x^{1/3}, 2x^{1/2}\right) \ll \max_{x^{1/3}<N\leqslant 2x^{1/2}} \left(\frac{x}{N^2} + \frac{y}{N} + \frac{y}{\sqrt{x}} + 1\right) \ll x^{1/3},$$

and thus

$$\sum_{x<n\leqslant x+y} \mu_2(n) = \frac{y}{\zeta(2)} + O\left(x^{1/3}\log x\right),$$

and hence this result does not improve on the trivial estimate of Remark 5.3 in that problem.

5.2.2   The Second Derivative Test

The next result, coming from a combinatorial argument, will enable us to pass from the first derivative to the second derivative of f. In this section, δ is supposed to verify the inequalities

$$0 < \delta < \tfrac{1}{8}.$$

5.2.2.1   The Reduction Principle

Lemma 5.7 *Let* $f : [N, 2N] \to \mathbb{R}$ *be any map,* $A \in \mathbb{R}$ *such that* $1 \leqslant A \leqslant N$ *and, for all integers* $a \in [1, A]$, *we define on* $[N, 2N - a]$ *the function* $\Delta_a f$ *by*

$$\Delta_a f(x) = f(x + a) - f(x).$$

Then

$$\mathcal{R}(f, N, \delta) \leqslant \frac{N}{A} + \sum_{a \leqslant A} \mathcal{R}(\Delta_a f, N, 2\delta) + 1.$$

Proof For all $a \in \mathbb{Z}_{\geqslant 1}$, define

$$S(a) = \{n \in [N, 2N] \cap \mathbb{Z} : n \text{ and } n + a \text{ are consecutive in } S(f, N, \delta)\}.$$

▷ We first prove that

$$\mathcal{R}(f, N, \delta) \leqslant \frac{N}{A} + \sum_{a \leqslant A} |S(a)| + 1. \tag{5.10}$$

Each integer of $S(f, N, \delta)$, except the largest one, has a successive element and then lies in only one subset $S(a)$, so that

$$\mathcal{R}(f, N, \delta) = \sum_{a=1}^{\infty} |S(a)| + 1 = \sum_{a \leqslant A} |S(a)| + \sum_{a > A} |S(a)| + 1$$

with $A \in \mathbb{R}$ satisfying $1 \leqslant A \leqslant N$. Now if $S(f, N, \delta) = \{n_1 \leqslant n_2 \leqslant \cdots \leqslant n_k\}$ and if we set

$$d_1 = n_2 - n_1, \ d_2 = n_3 - n_2, \ldots, d_{k-1} = n_k - n_{k-1},$$

then, for all $a \in \mathbb{Z}_{\geqslant 1}$, $|S(a)|$ is the number of indexes $j \in \{1, \ldots, k - 1\}$ such that $d_j = a$, and thus

$$\sum_{a=1}^{\infty} a \, |S(a)| = \sum_{j=1}^{k-1} d_j = \sum_{j=1}^{k-1} (n_{j+1} - n_j) = n_k - n_1 \leqslant N.$$

Therefore

$$N \geqslant \sum_{a=1}^{\infty} a \, |S(a)| \geqslant \sum_{a > A} a \, |S(a)| \geqslant A \sum_{a > A} |S(a)|,$$

which implies (5.10).
▷ Now let $n \in S(a)$. Then n and $n + a$ are consecutive in $S(f, N, \delta)$, so that

$$\|\Delta_a f(n)\| = \|f(n + a) - f(n)\| \leqslant \|f(n + a)\| + \|f(n)\| < 2\delta,$$

and hence $n \in \mathcal{S}(\Delta_a f, N, 2\delta)$, which gives

$$|\mathcal{S}(a)| \leqslant \mathcal{R}(\Delta_a f, N, 2\delta).$$

Inserting this bound in (5.10) gives the asserted result.   □

5.2.2.2 The Main Result

If $f \in C^2$ satisfies $|f''| \asymp \lambda_2$, then, by the mean value theorem, we have $|(\Delta_a f)'(x)| \asymp a\lambda_2$. We may apply Theorem 5.3 to $\Delta_a f$ and use the previous lemma to go back to f. These ideas provide the following criterion.

Theorem 5.4 (Second Derivative Test) *Let $f \in C^2[N, 2N]$ such that there exist $\lambda_2 > 0$ and $c_2 \geqslant 1$ such that, for all $x \in [N, 2N]$, we have*

$$\lambda_2 \leqslant |f''(x)| \leqslant c_2 \lambda_2 \tag{5.11}$$

and

$$N\lambda_2 \geqslant c_2^{-1}. \tag{5.12}$$

Then

$$\mathcal{R}(f, N, \delta) \leqslant 6 \left\{ (3c_2)^{1/3} \, N\lambda_2^{1/3} + (12c_2)^{1/2} \, N\delta^{1/2} + 1 \right\}.$$

In practice, one may use this result in the following way. If

$$|f''(x)| \asymp \lambda_2 \quad \text{and} \quad N\lambda_2 \gg 1,$$

then

$$\mathcal{R}(f, N, \delta) \ll N\lambda_2^{1/3} + N\delta^{1/2} + 1.$$

Proof If $\lambda_2 \geqslant (3c_2)^{-1}$, then $(3c_2)^{1/3} \, N\lambda_2^{1/3} \geqslant N + 1 \geqslant \mathcal{R}(f, N, \delta)$ by (5.4). Similarly, if $\delta \geqslant (12c_2)^{-1}$, then $(12c_2)^{1/2} \, N\delta^{1/2} + 1 \geqslant N + 1$. Henceforth, we assume that

$$0 < \lambda_2 < (3c_2)^{-1} \quad \text{and} \quad 0 < \delta < (12c_2)^{-1}. \tag{5.13}$$

Let $A \in \mathbb{R}$ such that $1 \leqslant A \leqslant N$. For all $x \in [N, 2N]$ and all $a \in [1, A] \cap \mathbb{Z}$ such that $x + a \in [N, 2N]$, the mean value theorem gives

$$(\Delta_a f)'(x) = f'(x + a) - f'(x) = af''(t),$$

for some $t \in (x, x + a)$. Since $(x, x + a) \subseteq [N, 2N]$, (5.11) implies that, for all $x \in [N, 2N]$ and all $a \in [1, A] \cap \mathbb{Z}$ such that $x + a \in [N, 2N]$, we have

$$a\lambda_2 \leqslant \left|(\Delta_a f)'(x)\right| \leqslant c_2 a\lambda_2.$$

Therefore, using Lemma 5.7 and Theorem 5.3, we get

$$\mathcal{R}(f, N, \delta) \leqslant \frac{N}{A} + \sum_{a \leqslant A} \left(2c_2 N a\lambda_2 + 8c_2 N\delta + \frac{4\delta}{a\lambda_2} + 1 \right) + 1,$$

and (5.12) implies that $1 \leqslant c_2 N a\lambda_2$ and

$$4c_2 N\delta \geqslant 4\delta\lambda_2^{-1} \geqslant 4\delta(a\lambda_2)^{-1},$$

for all $a \geqslant 1$, so that

$$\mathcal{R}(f, N, \delta) \leqslant \frac{N}{A} + \sum_{a \leqslant A} (3c_2 N a\lambda_2 + 12c_2 N\delta) + 1$$

$$\leqslant \frac{N}{A} + 3c_2 A^2 N\lambda_2 + 12c_2 N A\delta + 1.$$

Now Lemma 5.6 implies that

$$\mathcal{R}(f, N, \delta) \leqslant 3N \left\{ (3c_2\lambda_2)^{1/3} + 3c_2\lambda_2 + (12c_2\delta)^{1/2} + 12c_2\delta \right\} + 6,$$

and the proof is achieved with the use of (5.13).   □

Example 5.1 We use Lemma 5.3 with $A = 2x^{1/4}$ and $B = x^{1/3}$. By (5.3), we have $A > 2y^{1/2}$ and hence

$$\sum_{x < n \leqslant x+y} \mu_2(n) = \frac{y}{\zeta(2)} + O\left\{ R_1\left(2x^{1/4}, x^{1/3}\right) \log x + R_2\left(x^{1/3}\right) \log x + x^{1/4} \right\}.$$

The condition (5.12) is satisfied by the two functions in their respective ranges of summation, and Theorem 5.8 gives

$$R_1\left(2x^{1/4}, x^{1/3}\right) \ll \max_{2x^{1/4} < N \leqslant x^{1/3}} \left(\left(xN^{-1}\right)^{1/3} + y^{1/2} \right) \ll x^{1/4}$$

$$R_2\left(x^{1/3}\right) \ll \max_{N \leqslant 2x^{1/3}} \left((Nx)^{1/6} + y^{1/2}\left(N^3 x^{-1}\right)^{1/4}\right) \ll x^{2/9} + y^{1/2},$$

and thus

$$\sum_{x < n \leqslant x+y} \mu_2(n) = \frac{y}{\zeta(2)} + O\left(x^{1/4} \log x\right).$$

Thus there exists a constant $c_0 > 0$ such that, if $c_0 x^{1/4} \log x \leqslant y < \frac{1}{4} x^{1/2}$, the interval $(x, x + y]$ contains a squarefree number.

5.2.3 The kth Derivative Test

We now intend to generalize the results of the previous sections. To this end, we need a generalization of the mean value theorem. The first idea is to use the Taylor–Lagrange formula, which generalizes Theorem 5.2 by providing more terms if f has higher derivatives. This was done by Konyagin in [11] where the author also used properties of lattices. In what follows, we rather focus on a generalization of the mean value theorem in the number of points that a function may interpolate. Thus, the divided differences are the main tools in this section.

We first start with a very easy lemma.

Lemma 5.8 *Let $n = x + y \in \mathbb{Z}$ with $x, y \in \mathbb{R}$ such that $|y| < |x|$. Then $|x| > \frac{1}{2}$.*

Proof Since $|y| < |x|$, we have $n \neq 0$ and hence $1 \leqslant |n| \leqslant |x| + |y| < 2|x|$ as asserted. ☐

We now are in a position to show the main result of this section.

Proposition 5.1 *Let $k \geqslant 1$ be an integer and $f \in C^k[N, 2N]$ such that there exist $\lambda_k > 0$ and $c_k \geqslant 1$ such that, for all $x \in [N, 2N]$, we have*

$$\lambda_k \leqslant \left|f^{(k)}(x)\right| \leqslant c_k \lambda_k. \tag{5.14}$$

Assume also that

$$(k + 1)! \, \delta < \lambda_k. \tag{5.15}$$

If $\alpha_k = 2k \, (2c_k)^{\frac{2}{k(k+1)}}$, then

$$\mathcal{R}(f, N, \delta) \leqslant \alpha_k N \lambda_k^{\frac{2}{k(k+1)}} + 2k.$$

As usual, this result is mostly used in the following form. If

$$\left| f^{(k)}(x) \right| \asymp \lambda_k \quad \text{and} \quad \delta \ll \lambda_k,$$

then

$$\mathcal{R}(f, N, \delta) \ll N \lambda_k^{\frac{2}{k(k+1)}} + 1.$$

Proof If $\lambda_k \geqslant \frac{1}{2}$, then

$$\alpha_k N \lambda_k^{\frac{2}{k(k+1)}} + 1 \geqslant N + 1 \geqslant \mathcal{R}(f, N, \delta),$$

so that we may suppose that $\lambda_k < \frac{1}{2}$. We generalize the proof of Theorem 5.3 in the following way. Take $k + 1$ consecutive points $n < n + a_1 < n + a_2 < \cdots < n + a_k$ in $\mathcal{S}(f, N, \delta)$, and we will prove that

$$a_k \geqslant 2k\,\alpha_k^{-1} \lambda_k^{-\frac{2}{k(k+1)}}. \tag{5.16}$$

Assume first that (5.16) is true. Taking each $(k + 1)$th element of $\mathcal{S}(f, N, \delta)$, we may construct a subset T of $\mathcal{S}(f, N, \delta)$ such that any two distinct elements of T differ by more than $d_k = 2k\,\alpha_k^{-1}\lambda_k^{-\frac{2}{k(k+1)}}$ and therefore

$$\mathcal{R}(f, N, \delta) \leqslant (k + 1)\,(|T| + 1) \leqslant 2k \left(\frac{N}{d_k} + 1 \right),$$

giving the asserted result. *The rest of the text is devoted to the proof of* (5.16). By definition, there exist integers $m_0, \ldots, m_k$ and real numbers $\delta_0, \ldots, \delta_k$ such that

$$f(n + a_j) = m_j + \delta_j$$

with $a_0 = 0$ and $\left| \delta_j \right| < \delta$ for all $j \in \{0, \ldots, k\}$. Using (5.6), there exists $t \in (n, n + a_k)$ such that

$$\sum_{j=0}^{k} \frac{m_j + \delta_j}{\prod\limits_{\substack{0 \leqslant i \leqslant k \\ i \neq j}} (a_j - a_i)} = \frac{f^{(k)}(t)}{k!} \tag{5.17}$$

and by (5.5) and what follows, if $P = b_k X^k + \cdots + b_0$ is the Lagrange polynomial interpolating the points $(n + a_j, m_j)$, then we have

$$b_k = \sum_{j=0}^{k} \frac{m_j}{\prod_{\substack{0 \leqslant i \leqslant k \\ i \neq j}} (a_j - a_i)} = \frac{A_k}{D_k}, \tag{5.18}$$

for some $A_k \in \mathbb{Z}$ and where $D_k = \prod_{0 \leqslant i < j \leqslant k} (a_j - a_i) > 0$. Hence by (5.17), we get

$$b_k = \frac{f^{(k)}(t)}{k!} - \sum_{j=0}^{k} \frac{\delta_j}{\prod_{\substack{0 \leqslant i \leqslant k \\ i \neq j}} (a_j - a_i)},$$

and then

$$k! \, A_k = k! \, D_k \, b_k = D_k f^{(k)}(t) - k! \, D_k \sum_{j=0}^{k} \frac{\delta_j}{\prod_{\substack{0 \leqslant i \leqslant k \\ i \neq j}} (a_j - a_i)} = x + y.$$

Now the condition $\left| \delta_j \right| < \delta$ implies

$$|y| < k! \, D_k \, \delta \sum_{j=0}^{k} \frac{1}{\prod_{\substack{0 \leqslant i \leqslant k \\ i \neq j}} |a_j - a_i|} \leqslant (k+1)! \, D_k \, \delta$$

where we used the fact that $|a_j - a_i| \geqslant 1$ for all $i \neq j$, and using (5.14) and (5.15), we get

$$|y| < D_k \lambda_k \leqslant D_k \left| f^{(k)}(t) \right| = |x|.$$

Lemma 5.8 gives $|x| > \frac{1}{2}$ and the bounds $D_k \leqslant a_k^{\frac{k(k+1)}{2}}$ and (5.14) imply

$$\frac{1}{2} < D_k \left| f^{(k)}(t) \right| \leqslant a_k^{\frac{k(k+1)}{2}} c_k \lambda_k = \frac{1}{2} \lambda_k \left((2k)^{-1} a_k \alpha_k \right)^{\frac{k(k+1)}{2}},$$

which implies (5.16), concluding the proof.   □

This result does generalize Theorem 5.3, but the proof highlights the following weakness. The lower bound $|a_j - a_i| \geqslant 1$ may probably be improved in many special cases. This would enable us to decrease the upper bound for $|y|$ and then to have a condition less restrictive than (5.15). For instance, in the squarefree number problem, (5.15) requires to estimate R_1 and R_2 of Lemma 5.3 in the range

$$N < 8^{-1} (x/y)^{1/3},$$

which does not allow us to cover all the range of summation.

5.3   The Theorem of Huxley and Sargos

By the remark above, establishing a kth derivative criterion free from condition (5.15) arises naturally. This was done by Huxley and Sargos in [8] who proved the important result we shall see in this section. This turns out to be rather difficult, and several tools are needed in the proof. We first state the main result and then provide the proof in several steps. We shall essentially follow the line of [8], but our exposition may differ in certain minor points.[1]

Theorem 5.5 (Huxley–Sargos) *Let $k \geqslant 3$ be an integer and $f \in C^k[N, 2N]$ such that there exist $\lambda_k > 0$ and $c_k \geqslant 1$ such that, for all $x \in [N, 2N]$, we have*

$$\lambda_k \leqslant \left| f^{(k)}(x) \right| \leqslant c_k \lambda_k. \tag{5.19}$$

Let δ be a real number satisfying (5.2). Then

$$\mathcal{R}(f, N, \delta) \leqslant \alpha_k N \lambda_k^{\frac{2}{k(k+1)}} + \beta_k N \delta^{\frac{2}{k(k-1)}} + 8k^3 \left(\frac{\delta}{\lambda_k} \right)^{1/k} + 2k^2 \left(5e^3 + 1 \right),$$

where

$$\alpha_k = 2k^2 c_k^{\frac{2}{k(k+1)}} \quad \text{and} \quad \beta_k = 4k^2 \left(5e^3 c_k^{\frac{2}{k(k-1)}} + 1 \right).$$

As for the previous results, this inequality is used as

$$\mathcal{R}(f, N, \delta) \ll N \lambda_k^{\frac{2}{k(k+1)}} + N \delta^{\frac{2}{k(k-1)}} + \left(\frac{\delta}{\lambda_k} \right)^{1/k} + 1$$

---

[1]For instance, the authors did not make use of Gorny's inequality but proved a Landau–Hadamard–Kolmogorov-like result similar to (5.20) using divided differences.

under the sole hypothesis (5.19), where the implied constants depend only on k and c_k. We shall make use of the following additional notation:

$$C_\delta = \{(x, y) \in [N, 2N] \times \mathbb{R} : |y - f(x)| < \delta\}.$$

5.3.1 Preparatory Lemmas

The first tool is an easy enumeration principle.

Lemma 5.9 *Let S be a finite set of integers with length $\leqslant N$. If one can recover S by pairwise distinct intervals $\mathcal{I}$ and if $L(\mathcal{I})$ is the length of $\mathcal{I}$, then*

$$|S| \leqslant N \max_{\mathcal{I}} \left(\frac{|S \cap \mathcal{I}|}{L(\mathcal{I})} \right) + 2 \max_{\mathcal{I}} |S \cap \mathcal{I}|.$$

Proof Let $\mathcal{I}_1, \ldots, \mathcal{I}_J$ be such a covering of S where we suppose that $h \neq j \Rightarrow \mathcal{I}_h \cap \mathcal{I}_j = \varnothing$ and, if $2 \leqslant j \leqslant J - 1$, then $\mathcal{I}_j \subseteq S$. We have

$$|S| \leqslant \sum_{j=1}^{J} |S \cap \mathcal{I}_j| = \sum_{j=2}^{J-1} \left\{ L(\mathcal{I}_j) \times \frac{|S \cap \mathcal{I}_j|}{L(\mathcal{I}_j)} \right\} + |S \cap \mathcal{I}_1| + |S \cap \mathcal{I}_J|$$

$$\leqslant \max_{1 \leqslant j \leqslant J} \left(\frac{|S \cap \mathcal{I}_j|}{L(\mathcal{I}_j)} \right) \sum_{j=2}^{J-1} L(\mathcal{I}_j) + 2 \max_{1 \leqslant j \leqslant J} |S \cap \mathcal{I}_j|$$

$$\leqslant N \max_{1 \leqslant j \leqslant J} \left(\frac{|S \cap \mathcal{I}_j|}{L(\mathcal{I}_j)} \right) + 2 \max_{1 \leqslant j \leqslant J} |S \cap \mathcal{I}_j|,$$

where in the last inequality we used the fact that the intervals are pairwise distinct.
$\square$

The next tool belongs to a certain class of inequalities, called the *Landau–Hadamard–Kolmogorov inequalities*. In 1913, E. Landau proved that, if I is an interval with length $\geqslant 2$ and $f \in C^2(I)$ satisfies the conditions $|f(x)| \leqslant 1$ and $|f''(x)| \leqslant 1$ on I, then $|f'(x)| \leqslant 2$ and the constant 2 is the best possible. This was generalized by Hadamard who showed that, if $a \in \mathbb{R}$ and $L > 0$, then

$$\sup_{a \leqslant x \leqslant a+L} |f'(x)| \leqslant \frac{2}{L} \sup_{a \leqslant x \leqslant a+L} |f(x)| + \frac{L}{2} \sup_{a \leqslant x \leqslant a+L} |f''(x)|.$$

The generalization to higher orders of derivative was established by several mathematicians. In 1930, L. Neder proved that, if $a \in \mathbb{R}$, $L > 0$ and $f \in C^k[a, a + L]$, then, for all $j \in \{1, \ldots, k - 1\}$, we have

$$\sup_{a \leqslant x \leqslant a+L} |f^{(j)}(x)| \leqslant (2k)^{2k} L^{-j} \sup_{a \leqslant x \leqslant a+L} |f(x)| + L^{k-j} \sup_{a \leqslant x \leqslant a+L} |f^{(k)}(x)|.$$

(5.20)

This result would be sufficient for our proof of Theorem 5.5, but, for the sake of completeness, we mention the following improvement due to Gorny [4].

Lemma 5.10 (Gorny) *Let $k \geqslant 2$ be an integer, $a \in \mathbb{R}$, $L > 0$ and $f \in C^k[a, a + L]$ such that, for all $x \in [a, a + L]$, we have*

$$|f(x)| \leqslant M_0 \quad and \quad |f^{(k)}(x)| \leqslant M_k,$$

with $M_0 < \infty$ and $M_k < \infty$. Then, for all $x \in [a, a + L]$ and $j \in \{1, \ldots, k - 1\}$, we have

$$\left| f^{(j)}(x) \right| < 4 \left(e^2 k/j \right)^j M_0^{1-j/k} \left\{ \max \left(M_k, \, k! \, M_0 L^{-k} \right) \right\}^{j/k}.$$

In practice, we will use this result in the following form:

$$\left| f^{(j)}(x) \right| < 4e \, (ek/j)^j \left\{ k^{j+1} M_0 L^{-j} + e^{j-1} M_0^{1-j/k} M_k^{j/k} \right\}. \tag{5.21}$$

Let us finally mention the following elegant version of Kolmogorov's inequality, which can be found in [13, Théorème 6.3.III].

Let $f \in C^k(\mathbb{R})$ such that $M_0 = \sup\limits_{x \in \mathbb{R}} |f(x)| < \infty$ and $M_k = \sup\limits_{x \in \mathbb{R}} |f^{(k)}(x)| < \infty$.
Then, for all $j \in \{0, \ldots, k\}$ and all $x \in \mathbb{R}$, we have

$$\left| f^{(j)}(x) \right| \leqslant 2 M_0^{1-j/k} M_k^{j/k}.$$

To end this section, we shall need the next technical tool to calculate the implied constants appearing in Theorem 5.5.

Lemma 5.11 *Let $k \geqslant 3$ be an integer and $a > e(k - 1)$ be a real number. Then*

$$\sum_{j=1}^{k-1} \left(\frac{a}{j} \right)^{2j} \leqslant e^2 k \left(\frac{a}{k} \right)^{2k-2}.$$

Proof Since $a > e(k-1)$, the function $x \mapsto (a/x)^{2x}$ is increasing as soon as $1 \leqslant x \leqslant k-1$ and then

$$\sum_{j=1}^{k-1} \left(\frac{a}{j}\right)^{2j} \leqslant (k-1) \left(\frac{a}{k-1}\right)^{2k-2} \leqslant e^2 k \left(\frac{a}{k}\right)^{2k-2}$$

as asserted. $\square$

5.3.2 Major Arcs

In this section, we take the notation above. Besides, we recall that $\lfloor x \rceil$ is the nearest integer to x.

Definition 5.3

▷ A *major arc* associated to $f^{(k)}$ is a maximal set $\mathcal{A} = \{n_1, \ldots, n_J\}$ of consecutive points of $\mathcal{S}(f, N, \delta)$, where $J \geqslant k^2 + 1$ is an integer, such that, for all $j \in \{1, \ldots, J\}$, we have

$$\lfloor f(n_j) \rceil = P(n_j),$$

where $P \in \mathbb{Q}[X]$ is the Lagrange polynomial of degree $< k$ interpolating the points $(n_j, \lfloor f(n_j) \rceil)$. The equation $y = P(x)$ is called the *equation* of $\mathcal{A}$. We set C_P the curve with equation $y = P(x)$.

▷ Let q be the smallest positive integer such that $P \in \frac{1}{q}\mathbb{Z}[X]$. Then q is called the *denominator* of $\mathcal{A}$.

The first result gives a bound for the number of connected components of the set $C_\delta \cap C_P$.

Lemma 5.12 *The set $C_\delta \cap C_P$ has at most k connected components.*

Proof Let $\widetilde{f} \in C^k(\mathbb{R})$ defined by $\widetilde{f}(x) = f(x)$ for $x \in [N, 2N]$, $\widetilde{f}^{(k)}(x) = f^{(k)}(N)$ if $x \leqslant N$ and $\widetilde{f}^{(k)}(x) = f^{(k)}(2N)$ if $x \geqslant 2N$, so that

$$\left| \widetilde{f}^{(k)}(x) \right| \asymp \lambda_k$$

for all $x \in \mathbb{R}$. Similarly, the set $\widetilde{C}_\delta$ is the analogue of C_δ for the function $\widetilde{f}$. Since the behaviours of $\widetilde{f}$ and P at infinity are different, the set $\widetilde{C}_\delta \cap C_P$ is bounded, and hence each connected component of this set, not reduced to a singleton, has two extremities satisfying $P(x) = f(x) \pm \delta$. Now assume that there exist $k+1$ connected components of $\widetilde{C}_\delta \cap C_P$, not reduced to a singleton. Then the equation

$P(x) = f(x) + \delta$, say, has $k + 1$ solutions $x_0 < \cdots < x_k$. Since the polynomials $P(X)$ and $P(X) - \delta$ have the same leading coefficient, we get by (5.6)

$$0 = \frac{P^{(k)}(t_1)}{k!} = \sum_{j=0}^{k} \frac{P(x_j)}{\prod_{\substack{0 \leqslant i \leqslant k \\ i \neq j}} (x_j - x_i)} = \sum_{j=0}^{k} \frac{P(x_j) - \delta}{\prod_{\substack{0 \leqslant i \leqslant k \\ i \neq j}} (x_j - x_i)}$$

$$= \sum_{j=0}^{k} \frac{f(x_j)}{\prod_{\substack{0 \leqslant i \leqslant k \\ i \neq j}} (x_j - x_i)} = \frac{f^{(k)}(t_2)}{k!} \neq 0,$$

for some $t_1, t_2 \in (x_0, x_k)$, giving a contradiction. $\square$

This result leads to the following slight refinement of Definition 5.3.

Definition 5.4 Among the connected components of the set $C_\delta \cap C_P$, choose the one having the largest number of points $(n_{h+j}, \lfloor f(n_{h+j}) \rceil)$ for all $j \in \{1, \ldots, \ell\}$ with $\ell > k$. Then the set

$$\overline{\mathcal{A}} = \{n_{h+1}, \ldots, n_{h+\ell}\}$$

is called a *proper major arc* extracted from $\mathcal{A}$. The *length* of $\overline{\mathcal{A}}$ is the number

$$L = n_{h+\ell} - n_{h+1}.$$

For convenience, we introduce the following numbers:

$$r_k = 36e^{-2}k\,(2e^3)^k c_k \quad \text{and} \quad s_k = 20e^3\,k^2 c_k^{\frac{2}{k(k-1)}} \tag{5.22}$$

so that $\beta_k = s_k + 4k^2$. The next result summarizes the basic properties of the major arcs.

Lemma 5.13 *Let $\mathcal{A}$ be a major arc associated to $f^{(k)}$ and $\overline{\mathcal{A}}$ be the proper major arc taken from $\mathcal{A}$ with denominator q, length L and equation $y = P(x)$.*

▷ *We have*

$$L \leqslant 2k \left(\frac{\delta}{\lambda_k}\right)^{1/k}.$$

▷ *We have*

$$|\overline{\mathcal{A}}| \leqslant 2kLq^{-\frac{2}{k(k-1)}}.$$

▷ If $q \leqslant (r_k \delta)^{-1}$, then the distance d between each point of $\mathcal{S}(f, N, \delta) \setminus \overline{\mathcal{A}}$ and $\overline{\mathcal{A}}$ satisfies

$$d > L \, (r_k q \delta)^{-1/k} \, ,$$

where r_k is defined in (5.22).

Proof

▷ Set $\overline{\mathcal{A}} = \{n_h, \dots, n_h + L\}$ and define $g(x) = f(x) - P(x)$. Let $x_j = n_h + jL/k$ for $j \in \{0, \dots, k\}$. Using (5.6), we get

$$\frac{f^{(k)}(t)}{k!} = \frac{g^{(k)}(t)}{k!} = \sum_{j=0}^{k} \frac{g(x_j)}{\prod_{\substack{0 \leqslant i \leqslant k \\ i \neq j}} (x_j - x_i)} = \left(\frac{k}{L}\right)^k \sum_{j=0}^{k} \frac{g(x_j)}{\prod_{\substack{0 \leqslant i \leqslant k \\ i \neq j}} (j - i)}$$

for some $t \in (x_0, x_k)$, and taking account of the bound $|g(x_j)| \leqslant \delta$ and (5.19), we obtain

$$\frac{\lambda_k}{k!} \leqslant \delta \left(\frac{k}{L}\right)^k \sum_{j=0}^{k} \frac{1}{(k-j)! \, j!} = \frac{\delta}{k!} \left(\frac{k}{L}\right)^k \sum_{j=0}^{k} \binom{k}{j} = \frac{2^k \delta}{k!} \left(\frac{k}{L}\right)^k,$$

which gives the required bound.

▷ Let $n_1 < \cdots < n_k$ be k points lying in $\overline{\mathcal{A}}$. By the Lagrange interpolation formula of Remark 5.4, we have

$$P(x) = \sum_{j=1}^{k} \left(\prod_{\substack{i=1 \\ i \neq j}}^{k} \frac{x - n_i}{n_j - n_i} \right) P(n_j),$$

and hence q divides $\displaystyle\prod_{1 \leqslant i < j \leqslant k} (n_j - n_i) \leqslant (n_k - n_1)^{\frac{k(k-1)}{2}}$ so that

$$L \geqslant n_k - n_1 \geqslant q^{\frac{2}{k(k-1)}},$$

implying the asserted estimate.

▷ Take $n \in \mathcal{S}(f, N, \delta) \setminus \overline{\mathcal{A}}$ and $n_0 \in \overline{\mathcal{A}}$. Without loss of generality, one may assume that $n > n_0$ and set $d = n - n_0$ and $m = \lfloor f(n) \rceil$. Since $m \neq P(n)$, we have

$$|P(n) - m| \geqslant \frac{1}{q} \geqslant \frac{1}{3q} + 2\delta,$$

since $q \leqslant (r_k \delta)^{-1} \leqslant (3\delta)^{-1}$, and thus, taking up again the function $g(x) = f(x) - P(x)$ defined above, we get

$$|g(n) - g(n_0)| \geqslant |g(n)| - |g(n_0)| \geqslant |g(n)| - \delta$$

$$= |P(n) - f(n)| - \delta$$

$$\geqslant |P(n) - m| - |f(n) - m| - \delta$$

$$\geqslant \frac{1}{3q} + 2\delta - \delta - \delta = \frac{1}{3q}.$$

On the other hand, the Taylor–Lagrange formula yields

$$g(n) - g(n_0) = \sum_{j=1}^{k-1} g^{(j)}(n_0) \frac{d^j}{j!} + g^{(k)}(t) \frac{d^k}{k!}$$

for some $t \in (n_0, n)$, and hence, using (5.19) and (5.21) applied to the function g with $M_0 = \delta$ and $M_k = c_k \lambda_k$, we get

$$|g(n) - g(n_0)| < 4e\delta \sum_{j=1}^{k-1} \left(\frac{ke}{j} \right)^j \frac{d^j}{j!} \left\{ k^{j+1} L^{-j} + e^{j-1} \left(\frac{c_k \lambda_k}{\delta} \right)^{j/k} \right\} + \frac{c_k \lambda_k d^k}{k!},$$

and using the previous estimate in the form

$$\left(\frac{\lambda_k}{\delta} \right)^{1/k} \leqslant \frac{2k}{L}$$

along with the easy inequality $j! > e(j/e)^j$ implies that $|g(n) - g(n_0)|$ does not exceed

$$< 4e\delta \sum_{j=1}^{k-1} \left(\frac{ke}{j} \right)^j \frac{1}{j!} \left\{ k^{j+1} + c_k^{j/k} e^{-1} (2ek)^j \right\} \left(dL^{-1} \right)^j + \frac{(2k)^k c_k \delta d^k}{k! L^k}$$

$$\leqslant 4\delta \left(dL^{-1} + \left(dL^{-1} \right)^{k-1} \right) \sum_{j=1}^{k-1} \left(\frac{ke}{j} \right)^{2j} \left\{ k + c_k^{j/k} e^{-1} (2e)^j \right\} + (2e)^k e^{-1} c_k \delta \left(dL^{-1} \right)^k$$

$$\leqslant 4\delta \left(dL^{-1} + \left(dL^{-1} \right)^{k-1} \right) \left(k + c_k 2^{k-1} e^{k-2} \right) \sum_{j=1}^{k-1} \left(\frac{ke}{j} \right)^{2j} + (2e)^k e^{-1} c_k \delta \left(dL^{-1} \right)^k$$

$$< 2^{k+2} e^{k-2} c_k \delta \left(dL^{-1} + \left(dL^{-1} \right)^{k-1} \right) \sum_{j=1}^{k-1} \left(\frac{ke}{j} \right)^{2j} + (2e)^k e^{-1} c_k \delta \left(dL^{-1} \right)^k,$$

where we used $c_k 2^{k-1} e^{k-2} \geqslant 2^{k-1} e^{k-2} > k$. Now Lemma 5.11 implies that

$$|g(n) - g(n_0)| < 4e^{-2} k \, (2e^3)^k c_k \, \delta \left(dL^{-1} + \left(dL^{-1} \right)^{k-1} \right) + (2e)^k e^{-1} c_k \, \delta \left(dL^{-1} \right)^k$$

$$= 9^{-1} r_k \, \delta \left(dL^{-1} + \left(dL^{-1} \right)^{k-1} \right) + (2e)^k e^{-1} c_k \, \delta \left(dL^{-1} \right)^k$$

$$\leqslant 9^{-1} r_k \, \delta \left(dL^{-1} + \left(dL^{-1} \right)^{k-1} + \left(dL^{-1} \right)^k \right).$$

Combining with the above inequality, we then get

$$q^{-1} < 3^{-1} r_k \, \delta \left(dL^{-1} + \left(dL^{-1} \right)^{k-1} + \left(dL^{-1} \right)^k \right),$$

so that

$$1 < r_k \, q \delta \max \left(dL^{-1}, \left(dL^{-1} \right)^{k-1}, \left(dL^{-1} \right)^k \right),$$

and hence

$$d > L \min \left\{ (r_k q \delta)^{-1}, \, (r_k q \delta)^{-\frac{1}{k-1}}, \, (r_k q \delta)^{-1/k} \right\},$$

and the inequality $(q\delta)^{-1} \geqslant r_k$ implies the statement of the lemma. $\qquad \square$

We now are in a position to estimate the contribution of the points coming from the major arcs.

Lemma 5.14 *Let R_0 be the contribution of the points coming from the major arcs associated to $f^{(k)}$ to the number $\mathcal{R}(f, N, \delta)$. Then*

$$R_0 \leqslant s_k N \delta^{\frac{2}{k(k-1)}} + 8k^3 \left(\frac{\delta}{\lambda_k} \right)^{1/k} + 10e^3 k^2,$$

where s_k is defined in (5.22).

Proof Let $\mathcal{M}_0$ be the set of major arcs and $Q_k > 0$ be the real number defined by

$$Q_k = (r_k \delta)^{-1},$$

where r_k is given in (5.22). Write $\mathcal{M}_0 = \mathcal{M}_1 \cup \mathcal{M}_2$, where $\mathcal{M}_1$ is the set of major arcs with denominator $> Q_k$ and $\mathcal{M}_2 = \mathcal{M}_0 \setminus \mathcal{M}_1$. For $i \in \{1, 2\}$, let

$$S_i = \bigcup_{\mathcal{A} \in \mathcal{M}_i} \mathcal{A} \quad \text{and} \quad R_i = |S_i|.$$

Using Lemma 5.12, we infer that $R_0 \leqslant k \, (R_1 + R_2)$.

▷ **Estimate of R_1.** Take $2k-1$ consecutive points of S_1. One may take k consecutive points $n_1, \ldots, n_k$ from the same proper major arc with denominator $q > Q_k$. As in Lemma 5.13, we deduce that $n_k - n_1 \geqslant q^{\frac{2}{k(k-1)}} > Q_k^{\frac{2}{k(k-1)}}$ so that

$$R_1 \leqslant 2k \left(N Q_k^{-\frac{2}{k(k-1)}} + 1 \right) < 10 e^3 k \left(N \, (c_k \delta)^{\frac{2}{k(k-1)}} + 1 \right) = s_k (2k)^{-1} N \delta^{\frac{2}{k(k-1)}} + 10 e^3 k.$$

▷ **Estimate of R_2.** Without loss of generality, we may assume that $Q_k \geqslant 1$, otherwise $S_2 = \varnothing$. Let $\overline{\mathscr{A}}_1, \ldots, \overline{\mathscr{A}}_J$ be the ordered sequence of proper major arcs with denominator $q_j \leqslant Q_k$. For each proper major arc $\overline{\mathscr{A}}_j$, we set n_j and L_j its first point and length and define

$$d_j = L_j \left(r_k q_j \delta \right)^{-1/k}$$

and $\mathcal{I}_j = [n_j, n_j + d_j]$. We claim that $\mathcal{I}_j$ contains $\overline{\mathscr{A}}_j$ and does not contain $\overline{\mathscr{A}}_{j+1}$. Indeed, observe that:

◇ $\overline{\mathscr{A}}_j$ is lying in an interval of length L_j and since $q_j \leqslant Q_k$, we have

$$d_j \geqslant L_j \left(r_k Q_k \delta \right)^{-1/k} = L_j$$

so that $\mathcal{I}_j$ contains $\overline{\mathscr{A}}_j$.
◇ Assume that there exists an element of $\overline{\mathscr{A}}_{j+1}$ belonging to $\mathcal{I}_j$. Then the distance d between this element and $\overline{\mathscr{A}}_j$ satisfies $d \leqslant d_j$, contradicting Lemma 5.13.

Therefore, the intervals $\mathcal{I}_j$ are pairwise distinct, and using Lemma 5.9 with $S = S_2$, we get

$$R_2 \leqslant N \max_j \frac{|\overline{\mathscr{A}}_j|}{d_j} + 2 \max_j |\overline{\mathscr{A}}_j|.$$

Now by Lemma 5.13 and the choice of d_j, we have

$$\frac{|\overline{\mathscr{A}}_j|}{d_j} \leqslant 2k L_j q_j^{-\frac{2}{k(k-1)}} L_j^{-1} (r_k q_j \delta)^{1/k} = 2k \, (r_k \delta)^{1/k} \, q_j^{\frac{k-3}{k(k-1)}},$$

and since $q_j \leqslant Q_k = (r_k \delta)^{-1}$ and $k \geqslant 3$, we obtain

$$\frac{|\overline{\mathscr{A}}_j|}{d_j} \leqslant 2k \, (r_k \delta)^{\frac{2}{k(k-1)}} < 10 e^3 k \, (c_k \delta)^{\frac{2}{k(k-1)}} = s_k (2k)^{-1} \delta^{\frac{2}{k(k-1)}}$$

for $j \in \{1, \ldots, J\}$, and therefore

$$R_2 \leqslant s_k(2k)^{-1} N\delta^{\frac{2}{k(k-1)}} + 4k \max_j L_j \leqslant s_k(2k)^{-1} N\delta^{\frac{2}{k(k-1)}} + 8k^2 \left(\frac{\delta}{\lambda_k}\right)^{1/k}$$

by Lemma 5.13. The proof is complete.   □

5.3.3 The Proof of Theorem 5.5

We first need to estimate the contribution of the points of $S(f, N, \delta)$, which do not come from major arcs. The proof of the next result, which provides such an estimate, is similar to that of Theorem 5.1.

Lemma 5.15 *Let $N \leqslant n_0 < \cdots < n_k \leqslant 2N$ be $k + 1$ points of $S(f, N, \delta)$, which do not lie on the same algebraic curve of degree $< k$. Then*

$$n_k - n_0 > \min\left((c_k\lambda_k)^{-\frac{2}{k(k+1)}},\ 2^{-1}\delta^{-\frac{2}{k(k-1)}}\right).$$

Proof As in the proof of Theorem 5.1, there exist integers $m_0, \ldots, m_k$ and real numbers $\delta_0, \ldots, \delta_k$ such that

$$f(n_j) = m_j + \delta_j,$$

with $|\delta_j| < \delta$ for all $j \in \{0, \ldots, k\}$. We take up again the number

$$D_k = \prod_{0 \leqslant h < i \leqslant k} (n_i - n_h) > 0,$$

and if $P = b_k X^k + \cdots + b_0$ is the Lagrange polynomial interpolating the points $\langle n_j, m_j \rangle$, then we have

$$b_k = \sum_{j=0}^{k} \frac{m_j}{\prod\limits_{\substack{0 \leqslant i \leqslant k \\ i \neq j}} (n_j - n_i)} = \frac{A_k}{D_k},$$

where $A_k \in \mathbb{Z}$ is analogous to the number in (5.18). Reasoning exactly in the same way as in the proof of Theorem 5.1, we get

$$k!\, A_k = D_k f^{(k)}(t) - k!\, D_k \sum_{j=0}^{k} \frac{\delta_j}{\prod\limits_{\substack{0 \leqslant i \leqslant k \\ i \neq j}} (n_j - n_i)}.$$

Now since the points $\langle n_j, m_j \rangle$ do not all lie on the same algebraic curve of degree $< k$, we have $b_k \neq 0$ and hence $|A_k| \geqslant 1$, and using $|\delta_j| < \delta$ and (5.19), we get

$$k! \leqslant k! |A_k| < c_k \lambda_k \, D_k + k! \, \delta \, D_k \sum_{j=0}^{k} \frac{1}{\prod_{\substack{0 \leqslant i \leqslant k \\ i \neq j}} |n_j - n_i|}$$

$$= c_k \lambda_k \, D_k + k! \, \delta \sum_{j=0}^{k} \prod_{\substack{0 \leqslant h < i \leqslant k \\ h \neq j, \, i \neq j}} (n_i - n_h)$$

$$\leqslant c_k \lambda_k \, (n_k - n_0)^{\frac{k(k+1)}{2}} + (k+1)! \, \delta \, (n_k - n_0)^{\frac{k(k-1)}{2}},$$

implying that

$$n_k - n_0 > \min \left(\left(\frac{k!}{2} \right)^{\frac{2}{k(k+1)}} (c_k \lambda_k)^{-\frac{2}{k(k+1)}} , \; (2k+2)^{-\frac{2}{k(k-1)}} \delta^{-\frac{2}{k(k-1)}} \right),$$

which is slightly better than the asserted lower bound.   □

Proof (Proof of Theorem 5.5) Let S_0 be the set of the points of $S(f, N, \delta)$ coming from the major arcs and $T_0 = S(f, N, \delta) \setminus S_0$. By Lemma 5.14, we have

$$|S_0| = R_0 \leqslant s_k N \delta^{\frac{2}{k(k-1)}} + 8k^3 \left(\frac{\delta}{\lambda_k} \right)^{1/k} + 10 e^3 k^2,$$

where s_k is given in (5.22). Now let $G = \{n_0, \ldots, n_{k^2}\}$ be a set of $k^2 + 1$ consecutive ordered points of T_0. Since G is not contained in any major arc, one may find an integer $j \in \{k, \ldots, k^2\}$ such that the $j + 1$ points $\langle n_i, m_i \rangle$ do not lie on the same algebraic curve of degree $< k$. By Lemma 5.15, we have

$$n_{k^2} - n_0 \geqslant n_j - n_0 \geqslant n_k - n_0 > \min \left((c_k \lambda_k)^{-\frac{2}{k(k+1)}} , \; 2^{-1} \delta^{-\frac{2}{k(k-1)}} \right),$$

implying that

$$|T_0| \leqslant 2k^2 \left(N (c_k \lambda_k)^{\frac{2}{k(k+1)}} + 2N \delta^{\frac{2}{k(k-1)}} + 1 \right),$$

and we conclude the proof using $\mathcal{R}(f, N, \delta) \leqslant |T_0| + |S_0|$.   □

5.3.4 Application

We return to the squarefree number problem and intend to use Theorem 5.5 in order to bound the sum S_2. To this end, suppose first that $y \leqslant x^{4/9}$ and we use Lemma 5.3 with $A = 2x^{2/9}$ and $B = x^{1/3}$. We take up again the estimate of $R_2\left(x^{1/3}\right)$ we obtained in Example 5.1, namely

$$R_2\left(x^{1/3}\right) \ll x^{2/9} + yx^{-2/9}.$$

Furthermore, we use Huxley–Sargos's result with $k = 3$ for $R_1\left(2x^{2/9}, x^{1/3}\right)$, which gives

$$R_1\left(2x^{2/9}, x^{1/3}\right) \ll \max_{2x^{2/9} < N \leqslant x^{1/3}} \left\{ (Nx)^{1/6} + (Ny)^{1/3} + N\left(yx^{-1}\right)^{1/3} \right\}$$

$$\ll x^{2/9} + x^{1/9}y^{1/3}.$$

Thus, if $y \leqslant x^{1/3}$, we get

$$\sum_{x < n \leqslant x+y} \mu_2(n) = \frac{y}{\zeta(2)} + O\left(x^{2/9} \log x\right),$$

so that there exists a constant $c_0 > 0$ such that, if $c_0 x^{2/9} \log x \leqslant y < \frac{1}{4}x^{1/2}$, the interval $(x, x + y]$ contains a squarefree number.

5.3.5 Refinements

The refinements of Theorem 5.5 have been made in several directions. First, it has been shown that this result also holds for $k = 2$ (see [1, 8]).

Theorem 5.6 (Branton–Sargos) *Let $f \in C^2[N, 2N]$ such that there exist $\lambda_2 > 0$ such that, for all $x \in [N, 2N]$, we have*

$$\left| f''(x) \right| \asymp \lambda_2.$$

Then

$$\mathcal{R}(f, N, \delta) \ll N\lambda_2^{1/3} + N\delta + \left(\frac{\delta}{\lambda_2}\right)^{1/2} + 1.$$

(continued)

Theorem 5.6 (continued)
If in addition there exists $\lambda_1 > 0$ such that $\left|f'(x)\right| \asymp \lambda_1$ for all $x \in [N, 2N]$, then

$$\mathcal{R}(f, N, \delta) \ll N\lambda_2^{1/3} + N\delta + \lambda_1 \left(\frac{\delta}{\lambda_2}\right)^{1/2} + \frac{\delta}{\lambda_1} + 1.$$

In Theorem 5.5, $N\lambda_k^{\frac{2}{k(k+1)}}$ is the main term, also sometimes called the *smoothness term*, and the others are the secondary terms. It is very difficult to improve on the main term, and the quantity $\left(\delta\lambda_k^{-1}\right)^{1/k}$ is quasi-optimal. Thus, one may wonder whether the term $N\delta^{\frac{2}{k(k-1)}}$ may be improved, since, when δ is small, it increases rapidly as k grows. In [9], the authors dealt with this problem. By generalizing the method of [1], using a k-dimensional version of the reduction principle[2] and a new divisibility relation on the divided differences discovered by Filaseta and Trifonov [3], they proved the following result.

Theorem 5.7 (Huxley–Sargos) *Let $k \geqslant 3$ be an integer and $f \in C^k[N, 2N]$ such that there exist $\lambda_{k-1} > 0$ and $\lambda_k > 0$ such that, for all $x \in [N, 2N]$, we have*

$$\left|f^{(k-1)}(x)\right| \asymp \lambda_{k-1}, \quad \left|f^{(k)}(x)\right| \asymp \lambda_k \quad and \quad \lambda_{k-1} = N\lambda_k. \tag{5.23}$$

Then the following upper bounds hold.

▷ *For all $k \geqslant 3$, we have*

$$\mathcal{R}(f, N, \delta) \ll N\lambda_k^{\frac{2}{k(k+1)}} + N\delta^{\frac{2}{(k-1)(k-2)}} + N\left(\delta\lambda_{k-1}\right)^{\frac{2}{k^2-k+2}} + \left(\frac{\delta}{\lambda_{k-1}}\right)^{\frac{1}{k-1}} + 1.$$

▷ *For $k = 3$, we have*

$$\mathcal{R}(f, N, \delta) \ll N\lambda_3^{1/6} + N\delta^{2/3} + N\left(\delta^3\lambda_3\right)^{1/12} + \left(\frac{\delta}{\lambda_2}\right)^{1/2} + 1.$$

(continued)

---

[2]See Lemma 5.7.

Theorem 5.7 (continued)

▷ *For all $k \geqslant 4$ and $\varepsilon > 0$, we have*

$$\mathcal{R}(f, N, \delta) \ll \left\{ N\lambda_k^{\frac{2}{k(k+1)}} + N\left(\delta\lambda_{k-1}\right)^{\frac{2}{k^2-k+2}} + N\delta^{\frac{4}{k^2-3k+6}} \right.$$

$$\left. + N\left(\delta^2 N^{-1}\lambda_{k-1}^{-1}\right)^{\frac{2}{k^2-3k+4}} \right\} N^\varepsilon + \left(\frac{\delta}{\lambda_{k-1}}\right)^{\frac{1}{k-1}} + 1.$$

▷ *For all $k \geqslant 5$, we have*

$$\mathcal{R}(f, N, \delta) \ll N\lambda_k^{\frac{2}{k(k+1)}} + N\delta^{\frac{2}{(k-1)(k-2)}} + \left(\frac{\delta}{\lambda_{k-1}}\right)^{\frac{1}{k-1}} + 1.$$

It can easily be seen that Theorem 5.1 is a simple consequence of Theorem 5.5, since the conditions $\delta \ll \lambda_k \ll 1$ imply that the main term dominates all the others. The purpose of the next result is to provide an estimate analogous to that of Theorem 5.1 but with a hypothesis more flexible than (5.15).

Proposition 5.2 *Let $f \in C^\infty[N, 2N]$ such that there exists $T \geqslant 1$ such that, for all $x \in [N, 2N]$ and all $j \in \mathbb{Z}_{\geqslant 0}$, we have*

$$\left| f^{(j)}(x) \right| \asymp \frac{T}{N^j} \tag{5.24}$$

and

$$N\delta \leqslant T \leqslant \delta^{-1}. \tag{5.25}$$

Then, for all $k \geqslant 1$, we have

$$\mathcal{R}(f, N, \delta) \ll T^{\frac{2}{k(k+1)}} N^{\frac{k-1}{k+1}}.$$

Note that, using (5.24), Huxley–Sargos's result may be stated as

$$\mathcal{R}(f, N, \delta) \ll T^{\frac{2}{k(k+1)}} N^{\frac{k-1}{k+1}} + N\delta^{\frac{2}{k(k-1)}} + N\left(\delta T^{-1}\right)^{1/k} \tag{5.26}$$

for all $k \geqslant 2$, and note that the term $N\left(\delta T^{-1}\right)^{1/k}$ is dominated by the term $N\delta^{\frac{2}{k(k-1)}}$ as soon as $k \geqslant 3$. Hence, Proposition 5.2 shows that the conditions (5.25) are sufficient to remove this term.

Proof We use induction, the case $k = 1$ being clearly true by Theorem 5.3 in which the conditions (5.25) enable us to eliminate the secondary terms. Now suppose that the result is true for some $k \geqslant 1$. By induction hypothesis and (5.26) used with $k+1$ instead of k, we get

$$\mathcal{R}(f, N, \delta) \ll \min\left(E, \ T^{\frac{2}{k(k+1)}} N^{\frac{k-1}{k+1}}\right),$$

where

$$E = \max\left(T^{\frac{2}{(k+1)(k+2)}} N^{\frac{k}{k+2}}, \ N\delta^{\frac{2}{k(k+1)}}, \ N\left(\delta T^{-1}\right)^{\frac{1}{k+1}}\right) = \max(e_1, e_2, e_3)$$

say. The result follows at once if $E = e_1$. The cases $E = e_2$ and $E = e_3$ are treated using the following inequality for means: *if $x, y \geqslant 0$ and $0 \leqslant a \leqslant 1$, then*

$$\min(x, y) \leqslant x^a y^{1-a}.$$

▷ *Case $E = e_2$.* We choose $a = \frac{1}{k+2}$, which gives

$$\min\left(e_2, \ T^{\frac{2}{k(k+1)}} N^{\frac{k-1}{k+1}}\right) \leqslant T^{\frac{2}{(k+1)(k+2)}} N^{\frac{k}{k+2}} (T\delta)^{\frac{2}{k(k+1)(k+2)}} \leqslant T^{\frac{2}{(k+1)(k+2)}} N^{\frac{k}{k+2}}$$

by (5.25).

▷ *Case $E = e_3$.* We choose $a = \frac{2}{k+2}$, which gives

$$\min\left(e_3, \ T^{\frac{2}{k(k+1)}} N^{\frac{k-1}{k+1}}\right) \leqslant T^{\frac{2}{(k+1)(k+2)}} N^{\frac{k}{k+2}} (N\delta T^{-1})^{\frac{2}{(k+1)(k+2)}} \leqslant T^{\frac{2}{(k+1)(k+2)}} N^{\frac{k}{k+2}}$$

by (5.25). The proof is complete.   □

Example 5.2 We return to the squarefree number problem. Taking account of (5.25), we get if $y \leqslant x^{1/3}$

$$\max_{(xy)^{1/4} < N \leqslant x^{1/3}} \mathcal{R}\left(\frac{x}{n^2}, N, \frac{y}{N^2}\right) + \max_{y < N \leqslant 2x^{1/3}} \mathcal{R}\left(\sqrt{\frac{x}{n}}, N, \frac{y}{\sqrt{Nx}}\right) \ll x^{\frac{k^2-k+2}{3k(k+1)}}$$

for all $k \geqslant 3$. With $k = 3$, we get the bound $\ll x^{2/9}$.

The main term in the case $k = 2$ was improved by Huxley [5], Huxley and Trifonov [10] and then Trifonov [17], who extended an earlier work by Swinnerton-Dyer that we will see in Sect. 5.4. The basic idea in this method is that the integer points close to the curve form a convex polygonal line. The Dirichlet pigeon-hole principle is then applied to the determinant formed with the coordinates of consecutive vertices. For an exhaustive exposition of Swinnerton-Dyer's method, the reader may refer to [5]. We provide below one of the many versions of the theorem proved by the author [6].

Theorem 5.8 (Huxley) *Let $f \in C^3[N, 2N]$ such that there exist $C \geqslant 1$, $0 < \lambda_2 \leqslant C^{-1}$ and $\lambda_3 > 0$ such that, for all $x \in [N, 2N]$, we have*

$$C^{-1}\lambda_2 \leqslant |f''(x)| \leqslant C\lambda_2, \quad C^{-1}\lambda_3 \leqslant |f'''(x)| \leqslant C\lambda_3 \quad and \quad \lambda_2 = N\lambda_3. \tag{5.27}$$

Then

$$\mathcal{R}(f, N, \delta) \ll \left\{ N^{9/10}\lambda_2^{3/10} + N^{4/5}\lambda_2^{1/5} + N\lambda_2^{3/8}\delta^{1/8} + N^{7/8}\lambda_2^{1/4}\delta^{1/8} \right.$$
$$\left. + N^{6/7}(\lambda_2\delta)^{1/7} + N\lambda_2^{1/5}\delta^{2/5} \right\} (\log N)^{2/5} + N\delta + \left(\delta\lambda_2^{-1}\right)^{1/2} + 1.$$

The implied constant depends only on C.

Example 5.3 Assume $y \leqslant x^{2/5}$.

▷ Theorem 5.8 implies for all $N \geqslant 2x^{1/4}$ that

$$(\log N)^{-2/5}\mathcal{R}\left(\frac{x}{n^2}, N, \frac{y}{N^2}\right) \ll \left(xN^{-1}\right)^{3/10} + x^{1/5} + \left(x^3yN^{-6}\right)^{1/8} + (xy)^{1/7}$$
$$+ \left(xy^2N^{-3}\right)^{1/5} + yN^{-1} + N\left(yx^{-1}\right)^{1/2},$$

so that

$$\max_{x^{2/7} < N \leqslant x^{1/3}} \mathcal{R}\left(\frac{x}{n^2}, N, \frac{y}{N^2}\right) \ll x^{3/14}(\log x)^{2/5}.$$

Furthermore, using Theorem 5.7, we get

$$\mathcal{R}\left(\frac{x}{n^2}, N, \frac{y}{N^2}\right) \ll (Nx)^{1/6} + \left(y^2N^{-1}\right)^{1/3} + \left(xy^3N\right)^{1/12} + N\left(yx^{-1}\right)^{1/2},$$

so that

$$\max_{4\sqrt{y} < N \leqslant x^{2/7}} \mathcal{R}\left(\frac{x}{n^2}, N, \frac{y}{N^2}\right) \ll x^{3/14}.$$

Therefore

$$\max_{4\sqrt{y} < N \leqslant x^{1/3}} \mathcal{R}\left(\frac{x}{n^2}, N, \frac{y}{N^2}\right) \ll x^{3/14}(\log x)^{2/5}. \tag{5.28}$$

▷ Theorem 5.8 implies for all $N \geqslant 3x^{1/5}$ that

$$(\log N)^{-2/5} \mathcal{R}\left(\sqrt{\frac{x}{n}}, N, \frac{y}{\sqrt{Nx}}\right) \ll (Nx)^{3/20} + \left(N^3 x\right)^{1/10} + y(Nx^{-1})^{1/2}$$

$$+ (xy)^{1/8} + \left(xy^2 N^3\right)^{1/16} + \left(N^3 y\right)^{1/7}$$

$$+ \left(x^{-1} y^4 N^3\right)^{1/10} + N\left(yx^{-1}\right)^{1/2},$$

so that

$$\max_{3x^{1/5} < N \leqslant 8x^{1/3}} \mathcal{R}\left(\sqrt{\frac{x}{n}}, N, \frac{y}{\sqrt{Nx}}\right) \ll x^{1/5}(\log x)^{2/5},$$

since $y \leqslant x^{2/5}$. Furthermore, using the trivial estimate (5.4), we have

$$\max_{N \leqslant 3x^{1/5}} \mathcal{R}\left(\sqrt{\frac{x}{n}}, N, \frac{y}{\sqrt{Nx}}\right) \ll x^{1/5}.$$

Therefore

$$\max_{N \leqslant 8x^{1/3}} \mathcal{R}\left(\sqrt{\frac{x}{n}}, N, \frac{y}{\sqrt{Nx}}\right) \ll x^{1/5}(\log x)^{2/5}. \tag{5.29}$$

Using Lemma 5.3 with $A = 4\sqrt{y}$ and $B = x^{1/3}$, we then get assuming $y \leqslant x^{2/5}$

$$\sum_{x < n \leqslant x+y} \mu_2(n) = \frac{y}{\zeta(2)} + O\left(x^{3/14}(\log x)^{7/5}\right).$$

We infer that there exists a constant $c_0 > 0$ such that, if $c_0 x^{3/14}(\log x)^{7/5} \leqslant y < \frac{1}{4}x^{1/2}$, the interval $(x, x + y]$ contains a squarefree number.

A slight improvement of this estimate can be obtained via the following result due to Trifonov [17].

Theorem 5.9 (Trifonov) *Let $f \in C^3[N, 2N]$ such that there exist $C \geqslant 1$, $0 < \lambda_2 \leqslant 1$ and $\lambda_3 > 0$ such that, for all $x \in [N, 2N]$, we have*

$$C^{-1}\lambda_2 \leqslant |f''(x)| \leqslant C\lambda_2, \quad C^{-1}\lambda_3 \leqslant |f'''(x)| \leqslant C\lambda_3 \quad \text{and} \quad \lambda_2 = N\lambda_3 \tag{5.30}$$

(continued)

Theorem 5.9 (continued)
and

$$N\lambda_2 \geqslant 1 \quad and \quad N\delta^2 \leqslant C^{-1}. \tag{5.31}$$

Then, for all $\varepsilon > 0$, we have

$$\mathcal{R}(f, N, \delta) \ll \left\{ N^{43/54}\lambda_2^{4/27} + N^{4/5}\lambda_2^{4/25} + N^{9/10}\delta^{4/15} + N^{12/13}\delta^{4/13} \right.$$
$$\left. + N^{6/7}\lambda_2^{2/7} + N\lambda_2 + N\left(\lambda_2\delta\right)^{1/4} \right\} N^\varepsilon + \lambda_2 \left(N\delta\right)^{5/2}.$$

The implied constant depends only on C and ε.

Example 5.4 We apply this result to the function $f : u \mapsto x/u^2$ of the squarefree number problem. First, the conditions (5.31) are fulfilled as soon as

$$\max\left(x^{1/4}, 4y^{2/3}\right) \leqslant N \leqslant 2^{-1}x^{1/3}.$$

We assume then that $y \leqslant 8^{-1}x^{3/8}$ and use Lemma 5.3 with $A = 4\sqrt{y}$ and $B = 2^{-1}x^{1/3}$. We split the range of R_1 into three parts.

▷ For $4\sqrt{y} < N \leqslant x^{13/48}$, we use Theorem 5.7 with $k = 4$ giving

$$\mathcal{R}\left(\frac{x}{n^2}, N, \frac{y}{N^2}\right) \ll x^{1/10}N^{2/5} + (xy)^{1/7} + (Ny^2)^{1/5} + N\left(y^2x^{-1}\right)^{1/4} + N\left(yx^{-1}\right)^{1/3},$$

and hence

$$\max_{4\sqrt{y}<N\leqslant x^{13/48}} \mathcal{R}\left(\frac{x}{n^2}, N, \frac{y}{N^2}\right) \ll x^{5/24}$$

as long as $y \leqslant 8^{-1}x^{3/8}$.

▷ For $x^{13/48} < N \leqslant x^{41/136}$, we use Theorem 5.9 giving for all $\varepsilon > 0$

$$\mathcal{R}\left(\frac{x}{n^2}, N, \frac{y}{N^2}\right) \ll \left\{ N^{11/54}x^{4/27} + (Nx)^{4/25} + N^{11/30}y^{4/15} + (Ny)^{4/13} \right.$$
$$\left. + \left(xN^{-1}\right)^{2/7} + xN^{-3} + \left(xyN^{-2}\right)^{1/4} \right\} N^\varepsilon + x\left(y^5N^{-13}\right)^{1/2},$$

and hence

$$\max_{x^{13/48}<N\leqslant x^{41/136}} \mathcal{R}\left(\frac{x}{n^2}, N, \frac{y}{N^2}\right) \ll x^{57/272+\varepsilon}$$

as long as $y \leqslant x^{101/272}$.

▷ For $x^{41/136} < N \leqslant 2^{-1}x^{1/3}$, we use Theorem 5.8 giving as in Example 5.3

$$\max_{x^{41/136}<N\leqslant x^{1/3}} \mathcal{R}\left(\frac{x}{n^2}, N, \frac{y}{N^2}\right) \ll x^{57/272}(\log x)^{2/5}$$

as long as $y \leqslant x^{127/272}$.

Hence, we get for all $\varepsilon > 0$

$$\max_{4\sqrt{y}\leqslant N\leqslant 2^{-1}x^{1/3}} \mathcal{R}\left(\frac{x}{n^2}, N, \frac{y}{N^2}\right) \ll x^{57/272+\varepsilon}, \tag{5.32}$$

and taking account of (5.29) and Lemma 5.3, we infer that

$$\sum_{x<n\leqslant x+y} \mu_2(n) = \frac{y}{\zeta(2)} + O\left(x^{57/272+\varepsilon}\right)$$

if $y \leqslant x^{101/272}$. Hence, there exists a constant $c_0 > 0$ such that, if $c_0 x^{57/272+\varepsilon} \leqslant y < \frac{1}{4}x^{1/2}$, the interval $(x, x + y]$ contains a squarefree number.

5.4   The Method of Filaseta and Trifonov

The exponent $\frac{57}{272} \approx 0.209558\dots$ obtained in Example 5.4 is a very good result that is not attainable by current exponential sum results.[3] The bound (5.29) shows that any improvement must come from estimates of lattice points close to the curve of the function $f : u \mapsto x/u^2$. Using divided differences and the polynomial identity (5.34), which takes the particular structure of f into account, we will prove the following theorem, due to Filaseta and Trifonov [2, 3], which supersedes each previous result.

Theorem 5.10 (Filaseta–Trifonov) *Let $x \geqslant 1$ and δ satisfying (5.2) be real numbers and N be an integer such that $4 \leqslant N \leqslant x^{1/2}$. Then there exists a small real number $c_0 > 0$ such that*

$$N\delta \leqslant c_0 \tag{5.33}$$

(continued)

---

[3] see Chap. 6.

Theorem 5.10 (continued)
and, for x sufficiently large

$$\mathcal{R}\left(\frac{x}{n^2}, N, \delta\right) \ll x^{1/5} + x^{1/15}\delta N^{5/3}.$$

The authors combined the divided difference techniques with the following identity:

$$(X + Y)^2 P(X, Y) - X^2 Q(X, Y) = Y^3, \tag{5.34}$$

where $P, Q \in \mathbb{Z}[X, Y]$ are the homogeneous polynomials in two variables defined by

$$P(X, Y) = -2X + Y \quad \text{and} \quad Q(X, Y) = -2X - 3Y. \tag{5.35}$$

Note that the identity (5.34) has to be used with Y being small compared to X.

In what follows, x, δ and N are as stated in Theorem 5.10 and f is the function supported on $[N, 2N]$ and defined by $f(u) = x/u^2$. The constant c_0 appearing in (5.33) is sufficiently small, say $c_0 < 600^{-1}$. When $\mathcal{I}$ is an interval of $\mathbb{R}$, the notation $|\mathcal{I}|$ always means $|\mathcal{I} \cap \mathbb{Z}|$. Finally, one may assume that (5.33) is always satisfied.

The next tool is similar to Lemma 5.8, and the proof, which is left to the reader, is analogous.

Lemma 5.16 *Let $n = x + y \in \mathbb{Z}$ with $x, y \in \mathbb{R}$. Then*

▷ *If $|x| \geqslant \frac{1}{2}$ and $|y| < \frac{1}{2}$, then $n \neq 0$.*
▷ *If $|x| \leqslant \frac{1}{2}$ and $|y| < \frac{1}{2}$, then $n = 0$.*

5.4.1   Preparatory Lemma

The first step in Filaseta–Trifonov's method is the following construction of a subset of $\mathcal{S}(f, N, \delta)$ in which the elements are not too close to each other.

Lemma 5.17 *There exists a subset T of $\mathcal{S}(f, N, \delta)$ satisfying the following two properties:*

▷ $\mathcal{R}(f, N, \delta) \leqslant 4\,(|T| + 1)$.
▷ *Any two consecutive elements of T differ by $> (2x)^{-1/3}N^{4/3}$.*

Proof If $N < x^{1/4}$, then one may take $T = S(f, N, \delta)$ because we have in this case $x^{-1/3}N^{4/3} < 1$. Now suppose that $N \geqslant x^{1/4}$ and let $a, b \in \mathbb{Z}_{\geqslant 1}$ and n, $n + a$ and $n + a + b$ be three consecutive elements of $S(f, N, \delta)$ such that

$$1 \leqslant a, b \leqslant (2x)^{-1/3}N^{4/3}.$$

We will show that there are only *two* possibilities for the choice of b. The result will then follow by taking each 4th element of $S(f, N, \delta)$.

By definition, there exist non-zero integers m_i and real numbers δ_i such that

$$f(n) = m_1 + \delta_1$$
$$f(n + a) = m_2 + \delta_2$$
$$f(n + a + b) = m_3 + \delta_3$$

with $|\delta_i| < \delta$ for $i \in \{1, 2, 3\}$. In fact, each integer m_i is positive since, for all $u \in [N, 2N]$, we have $f(u) \geqslant x/(4N^2) \geqslant 1/4$ and $\delta \leqslant c_0 N^{-1}$. Also note that, by (5.34), we get

$$f(n)P(n, a) - f(n + a)Q(n, a) = \frac{xa^3}{n^2(n+a)^2} \leqslant xa^3 N^{-4} \leqslant \tfrac{1}{2}.$$

On the other hand, we have

$$f(n)P(n, a) - f(n + a)Q(n, a) = m_1 P(n, a) - m_2 Q(n, a) + \varepsilon$$

with

$$|\varepsilon| < \delta\left(|P(n, a)| + |Q(n, a)|\right) \leqslant 4\delta(n + a) \leqslant 8N\delta \leqslant 8c_0 < \tfrac{1}{2}.$$

Hence by Lemma 5.16, we obtain

$$m_1 P(n, a) - m_2 Q(n, a) = 0. \tag{5.36}$$

Similarly, we also have

$$m_2 P(n + a, b) - m_3 Q(n + a, b) = 0$$
$$m_1 P(n, a + b) - m_3 Q(n, a + b) = 0,$$

and eliminating m_3, we get

$$m_2 P(n + a, b)Q(n, a + b) - m_1 P(n, a + b)Q(n + a, b) = 0,$$

which gives

$$3b^2(m_1 - m_2) + \kappa_1 b + 2\kappa_2 = 0, \tag{5.37}$$

where

$$\kappa_1 = a\,(5m_1 + 3m_2) - 4n(m_1 - m_2),$$

$$\kappa_2 = a^2\,(m_1 + 3m_2) - an\,(m_1 - 5m_2) - 2n^2(m_1 - m_2).$$

If $m_1 = m_2$, then by (5.36) we have $P(n, a) = Q(n, a)$ and then $-2n + a = -2n - 3a$, so that $a = 0$, which is impossible since $a \geqslant 1$. Therefore, $m_1 \neq m_2$ and (5.37) is a quadratic equation in b, concluding the proof. $\qquad\square$

We deduce at once the following result.

Corollary 5.1 *There exists a small constant $c_0 > 0$ such that $N\delta \leqslant c_0$ and*

$$\mathcal{R}\left(\frac{x}{n^2}, N, \delta\right) \leqslant 10 \left(\frac{x}{N}\right)^{1/3}.$$

Furthermore, if $x^{2/5} \leqslant N \leqslant x^{1/2}$, then

$$\mathcal{R}\left(\frac{x}{n^2}, N, \delta\right) \leqslant 10\,x^{1/5}.$$

Note that this result is equivalent to Proposition 5.2 used with $k = 2$ except with the condition $N\delta \ll 1$ instead of (5.25).

5.4.2 Higher Divided Differences

By Corollary 5.1, it is now sufficient to assume

$$2^{2/3}x^{1/5} \leqslant N < x^{2/5}, \tag{5.38}$$

and we set

$$R = \left(2^{-1}N^4 x^{-1}\right)^{1/3} \quad \text{and} \quad A = 2^{-1/3}N x^{-1/5}.$$

By (5.38), we have $R < A \leqslant N^{1/2}$. We take up again the subset T of Lemma 5.17 in which any two consecutive elements differ by $> R$. As in Lemma 5.7, we will use the following subsets of T. For $a \in (R, A] \cap \mathbb{Z}$, we define

$$T(a) = \left\{n \in \mathbb{Z}_{\geqslant 1} : n \text{ and } n + a \text{ are consecutive in } T\right\}.$$

By (5.10), we have

$$|T| \ll \frac{N}{A} + \sum_{R < a \leqslant A} |T(a)|. \tag{5.39}$$

The next tool will enable us to get an upper bound for the sets $T(a)$.

Lemma 5.18 *Let $a \in (R, A) \cap \mathbb{Z}$ and I be a subinterval of $[N, 2N]$ satisfying*

$$|I| \leqslant 16^{-1} N^5 x^{-1} a^{-3}.$$

Then of each three consecutive elements in $T(a) \cap I$, there are two consecutive elements that differ by

$$> 6^{-1} (ax)^{-1/3} N^{5/3}.$$

Proof Let n, $n + b$ and $n + b + d$ be three consecutive elements of $T(a) \cap I$ with $b, d \in \mathbb{Z}_{\geqslant 1}$. Hence, the six integers n, $n + a$, $n + b$, $n + a + b$, $n + b + d$ and $n + a + b + d$ are all lying in $[N, 2N]$. Since $a > R$, observe also that

$$|I| \leqslant 16^{-1} N^5 x^{-1} a^{-3} < 16^{-1} N^5 x^{-1} R^{-3} = \tfrac{1}{8} N.$$

Consider the six integers $m_1, \ldots, m_6$ and real numbers $\delta_1, \ldots, \delta_6$ satisfying $|\delta_i| < \delta$ and defined by

$$f(n) = m_1 + \delta_1,$$
$$f(n + a) = m_2 + \delta_2,$$
$$f(n + a + b) = m_3 + \delta_3,$$
$$f(n + b) = m_4 + \delta_4,$$
$$f(n + b + d) = m_5 + \delta_5,$$
$$f(n + a + b + d) = m_6 + \delta_6,$$

and we set $F_a(n) = -\Delta_a f(n) = f(n) - f(n + a)$ and

$$D_1 = d F_a(n) - (b + d) F_a(n + b) + b F_a(n + b + d).$$

Observe that, with the notation of Remark 5.4, we have

$$\frac{D_1}{bd(b + d)} = F_a[n, n + b, n + b + d],$$

so that by (5.6) there exists a real number $t \in (n, n + b + d) \subseteq [N, 2N]$ such that

$$D_1 = \frac{bd(b+d)}{2!} F_a''(t) = bd(b+d) \frac{3ax(2t+a)(2t^2 + 2at + a^2)}{t^4(t+a)^4}.$$

Hence, $D_1 > 0$ and, since $t, t + a \in [N, 2N]$, we get

$$D_1 \geqslant bd(b+d) \frac{3ax(2N+a)(2N^2 + 2aN + a^2)}{(2N)^8} \geqslant bd(b+d) \frac{ax}{25N^5}.$$

Recall that $n + b$ and $n + b + d$ are consecutive in $T(a) \cap I$ and hence $\min(b, d) \geqslant a > R$ and then

$$D_1 \geqslant (b+d) \frac{R^3 x}{25N^5} = \frac{b+d}{50N},$$

so that

$$b + d \leqslant 50 \, N D_1. \tag{5.40}$$

Now set

$$E_1 = d(m_1 - m_2) - (b+d)(m_4 - m_3) + b(m_5 - m_6) \in \mathbb{Z}.$$

Using the definition of the integers $m_1, \ldots, m_6$ and the function F_a, we get

$$E_1 = d(F_a(n) + \delta_1 - \delta_2) - (b+d)(F_a(n+b) + \delta_4 - \delta_3) + b(F_a(n+b+d) + \delta_5 - \delta_6)$$

$$= dF_a(n) - (b+d)F_a(n+b) + bF_a(n+b+d) + d(\delta_1 - \delta_2) - (b+d)(\delta_4 - \delta_3) + b(\delta_5 - \delta_6)$$

$$= D_1 + R_1,$$

with $|R_1| < 4\delta(b+d)$. By (5.40) and using (5.33) with $c_0 < 600^{-1}$, we obtain

$$4\delta(b+d) \leqslant 200N\delta D_1 < \tfrac{1}{3}D_1,$$

and therefore

$$|E_1 - D_1| < \tfrac{1}{3}D_1,$$

which implies that $E_1 \neq 0$. Since $E_1 \in \mathbb{Z}$, we infer that $|E_1| \geqslant 1$, and since

$$|E_1| \leqslant D_1 + |R_1| < \tfrac{4}{3}D_1,$$

we get

$$D_1 > \tfrac{3}{4}.$$

One may obtain an upper bound for D_1 in a similar way, since $a \leqslant A \leqslant N^{1/2} \leqslant \tfrac{1}{2}N$ and $t \in [N, 2N]$ imply

$$D_1 \leqslant bd(b+d)\, 3ax(4N + N/2)(8N^2 + 2 \times 2N^2)N^{-8} = 162\, bd(b+d)\, ax N^{-5}.$$

Combining both of these estimates, we get

$$bd(b+d) > \frac{N^5}{216\, ax},$$

so that either b or d is $> 6^{-1}(ax)^{-1/3}N^{5/3}$ as asserted.   □

5.4.3   Proof of the Main Result

We now are in a position to prove Theorem 5.10. To this end, we pick up from Lemma 5.18 the interval $\mathcal{I}$ and the integers $m_1, \ldots, m_4$ and define

$$G_a(n) = P(n, a)f(n) - Q(n, a)f(n+a)$$
$$D_2 = -\Delta_b G_a(n) = G_a(n) - G_a(n+b)$$
$$E_2 = P(n, a)m_1 - Q(n, a)m_2 - P(n+b, a)m_4 + Q(n+b, a)m_3 \in \mathbb{Z}.$$

As in the proof of Lemma 5.18, we have $D_2 = E_2 + R_2$, and using (5.33), we have again $|R_2| \leqslant 20N\delta < \tfrac{1}{2}$. By (5.6), there exists a real number $t \in (n, n+b)$ such that

$$D_2 = -b\, G_a'(t) = 2a^3 xb \left(t^{-2}(t+a)^{-3} + t^{-3}(t+a)^{-2} \right).$$

Since $t, t+a \in [N, 2N]$, we obtain as above

$$16^{-1}ba^3 x N^{-5} \leqslant |D_2| \leqslant 4ba^3 x N^{-5}, \tag{5.41}$$

and since $b \leqslant |\mathcal{I}| \leqslant 16^{-1}N^5 x^{-1}a^{-3}$, we get

$$|D_2| \leqslant \tfrac{1}{4}.$$

By Lemma 5.31, we infer that $E_2 = 0$, which implies that $|D_2| = |R_2| \leqslant 20N\delta$, and using (5.41), we get

$$b \ll a^{-3}x^{-1}N^5|D_2| \ll a^{-3}x^{-1}N^6\delta.$$

Thus, any two elements of $T(a) \cap \mathcal{I}$ are lying in a subinterval $\mathcal{J}$ of $\mathcal{I}$ satisfying

$$|\mathcal{J}| \ll a^{-3}x^{-1}N^6\delta.$$

Now by Lemma 5.18, of each three consecutive elements in $T(a) \cap \mathcal{I}$, there are two consecutive elements that differ by $\gg (ax)^{-1/3}N^{5/3}$ so that

$$|T(a) \cap \mathcal{I}| \ll \frac{|\mathcal{J}|}{(ax)^{-1/3}N^{5/3}} + 1 \ll x^{-2/3}\delta a^{-8/3}N^{13/3} + 1.$$

We saw that $|\mathcal{I}| \leqslant \frac{1}{8}N$. Subdividing $[N, 2N]$ into $s = \lfloor N/(16^{-1}N^5x^{-1}a^{-3})\rfloor + 1$ distinct subintervals $\mathcal{I}_1, \ldots, \mathcal{I}_s$ with lengths $\leqslant 16^{-1}N^5x^{-1}a^{-3}$, we get

$$
\begin{aligned}
|T(a)| &\ll \left(\frac{N}{N^5x^{-1}a^{-3}} + 1 \right) \left(x^{-2/3}\delta a^{-8/3}N^{13/3} + 1 \right) \\
&\ll \left(xa^3N^{-4} + 1 \right) \left(x^{-2/3}\delta a^{-8/3}N^{13/3} + 1 \right) \\
&\ll xa^3N^{-4} + (Nax)^{1/3}\delta
\end{aligned}
$$

where we used the fact that $a > R = (2x)^{-1/3}N^{4/3}$, implying that $1 < 2xa^3N^{-4}$, in the last inequality. Using (5.39), we obtain

$$
\begin{aligned}
|T| &\ll NA^{-1} + \sum_{R < a \leqslant A} \left(xa^3N^{-4} + (Nax)^{1/3}\delta \right) \\
&\ll NA^{-1} + x\left(AN^{-1} \right)^4 + \left(NA^4x \right)^{1/3}\delta,
\end{aligned}
$$

and the choice of $A = 2^{-1/3}Nx^{-1/5}$ gives

$$|T| \ll x^{1/5} + x^{1/15}\delta N^{5/3}. \tag{5.42}$$

Let us summarize all the results obtained above.

▷ If $4 \leqslant N \leqslant 2^{2/3}x^{1/5}$, we use the trivial bound (5.4) giving

$$\mathcal{R}\left(\frac{x}{n^2}, N, \delta \right) \leqslant N + 1 \ll x^{1/5}.$$

▷ If $2^{2/3}x^{1/5} < N < x^{2/5}$, Lemma 5.17 and (5.42) give

$$\mathcal{R}\left(\frac{x}{n^2}, N, \delta\right) \ll |T| + 1 \ll x^{1/5} + x^{1/15}\delta N^{5/3}.$$

▷ If $x^{2/5} \leqslant N \leqslant x^{1/2}$, Corollary 5.1 provides

$$\mathcal{R}\left(\frac{x}{n^2}, N, \delta\right) \ll x^{1/5}.$$

The proof of Theorem 5.10 is complete. □

5.4.4 Application

Applying Theorem 5.10 to the gaps between squarefree integers gives the following consequence.

Corollary 5.2 *Let x, y be real numbers satisfying (5.3). Then we have*

$$\sum_{x < n \leqslant x+y} \mu_2(n) = \frac{y}{\zeta(2)} + O\left\{\left(x^{1/5} + yx^{-1/60}\right)\log x + y^{1/2}\right\}.$$

Proof Let c_0 be the constant in (5.33) and set $c_1 = c_0^{-1}$. We consider two cases.

▷ *Case* $16 \leqslant y \leqslant x^{1/4}$. We may write

$$\sum_{2\sqrt{y} < n \leqslant \sqrt{x}} \left(\left\lfloor \frac{x+y}{n^2} \right\rfloor - \left\lfloor \frac{x}{n^2} \right\rfloor\right) = \left(\sum_{2\sqrt{y} < n \leqslant c_1 x^{1/4}} + \sum_{c_1 x^{1/4} < n \leqslant \sqrt{x}}\right)\left(\left\lfloor \frac{x+y}{n^2} \right\rfloor - \left\lfloor \frac{x}{n^2} \right\rfloor\right)$$

$$= \Sigma_1 + \Sigma_2.$$

We use Theorem 5.7 with $k = 4$ for Σ_1 and Theorem 5.10 for Σ_2, which gives

$$\Sigma_1 \ll \max_{2\sqrt{y} < N \leqslant c_1 x^{1/4}} \left(x^{1/10}N^{2/5} + (Ny)^{1/3} + (xy)^{1/7} + N\left(yx^{-1}\right)^{1/3}\right)\log x$$

$$\ll \left(x^{1/5} + x^{1/12}y^{1/3} + (xy)^{1/7}\right)\log x \ll x^{1/5}\log x$$

since $y \leqslant x^{1/4}$, and

$$\Sigma_2 \ll \max_{c_1 x^{1/4} < n \leqslant \sqrt{x}} \left(x^{1/5} + x^{1/15}yN^{-1/3}\right)\log x \ll \left(x^{1/5} + yx^{-1/60}\right)\log x.$$

▷ *Case* $x^{1/4} < y < \frac{1}{4}x^{1/2}$. We now have

$$\sum_{2\sqrt{y}<n\leqslant\sqrt{x}}\left(\left\lfloor\frac{x+y}{n^2}\right\rfloor - \left\lfloor\frac{x}{n^2}\right\rfloor\right) = \left(\sum_{2\sqrt{y}<n\leqslant c_1 y} + \sum_{c_1 y<n\leqslant\sqrt{x}}\right)\left(\left\lfloor\frac{x+y}{n^2}\right\rfloor - \left\lfloor\frac{x}{n^2}\right\rfloor\right),$$

and Theorem 5.7 with $k = 4$ applied to the first sum implies that it contributes

$$\ll \left(x^{1/10}y^{2/5} + (xy)^{1/7} + y^{2/3} + y^{4/3}x^{-1/3}\right)\log x \ll yx^{-1/60}\log x,$$

where we used the fact that $x^{1/4} < y < \frac{1}{4}x^{1/2}$, and the second sum contributes

$$\ll \left(x^{1/5} + x^{1/15}y^{2/3}\right)\log x \ll \left(x^{1/5} + yx^{-1/60}\right)\log x,$$

since $y > x^{1/4}$. Applying Lemma 5.3 with $A = 2\sqrt{y}$ and $B = \sqrt{x}$ concludes the proof.   □

5.4.5   *Generalization*

The computations made above may be generalized to the case of the function $u \mapsto x/u^r$, where $r \geqslant 2$ is a fixed integer, which enables us to treat the r-free number problem. To this end, we need to have at our disposal two homogeneous integer polynomials $P_{r-1}(X, Y)$ and $Q_{r-1}(X, Y)$ with total degree $r-1$ analogous to (5.35) in order to generalize (5.34). It may be proved [3] that the polynomials

$$P_{r-1}(X, Y) = r\binom{2r-1}{r}\sum_{j=0}^{r-1}(-1)^{r-1+j}\binom{r-1}{j}\frac{(X+Y)^{r-1-j}Y^j}{2r-j-1}$$

$$Q_{r-1}(X, Y) = r\binom{2r-1}{r}\sum_{j=0}^{r-1}(-1)^{r-1}\binom{r-1}{j}\frac{X^{r-1-j}Y^j}{2r-j-1},$$

with $r \geqslant 1$, have the desired properties and satisfy the identity

$$(X + Y)^r P_{r-1}(X, Y) - X^r Q_{r-1}(X, Y) = Y^{2r-1}. \tag{5.43}$$

Similarly, in order to prove the analogue of Lemma 5.34 using divided differences of the second order, the following polynomials

$$P_{r-2}(X, Y) = (r - 1)\binom{2r - 2}{r - 1} \sum_{j=0}^{r-2}(-1)^{r+j}\binom{r - 2}{j} \frac{(X + Y)^{r-2-j}Y^j}{(2r - j - 3)(2r - j - 2)}$$

$$Q_{r-2}(X, Y) = (r - 1)\binom{2r - 2}{r - 1} \sum_{j=0}^{r-2}(-1)^{r}\binom{r - 2}{j} \frac{X^{r-2-j}Y^j}{(2r - j - 3)(2r - j - 2)},$$

with $r \geqslant 2$, are homogeneous integer polynomials of total degree $r - 2$ and satisfy the identity

$$(X + Y)^r P_{r-2}(X, Y) - X^r Q_{r-2}(X, Y) = Y^{2r-3}(2X + Y). \tag{5.44}$$

The polynomial identities (5.43) and (5.44) are very difficult to show and may be replaced by the theory of *Padé approximants* such as in [3], which allows the authors to take r as a rational number. Adapting the proof above to the general case, Filaseta and Trifonov proved the following result.

Theorem 5.11 (Filaseta–Trifonov) *Let $r \geqslant 2$ be an integer, $x \geqslant 1$ and δ satisfying (5.2) be real numbers and N be an integer such that $4 \leqslant N \leqslant x^{1/r}$. Then there exists a small real number $c_r > 0$, depending only on r, such that*

$$N^{r-1}\delta \leqslant c_r$$

and

$$\mathcal{R}\left(\frac{x}{n^r}, N, \delta\right) \ll x^{\frac{1}{2r+1}} + x^{\frac{1}{6r+3}} \delta N^{r-1/3}.$$

We shall see in Exercise 106 that the method may be adapted to the function $u \mapsto \sqrt{x/u}$ and, more generally, to the function $u \mapsto (x/u)^{1/r}$ where $r \geqslant 2$. The reader can check that the following dual result holds.

Theorem 5.12 *Let* $r \geqslant 2$ *be an integer,* $x \geqslant 1$ *and* δ *satisfying* (5.2) *be real numbers and* N *be an integer such that* $4 \leqslant N \leqslant x$. *Then there exists a small real number* $c_r > 0$, *depending only on* r, *such that*

$$N^{r-1}\delta \leqslant c_r$$

and

$$\mathcal{R}\left(\left(\frac{x}{n}\right)^{1/r}, N, \delta\right) \ll \left(N^{r^2-1}x\right)^{\frac{1}{r(2r+1)}} + x^{\frac{1}{3r(2r+1)}} \delta \, N^{r-1/6 - \frac{r+2}{6r(2r+1)}}.$$

5.5   Recent Results

5.5.1   Smooth Curves

Using Theorem 5.1 and letting $\delta \to 0$, we infer that the number of integer points *lying on* the arc of the curve $y = f(x)$ with $N < x \leqslant 2N$ is

$$\ll N\lambda_k^{\frac{2}{k(k+1)}} + 1.$$

Historically, this number was first investigated by Jarnik who proved that a strictly convex arc $y = f(x)$ with length L has at most

$$\leqslant \frac{3}{(2\pi)^{1/3}} L^{2/3} + O\left(L^{1/3}\right)$$

integer points, and this is a nearly best possible result under the sole hypothesis of convexity. However, Swinnerton-Dyer and Schmidt proved independently that if $f \in C^3[0, N]$ is such that $|f(x)| \leqslant N$ and $f'''(x) \neq 0$ for all $x \in [0, N]$, then the number of integer points on the arc $y = f(x)$ with $0 \leqslant x \leqslant N$ is $\ll N^{3/5+\varepsilon}$. This result was generalized by Bombieri and Pila who showed the following result.

Proposition 5.3 (Bombieri-Pila) *Let* $N \geqslant 1$, $k \geqslant 4$ *be integers and set* $K = \binom{k+2}{2}$. *Let* $\mathcal{I}$ *be an interval with length* N *and* $f \in C^K(\mathcal{I})$ *satisfying* $|f'(x)| \leqslant 1$, $f''(x) > 0$ *and such that the number of solutions of the equation* $f^{(K)}(x) = 0$ *is* $\leqslant m$. *Then there exists a constant* $c_0 = c_0(k) > 0$ *such that the number of integer points on the arc* $y = f(x)$ *with* $x \in \mathcal{I}$ *is*

$$\leqslant c_0(m+1)N^{1/2+3/(k+3)}.$$

The ideas of Bombieri and Pila have been extended by Huxley [7] to counting the number of integer points, which are very close to regular curves. The function is supposed to be C^5 and, along with the usual non-vanishing conditions of the derivatives on $[N, 2N]$, the proof also requires lower bounds of the following determinants:

$$D_1(f; x) = \begin{vmatrix} f'''(x) & 3f''(x) \\ f^{(4)}(x) & 4f'''(x) \end{vmatrix}$$

$$D_2(f; x) = \frac{1}{2f''(x)} \begin{vmatrix} f'''(x) & 3f''(x) & 0 \\ f^{(4)}(x) & 4f^{(3)}(x) & 6f''(x)^2 \\ f^{(5)}(x) & 5f^{(4)}(x) & 20f''(x)f'''(x) \end{vmatrix}.$$

Proposition 5.4 (Huxley) *Assume that $f \in C^5[N, 2N]$ such that there exist real numbers $C, T \geqslant 1$ such that*

$$\left| f^{(j)}(x) \right| \leqslant C^{j+1} j! \times \frac{T}{N^j} \quad (j = 1, \ldots, 5), \quad \left| f^{(j)}(x) \right| \geqslant \frac{j!}{C^{j+1}} \times \frac{T}{N^j} \quad (j = 2, 3),$$

$$|D_1(f; x)| \geqslant 144 \, C^{-8} \times \frac{T^2}{N^6}, \quad |D_2(f; x)| \geqslant 4320 \, C^{-12} \times \frac{T^3}{N^9}.$$

Let δ be a real number satisfying (5.2). Then we have

$$\mathcal{R}(f, N, \delta) \ll (NT)^{4/15} + N \left(\delta^{11} T^9 \right)^{1/75}.$$

The implied constant depends only on C.

The main term is a very good result, since Theorem 5.4 only yields $(NT)^{1/3}$ and Theorem 5.8 provides the bound $(NT)^{3/10}$. On the other hand, the secondary term is too large, and thus useless in many applications. In this direction, the author proved that this term may be improved subject to some additional non-vanishing conditions of certain quite complicated determinants.

5.5.2 Polynomials

An improvement of the smooth term appeared in [12] in the case of a polynomial.

Proposition 5.5 *Let* $\delta \in \left[0, \frac{1}{4}\right]$, $N, k \in \mathbb{Z}_{\geqslant 1}$ *and* $P = \alpha_k X^k + \cdots + \alpha_0 \in \mathbb{R}[X]$. *If there exist* $a \in \mathbb{Z}$ *and* $q \in \left[1, N^k\right] \cap \mathbb{Z}$ *such that* $(a, q) = 1$ *and* $\left|\alpha_k - \frac{a}{q}\right| \leqslant \frac{1}{q^2}$, *then, for all* $\varepsilon > 0$

$$\mathcal{R}(P, N, \delta) \ll Nq^{-1/k} + N\delta^{\frac{2}{k(k+1)}} + N^{\varepsilon},$$

the term N^{ε} *being replaced by* 1 *if* $k = 1$.

If $\alpha_k \in (0, 1]$, then one may take $a = 1$ and $q = \left\lfloor \alpha_k^{-1} \right\rfloor$, and hence, if we assume furthermore that $\alpha_k \in \left[N^{-k}, 1\right]$, then we derive

$$\mathcal{R}(P, N, \delta) \ll N\alpha_k^{1/k} + N\delta^{\frac{2}{k(k+1)}} + N^{\varepsilon}.$$

If $\alpha_k > 1$, then this bound is trivial, and therefore we derive

$$\mathcal{R}(P, N, \delta) \ll N\alpha_k^{1/k} + N\delta^{\frac{2}{k(k+1)}} + N^{\varepsilon}$$

provided that $\alpha_k \geqslant N^{-k}$. Now if $\alpha_k < N^{-k}$, then $N < \alpha_k^{-1/k}$, and thus we may state the following corollary.

Corollary 5.3 *Let* $\delta \in \left[0, \frac{1}{4}\right]$, $N, k \in \mathbb{Z}_{\geqslant 1}$ *and* $P = \alpha_k X^k + \alpha_{k-1} X^{k-1} + \cdots + \alpha_0 \in \mathbb{R}[X]$ *with* $\alpha_k > 0$. *Then, for all* $\varepsilon > 0$

$$\mathcal{R}(P, N, \delta) \ll N\alpha_k^{1/k} + N\delta^{\frac{2}{k(k+1)}} + \alpha_k^{-1/k} + N^{\varepsilon}$$

the term N^{ε} *being replaced by* 1 *if* $k = 1$.

Exercises

102   Show that the function $x \longmapsto \|x\|$ is bounded, even, periodic of period 1 and that we have

$$\left| \|x\| - \|y\| \right| \leqslant |x - y| \quad \text{and} \quad \|x + y\| \leqslant \|x\| + \|y\|.$$

103   Prove that, for all $x \in \mathbb{R}$, we have $2\|x\| \leqslant |\sin(\pi x)| \leqslant \pi \|x\|$.

104   Let $m, n \in \mathbb{Z}_{\geqslant 1}$ such that $n \nmid m$. Show that $\left\| \frac{m}{n} \right\| \geqslant \frac{1}{n}$.

105 (Reflection Principle). Let $f \in C^1[N, 2N]$ be a monotone real-valued function, such that there exist $\lambda_1 > 0$ and $0 < c_1 < c_2$ such that, for all $x \in [N, 2N]$

$$c_1\lambda_1 \leqslant |f'(x)| \leqslant c_2\lambda_1 \quad \text{and} \quad \delta \leqslant \tfrac{1}{4}c_1\lambda_1.$$

Prove that

$$\mathcal{R}(f, N, \delta) \leqslant \mathcal{R}\left(f^{-1}, M, \frac{\delta}{c_1\lambda_1}\right) + \frac{2\delta}{c_1\lambda_1} + 2,$$

where $M \in \mathbb{Z}_{\geqslant 1}$ is such that $c_1 N\lambda_1 \leqslant M \leqslant c_2 N\lambda_1$.

106 *This exercise is the analogue of Lemma 5.17 for the function $f : u \mapsto (x/u)^{1/2}$.* Let $x \geqslant 1$ and δ satisfying (5.2) and N be an integer such that $4 \leqslant N \leqslant x$. The aim is to show that there exists a small real number $c_0 > 0$ such that $N\delta \leqslant c_0$ and

$$\mathcal{R}\left(\sqrt{\frac{x}{n}}, N, \delta\right) \ll (Nx)^{1/6}.$$

One may use the polynomials

$$P(X, Y) = 4X + Y \quad \text{and} \quad Q(X, Y) = 4X + 3Y.$$

(a) Show that there exists a subset T of $\mathcal{S}(f, N, \delta)$ such that $\mathcal{R}(f, N, \delta) \leqslant 4(|T| + 1)$ and any two consecutive elements of T differ by $> 2^{2/3}x^{-1/6}N^{5/6}$.
(b) Deduce the desired result.

107 Using Theorem 5.10, show that for $16 \leqslant y < \tfrac{1}{4}x^{1/2}$, the following asymptotic formula holds.

$$\sum_{x < n \leqslant x+y} \mu_2(n) = \frac{y}{\zeta(2)} + O\left(x^{1/15}y^{2/3}\log x\right).$$

108 Let $r \geqslant 2$ be an integer and x, y be real numbers such that $4^r \leqslant y < x^{1/r}$. By adapting the proof of Corollary 5.2 with the use of Theorem 5.11 instead of Theorem 5.10, show that

$$\sum_{x < n \leqslant x+y} \mu_r(n) = \frac{y}{\zeta(r)} + O\left\{\left(x^{\frac{1}{2r+1}} + yx^{-\frac{1}{6r(2r+1)}}\right)\log x + y^{1/r}\right\}.$$

109   This exercise deals with the problem of square-full numbers in short intervals. Let $x \geqslant 10^{35}$ be a large real number and y be a real number satisfying

$$16\,x^{1/2}(\log x)^3 \leqslant y \leqslant 4^{-3}x(\log x)^{-1}. \tag{5.45}$$

Set $L = L(x, y) = y(x \log x)^{-1/2}$.

(a) Splitting the sum into two subsums, show that

$$\sum_{L<b\leqslant(x+y)^{1/3}} \left(\left\lfloor \sqrt{\frac{x+y}{b^3}} \right\rfloor - \left\lfloor \sqrt{\frac{x}{b^3}} \right\rfloor \right) \ll (R_1 + R_2) \log x + L$$

, where

$$R_1 = \max_{L<B\leqslant(2x)^{1/5}} \mathcal{R}\left(\sqrt{\frac{x}{b^3}}, B, \frac{y}{\sqrt{xB^3}} \right)$$

and

$$R_2 = \max_{L<A\leqslant(2x)^{1/5}} \mathcal{R}\left(\left(\frac{x}{a^2}\right)^{1/3}, A, \frac{y}{(Ax)^{2/3}} \right).$$

(b) Deduce that we have

$$\sum_{x<n\leqslant x+y} s_2(n) = \frac{\zeta(3/2)}{2\zeta(3)} \frac{y}{x^{1/2}} + O\left\{ (R_1 + R_2)\log x + L \right\}.$$

(c) Using Theorems 5.6 and 5.7, show that

$$\sum_{x<n\leqslant x+y} s_2(n) = \frac{\zeta(3/2)}{2\zeta(3)} \frac{y}{x^{1/2}} + O\left\{ x^{2/15}\log x + \frac{y}{(x\log x)^{1/2}} \right\}.$$

(d) Using Theorem 5.8 to estimate R_1 and R_2 in some specific ranges and using Theorem 5.7 in the complementary ranges, show that, if (5.45) is replaced by

$$x^{37/60}(\log x)^3 \leqslant y \leqslant 4^{-3}x(\log x)^{-1}, \tag{5.46}$$

then we have the following improvement:

$$\sum_{x<n\leqslant x+y} s_2(n) = \frac{\zeta(3/2)}{2\zeta(3)} \frac{y}{x^{1/2}} + O\left\{ x^{1/8}(\log x)^{7/5} + \frac{y}{(x\log x)^{1/2}} \right\}.$$

110   Let $x \geqslant 1$ and δ satisfying (5.2) be real numbers, $r \geqslant 2$ and $N \geqslant 4$ be integers satisfying

$$x^\varepsilon \leqslant 2^{r+1} N^r \delta \leqslant \tfrac{2}{3} x$$

for all $\varepsilon > 0$. Using Theorem 4.17 or 4.18, show that

$$R\left(\frac{x}{n^r}, N, \delta\right) \ll_{r,\varepsilon} N^r \delta.$$

111   This exercise provides a variant of Theorem 5.10, from which we take up all the notation. Furthermore, assume that

$$N\delta \leqslant c_0, \quad N < (c_0^{-1}x)^{1/3} \quad \text{and} \quad N^2\delta \geqslant 4x^{-1}. \tag{5.47}$$

We set

$$A = 2^{-1/3} N^{2/3} (x\delta)^{-1/6}. \tag{5.48}$$

(a) Check that this choice of A is admissible in the proof of Theorem 5.10, i.e. using (5.47) check that

$$R < A \leqslant \tfrac{1}{2} N.$$

(b) Deduce that, if (5.47) holds, then

$$R\left(\frac{x}{n^2}, N, \delta\right) \ll x^{1/5} + \left(N^2\delta x\right)^{1/6} + \left(x\delta^{-2}N^{-4}\right)^{1/3} + \left(x\delta^7 N^{11}\right)^{1/9}.$$

(c) Prove that, if $x^{1/5} \leqslant y \leqslant x^{1/3}$, then

$$\sum_{x < n \leqslant x+y} \mu_2(n) = \frac{y}{\zeta(2)} + O\left(x^{1/9}y^{4/9}(\log x)^{7/5}\right)$$

and check that this error term is better than that of Exercise 107 apart from the logarithmic power.

112   Let x, y be real numbers satisfying $x^{1/3} \leqslant y \leqslant x$. Using Theorem 5.6 or otherwise, prove that

$$\max_{4y < N \leqslant x} \sum_{N < n \leqslant 2N} \left(\left\lfloor \frac{x+y}{n} \right\rfloor - \left\lfloor \frac{x}{n} \right\rfloor\right) \ll y.$$

This is a slight improvement of (4.43) but in the restricted range $x^{1/3} \leqslant y \leqslant x$.

113 Let $a \geqslant 2$ be an integer and $f \in C^\infty[N, 2N]$ such that there exists $T \geqslant 1$ such that, for all $x \in [N, 2N]$ and all $j \in \mathbb{Z}_{\geqslant 0}$, we have

$$\left| f^{(j)}(x) \right| \asymp T N^{-j} \tag{5.49}$$

and

$$N\delta \leqslant T \leqslant N^a. \tag{5.50}$$

By mimicking the proof of Proposition 5.2, show that, for all $k \geqslant 2a$, we have

$$\mathcal{R}(f, N, \delta) \ll T^{\frac{2}{k(k+1)}} N^{\frac{k-1}{k+1}} + N\delta^{\frac{1}{a(2a-1)}}.$$

This implies in particular that, if $k \geqslant 4$ and $N\delta \leqslant T \leqslant N^2$, then

$$\mathcal{R}(f, N, \delta) \ll T^{\frac{2}{k(k+1)}} N^{\frac{k-1}{k+1}} + N\delta^{1/6}.$$

References

1. Branton, M., Sargos, P.: Points entiers au voisinage d'une courbe plane à très faible courbure. Bull. Sci. Math. **118**, 15–28 (1994)
2. Filaseta, M., Trifonov, O.: On gaps between squarefree numbers II. J. Lond. Math. Soc. **45**, 215–221 (1992)
3. Filaseta, M., Trifonov, O.: The distribution of fractional parts with application to gap results in number theory. Proc. Lond. Math. Soc. **73**, 241–278 (1996)
4. Gorny, A.: Contribution à l'étude des fonctions dérivables d'une variable réelle. Acta Math. **71**, 317–358 (1939)
5. Huxley, M.N.: Area, Lattice Points and Exponential Sums. Oxford Science Pub., Oxford (1996)
6. Huxley, M.N.: The integer points close to a curve III. In: Győry et al. (eds.) Number Theory in Progress, vol. 2, pp. 911–940. de Gruyter, Berlin (1999)
7. Huxley, M.N.: The integer points in a plane curve. Funct. Approx. **37**, 213–231 (2007)
8. Huxley, M.N., Sargos, P.: Points entiers au voisinage d'une courbe plane de classe C^n. Acta Arith. **69**, 359–366 (1995)
9. Huxley, M.N., Sargos, P.: Points entiers au voisinage d'une courbe plane de classe C^n, II. Funct. Approx. **35**, 91–115 (2006)
10. Huxley, M.N., Trifonov, O.: The square-full numbers in an interval. Math. Proc. Camb. Phil. Soc. **119**, 201–208 (1996)
11. Konyagin, S.: Estimates for the least prime factor of a binomial coefficient. Mathematika **45**, 41–55 (1998)
12. Letendre, P.: The number of integer points close to a polynomial, 12 pp. Preprint (2019). https://arxiv.org/pdf/1901.10004v1.pdf
13. Mandelbrojt, S.: Séries Adhérentes, Régularisation des Suites, Applications. Gauthiers-Villars, Paris (1952)
14. Matomäki, K., Teräväinen, J.: On the Möbius function in all short intervals, 17 pp. Preprint (2019). https://arxiv.org/pdf/1911.09076.pdf

15. Shiu, P.: On the number of square-full integers between successive squares. Mathematika **27**, 171–178 (1980)
16. Srinivasan, B.R.: On Van der Corput's and Nieland's results on the Dirichlet's divisor problem and the circle problem. Proc. Natl. Inst. Sci. Part A **28**, 732–742 (1962)
17. Trifonov, O.: Lattice points close to a smooth curve and squarefull numbers in short intervals. J. Lond. Math. Soc. **65**, 309–319 (2002)

Chapter 6
Exponential Sums

6.1 The ψ-Function

6.1.1 Back to the Divisor Problem

The function $\psi : x \mapsto \psi(x) = x - \lfloor x \rfloor - \frac{1}{2}$ appears in most problems in number theory. The reason is quite simple: when we look at asymptotics for average orders of classical arithmetic functions, the summations are often taken over some subsets of the set of divisors of some integer n, and when interchanging the order of the summations, integer parts, and hence the function ψ, arise. For instance, let us consider the number $\tau(n)$ of divisors of n. By the Dirichlet hyperbola principle[1] or equivalently using (7.48), we first derive

$$\sum_{n \leqslant x} \tau(n) = 2 \sum_{d \leqslant \sqrt{x}} \left\lfloor \frac{x}{d} \right\rfloor - \lfloor \sqrt{x} \rfloor^2.$$

Now replacing each integer part by ψ and using the estimate

$$\sum_{d \leqslant x} \frac{1}{d} = \log x + \gamma - \frac{\psi(x)}{x} + O\left(\frac{1}{x^2}\right)$$

---

[1] See also the proof of Theorem 4.28.

© Springer Nature Switzerland AG 2020
O. Bordellès, *Arithmetic Tales*, Universitext,
https://doi.org/10.1007/978-3-030-54946-6_6

obtained in Example 1.1, we get

$$\sum_{n \leqslant x} \tau(n) = 2 \sum_{d \leqslant \sqrt{x}} \left(\frac{x}{d} - \psi \left(\frac{x}{d} \right) - \frac{1}{2} \right) - \left(\sqrt{x} - \psi \left(\sqrt{x} \right) - \frac{1}{2} \right)^2$$

$$= 2x \sum_{d \leqslant x} \frac{1}{d} - 2 \sum_{d \leqslant \sqrt{x}} \psi \left(\frac{x}{d} \right) - x - \psi \left(\sqrt{x} \right)^2 + \frac{1}{4} + 2\psi \left(\sqrt{x} \right) \sqrt{x}$$

$$= x (\log x + 2\gamma - 1) - 2 \sum_{d \leqslant \sqrt{x}} \psi \left(\frac{x}{d} \right) + O(1).$$

The error term in this problem is usually denoted by $\Delta(x)$, so that

$$\Delta(x) = \sum_{n \leqslant x} \tau(n) - x(\log x + 2\gamma - 1) = -2 \sum_{d \leqslant \sqrt{x}} \psi \left(\frac{x}{d} \right) + O(1).$$

In fact, with a little more work, one can prove [5] that

$$|\Delta(x)| \leqslant 2 \left| \sum_{d \leqslant \sqrt{x}} \psi (x/d) \right| + \tfrac{1}{2}. \tag{6.1}$$

Using the trivial bound $|\psi(x)| \leqslant \frac{1}{2}$ enables us to recover Theorem 4.28. The *Dirichlet divisor problem* is about the smallest exponent $\theta \in [0, 1)$ for which the following estimate

$$\sum_{n \leqslant x} \tau(n) = x(\log x + 2\gamma - 1) + O \left(x^\theta (\log x)^\beta \right)$$

holds for some $\beta \geqslant 0$. By Theorem 4.28 or the computations made above, the pair $(\theta, \beta) = \left(\frac{1}{2}, 0 \right)$ is admissible. Hardy showed in [30] that $\Delta(x)$ cannot be a $o(x^{1/4})$, in other words we necessarily have $\theta \geqslant \frac{1}{4}$ with some $\beta > 0$.

It is surmised that $\theta = \frac{1}{4}$ is the right value in the Dirichlet divisor problem, which is supported by the following mean value result

$$\int_1^T \Delta(x)^2 dx = \frac{\zeta(3/2)^4}{36\zeta(2)\zeta(3)} T^{3/2} + O \left(T(\log T)^5 \right)$$

due to Tong [82]. However, no one has been in a position to prove it until now. The best result up to date is due to Bourgain and Watt [7] who showed that

$$\theta \leqslant \frac{517}{1\,648}.$$

Note that $\frac{517}{1648} \approx 0,3137\ldots$ so that we are still far from the conjectured value, but this problem gradually gets harder and harder.

6.1.2 Vaaler's and Steckin's Inequalities

Now let us return to the function ψ. This function is odd and 1-periodic, and hence permits a Fourier series expansion. By Proposition 1.5, the series

$$-\sum_{h=1}^{\infty} \frac{\sin(2\pi hx)}{\pi h} = -\sum_{h\in\mathbb{Z}\setminus\{0\}} \frac{e(hx)}{2\pi ih}$$

converges to $\psi(x)$ if $x \in \mathbb{R} \setminus \mathbb{Z}$ and to 0 if $x \in \mathbb{Z}$. Unfortunately, the convergence is not uniform, so that it would be better to work with partial sums of this series, or more generally with trigonometric polynomials. The following result, due to Vaaler [83], is a useful answer to this question.

Theorem 6.1 (Vaaler) *For all real numbers $x \geqslant 1$ and all integers $H \geqslant 1$, we have*

$$\psi(x) = -\sum_{0<|h|\leqslant H} \Phi\left(\frac{h}{H+1}\right) \frac{e(hx)}{2\pi ih} + \mathcal{R}_H(x),$$

where $\Phi(t) = \pi t(1 - |t|)\cot(\pi t) + |t|$ for $0 < |t| < 1$ and

$$|\mathcal{R}_H(x)| \leqslant \frac{1}{2H+2} \sum_{|h|\leqslant H} \left(1 - \frac{|h|}{H+1}\right) e(hx).$$

Note that $0 < \Phi(t) < 1$ for $0 < |t| < 1$ and, using Exercise 118, we have

$$\sum_{|h|\leqslant H} \left(1 - \frac{|h|}{H+1}\right) e(hx) = \frac{1}{H+1} \left|\sum_{h=0}^{H} e(hx)\right|^2$$

so that the sum in the error term is a non-negative real number. Using this with $x = f(n)$ and summing over $(N, 2N]$ we get the following inequality.

Corollary 6.1 *Let N be a positive integer and $f : [N, 2N] \to \mathbb{R}$ be any function. Then, for any $H \in \mathbb{Z}_{\geqslant 1}$*

$$\left|\sum_{N<n\leqslant 2N} \psi(f(n))\right| \leqslant \frac{N}{2H+2} + \left(1 + \frac{1}{\pi}\right) \sum_{h=1}^{H} \frac{1}{h} \left|\sum_{N<n\leqslant 2N} e(hf(n))\right|.$$

Proof Using Theorem 6.1 we get

$$\sum_{N<n\leqslant 2N} \psi\left(f(n)\right) = -\sum_{0<|h|\leq H} \Phi\left(\frac{h}{H+1}\right)\frac{1}{2\pi ih}\sum_{N<n\leqslant 2N} e\left(hf(n)\right) + \sum_{N<n\leqslant 2N} \mathcal{R}_H(f(n)),$$

where Φ and $\mathcal{R}_H$ are defined in Theorem 6.1, and hence

$$\left|\sum_{N<n\leqslant 2N} \psi\left(f(n)\right)\right| \leqslant |\Sigma_1| + |\Sigma_2|$$

with

$$|\Sigma_1| \leqslant \sum_{0<|h|\leqslant H} \frac{1}{2\pi|h|}\left|\sum_{N<n\leqslant 2N} e\left(hf(n)\right)\right| = \frac{1}{\pi}\sum_{h=1}^{H}\frac{1}{h}\left|\sum_{N<n\leqslant 2N} e\left(hf(n)\right)\right|$$

and

$$|\Sigma_2| \leqslant \frac{1}{2H+2}\sum_{|h|\leqslant H}\left(1-\frac{|h|}{H+1}\right)\sum_{N<n\leqslant 2N} e\left(hf(n)\right)$$

$$= \frac{N}{2H+2} + \frac{1}{H+1}\sum_{h=1}^{H}\left(1-\frac{h}{H+1}\right)\mathrm{Re}\left(\sum_{N<n\leqslant 2N} e\left(hf(n)\right)\right)$$

$$\leqslant \frac{N}{2H+2} + \sum_{h=1}^{H}\frac{1}{h}\left|\sum_{N<n\leqslant 2N} e\left(hf(n)\right)\right|$$

which implies the asserted result. □

Remark 6.1 One can also work directly with the partial sums of the Fourier series of the function ψ with the following asymptotic formula. For all positive integers H, we have (see [64], for instance)

$$\psi(x) = -\sum_{0<|h|\leqslant H}\frac{e(hx)}{2\pi ih} + O\left\{\min\left(1,\frac{1}{H\|x\|}\right)\right\},$$

where $\|x\|$ is the distance from x to its nearest integer. The two results are equivalent, but Vaaler's theorem is often more useful in applications.

Corollary 6.1 is a particular case of the following more general result due to Stečkin [81].[2]

---

[2]I thank Ognian Trifonov for pointing out this result to me and writing a proof of it.

Proposition 6.1 (Stečkin) *Let g be a 1-periodic function of bounded variation on the interval $[0, 1]$. Then for any function f, any integers a, b with $b - a > 1$, any set S of positive integers, and any integer $H \geqslant 1$, we have*

$$\sum_{\substack{a < n \leqslant b \\ n \in S}} g\left(f(n)\right) = \left(\sum_{\substack{a < n \leqslant b \\ n \in S}} 1\right) \int_0^1 g(t)\, dt + O\left\{\frac{1}{H} \sum_{\substack{a < n \leqslant b \\ n \in S}} 1 + \sum_{h=1}^{H} \frac{1}{h} \left| \sum_{\substack{a < n \leqslant b \\ n \in S}} e\left(hf(n)\right) \right| \right\}.$$

Proof Let $H \in \mathbb{Z}_{\geqslant 1}$ and define $t_H(x) = \sum_{|h| \leqslant H} c_h e(hx)$ and $T_H(x) = \sum_{|h| \leqslant H} C_h e(hx)$ to be the trigonometric polynomials of degree H of best one-sided L^1 approximation of g. Thus, t_H and T_H are real-valued and, for all $x \in \mathbb{R}$, we have $t_H(x) \leqslant g(x) \leqslant T_H(x)$, so that

$$\sum_{\substack{a < n \leqslant b \\ n \in S}} \sum_{|h| \leqslant H} c_h e\left(hf(n)\right) \leqslant \sum_{\substack{a < n \leqslant b \\ n \in S}} g\left(f(n)\right) \leqslant \sum_{\substack{a < n \leqslant b \\ n \in S}} \sum_{|h| \leqslant H} C_h e\left(hf(n)\right).$$

Therefore

$$\left(\sum_{\substack{a < n \leqslant b \\ n \in S}} 1\right)\left(c_0 - \int_0^1 g(t)\, dt\right) + \sum_{\substack{a < n \leqslant b \\ n \in S}} \sum_{0 < |h| \leqslant H} c_h e\left(hf(n)\right)$$

$$\leqslant \sum_{\substack{a < n \leqslant b \\ n \in S}} g\left(f(n)\right) - \left(\sum_{\substack{a < n \leqslant b \\ n \in S}} 1\right) \int_0^1 g(t)\, dt$$

$$\leqslant \left(\sum_{\substack{a < n \leqslant b \\ n \in S}} 1\right)\left(C_0 - \int_0^1 g(t)\, dt\right) + \sum_{\substack{a < n \leqslant b \\ n \in S}} \sum_{0 < |h| \leqslant H} C_h e\left(hf(n)\right).$$

Now using [25, Theorems II and III], we get

$$\left| c_0 - \int_0^1 g(t)\, dt \right| = \|g - t_H\|_1 \ll \frac{V(g)}{H},$$

where $V(g)$ is the total variation of g over a period. Similarly

$$\left| C_0 - \int_0^1 g(t)\, dt \right| = \|g - T_H\|_1 \ll \frac{V(g)}{H}.$$

Furthermore, for $1 \leqslant h \leqslant H$, we have

$$c_h = \int_0^1 t_H(x) e(-hx)\, dx = \int_0^1 \left(t_H(x) - g(x)\right) e(-hx)\, dx + \int_0^1 g(x) e(-hx)\, dx$$

and hence, using Lemma 6.7, we derive

$$\left| c_h \right| \leqslant \| g - t_H \|_1 + \left| \int_0^1 g(x) e(-hx) \, dx \right| \ll V(g) \left(\frac{1}{H} + \frac{1}{h} \right) \ll \frac{1}{h}$$

and similarly $|C_h| \ll h^{-1}$, completing the proof.   □

6.2   Basic Inequalities

6.2.1   Cauchy–Schwarz

Cauchy–Schwarz's inequality, which plays an important role in many branches of modern mathematics, is one of the most famous results in inequalities theory. In analytic number theory, this inequality is often used to separate certain functions for which better results are obtained when they are summed individually. For a survey of this inequality and related results, the reader should refer to [21].

Lemma 6.1 (Cauchy–Schwarz's Inequality) *Let M be a positive integer and $a_1, \ldots, a_M$ and $b_1, \ldots, b_M$ be two finite sequences of arbitrary complex numbers. Then we have*

$$\left| \sum_{m=1}^{M} a_m b_m \right|^2 \leqslant \sum_{m=1}^{M} |a_m|^2 \sum_{m=1}^{M} |b_m|^2.$$

Proof A possible proof of this inequality rests on Cauchy–Binet's identity for complex numbers

$$\sum_{m=1}^{M} x_m y_m \sum_{m=1}^{M} z_m t_m - \sum_{m=1}^{M} x_m t_m \sum_{m=1}^{M} z_m y_m$$

$$= \frac{1}{2} \sum_{m=1}^{M} \sum_{n=1}^{M} (x_m z_n - x_n z_m)(y_m t_n - y_n t_m)$$

which is a generalization of Lagrange's identity, applied with $x_m = \overline{a_m}$, $y_m = a_m$, $z_m = b_m$ and $t_m = \overline{b_m}$ (for $m = 1, \ldots, M$), so that we get at once

$$\sum_{m=1}^{M} |a_m|^2 \sum_{m=1}^{M} |b_m|^2 - \left| \sum_{m=1}^{M} a_m b_m \right|^2 = \frac{1}{2} \sum_{m=1}^{M} \sum_{n=1}^{M} |\overline{a_m} \, b_n - \overline{a_n} \, b_m|^2 \geqslant 0$$

as required.   □

6.2.2 Weyl's Shift

The next tool is a device based upon the following observation. Suppose that N is a positive integer and $a_{N+1}, \ldots, a_{2N}$ are N arbitrary complex numbers. If we set for all $n \in \mathbb{Z}$

$$\alpha_n = \begin{cases} a_n, & \text{if } n \in \{N+1, \ldots, 2N\} \\ 0, & \text{otherwise,} \end{cases} \tag{6.2}$$

then we have for all integers h

$$\sum_{N < n \leqslant 2N} a_n = \sum_{n \in \mathbb{Z}} \alpha_n = \sum_{n \in \mathbb{Z}} \alpha_{n+h}. \tag{6.3}$$

Equality (6.3) is called *Weyl's shift*. It allows a great flexibility with the indices of the sums. For instance, for all positive integers H, we can write

$$\sum_{N < n \leqslant 2N} a_n = \frac{1}{H} \sum_{h=1}^{H} \sum_{n \in \mathbb{Z}} \alpha_{n+h}$$

and interchanging the summations we infer that

$$\sum_{N < n \leqslant 2N} a_n = \frac{1}{H} \sum_{n=N-H+1}^{2N-1} \sum_{h=1}^{H} \alpha_{n+h}. \tag{6.4}$$

As a first application, we derive the following useful lemma.

Lemma 6.2 *Let $N, N_1 \in \mathbb{Z}_{\geqslant 1}$ such that $N < N_1 \leqslant 2N$ and $a_{N+1}, \ldots, a_{N_1} \in \mathbb{C}$ satisfying $|a_n| \leqslant 1$. Then, for any $H \in \{1, \ldots, N-1\}$*

$$\left| \sum_{N < n \leqslant N_1} a_n \right| \leqslant \frac{1}{H} \left| \sum_{N < n \leqslant N_1 - H} \sum_{h \leqslant H} a_{n+h} \right| + 2H.$$

Proof Define

$$\alpha_n := \begin{cases} a_n, & \text{if } N < n \leqslant N_1 \\ 0, & \text{otherwise.} \end{cases}$$

Then

$$\sum_{N < n \leqslant N_1} a_n = \frac{1}{H} \sum_{h \leqslant H} \sum_{n \in \mathbb{Z}} \alpha_n = \frac{1}{H} \sum_{h \leqslant H} \sum_{n \in \mathbb{Z}} \alpha_{n+h} = \frac{1}{H} \sum_{h \leqslant H} \sum_{N-h < n \leqslant N_1 - h} \alpha_{n+h}$$

and interverting the sums yields

$$\sum_{N<n\leqslant N_1} a_n = \frac{1}{H} \sum_{N-H<n\leqslant N_1-1} \sum_{h\leqslant H} \alpha_{n+h}.$$

Now splitting the outer sum into three subsums as

$$\sum_{N<n\leqslant N_1} a_n = \frac{1}{H} \left(\sum_{N-H<n\leqslant N} + \sum_{N<n\leqslant N_1-H} + \sum_{N_1-H<n\leqslant N_1-1} \right) \sum_{h\leqslant H} \alpha_{n+h}$$

and estimating the 1st and the 3rd sum trivially, we derive

$$\left| \sum_{N<n\leqslant N_1} a_n \right| \leqslant \frac{1}{H} \left| \sum_{N<n\leqslant N_1-H} \sum_{h\leqslant H} \alpha_{n+h} \right| + \frac{2H^2}{H}$$

completing the proof since, in the range of the sum, we have $\alpha_{n+h} = a_{n+h}$. □

6.2.3 Van der Corput's A-Process

The following inequality, discovered by van der Corput, is a very clever application of Weyl's shift and Cauchy–Schwarz's inequality. Often called the *A-process*, it is a first result given by van der Corput by which the problem of estimating a given exponential sum with a function $f(n)$ is replaced by estimating another exponential sum with the upper finite difference $\Delta_h f(n) = f(n+h) - f(n)$ of f. Since $\Delta_h f(n)$ is of the same order of magnitude as $hf'(n)$, one may expect some improvement in the estimates.

Proposition 6.2 *Let N be a positive integer and $a_{N+1}, \ldots, a_{2N}$ be arbitrary complex numbers. For all integers $H \in \{1, \ldots, N\}$, we have*

$$\left| \sum_{N<n\leqslant 2N} a_n \right|^2 \leqslant \frac{2N}{H+1} \sum_{N<n\leqslant 2N} |a_n|^2 + \frac{4N}{H+1} \sum_{h=1}^{H} \left| \sum_{N<n\leqslant 2N-h} \overline{a_n}\, a_{n+h} \right|.$$

Proof Let S_N be the sum on the left-hand side. By (6.4), we have

$$|S_N| = \frac{1}{H+1} \left| \sum_{n=N-H}^{2N-1} \sum_{h=1}^{H+1} \alpha_{n+h} \right|,$$

where the numbers α_n are defined in (6.2), and by Cauchy–Schwarz's inequality, we infer that

$$|S_N|^2 \leqslant \frac{N+H}{(H+1)^2} \sum_{n=N-H}^{2N-1} \left| \sum_{h=1}^{H+1} \alpha_{n+h} \right|^2.$$

Squaring out the modulus we obtain

$$|S_N|^2 \leqslant \frac{N+H}{(H+1)^2} \sum_{n=N-H}^{2N-1} \sum_{i=1}^{H+1} \sum_{j=1}^{H+1} \alpha_{n+i} \, \overline{\alpha_{n+j}}.$$

We now change the variables in the following way. Set $i = j+h$ with $1 \leqslant i \leqslant H+1$, so that $1 - h \leqslant j \leqslant H + 1 - h$ with $|h| = |i - j| \leqslant H$. This gives

$$|S_N|^2 \leqslant \frac{N+H}{(H+1)^2} \sum_{n=N-H}^{2N-1} \sum_{|h| \leqslant H} \sum_{j=1-h}^{H+1-h} \alpha_{n+h+j} \, \overline{\alpha_{n+j}}$$

and using Weyl's shift (6.3) we get

$$|S_N|^2 \leqslant \frac{N+H}{(H+1)^2} \sum_{n \in \mathbb{Z}} \sum_{|h| \leqslant H} \sum_{j=1-h}^{H+1-h} \alpha_{n+h} \, \overline{\alpha_n}$$

$$= \frac{N+H}{(H+1)^2} \sum_{|h| \leqslant H} \sum_{N<n \leqslant 2N-h} (H+1-|h|) \, \alpha_{n+h} \, \overline{\alpha_n}$$

$$= \frac{N+H}{H+1} \sum_{|h| \leqslant H} \left(1 - \frac{|h|}{H+1} \right) \sum_{N<n \leqslant 2N-h} \alpha_{n+h} \, \overline{\alpha_n}.$$

We then separate the case $h = 0$ from the other values, which gives

$$|S_N|^2 \leqslant \frac{N+H}{H+1} \sum_{N<n \leqslant 2N} |\alpha_n|^2 + \frac{N+H}{H+1} \sum_{\substack{|h| \leqslant H \\ h \neq 0}} \left(1 - \frac{|h|}{H+1} \right) \sum_{N<n \leqslant 2N-h} \alpha_{n+h} \, \overline{\alpha_n}$$

$$= \frac{N+H}{H+1} \sum_{N<n \leqslant 2N} |\alpha_n|^2 + \frac{N+H}{H+1} \left\{ \sum_{h=1}^{H} \left(1 - \frac{h}{H+1} \right) \sum_{N<n \leqslant 2N-h} \alpha_{n+h} \, \overline{\alpha_n} \right.$$

$$+ \sum_{h=1}^{H} \left(1 - \frac{h}{H+1} \right) \sum_{N<n \leqslant 2N+h} \alpha_{n-h} \, \overline{\alpha_n} \bigg\}$$

and the use of Weyl's shift (6.4) again on the last sum gives

$$
|S_N|^2 \leqslant \frac{N+H}{H+1} \sum_{N<n\leqslant 2N} |\alpha_n|^2 + \frac{N+H}{H+1} \left\{ \sum_{h=1}^{H} \left(1 - \frac{h}{H+1}\right) \sum_{N<n\leqslant 2N-h} \alpha_{n+h}\,\overline{\alpha_n} \right.
$$

$$
\left. + \sum_{h=1}^{H} \left(1 - \frac{h}{H+1}\right) \sum_{N<n\leqslant 2N-h} \alpha_n\,\overline{\alpha_{n+h}} \right\}
$$

$$
= \frac{N+H}{H+1} \sum_{N<n\leqslant 2N} |\alpha_n|^2 + \frac{2(N+H)}{H+1} \operatorname{Re}\left\{ \sum_{h=1}^{H} \left(1 - \frac{h}{H+1}\right) \sum_{N<n\leqslant 2N-h} \alpha_{n+h}\,\overline{\alpha_n} \right\}
$$

$$
\leqslant \frac{N+H}{H+1} \sum_{N<n\leqslant 2N} |a_n|^2 + \frac{2(N+H)}{H+1} \sum_{h=1}^{H} \left| \sum_{N<n\leqslant 2N-h} a_{n+h}\,\overline{a_n} \right|
$$

and the bound $H \leqslant N$ gives the asserted result.   □

6.3 Exponential Sums Estimates

6.3.1 The First Derivative Theorem

The first result in exponential sums is the following estimate which was of interest to mathematicians including van der Corput, Kusmin, Landau, Karamata and Tomic (see [67]). We begin by proving the original result, and then we provide some more practical versions we shall be able to use later.

We first start with a technical lemma.

Lemma 6.3 *Let* $M \in \mathbb{N}$ *and* $x_1, \ldots, x_M$ *be pairwise distinct complex numbers. Then*

$$
2 \sum_{n=1}^{M} x_n = x_1 \left(\frac{x_1 + x_2}{x_1 - x_2} + 1 \right)
$$

$$
+ \sum_{n=2}^{M-1} x_n \left(\frac{x_n + x_{n+1}}{x_n - x_{n+1}} - \frac{x_{n-1} + x_n}{x_{n-1} - x_n} \right) + x_M \left(1 - \frac{x_{M-1} + x_M}{x_{M-1} - x_M}\right).
$$

Proof If we set $x_0 = x_{M+1} = 0$ and $T_n = \dfrac{x_n + x_{n+1}}{x_n - x_{n+1}}$ for $n = 0, \ldots, M$, then the right-hand side above is equal to

$$
\sum_{n=1}^{M} x_n \left(T_n - T_{n-1}\right).
$$

Abel's summation yields

$$\sum_{n=1}^{M} x_n \left(T_n - T_{n-1}\right) = x_M \sum_{n=1}^{M} \left(T_n - T_{n-1}\right) - \sum_{n=1}^{M-1} \left(x_{n+1} - x_n\right) \sum_{k=1}^{n} \left(T_k - T_{k-1}\right)$$

$$= x_M \left(T_M - T_0\right) - \sum_{n=1}^{M-1} \left(x_{n+1} - x_n\right) \left(T_n - T_0\right)$$

and using $T_0 = -1$, $T_M = 1$ and $(x_{n+1} - x_n) T_n = -(x_n + x_{n+1})$ we get

$$\sum_{n=1}^{M} x_n \left(T_n - T_{n-1}\right) = 2x_M + \sum_{n=1}^{M-1} \left(x_n + x_{n+1}\right) - \sum_{n=1}^{M-1} \left(x_{n+1} - x_n\right) = 2 \sum_{n=1}^{M} x_n$$

as required.   □

We now are in a position to prove the main result of this section.

Theorem 6.2 (Kusmin–Landau's Inequality) *Let $M \in \mathbb{Z}_{\geq 1}$ and $f :$ $[1, M] \to \mathbb{R}$ be a function such that there exists $\lambda_1 \in \mathbb{R}$ such that*

$$0 < \lambda_1 \leqslant f(2) - f(1) \leqslant \cdots \leqslant f(M) - f(M-1) \leqslant 1 - \lambda_1. \qquad (6.5)$$

Then we have

$$\left| \sum_{n=1}^{M} e\left(\pm f(n)\right) \right| \leqslant \frac{2}{\pi \lambda_1}.$$

Proof Let $S_M := \sum_{n=1}^{M} e\left(f(n)\right)$. We use Lemma 6.3 with $x_n = e(f(n))$ and the fact that

$$\frac{e(x) + e(y)}{e(x) - e(y)} = \frac{1 + e(y - x)}{1 - e(y - x)} = i \cot \pi (y - x)$$

so that

$$2S_M = e(f(1)) \left\{1 + i \cot \pi \left(f(2) - f(1)\right)\right\} + e\left(f(M)\right) \left\{1 + i \cot \pi \left(f(M) - f(M-1)\right)\right\}$$

$$+ i \sum_{n=2}^{M-1} e\left(f(n)\right) \left\{\cot \pi \left(f(n+1) - f(n)\right) - \cot \pi \left(f(n) - f(n-1)\right)\right\}.$$

Now using (6.5) and the fact that the function $x \mapsto \cot(\pi x)$ is strictly decreasing in $(0, 1)$, we derive

$$
2\,|S_M| \leqslant \frac{1}{\sin \pi\,(f(2) - f(1))} + \frac{1}{\sin \pi\,(f(M) - f(M-1))}
$$

$$
+ \sum_{n=2}^{M-1} |\cot \pi\,(f(n+1) - f(n)) - \cot \pi\,(f(n) - f(n-1))|
$$

$$
= \frac{1}{\sin \pi\,(f(2) - f(1))} + \frac{1}{\sin \pi\,(f(M) - f(M-1))}
$$

$$
+ \cot \pi\,(f(2) - f(1)) - \cot \pi\,(f(M) - f(M-1))
$$

$$
= \cot\left(\tfrac{\pi}{2}\,(f(2) - f(1))\right) + \tan\left(\tfrac{\pi}{2}\,(f(M) - f(M-1))\right)
$$

$$
\leqslant \cot\left(\tfrac{1}{2}\pi\lambda_1\right) + \tan\left(\tfrac{\pi}{2}\,(1 - \lambda_1)\right) = 2\cot\left(\tfrac{1}{2}\pi\lambda_1\right)
$$

and we conclude the proof by using the inequality $\cot x \leqslant \frac{1}{x}$ valid for $0 < x \leqslant \frac{1}{2}\pi$. $\qquad\square$

Corollary 6.2 *Let $N < N_1 \leqslant 2N$ be positive integers and $f \in C^1[N, N_1]$ such that f' is non-decreasing and there exist real numbers $c_1 \geqslant 1$ and $0 < \lambda_1 \leqslant (c_1 + 1)^{-1}$ such that, for all $x \in [N, N_1]$, we have*

$$
\lambda_1 \leqslant f'(x) \leqslant c_1 \lambda_1. \tag{6.6}
$$

Then

$$
\left| \sum_{N < n \leqslant N_1} e\,(\pm f(n)) \right| \leqslant \frac{2}{\pi \lambda_1}.
$$

Proof First note that, since the hypothesis $0 < \lambda_1 \leqslant (c_1 + 1)^{-1}$ implies that $c_1 \lambda_1 \leqslant 1 - \lambda_1$, it suffices to show the Corollary with (6.6) replaced by

$$
\lambda_1 \leqslant f'(x) \leqslant 1 - \lambda_1.
$$

Define the functions g and h by

$$
g(x) = f(x + N) \qquad (1 \leqslant x \leqslant N_1 - N)
$$
$$
h(x) = g(x) - g(x - 1) \quad (2 \leqslant x \leqslant N_1 - N).
$$

Since $g'(x) = f'(x + N)$, the function g' is non-decreasing and we have $\lambda_1 \leqslant g'(x) \leqslant 1 - \lambda_1$. Furthermore, since $h'(x) = g'(x) - g'(x - 1) \geqslant 0$, the function h is non-decreasing. Hence we have

$$\lambda_1 \leqslant g(2) - g(1) \leqslant \cdots \leqslant g(N_1 - N) - g(N_1 - N - 1) \leqslant 1 - \lambda_1$$

so that g satisfies the hypothesis of Theorem 6.2, and we get

$$\left| \sum_{N < n \leqslant N_1} e\left(f(n)\right) \right| = \left| \sum_{n=1}^{N_1 - N} e\left(g(n)\right) \right| \leqslant \frac{2}{\pi \lambda_1}$$

which concludes the proof. □

Remark 6.2 Since $e(f(n)) = e(f(n) - kn)$ for any $k \in \mathbb{Z}$, the inequality of Corollary 6.2 is still true under the weaker assumption

$$k + \lambda_1 \leqslant f'(x) \leqslant k + 1 - \lambda_1 \quad (k \in \mathbb{Z}).$$

6.3.2 The Second Derivative Theorem

Kusmin–Landau's inequality is sharp under the sole hypothesis (6.6) but the condition $0 < \lambda_1 \leqslant (c_1 + 1)^{-1}$ is too restrictive to be really used efficiently in usual problems of number theory. In the 1920s, van der Corput established a very useful inequality which can be considered as the starting point of crucial theorems supplying estimates for exponential sums.

Theorem 6.3 (van der Corput) *Let $N < N_1 \leqslant 2N$ be positive integers and $f \in C^2[N, N_1]$ such that there exist real numbers $c_2 \geqslant 1$ and $\lambda_2 > 0$ such that, for all $x \in [N, N_1]$, we have*

$$\lambda_2 \leqslant f''(x) \leqslant c_2 \lambda_2. \tag{6.7}$$

Then

$$\left| \sum_{N < n \leqslant N_1} e\left(\pm f(n)\right) \right| \leqslant \frac{4}{\sqrt{\pi}} \left(c_2 N \lambda_2^{1/2} + 2\lambda_2^{-1/2} \right).$$

Remark 6.3 The condition (6.7) will prove to be much more useful than (6.6). It also should be mentioned that the proof of the theorem will make the following slightly stronger result appear.

Let $N < N_1 \leqslant 2N$ be positive integers and $f \in C^2[N, N_1]$ such that there exists a real number $\lambda_2 > 0$ such that, for all $x \in [N, N_1]$, we have $f''(x) \geqslant \lambda_2$. Then

$$\left| \sum_{N < n \leqslant N_1} e\left(\pm f(n)\right) \right| \leqslant \frac{4}{\sqrt{\pi \lambda_2}} \left(f'(N_1) - f'(N) + 2 \right).$$

In practice, we shall make no use of such an improvement.

Many methods can be used to show Theorem 6.3. The one we will introduce here is the discrete analogue of the proof of the second derivative test for integrals, also proved by van der Corput.[3] Another proof is supplied in Exercise 115. See also [26, 63].

We begin with an intermediate result.

Lemma 6.4 *Let $N < N_1 \leqslant 2N$ be positive integers and $f \in C^2[N, N_1]$ such that there exists a real number $\lambda_2 \in (0, \pi^{-1})$ such that, for all $x \in [N, N_1]$, we have $f''(x) \geqslant \lambda_2$. Assume also that, for all $x \in (N, N_1)$, we have $f'(x) \notin \mathbb{Z}$. Then*

$$\left| \sum_{N < n \leqslant N_1} e\left(\pm f(n)\right) \right| \leqslant \frac{4}{\sqrt{\pi \lambda_2}}.$$

Proof Let $t \in (0, 1)$ be a parameter at our disposal. Since $f'(x) \notin \mathbb{Z}$ for all $x \in (N, N_1)$, there exist $u, v \in \mathbb{R}$ and non-negative integers $M_1, M_2 \in (N, N_1)$ such that

$$u = f'(N) \quad \text{and} \quad v = f'(N_1)$$
$$f'(M_1) = \lfloor u \rfloor + t \text{ and } f'(M_2) = \lfloor u \rfloor + 1 - t.$$

We now split the sum into three subsums

$$\sum_{N < n \leqslant N_1} e\left(f(n)\right) = \sum_{N < n \leqslant M_1} e\left(f(n)\right) + \sum_{M_1 < n \leqslant M_2} e\left(f(n)\right) + \sum_{M_2 < n \leqslant N_1} e\left(f(n)\right)$$

---

[3] See Lemma 6.6.

and estimate the first and third sums trivially. The mean value theorem implies the
existence of a real number $c \in (N, 2N)$ such that

$$\left| \sum_{N < n \leqslant M_1} e\left(f(n)\right) \right| \leqslant \max\left(M_1 - N, 1\right) = \max\left(\frac{f'(M_1) - f'(N)}{f''(c)}, 1\right)$$

$$\leqslant \max\left(\frac{\lfloor u \rfloor + t - u}{\lambda_2}, 1\right) \leqslant \max\left(\frac{t}{\lambda_2}, 1\right).$$

Similarly, we have

$$\left| \sum_{M_2 < n \leqslant N_1} e\left(f(n)\right) \right| \leqslant \max\left(\frac{v - (\lfloor u \rfloor + 1) + t}{\lambda_2}, 1\right) \leqslant \max\left(\frac{t}{\lambda_2}, 1\right)$$

since $\lfloor u \rfloor \leqslant u \leqslant v \leqslant \lfloor u \rfloor + 1$. The second sum is estimated using Corollary 6.2.
Since f' is non-decreasing and, in the interval $[M_1, M_2]$, we have $\lfloor u \rfloor + t \leqslant$
$f'(x) \leqslant \lfloor u \rfloor + 1 - t$, Corollary 6.2 applies and yields

$$\left| \sum_{M_1 < n \leqslant M_2} e\left(f(n)\right) \right| \leqslant \frac{2}{\pi t}.$$

Hence we get

$$\left| \sum_{N < n \leqslant N_1} e\left(\pm f(n)\right) \right| \leqslant 2 \max\left(\frac{t}{\lambda_2}, 1\right) + \frac{2}{\pi t}$$

and the asserted estimate follows from taking $t = \sqrt{\lambda_2 \pi^{-1}}$ since $\lambda_2 < \pi^{-1}$.   □

We are now able to prove Theorem 6.3.

Proof (Proof of Theorem 6.3) If $\lambda_2 \geqslant \pi^{-1}$, then we have

$$\frac{4N\lambda_2^{1/2}}{\sqrt{\pi}} \geqslant \frac{4N}{\pi} > N \geqslant N_1 - N \geqslant \left| \sum_{N < n \leqslant N_1} e\left(f(n)\right) \right|$$

so that we may suppose that $\lambda_2 < \pi^{-1}$. We pick up the numbers u and v from the
proof of Lemma 6.4 and set

$$[u, v] \cap \mathbb{Z} = \{m + 1, \dots, m + K\},$$

where $m \in \mathbb{Z}$ and $K \in \mathbb{Z}_{\geqslant 1}$, and we define for all integers $k \in \{1, \ldots, K+1\}$ the intervals

$$J_k = (m + k - 1, m + k] \cap [u, v].$$

Lemma 6.4 implies that

$$\left| \sum_{N < n \leqslant N_1} e\left(f(n)\right) \right| \leqslant \sum_{k=1}^{K+1} \left| \sum_{n \in J_k \cap \mathbb{Z}} e\left(f(n)\right) \right| \leqslant \frac{4(K+1)}{\sqrt{\pi \lambda_2}}$$

and we have by the mean value theorem

$$K - 1 \leqslant v - u = f'(N_1) - f'(N) \leqslant c_2(N_1 - N)\lambda_2 \leqslant c_2 N \lambda_2$$

which concludes the proof.   □

When the function f is sufficiently smooth, one can improve the secondary term with the use of Corollary 6.2. The following result is the practical version of this inequality that we shall use in most of the applications.

Corollary 6.3 *Let $N < N_1 \leqslant 2N$ be positive integers and $f \in C^2[N, N_1]$ such that there exist real numbers $c_1, c_2 \geqslant 1$ and $\lambda_1, \lambda_2, s_1 > 0$ such that, for all $x \in [N, N_1]$, we have*

$$\lambda_1 \leqslant \left|f'(x)\right| \leqslant c_1 \lambda_1, \quad \lambda_2 \leqslant \left|f''(x)\right| \leqslant c_2 \lambda_2 \quad \textit{and} \quad \lambda_1 = s_1(N_1 - N)\lambda_2. \tag{6.8}$$

We set $c = 4\pi^{-1/2}\{c_2 + 2s_1(c_1 + 1)\}$. Then

$$\left| \sum_{N < n \leqslant N_1} e\left(f(n)\right) \right| \leqslant c N \lambda_2^{1/2} + \frac{2}{\pi \lambda_1}.$$

Proof If $\lambda_1 \leqslant (c_1 + 1)^{-1}$, then we apply Corollary 6.2. Suppose that $\lambda_1 > (c_1 + 1)^{-1}$. Since $\lambda_1 = s_1(N_1 - N)\lambda_2$, we can write

$$2s_1(c_1 + 1)N\lambda_2^{1/2} \geqslant 2s_1(c_1 + 1)(N_1 - N)\lambda_2^{1/2} = 2(c_1 + 1)\lambda_1 \lambda_2^{-1/2} > 2\lambda_2^{-1/2}$$

so that by Theorem 6.3 we get

$$\left| \sum_{N < n \leqslant N_1} e\left(f(n)\right) \right| \leqslant \frac{4}{\sqrt{\pi}} \left(c_2 N \lambda_2^{1/2} + 2\lambda_2^{-1/2} \right)$$

$$< \frac{4}{\sqrt{\pi}} \left(c_2 N \lambda_2^{1/2} + 2s_1(c_1 + 1) N \lambda_2^{1/2} \right) = c \, N \lambda_2^{1/2}$$

which implies the asserted result. $\square$

6.3.3   The Third Derivative Theorem

Van der Corput's A-process will allow to work with a function of the same order of magnitude of the derivative of the initial function. Applying Theorem 6.3 to this function may then give a criterion depending on the order of magnitude of the third derivative of the initial function. More precisely, we will show the following result.

Theorem 6.4 (Third Derivative Criterion) *Let* $N \in \mathbb{N}$ *and* $f \in C^3 [N, 2N]$ *such that there exist real numbers* $c_2, c_3 \geqslant 1$ *and* $\lambda_2, \lambda_3, s_2 > 0$ *such that, for all* $x \in [N, 2N]$, *we have*

$$\lambda_2 \leqslant |f''(x)| \leqslant c_2 \lambda_2, \quad \lambda_3 \leqslant |f'''(x)| \leqslant c_3 \lambda_3 \quad \text{and} \quad \lambda_2 = s_2 N \lambda_3.$$
$$\tag{6.9}$$

We set $c = 4\pi^{-1/2} \{c_3 + 2s_2(c_2 + 1)\}$ *and suppose that* $\pi s_2 c^{-2} \leqslant 8$. *Then we have*

$$\left| \sum_{N < n \leqslant 2N} e\left(f(n)\right) \right| \leqslant 6^{1/2} c^{1/3} N \lambda_3^{1/6} + 8^{1/2} c^{1/3} (s_2 \pi)^{-1/2} \lambda_3^{-1/3} (\log eN)^{1/2}.$$

Proof We set S_N the sum on the left-hand side. We first notice that if $\lambda_3 \geqslant c^{-2}$, then $c^{1/3} N \lambda_3^{1/6} \geqslant N$ and if $\lambda_3 \leqslant c^{-2} N^{-3}$, then

$$8^{1/2} c^{1/3} (s_2 \pi)^{-1/2} \lambda_3^{-1/3} \geqslant \left(8\pi^{-1} c^2 s_2^{-1} \right)^{1/2} N \geqslant N$$

because of $\frac{1}{8} \pi s_2 c^{-2} \leqslant 1$. Therefore we may suppose that

$$c^{-2} N^{-3} < \lambda_3 < c^{-2}.$$
$$\tag{6.10}$$

By van der Corput's A-process applied with $a_n = e(f(n))$, for all integers $H \in \{1, \ldots, N\}$, we have

$$|S_N|^2 \leqslant \frac{2N^2}{H+1} + \frac{4N}{H+1} \sum_{h=1}^{H} \left| \sum_{N < n \leqslant 2N-h} e(\Delta_h f(n)) \right|,$$

where we set $\Delta_h f(n) = f(n+h) - f(n)$. The conditions (6.9) imply that

$$\Lambda_1 \leqslant |(\Delta_h f)'(x)| \leqslant c_2 \Lambda_1, \quad \Lambda_2 \leqslant |(\Delta_h f)''(x)| \leqslant c_3 \Lambda_2 \quad \text{and} \quad \Lambda_1 = s_2 N \Lambda_2$$

with $\Lambda_j = h\lambda_{j+1}$ for $j \in \{1, 2\}$. We may apply Corollary 6.3 which yields

$$|S_N|^2 \leqslant \frac{2N^2}{H+1} + \frac{4N}{H+1} \sum_{h=1}^{H} \left(cN(h\lambda_3)^{1/2} + \frac{2}{\pi h \lambda_2} \right)$$

$$\leqslant \frac{2N^2}{H+1} + 4cN^2 (H\lambda_3)^{1/2} + \frac{8 \log eN}{\pi s_2 (H+1)\lambda_3},$$

where we used $\lambda_2 = s_2 N \lambda_3$ and the bound $H \leqslant N$. Now the choice of $H = \left\lfloor \left(c^2 \lambda_3 \right)^{-1/3} \right\rfloor$ gives the asserted result. Note that the conditions (6.10) ensure that $1 \leqslant H \leqslant N$. $\qquad\qquad\qquad\qquad\qquad\qquad\qquad\qquad\qquad\qquad\qquad\qquad\qquad\qquad\qquad\square$

As for Corollary 6.3, we may improve the secondary term in Theorem 6.4 if we suppose that f is sufficiently smooth, a condition that appears frequently in applications.

Corollary 6.4 *Let N be a positive integer and $f \in C^3[N, 2N]$ such that there exist real numbers $c_1, c_2, c_3 \geqslant 1$ and $\lambda_1, \lambda_2, \lambda_3, s_1, s_2 > 0$ such that, for all $x \in [N, 2N]$, we have*

$$\lambda_j \leqslant |f^{(j)}(x)| \leqslant c_j \lambda_j \quad (j = 1, 2, 3) \quad \text{and} \quad \lambda_j = s_j N \lambda_{j+1} \quad (j = 1, 2).$$

We set

$$c = 4\pi^{-1/2} \{c_3 + 2s_2(c_2 + 1)\}$$

$$\kappa_1 = 4\pi^{-1/2} \left\{ c_2 \left(s_1(c_1 + 1) \right)^{-1/2} + 2 \left(s_1(c_1 + 1) \right)^{1/2} \right\}$$

$$\kappa_2 = 6^{1/2} c^{1/3} + 8^{1/2} c^{1/3} \left(s_1 \pi^{-1}(c_1 + 1) \right)^{1/2}$$

and $\kappa = \max(\kappa_1, \kappa_2)$. *Suppose that* $\pi s_2 c^{-2} \leqslant 8$ *and* $N \geqslant 312 \{s_1 s_2 (c_1 + 1)\}^{5/3}$. *Then*

$$\left| \sum_{N < n \leqslant 2N} e(f(n)) \right| \leqslant \kappa \, N \lambda_3^{1/6} + \frac{2}{\pi \lambda_1}.$$

Proof First note that the condition $N \geqslant 312 \{s_1 s_2 (c_1 + 1)\}^{5/3}$ implies that

$$N \geqslant s_1 s_2 (c_1 + 1)(\log eN)^3.$$

Now if $\lambda_1 \leqslant (c_1 + 1)^{-1}$, then we apply Corollary 6.2. If

$$(c_1 + 1)^{-1} < \lambda_1 \leqslant (c_1 + 1)^{-1} \log(eN),$$

then we apply Corollary 6.3. In view of $\lambda_2 = s_1^{-1} N^{-1} \lambda_1$, we have

$$(s_1 (c_1 + 1))^{-1} N^{-1} < \lambda_2 \leqslant (s_1 (c_1 + 1))^{-1} N^{-1} \log(eN)$$

so that

$$\left| \sum_{N < n \leqslant 2N} e(f(n)) \right| \leqslant \left(4\pi^{-1/2} \{c_2 + 2 s_1 (c_1 + 1)\} \right) N \lambda_2^{1/2} + \frac{2}{\pi \lambda_1}$$

$$\leqslant \kappa_1 (N \log eN)^{1/2} + \frac{2}{\pi \lambda_1}.$$

But we also have $N^3 \lambda_3 = (s_1 s_2)^{-1} N \lambda_1 > (s_1 s_2 (c_1 + 1))^{-1} N \geqslant \log^3(eN)$ so that

$$(N \log eN)^{1/2} \leqslant N \lambda_3^{1/6}.$$

Finally, if $\lambda_1 > (c_1 + 1)^{-1} \log(eN)$, then the equality $\lambda_1 = s_1 s_2 N^2 \lambda_3$ implies that

$$\lambda_3^{-1/3} \log^{1/2}(eN) \leqslant (s_1 s_2 (c_1 + 1))^{1/2} N \lambda_3^{1/6}$$

and we use Theorem 6.4 to conclude the proof. $\qquad\square$

6.3.4 The kth Derivative Theorem

Theorems 6.3 and 6.4 can be generalized to functions having derivatives of higher orders. More precisely, one can prove by induction the following result.

Theorem 6.5 (kth Derivative Test) *Let $k \geqslant 2$ be an integer, $N < N_1 \leqslant 2N$ be positive integers and $f \in C^k [N, N_1]$ such that there exists a real number $\lambda_k > 0$ such that, for all $x \in [N, N_1]$, we have*

$$\left| f^{(k)}(x) \right| \asymp \lambda_k.$$

Then we have

$$\sum_{N < n \leqslant N_1} e(f(n)) \ll N \lambda_k^{1/(2^k - 2)} + N^{1 - 2^{2-k}} \lambda_k^{-1/(2^k - 2)}.$$

As in Corollaries 6.3 and 6.4, one may improve on the secondary term if f is sufficiently smooth with the help of Kusmin–Landau's inequality.

Corollary 6.5 (Improved kth Derivative Test) *Let $k \geqslant 2$, $N \geqslant 1$ be integers and $f \in C^k [N, 2N]$ such that there exist real numbers $T \geqslant 1$ and $1 \leqslant c_0 \leqslant \cdots \leqslant c_k$ such that $T \geqslant N$ and, for all $x \in [N, 2N]$ and all $j \in \{0, \ldots, k\}$, we have*

$$\frac{T}{N^j} \leqslant \left| f^{(j)}(x) \right| \leqslant c_j \frac{T}{N^j}. \tag{6.11}$$

Then we have

$$\sum_{N < n \leqslant 2N} e(f(n)) \ll c_k^{4/2^k} T^{1/(2^k - 2)} N^{1 - k/(2^k - 2)},$$

where the implied constants are absolute.

6.4   Applications to the ψ-Function

We will use the previous results to give explicit upper bounds for sums of the ψ-function.

6.4.1   The First Derivative Test

The first criterion is a consequence of the Kusmin–Landau's inequality.

Proposition 6.3 *Let $N \in \mathbb{Z}_{\geqslant 1}$ and $f \in C^1 [N, 2N]$ such that f' is monotonic and there exist $c_1 \geqslant 1$ and $0 < \lambda_1 \leqslant c_1^{-1} (c_1 + 1)^{-1}$ such that, for all $x \in [N, 2N]$, we have*

$$\lambda_1 \leqslant |f'(x)| \leqslant c_1 \lambda_1. \tag{6.12}$$

Set $\kappa = c_1 (c_1 + 1)$. Then

$$\left| \sum_{N < n \leqslant 2N} \psi (f(n)) \right| < \kappa N \lambda_1 + 2\lambda_1^{-1}.$$

Proof We assume without loss of generality that f is non-decreasing. Since $\lambda_1 \leqslant \kappa^{-1}$, the condition of Corollary 6.2 is fulfilled for the function $x \mapsto hf(x)$ for all $h \in \{1, \ldots, H\}$ whenever $H \leqslant (\kappa \lambda_1)^{-1}$. Hence, using Corollaries 6.1 and 6.2 we derive

$$\left| \sum_{N < n \leqslant 2N} \psi (f(n)) \right| \leqslant \frac{N}{2H + 2} + \left(1 + \frac{1}{\pi} \right) \sum_{h=1}^{H} \frac{1}{h} \left| \sum_{N < n \leqslant 2N} e (hf(n)) \right|$$

$$\leqslant \frac{N}{2H + 2} + \frac{2}{\pi} \left(1 + \frac{1}{\pi} \right) \sum_{h=1}^{H} \frac{1}{\lambda_1 h^2}$$

$$\leqslant \frac{N}{2H + 2} + \frac{\pi + 1}{3\lambda_1},$$

where we used $\sum_{h=1}^{H} h^{-2} \leqslant \zeta(2) = \frac{1}{6}\pi^2$. Now choosing $H = \lfloor (\kappa \lambda_1)^{-1} \rfloor$ yields the asserted result. $\qquad \square$

6.4.2   The Second Derivative Test

The condition (6.12) is very restrictive. Corollary 6.3 implies the following more useful estimate.

Proposition 6.4 *Let $N \in \mathbb{Z}_{\geqslant 1}$ and $f \in C^2 [N, 2N]$ such that there exist $c_1, c_2 \geqslant 1$ and $\lambda_1, \lambda_2, s_1 > 0$ such that, for all $x \in [N, 2N]$, we have*

$$\lambda_j \leqslant |f^{(j)}(x)| \leqslant c_j \lambda_j \quad (j = 1, 2) \quad \text{and} \quad \lambda_1 = s_1 N \lambda_2.$$

Set c as in Corollary 6.3 *and* $\kappa = \frac{3}{2}\left(2c\left(1 + \pi^{-1}\right)\right)^{2/3}$. *Then*

$$\left|\sum_{N < n \leqslant 2N} \psi\left(f(n)\right)\right| < \kappa\, N\lambda_2^{1/3} + 2\lambda_1^{-1}.$$

Proof If $\lambda_2 \geqslant \kappa^{-3}$, then $\kappa\, N\lambda_2^{1/3} \geqslant N$, so that we may suppose

$$0 < \lambda_2 < \kappa^{-3}. \tag{6.13}$$

As in Proposition 6.3 and using Corollary 6.3 instead, we get

$$\left|\sum_{N < n \leqslant 2N} \psi\left(f(n)\right)\right| \leqslant \frac{N}{2H + 2} + \left(1 + \frac{1}{\pi}\right)\left(cN\lambda_2^{1/2}\sum_{h=1}^{H}\frac{1}{h^{1/2}} + \frac{2}{\pi\lambda_1}\sum_{h=1}^{H}\frac{1}{h^2}\right)$$

$$< \frac{N}{2H + 2} + \left(\frac{2\kappa}{3}\right)^{3/2}N(H\lambda_2)^{1/2} + \frac{2}{\lambda_1},$$

where we used $\sum_{h=1}^{H}h^{-2} \leqslant \frac{1}{6}\pi^2$ once again. The choice of $H = \left\lfloor 3(2\kappa)^{-1}\lambda_2^{-1/3}\right\rfloor$ gives the desired result. Note that hypothesis (6.13) implies that $H \geqslant 1$. $\qquad\square$

6.4.3   The Third Derivative Test

Proposition 6.5 *Let* $N \in \mathbb{Z}_{\geqslant 1}$ *and* $f \in C^3[N, 2N]$ *such that there exist* $c_1, c_2, c_3 \geqslant 1$ *and* $\lambda_1, \lambda_2, \lambda_3, s_1, s_2 > 0$ *such that, for all* $x \in [N, 2N]$, *we have*

$$\lambda_j \leqslant \left|f^{(j)}(x)\right| \leqslant c_j\lambda_j \quad (j = 1, 2, 3) \quad \text{and} \quad \lambda_j = s_j N\lambda_{j+1} \quad (j = 1, 2). \tag{6.14}$$

Set c and κ *as in* Corollary 6.4, $\nu = \frac{7}{2}\left(2\kappa\left(1 + \pi^{-1}\right)\right)^{6/7}$ *and suppose that* $\pi s_2 c^{-2} \leqslant 8$ *and* $N \geqslant 312\{s_1 s_2(c_1 + 1)\}^{5/3}$. *Then*

$$\left|\sum_{N < n \leqslant 2N} \psi\left(f(n)\right)\right| < \nu\, N\lambda_3^{1/7} + 2\lambda_1^{-1}.$$

Proof The proof is exactly the same as in Proposition 6.4, except that we use Corollary 6.4 instead of Corollary 6.3. $\qquad\square$

6.4.4 The Dirichlet Divisor Problem

We intend to prove the following result.

Theorem 6.6 (Voronoï, van der Corput) *For all $x \geqslant 3$, we have*

$$\sum_{n \leqslant x} \tau(n) = x \left(\log x + 2\gamma - 1\right) + O^\star \left(18 x^{1/3} \log x\right).$$

Proof We first split the sum (6.1) into the ranges $\left[1, 6x^{1/3}\right]$ and $\left(6x^{1/3}, x^{1/2}\right]$, then estimate the sum trivially in the first one and use a splitting argument for the second sum. Since there are at most $\log x / \log 64$ subintervals of the form $(N, 2N]$ in $\left(6x^{1/3}, x^{1/2}\right]$, (6.1) becomes

$$|\Delta(x)| \leqslant 6x^{1/3} + \max_{6x^{1/3} < N \leqslant \sqrt{x}} \left| \sum_{N < n \leqslant 2N} \psi \left(\frac{x}{n}\right) \right| \frac{\log x}{\log 8} + \frac{1}{2}.$$

Proposition 6.4 may be applied with $\lambda_1 = x \left(4N^2\right)^{-1}$, $c_1 = 4$, $\lambda_2 = x \left(4N^3\right)^{-1}$, $c_2 = 8$, $s_1 = 1$ so that $c = 72\pi^{-1/2}$, and then $\kappa = 18 \times 12^{1/3}\pi^{-1}(1 + \pi)^{2/3} \approx 33,83$, which yields

$$|\Delta(x)| < 6x^{1/3} + \max_{6x^{1/3} < N \leqslant \sqrt{x}} \left(22 x^{1/3} + 8N^2x^{-1}\right) \frac{\log x}{\log 8} + \frac{1}{2}$$

$$< 6x^{1/3} + 11x^{1/3} \log x + 4 \log x + \tfrac{1}{2} \leqslant 18 x^{1/3} \log x.$$

The proof is complete. $\qquad\square$

Remark 6.4 Theorem 6.6 was rediscovered by van der Corput using his method of exponential sums, but it was first shown by Voronoï [87] in an elementary way, since theorems for exponential sums did not exist at the time. Voronoï improved on Dirichlet hyperbola principle by considering triangles instead of rectangles under the hyperbola $mn = x$ in Dirichlet's method. He was then able to prove the following estimate.

Lemma 6.5 (Voronoï) *For all real numbers $x, T \geqslant 1$, we have*

$$|\Delta(x)| \leqslant \frac{19}{12} \sum_{n \leqslant T} \tau(n) + \left(\frac{\sqrt{x}}{4T} + \frac{\sqrt{T}}{6}\right) \sum_{n \leqslant T} \frac{\tau(n)}{n^{1/2}} + \frac{3x^{1/4}}{4} \sum_{n \leqslant T} \frac{\tau(n)}{n^{3/4}} + \frac{T}{6} + \sqrt{\frac{x}{T}} + \frac{7}{4}.$$

Estimating the sums $\sum_{n \leqslant T} \tau(n) n^{-1/2}$ and $\sum_{n \leqslant T} \tau(n) n^{-3/4}$ by partial summation and choosing the parameter T optimally, Voronoï deduced that

$$|\Delta(x)| < 3 x^{1/3} \log x$$

for all $x \geqslant 308$. Unfortunately, his method did not seem to produce better exponents than $\frac{1}{3}$, whereas van der Corput's was to show great promise.

6.5   The Method of Exponent Pairs

6.5.1   Van der Corput's B-Process

The integral analogues of Theorem 6.2 and Lemma 6.4 exist and have also been proved by van der Corput. We put them together for the sake of clarity.

Lemma 6.6 (1st and 2nd Derivative Tests for Integrals) *Let f be a real-valued function defined on an interval $[a, b]$.*

▷ *Suppose $f \in C^1[a, b]$ such that f' is monotonic and there exists $\lambda_1 > 0$ such that, for all $x \in [a, b]$, we have $f'(x) \geqslant \lambda_1$. Then*

$$\left| \int_a^b e(f(x)) \, dx \right| \leqslant \frac{2}{\pi \lambda_1}.$$

▷ *Suppose $f \in C^2[a, b]$ such that there exists $\lambda_2 > 0$ such that, for all $x \in [a, b]$, we have $f''(x) \geqslant \lambda_2$. Then*

$$\left| \int_a^b e(f(x)) \, dx \right| \leqslant \frac{4\sqrt{2}}{\sqrt{\pi \lambda_2}}.$$

Proof

▷ We use the following result due to Ostrowski.[4]
  Let F be a real-valued monotone integrable function on $[a, b]$ and G be a complex-valued integrable function on $[a, b]$. Then we have

$$\left| \int_a^b F(x) G(x) \, dx \right| \leqslant |F(a)| \max_{a \leqslant t \leqslant b} \left| \int_a^t G(x) \, dx \right| + |F(b)| \max_{a \leqslant t \leqslant b} \left| \int_t^b G(x) \, dx \right|.$$

---

[4]See [9], for instance.

Applying this with $F(x) = 1/(2\pi f'(x))$ which is monotone and $G(x) = 2\pi f'(x)e(f(x))$, we get

$$\left| \int_a^b e(f(x))\, dx \right| \leqslant \frac{1}{2\pi \lambda_1} \left(\max_{a \leqslant t \leqslant b} \left| [ie(f(x))]_a^t \right| + \max_{a \leqslant t \leqslant b} \left| [ie(f(x))]_t^b \right| \right)$$

$$\leqslant \frac{1}{2\pi \lambda_1} \times 4 = \frac{2}{\pi \lambda_1}.$$

▷ Let $\delta > 0$ be a parameter to be chosen later and write $[a, b] = E \cup F$ where $E = \{x : |f'(x)| < \delta\}$ and $F = \{x : |f'(x)| \geqslant \delta\}$. Since $f'' > 0$, the set E consists of at most one interval and F consists of at most two intervals. If $E = [c, d]$, then $(d - c)\lambda_2 \leqslant |f'(d) - f'(c)| \leqslant 2\delta$ so that

$$\left| \int_E e(f(x))\, dx \right| \leqslant d - c \leqslant \frac{2\delta}{\lambda_2}$$

and using the previous bound we have

$$\left| \int_F e(f(x))\, dx \right| \leqslant \frac{4}{\pi \delta}.$$

Therefore

$$\left| \int_a^b e(f(x))\, dx \right| \leqslant \frac{2\delta}{\lambda_2} + \frac{4}{\pi \delta}$$

and the choice of $\delta = \left(2\pi^{-1}\lambda_2\right)^{1/2}$ gives the desired result. $\qquad\qquad\qquad\square$

Note that this lemma may be generalized in the case of weight functions of bounded variation. For a proof, see [41, Lemmas 5.1.2 and 5.1.3].

Lemma 6.7 (1st and 2nd Derivative Tests with Weight) *Let g be a function of bounded variation on $[a, b]$. Under the same assumptions for the function f as that of Lemma 6.6, we, respectively, have the bounds*

$$\left| \int_a^b g(x)e(f(x))\, dx \right| \leqslant \frac{V_a^b(g)}{\pi \lambda_1} \quad and \quad \left| \int_a^b g(x)e(f(x))\, dx \right| \leqslant \frac{4V_a^b(g)}{\sqrt{\pi \lambda_2}}.$$

Besides proving the A-process,[5] van der Corput provided another tool for estimating exponential sums called the B-process. This result relies on a truncated Poisson summation formula and the method of the stationary phase. The Poisson formula,

---

[5] See Lemma 6.2.

which is a useful tool in harmonic analysis, allows us to replace an exponential sum by an integral, which may be estimated by the stationary phase. We first state without proof one of the many versions of this theorem.[6]

Lemma 6.8 (Poisson Summation) *Let f be a continuous function on $\mathbb{R}$ such that $\int_{\mathbb{R}} |f(x)|\,dx$ exists and is finite. Then we have*

$$\sum_{k \in \mathbb{Z}} f(k) = \lim_{K \to \infty} \sum_{|k| \leqslant K} \widehat{f}(k),$$

where $\widehat{f}(t) = \int_{\mathbb{R}} f(x)\, e(-tx)\,dx$.

For our purpose, it will be more useful if we have at our disposal a truncated version of Lemma 6.8. We state here the following two results without proof.[7]

Lemma 6.9 (Truncated Poisson Summation) *Let $f \in C^1[a, b]$ such that f' is increasing on $[a, b]$ and put $\alpha = f'(a)$ and $\beta = f'(b)$. Then we have*

$$\sum_{a \leqslant n \leqslant b} e(f(n)) = \sum_{\alpha - 1 \leqslant k \leqslant \beta + 1} \int_a^b e(f(x) - kx)\,dx + O\left(\log(\beta - \alpha + 2)\right).$$

The range of summation in the sum of the right-hand side may be restricted to $\left[\alpha - \frac{1}{4}, \beta + \frac{1}{4}\right]$, say, without changing the error term, as may be seen in [41]. Lemmas 6.6 and 6.9 allow us to recover the van der Corput inequality. Indeed, using Lemma 6.6 in the integral above, we get with the hypothesis of Theorem 6.3

$$\sum_{N < n \leqslant 2N} e(f(n)) \ll \left(f'(2N) - f'(N) + 1\right) \lambda_2^{-1/2} + \log(f'(2N) - f'(N) + 2)$$

$$\ll N\lambda_2^{1/2} + \lambda_2^{-1/2} + \log(N\lambda_2 + 2) \ll N\lambda_2^{1/2} + \lambda_2^{-1/2}$$

since $N\lambda_2^{1/2} + \lambda_2^{-1/2} \geqslant 2N^{1/2} \gg \log(N\lambda_2 + 2)$ if N is sufficiently large.

The next result is an effective version of the method of the stationary phase.

Lemma 6.10 (Stationary Phase) *Let $g \in C^4[a, b]$ and assume there exists $x_0 \in (a, b)$ such that $g'(x_0) = 0$. We also suppose that there exist $\lambda_2, \lambda_3, \lambda_4 > 0$ such that, for all $x \in [a, b]$, we have*

$$g''(x) \geqslant \lambda_2, \quad |g'''(x)| \leqslant \lambda_3 \quad and \quad |g^{(4)}(x)| \leqslant \lambda_4.$$

---

[6]See [65], for instance.

[7]See [63].

Then we have

$$\int_a^b e(g(x))\,\mathrm{d}x = \frac{e(g(x_0) + 1/8)}{\sqrt{g''(x_0)}} + O\left(E_1(x_0) + E_2\right),$$

where

$$E_1(x_0) = \min\left(\frac{1}{\lambda_2(x_0 - a)}, \lambda_2^{-1/2}\right) + \min\left(\frac{1}{\lambda_2(b - x_0)}, \lambda_2^{-1/2}\right)$$

$$E_2 = (b - a)\left\{\lambda_2^{-2}\lambda_4 + \lambda_2^{-3}\lambda_3^2\right\}.$$

Combining Lemmas 6.9 and 6.10 gives van der Corput's *B*-process.

Theorem 6.7 (B-Process) *Let $N \in \mathbb{Z}_{\geqslant 1}$, $a \leqslant b \leqslant a + N$ and $f \in C^4[a, b]$ such that there exist $c_2 \geqslant 1$ and $\lambda_2, \lambda_3, \lambda_4 > 0$ such that, for all $x \in [a, b]$, we have*

$$\lambda_2 \leqslant f''(x) \leqslant c_2\lambda_2, \quad \left|f'''(x)\right| \leqslant \lambda_3 \quad and \quad \left|f^{(4)}(x)\right| \leqslant \lambda_4.$$

We set $\alpha = f'(a)$ and $\beta = f'(b)$ and for each integer $k \in [\alpha, \beta]$, let x_k be defined by $f'(x_k) = k$. Then we have

$$\sum_{a \leqslant n \leqslant b} e(f(n)) = \sum_{\alpha \leqslant k \leqslant \beta} \frac{e(f(x_k) - kx_k + 1/8)}{\sqrt{f''(x_k)}} + O\left(R_1 + R_2\right),$$

where

$$R_1 = \lambda_2^{-1/2} + \log(N\lambda_2 + 2) \quad and \quad R_2 = N^2\left\{\lambda_2^{-1}\lambda_4 + \lambda_2^{-2}\lambda_3^2\right\}.$$

Proof If $\lambda_2 \leqslant N^{-2}$, then $R_1 \gg N$, so that we may suppose that $\lambda_2 > N^{-2}$. By Lemma 6.6 we have as above

$$\int_a^b e(f(x) - kx)\,\mathrm{d}x \ll \lambda_2^{-1/2} \tag{6.15}$$

uniformly in k and $\beta - \alpha = f'(b) - f'(a) \leqslant c_2 N\lambda_2$ so that by Lemma 6.9 we get

$$\sum_{a \leqslant n \leqslant b} e(f(n)) = \sum_{\alpha \leqslant k \leqslant \beta} \int_a^b e(f(x) - kx)\,\mathrm{d}x + O\left(R_1\right).$$

If $\beta - \alpha \leqslant 1$, then the theorem is proved by (6.15). If $\beta - \alpha > 1$, then the sum on the right-hand side is non-empty. We may use Lemma 6.10 to estimate each integral, which gives

$$\sum_{a \leqslant n \leqslant b} e(f(n)) = \sum_{\alpha \leqslant k \leqslant \beta} \frac{e(f(x_k) - kx_k + 1/8)}{\sqrt{f''(x_k)}} + O\left\{\sum_{\alpha \leqslant k \leqslant \beta} (E_1(x_k) + E_2)\right\} + O(R_1).$$

Now $k - \alpha = f'(x_k) - f'(a) \leqslant c_2\lambda_2 (x_k - a)$ and similarly $\beta - k \leqslant c_2\lambda_2 (b - x_k)$ so that

$$E_1(x_k) \leqslant \min\left(\frac{c_2}{k - \alpha}, \lambda_2^{-1/2}\right) + \min\left(\frac{c_2}{\beta - k}, \lambda_2^{-1/2}\right).$$

Furthermore, since $\sum_{\alpha \leqslant k \leqslant \beta} 1 = \lfloor \beta - \alpha \rfloor$ or $\lfloor \beta - \alpha \rfloor + 1$ and since $\beta - \alpha > 1$, we have

$$\frac{N\lambda_2}{2} \leqslant \frac{\beta - \alpha}{2} \leqslant \sum_{\alpha \leqslant k \leqslant \beta} 1 \leqslant \beta - \alpha + 1 \leqslant 2c_2N\lambda_2$$

and therefore

$$\sum_{\alpha \leqslant k \leqslant \beta} (E_1(x_k) + E_2) \ll \lambda_2^{-1/2} + \sum_{\alpha+1 \leqslant k \leqslant \beta} \frac{1}{k - \alpha} + \sum_{\alpha \leqslant k \leqslant \beta-1} \frac{1}{\beta - k} + \sum_{\alpha \leqslant k \leqslant \beta} E_2$$

$$\ll \lambda_2^{-1/2} + \log(\beta - \alpha) + N\lambda_2 E_2 \ll R_1 + R_2$$

as required.   □

Note that if we estimate the exponential sum in Theorem 6.7 trivially, we recover Theorem 6.3. Thus, van der Corput's inequality follows from the B-process and the triangle inequality.

One may state the B-process in another form if we introduce the function s_f defined by

$$s_f(y) = f \circ (f')^{-1}(y) - y(f')^{-1}(y) \tag{6.16}$$

for all functions f sufficiently smooth and monotone on $[a, b]$ and all $y \in [\alpha, \beta]$, where $\alpha = f'(a)$ and $\beta = f'(b)$. With this notation, it is easy to see that, for all integers k, we have $s_f(k) = f(x_k) - kx_k$ where x_k is defined in Theorem 6.7, so that van der Corput's B-process may be rewritten in the following form.

Corollary 6.6 (B-Process Version 2) *Let $N \in \mathbb{Z}_{\geqslant 1}$, $a \leqslant b \leqslant a + N$ and $f \in C^4[a, b]$ satisfying the hypotheses of Theorem 6.7. We set $\alpha = f'(a)$ and $\beta = f'(b)$ and for each integer $k \in [\alpha, \beta]$, let x_k be defined by $f'(x_k) = k$. The function*

s_f being defined in (6.16), we have

$$\sum_{a \leqslant n \leqslant b} e(f(n)) = e\left(\frac{1}{8}\right) \sum_{\alpha \leqslant k \leqslant \beta} \frac{e(s_f(k))}{\sqrt{f'' \circ (f')^{-1}(k)}} + O\left(R_1 + R_2\right),$$

where the remainder terms R_i are given in Theorem 6.7.

It could be interesting to have some properties of the function s_f. For instance, one can see that, if $f \in C^k[a, b]$ for $k \in \{1, 2, 3\}$ and f' is monotone, then $s_f \in C^k[\alpha, \beta]$ and if $f'' \neq 0$ when $k \geqslant 2$, we have

$$(s_f)'(y) = -(f')^{-1}(y),$$

$$(s_f)''(y) = \frac{-1}{f'' \circ (f')^{-1}(y)} \quad \text{and} \quad (s_f)'''(y) = \frac{f''' \circ (f')^{-1}(y)}{\left(f'' \circ (f')^{-1}(y)\right)^3}.$$

6.5.2 Exponent Pairs

The A- and B-processes can be systematized in an algorithmic type procedure which was developed by van der Corput and later simplified by Phillips. To make things more accurate, we start by giving a definition of a class of function on which the process may apply. In practice, this class contains almost all functions that we may encounter in the usual problems of number theory. Roughly speaking, we need to work with functions satisfying hypotheses of the type (6.11) and possibly the logarithm function for the problem of the Riemann zeta-function in the critical strip.

Definition 6.1 Let $N \geqslant 1$ and $r \geqslant 2$ be integers, $T \geqslant 1$, $\sigma > -1$ and $0 < \varepsilon < \frac{1}{2}$ be real numbers. A function f belongs to the class $\mathcal{F}(T, N; r, \sigma, \varepsilon)$ if $f \in C^r(\mathcal{I})$ such that $\mathcal{I} \subseteq [N, 2N]$ and, for all $x \in \mathcal{I}$, we have

$$f(x) = \begin{cases} \pm \dfrac{T N^\sigma}{x^\sigma} + u(x), & \text{if } \sigma \neq 0 \\[2mm] \pm T \log x + u(x), & \text{if } \sigma = 0, \end{cases}$$

where $u \in C^r(\mathcal{I})$ is a *remainder function* satisfying

$$|u^{(j)}(x)| \leqslant \begin{cases} (\sigma)_j \, \varepsilon \, T N^{-j}, & \text{if } \sigma \neq 0 \\ (j-1)! \, \varepsilon \, T N^{-j}, & \text{if } \sigma = 0 \end{cases}$$

for all $x \in I$ and $j \in \{1, \ldots, r\}$, and where the *Pochhammer symbol* $(\sigma)_j$ is defined by

$$(\sigma)_j = \begin{cases} 1, & \text{if } j = 0 \\ \sigma(\sigma + 1) \cdots (\sigma + j - 1), & \text{if } j \geq 1. \end{cases}$$

A function belonging to the class $\mathcal{F}(T, N; r, \sigma, \varepsilon)$ with $\sigma \neq 0$ is called a *monomial function*.

We summarize in the next result the basic properties of this class of functions.

Lemma 6.11 *Let $N \geq 1$ and $r \geq 2$ be integers, $T \geq 1$, $\sigma > -1$ and $\varepsilon > 0$ be real numbers.*

▷ *Let $f \in \mathcal{F}(T, N; r, \sigma, \varepsilon)$, $1 \leq h \leq \frac{2\varepsilon N}{\sigma + r}$ be an integer and set $\Delta_h f(x) = f(x + h) - f(x)$. Then*

$$\Delta_h f \in \mathcal{F}\left(hTN^{-1}, N; r - 1, \sigma + 1, 2\varepsilon\right).$$

▷ *If $g_\sigma(x) = -T(Nx^{-1})^\sigma$ with $\sigma \neq 0$ or if $g_0(x) = T \log x$, then, for all $\varepsilon > 0$ and all integers $r \geq 2$, we have*

$$s_{g_\sigma} \in \mathcal{F}\left(T, TN^{-1}; r, -\frac{\sigma}{\sigma+1}, \varepsilon\right).$$

Proof We will make use of the following result which is a particular case of an inequality by Ostrowski.[8]
Let $a < b$ be real numbers and $f \in C^1[a, b]$ such that there exists $\lambda_1 > 0$ such that, for all $x \in [a, b]$, we have $\left|f'(x)\right| \leq \lambda_1$. Then

$$\left|(b - a)f(a) - \int_a^b f(t)\,dt\right| \leq \tfrac{1}{2}(b - a)^2 \lambda_1.$$

▷ Let $f \in \mathcal{F}(T, N; r, \sigma, \varepsilon)$ and we set $\mathbb{T} = hTN^{-1}$. We have

$$\Delta_h f(x) = \pm \frac{hTN^\sigma}{x^{\sigma+1}} + \mathcal{R}_{h,\sigma}(x) = \pm \frac{\mathbb{T}N^{\sigma+1}}{x^{\sigma+1}} + \mathcal{R}_{h,\sigma}(x),$$

where

$$\mathcal{R}_{h,\sigma}(x) = \pm TN^\sigma \left(\int_x^{x+h} \frac{dt}{t^{\sigma+1}} - \frac{h}{x^{\sigma+1}}\right) + \Delta_h u(x)$$

---

[8] See [68], for instance.

so that for all $j \in \{1, \ldots, r - 1\}$ we have, using Ostrowski's inequality

$$\left| R_{h,\sigma}^{(j)}(x) \right| \leqslant (\sigma + 1)_j T N^\sigma \left| \int_x^{x+h} \frac{dt}{t^{\sigma+j+1}} - \frac{h}{x^{\sigma+j+1}} \right| + \int_x^{x+h} |u^{(j+1)}(t)| \, dt$$

$$\leqslant \frac{(\sigma + 1)_j}{2} h T N^{-j-1}(\sigma + j + 1)(hN^{-1}) + \varepsilon (\sigma + 1)_j h T N^{-j-1}$$

$$\leqslant \varepsilon (\sigma + 1)_j \, \mathbb{T}N^{-j} \left(\frac{\sigma + j + 1}{\sigma + r} + 1 \right) \leqslant 2\varepsilon (\sigma + 1)_j \, \mathbb{T}N^{-j}.$$

▷ Easy computations from the definition (6.16) give

$$s_{g_\sigma}(y) = - \left(\sigma^{\frac{1}{\sigma+1}} + \sigma^{-\frac{\sigma}{\sigma+1}} \right) \frac{T(TN^{-1})^{-\frac{\sigma}{\sigma+1}}}{y^{-\frac{\sigma}{\sigma+1}}}$$

and

$$s_{g_0}(y) = -T \log \left(e y T^{-1} \right).$$

The proof is complete.   □

We are now in a position to define the exponent pairs.

Definition 6.2 (Exponent Pairs) Let k and ℓ be real numbers satisfying

$$0 \leqslant k \leqslant \tfrac{1}{2} \leqslant \ell \leqslant 1.$$

The pair (k, ℓ) is called an *exponent pair* if, for all $\sigma > -1$, there exist an integer $r \geqslant 2$ and $0 < \varepsilon < \frac{1}{2}$ such that, for all $N \in \mathbb{Z}_{\geqslant 1}$, all real numbers $T \geqslant 1$ and all $f \in \mathcal{F}(T, N; r, \sigma, \varepsilon)$, the estimate

$$\sum_{a < n \leqslant b} e(f(n)) \ll T^k N^{\ell-k} + NT^{-1}$$

holds, with $[a, b] \subseteq [N, 2N]$.

Note that the pair $(0, 1)$ is an exponent pair by the triangle inequality and is called the *trivial exponent pair*. By Corollary 6.3, the pair $\left(\frac{1}{2}, \frac{1}{2} \right)$ is an exponent pair, and by Corollary 6.4 the pair $\left(\frac{1}{6}, \frac{2}{3} \right)$ is also an exponent pair. Furthermore, there is no need to ever take $\ell > 1$ in an exponent pair, otherwise we would get an estimate worse than the one provided by the triangle inequality. Similarly, there is no need to take $k > \frac{1}{2}$, for then (k, ℓ) would provide a weaker result than the pair $\left(\frac{1}{2}, \frac{1}{2} \right)$. More generally, the reader should refer to [26] where an interesting explanation is given about the hypotheses $0 \leqslant k \leqslant \frac{1}{2} \leqslant \ell \leqslant 1$ appearing in this definition.

The term NT^{-1}, which comes from Kusmin–Landau's inequality, is here to prevent the case where the order of magnitude of the function in the interval $[N, 2N]$ is very small compared to N. However, we shall mostly have $T \gg N$, so that NT^{-1} is absorbed by the main term $T^k N^{\ell-k}$. Thus, the exponent pair $\left(\frac{1}{2}, \frac{1}{2}\right)$ tells us that the exponential sum is bounded by the square-root of the order of magnitude of the function in $[N, 2N]$, which is an interesting result. But it would be more efficient to have the bound $\sum_{a < n \leqslant b} e(f(n)) \ll N^{1/2+\varepsilon}$ instead of $T^{1/2}$. Unfortunately, this conjecture, called the *exponent pair conjecture*, still remains unproven. Note that this conjecture would solve Dirichlet's divisor problem.

6.5.3 Applications

Using Corollary 6.1 along with exponent pairs gives the following useful result.

Corollary 6.7 *Let $f \in \mathscr{F}(T, N; r, \sigma, \varepsilon)$ and (k, ℓ) be an exponent pair. Then we have*

$$\sum_{a < n \leqslant b} \psi(f(n)) \ll \left(T^k N^\ell\right)^{\frac{1}{k+1}} + NT^{-1}$$

with $[a, b] \subseteq [N, 2N]$.

Proof If $N \leqslant T^{\frac{k}{1+k-\ell}}$, then we have

$$\left(T^k N^\ell\right)^{\frac{1}{k+1}} \geqslant N \geqslant b - a \geqslant \left| \sum_{a < n \leqslant b} \psi(f(n)) \right|$$

so that we may suppose that $N > T^{\frac{k}{1+k-\ell}}$. Using Corollary 6.1 and Definition 6.2, we get for any positive integer H

$$\sum_{a < n \leqslant b} \psi(f(n)) \ll NH^{-1} + (HT)^k N^{\ell-k} + NT^{-1}$$

and the choice of $H = \left\lfloor \left(T^{-k} N^{1+k-\ell}\right)^{\frac{1}{k+1}} \right\rfloor$ gives the asserted result. $\qquad \square$

Lemma 6.11 implies that in a certain sense the exponent pairs are compatible with the A- and B-processes. As shown by the next result, this compatibility has the advantage of producing some new exponent pairs from older ones. For the proof, we refer the reader to [26, 63].

Theorem 6.8 (Exponent Pairs by A- and B-Processes)   *Let (k, ℓ) be an exponent pair. Then*

$$A(k, \ell) = \left(\frac{k}{2k+2}, \frac{1}{2} + \frac{\ell}{2k+2}\right) \quad and \quad B(k, \ell) = \left(\ell - \frac{1}{2}, k + \frac{1}{2}\right)$$

are exponent pairs.

This theorem shows the great usefulness of the exponent pairs, for one may combine the A- and B-processes successively to produce exponent pairs from the pair $(0, 1)$. For instance, setting A^j and B^j to symbolize the A- and B-processes applied j times, respectively, we have

$$BA^3 \left(BA^2\right)^2 B(0, 1) = \left(\frac{97}{251}, \frac{132}{251}\right) \tag{6.17}$$

so that $\left(\frac{97}{251}, \frac{132}{251}\right)$ is an exponent pair. Note that $B^2(k, \ell) = (k, \ell)$, which is not surprising since the Poisson summation formula is one of the ingredients of the B-process.

Using (6.17) in Corollary 6.7 gives the following improvement in the Dirichlet divisor problem.

$$\sum_{N<n\leqslant 2N} \psi\left(\frac{x}{n}\right) \ll x^{97/348} N^{35/348} + N^2 x^{-1}$$

so that, for x sufficiently large, we have

$$\sum_{n\leqslant\sqrt{x}} \psi\left(\frac{x}{n}\right) \ll x^{229/696} \log x.$$

Note that $\frac{229}{696} \approx 0.329022988\ldots$ Compared to the conjectured value $\frac{1}{4}$, this estimate represents a saving of a little more than 5% in comparison to the bound provided by Theorem 6.6.

The search for the best exponent pair to a given estimate implies a difficult optimization problem which was investigated by Graham and Kolesnik in [26]. For the Dirichlet divisor problem, the authors proved that the best exponent accessible by this method is $0.32902135685\ldots$ which is slightly better than the exponent above.

6.5.4   A New Third Derivative Theorem

The question whether the exponents $\frac{1}{3}$ and $\frac{1}{7}$ in Corollaries 6.4 and 6.5 can be increased arises naturally. Concerning the first one, a counterexample, given by Grekos [29], shows that this exponent cannot be improved. In [75], the authors investigate the problem of the second exponent. Their idea is to estimate the following exponential sum with a parameter

$$\frac{1}{H} \sum_{H < h \leqslant 2H} \left| \sum_{n \in I_h} e\left(\frac{h}{H} f(n) \right) \right|,$$

where H, N are large positive integers and I_h is any subinterval of $(N, 2N]$ which may depend on h for all $h \in [H, 2H]$. The non-trivial treatment of this sum rests on bounds of quadruple exponential sums and the use of the so-called *spacing lemmas*, which, in Huxley's terminology, are tools for estimating certain Diophantine systems. Their main result may be stated as follows.

Theorem 6.9 *Let H, N be large positive integers, I_h be any subinterval of $(N, 2N]$ which may depend on h for all $h \in [H, 2H]$ and $f \in C^3[N, 2N]$ such that there exists a real number $\lambda_3 > 0$ such that, for all $x \in [N, 2N]$ we have*

$$\left| f'''(x) \right| \asymp \lambda_3.$$

Then, for all $\varepsilon > 0$, we have

$$\frac{1}{H} \sum_{H < h \leqslant 2H} \left| \sum_{n \in I_h} e\left(\frac{h}{H} f(n) \right) \right| \ll N^{1+\varepsilon} \lambda_3^{1/6} H^{-1/9} + N^{1+\varepsilon} \lambda_3^{1/5} + N^{3/4+\varepsilon} + \lambda_3^{-1/3}.$$

We now use this tool along with Corollary 6.1 in the following way. Suppose first that $N^{-19/3} \leqslant \lambda_3 < 1$. Splitting the interval $[1, H]$ into $O(\log H)$ subintervals of the form $[H_1, 2H_1]$, we get

$$\sum_{N < n \leqslant 2N} \psi(f(n)) \ll \frac{N}{H} + \sum_{h=1}^{H} \frac{1}{h} \left| \sum_{N < n \leqslant 2N} e(hf(n)) \right|$$

$$\ll \frac{N}{H} + \max_{1 \leqslant H_1 \leqslant H} \sum_{H_1 < h \leqslant 2H_1} \frac{1}{h} \left| \sum_{N < n \leqslant 2N} e(hf(n)) \right| \log H$$

$$\ll \frac{N}{H} + \max_{1 \leqslant H_1 \leqslant H} \frac{1}{H_1} \sum_{H_1 < h \leqslant 2H_1} \left| \sum_{N < n \leqslant 2N} e(hf(n)) \right| \log H$$

$$\ll \frac{N}{H} + \max_{1 \leqslant H_1 \leqslant H} \left(N^{1+\varepsilon} (H_1 \lambda_3)^{1/6} H_1^{-1/9} + N^{1+\varepsilon} (H_1 \lambda_3)^{1/5} \right.$$

$$\left. + N^{3/4+\varepsilon} + (H_1 \lambda_3)^{-1/3} \right) \log H$$

$$\ll \frac{N}{H} + \left(N^{1+\varepsilon} (H\lambda_3^3)^{1/18} + N^{1+\varepsilon} (H\lambda_3)^{1/5} + N^{3/4+\varepsilon} + \lambda_3^{-1/3} \right) \log H$$

and choosing $H = \left\lfloor \lambda_3^{-3/19} \right\rfloor$ gives

$$\sum_{N < n \leqslant 2N} \psi(f(n)) \ll N^\varepsilon \left(N\lambda_3^{3/19} + N^{3/4} + \lambda_3^{-1/3} \right)$$

the second term being absorbed by the first one, and $\log H \leqslant \log N \ll N^\varepsilon$ since $\lambda_3 \geqslant N^{-19/3}$. If $\lambda_3 \geqslant 1$, then $N\lambda_3^{3/19} \gg N$ and if $\lambda_3 < N^{-19/3}$, then $\lambda_3^{-1/3} > N^{19/9}$ so that we may state the following result.

Corollary 6.8 *Let $f \in C^3[N, 2N]$ such that there exists a real number $\lambda_3 > 0$ such that, for all $x \in [N, 2N]$ we have $|f'''(x)| \asymp \lambda_3$. Then, for all $\varepsilon > 0$, we have*

$$N^{-\varepsilon} \sum_{N < n \leqslant 2N} \psi(f(n)) \ll N\lambda_3^{3/19} + N^{3/4} + \lambda_3^{-1/3}.$$

Applied to the Dirichlet divisor problem, this result implies that

$$\sum_{N < n \leqslant 2N} \psi\left(\frac{x}{n}\right) \ll N^\varepsilon \left(x^{3/19} N^{7/19} + N^{3/4} + N^{4/3} x^{-1/3} \right)$$

so that, for x sufficiently large, we get

$$\sum_{n \leqslant x^{2/5}} \psi\left(\frac{x}{n}\right) \ll x^{29/95+\varepsilon}$$

which is better than the previous result obtained by the third derivative test, but once again this does not allow us to recover the whole range of summation.

6.6   Character Sums

6.6.1   Additive Characters and Gauss Sums

6.6.1.1   Additive Characters

The Dirichlet characters modulo q are characters from the multiplicative groups $(\mathbb{Z}/q\mathbb{Z})^\times$ to the multiplicative group $\mathbb{C}^*$ of non-zero complex numbers, and are thus called *multiplicative characters*. The dual concept of additive characters is defined in a similar way. Let a, q be positive coprime integers with $1 \leqslant a \leqslant q$ and consider the primitive qth roots of unity $e_q(a)$. From the well-known identity

$$\sum_{k \,(\mathrm{mod}\,q)} (e_q(a))^k = 0$$

we get

$$\frac{1}{q} \sum_{k \,(\mathrm{mod}\,q)} e_q(-ka)e_q(kn) = \begin{cases} 1, & \text{if } n \equiv a \pmod{q} \\ 0, & \text{otherwise} \end{cases} \tag{6.18}$$

which is the analogue of Proposition 3.13. Thus the characteristic function of the integers $n \equiv a \pmod{q}$ may be written as a linear combination of the sequence $(e_q(kn))$. These functions are called *additive characters* since they are characters from the additive groups $(\mathbb{Z}/q\mathbb{Z}, +)$ to the multiplicative group $\mathbb{C}^*$.

Let f be an arithmetic function of period q. Multiplying both sides of (6.18) by $f(n)$ and summing over n running through a complete residue system modulo q, we obtain

$$f(a) = \sum_{k \,(\mathrm{mod}\,q)} \frac{e_q(ka)}{q} \sum_{n \,(\mathrm{mod}\,q)} f(n)e_q(-kn).$$

The function

$$\widehat{f}(k) = \frac{1}{q} \sum_{n \,(\mathrm{mod}\,q)} f(n)e_q(-kn)$$

is called the *finite Fourier transform* of f, and we thus have

$$f(n) = \sum_{k \,(\mathrm{mod}\,q)} \widehat{f}(k)e_q(kn).$$

For all q-periodic arithmetic function f, we have

$$\frac{1}{q}\sum_{n=1}^{q}|f(n)|^2 = \sum_{k=1}^{q}|\widehat{f}(k)|^2$$

which is the analogue of Plancherel's formula for functions $f \in L^2(\mathbb{R})$ or of Parserval's formula for functions $f \in L^2(\mathbb{R}/\mathbb{Z})$.

6.6.1.2   Ramanujan Sum

Definition 6.3 Let $q \in \mathbb{Z}_{\geqslant 1}$ and $n \in \mathbb{Z}$. The *Ramanujan sum* is defined by

$$c_q(n) := \sum_{\substack{a=1 \\ (a,q)=1}}^{q} e_q(na).$$

In other words, the Ramanujan sum is the sum of all *primitive* additive characters modulo q. This sum plays a key part in number theory and possesses some interesting properties.

Proposition 6.6 *Let $q \in \mathbb{Z}_{\geqslant 1}$ and $n \in \mathbb{Z}$, and set $g := (q, n)$. Then*

$$c_q(n) = \sum_{d|g} d\mu(q/d).$$

Furthermore, the arithmetic function $q \mapsto c_q(n)$ is multiplicative and, for all $q \in \mathbb{Z}_{\geqslant 1}$ and $n \in \mathbb{Z}$

$$c_q(n) = \varphi(q)\mu(q/g)\varphi(q/g)^{-1}.$$

Proof Using Corollary 4.6, we get

$$c_q(n) = \sum_{d|q}\mu(d)\sum_{k\leqslant q/d}e_q(kdn) = q\sum_{d|q}\frac{\mu(d)}{d}\left(\frac{d}{q}\sum_{k\leqslant q/d}e_{q/d}(kn)\right)$$

and, by (6.18), the sum in brackets is equal to 1 if $(q/d) \mid n$ and 0 otherwise. Hence

$$c_q(n) = q\sum_{\substack{d|q \\ (q/d)|n}}\frac{\mu(d)}{d} = \sum_{d|q,\, d|n} d\mu(q/d)$$

implying the first identity. Now define $\mathrm{id}_n(d) = d$ if $d \mid n$, 0 otherwise. The previous result may be written as

$$c_q(n) = (\mu \star \mathrm{id}_n)(q)$$

and since the functions μ and id_n are multiplicative, we infer that the function $q \mapsto c_q(n)$ is also multiplicative by Theorem 4.1. For the last part of the proposition, we note that both sides are multiplicative functions of the variable q, so that it suffices to prove the identity for prime powers p^α. If $p \nmid n$, then both sides of the identity are equal to $\mu(p^\alpha)$. Now assume $p \mid n$ and set $\beta := v_p(n) \geqslant 1$. If $\beta \geqslant \alpha$, then both sides of the identity are equal to $\varphi(p^\alpha)$. If $\beta < \alpha$, then both sides of the identity are equal to $-p^{\alpha-1}$ if $\alpha - \beta = 1$, 0 if $\alpha - \beta > 1$, completing the proof.   □

6.6.1.3 Gauss Sums

It may be interesting to deal with a discrete inner product of the multiplicative character χ and the additive character e_q. To this end, we define the sum

$$\tau(\chi) = \sum_{a\,(\mathrm{mod}\,q)} \chi(a)e_q(a). \tag{6.19}$$

This is called the *Gauss sum* of χ and is an important tool connecting additive and multiplicative number theory. One of its most important properties is to express any primitive Dirichlet character as a linear combination of additive characters [65, Theorem 9.5, Corollary 9.8].

Proposition 6.7 *Let χ be a Dirichlet character modulo q. If $(n, q) = 1$, then*

$$\chi(n)\tau(\overline{\chi}) = \sum_{a\,(\mathrm{mod}\,q)} \overline{\chi}(a)e_q(na). \tag{6.20}$$

In particular, if χ is a primitive Dirichlet character modulo q, then for all positive integers n, we have

$$\chi(n) = \frac{1}{\tau(\overline{\chi})} \sum_{a\,(\mathrm{mod}\,q)} \overline{\chi}(a)e_q(na).$$

Proof When $(n, q) = 1$, the map $a \mapsto an$ is a one-to-one in $\mathbb{Z}/q\mathbb{Z}$, so that by (6.19) we get

$$\tau(\chi) = \sum_{a\,(\mathrm{mod}\,q)} \chi(an)e_q(an) = \chi(n) \sum_{a\,(\mathrm{mod}\,q)} \chi(a)e_q(an).$$

The first part of the theorem follows on replacing χ by $\overline{\chi}$. For the second part, assume χ primitive, $(n, q) > 1$ and choose m and d such that $m/d = n/q$ and $(m, d) = 1$. Then

$$\sum_{a \,(\mathrm{mod}\, q)} \overline{\chi}(a) e_q(na) = \sum_{h=1}^{d} e_d(mh) \sum_{\substack{a=1 \\ a \equiv h \,(\mathrm{mod}\, d)}}^{q} \overline{\chi}(a)$$

and the inner sum vanishes by [65, Theorem 9.4] since χ is primitive and $d \mid q$, $d < q$. Hence (6.20) is true for all $n \in \mathbb{Z}_{\geqslant 1}$ in this case, and the result follows by using $\tau(\overline{\chi}) \neq 0$ when χ is primitive. $\square$

The modulus of the Gauss sum is also an important fact. Proposition 6.7, along with the next result, may be used to estimate twisted exponential sums of the form

$$\sum_{N < n \leqslant 2N} \chi(n) e(f(n))$$

arising in many problems in number theory.

Proposition 6.8 *If χ is a primitive Dirichlet character modulo q, then $|\tau(\chi)| = \sqrt{q}$.*

Proof By (6.19) we get

$$|\tau(\chi)|^2 = \sum_{a,b \,(\mathrm{mod}\, q)} \chi(a) \overline{\chi}(b) e_q(a - b)$$

$$= \sum_{a \,(\mathrm{mod}\, q)} \chi(a) \sum_{\substack{b \,(\mathrm{mod}\, q) \\ (b,q)=1}} e_q(b(a - 1))$$

$$= \sum_{a \,(\mathrm{mod}\, q)} \chi(a) c_q(a - 1)$$

and from Proposition 6.6 we derive

$$|\tau(\chi)|^2 = \sum_{a \,(\mathrm{mod}\, q)} \chi(a) \sum_{d \mid (q, a-1)} d\mu(q/d) = \sum_{d \mid q} d\mu(q/d) \sum_{\substack{a \,(\mathrm{mod}\, q) \\ a \equiv 1 \,(\mathrm{mod}\, d)}} \chi(a)$$

and since χ is primitive and $d \mid q$, the inner sum vanishes when $d < q$ by [65, Theorem 9.4], and is equal to 1 if $d = q$, so that $|\tau(\chi)|^2 = q$ as required. $\square$

Remark 6.5 When χ is not primitive, we have the following important generalization [65, Theorem 9.10].

Proposition 6.9 *Let χ be a Dirichlet character modulo q that is induced by the primitive character $\chi^\star$ modulo $q^\star$. Set $q = kq^\star$. Then*

$$\tau(\chi) = \mu(k)\chi^\star(k)\tau\left(\chi^\star\right).$$

6.6.2 Incomplete Character Sums

In many applications the bound of Proposition 3.14 may be not sufficient. The next result provides a nearly best possible estimate for this character sum.

Theorem 6.10 (Pólya–Vinogradov Inequality)   *For all non-principal Dirichlet characters χ modulo q and all integers M and N with $N \geqslant 1$, we have*

$$\left| \sum_{M < n \leqslant M+N} \chi(n) \right| \leqslant \sqrt{3q} \log q.$$

Proof Assume first that χ is primitive. From Proposition 6.7, we derive

$$\sum_{M < n \leqslant M+N} \chi(n) = \frac{1}{\tau(\overline{\chi})} \sum_{a=1}^{q} \overline{\chi}(a) \sum_{M < n \leqslant M+N} e_q(na)$$

$$= \frac{1}{\tau(\overline{\chi})} \sum_{a=1}^{q-1} \overline{\chi}(a) e_q\left((M + \tfrac{1}{2}(N+1))a \right) \frac{\sin \pi Na/q}{\sin \pi a/q}$$

and Proposition 6.8 and Jensen's inequality applied to the convex function $x \mapsto (\sin \pi x)^{-1}$ yield

$$\left| \sum_{M < n \leqslant M+N} \chi(n) \right| \leqslant \frac{1}{\sqrt{q}} \sum_{a=1}^{q-1} \frac{1}{|\sin \pi a/q|} \leqslant \frac{1}{\sqrt{q}} \times q \sum_{a=1}^{q-1} \int_{a/q-1/(2q)}^{a/q+1/(2q)} \frac{dx}{\sin \pi x}$$

$$= \sqrt{q} \int_{1/(2q)}^{1-1/(2q)} \frac{dx}{\sin \pi x} \leqslant \sqrt{q} \int_{1/(2q)}^{1/2} \frac{dx}{x} \leqslant \sqrt{q} \log q.$$

Now if χ is a non-principal Dirichlet character modulo q induced by the primitive character $\chi^\star$ of conductor $q^\star$, then $q^\star \mid q$ and writing $q = kq^\star$ we get using Corollary 4.6

$$\sum_{M<n\leqslant M+N} \chi(n) = \sum_{\substack{M<n\leqslant M+N \\ (n,k)=1}} \chi^\star(n)$$

$$= \sum_{d\mid k} \mu(d) \sum_{\frac{M}{d}<m\leqslant \frac{M+N}{d}} \chi^\star(md)$$

$$= \sum_{d\mid k} \mu(d)\chi^\star(d) \sum_{\frac{M}{d}<m\leqslant \frac{M+N}{d}} \chi^\star(m)$$

and the estimate above yields

$$\left| \sum_{M<n\leqslant M+N} \chi(n) \right| \leqslant \sqrt{q^\star} \log q^\star \sum_{d\mid k} \mu(d)^2 = 2^{\omega(q/q^\star)}\sqrt{q^\star} \log q^\star$$

and the bound $2^{\omega(n)} \leqslant \sqrt{3n}$ completes the proof.   □

This result, along with Corollary 1.1, immediately improves Corollary 3.8.

Corollary 6.9   *Let $F \in C^1[1, +\infty[$ be a decreasing function such that $F > 0$ and $F(x) \to 0$ as $x \to \infty$. For all non-principal Dirichlet characters χ modulo q and all real numbers $x \geqslant 1$, we have*

$$\left| \sum_{n>x} \chi(n)F(n) \right| \leqslant 2F(x)\sqrt{3q} \log q.$$

If $N \ll q^{1/2}$, the Pólya–Vinogradov inequality is weaker than the trivial bound provided by the triangle inequality. In the early 1960s, Burgess [10, 11] discovered another method to estimate character sums whose bound depends on the length of the range of summation.

Theorem 6.11 (Burgess)   *For all non-principal* primitive *Dirichlet characters χ modulo q, all real numbers $0 < \varepsilon < 1$ and all integers M and N with $N \geqslant 1$, we have*

$$\sum_{M<n\leqslant M+N} \chi(n) \ll N^{1-1/r} q^{\frac{r+1}{4r^2}+\varepsilon}$$

for $r \in \{1, 2, 3\}$.

The proof uses in a crucial way the Riemann hypothesis for curves over finite fields which was proved by Weil[9] [89], implying that for prime p and Dirichlet character χ modulo p of order d, then

$$\left| \sum_{x=1}^{p} \chi\left(P(x)\right) \right| \leqslant (m-1)\sqrt{p},$$

where $P \in \mathbb{F}_p[X]$ is not a dth power of a polynomial and m is the number of distinct roots of P.

For incomplete Gauss sums, Burgess [12] obtained the following result.

Proposition 6.10 (Burgess) *For all non-principal Dirichlet characters χ modulo q and all integers $a \geqslant 0$, M and N with $1 \leqslant N < q$, we have*

$$\sum_{M < n \leqslant M+N} \chi(n) e_q(an) \ll N^{2/3} q^{1/8} (\log q)^2.$$

6.6.3   Kloosterman Sums

6.6.3.1   Definition

Definition 6.4 Let $a, b \in \mathbb{Z}$ et $q \in \mathbb{Z}_{\geqslant 1}$. A *Kloosterman sum*, resp. *incomplete Kloosterman sum*, is the exponential sum

$$S(a,b,q) := \sum_{\substack{n=1 \\ (n,q)=1}}^{q} e_q\left(an + b\bar{n}\right) \quad \text{resp.} \quad S(a,b,q,I) := \sum_{\substack{n \in I \\ (n,q)=1}} e_q\left(an + b\bar{n}\right),$$

where $\bar{n}$ is any integer for which $n \times \bar{n} \equiv 1 \pmod{q}$[10] and I is any finite real interval such that Long $I \leqslant q$.

These sums have played an increasingly important role in the analytic theory of numbers. It was firstly studied by Kloosterman [50] in his refinement of the Hardy–Littlewood method in 1926, which arised from the problem of representing integers by a positive definite integral quadratic forms in four variables, and gave the non-trivial upper bound, if $q = p > 2$ is prime

$$|S(a,b,p)| < 3^{1/4} p^{3/4} \quad \left(p \nmid b\right).$$

---

[9] See Theorem 6.34.

[10] We follow the convention that values where $\bar{n}$ is not defined are omitted.

Afterwards, Salié [77] and Davenport [20] independently improved this upper bound to

$$S(a, b, p) \ll p^{2/3}$$

when $q = p > 2$ is prime again.

6.6.3.2 Weil's Bound

As mentioned above, the first breakthrough in this problem came from Weil[89] who obtained the best possible bound from his famous proof of the Riemann hypothesis for the algebraic curves over finite fields.

Theorem 6.12 (Weil) *Let* $a, b \in \mathbb{Z}$, $q \in \mathbb{Z}_{\geq 1}$. *Then*

$$|S(a, b, q)| \leq q^{1/2}(a, b, q)^{1/2}\tau(q).$$

This result is a derivation from the classical Weil's Theorem 6.34, who showed it in the case where $q = p > 2$ is prime, and is contained in the paper of Estermann [22].

6.6.3.3 Incomplete Sums

A basic idea in analytic number theory, with countless applications, is the completion technique, which gives estimates for sums over intervals of integers in terms of longer sums. When f is a periodic function, one can use Fourier theory to handle such a completion. This yields the next result.

Theorem 6.13 (Completion of Periodic Functions) *Let* $M < N \in \mathbb{Z}$, $q \in \mathbb{Z}_{\geq 1}$ *and* $f : \mathbb{Z} \to \mathbb{C}$ *a* q-*periodic function. Then*

$$\left| \sum_{M < n \leq N} f(n) - \frac{M - N}{q} \sum_{n=1}^{q} f(n) \right| \leq \max_{1 \leq m < q} \left| \sum_{n=1}^{q} f(n) e_q(mn) \right| \log q.$$

In general, the sum on the right-hand side is expected to be small for $1 \leq m < q$, so that the partial sum of the $f(n)$ is approximately described by its average value. Applied to Kloosterman sums, we derive the following bound.

Corollary 6.10 *Let* $a, b \in \mathbb{Z}$, $q \in \mathbb{Z}_{\geqslant 1}$ *and* I *be a finite real interval such that* Long $I \leqslant q$. *Then*

$$|S(a, b, q, I)| \leqslant \max_{1 \leqslant m \leqslant q} |S(a + m, b, q)| (1 + \log q).$$

Combining Theorem 6.34 and the previous corollary, we derive

$$|S(a, b, q, I)| \leqslant 2q^{1/2}(b, q)^{1/2}\tau(q) \log q \tag{6.21}$$

whenever $q \geqslant 3$ and $1 \leqslant$ Long $I < q$. On the other hand, it can be proved that

$$\sum_{a \,(\mathrm{mod}\, q)} |S(a, b, q, I)|^2 \ll q\,(|I| + 1),$$

where $|I|$ is the number of integers contained in I. It is then reasonable to surmise that

$$S(a, b, q, I) \ll \left(|I|^{1/2+\varepsilon} + 1\right)(a, b, q)^{1/2}.$$

In [39], Hooley stated the R^*-*conjecture*, still unproven to date, namely

$$S(a, b, q, I) \ll \left(|I|^{1/2} + 1\right)(b, q)^{1/2}q^{\varepsilon}.$$

When $I = [1, x]$ with $x > e^{\frac{c \log q}{\log \log q}}$ for some $c > \log 2$, Korolev [51] proved that there exists $\delta > 0$ such that

$$S\,(a, b, q, [1, x]) \ll x(\log q)^{-\delta}.$$

6.6.3.4 Twisted Incomplete Sums

In recent years, the previous results have been extended in various directions. For instance, a great deal of efforts have been made to get bounds for the twisted sums

$$S_f(a, b, q, x) := \sum_{\substack{n \leqslant x \\ (n,q)=1}} f(n)e_q\,(an + b\overline{n}),$$

where $\sqrt{q} \log q \ll x \leqslant q$ is any fixed real number, $(a, q) = 1$ and f is a complex-valued multiplicative function satisfying $|f(n)| \leqslant 1$. When q is sufficiently large, Korolev [52] showed the next estimate.

Theorem 6.14 *Let $\varepsilon \in \left(0, \frac{1}{2}\right)$ be any fixed real number, $q \geqslant q_0(\varepsilon)$ be a sufficiently large integer, and assume that $q^{1/2+\varepsilon} \ll x \leqslant q$. Let f be a multiplicative function satisfying $|f(n)| \leqslant 1$. Then*

$$\left| S_f(a, b, q, x) \right| \leqslant 562 \, x \frac{\log \log q}{\varepsilon \log q}.$$

6.7 The Hardy–Littlewood Method

6.7.1 The Circle Method

For an exhaustive exposition on this subject, see Vaughan's monograph [84] or the survey [94].

6.7.1.1 Additive Prime Number Theory

Euclid established the basic multiplicative structure of primes. However, ever since their discovery, any attempt to understand the *additive* nature of primes has been proved to be a very hard task. As a result, two landmark problems have risen, namely the strong and weak Goldbach Conjectures.

Conjecture 6.1 (Strong Goldbach Conjecture) Every even number $n \geqslant 4$ can be written as the sum of two primes.

The second part of the Goldbach Conjecture, also known as his Odd Conjecture or his Ternary Conjecture, can also easily be described.

Conjecture 6.2 (Weak Goldbach Conjecture) Every odd number $n \geqslant 7$ can be written as the sum of three primes.

Early attempts were largely unsuccessful, and many statements were made without proof. For instance, Euler in 1780 stated that every number of the form $4n + 2$ is the sum of two primes of the form $4k + 1$ and checked this for all $4n + 2 \leqslant 110$.

As pointed out in [94], a general additive problem can be expressed in the following form: given $n \in \mathbb{Z}_{\geqslant 1}$, $s \in \mathbb{Z}_{\geqslant 2}$ and subsets $A_1, \ldots, A_s$ of $\mathbb{Z}_{\geqslant 1}$ not necessarily distinct, consider the Diophantine equation

$$n = a_1 + \cdots + a_s \tag{6.22}$$

with $a_i \in A_i$. We then ask for either the number of solutions of (6.22), or for a proof that this equation has at least one solution for "sufficiently large" n. As seen in Proposition 2.3, we cannot expect, in general, that for very small n there will be a solution of Eq. (6.22). Also, depending on the nature of the sets A_i, there may be some constraints on those n written in the shape (6.22).

6.7.1.2 Hardy, Littlewood and Ramanujan's Method

The method that we are going to describe is certainly one of the most fundamental methods in the study of Additive Prime Number Theory. It was first conceptualized in [34] by G.H. Hardy and S. Ramanujan in 1918, approximately 20 years after the proof of the Prime Number Theorem. However, the method was not formalized for general use until the paper [31].

We illustrate the basic ideas behind the circle method to Eq. (6.22), where we assume, for the sake of clearness, that $A_1 = \cdots = A_s = A$ is an infinite set. Define $r(n; A)$ to be the number of solutions of (6.22) and set

$$F(z) := \sum_{a \in A} z^a.$$

Then, by Cauchy's rule for the product of absolutely convergent power series, we have for $|z| < 1$

$$F(z)^s = \sum_{n=0}^{\infty} \left(\sum_{\substack{n=a_1+\cdots+a_s \\ a_j \in A}} 1 \right) z^n = \sum_{n=0}^{\infty} r(n; A) z^n$$

so that, using Cauchy's formula, we derive for any $0 < \rho < 1$

$$r(n; A) = \frac{1}{2\pi i} \oint_{|z|=\rho} F(z)^s z^{-n-1} \, dz. \qquad (6.23)$$

This is why this method is called "Circle Method". In general, this identity is used to establish bounds for the integral by taking $\rho = \rho(n)$ as a function depending on n such that $\rho(n) \to 1$ as $n \to \infty$, and to show that there exists a constant N such that, for all $n > N$, the integral is greater than or equal to one. This is done by splitting the integral into what are called *major and minor arcs*, and is the strategy followed by Hardy and Littlewood in [32, 33] for the weak Goldbach conjecture, in which they however needed the Generalized Riemann Hypothesis to conclude their work.

6.7.1.3 Vinogradov's Approach

In [85], Vinogradov was able to eliminate the dependency on the generalized Riemann hypothesis. Let us sketch how Vinogradov's ideas lead to the following result in the Goldbach problem, where we set

$$R_3(n) := \sum_{a+b+c=n} \Lambda(a)\Lambda(b)\Lambda(c) \quad \left(n \in \mathbb{Z}_{\geq 1} \right).$$

Theorem 6.15 (Vinogradov) *For any $A \geqslant 0$ and any n sufficiently large*

$$R_3(n) = \tfrac{1}{2}\mathfrak{S}(n)n^2 + O_A\left(n^2(\log n)^{-A}\right),$$

where

$$\mathfrak{S}(n) := \prod_{p|n}\left(1 - \frac{1}{(p-1)^2}\right)\prod_{p\nmid n}\left(1 + \frac{1}{(p-1)^3}\right). \qquad (6.24)$$

Remark 6.6 As in the Prime Number Theorem, a *weighted* number of representations of n as the sum of three *prime powers* is studied, rather than the quantity

$$r_3(n) := \sum_{p_1+p_2+p_3=n} 1$$

itself. However, it is not really difficult to recover the weak Goldbach conjecture from Theorem 6.15.

Corollary 6.11 *Every sufficiently large odd integer is the sum of three prime numbers.*

Proof If n is odd, then

$$\mathfrak{S}(n) \geqslant \prod_{p>2}\left(1 - \frac{1}{(p-1)^2}\right) \geqslant \frac{1}{2}$$

so that, by Theorem 6.15, if n is sufficiently large

$$n^2 \ll R_3(n) = \sum_{p_1+p_2+p_3=n} \log p_1 \log p_2 \log p_3 + \sum_{\substack{p_1^{\alpha_1}+p_2^{\alpha_2}+p_3^{\alpha_3}=n \\ \alpha_j \geqslant 2}} \log p_1 \log p_2 \log p_3$$

$$\ll r_3(n)(\log n)^3 + \text{error terms}$$

and therefore $r_3(n) \gg n^2(\log n)^{-3} + \text{error}$, which is positive if n is sufficiently large. $\square$

In order to prove Theorem 6.15, Vinogradov introduced some refinements, one of which was to replace the infinite power series $F(z)$ by a finite trigonometric polynomial as in (4.59), namely

$$S_n(\alpha) := \sum_{k=1}^n \Lambda(k)e(\alpha k) \quad (\alpha \in \mathbb{R}/\mathbb{Z}).$$

As in the previous section using (4.62), we derive

$$S_n(\alpha)^3 = \sum_{k_1,k_2,k_3 \leqslant n} \Lambda(k_1)\Lambda(k_2)\Lambda(k_3) e(\alpha(k_1+k_2+k_3))$$

$$= \sum_{k \leqslant 3n} e(\alpha k) \sum_{\substack{a+b+c=k \\ a,b,c \leqslant n}} \Lambda(a)\Lambda(b)\Lambda(c)$$

and using (4.62), we obtain

$$\int_0^1 S_n(\alpha)^3 e(-\alpha n)\, d\alpha = \sum_{k \leqslant 3n} \sum_{\substack{a+b+c=k \\ a,b,c \leqslant n}} \Lambda(a)\Lambda(b)\Lambda(c) \int_0^1 e(\alpha(k-n))\, d\alpha$$

$$= \sum_{a+b+c=n} \Lambda(a)\Lambda(b)\Lambda(c)$$

so that

$$R_3(n) = \int_0^1 S_n(\alpha)^3 e(-\alpha n)\, d\alpha. \tag{6.25}$$

Let us first have a glance at the case $\alpha = \frac{a}{q}$ where $1 \leqslant a \leqslant q$ are coprime integers. For $t \in (\sqrt{n}, n]$, set

$$S(a/q; t) = \sum_{k \leqslant t} \Lambda(k) e_q(ak).$$

Splitting the sum according to $(k,q) = 1$ or not, considering in the first case a reduced residue system modulo q and taking the q-periodicity of the function $k \mapsto e_q(ak)$ into account, we derive

$$S(a/q; t) = \sum_{\substack{\ell \leqslant q \\ (\ell,q)=1}} \Psi(t; q, \ell) e_q(a\ell) + \sum_{\substack{k \leqslant t \\ (k,q)>1}} \Lambda(k) e_q(ak),$$

where the function $\Psi(t; q, \ell)$ is given in (3.55). Now the absolute value of the second sum is easily seen to be $\leqslant \omega(q) \log t$, and the Siegel–Walfisz theorem (3.56) yields, for any $B > 0$

$$S(a/q; t) = \frac{t}{\varphi(q)} \sum_{\substack{\ell \leqslant q \\ (\ell,q)=1}} e_q(a\ell) + O_B\left(\varphi(q)ne^{-c_0\sqrt{\log n}}\right)$$

$$= \frac{t c_q(a)}{\varphi(q)} + O_B\left(ne^{-c_1(B)\sqrt{\log n}}\right) = \frac{t\mu(q)}{\varphi(q)} + O_B\left(ne^{-c_1(B)\sqrt{\log n}}\right) \tag{6.26}$$

whenever $q \leqslant (\log n)^B$, and where we used Proposition 6.6 in the last equality. Also note that (6.26) is trivial when $t \leqslant \sqrt{n}$. □

These heuristics lead naturally to the idea that splitting the interval $[0, 1]$ of integration in (6.25) into subintervals, depending on whether α is close to rational numbers with small denominator or not, could be relevant.

Definition 6.5 Let $R > 0$, $Q \in \mathbb{Z}_{\geqslant 1}$ such that $Q \leqslant \frac{1}{2}R$ and, for $1 \leqslant a \leqslant q \leqslant Q$ such that $(a, q) = 1$, define

$$\mathfrak{M}(q, a) := \left[\frac{a}{q} - \frac{1}{R}, \frac{a}{q} + \frac{1}{R} \right].$$

These intervals are called *major arcs* and it is customary to define

$$\mathfrak{M} = \bigcup_{\substack{q \leqslant Q \\ (a,q)=1}} \mathfrak{M}(q, a).$$

The set $\mathfrak{m} := [0, 1] \setminus \mathfrak{M}$ forms the *minor arcs*.

Note that, if $a_1/q_1 \neq a_2/q_2$ are such that $q_1, q_2 \leqslant Q \leqslant \frac{1}{2}R$, then

$$\left| \frac{a_1}{q_1} - \frac{a_2}{q_2} \right| \geqslant \frac{1}{q_1 q_2} \geqslant \left(\frac{1}{q_1} + \frac{1}{q_2} \right) \frac{1}{2Q} \geqslant \left(\frac{1}{q_1} + \frac{1}{q_2} \right) \frac{1}{R}$$

so that the major arcs $\mathfrak{M}(q, a)$ are pairwise disjoint.

6.7.1.4 The Major and Minor Arcs

From now on, we choose

$$A > 0, \quad B := 2A + 5, \quad Q := (\log n)^B \quad \text{and} \quad R := n(\log n)^{-B}. \tag{6.27}$$

Lemma 6.12 *For* $x \in \mathbb{R}$, *let*

$$T_n(x) := \sum_{k=1}^{n} e(kx).$$

Then there exists a constant $c_2 > 0$ *such that, for* $1 \leqslant a \leqslant q \leqslant Q$, $(a, q) = 1$, $\alpha \in \mathfrak{M}(q, a)$, *one has*

$$S_n(\alpha) = \frac{\mu(q)}{\varphi(q)} T_n\left(\alpha - \frac{a}{q} \right) + O\left(n e^{-c_2\sqrt{\log n}} \right).$$

Proof For simplicity, set $\beta := \alpha - \frac{a}{q}$ and note that $|\beta| \leqslant R^{-1} = n^{-1}(\log n)^B$ since $\alpha \in \mathfrak{M}(q, a)$. By partial summation

$$S_n(\alpha) = \sum_{k \leqslant n} \Lambda(k) e_q(ka) e(k\beta) = e(n\beta) S\left(\frac{a}{q}; n\right) - 2\pi i \beta \int_1^n e(t\beta) S\left(\frac{a}{q}; t\right) dt$$

and similarly

$$T_n(\beta) = ne(n\beta) - 2\pi i \beta \int_1^n \lfloor t \rfloor e(t\beta) \, dt$$

so that, using (6.26), the inequality $q \leqslant Q$ and (6.27)

$$\left| S_n(\alpha) - \frac{\mu(q)}{\varphi(q)} T_n(\beta) \right| \leqslant \left| S\left(\frac{a}{q}; n\right) - \frac{n\mu(q)}{\varphi(q)} \right| + 2\pi |\beta| \int_1^n \left| S\left(\frac{a}{q}; t\right) - \frac{\lfloor t \rfloor \mu(q)}{\varphi(q)} \right| dt$$

$$\ll n e^{-c_1 \sqrt{\log n}} (1 + n|\beta|) \ll n e^{-c_2 \sqrt{\log n}}$$

as required.   □

We are now in a position to prove the main result of this section.

Proposition 6.11 *With $\mathfrak{S}(n)$ given in (6.24), we have*

$$\int_{\mathfrak{M}} S_n(\alpha)^3 e(-\alpha n) \, d\alpha = \tfrac{1}{2} \mathfrak{S}(n) n^2 + O\left(n^2 (\log n)^{-B}\right).$$

Proof With Q and R defined in (6.27) and using Lemma 6.12, we derive

$$\int_{\mathfrak{M}} S_n(\alpha)^3 e(-\alpha n) \, d\alpha$$

$$= \sum_{q \leqslant Q} \sum_{\substack{a=1 \\ (a,q)=1}}^q \int_{\mathfrak{M}(q,a)} \left\{ \frac{\mu(q)}{\varphi(q)^3} T_n(\beta)^3 e(-\alpha n) + O\left(n^3 e^{-c_3 \sqrt{\log n}}\right) \right\} d\alpha$$

$$= \sum_{q \leqslant Q} \sum_{\substack{a=1 \\ (a,q)=1}}^q \int_{-1/R}^{1/R} \frac{\mu(q)}{\varphi(q)^3} T_n(\beta)^3 e(-\beta n) e_q(-an) \, d\beta + O\left(Q^2 R^{-1} n^3 e^{-c_3 \sqrt{\log n}}\right)$$

$$= \mathfrak{S}(n; Q) \int_{-1/R}^{1/R} T_n(\beta)^3 e(-\beta n) \, d\beta + O\left(n^2 (\log n)^{-B}\right),$$

where

$$\mathfrak{S}(n; Q) := \sum_{q \leqslant Q} \sum_{\substack{a=1 \\ (a,q)=1}}^{q} \frac{\mu(q)}{\varphi(q)^3} e_q(-an)$$

$$= \sum_{q=1}^{\infty} \frac{\mu(q)}{\varphi(q)^3} \sum_{\substack{a=1 \\ (a,q)=1}}^{q} e_q(-an) + O\left(\sum_{q>Q} \frac{1}{\varphi(q)^2}\right)$$

$$= \sum_{q=1}^{\infty} \frac{\mu(q)c_q(n)}{\varphi(q)^3} + O\left(Q^{-1}\right)$$

$$= \prod_{p} \left(1 - \frac{c_p(n)}{\varphi(p)^3}\right) + O\left((\log n)^{-B}\right)$$

and by Proposition 6.6, we get

$$c_p(n) = \begin{cases} \varphi(p), & \text{if } p \mid n \\ \mu(p), & \text{if } p \nmid n \end{cases}$$

so that

$$\mathfrak{S}(n; Q) = \mathfrak{S}(n) + O\left((\log n)^{-B}\right).$$

Now observe that, if $\beta \notin \mathbb{Z}$, then $T_n(\beta) \leqslant \|\beta\|^{-1}$ by Exercise 114, so that one can replace the interval $\left[-\frac{1}{R}, \frac{1}{R}\right]$ by $\left[-\frac{1}{2}, \frac{1}{2}\right]$, and the error does not exceed

$$\ll \sum_{q \leqslant Q} \frac{1}{\varphi(q)^3} \sum_{\substack{a=1 \\ (a,q)=1}}^{q} \int_{1/R}^{1/2} \frac{d\beta}{\|\beta\|^3} \ll \sum_{q \leqslant Q} \frac{1}{\varphi(q)^2} \int_{1/R}^{1/2} \frac{d\beta}{\beta^3} \ll R^2 \ll n^2 (\log n)^{-2B}$$

and a computation as above using (4.62) again yields

$$\int_{-1/2}^{1/2} T_n(\beta)^3 e(-\beta n) \, d\beta = \int_{-1/2}^{1/2} \sum_{k \leqslant 3n} e(k\beta) \left(\sum_{a+b+c=k} 1\right) e(-\beta n) \, d\beta$$

$$= \sum_{k \leqslant 3n} \left(\sum_{a+b+c=k} 1\right) \int_{-1/2}^{1/2} e\left(\beta(k-n)\right) \, d\beta$$

$$= \sum_{a+b+c=n} 1 = \binom{n-1}{2} = \tfrac{1}{2}(n-1)(n-2),$$

where we used Proposition 2.8. Putting all together gives the asserted estimate.   □

Now we turn to bound the contribution of the minor arcs to the integral. First note that, by Lemma 4.15 and Corollary 3.3, we have

$$\int_0^1 |S_n(\alpha)|^2 \, d\alpha = \sum_{k \leqslant n} \Lambda(k)^2 \leqslant \log n \sum_{k \leqslant n} \Lambda(k) \leqslant 2n \log n$$

and hence

$$\left| \int_{\mathfrak{m}} S_n(\alpha)^3 e(-n\alpha) \, d\alpha \right| \leqslant \sup_{\alpha \in \mathfrak{m}} |S_n(\alpha)| \int_0^1 |S_n(\alpha)|^2 \, d\alpha \leqslant 2n \log n \cdot \sup_{\alpha \in \mathfrak{m}} |S_n(\alpha)|$$

$$(6.28)$$

so that it remains to prove a pointwise estimate for $|S_n(\alpha)|$ for α far away from rationals with small denominators. This is done by using (6.39) as in Theorem 4.51 to derive a Vaughan type inequality for $|S_n(\alpha)|$. The next result is an explicit version of such a bound [19].

Proposition 6.12 *Let $x \geqslant 1$ and $\alpha > 0$ such that there exist positive coprime integers $a \leqslant q \leqslant x$ satisfying $\left| \alpha - \frac{a}{q} \right| \leqslant \frac{1}{q^2}$. Then*

$$\left| \sum_{k \leqslant x} \Lambda(k) e(k\alpha) \right| \leqslant x (\log x)^{3/4} L(x) \left(14.86 \left(\frac{1}{q} + \frac{q \log 4q}{\pi x} \right)^{1/2} + 6.45 \sqrt{\log x} \, e^{-\frac{1}{2}\sqrt{\log x}} \right),$$

where $L(x) := \sqrt{\log \log x + \frac{1}{2}}$.

This result and the argument above lead to the following estimates for the minor arcs.

Proposition 6.13 *Assume $\log n \geqslant B^B$. Then*

$$\left| \int_{\mathfrak{m}} S_n(\alpha)^3 e(-n\alpha) \, d\alpha \right| < 68n^2 (\log n)^{(5-B)/2}.$$

Proof The condition $\log n \geqslant B^B$ ensures that

$$(\log n)^{5/4} L(n) \, e^{-\frac{1}{2}\sqrt{\log n}} < (\log n)^{(3-B)/2}.$$

Now let $\alpha \in \mathfrak{m}$ and take Q and R as in (6.27). From Dirichlet's approximation theorem, there exist positive coprime integers $a \leqslant q \leqslant R$ such that

$$\left| \alpha - \frac{a}{q} \right| \leqslant \frac{1}{qR} \leqslant \frac{1}{q^2}.$$

Hence $\left|\alpha - \frac{a}{q}\right| \leqslant R^{-1}$, and since $\alpha \in \mathfrak{m}$, this implies that $q > Q$, and therefore, by Proposition 6.12

$$|S_n(\alpha)| \leqslant 14.86\, n \log n \left(\frac{1}{Q^{1/2}} + \frac{R^{1/2}(\log 4R)^{1/2}}{(\pi n)^{1/2}} \right) + 6.45n(\log n)^{(3-B)/2}$$

$$\leqslant 14.86\, n \log n \left((\log n)^{-B/2} + \sqrt{\tfrac{2}{\pi}}(\log n)^{(1-B)/2} \right) + 6.45n(\log n)^{(3-B)/2}$$

$$< 34n(\log n)^{(3-B)/2}$$

and inserting this bound in (6.28) yields the asserted estimate. $\qquad\square$

6.7.1.5 Conclusion

From Propositions 6.11 and 6.13, and also from (6.25) and (6.27), we derive

$$R_3(n) = \int_{\mathfrak{M}} S(\alpha)^3 e(-n\alpha)\, d\alpha + \int_{\mathfrak{m}} S(\alpha)^3 e(-n\alpha)\, d\alpha$$

$$= \tfrac{1}{2}\mathfrak{S}(n)n^2 + O\left(n^2(\log n)^{(5-B)/2}\right)$$

$$= \tfrac{1}{2}\mathfrak{S}(n)n^2 + O\left(n^2(\log n)^{-A}\right)$$

completing the proof of Theorem 6.15.

Now let us point out several remarks.

Remark 6.7

▷ As seen in the proof above, Vinogradov's theorem from 1937 relies heavily on the Siegel–Walfisz theorem. Since all of the implied constants in Siegel–Walfisz are ineffective, this proof does not either yield any computable constant C such that the weak Goldbach conjecture holds for all $n \geqslant C$, or gives no way to determine what the constant C might be.

▷ A few years later, Mardzhanishvili [62] improved on Vinogradov's result so that he already had the effect of making the result *effective*, i.e. he showed a way that makes it clear that a constant C could in principle be determined following and reworking the proof with great care.

▷ Subsequently, some authors made Vinogradov's work *explicit*, giving a value to C. For instance, it was proved by Liu and Wang [59] that the value $C = 2 \times 10^{1346}$ is admissible. Although completely explicit, this value is not really *efficient*, meaning that it is too large to be attacked by any foreseeable computational means within our physical universe.

▷ Finally, by careful analysis of the previous work and, as the author points out, using the large sieve "not just as a tool [...] but as a source for ideas on how to apply the circle method more effectively", Helfgott [37] proved the weak Goldbach conjecture for all $n \geqslant 10^{27}$, which really is an efficient value since the conjecture had previously been verified computationally up to 8.8×10^{30} by Helfgott and Platt [38]. The ternary Goldbach conjecture is henceforth a theorem.

> **Theorem 6.16** *Every odd integer $n \geqslant 7$ can be written as the sum of three primes.*

Remark 6.8

▷ Note that if n is even, then $\mathfrak{G}(n) = 0$, so Vinogradov's theorem does not tell us anything. Indeed, if a sufficiently large even integer n was the sum of three primes, then one of these primes must be 2. Thus, $n - 2$ could be written as the sum of two primes and the strong Goldbach conjecture would follow for sufficiently large integers.

▷ On the other hand, Vinogradov's strategy cannot be applied to the binary Goldbach conjecture. Indeed, with obvious notation, we now have

$$R_2(n) = \int_0^1 S_n(\alpha)^2 e(-n\alpha) \, d\alpha$$

and the contribution of the minor arcs to the integral is greater than the main term given by the major arcs, since

$$\int_0^1 |S_n(\alpha)|^2 \, d\alpha \ll N \log N \quad \text{and} \quad \sup_{\alpha \in \mathfrak{m}} |S_n(\alpha)|^2 \ll N^2 (\log N)^{-\kappa(A)}.$$

▷ As Helfgott pointed out in [37, p. 6], the Circle Method is an application of Fourier analysis over $\mathbb{Z}$, since exponential sums could be viewed as Fourier transforms over $\mathbb{Z}$. The method underlines the duality between $\mathbb{Z}$ and the "circle" $\mathbb{R}/\mathbb{Z}$.

6.7.2   The Discrete Circle Method

The idea to adapt Hardy–Littlewood's circle method to exponential sums first appeared in [6] in which the authors obtained an improvement in the problem of bounding $\zeta\left(\frac{1}{2} + it\right)$. The method was later used by Iwaniec and Mozzochi [45] to treat the Dirichlet divisor problem and the Gauss circle problem, which are similar by nature. This method is one of the trickiest in exponential sum theory. The proof is far beyond the scope of this book,[11] but one can systematize the ideas in the following six steps.

In what follows, let $f \in C^r \left[\frac{1}{2} N, 2N \right]$ with $r \geq 4$ and the derivatives satisfying van der Corput type hypotheses as in (6.11) so that there exist $c_i \geq 1$ and $T \geq 1$ such that, for all $x \in \left[\frac{1}{2} N, 2N \right]$, we have

$$\frac{T}{N^j} \leq \left| f^{(j)}(x) \right| \leq c_j \frac{T}{N^j} \quad (j = 1, \ldots, r).$$

We also suppose that $c_3 T \leq N^3$. The purpose is to get an estimate for the exponential sum

$$S = \sum_{N \leq n \leq 2N} e(f(n)).$$

6.7.2.1   Step 1: Major and Minor Arcs

As in the original circle method, the interval $[N, 2N]$ is covered by $\lfloor N/M \rfloor + 1$ intervals $I_k = [N + (k-1)M, N + kM)$ where M is a positive integer chosen so that $M \leq \frac{1}{10} N$, $M^2 T \leq 2c_3 N^3$ and $M^4 T \leq N^4$. Let R be the integer such that

$$(R - 1)^2 MT < 2c_3 N^3 \leq R^2 MT$$

so that $R^2 \geq M$ and suppose that $R \leq M$, since this is the range where this method really works. The role of R is to give the right order of magnitude of q when we approximate the real number $\frac{1}{2} f''(x)$ by a rational number $\frac{a}{q}$. With the two parameters defined above, we have ensured that, for all $x \in \left[\frac{1}{2} N, 2N \right]$,

$$M^{-3} \ll f'''(x) \ll M^{-2} \quad \text{and} \quad f^{(4)}(x) \ll M^{-4}.$$

The problem is then to find a point in I_k such that the derivatives of f at this point are rational numbers, in order to expand $f(x)$ in a Taylor series about this point. Since I_k has length M, as x runs through I_k, $\frac{1}{2} f''(x)$ runs through an interval J_k of length $\asymp MTN^{-3} \asymp R^{-2}$. Choose a rational $\frac{a}{q} = \frac{a_k}{q_k}$ in J_k with $(a, q) = 1$ for which q is minimal. If $q \leq R^2 M^{-1}$, we call I_k or J_k a *major arc*, otherwise a *minor arc*. Having chosen $\frac{a}{q}$, we pick the integer $n_0 = n_0^{(k)} \in I_k$ for which $\frac{1}{2} f''(n_0)$ is the closest to $\frac{a}{q}$. Since f'' is strictly monotone, we have $|n_0 - x_0| \leq 1$ where x_0 is the

---

[11]For a complete exposition of this subject, the reader should refer to [26, 41].

unique solution of the equation $\frac{1}{2} f''(x) = \frac{a}{q}$. Hence we have

$$\left| \frac{f''(n_0)}{2} - \frac{a}{q} \right| \leqslant \frac{c_3 T}{2 N^3} \leqslant \frac{4 c_3^2}{M^2}.$$

Next we choose a rational approximation $\frac{b}{s}$ to $q f'(n_0)$ such that

$$\left| \frac{b}{s} - q f'(n_0) \right| \leqslant q M^{-1} \quad \text{and} \quad s \leqslant \max(1, M q^{-1})$$

which is always possible by Dirichlet's approximation theorem, and set

$$g(x) = f(n_0) + \frac{bx}{qs} + \frac{ax^2}{q} + \mu x^3,$$

where $\mu = \frac{1}{6} f'''(n_0)$. This polynomial $g(x)$ is the function which is used to approximate $f(x + n_0)$. The values of m such that $n_0 + m$ lies in I_k belong to an interval $[M_1, M_2]$ such that $M_1 \geqslant M$, $M_2 \leqslant 3M$ and $M_2 - M_1 \leqslant M$, and writing

$$\sum_{M_1 \leqslant m \leqslant M_2} e(f(n_0 + m)) = \sum_{M_1 \leqslant m \leqslant M_2} e(f(n_0 + m) - g(m)) e(g(m))$$

and using the fact that

$$f'(n_0 + x) - g'(x) = f'(n_0) - \frac{b}{qs} + x \left(f''(n_0) - \frac{2a}{q} \right) + \frac{x^3}{6} f^{(4)}(n_0 + \theta x),$$

where $0 < \theta < 1$, we get using the estimates above

$$f'(n_0 + x) - g'(x) \ll M^{-1}$$

for all $x \ll M$ such that $n_0 + x \in \left[\frac{1}{2} N, 2N \right]$, so that by partial summation we get

$$\sum_{M_1 \leqslant m \leqslant M_2} e(f(n_0 + m)) \ll \max_{M_1 \leqslant M_3 \leqslant M_2} \left| \sum_{M_3 \leqslant m \leqslant M_2} e(g(m)) \right|.$$

For each k we have a different polynomial $g(m) = g_k(m)$ and different limits $M_3^{(k)}, M_2^{(k)}$. Also note that we have bounded S by the sum of the moduli of $\ll NM^{-1}$ shorter sums of length $\asymp M$. Since the best result one can hope for these sums is $\ll M^{1/2}$, the best estimate we could obtain for S would be

$$S \ll \left(\frac{N}{M}\right) M^{1/2} \ll \frac{N}{M^{1/2}}. \tag{6.29}$$

6.7.2.2   Step 2: Gauss Sums

The sum

$$\sum_{M_3 \leqslant m \leqslant M_2} e(g(m)) = e(f(n_0)) \sum_{M_3 \leqslant m \leqslant M_2} e\left(\mu m^3 + \frac{am^2}{q} + \frac{bm}{qs}\right)$$

may be viewed as an incomplete quadratic Gauss sum perturbed by the factor $e(\mu m^3)$. Gauss sums are usually defined as sums of additive or multiplicative characters. In our problem, the quadratic Gauss sum

$$G(a, b; q) = \sum_{m \;(\mathrm{mod}\, q)} e_q(am^2 + bm)$$

plays a key part since

$$\sum_{M_3 \leqslant m \leqslant M_2} e\left(\mu m^3 + \frac{am^2}{q} + \frac{bm}{qs}\right) = \frac{1}{qs} \sum_{k \;(\mathrm{mod}\, qs)} G(as, b+k; qs) \sum_{M_3 \leqslant m \leqslant M_2} e\left(\mu m^3 - \frac{km}{qs}\right).$$

Quadratic Gauss sums have some interesting properties. For instance, if $b + k = ds$ for some integer d and if q is odd, then

$$G(as, b+k; qs) = s\, e_q(\overline{-4a}\, d^2)\, G(a, 0; q)$$

otherwise the sum is zero. If q is even, there is a slight variation in this identity which does not really affect the result. The notation $\overline{\alpha}$ means the integer β satisfying $\alpha\beta \equiv 1 \pmod{q}$. Since $|G(a, 0; q)| = \sqrt{q}$ if q is odd and $(a, q) = 1$, we get by [41, Lemma 5.4.5]

$$\left| \sum_{M_3 \leqslant m \leqslant M_2} e(g(m)) \right| = \frac{1}{\sqrt{q}} \left| \sum_{d \;(\mathrm{mod}\, q)} e_q(\overline{-4a}\, d^2) \sum_{M_3 \leqslant m \leqslant M_2} e\left(\mu m^3 - \frac{m}{q}\left(d - \frac{b}{s}\right)\right) \right|$$

if q is odd, which we shall suppose from now on.

6.7.2.3   Step 3: Poisson Summation

One may apply the truncated Poisson summation formula of Lemma 6.9 along with a computation about a special integral called the *Airy integral* and the use of Lemma 6.6. This process replaces the sums over d and m above by

$$\left| \sum_{M_3 \leqslant m \leqslant M_2} e(g(m)) \right| = \frac{1}{\sqrt{q}} \left| \sum_h \left(\frac{q}{12 h_1 \mu} \right)^{1/4} e_q(\overline{-4a}\, h^2)\, e\left(-\frac{2}{\mu^{1/2}} \left(\frac{h_1}{3q} \right)^{3/2} \right) \right|$$

$$+ O\left(\frac{q + R}{\sqrt{q}} \log M \right)$$

for $\mu > 0$, where h is an integer variable, $h_1 = h - \frac{b}{s}$ and the sum is such that

$$3\mu q M_3^2 \leqslant h_1 \leqslant 3\mu q M_2^2.$$

For $\mu < 0$ we use $-g(m)$. Since $R^2 \asymp N^3 (MT)^{-1}$, the size of each summand is

$$\ll \left(\frac{q}{\mu h_1} \right)^{1/4} \ll \left(\frac{q}{\mu^2 M^2 q} \right)^{1/4} \asymp (\mu M)^{-1/2} \asymp R$$

and the length of the range of summation is $\leqslant 18\mu q M^2 < 144 c_3^2 q M R^{-2}$. Hence the trivial estimate on the sum over h is $\ll Rq^{-1/2}$ times the number of terms. On major arcs, we have $q M R^{-2} \leqslant 1$ and one can prove [43] that the number of major arcs with $Q \leqslant q < 2Q$ is $\ll N Q^2 (MR^2)^{-1} + 1$. Thus, dividing the major arcs into ranges $Q \leqslant q < 2Q$ where Q is a power of 2 such that $Q \leqslant R^2/M$, we infer that the major arcs contribute

$$\ll \sum_{Q=2^\alpha \leqslant R^2/M} \frac{R}{Q^{1/2}} \left(\frac{NQ^2}{MR^2} + 1 \right) \log M$$

$$\ll \left(\frac{N}{MR} \cdot \frac{R^3}{M^{3/2}} + R \right) \log M$$

$$\ll \left(\frac{N}{M^{1/2}} + R \right) \log M$$

to the sum S, where we used the inequality $R \leqslant M$. Taking (6.29) into account, this estimate is nearly a best possible result.

6.7.2.4 Step 4: The Large Sieve on Minor Arcs

We now look for the contribution of the minor arcs for which $q > R^2 M^{-1}$. The use of the large sieve corresponds to Vinogradov's mean value theorem in the continuous case.[12] The sum over h above is first tidied with some analytic tools like partial sums by Fourier transforms, Hölder's inequality, etc. Sums of bilinear forms of the type

$$\sum_{M_3 \leqslant m \leqslant M_2} e(g(m)) \ll Rq^{-1/2} \sum_H \int_{-1/2}^{1/2} \mathcal{K}(t) \left| \sum_{H \leqslant h < 2H} e(-\mathbf{x}^{(h)} \cdot \mathbf{y}(t)) \right| dt + q^{1/2} \log M$$

thus appear, where H runs over all powers of 2 covering the range of h in the summation above, where $\mathcal{K}(t)$ is a kernel function satisfying

$$\int_{-1/2}^{1/2} \mathcal{K}(t)\, dt \ll \log M$$

and where the vectors $\mathbf{x}^{(h)}$ and $\mathbf{y}(t)$ have, respectively, integer and rational coordinates in four dimensions, the first one corresponding to h and the second one to the minor arc J_k. After some rearrangements, the sum is treated with the following inequality.

Lemma 6.13 (Double Large Sieve Inequality) *Let* $\mathbf{x}^{(1)}, \ldots, \mathbf{x}^{(K)}$ *and* $\mathbf{y}^{(1)}, \ldots,$ $\mathbf{y}^{(L)}$ *be real vectors in d dimensions satisfying*

$$\left| \mathbf{x}_i^{(k)} \right| \leqslant \tfrac{1}{2} X_i \quad \text{and} \quad \left| \mathbf{y}_i^{(\ell)} \right| \leqslant \tfrac{1}{2} Y_i$$

for $i = 1, \ldots, d$, $k = 1, \ldots, K$ *and* $\ell = 1, \ldots, L$. *Then*

$$\left| \sum_{k=1}^K \sum_{\ell=1}^L e\left(\mathbf{x}^{(k)} \cdot \mathbf{y}^{(\ell)} \right) \right|^2 \leqslant \left(\frac{\pi}{2} \right)^{4d} A(\mathbf{a}) B(\mathbf{b}) \prod_{i=1}^d (X_i Y_i + 1),$$

where

$$A(\mathbf{a}) = \sum_k \sum_j \prod_{i=1}^d \max\left(0, 1 - \left| \frac{\mathbf{x}_i^{(k)} - \mathbf{x}_i^{(j)}}{a_i} \right| \right)$$

$$B(\mathbf{b}) = \sum_\ell \sum_j \prod_{i=1}^d \max\left(0, 1 - \left| \frac{\mathbf{y}_i^{(\ell)} - \mathbf{y}_i^{(j)}}{b_i} \right| \right)$$

and $a_i = X_i(1 + X_i Y_i)^{-1}$ *and* $b_i = X_i^{-1}$.

---

[12] See Lemma 6.14.

6.7.2.5 Step 5: Semicubical Powers of Integers

This is the trickiest point of the method. The numbers $A(\mathbf{a})$ and $B(\mathbf{b})$ of Lemma 6.13 require to count the number of solutions of a system of four Diophantine inequalities involving semicubical powers of eight integer unknowns. This is done using very complicated combinatorial arguments [6, 26].

6.7.2.6 Step 6: Final Step

The final step rests on counting the points indexed by the minor arcs. One has to count the number of coincidences of the points $\mathbf{y}$, i.e. to count the number of Diophantine inequalities with rational unknowns. The numbers M and R are then chosen optimally.

6.7.2.7 Conclusion

This method enables us to get some new exponent pairs which are not attainable by any successive A- and B-processes. In [43], Huxley and Watt were able to get the exponent pair $\left(\frac{9}{56} + \varepsilon, \frac{1}{2} + \frac{9}{56} + \varepsilon \right)$. They gradually improved their estimates by getting the pairs

$$
(k, \ell) =
\begin{cases}
\left(\frac{2}{13} + \varepsilon, \frac{35}{52} + \varepsilon \right) & (1987) \\[2ex]
\left(\frac{89}{570} + \varepsilon, \frac{1}{2} + \frac{89}{570} + \varepsilon \right) & (1996) \\[2ex]
\left(\frac{32}{205} + \varepsilon, \frac{1}{2} + \frac{32}{205} + \varepsilon \right) & (2005)
\end{cases}
$$

the last two exponent pairs being obtained by Huxley [41, 42].

In [42], Huxley used the ideas exposed above and added some refinements, notably bounding the number of integer points near a specific smooth curve as seen in Chap. 5, to estimate the following double exponential sum

$$
\sum_{H < h \leqslant 2H} a_h \sum_{N < n \leqslant 2N} b_n e(hf(n))
$$

where (a_h) and (b_n) are of bounded variation. Using Corollary 6.1, he then deduced the following result.

Theorem 6.17 (Huxley) *Let $r \geqslant 5$, $N \geqslant 1$ be integers and $f \in C^r [N, 2N]$ such that there exist real numbers $T \geqslant N$ and $1 \leqslant c_0 \leqslant \cdots \leqslant c_r$ such that, for all $x \in [N, 2N]$ and all $j \in \{0, \ldots, r\}$, we have*

$$\frac{T}{N^j} \leqslant \left| f^{(j)}(x) \right| \leqslant c_j \frac{T}{N^j}.$$

Then we have

$$\sum_{N < n \leqslant 2N} \psi \left(f(n) \right) \ll (NT)^{131/416} (\log NT)^{18\,627/8\,320}.$$

Applied to the Dirichlet divisor problem, this gives the best known result up to now namely

$$\Delta(x) \ll x^{131/416} (\log x)^{26\,947/8\,320}.$$

Corollary 6.4 with van der Corput's theory gives under the same hypotheses

$$\sum_{N < n \leqslant 2N} \psi \left(f(n) \right) \ll (NT)^{1/3}$$

if $T \geqslant N$. We have $1/3 \approx 0.33 \ldots$ and $131/416 \approx 0.31490384$ which shows the extreme difficulty in getting some new improvements—this is actually the case with every problem in analytic number theory.

Huxley's result was recently slightly superseded by Bourgain and Watt [7, (7.4)] who combined Huxley's method with the so-called *decoupling principle* in harmonic analysis. This yields some improvement in the first spacing problem of Huxley's method, and allows the authors to derive the following improvement of Theorem 6.17 in the case where $f(n) = x/n$.

Theorem 6.18 *Let $a, b \in \mathbb{Z}$ be integers such that $|a| + |b| \leqslant 1$, $N \in \mathbb{Z}_{\geqslant 3}$ and $x \in \mathbb{R}_{\geqslant 1}$ satisfying $3 \leqslant N \leqslant x^{1/2}$. Let $\varepsilon \in \left(0, \frac{1}{2} \right]$. Then*

$$\sum_{N < n \leqslant 2N} \psi \left(\frac{4x}{4n + a} + \frac{b}{4} \right) \ll x^{\frac{517}{1648} + \varepsilon}.$$

As a corollary, this result immediately provides the following new estimates in the Dirichlet divisor problem and the Gauss circle problem.

Corollary 6.12 (Dirichlet and Gauss Problems) *For x sufficiently large*

$$\sum_{n \leqslant x} \tau(n) = x \log x + x(2\gamma - 1) + O\left(x^{517/1\,648+\varepsilon}\right)$$

and

$$\sum_{n \leqslant x} r(n) = \pi x + O\left(x^{517/1\,648+\varepsilon}\right).$$

6.8 Vinogradov's Method

6.8.1 Introduction

Let us move on now to another divisor problem and suppose we want to estimate the sum $\sum_{n \leqslant x} \sigma(n)$. As in Chap. 4, we have

$$\sum_{n \leqslant x} \sigma(n) = \frac{1}{2} \sum_{d \leqslant x} \left\lfloor \frac{x}{d} \right\rfloor \left(\left\lfloor \frac{x}{d} \right\rfloor + 1 \right)$$

$$= \frac{1}{2} \sum_{d \leqslant x} \left(\frac{x}{d} - \psi\left(\frac{x}{d}\right) - \frac{1}{2} \right) \left(\frac{x}{d} - \psi\left(\frac{x}{d}\right) + \frac{1}{2} \right)$$

$$= \frac{1}{2} \sum_{d \leqslant x} \left(\frac{x^2}{d^2} - \frac{2x}{d} \psi\left(\frac{x}{d}\right) + \psi\left(\frac{x}{d}\right)^2 - \frac{1}{4} \right)$$

$$= \frac{x^2}{2} \sum_{d \leqslant x} \frac{1}{d^2} - x \sum_{d \leqslant x} \frac{1}{d} \psi\left(\frac{x}{d}\right) + O(x)$$

$$= \frac{x^2 \zeta(2)}{2} - \frac{x^2}{2} \sum_{d > x} \frac{1}{d^2} - x \sum_{d \leqslant x} \frac{1}{d} \psi\left(\frac{x}{d}\right) + O(x)$$

$$= \frac{x^2 \zeta(2)}{2} - x \sum_{d \leqslant \sqrt{x}} \frac{1}{d} \psi\left(\frac{x}{d}\right) - x \sum_{\sqrt{x} < d \leqslant x} \frac{1}{d} \psi\left(\frac{x}{d}\right) + O(x).$$

We apply (1.2) to the second sum with $f(t) = t^{-1}\psi(x/t)$, $a = \sqrt{x}$ and $b = x$. Since $V_a^b(f) \ll 1 + x^{-1/2}$ we have

$$\sum_{\sqrt{x}<d\leqslant x} \frac{1}{d}\psi\left(\frac{x}{d}\right) = \int_{\sqrt{x}}^x \frac{1}{t}\psi\left(\frac{x}{t}\right)dt + \frac{\psi(x)}{2x} + \frac{\psi(\sqrt{x})^2}{\sqrt{x}} + O(1)$$

$$= \int_1^{\sqrt{x}} \frac{\psi(u)}{u}du + O(1) \ll 1$$

so that

$$\sum_{n\leqslant x}\sigma(n) = \frac{x^2\zeta(2)}{2} - x\sum_{n\leqslant\sqrt{x}}\frac{1}{n}\psi\left(\frac{x}{n}\right) + O(x).$$

The trivial estimate on the sum gives

$$\left|\sum_{n\leqslant\sqrt{x}}\frac{1}{n}\psi\left(\frac{x}{n}\right)\right| \leqslant \frac{1}{2}\sum_{n\leqslant\sqrt{x}}\frac{1}{n} \ll \log x \tag{6.30}$$

so that we get the asymptotic formula

$$\sum_{n\leqslant x}\sigma(n) = \frac{x^2\zeta(2)}{2} + O(x\log x).$$

If we intend to improve the error term, the use of previous results such as Corollary 6.7 can help to reduce the interval of summation. Indeed, using Abel's summation and Corollary 6.7 with the exponent pair (6.17), we get

$$\sum_{a(x)<n\leqslant\sqrt{x}}\frac{1}{n}\psi\left(\frac{x}{n}\right) \ll \max_{a(x)<N\leqslant\sqrt{x}}\left|\sum_{N<n\leqslant 2N}\frac{1}{n}\psi\left(\frac{x}{n}\right)\right|\log x$$

$$\ll \max_{a(x)<N\leqslant\sqrt{x}}N^{-1}\max_{N\leqslant N_1\leqslant 2N}\left|\sum_{N<n\leqslant N_1}\psi\left(\frac{x}{n}\right)\right|\log x$$

$$\ll \max_{a(x)<N\leqslant\sqrt{x}}N^{-1}\max_{N\leqslant N_1\leqslant 2N}\left(x^{97/348}N^{35/348} + N^2x^{-1}\right)\log x$$

$$\ll \max_{a(x)<N\leqslant\sqrt{x}}\left(x^{97/348}N^{-313/348} + Nx^{-1}\right)\log x \ll 1,$$

where $a(x) = x^{97/313}(\log x)^{348/313} \approx x^{0.279}(\log x)^{1.112}$. However, this method does not enable us to cover the whole range of summation.

6.8.2 Vinogradov's Mean Value Theorem

From 1934 to 1937, Vinogradov created another tool to deal with exponential sums which may supersede Weyl–van der Corput's method when the orders of the derivatives of the function are large. We sketch below the ideas underlying this method, but we omit the proofs. For more information about Vinogradov's work, see [23, 63, 86].

Let k be a large integer, say $k \geqslant 5$, and $f \in C^{k+1}[N, 2N]$ such that there exists a real number $\lambda_{k+1} \in \left[N^{-2}, N^{-1} \right]$ such that, for all $x \in [N, 2N]$, we have

$$\left| \frac{f^{(k+1)}(x)}{(k+1)!} \right| \asymp \lambda_{k+1}.$$

Let H be a positive integer satisfying $H \leqslant \frac{1}{2} N$ and $H^{k+1}\lambda_{k+1} \ll 1$. Using Weyl's shift (6.4) with $a_n = e(f(n))$ and Taylor's series expansion we get

$$\left| \sum_{N < n \leqslant 2N} e(f(n)) \right| \leqslant \frac{1}{H} \sum_{N-H < n < 2N} \left| \sum_{h \leqslant H} e(f(n+h)) \right| + 2H$$

$$= \frac{1}{H} \sum_{N-H < n < 2N} \left| \sum_{h \leqslant H} e(Q_n(h) + u_n(h)) \right| + 2H,$$

where

$$Q_n(h) = hf'(n) + \cdots + \frac{h^k}{k!} f^{(k)}(n) \quad \text{and} \quad u_n(h) = f(n) + \frac{1}{k!} \int_0^h (h-t)^k f^{(k+1)}(n+t)\, \mathrm{d}t.$$

Since $H^{k+1}\lambda_{k+1} \ll 1$, the total variation of $u_n(h)$ is $\ll 1$, so that using Abel's summation we get

$$\sum_{N < n \leqslant 2N} e(f(n)) \ll \frac{1}{H} \sum_{n \sim N} \left| \sum_{h \sim H} e(Q_n(h)) \right| + H$$

and writing $f^{(j)}(n)/j! = \left\lfloor f^{(j)}(n)/j! \right\rfloor + \left\{ f^{(j)}(n)/j! \right\}$ we have

$$\sum_{N < n \leqslant 2N} e(f(n)) \ll \frac{1}{H} \sum_{n \sim N} \left| \sum_{h \sim H} e(P_n(h)) \right| + H,$$

where $P_n(h) = h\left\{f'(n)\right\} + h^2\left\{f''(n)/2\right\} + \cdots + h^k\left\{f^{(k)}(n)/k!\right\}$. The first step in Vinogradov's method is to prove the following estimate

$$\sum_{N<n\leqslant 2N} e(f(n)) \ll NH^{-1}\left(H^{k(k+1)/2}\mathcal{B}(H)J_{s,k}(H)N^{-2}\right)^{\frac{1}{2s}} + H, \qquad (6.31)$$

where s is a fixed positive integer, $\mathcal{B}(H)$ is the number of pair (n, n') such that

$$\left|\frac{f^{(k)}(n)}{k!} - \frac{f^{(k)}(n')}{k!}\right| \leqslant H^{-k} \quad \text{and} \quad \left|\left\{\frac{f^{(k-1)}(n)}{(k-1)!}\right\} - \left\{\frac{f^{(k-1)}(n')}{(k-1)!}\right\}\right| \leqslant H^{1-k}$$

and $J_{s,k}(H)$ is the *Vinogradov integral* defined by

$$J_{s,k}(H) = \int_0^1 \cdots \int_0^1 \left|\sum_{h\leqslant H} e\left(a_1 h + \cdots + a_k h^k\right)\right|^{2s} da_1 \cdots da_k.$$

One can get an upper bound for $\mathcal{B}(H)$ in the following way. Set $n' = n + \ell$ with $|\ell| \leqslant L \ll N$ for some $L \in \mathbb{Z}_{>0}$. Then the first condition implies that $(k + 1)|\ell|\lambda_{k+1} \leqslant H^{-k}$. Taking $L \asymp \left(\lambda_{k+1}H^k\right)^{-1}$ and taking the inequality $|\{a\} - \{b\}| \geqslant \|a - b\|$ into account, we get

$$\mathcal{B}(H) \ll \text{Card}\left\{(n, \ell) : 0 \leqslant |\ell| \leqslant L, \tfrac{1}{2}N < n \leqslant 2N, \left\|\frac{\Delta_\ell f^{(k-1)}(n)}{(k-1)!}\right\| \leqslant H^{1-k}\right\}$$

$$\ll N + \sum_{0<|\ell|\leqslant L} \mathcal{R}\left(\frac{\Delta_\ell f^{(k-1)}(n)}{(k-1)!}, N, 2H^{1-k}\right)$$

$$\ll N + \sum_{0<|\ell|\leqslant L} \left(N|\ell|\lambda_{k+1} + NH^{1-k} + \left(|\ell|\lambda_{k+1}H^{k-1}\right)^{-1} + 1\right),$$

where we used Theorem 5.3. Now since $NL\lambda_{k+1} \ll NH^{1-k}$ and $\left(|\ell|\lambda_{k+1}H^{k-1}\right)^{-1} \gg H$ we get

$$\mathcal{B}(H) \ll N + \frac{N}{H^{2k-1}\lambda_{k+1}} + \frac{\log N}{\lambda_{k+1}H^{k-1}}$$

and the choice of $H = \left\lfloor\lambda_{k+1}^{-1/(2k-1)}\right\rfloor$ gives

$$\mathcal{B}(H) \ll N \log N, \qquad (6.32)$$

where we used the fact that $N^2\lambda_{k+1} \geqslant 1$. Note that from the hypotheses above we have

$$N^{\frac{1}{2k-1}} \ll H \ll N^{\frac{2}{2k-1}} \quad \text{and} \quad H^{k+1}\lambda_{k+1} \ll N^{-\frac{k-2}{2k-1}}.$$

The next result, which gives a non-trivial upper bound for $J_{s,k}(H)$, lies at the heart of Vinogradov's method.

Lemma 6.14 (Vinogradov's Mean Value Theorem)   *Set* $\varepsilon = e^{-s/k^2}$. *Then*

$$J_{s,k}(H) \ll H^{2s-(1-\varepsilon)k(k+1)/2}.$$

The proof relies on the observation that $J_{s,k}(H)$ counts the number of integer solutions of the following system of Diophantine equations

$$\begin{cases} h_1 + \cdots + h_s = h_1' + \cdots + h_s' \\ h_1^2 + \cdots + h_s^2 = h_1'^2 + \cdots + h_s'^2 \\ \qquad \vdots \qquad \qquad \vdots \qquad \qquad \vdots \\ h_1^k + \cdots + h_s^k = h_1'^k + \cdots + h_s'^k \end{cases}$$

with $1 \leqslant h_i, h_i' \leqslant H$. Indeed, expanding the power in the integrand of $J_{s,k}(H)$ and exchanging the order of summation and integration, we get

$$J_{s,k}(H) = \sum_{\mathbf{h},\mathbf{h}'} \prod_{j=1}^{k} \left(\int_0^1 e\left\{ \left(h_1^j + \cdots + h_s^j - h_1'^j - \cdots - h_s'^j \right) a_j \right\} da_j \right),$$

where $\mathbf{h} = (h_1, \ldots, h_s)$ and $\mathbf{h}' = (h_1', \ldots, h_s')$ run independently over $\{1, \ldots, H\}^s$. Hence the integral vanishes unless $h_1^j + \cdots + h_s^j = h_1'^j + \cdots + h_s'^j$ for all $1 \leqslant j \leqslant k$.

Note that Lemma 6.14 is nearly the best possible result since one can show that

$$J_{s,k}(H) \gg H^{2s-k(k+1)/2}$$

and it is surmised that, for all $\varepsilon > 0$,

$$J_{s,k}(H) \ll H^\varepsilon \left(H^s + H^{2s-k(k+1)/2} \right). \tag{6.33}$$

This conjecture is usually called *the main conjecture* in Vinogradov's mean value theorem.

Now putting (6.32) and Lemma 6.14 into (6.31) we get

$$\sum_{N<n\leqslant 2N} e(f(n)) \ll N^{1-\frac{1}{2s}} H^{\frac{\varepsilon k(k+1)}{4s}} \log N + H$$

$$\ll N^{1-\frac{1}{2s}} \lambda_{k+1}^{-\frac{\varepsilon k(k+1)}{4s(2k-1)}} \log N + \lambda_{k+1}^{-\frac{1}{2k-1}}$$

$$\ll N^{1-\frac{r(s,k)}{2s}} \log N + N^{\frac{2}{2k-1}},$$

where

$$r(s,k) := 1 - e^{-s/k^2} \frac{k(k+1)}{2k-1}.$$

Taking $s = k^2 \log k$ with $k \geqslant 5$ then gives

$$\sum_{N<n\leqslant 2N} e(f(n)) \ll N^{1-\frac{1}{6k^2 \log k}} \log N.$$

We can state the following result.

Theorem 6.19 (Vinogradov) *Let $k \geqslant 5$ be an integer and $f \in C^{k+1}[N, 2N]$ such that there exists a real number $\lambda_{k+1} \in [N^{-2}, N^{-1}]$ such that, for all $x \in [N, 2N]$, we have*

$$\left| \frac{f^{(k+1)}(x)}{(k+1)!} \right| \asymp \lambda_{k+1}.$$

Then

$$\sum_{N<n\leqslant 2N} e(f(n)) \ll N^{1-\frac{1}{6k^2 \log k}} \log N.$$

6.8.3 Proving the Main Conjecture

Despite 80 years of intense investigation, it was only recently that the main Conjecture (6.33) in Vinogradov's mean value theorem has been established, first in the case $k = 3$ by Wooley [91], and then by a completely different method[13] in the case $k \geqslant 4$ in [8]. The cases $k = 1$ and $k = 2$ can be proved by elementary methods, so that we may now state the following important result.

Theorem 6.20 *For all $\varepsilon > 0$*

$$J_{s,k}(H) \ll_{s,k,\varepsilon} H^{\varepsilon} \left(H^s + H^{2s-k(k+1)/2} \right)$$

the term H^{ε} being omitted if $s > \frac{1}{2}k(k+1)$.

As an application, Heath-Brown, building his work on an earlier idea of Sargos, improved significantly the main term of van der Corput's estimate of Theorem 6.5.

Theorem 6.21 *Assume $m \geqslant 3$ and let $\varepsilon > 0$ and $f \in C^m ([N, 2N])$ such that there exist $\lambda_m > 0$ and $c_m \geqslant 1$ such that, for all $x \in [N, 2N]$, we have $\lambda_m \leqslant f^{(m)}(x) \leqslant c_m \lambda_m$. Then*

$$\sum_{N < n \leqslant 2N} e\left(f(n) \right) \ll N^{\varepsilon} \left(N \lambda_m^{\frac{1}{m(m-1)}} + N^{1 - \frac{1}{m(m-1)}} + N^{1 - \frac{2}{m(m-1)}} \lambda_m^{-\frac{2}{m^2(m-1)}} \right).$$

6.8.4 Walfisz's Estimates

In [88], Walfisz was able to get a non-trivial upper bound for the sum (6.30). For large N, he used a van der Corput type estimate as in Corollary 6.5, and for smaller ones he used a Vinogradov type result. Afterwards, Karacuba [48, 70] obtained a refined estimate using similar ideas as in Vinogradov's work which contains both methods. Applying this result to the function $f(n) = T/n$, Pétermann and Wu showed in [71] that, if $e^{200} \leqslant N < N_1 \leqslant 2N$ and $T \geqslant N^2$, then there exists $c_0 > 0$ such that we have

$$\sum_{N < n \leqslant N_1} e\left(\frac{T}{n} \right) \ll N \exp\left(-c_0 \frac{(\log N)^3}{(\log T N^{-1})^2} \right).$$

---

[13]One of the key tools in the proof is the use of the decoupling principle previously seen in Theorem 6.18.

Applying this in Corollary 6.1 we obtain with $e^{200} \leqslant N < N_1 \leqslant 2N$ and $N \leqslant x^{1/2}$

$$\sum_{N < n \leqslant N_1} \psi\left(\frac{x}{n}\right) \ll NH^{-1} + N \exp\left(-c_0 \frac{(\log N)^3}{(\log HxN^{-1})^2}\right) \log H.$$

Set $c = 15(4c_0)^{-1/3}$ and $w(x) = e^{c(\log x)^{2/3}}$. Choosing $H = \lfloor \exp\{(\log N)^3 / (\log x)^2\} \rfloor$ gives for all $w(x) \leqslant N \leqslant x^{1/2}$

$$\sum_{N < n \leqslant N_1} \psi\left(\frac{x}{n}\right) \ll \frac{N(\log N)^3}{(\log x)^2} \exp\left(-c_1 \frac{(\log N)^3}{(\log x)^2}\right)$$

with $c_1 = 64c_0/81$ and where we have used the bounds $N \geqslant w(x)$ and $H \leqslant x^{1/8}$. An application of Abel's summation yields

$$\sum_{N < n \leqslant N_1} \frac{1}{n} \psi\left(\frac{x}{n}\right) \ll \frac{(\log N)^3}{(\log x)^2} \exp\left(-c_1 \frac{(\log N)^3}{(\log x)^2}\right)$$

for $w(x) \leqslant N \leqslant x^{1/2}$ and a similar argument to that used in the proof of [71, Lemma 2.3] finally gives

$$\sum_{w(x) < n \leqslant x^{1/2}} \frac{1}{n} \psi\left(\frac{x}{n}\right) \ll (\log x)^{2/3}.$$

Since we have trivially

$$\sum_{n \leqslant w(x)} \frac{1}{n} \psi\left(\frac{x}{n}\right) \ll \sum_{n \leqslant w(x)} \frac{1}{n} \ll (\log x)^{2/3}$$

we may state the following result.[14]

Theorem 6.22 (Walfisz)   *For all $x \geqslant 2$ sufficiently large, we have*

$$\sum_{n \leqslant x} \sigma(n) = \frac{x^2 \zeta(2)}{2} + O\left(x(\log x)^{2/3}\right).$$

---

[14]Compare to Corollary 4.11.

A similar method was used to get the following improvement for the average order of Euler's totient function.[15]

Theorem 6.23 (Walfisz) *For all $x \geqslant 3$ sufficiently large, we have*

$$\sum_{n \leqslant x} \varphi(n) = \frac{x^2}{2\zeta(2)} + O\left(x(\log x)^{2/3}(\log\log x)^{1/3}\right).$$

6.9   Vaughan's Identity

6.9.1   Introduction

The sums $\sum_{p \leqslant x} e(f(p))$ and $\sum_{n \leqslant x} \mu(n)e(f(n))$ frequently arise in number theory. For instance, if we intend to get an estimate for $\sum_{N < n \leqslant 2N} \mu(n)\psi(f(n))$, we may use the asymptotic formula of Remark 6.1 which implies that, for any positive integer H, we have

$$\sum_{N < n \leqslant 2N} \mu(n)\psi(f(n)) = -\frac{1}{2\pi i} \sum_{0 < |h| \leqslant H} \frac{1}{h} \sum_{N < n \leqslant 2N} \mu(n)e(hf(n))$$

$$+ O\left\{ \sum_{N < n \leqslant 2N} \min\left(1, \frac{1}{H\|f(n)\|}\right) \right\} \qquad (6.34)$$

so that we need to have at our disposal bounds for sums of the type

$$\sum_{N < n \leqslant 2N} \mu(n)e(hf(n)). \qquad (6.35)$$

For the error term of (6.34), it can be shown [64] that the minimum admits a Fourier series expansion. Nevertheless, it is sometimes simpler to use the following lemma.

Lemma 6.15 *Let $N \geqslant 1$ and $H \geqslant 4$ be integers, and $f : [N, 2N] \longrightarrow \mathbb{R}$ be any function. We set $K = \lceil \log H / \log 2 \rceil$. Then we have*

$$\sum_{N \leqslant n \leqslant 2N} \min\left(1, \frac{1}{H\|f(n)\|}\right) < 24NH^{-1} + 2\sum_{k=0}^{K-2} 2^{-k}\mathcal{R}\left(f, N, 2^k H^{-1}\right).$$

---

[15]The exponent of the log log x has recently been ameliorated by Liu [57], improving the previous result of $\frac{4}{3}$ obtained by Walfisz [88, Satz 1 p. 144].

Proof We have

$$
\sum_{N \leqslant n \leqslant 2N} \min\left(1, \frac{1}{H\|f(n)\|}\right) = \sum_{\substack{N \leqslant n \leqslant 2N \\ \|f(n)\| < H^{-1}}} 1 + \frac{1}{H} \sum_{\substack{N \leqslant n \leqslant 2N \\ \|f(n)\| \geqslant H^{-1}}} \frac{1}{\|f(n)\|}
$$

$$
= \mathcal{R}(f, N, H^{-1}) + \frac{1}{H} \sum_{\substack{N \leqslant n \leqslant 2N \\ \|f(n)\| \geqslant H^{-1}}} \frac{1}{\|f(n)\|}.
$$

Since

$$
\left\{ n \in [N, 2N] \cap \mathbb{Z} : \|f(n)\| \geqslant H^{-1} \right\}
$$

$$
\subseteq \bigcup_{k=1}^{K} \left\{ n \in [N, 2N] \cap \mathbb{Z} : 2^{k-1} H^{-1} \leqslant \|f(n)\| < 2^k H^{-1} \right\}
$$

we get

$$
\sum_{\substack{N \leqslant n \leqslant 2N \\ \|f(n)\| \geqslant H^{-1}}} \frac{1}{\|f(n)\|} \leqslant \sum_{k=1}^{K} \sum_{\substack{N \leqslant n \leqslant 2N \\ 2^{k-1} H^{-1} \leqslant \|f(n)\| < 2^k H^{-1}}} \frac{1}{\|f(n)\|}
$$

$$
\leqslant (N+1)\left(2^{1-K} + 2^{2-K}\right) H + \sum_{k=1}^{K-2} \sum_{\substack{N \leqslant n \leqslant 2N \\ 2^{k-1} H^{-1} \leqslant \|f(n)\| < 2^k H^{-1}}} \frac{1}{\|f(n)\|}
$$

$$
\leqslant 6 \times 2^{-K}(N+1)H + 2H \sum_{k=1}^{K-2} 2^{-k} \sum_{\substack{N \leqslant n \leqslant 2N \\ \|f(n)\| < 2^k H^{-1}}} 1
$$

$$
< 12(N+1) + 2H \sum_{k=1}^{K-2} 2^{-k} \mathcal{R}\left(f, N, 2^k H^{-1}\right)
$$

since $2^{-K} < 2H^{-1}$. Thus we get

$$
\sum_{N \leqslant n \leqslant 2N} \min\left(1, \frac{1}{H\|f(n)\|}\right) < \mathcal{R}(f, N, H^{-1}) + 24NH^{-1} + 2\sum_{k=1}^{K-2} 2^{-k}\mathcal{R}\left(f, N, 2^k H^{-1}\right)
$$

which implies the desired result. $\square$

In the case of the function $f(n) = x/n$ with $x \geqslant 1$ being a large real number, Shiu's result may be used to get a quasi-optimal estimate.[16]

Lemma 6.16 *Let $x \geqslant 1$ be a real number, $N \geqslant 1$ and $4 \leqslant H \leqslant x$ be integers. Then, for all $\varepsilon > 0$, we have*

$$\sum_{N \leqslant n \leqslant 2N} \min\left(1, \frac{1}{H \|x/n\|}\right) \ll_\varepsilon N H^{-1} x^\varepsilon.$$

Proof Using Lemma 6.15 we get

$$\sum_{N \leqslant n \leqslant 2N} \min\left(1, \frac{1}{H \|x/n\|}\right) \ll N H^{-1} + \sum_{k=0}^{[\log H/\log 2]-2} 2^{-k} \mathcal{R}\left(\frac{x}{n}, N, \frac{2^k}{H}\right)$$

and interchanging the summations we obtain using Lemma 5.2

$$\mathcal{R}\left(\frac{x}{n}, N, \frac{2^k}{H}\right) \leqslant \sum_{N \leqslant n \leqslant 2N} \left(\left[\frac{x}{n} + \frac{2^k}{H}\right] - \left[\frac{x}{n} - \frac{2^k}{H}\right]\right)$$

$$\leqslant \sum_{x - 2^{k+1} N H^{-1} < m \leqslant x + 2^{k+1} N H^{-1}} \sum_{\substack{d \mid m \\ N \leqslant d \leqslant 2N}} 1$$

$$\leqslant \sum_{x - 2^{k+1} N H^{-1} < m \leqslant x + 2^{k+1} N H^{-1}} \tau(m)$$

$$\ll_\varepsilon 2^k N H^{-1} x^\varepsilon,$$

where we used Theorem 4.17 in the last estimate. This implies the asserted result. □

It is obvious that the equality (6.34) still holds when replacing $\mu(n)$ by any complex-valued sequence (a_n). Furthermore, with $f(n) = x/n$, if $a_n \ll n^\varepsilon$ in (6.34), then the error term of Lemma 6.16 is only affected by a factor x^ε if $N \leqslant x$. We may therefore state the following useful result which generalizes Corollary 6.1 in the case of the function $f(n) = x/n$.

---

[16]One may slightly improve on this result by using Δ-Hooley's divisor function instead of the τ-function, and making use of Theorem 4.18 instead of Shiu's theorem.

Proposition 6.14 *Let $\varepsilon > 0$, $x \geqslant 1$ be real numbers, $N \geqslant 1$, $H \geqslant 4$ be integers satisfying $\max(N, H) \leqslant x$, and (a_n) be a complex-valued sequence supported on $[N, 2N]$ such that, for all $n \in \{N, \ldots, 2N\}$, we have $a_n \ll n^\varepsilon$. Then we have*

$$\sum_{N < n \leqslant 2N} a_n \psi\left(\frac{x}{n}\right) \ll NH^{-1}x^{2\varepsilon} + \sum_{h \leqslant H} \frac{1}{h} \left| \sum_{N < n \leqslant 2N} a_n e\left(\frac{hx}{n}\right) \right|.$$

6.9.2 Prime Numbers in Intervals

Another long-standing problem in number theory is to ask whether there is a prime number in the interval $I = \left(x, x + x^{1/2}\right]$ for large x. Even if we assume the Riemann hypothesis, it seems to be extremely difficult to answer this question. As an approximation, Ramachandra [73] suggested the problem of showing that there is $n \in I$ having a large prime factor p with $p > x^\phi$ and ϕ being as large as possible. We may proceed as follows. Let $P(n)$ be the greatest prime factor of n and, for all positive integers d, set

$$N(d) = \sum_{\substack{n \in I \\ d \mid n}} 1.$$

The starting point is the following easy estimate. Interchanging the summations and using the convolution identity $\Lambda \star \mathbf{1} = \log$, we get

$$\sum_{d \leqslant x} N(d)\Lambda(d) = \sum_{x < n \leqslant x + x^{1/2}} \sum_{d \mid n} \Lambda(d) = \sum_{x < n \leqslant x + x^{1/2}} \log n = x^{1/2} \log x + O(x^{1/2}).$$

Now note that $N(d) \leqslant 1$ for all $d > x^{1/2}$ so that, using Chebyshev's estimate of Lemma 3.8, we get for all $\varepsilon > 0$

$$\sum_{\substack{x^{3/5-\varepsilon} < d \leqslant x \\ d \text{ not prime}}} N(d)\Lambda(d) = \sum_{\alpha \geqslant 2} \sum_{x^{3/5-\varepsilon} < p^\alpha \leqslant x} N(p^\alpha) \log p$$

$$\leqslant \sum_{p \leqslant x^{1/2}} \log p + \sum_{\alpha \geqslant 3} \sum_{p^\alpha \leqslant x} \log p$$

$$\leqslant x^{1/2} \log 4 + x^{1/3} \log x \ll x^{1/2}$$

for large x. Similarly, using $N(d) = x^{1/2}d^{-1} + O(1)$, we get

$$\sum_{d \leqslant x^{1/2-\varepsilon}} N(d)\Lambda(d) = x^{1/2} \sum_{d \leqslant x^{1/2-\varepsilon}} \frac{\Lambda(d)}{d} + O\left(\Psi(x^{1/2-\varepsilon})\right) = \left(\tfrac{1}{2} - \varepsilon\right) x^{1/2} \log x + O(x^{1/2}),$$

where we used Chebyshev's estimates of Corollary 3.3 and Lemma 3.8. We then have at this step

$$x^{1/2} \log x + O(x^{1/2}) = \sum_{d \leqslant x} N(d)\Lambda(d)$$

$$= \sum_{d \leqslant x^{1/2-\varepsilon}} N(d)\Lambda(d) + \sum_{x^{1/2-\varepsilon} < d \leqslant x^{3/5-\varepsilon}} N(d)\Lambda(d) + \sum_{\substack{x^{3/5-\varepsilon} < d \leqslant x \\ d \text{ not prime}}} N(d)\Lambda(d)$$

$$+ \sum_{x^{3/5-\varepsilon} < p \leqslant x^{\phi}} N(p) \log p + \sum_{x^{\phi} < p \leqslant x} N(p) \log p$$

$$= \left(\tfrac{1}{2} - \varepsilon\right) x^{1/2} \log x + \Sigma_2 + \Sigma_3 + \sum_{x^{\phi} < p \leqslant x} N(p) \log p + O(x^{1/2}),$$

where

$$\Sigma_2 = \sum_{x^{1/2-\varepsilon} < d \leqslant x^{3/5-\varepsilon}} N(d)\Lambda(d) \quad \text{and} \quad \Sigma_3 = \sum_{x^{3/5-\varepsilon} < p \leqslant x^{\phi}} N(p) \log p$$

so that

$$\sum_{x^{\phi} < p \leqslant x} N(p) \log p = \left(\tfrac{1}{2} + \varepsilon\right) x^{1/2} \log x - \Sigma_2 - \Sigma_3 + O(x^{1/2}).$$

The next step is to show the following estimate

$$\Sigma_2 = \tfrac{1}{10} x^{1/2} \log x + O(x^{1/2}) \tag{6.36}$$

and to find the largest exponent ϕ such that the upper bound

$$\Sigma_3 < \tfrac{2}{5} x^{1/2} \log x \tag{6.37}$$

holds, so that we shall have

$$\sum_{x^{\phi} < p \leqslant x} N(p) \log p > \tfrac{1}{2} \varepsilon x^{1/2} \log x$$

for x sufficiently large, which yields the existence of a prime p satisfying $p > x^{\phi}$ such that $N(p) = 1$, so that there exists an integer $n \in I$ such that $P(n) > x^{\phi}$.

The estimate (6.36) has been proved in [54, 35] using the following arguments. Since

$$N(d) = \frac{x^{1/2}}{d} - \psi\left(\frac{x + x^{1/2}}{d}\right) + \psi\left(\frac{x^{1/2}}{d}\right)$$

it is sufficient to show that, for all positive integers N such that $x^{1/2} < N \leqslant x^{3/5-\varepsilon}$, we have

$$\sum_{N < d \leqslant 2N} \Lambda(d) f(d) \ll \frac{x^{1/2}}{\log x}, \tag{6.38}$$

where

$$f(d) = \psi\left(\frac{x + x^{1/2}}{d}\right) - \psi\left(\frac{x^{1/2}}{d}\right).$$

The estimate (6.37) is shown in [4] with $\phi = 0,742\,8$, and also in [35] with the slightly weaker value $\phi = 0,74$, by using sieve techniques, and more precisely the Rosser-Iwaniec sieve and an alternative sieve.

6.9.3 The von Mangoldt Function

To treat the sum (6.35), one can make use of Vaughan's ingenious combinatorial identities based upon some decompositions of certain formulae involving the Riemann zeta-function [64]. Vaughan's Identity can take several forms, each of which based upon the next equality, namely for $\sigma > 1$

$$-\frac{\zeta'(s)}{\zeta(s)} = F(s) - \zeta(s)F(s)G(s) - \zeta'(s)G(s)$$

$$+ \left(-\frac{\zeta'(s)}{\zeta(s)} - F(s)\right)(1 - \zeta(s)G(s)).$$

Expanding these functions as Dirichlet series and using Proposition 4.12, we get

$$\Lambda(n) = a_1(n) + a_2(n) + a_3(n) + a_4(n),$$

where

$$a_1(n) = \begin{cases} \Lambda(n), \text{ if } n \leqslant U \\ \quad 0, \text{ otherwise} \end{cases} \qquad a_2(n) = - \sum_{\substack{mdr=n \\ m,d \leqslant U}} \Lambda(m)\mu(d)$$

$$a_3(n) = \sum_{\substack{kd=n \\ d \leqslant U}} \mu(d) \log k \quad \text{and} \quad a_4(n) = - \sum_{\substack{mk=n \\ m>U, k>U}} \Lambda(m) \left(\sum_{\substack{d|k \\ d \leqslant U}} \mu(d) \right).$$

$$(6.39)$$

Choosing $U = R_1^{1/3}$, multiplying throughout by $f(n)$ and interchanging the summations, we derive the important following result.

Theorem 6.24 (Vaughan) *There exist 6 arithmetic functions $\alpha_1, \ldots, \alpha_6$ satisfying*

$$|\alpha_k(n)| \leqslant \tau(n) \log 2n \quad (k = 1, \ldots, 6)$$

such that, for all integers $R < R_1 \leqslant 2R$, $R \geqslant 100$, and every arithmetic function $f : \mathbb{Z}_{\geqslant 1} \to \mathbb{C}$, we have

$$\sum_{R < n \leqslant R_1} \Lambda(n) f(n) = S_1 + S_2 + S_3 + S_4$$

with

$$S_1 = \sum_{m \leqslant R_1^{1/3}} \alpha_1(m) \sum_{R < mn \leqslant R_1} f(mn)$$

$$S_2 = \sum_{m \leqslant R_1^{1/3}} \alpha_2(m) \sum_{R < mn \leqslant R_1} f(mn) \log n$$

$$S_3 = \sum_{\substack{R_1^{1/3} \leqslant m \leqslant R_1^{2/3} \; R_1^{1/3} \leqslant n \leqslant R_1^{2/3} \\ R < mn \leqslant R_1}} \alpha_3(m)\alpha_4(n) f(mn)$$

$$S_4 = \sum_{\substack{R_1^{1/3} \leqslant m \leqslant R_1^{2/3} \; R_1^{1/3} \leqslant n \leqslant R_1^{2/3} \\ R < mn \leqslant R_1}} \alpha_5(m)\alpha_6(n) f(mn).$$

Remark 6.9 In Vaughan's terminology, the sums S_1 and S_2 are called *sums of type* I, while S_3 and S_4 are *sums of type* II. It should also be mentioned that these identities do not work if f is multiplicative. Indeed, we have in this case, supposing f completely multiplicative and neglecting the multiplicative condition $R < mn \leqslant R_1$ for the sake of simplicity,

$$S_{II} = \left(\sum_{M < m \leqslant 2M} a_m f(m) \right) \left(\sum_{N < n \leqslant 2N} b_n f(n) \right)$$

so that the new sums are not easier to deal with than the original sum. In fact, one generally uses these identities with $f(n) = e(g(n))$ or $f(n) = \sum_{h \sim H} e(g(h, n))$ for some real-valued function g. Furthermore, for the sums S_3 and S_4, one may exchange the role of the variables m and n by symmetry, which enables us to reduce the range of summation to $\left[\frac{1}{2}, \frac{2}{3} \right]$.

As an example, let us prove the following skilful estimate, which is a simplified version of the main result of [61].

Theorem 6.25 *Let $\delta \in [0, 1]$, $x \geqslant 1$ be a large real number, R, R_1 be positive integers such that $1 \leqslant R < R_1 \leqslant 2R \leqslant x^{2/3}$. Then, for all $\varepsilon \in \left(0, \frac{1}{2} \right]$*

$$x^{-\varepsilon} \sum_{R < n \leqslant R_1} \Lambda(n) \psi \left(\frac{x}{n+\delta} \right) \ll \left(x^2 R^{33} \right)^{1/38} + \left(x^2 R^{19} \right)^{1/24}$$

$$+ \left(x^3 R^2 \right)^{1/9} + \left(x^3 R^{-1} \right)^{1/6} + R^{5/6}.$$

We start with an easy lemma giving the order of magnitude of the first three derivatives of certain functions we that we will encounter in the proof of Theorem 6.25.

Lemma 6.17 *Let $\delta \in [0, 1]$, $x \geqslant 1$ and H, M, T be positive integers.*

▷ *Given integers $h, h', n, n' \geqslant 1$ and $k := hn' - h'n$, define*

$$f : t \in (M, 2M] \mapsto x \frac{kt + \delta (h - h')}{(nt + \delta)(n't + \delta)}. \tag{6.40}$$

Assume $\max(n, n') \leqslant 4 \min(n, n')$ and $h, h' \leqslant c_0 M$ where $0 < c_0 \leqslant 2020^{-1}$ is a small absolute constant. Then, we have for all $t \in (M, 2M]$ and $j \in \{1, 2, 3\}$

$$\frac{1}{2^{j+2}} \frac{j! x |k|}{nn' M^{j+1}} \leqslant \left| f^{(j)}(t) \right| \leqslant \frac{3}{2} \frac{j! x |k|}{nn' M^{j+1}}.$$

▷ *Given integers $h, h', m \geqslant 1$, define*

$$g : t \in (T, 2T] \mapsto \frac{hx}{mt + \delta}. \tag{6.41}$$

Then, we have for all $t \in (T, 2T]$ and $j \in \{1, 2, 3\}$

$$\frac{1}{2^{j+2}} \frac{j!hx}{mT^{j+1}} \leqslant \left| g^{(j)}(t) \right| \leqslant \frac{3}{2} \frac{j!hx}{mT^{j+1}}.$$

Proof The two estimates are similar, so that we only prove the first one. With all the parameters given in the lemma, we have

$$f^{(j)}(t) = (-1)^j j! x \left(\frac{n^j h}{(nt + \delta)^{j+1}} - \frac{n'^j h}{(n't + \delta)^{j+1}} \right),$$

where we used the fact that $k = hn' - h'n$. Therefore

$$\left| f^{(j)}(t) - (-1)^j j! \frac{xk}{nn't^{j+1}} \right|$$

$$= \frac{j!x}{t^{j+1}} \left| \frac{h}{n} \left(1 + \frac{\delta}{nt} \right)^{-j-1} - \frac{h'}{n'} \left(1 + \frac{\delta}{n't} \right)^{-j-1} - \frac{k}{nn'} \right|$$

$$= \frac{j!x}{t^{j+1}} \left| \frac{hn' - h'n - k}{nn'} + \sum_{\ell=1}^{\infty} (-1)^\ell \binom{j + \ell}{j} \left(\frac{\delta}{t} \right)^\ell \left(\frac{h}{n^{\ell+1}} - \frac{h'}{n'^{\ell} + 1} \right) \right|.$$

Now since $k = hn' - h'n$, the first term vanishes, and since

$$\left| \frac{h}{n^{\ell+1}} - \frac{h'}{n'^{\ell} + 1} \right| \leqslant \frac{h}{n^{\ell+1}} + \frac{h'}{n'^{\ell} + 1} < \frac{M}{1\,010 N^{\ell+1}},$$

where $N := \min(n, n')$, we derive

$$\left| f^{(j)}(t) - (-1)^j j! \frac{xk}{nn't^{j+1}} \right| \leqslant \frac{j!x}{t^{j+1}} \sum_{\ell=1}^{\infty} \binom{j + \ell}{j} t^{-\ell} \left| \frac{h}{n^{\ell+1}} - \frac{h'}{n'^{\ell} + 1} \right|$$

$$\leqslant \frac{j!x}{t^{j+1}} \frac{M}{1\,010 N} \sum_{\ell=1}^{\infty} \binom{j + \ell}{j} (MN)^{-\ell}$$

$$= \frac{j!x}{t^{j+1}} \frac{M}{1\,010 N} \left\{ \left(1 - \frac{1}{MN} \right)^{-j-1} - 1 \right\}$$

$$\leqslant \left(2^{j+2} - 2\right) \frac{j!x}{t^{j+1}} \frac{1}{1010N^2} < \frac{2^{j+6} - 32}{1010} \frac{j!x}{nn't^{j+1}}$$

$$< \frac{1}{2} \frac{j!x}{nn't^{j+1}}$$

and hence

$$\frac{1}{2} \frac{j!x|k|}{nn't^{j+1}} \leqslant \left| f^{(j)}(t) \right| \leqslant \frac{3}{2} \frac{j!x|k|}{nn't^{j+1}}$$

implying the asserted bounds since $M < t \leqslant 2M$.   □

We are now in a position to prove Theorem 6.25.

Proof ***(Proof of Theorem 6.25)*** Let S_R be the sum of the theorem. Applying (6.39), we get

$$S_R = S_1 + S_2 + S_3 + S_4 \tag{6.42}$$

with

$$S_1 = \sum_{m \leqslant R_1^{1/3}} \alpha_1(m) \sum_{R < mn \leqslant R_1} \psi\left(\frac{x}{mn + \delta}\right)$$

$$S_2 = \sum_{m \leqslant R_1^{1/3}} \alpha_2(m) \sum_{R < mn \leqslant R_1} \psi\left(\frac{x}{mn + \delta}\right) \log n$$

$$S_3 = \sum_{\substack{R_1^{1/3} \leqslant m \leqslant R_1^{2/3} \; R_1^{1/3} \leqslant n \leqslant R_1^{2/3} \\ R < mn \leqslant R_1}} \alpha_3(m)\alpha_4(n)\psi\left(\frac{x}{mn + \delta}\right)$$

$$S_4 = \sum_{\substack{R_1^{1/3} \leqslant m \leqslant R_1^{2/3} \; R_1^{1/3} \leqslant n \leqslant R_1^{2/3} \\ R < mn \leqslant R_1}} \alpha_5(m)\alpha_6(n)\psi\left(\frac{x}{mn + \delta}\right).$$

We first estimate S_3 and S_4. The two sums are treated equivalently, so we only estimate S_3. Using Theorem 6.1 we derive

$$S_3 = -\frac{1}{2\pi i}\left(S_{31} + \overline{S_{31}}\right) + S_{32},$$

where

$$S_{31} := \sum_{\substack{R_1^{1/3} \leqslant m \leqslant R_1^{2/3} \\ R < mn \leqslant R_1}} \sum_{R_1^{1/3} \leqslant n \leqslant R_1^{2/3}} \alpha_3(m)\alpha_4(n) \sum_{h \leqslant H} \frac{1}{h} \Phi\left(\frac{h}{H+1}\right) e\left(\frac{hx}{mn+\delta}\right);$$

$$S_{32} := \sum_{\substack{R_1^{1/3} \leqslant m \leqslant R_1^{2/3} \\ R < mn \leqslant R_1}} \sum_{R_1^{1/3} \leqslant n \leqslant R_1^{2/3}} \alpha_3(m)\alpha_4(n)\mathcal{R}_H\left(\frac{x}{mn+\delta}\right).$$

Splitting into dyadic intervals and using the symmetry of the variables m and n as explained in Remark 6.9, we are led to prove

$$S_3 + S_4 \ll \max_{R^{1/2} \leqslant M \leqslant R^{2/3}} |S_{31}(M) + S_{32}(M)| \log R, \tag{6.43}$$

where

$$S_{31}(M) := \sum_{M < m \leqslant 2M} \alpha_3(m) \sum_{\frac{R}{m} \leqslant n \leqslant \frac{R_1}{m}} \alpha_4(n) \sum_{h \leqslant H} \frac{1}{h} \Phi\left(\frac{h}{H+1}\right) e\left(\frac{hx}{mn+\delta}\right);$$

$$S_{32}(M) := \sum_{M < m \leqslant 2M} \alpha_3(m) \sum_{\frac{R}{m} \leqslant n \leqslant \frac{R_1}{m}} \alpha_4(n)\mathcal{R}_H\left(\frac{x}{mn+\delta}\right).$$

By Cauchy–Schwarz's inequality, we get

$$|S_{31}(M)|^2 \ll x^\varepsilon M \sum_{M < m \leqslant 2M} \left| \sum_{\frac{R}{m} \leqslant n \leqslant \frac{R_1}{m}} \alpha_4(n) \sum_{h \leqslant H} \frac{1}{h} \Phi\left(\frac{h}{H+1}\right) e\left(\frac{hx}{mn+\delta}\right) \right|^2$$

$$\ll x^\varepsilon M \sum_{M < m \leqslant 2M} \sum_{\frac{R}{m} \leqslant n,n' \leqslant \frac{R_1}{m}} \sum_{h,h' \leqslant H} c_{n,n',h,h'}\, e\left(\frac{hx}{mn+\delta} - \frac{h'x}{mn'+\delta}\right)$$

with

$$c_{n,n',h,h'} := \frac{\alpha_4(n)\alpha_4(n')}{hh'} \Phi\left(\frac{h}{H+1}\right) \Phi\left(\frac{h'}{H+1}\right).$$

Note that $0 < c_{n,n',h,h'} \ll x^\varepsilon \left(hh'\right)^{-1}$, so that

$$|S_{31}(M)|^2 \ll x^\varepsilon M \sum_{h,h' \leqslant H} \sum_{N \leqslant n,n' \leqslant 4N} \sum_{m \in I_n} c_{n,n',h,h'} \, e\left(\frac{(hn' - nh')m + \delta(h - h')}{(mn + \delta)(mn' + \delta)}x\right)$$

$$\ll x^\varepsilon M \sum_{0 \leqslant |k| \leqslant 4NH} S_{31}(M, k),$$

where $N = R/(2M)$, $I_n = \bigl(\max(M, R/n, R/n'), \ \min(2M, 2R/n, 2R/n')\bigr]$ and

$$S_{31}(M, k) := \sum_{\substack{N < n,n' \leqslant 4N \\ hn' - nh' = k}} \sum_{h,h' \leqslant H} \frac{1}{hh'} \left|\sum_{m \in I_n} e\left(f(m)\right)\right|,$$

where the function f is given in (6.40). Now

$$S_{31}(M, 0) \ll M \sum_{N < n,n' \leqslant 4N} \sum_{h,h' \leqslant H} \frac{1}{hh'} \ll MN^2 \sum_{N \leqslant d \leqslant 4NH} \frac{\tau(d)^2}{d^2} \ll Rx^\varepsilon$$

and, for $k \geqslant 1$ or $k \leqslant -1$, we may apply Corollary 6.4 to the exponential sum over m by Lemma 6.17, yielding

$$S_{31}(M, k) \ll \sum_{\substack{N < n,n' \leqslant 4N \\ hn' - nh' = k}} \sum_{h,h' \leqslant H} \frac{1}{hh'} \left\{\left(\frac{x|k|M^2}{nn'}\right)^{1/6} + \frac{nn'M^2}{x|k|}\right\}$$

$$\ll \left\{\left(\frac{x|k|M^2}{N^2}\right)^{1/6} + \frac{N^2M^2}{x|k|}\right\} \sum_{\substack{N < n,n' \leqslant 4N \\ hn' - nh' = k}} \sum_{h,h' \leqslant H} \frac{1}{hh'}$$

$$\ll \left\{N^{5/3}\left(x|k|M^2\right)^{1/6} + \frac{N^4M^2}{x|k|}\right\} \sum_{N \leqslant d \leqslant 4NH} \frac{\tau(d)\tau(d + k)}{d(d + k)}$$

$$\ll x^\varepsilon \left\{N^{2/3}\left(x|k|M^2\right)^{1/6} + \frac{N^3M^2}{x|k|}\right\}$$

$$\ll x^\varepsilon \left\{\left(\frac{x|k|R^4}{M^2}\right)^{1/6} + \frac{NR^2}{x|k|}\right\}$$

provided $H \leqslant c_0 M$, and hence

$$|S_{31}(M)|^2 \ll x^\varepsilon M \left\{ R + \left(\frac{x R^4}{M^2} \right)^{1/6} \sum_{0 < |k| \leqslant 4NH} |k|^{1/6} + \frac{NR^2}{x} \sum_{0 < |k| \leqslant 4NH} \frac{1}{|k|} \right\}$$

$$\ll x^\varepsilon \left\{ MR + \left(x R^4 M^4 N^7 H^7 \right)^{1/6} + R^3 x^{-1} \right\}$$

$$\ll x^\varepsilon \left\{ MR + \left(x R^{11} M^{-3} H^7 \right)^{1/6} \right\}$$

since the conditions $R \leqslant x^{2/3}$ and $R^{1/2} \leqslant M \leqslant R^{2/3}$ entail that $MR \geqslant R^3 x^{-1}$. Therefore

$$S_{31}(M) \ll x^\varepsilon \left\{ (MR)^{1/2} + \left(x R^{11} M^{-3} H^7 \right)^{1/12} \right\}$$

and similarly we also derive

$$S_{32}(M) \ll x^\varepsilon \left\{ RH^{-1} + (MR)^{1/2} + \left(x R^{11} M^{-3} H^7 \right)^{1/12} \right\}$$

provided $H \leqslant c_0 M$ and $R^{1/2} \leqslant M \leqslant R^{2/3}$. Now we use Lemma 5.6 to optimize H over $[1, c_0 M]$, yielding

$$S_{31}(M) + S_{32}(M) \ll x^\varepsilon \left\{ \left(x R^{18} M^{-3} \right)^{1/19} + \left(x R^{11} M^{-3} \right)^{1/12} + (MR)^{1/2} \right\},$$

where the term RM^{-1} has been absorbed by the term $(MR)^{1/2}$. Finally, inserting these estimates in (6.43), we derive

$$S_3 + S_4 \ll x^\varepsilon \left\{ \left(x^2 R^{33} \right)^{1/38} + \left(x^2 R^{19} \right)^{1/24} + R^{5/6} \right\}. \tag{6.44}$$

The factor $\log n$ of S_2 can be removed by partial summation, so that it suffices to estimate S_1. As above, but in a simpler way

$$S_1 + S_2 \ll x^\varepsilon (S_{11} + S_{12}) \tag{6.45}$$

with

$$S_{11} := \sum_{m \leqslant R_1^{1/3}} \sum_{h \leqslant H} \frac{1}{h} \left| \sum_{\frac{R}{m} < n \leqslant \frac{R_1}{n}} e(g(n)) \right|, \quad S_{12} := \sum_{m \leqslant R_1^{1/3}} \sum_{\frac{R}{m} < n \leqslant \frac{R_1}{n}} |\mathcal{R}_H(g(n))|,$$

where the function g is defined in (6.41). By Lemma 6.17, we may apply Corollary 6.3 to the exponential sum over n in S_{11}, so that

$$S_{11} \ll \sum_{m \leqslant R_1^{1/3}} \sum_{h \leqslant H} \frac{1}{h} \left\{ \frac{R}{m} \left(\frac{hx}{m(Rm^{-1})^3} \right)^{1/2} + \frac{m(Rm^{-1})^2}{hx} \right\}$$

$$\ll \sum_{m \leqslant R_1^{1/3}} \sum_{h \leqslant H} \left\{ \left(\frac{x}{Rh} \right)^{1/2} + \frac{R^2}{h^2 mx} \right\}$$

$$\ll \sum_{m \leqslant R_1^{1/3}} \left\{ \left(\frac{Hx}{R} \right)^{1/2} + \frac{R^2}{mx} \right\}$$

$$\ll (Hx)^{1/2} R^{-1/6} + R^2 x^{-1} \log R$$

and similarly

$$S_{12} \ll RH^{-1} + (Hx)^{1/2} R^{-1/6} + R^2 x^{-1} \log R$$

for all $H \geqslant 1$. Optimizing H over $[1, \infty)$ with the help of Lemma 5.6 and inserting in (6.45) yields

$$S_1 + S_2 \ll x^\varepsilon \left\{ \left(x^3 R^2 \right)^{1/9} + \left(x^3 R^{-1} \right)^{1/6} + R^2 x^{-1} \right\} \tag{6.46}$$

and the result follows from (6.42), (6.44), (6.46) and the bound $R^2 x^{-1} \leqslant R^{5/6}$.   □

6.9.4   The Möbius Function

The sum (6.38) may be estimated similarly. Using

$$\frac{1}{\zeta(s)} = 2G(s) - G(s)^2 \zeta(s) + \left(\frac{1}{\zeta(s)} - G(s) \right) (1 - \zeta(s) G(s))$$

we derive the following result analog to Theorem 6.24.

Theorem 6.26 (Vaughan) *There exist 2 arithmetic functions β_1, β_2 satisfying*

$$|\beta_k(n)| \leqslant \tau(n)\log n \quad (k = 1, 2)$$

such that, for all integers $R < R_1 \leqslant 2R$, $R \geqslant 100$, and every arithmetic function $f : \mathbb{Z}_{\geqslant 1} \to \mathbb{C}$, we have

$$\sum_{R < n \leqslant R_1} \mu(n)f(n) = S_1 + S_2 + S_3 \tag{6.47}$$

with

$$S_1 = -\sum_{m \leqslant R_1^{1/3}} \beta_1(m) \sum_{R < mn \leqslant R_1} f(mn)$$

$$S_2 = -\sum_{R_1^{1/3} \leqslant m \leqslant R_1^{2/3}} \beta_1(m) \sum_{R < mn \leqslant R_1} f(mn)$$

$$S_3 = -\sum_{\substack{m > R_1^{1/3} \; n > R_1^{1/3} \\ R < mn \leqslant R_1}} \mu(m)\beta_2(n)f(mn).$$

6.9.5   Heath-Brown's Refinement

Subsequently, Heath-Brown [36] generalized Vaughan's identities by providing some formulae which are more flexible.

Theorem 6.27 (Heath-Brown) *Let $3 \leqslant u < v < z < R$, with $z \in \mathbb{Z}_{\geqslant 1} + \frac{1}{2}$, $z \geqslant 4u^2$, $R \geqslant 64z^2 u$ and $v^3 \geqslant 32R$. Let f be an arithmetic function defined over $\left] \frac{1}{2}R, R \right]$ such that $|f(n)| \leqslant A$. We extend f over $\mathbb{R}$ by setting $f(n) = 0$ if $n \leqslant \frac{1}{2}R$ or $n > R$. Then*

$$\sum_{\frac{1}{2}R < n \leqslant R} \Lambda(n)f(n) = S_1 + S_1' - S_2 - S_2' - S_3 - S_3'$$

(continued)

Theorem 6.27 (continued)
with

$$S_1, S_1', S_2, S_2' \ll \max_N \sum_{m=1}^{\infty} \tau_3(m) \left| \sum_{z<n\leqslant N} f(mn) \right| \log R$$

$$S_3, S_3' \ll A + \sup_g \sum_{m=1}^{\infty} \tau_4(m) \left| \sum_{u<n\leqslant v} g(n)f(mn) \right| (\log R)^8,$$

where the supremum is taken over all arithmetic functions g such that $|g(n)| \leqslant \tau_3(n)$.

By using Heath-Brown's identities, one may prove the following useful result [74]. To simplify the notation, we set

$$S_I := \sum_{\substack{M<m\leqslant 2M \ N<n\leqslant 2N \\ R<mn\leqslant R_1}} a_m f(mn) ; \tag{6.48}$$

$$S_{II} := \sum_{\substack{M<m\leqslant 2M \ N<n\leqslant 2N \\ R<mn\leqslant R_1}} a_m b_n f(mn), \tag{6.49}$$

where (a_m) and (b_n) are any sequences of complex numbers satisfying $|a_m| \leqslant 1$ and $|b_n| \leqslant 1$.

Proposition 6.15 Let $k \geqslant 4$ be an integer, $(2k)^{-1} \leqslant \alpha \leqslant \frac{1}{6}$ be a real number and $S > 0$ be a positive real number. Suppose that $MN \asymp R$ and that the following estimates

$$S_I \ll S \quad \text{for} \quad N \geqslant R^{(1-\alpha)/2} \quad \text{and for} \quad R^{2\alpha} < N \leqslant R^{1/3}$$

$$S_{II} \ll S \quad \text{for} \quad R^{\alpha} \leqslant N \leqslant R^{2\alpha}$$

hold for all sums of type I and type II given in (6.48) and (6.49). Then we have for all $\varepsilon > 0$

$$\sum_{R<n\leqslant R_1} \Lambda(n)f(n) \ll_{k,\varepsilon} SR^{\varepsilon}.$$

Remark 6.10 One can prove that the condition $R < mn \leqslant R_1$ can be removed from sums $\mathcal{S}_I$ and $\mathcal{S}_{II}$ at a cost of a factor $\log R$.[17] Over the last two decades, many authors have provided some non-trivial bounds for sums of type I and type II.[18]

A similar result also holds for the Möbius function as shown in [3].

Proposition 6.16 *Let $S > 0$ be a positive real number and suppose that the following estimates*

$$S_I \ll S \quad for \quad N \gg R^{1/2}$$

$$S_{II} \ll S \quad for \quad R^{1/3} \ll N \ll R^{1/2}$$

hold for all sums of type I and type II given in (6.48) and (6.49). Then we have

$$\sum_{R < n \leqslant R_1} \mu(n) f(n) \ll S (\log 3R)^5 .$$

To end this section, let us prove an analogue of the third derivative theorem for sums of type II.

Proposition 6.17 *Let $f \in C^3 ([M, 2M] \times [N, 2N])$ such that there exists $\lambda_3 > 0$ such that, for all $(x, y) \in [M, 2M] \times [N, 2N]$, we have*

$$\left| \frac{\partial}{\partial x} \frac{\partial^2}{\partial y^2} f(x, y) \right| \asymp \lambda_3.$$

Let (a_m) and (b_n) be two complex-valued sequences, respectively, supported on $[M, 2M]$ and $[N, 2N]$ satisfying $|a_m| \leqslant 1$ and $|b_n| \leqslant 1$. Then we have

$$\sum_{M < m \leqslant 2M} \sum_{N < n \leqslant 2N} a_m b_n e(f(m, n)) \ll MN\lambda_3^{1/6} + MN^{3/4} + M^{1/2}N + M^{3/4}N^{1/2}\lambda_3^{-1/4}.$$

Proof We may suppose $\lambda_3 \ll 1$ otherwise the estimate is trivial. Let $S_{M,N}$ be the sum on the left-hand side and

$$S_M(n) = \sum_{M < m \leqslant 2M} a_m e(f(m, n))$$

so that

$$S_{M,N} = \sum_{N < n \leqslant 2N} b_n S_M(n).$$

---

[17]See [1, Lemma 15] or [74], for instance.
[18]See [1, 2, 3, 13, 14, 15, 24, 28, 53, 55, 56, 58, 76, 78, 92, 93, 95] among a lot of references.

Let H be a positive integer such that $H \leqslant M$. By Cauchy–Schwarz's inequality we have

$$|S_{M,N}|^2 \leqslant N \sum_{N < n \leqslant 2N} |b_n S_M(n)|^2 \leqslant N \sum_{N < n \leqslant 2N} |S_M(n)|^2$$

and van der Corput's A-process[19] gives

$$|S_M(n)|^2 \leqslant \frac{2M^2}{H} + \frac{4M}{H} \operatorname{Re} \left\{ \sum_{h \leqslant H} \left(1 - \frac{h}{H}\right) \sum_{M < m \leqslant 2M - h} a_{m+h} \overline{a_m} e\left(\Delta_h f(m, n)\right) \right\},$$

where

$$\Delta_h f(m, n) = f(m + h, n) - f(m, n).$$

We infer that

$$\left|S_{M,N}\right|^2 \ll \frac{(MN)^2}{H} + \frac{MN}{H} \sum_{h \leqslant H} \sum_{M < m \leqslant 2M} \left| \sum_{N < n \leqslant 2N} e\left(\Delta_h f(m, n)\right) \right|$$

and since

$$\left| \frac{\partial^2}{\partial y^2} \Delta_h f(x, y) \right| \asymp h\lambda_3$$

for all $(x, y) \in [M, 2M] \times [N, 2N]$, using van der Corput's inequality[20] we get

$$|S_{M,N}|^2 \ll \frac{(MN)^2}{H} + \frac{MN}{H} \sum_{h \leqslant H} \sum_{M < m \leqslant 2M} \left\{ N(h\lambda_3)^{1/2} + (h\lambda_3)^{-1/2} \right\}$$

$$\ll \frac{(MN)^2}{H} + (MN)^2 (H\lambda_3)^{1/2} + M^2 N (H\lambda_3)^{-1/2}$$

so that

$$\left|S_{M,N}\right| \ll MNH^{-1/2} + MN(H\lambda_3)^{1/4} + MN^{1/2}(H\lambda_3)^{-1/4}$$

and Lemma 5.6 gives the asserted result plus a secondary term $MN\lambda_3^{1/4}$ which is absorbed by the main term since $\lambda_3 \ll 1$.   □

---

[19] See Lemma 6.2.
[20] See Theorem 6.3.

6.9.6 A Variant of Vaughan's Identity

Exploiting the identity

$$\Lambda(n) = \log n - \sum_{\substack{\ell m = n \\ \ell, m > 1}} \Lambda(\ell)$$

and summing over the integers n such that $P(n) \leqslant y$, Granville [27] derived the following Vaughan-like identity.

Theorem 6.28 *For any arithmetic function* $f : \mathbb{Z}_{\geqslant 1} \to \mathbb{C}$ *and any* $1 \leqslant y \leqslant x$, *we have*

$$\sum_{\substack{n \leqslant x \\ p(n) > y}} \Lambda(n) f(n) \ll S_I(x, y) \log x + \sqrt{S_{II}(x, y) x (\log x)^5},$$

where $S_I(x, y)$ *is the Type I sum given by*

$$S_I(x, y) := \max_{t \leqslant x} \left| \sum_{\substack{n \leqslant t \\ p(n) > y}} f(n) \right| \leqslant \max_{t \leqslant x} \sum_{\substack{d \geqslant 1 \\ P(d) \leqslant y}} \left| \sum_{m \leqslant t/d} f(md) \right|$$

and $S_{II}(x, y)$ *is the Type II sum given by*

$$S_{II}(x, y) := \max_{\substack{y < L \leqslant x/y \\ y < m \leqslant 2x/L}} \sum_{\frac{1}{2}m < n \leqslant 2m} \left| \sum_{\substack{L < \ell \leqslant 2L \\ \ell \leqslant \min(\frac{x}{m}, \frac{x}{n})}} f(\ell m) \overline{f(\ell n)} \right|.$$

As an example, Granville established a slightly stronger version of the Bombieri–Vinogradov theorem than that of Theorem 4.52.

6.10 The Chowla–Walum Conjecture

6.10.1 Genesis of the Conjecture

Let $a > -\frac{1}{2}$ be a fixed real number, $k \geqslant 1$ be a fixed integer and define

$$G_{a,k}(x) := \sum_{n \leqslant \sqrt{x}} n^a B_k \left(\left\{ \frac{x}{n} \right\} \right),$$

where $B_k(\{x\})$ is the kth Bernoulli function defined in Chap. 1. Note that when $a = 0$ and $k = 1$, we recover the sum of the ψ-function appearing in the Dirichlet divisor problem. Recall that it is surmised that

$$G_{0,1}(x) \ll x^{1/4+\varepsilon}.$$

As an extension of this problem, Chowla and Walum [16] stated the following conjecture.

Conjecture 6.3 (Chowla–Walum) For x large

$$G_{a,k}(x) \ll x^{a/2+1/4+\varepsilon}.$$

Chowla and Walum proved the conjecture for $(a, k) = (1, 2)$. Later, Kanemitsu and Sita Rama Chandra Rao [47] proved this conjecture in the case $a \geqslant \frac{1}{2}$ and $k \geqslant 2$.

6.10.2   Related Results

To date, further results have been obtained towards this conjecture. In [47], the author derived the next estimates.

Theorem 6.29 *The Chowla–Walum Conjecture is true if $k \in \mathbb{Z}_{\geqslant 2}$ and $a \geqslant \frac{1}{2}$, the ε being omitted if $a > \frac{1}{2}$ and being replaced by $\log x$ if $a = \frac{1}{2}$. Furthermore, if $k \in \mathbb{Z}_{\geqslant 2}$ and $0 \leqslant a < \frac{1}{2}$, then*

$$\sum_{n \leqslant \sqrt{x}} n^a B_k\left(\left\{\frac{x}{n}\right\}\right) \ll x^{2a/5+3/10}.$$

They also studied the mean square of $G_{a,k}(x)$ and proved the following bound which supports the Chowla–Walum Conjecture.

Theorem 6.30 *Assume $|a| \leqslant \frac{1}{2}$. Then*

$$\int_1^T G_{a,k}(x)^2\, dx \ll T^{a+3/2+\varepsilon}.$$

Let us also state the next result dealing with a Chowla–Walum type summation [69].

Theorem 6.31 *Let $k \in \mathbb{Z}_{\geqslant 2}$ and $a, b \geqslant 1$ be real numbers such that $2a \geqslant b + 1$. Then*

$$\sum_{n \leqslant x^{1/3}} \sum_{n < m \leqslant (x/n)^{1/2}} n^a m^b B_k\left(\left\{\frac{x}{mn}\right\}\right) \ll_\varepsilon x^{(a+b+1)/3}(\log x)^3.$$

There also are some estimates dealing with sums in shorter ranges. For instance, using Vinogradov's results on exponential sums, Kanemitsu [46] proved the following bound.

Theorem 6.32 *Let $x \geqslant 64$ and $t \geqslant 1$ such that $c_0 x^{1/3} \leqslant t \leqslant \sqrt{2x}$ for some $c_0 \geqslant 4$. Then*

$$\sum_{n \leqslant t} n \psi \left(\frac{x}{n} \right)^2 = \frac{t^2}{24} + O \left(x^{2/3} + (xt)^{1/2} \right).$$

6.11   Exponential Sums over a Finite Field

6.11.1   Main Result

In what follows, p is a prime and $f(x) = a_n x^n + \cdots + a_1 x \in \mathbb{Z}[x]$ is a polynomial such that $(a_1, \ldots, a_n, p) = 1$. In this section, we describe a method, due to Mordell [66], to estimate the exponential sum

$$S_p(f) = \sum_{x=1}^{p} e_p(f(x)).$$

Theorem 6.33 *Let $n \in \mathbb{Z}_{\geqslant 2}$, $p > n$ be a prime and $f(x) = a_n x^n + \cdots + a_1 x \in \mathbb{Z}[x]$ be a polynomial such that $(a_1, \ldots, a_n, p) = 1$. Then*

$$\left| \sum_{x=1}^{p} e_p(f(x)) \right| < n^{1/2} p^{1 - 1/n}.$$

6.11.2   An Overview of Mordell's Method

We make a first observation. Assume Theorem 6.33 is true under the conditions $(a_m, p) = 1$ and $a_{m+1} \equiv \cdots \equiv a_n \equiv 0 \pmod{p}$ for some $1 \leqslant m \leqslant n$. Then

$$\left| S_p(f) \right| = \left| \sum_{x=1}^{p} e_p(a_m x^m + \cdots + a_1 x) \right| \leqslant m^{1/2} p^{1 - 1/m} \leqslant n^{1/2} p^{1 - 1/n}.$$

so that it suffices to show the estimate assuming $(a_n, p) = 1$, which we will assume from now on. Also note that the theorem holds in the case $n = 2$ since, by Iwaniec and Kowalski [44, (8.7)], we have

$$\left| \sum_{x=1}^{p} e_p \left(a_2 x^2 + a_1 x \right) \right| = p^{1/2}$$

as soon as $(2a_2, p) = 1$, so that we may suppose $n \geqslant 3$.

The starting point of Mordell's method looks like a Weyl's schift. Assume that k, ℓ are positive integers such that $1 \leqslant k \leqslant p - 1$ and $1 \leqslant \ell \leqslant p$. We may write

$$f(kx + \ell) = b_0(k, \ell) + b_1(k, \ell)x + \cdots + b_n(k, \ell)x^n$$

with $b_n(k, \ell) = a_n k^n$ and $b_{n-1}(k, \ell) = (na_n\ell + a_{n-1})k^{n-1}$. Since $(k, p) = 1$, the linear function $kx + \ell$ runs through a complete set of residues modulo p when x runs through a complete residue set modulo p, hence, summing with respect to k and ℓ, we derive

$$|S_p(f)|^{2n} = \left| \sum_{x=1}^{p} e_p \left(f(kx + \ell) \right) \right|^{2n}$$

$$= \frac{1}{p(p-1)} \sum_{k=1}^{p-1} \sum_{\ell=1}^{p} \left| \sum_{x=1}^{p} e_p \left(b_1(k, \ell)x + \cdots + b_n(k, \ell)x^n \right) \right|^{2n}$$

$$= \frac{1}{p(p-1)} \sum_{b_1,\ldots,b_n=1}^{p} N(b_1, \ldots, b_n) \left| \sum_{x=1}^{p} e_p \left(b_1 x + \cdots + b_n x^n \right) \right|^{2n},$$

where $N(b_1, \ldots, b_n)$ is the number of solutions of the system

$$\begin{cases} b_1(k, \ell) \equiv b_1 \pmod{p} \\ \vdots \quad \vdots \quad \vdots \\ b_n(k, \ell) \equiv b_n \pmod{p}. \end{cases}$$

$N(b_1, \ldots, b_n)$ does not exceed the number of solutions of the system made up of the last two congruences

$$\begin{cases} (na_n\ell + a_{n-1})k^{n-1} \equiv b_{n-1} \pmod{p} \\ a_n k^n \equiv b_n \pmod{p} \end{cases}$$

and hence, since $(na_n, p) = (k, p) = 1$, we derive $N(b_1, \ldots, b_n) \leqslant n$. Inserting this bound above yields

$$\left| S_p(f) \right|^{2n} \leqslant \frac{n}{p(p-1)} \sum_{b_1,\ldots,b_n=1}^{p} \left| \sum_{x=1}^{p} e_p \left(b_1 x + \cdots + b_n x^n \right) \right|^{2n}. \tag{6.50}$$

6.11.3 The Heart of Mordell's Method

By (6.50), the problem goes back to finding non-trivial upper bound for the multiple exponential sum

$$T_{n,p} = \sum_{b_1,\ldots,b_n=1}^{p} \left| \sum_{x=1}^{p} e_p \left(b_1 x + \cdots + b_n x^n \right) \right|^{2n}.$$

Mordell's idea rests on the use of (6.18) to transfer this exponential sum to a system of congruences that we may more easily estimate. Indeed, expanding the square and using (6.18) with $q = p$ and $a = 0$, we derive

$$p^{-n} T_{n,p} = p^{-n} \sum_{b_1,\ldots,b_n=1}^{p} \sum_{x_1,\ldots,x_n=1}^{p} e_p \left(a_1(x_1 + \cdots + x_n - y_1 \cdots - y_n) \right) \cdots$$

$$\cdots e_p \left(a_n(x_1^n + \cdots + x_n^n - y_1^n \cdots - y_n^n) \right)$$

$$= N(n, p),$$

where $N(n, p)$ is the number of solutions of the system

$$(S): \begin{cases} x_1 + \cdots + x_n \equiv y_1 + \cdots + y_n \pmod{p} \\ \vdots \qquad \vdots \qquad \vdots \\ x_1^n + \cdots + x_n^n \equiv y_1^n + \cdots + y_n^n \pmod{p} \end{cases}$$

with $1 \leqslant x_i, y_i \leqslant p$. Now Theorem 6.33 follows from the following lemma.

Lemma 6.18 Let p be a prime and n be a positive integer such that $p > n$. Then

$$N(n, p) \leqslant n! \, p^n.$$

Indeed, assume the lemma is true. Then by above and (6.50)

$$\left| S_p(f) \right|^{2n} \leqslant \frac{n}{p(p-1)} n! \, p^{2n}.$$

Now since $p > n \geqslant 3$, we have $\frac{1}{p-1} \leqslant \frac{5}{4p}$ and using $\frac{5}{4}n \times n! < n^n$ as soon as $n \geqslant 3$, we derive the asserted bound.   □

Proof (Proof of Lemma 6.18) Let $(x_1, \ldots, x_n, y_1, \ldots, y_n)$ be a solution of the system (S) and define for $k = 1, \ldots, n$

$$s_k = s_k(x_1, \ldots, x_n) = \sum_{j=1}^{n} x_j^k \quad \text{and} \quad e_k = e_k(x_1, \ldots, x_n) = \sum_{1 \leqslant j_1 < \cdots < j_k \leqslant n} x_{j_1} \cdots x_{j_k}$$

the kth power sum and the kth elementary symmetric polynomial, respectively. Using Newton's identities, we derive for $k = 1, \ldots, n$

$$s_k = (-1)^{k-1} k e_k + \sum_{j=1}^{k-1} (-1)^{k-1+j} e_{k-j} s_j$$

and since $p > n$, we have $(k, p) = 1$ so that the system (S) implies that, for $k = 1, \ldots, n$

$$e_k(x_1, \ldots, x_n) \equiv e_k(y_1, \ldots, y_n) \pmod{p}.$$

Now since

$$\prod_{j=1}^{n} (X - x_j) = \sum_{k=0}^{n} (-1)^k e_k(x_1, \ldots, x_n) X^{n-k}$$

we get

$$\prod_{j=1}^{n} (X - x_j) \equiv \prod_{j=1}^{n} (X - y_j) \pmod{p}.$$

Thus, the $y_1, \ldots, y_n$ are obtained by permutations of $x_1, \ldots, x_n \pmod{p}$, yielding the asserted bound.   □

Remark 6.11 This result is a particular case of [90, Theorem 1.1], where the author used modern tools from algebraic geometry, whereas Mordell's proof involves notions of elimination theory.

6.11.4   Further Results

In this section, for any prime p, we denote $d_p(f)$ to be the degree of f reduced mod p.

Mordell's bound was superseded by the following crucial result due to Weil [89].

Theorem 6.34 (Weil) *Let p be a prime number. For any polynomial f of degree n such that $d_p(f) \geqslant 1$, there exists a set of $n - 1$ complex numbers $\omega_1, \ldots, \omega_{n-1}$ such that, for $k \in \{1, \ldots, n - 1\}$, $|\omega_k| \leqslant \sqrt{p}$ and*

$$\sum_{x=1}^{p} e_p(f(x)) = -\sum_{k=1}^{n-1} \omega_k.$$

In particular

$$\left| \sum_{x=1}^{p} e_p(f(x)) \right| \leqslant (n - 1)\sqrt{p}.$$

Note that this result implies that we can replace $n^{1/2}$ in Theorem 6.33 by an absolute constant. Indeed, it can easily be verified that

$$np^{1/2} \leqslant 2p^{1-1/n}$$

so that, by Theorem 6.34, we derive if $(n, p) \neq (2, 2)$

$$\left| \sum_{x=1}^{p} e_p(f(x)) \right| \leqslant 2p^{1-1/n}$$

and this bound is trivial if $(n, p) = (2, 2)$. In [17], the authors improved the constant by showing that 1.75 is admissible.

Weil's bound is essentially best possible, in the sense that it can be proved that, for any n and $\varepsilon > 0$, there exists an infinite family of pairs (p, a) with $p \nmid a$ such that

$$\left| S_p\left(ax^n\right) \right| \geqslant (n - 1 - \varepsilon)\sqrt{p}.$$

For composite moduli $q > 1$, first note that, by Hua [40, Lemma 1.3], the general sum

$$S_q(f) = \sum_{x=1}^{q} e_q(f(x))$$

satisfies the *multiplicative property*

$$S_{q_1 q_2}(f) = S_{q_1}\left(\frac{1}{q_2}f(q_2 x)\right) S_{q_2}\left(\frac{1}{q_1}f(q_1 x)\right)$$

whenever $(q_1, q_2) = 1$ and $f(0) = 0$, from which we derive

$$S_q(f) = \prod_{p^\alpha \| q} S_{p^\alpha}\left(\frac{p^\alpha}{q} f\left(\frac{qx}{p^\alpha}\right)\right).$$

This property enables us to reduce the evaluation of $S_q(f)$ to the case of prime power moduli. In [40, Lemma 1.6], Hua derived the following very precise estimate

$$\left|S_{p^\alpha}(f)\right| \leqslant n^3 p^{\alpha(1-1/n)}$$

for any polynomial f of degree n such that $d_p(f) \geqslant 1$. The exponent is best possible in view of the Gauss sum

$$S_{p^\alpha}\left(ax^n\right) = p^{\alpha(1-1/n)}$$

if $n \mid \alpha$, $p \nmid a$ and $p > n \geqslant 2$, as can be seen in [18]. Cochrane and Zheng [17, Theorem 1.1] improved Hua's result by replacing the constant n^3 by the following absolute constant

$$\left|S_{p^\alpha}(f)\right| \leqslant 4.41 p^{\alpha(1-1/n)}.$$

For general moduli, we have the bound [72]

$$\left|S_q(f)\right| \leqslant e^{1.74n} q^{1-1/n}$$

for any polynomial f such that $d_p(f) \geqslant 1$ for any prime $p \mid q$. As pointed out in [17], it can be observed that in order to get any further improvement, one must first derive a non-trivial upper bound for $S_p(f)$ in the range $p < (n-1)^2$, the interval where Weil's bound is worse than the trivial bound. The only significant progress in this direction appears to be for the case of *sparse* polynomials, i.e. polynomials whose zero coefficients are not explicitly stored. For instance, it is shown in [79, p. 88] that

$$\left|S_p\left(ax^n + bx\right)\right| \leqslant n^{1/4} p^{3/4}$$

provided that $p \nmid ab$, which is better than Theorem 6.34 whenever $p < n^{-1}(n-1)^4$. For more recent results and a large amount of reference on this topic, see [80].

To take another example, let $M_n = 2^n - 1$ be the nth Mersenne number. The following estimate is shown in [49].

Theorem 6.35 *Let q be a fixed odd prime, $\ell \in \mathbb{Z}_{\geqslant 1}$ and $A > 0$. Then there exists a constant $c_{q,A} > 0$ depending only on q and A and a constant $\delta_A > 0$ depending only on A such that, uniformly for all $x \geqslant 2$ and integers $a \geqslant 1$ such that $(a, q) = 1$, we have*

$$\left| \sum_{p \leqslant x} e_{q^{\ell}} \left(a M_p \right) \right| \leqslant c_{q,A} \left\{ x^{1 - \delta_A \rho_{\ell}^2} \log x + x q^{-\ell \delta_A} \right\} \tag{6.51}$$

provided that

$$x \leqslant q^{A\ell} \tag{6.52}$$

and where $\rho_{\ell} := \frac{\log x}{\ell \log q}$.

Surprisingly, the authors deduce from this bound the following consequence. Let q be a fixed odd prime, $\sigma = (a_{s-1}, \ldots, a_0) \in \{0, \ldots, q-1\}^s$ be a string of $s \geqslant 1$ digits to base q, and define $A_r(x, \sigma)$ to be the number of primes $p \leqslant x$ such that M_p written in base q has σ as the string on s consecutive digits on positions $r + s - 1, \ldots, r$, counting from the right to the left. The authors then observe in [49, p. 25] that this is equivalent to the property

$$\left\{ q^{-r-1} M_p \right\} \in I_{q,s} := \left[\overline{\sigma} q^{-s} , \, (1 + \overline{\sigma}) q^{-s} \right),$$

where $\overline{\sigma} := a_{s-1} q^{s-1} + \cdots + a_0$ is the integer which q-ary digits are given by σ. Now using Proposition 6.1 with S to be the set of primes, $a = 1, b = x, f(n) = q^{-r-1} M_n$ and

$$g(t) = \begin{cases} 1, & \text{if } \{t\} \in I_{q,s} \, ; \\ 0, & \text{otherwise} \, ; \end{cases}$$

we derive for all $H \in \mathbb{Z}_{\geqslant 1}$ and $x > 2$

$$A_r(x, \sigma) = \frac{\pi(x)}{q^s} + O\left(\frac{\pi(x)}{H} + \sum_{h=1}^{H} \frac{1}{h} \left| \sum_{p \leqslant x} e_{q^{r+1}} \left(h M_p \right) \right| \right)$$

$$= \frac{\pi(x)}{q^s} + O\left(\frac{\pi(x)}{H} + \frac{\pi(x) \log x}{q^r} + \sum_{\substack{h=1 \\ q^r \nmid h}}^{H} \frac{1}{h} \left| \sum_{p \leqslant x} e_{q^{r+1}} \left(h M_p \right) \right| \right),$$

where we estimated the exponential sum trivially in the case $q^r \mid h$. Now fix $\varepsilon \in (0, 1)$, assume $r \geqslant \varepsilon \log x$ and choose $H = \lfloor x^{\varepsilon/2} \rfloor$, implying that

$$\frac{\pi(x)}{H} + \frac{\pi(x)\log x}{q^r} \ll \frac{\pi(x)}{x^{\varepsilon/2}}.$$

For all $h \in \{1, \ldots, H\}$ such that $q^r \nmid h$, write $h = q^\beta h_1$ with $q \nmid h_1$ and $0 \leqslant \beta < r$, and set $\ell := r + 1 - \beta$. Hence

$$A_r(x, \sigma) = \frac{\pi(x)}{q^s} + O\left(\frac{\pi(x)}{x^{\varepsilon/2}} + \sum_{0 \leqslant \beta < r} \frac{1}{q^\beta} \sum_{h_1 \leqslant x^{\varepsilon/2}/q^\beta} \frac{1}{h_1} \left| \sum_{p \leqslant x} e_{q^\ell}(h_1 M_p) \right| \right).$$

We use (6.51) with $A = 2\varepsilon^{-1}$. Notice that

$$q^{A\ell} = q^{\frac{2}{\varepsilon}(r+1-\beta)} \geqslant \left(q^{r+1} h^{-1} \right)^{2/\varepsilon} > \left(e^r H^{-1} \right)^{2/\varepsilon} \geqslant x$$

so that (6.52) is satisfied. We then derive

$$A_r(x, \sigma) = \frac{\pi(x)}{q^s} + O_{\varepsilon, q} \left\{ \frac{\pi(x)}{x^{\varepsilon/2}} + \sum_{0 \leqslant \beta < r} \frac{1}{q^\beta} \left(x^{1-\delta_\varepsilon \rho_\ell^2}(\log x)^2 + xq^{-\ell\delta_\varepsilon}\log x \right) \right\}$$

for some $\delta_\varepsilon > 0$ depending only on ε. Since $2 \leqslant \ell \leqslant r + 1$ and $q^\ell \geqslant x^{\varepsilon/2}$ as seen above, we get

$$A_r(x, \sigma) = \frac{\pi(x)}{q^s} + O_{\varepsilon, q} \left(x^{1-\delta_\varepsilon \rho_{r+1}^2}(\log x)^2 + x^{1-\varepsilon_1} \right)$$

with $\varepsilon_1 := \frac{1}{2}\varepsilon(1 + \delta_\varepsilon)$, and where the term $\pi(x)x^{-\varepsilon/2}$ has been absorbed by the term $x^{1-\varepsilon_1}$. Finally, let us suppose that $r \leqslant (\log x)^{3/2-2\varepsilon}$, so that $r + 1 \leqslant (\log x)^{3/2-\varepsilon}$ as soon as $x \geqslant \exp\left(2^{1/\varepsilon} \right)$. We get $\rho_{r+1}^2 \geqslant (\log x)^{-1+\varepsilon}(\log q)^{-2}$, and therefore

$$A_r(x, \sigma) = \frac{\pi(x)}{q^s} + O_{\varepsilon, q} \left(x \exp\left(-\delta_{q, \varepsilon}(\log x)^\varepsilon \right) (\log x)^2 + x^{1-\varepsilon_1} \right),$$

where $\delta_{q, \varepsilon} := \delta_\varepsilon (\log q)^{-2}$. This result implies that, for any fixed odd prime q, any $\varepsilon \in (0, 1)$ and integer $s \geqslant 1$, for any integer $\varepsilon \log x \leqslant r \leqslant (\log x)^{3/2-2\varepsilon}$, on rightmost q-ary position $r, \ldots, r - s + 1$ of M_p, $p \leqslant x$, any block of q-ary digits of length s appears asymptotically the same number of times. $\square$

In another direction, it could be interesting to derive some estimates for sums of type I or II. To do this, let $q > 1$ be a positive *squarefree* integer and $I \subset \mathbb{Z}/q\mathbb{Z}$ be a set of N consecutive residues of $\mathbb{Z}/q\mathbb{Z}$, with $q - 1$ followed by 0. For any complex weight $\alpha = (\alpha_m)_{m \in \mathbb{Z}/q\mathbb{Z}}$, define the norms

$$\|\alpha\|_p = \left(\sum_{m \in \mathbb{Z}/q\mathbb{Z}} |\alpha_m|^p \right)^{1/p}$$

with $p > 0$, and $\|\alpha\|_\infty = \max_{m \in \mathbb{Z}/q\mathbb{Z}} |\alpha_m|$. In [60, Theorem 2.1] used with $\nu = 2$, the authors derived the following bound.

Theorem 6.36 *Let k, ℓ be fixed non-zero integers such that k/ℓ is not a positive integer. With the previous notation, we have*

$$\sum_{m \in \mathbb{Z}/q\mathbb{Z}} \alpha_m \sum_{n \in I} \sum_{x \in (\mathbb{Z}/q\mathbb{Z})^\times} e_q \left(mx^k + nx^\ell \right)$$

$$\ll \min \left\{ \|\alpha\|_2 N^{1/2} q , \ (\|\alpha\|_1 \|\alpha\|_2)^{1/2} \, q^{o(1)} \left(q^{11/12} N^{1/3} + q \right) \right\}.$$

For instance, if we take α_m to be the characteristic function of a set J of M consecutive residues of $\mathbb{Z}/q\mathbb{Z}$ and interchanging the roles of M and N, this result yields

$$\sum_{m \in J} \sum_{n \in I} \sum_{x \in (\mathbb{Z}/q\mathbb{Z})^\times} e_q \left(mx^k + nx^\ell \right) \ll q^{o(1)} N^{3/4} \left(q^{11/12} M^{1/3} + q \right).$$

As pointed out by the authors, such sums with $(k, \ell) = (2, -1)$ can be applied to estimate squarefree numbers in arithmetic progressions.

Exercises

114   Let $x, y, \alpha, \beta \in \mathbb{R}$ and $M < N$ be integers.

(a) Show that $4\|x - y\| \leqslant |e(x) - e(y)| \leqslant 2\pi \|x - y\|$.
(b) Show that

$$\left| \sum_{n=M+1}^{N} e(\alpha n + \beta) \right| \leqslant \min \left(N - M, \frac{1}{2\|\alpha\|} \right).$$

115 Let $N < N_1 \leqslant 2N$ be large integers and $f : [N, N_1] \to \mathbb{R}$ be a function satisfying the hypotheses of Theorem 6.3 with $\lambda_2 \leqslant 10^{-2}$ and let $\delta \in \left(0, \frac{1}{10}\right]$ be a small real number. Suppose also that f' is non-decreasing.

(a) Splitting the sum into two subsums, show that

$$\sum_{N < n \leqslant N_1} e(\pm f(n)) \ll \mathcal{R}(f', N, \delta) + (N\lambda_2 + 1)\,\delta^{-1}.$$

(b) Using Theorem 5.3 and choosing δ optimally, deduce another proof of van der Corput's inequality.

116 Prove

$$\sum_{(20x)^{1/2} < n \leqslant x^{3/4}} \psi\left(\frac{x}{n}\right) \ll x^{1/2} \log x.$$

117 Let $N \in \mathbb{Z}_{\geqslant 1}$, $U > 0$, and f and u be two real-valued functions defined on $[N, 2N]$, with u satisfying

$$\sum_{N < n \leqslant 2N-1} |u(n+1) - u(n)| \leqslant U.$$

Prove that, for any $H \in \mathbb{Z}_{\geqslant 1}$

$$\sum_{N < n \leqslant 2N} \psi\left(f(n) + u(n)\right) \ll \frac{N}{H} + \sum_{h \leqslant H} \left(\frac{1}{h} + U\right) \max_{N \leqslant N_1 \leqslant 2N} \left|\sum_{N \leqslant n \leqslant N_1} e\left(hf(n)\right)\right|.$$

118 Let $a \in \mathbb{R}$ and $H \in \mathbb{Z}_{\geqslant 1}$. Show that

$$\left|\sum_{h=0}^{H-1} e(ha)\right|^2 = \sum_{|h| \leqslant H-1} (H - |h|)\, e(ha).$$

119 Let $f : [N, 2N] \to \mathbb{R}$ be any function and $\delta \in (0, 1/4)$. Set $K = \lfloor(8\delta)^{-1}\rfloor + 1$. The purpose of this exercise is to prove that, for all positive integers $H \leqslant K$, we have

$$\mathcal{R}(f, N, \delta) \leqslant \frac{4N}{H} + \frac{4}{H}\sum_{h=1}^{H-1}\left|\sum_{N \leqslant n \leqslant 2N} e(hf(n))\right|. \tag{6.53}$$

(a) Let $n \in [N, 2N] \cap \mathbb{Z}$ such that $\|f(n)\| < \delta$. Prove that, for all integers h such that $|h| < H$, we have

$$\mathrm{Re}\,\{e(hf(n))\} \geqslant \tfrac{\sqrt{2}}{2}.$$

(b) Prove that

$$\mathcal{R}(f, N, \delta) \leqslant \frac{2}{H^2} \sum_{\substack{N \leqslant n \leqslant 2N \\ \|f(n)\| < \delta}} \left| \sum_{h=0}^{H-1} e(hf(n)) \right|^2$$

and show (6.53) by using Exercise 118.

120   Let f be a function satisfying the hypotheses of Definition 6.1 and let (k, ℓ) be an exponent pair. Suppose that $N \leqslant 8T$. Prove

$$\mathcal{R}(f, N, \delta) \ll N\delta + \left(T^k N^\ell\right)^{\frac{1}{k+1}} + T^k N^{\ell-k}. \tag{6.54}$$

121   Apply (6.54) to the squarefree number and square-full number problems from Chap. 5.

122   Let $s = \sigma + it \in \mathbb{C}$ such that $\frac{1}{2} \leqslant \sigma \leqslant 1$ and $t \geqslant 3$, and $\zeta(s)$ be the Riemann zeta-function.

(a) Show that $\zeta(\sigma + it) \ll \left| \sum_{n \leqslant t} n^{-\sigma - it} \right| + t^{1-2\sigma} \log t$.

(b) Let (k, ℓ) be an exponent pair such that $\ell - k \leqslant \frac{1}{2}$. Prove that

$$\zeta(\sigma + it) \ll t^{\frac{k(1-\sigma)}{1+k-\ell}} \log t.$$

Deduce that, for all $\varepsilon > 0$, we have $\zeta\left(\frac{1}{2} + it\right) \ll t^{13/84+\varepsilon}$.

123   Let $M \in \mathbb{Z}_{\geqslant 0}$, $N \in \mathbb{Z}_{\geqslant 1}$, $H \in \mathbb{Z}_{\geqslant 4}$ and $\alpha > 0$. Prove

$$\sum_{M < n \leqslant M+N} \min\left(H, \frac{1}{\|n\alpha\|}\right) < 8HN\alpha + \frac{4}{\log 2}\left(7N + \alpha^{-1}\right) \log H + 4H.$$

124   Let $x \geqslant 1$, $K := \left\lfloor \sqrt{2x + \frac{1}{4}} - \frac{1}{2} \right\rfloor$ and set

$$S(x) := x \sum_{k \leqslant K} \frac{\left\lfloor \frac{2x}{k} \right\rfloor - 2\left\lfloor \frac{x}{k} \right\rfloor}{\left\lfloor \frac{x}{k} \right\rfloor \left(2\left\lfloor \frac{x}{k} \right\rfloor + 1\right)}.$$

(a) Prove that, for $x \geqslant 1$, $N \in \mathbb{Z}_{\geqslant 1}$ large and if (k, ℓ) is an exponent pair, then

$$\left| \sum_{N < n \leqslant 2N} \left(\psi \left(\frac{x}{n} \right) + \psi \left(\frac{x}{n} + \frac{1}{2} \right) \right) \right| \ll \left(x^k N^{\ell-k} \right)^{\frac{1}{k+1}} + N^2 x^{-1}.$$

(b) Deduce that

$$S(x) = \frac{\sqrt{2}}{6} x^{1/2} + O \left(x^{\frac{k+\ell}{2k+2}} \log x \right).$$

125 In this exercise, m, n, d are positive integers, and $(m \bmod d)$ designates the remainder of the Euclidean division of m by d. The aim is to obtain an asymptotic formula for the sum

$$\sum_{d=1}^{n} (m \bmod d)^2 \quad (m \geqslant n \geqslant 64).$$

(a) Let (k, ℓ) be an exponent pair. Show that

$$\sum_{d \leqslant n} d^2 \psi \left(\frac{m}{d} \right) \ll \left\{ \left(m^k n^{k+\ell+2} \right)^{\frac{1}{k+1}} + n^4 m^{-1} \right\} \log n.$$

(b) With the help of Theorem 6.32, prove that, for all $n, m \in \mathbb{Z}_{\geqslant 1}$, $n \geqslant 64$, such that $m^{1/3} \ll n \leqslant \sqrt{2m}$, then

$$\sum_{d \leqslant n} d^2 \psi \left(\frac{m}{d} \right)^2 = \frac{n^3}{36} + O \left(nm^{2/3} + (mn^3)^{1/2} \right).$$

(c) Deduce that, for all $m \geqslant n \in \mathbb{Z}_{\geqslant 64}$ such that $m^{1/3} \ll n \leqslant \sqrt{2m}$

$$\sum_{d=1}^{n} (m \bmod d)^2 = \frac{n^3}{9} + O \left(n^2 m^{1/3} \log n \right).$$

126 Let $r \in \mathbb{Z}_{\geqslant 2}$. Using Example 4.12 and Corollary 6.7, prove that, if (k, ℓ) is an exponent pair , then

$$\sum_{n \leqslant x} \left\lfloor \frac{x^r}{n^r} \right\rfloor = x^r \zeta(r) + x \zeta \left(\frac{1}{r} \right) + O_r \left(x^{\theta_r} \log x \right),$$

where

$$\theta_r = \theta_r(k, \ell) := \begin{cases} \dfrac{r(k+\ell)}{(k+1)(r+1)}, & \text{if } \ell \geqslant rk \\[3mm] \dfrac{rk}{k(r+1)+1-\ell}, & \text{if } \ell \leqslant rk. \end{cases}$$

127 Use the Circle Method to recover the equality of Proposition 2.7, namely

$$\mathcal{D}_{(1,\dots,1)}(n) = \binom{k+n-1}{n}.$$

128 Let $N \in \mathbb{Z}_{\geqslant 1}$ be a large integer. Using Theorem 6.21, derive a bound for
$\displaystyle\sum_{N < n \leqslant 2N} \psi(f(n))$ and $\mathcal{R}(f, N, \delta)$, where f is a function satisfying the conditions of Theorem 6.21.

References

1. Baker, R.C.: The greatest prime factor of the integers in an interval. Acta Arith. **47**, 193–231 (1986)
2. Baker, R.C.: The square-free divisor problem. Quart. J. Math. **45**, 269–277 (1994)
3. Baker, R.C.: Sums of two relatively prime cubes. Acta Arith. **129**, 103–146 (2007)
4. Baker, R.C., Harman, G.: Numbers with a large prime factor II. In: Chen, W.W.L., Gowers, W.T., Halberstam, H., Schmidt, W.M., Vaughan, R.C. (eds.) Analytic Number Theory, Essays in Honour of Klaus Roth. Cambridge University Press, Cambridge (2009)
5. Berkane, D., Bordellès, O., Ramaré, O.: Explicit upper bounds for the remainder term in the divisor problem. Math. Comp. **81**, 1025–1051 (2012)
6. Bombieri, E., Iwaniec, H.: On the order of $\zeta\left(\frac{1}{2} + it\right)$. Ann. Scuola Norm Sup. Pisa **13**, 449–472 (1986)
7. Bourgain, J., Watt, N.: Mean square of zeta function, circle problem and divisor problem revisited, 23 pp. Preprint (2017). https://arxiv.org/abs/1709.04340
8. Bourgain, J., Demeter, C., Guth, L.: Proof of the main conjecture in Vinogradov's mean value theorem for degrees higher than three. Ann. Math. **184**, 633–682 (2016)
9. Bullen, P.S.: A dictionary of inequalities. Pitman Monographs. Addison Wesley Longman, Boston (1998)
10. Burgess, D.A.: On character sums and L-series, II. Proc. Lond. Math. Soc. **13**, 524–536 (1963)
11. Burgess, D.A.: The character sum estimate with $r = 3$. J Lond. Math. Soc. II Ser. **33**, 219–226 (1986)
12. Burgess, D.A.: Partial Gaussian sums II. Bull. Lond. Math. Soc. **21**, 153–158 (1989)
13. Cao, X., Zhai, W.G.: The distribution of square-free numbers of the form $[n^c]$. J. Théor. Nombres Bordeaux **10**, 287–299 (1998)
14. Cao, X., Zhai, W.G.: On the number of coprime integer pairs within a circle. Acta Arith. **89**, 163–187 (1999)

15. Cao, X., Zhai, W.G.: Multiple exponential sums with monomials. Acta Arith. **92**, 195–213 (2000)
16. Chowla, S., Walum, H.: On the divisor problem. Skr. K. Nor. Vidensk. Selsk. **36**, 127–134 (1963). Proceedings, Symp. Pure Math. No. 8. pp. 138–143. Amer. Math. Sot., Providence, R.I., 1965
17. Cochrane, T., Zheng, Z.: On upper bounds of Chalk and Hua for exponential sums. Proc. Amer. Math. Soc. **129**, 2505–2516 (2001)
18. Cochrane, T., Zheng, Z.: A survey on pure and mixed exponential sums modulo prime powers. In: Bennett, M.A., et al. (eds.) Number Theory for the Millennium I, Urbana-Champaign, IL, May 21–26, 2000, pp. 273–300. A K Peters, Natick (2002)
19. Daboussi, H., Rivat, J.: Explicit upper bounds for exponential sums over primes. Math. Comp. **233**, 431–447 (2000)
20. Davenport, H.: On certain exponential sums. J. Reine Angew. Math. **169**, 158–176 (1933)
21. Dragomir, S.S.: A survey on Cauchy- Bunyakovsky -Schwarz type discrete inequalities. J. Inequal. Pure Appl. Math. **4**, Article 63 (2003)
22. Estermann, T.: On Kloosterman's sum. Mathematika **8**, 83–86 (1961)
23. Ford, K.: (2002) Recent progress on the estimation of Weyl sums. In: Modern Problems of Number Theory and Its Applications; Topical Problems Part II, Tula, pp. 48–66 (2001)
24. Fouvry, E., Iwaniec, H.: Exponential sums with monomials. J. Number Theory **33**, 311–333 (1989)
25. Ganelius, T.: On one-sided approximation by trigonometric polynomials. Math. Scandinavica **4**, 247–258 (1957)
26. Graham, S.W., Kolesnik, G.: Van der Corput's method of exponential sums. London Mathematical Society Lecture Note Series, vol. 126. Cambridge University Press, Cambridge (1991)
27. Granville, A.: An alternative to Vaughan identity, 3 pp. Preprint (2020). https://arxiv.org/abs/2001.07777v1
28. Granville, A., Ramaré, O.: Explicit bounds on exponential sums and the scarcity of squarefree binomial coefficients. Mathematika **45**, 73–107 (1996)
29. Grekos, G.: Sur le nombre de points entiers d'une courbe convexe. Bull. Sci. Math. **112**, 235–254 (1988)
30. Hardy, G.H.: On Dirichlet's divisor problem. Proc. Lond. Math. Soc. **15**, 1–25 (1916)
31. Hardy, G.H., Littlewood, J.E.: A new solution of Waring's problem. Quart. J. **48**, 272–293 (1919)
32. Hardy, G.H., Littlewood, J.E.: Some problems of "Partito Numerorum" III. On the expression of a number as a sum of primes. Acta Math. **44**, 1–70 (1923a)
33. Hardy, G.H., Littlewood, J.E.: Some problems of "Partito Numerorum" V. further contribution to the study of Goldbach's problem. Proc. Lond. Math. Soc. **17**, 75–115 (1923b)
34. Hardy, G.H., Ramanujan, S.: Asymptotic formulae in combinatory analysis. Proc. Lond. Math. Soc. **17**, 75–115 (1918)
35. Harman, G.: Prime-detecting sieves. London Mathematical Society Monographs. Princeton University Press, Princeton (2007)
36. Heath-Brown, D.R.: Prime numbers in short intervals and a generalized Vaughan identity. Canad. J. Math. **34**, 1365–1377 (1982)
37. Helfgott, H.A.: The ternary Goldbach problem, 327 pp. (2015). Preprint. https://arxiv.org/abs/1501.05438
38. Helfgott, H.A., Platt, D.J.: Numerical verification of the ternary Goldbach conjecture up to $8,875 \cdot 10^{30}$. Exp. Math. **22**, 406–409 (2013)
39. Hooley, C.: On the greatest prime factor of a cubic polynomial. J. Reine. Angew. Math. **303/304**, 21–50 (1978)
40. Hua, L.K.: Additive Theory of Prime Numbers, vol. 13. American Mathematical Society, Providence (2009). Reprinted by the AMS from the 1965 edition
41. Huxley, M.N.: Area, Lattice Points and Exponential Sums. Oxford Science Pub., Oxford (1996)

42. Huxley, M.N.: Exponential sums and lattice points III. Proc. Lond. Math. Soc. **87**, 591–609 (2003)
43. Huxley, M.N., Watt, N.: Exponential sums and the Riemann zeta function. Proc. Lond. Math. Soc. **57**, 1–24 (1988)
44. Iwaniec, H., Kowalski, E.: Analytic Number Theory, vol. 53. Colloquium Publications. American Mathematical Society, Providence (2004)
45. Iwaniec, H., Mozzochi, C.J.: On the divisor and circle problems. J. Number Theory **29**, 60–93 (1988)
46. Kanemitsu, S.: On a kind of divisor problem. Mem. Fac. Sci. Kyushu Univ. **32**, 81–89 (1978)
47. Kanemitsu, S., Rao, R.S.R.C.: On a conjecture of S. Chowla and of S. Chowla and H. Walum, I. J. Number Theory **20**, 255–261 (1985)
48. Karacuba, A.A.: Estimates for trigonometric sums by Vinogradov's method, and some applications. Proc. Steklov. Inst. Math. **112**, 251–265 (1971)
49. Kerr, B., Mérai, L., Shparlinski, I.E.: On digits of Mersenne numbers, 26 pp. (2020). Preprint. https://arxiv.org/abs/2001.03380
50. Kloosterman, H.D.: On the representation of numbers in the form $ax^2 + by^2 + cz^2 + dt^2$. Acta Math. **49**, 407–464 (1926)
51. Korolev, M.A.: Incomplete Kloosterman sums and their applications. Izv. Ross. Akad Nauk. Ser. Mat. **64**, 41–64 (2000). Translation in Izv. Math. **64**, 1129–1152 (2000)
52. Korolev, M.A.: Kloosterman sums with multiplicative coefficients. Izv. Math. **82**, 647–661 (2018)
53. Kowalski, E., Robert, O., Wu, J.: Small gaps in coefficients of L-functions and $\mathfrak{B}$-free numbers in short intervals. Rev. Mat. Iberoam **23**, 281–326 (2007)
54. Liu, H.Q.: The greatest prime factor of the integers in an interval. Acta Arith. **65**, 301–328 (1993)
55. Liu, H.Q.: The distribution of 4-full numbers. Acta Arith. **67**, 165–176 (1994)
56. Liu, H.Q.: Divisor problems of 4 and 3 dimensions. Acta Arith. **73**, 249–269 (1995)
57. Liu, H.Q.: On Euler's function. Proc. R. Soc. Edinb. A **146**, 769–775 (2016)
58. Liu, H.Q., Wu, J.: Numbers with a large prime factor. Acta Arith. **89**, 163–187 (1999)
59. Liu, M.C., Wang, T.: On the Vinogradov bound in the three primes Goldbach conjecture. Acta Arith. **105**, 133–175 (2002)
60. Liu, K., Shparlinski, I.E., Zhang, T.: Bilinear forms with exponential sums with binomials. J. Number Theory **188**, 172–185 (2018)
61. Ma, J., Wu, J.: On a sum involving the Mangoldt function. Period. Math. Hung. (2020). To appear
62. Mardzhanishvili, K.K.: On the proof of the Goldbach-Vinogradov theorem. C R (Doklady) Acad. Sci. URSS (NS) **30**, 681–684 (1941), in Russian
63. Montgomery, H.L.: Ten Lectures on the Interface Between Analytic Number Theory and Harmonic Analysis, vol. 84. The American Mathematical Monthly, cMBS (1994)
64. Montgomery, H.L., Vaughan, R.C.: The distribution of squarefree numbers. In: Halberstam, H., Hooley, C. (eds.) Recent Progress in Analytic Number Theory, vol. I. Academic Press, Cambridge (1981)
65. Montgomery, H.L., Vaughan, R.C.: Multiplicative number theory I. Classical theory. No. 97. In: Cambridge Series in Advanced Mathematics. Cambridge University Press, Cambridge (2007)
66. Mordell, L.J.: On a sum analogous to a Gauss's sum. Q. J. Oxford Ser. **3**, 161–167 (1932)
67. Mordell, L.J. On the Kusmin–Landau inequality for exponential sums. Acta Arith. **4**, 3–9 (1958)
68. Niculescu, C.P., Persson, L.E.: Old and new on the Hermite-Hadamard inequality. Real Anal. Exch. **29**, 663–685 (2003/2004)
69. Nowak, W.G.: An Analogue to a Conjecture of S. Chowla and H. Walum. J. Number Theory **19**, 254–262 (1984)
70. Pétermann, Y.F.S.: On an estimate of Walfisz and Saltykov for an error term related to the Euler function. J. Théor Nombres Bordeaux **10**, 203–236 (1998)

71. Pétermann, Y.F.S., Wu, J.: On the sum of exponential divisors of an integer. Acta Math. Hungar **77**, 159–175 (1997)
72. Qi, M., Ding, P.: Further estimate of complete trigonometric sums. J. Tsinghua Univ. **29**, 74–85 (1989)
73. Ramachandra, K.: A note on numbers with a large prime factor. J. Lond. Math. Soc. **1**, 303–306 (1969)
74. Rivat, J., Sargos, P.: Nombres premiers de la forme $[n^c]$. Canad. J. Math. **53**, 190–209 (2001)
75. Robert, O., Sargos, P.: A third derivative test for mean values of exponential sums with application to lattice point problems. Acta Arith. **106**, 27–39 (2003)
76. Robert, O., Sargos, P.: Three-dimensional exponential sums with monomials. J. Reine. Angew. Math. **591**, 1–20 (2006)
77. Salié, H.: Zur abschätzung der fourierkoeffizienten ganzer modulformen. Math. Z. **36**, 263–278 (1933)
78. Sargos, P., Wu, J.: Multiple exponential sums with monomials and their applications in number theory. Acta Math. Hung. **88**, 333–354 (2000)
79. Shparlinski, I.E.: Computational and Algorithmic Problems in Finite Fields. Kluwer Academic Pub., Boston (1992)
80. Shparlinski, I.E., Voloch, J.F.: Binomial exponential sums, 12 pp. (2018), Preprint. https://arxiv.org/abs/1811.00765
81. Stečkin, S.B. The Theory of Approximation of Functions. Nauka, Moscow (1983)
82. Tong, K.C.: On divisor problem III. Acta Math. Sinica. **6**, 515–541 (1956)
83. Vaaler, J.: Some extremal functions in Fourier analysis. Bull. Amer. Math. Soc. **12**, 183–216 (1985)
84. Vaughan, R.C.: The Hardy-Littlewood Method, 2nd edn. Cambridge University Press, Cambridge (1997)
85. Vinogradov, I.M. Representation of an odd number as the sum of three primes. Dokl. Akad. Nauk. SSR **15**, 291–294 (1937)
86. Vinogradov, I.M.: The Method of Trigonometric Sums in the Theory of Numbers. Interscience, Geneva (1954)
87. Voronoï, G.: Sur un problème du calcul des fonctions asymptotiques. J. Reine. Angew. Math. **126**, 241–282 (1903)
88. Walfisz, A.: Weylsche Exponentialsummen in der Neueren Zahlentheorie. VEB Deutscher Verlag, Berlin (1963)
89. Weil, A.: On some exponential sums. Proc. Natl. Acad. Sci. USA **34**, 204–207 (1948)
90. Wooley, T.D.: A note on simultaneous congruences. II: Mordell revised. J. Aust. Math. Soc. **88**, 261–275 (2010)
91. Wooley, T.D.: The cubic case of the main conjecture in Vinogradov's mean value theorem. Adv. Math. **294**, 532–561 (2016)
92. Wu, J.: Nombres $\mathcal{B}$-libres dans les petits intervalles. Acta Arith. **65**, 97–98 (1993)
93. Wu, J.: On the primitive circle problem. Monatsh Math. **135**, 69–81 (2002)
94. Zaccagnini, A.: Introduction to the Circle Method of Hardy, Ramanujan and Littlewood. Harish-Chandra Research Institute, Jhusi (2005)
95. Zhai, W.G.: On sums and differences of two coprime kth powers. Acta Arith. **91**, 233–248 (1999)

Chapter 7
Algebraic Number Fields

7.1 Introduction

Algebraic number theory came from the necessity to solve certain Diophantine equations for which the classical tools borrowed from arithmetic in $\mathbb{Z}$ were not sufficient enough to provide a satisfying answer. For instance, the Fermat equation $x^n + y^n = z^n$, where x, y and z are positive integers and $n \geqslant 3$ is an odd integer, can be factored using a nth root of unity $\zeta_n = e_n(1) = e^{2\pi i/n}$ as

$$z^n = \prod_{k=0}^{n-1} (x + \zeta_n^k y).$$

The right-hand side makes numbers of the form $x + \zeta_n^k y$ appear. These numbers do not belong to $\mathbb{Q}$ but lie in a larger set which may be viewed as an extension of $\mathbb{Q}$, contained in $\mathbb{C}$ and obtained by *adjoining* the number ζ_n to $\mathbb{Q}$. This new set is denoted by $\mathbb{Q}(\zeta_n)$, and its elements can be written in the form

$$\sum_{k=0}^{n-1} a_k \zeta_n^k \quad \text{with} \quad a_k \in \mathbb{Q}.$$

One can prove that such a set is a field,[1] which belongs to the sets named *algebraic number fields*. Note that ζ_n is a root of the algebraic polynomial $X^n - 1$ but not a root of $X^d - 1$ for any $d < n$.

Such a generalization of $\mathbb{Z}$ and $\mathbb{Q}$ requires a more careful approach. Take for instance the set $\mathbb{Q}(\sqrt{-5})$, whose elements are of the form $a + b\sqrt{-5}$ with $a, b \in \mathbb{Q}$.

---

[1] Called a *cyclotomic field*.

© Springer Nature Switzerland AG 2020
O. Bordellès, *Arithmetic Tales*, Universitext,
https://doi.org/10.1007/978-3-030-54946-6_7

This set is also a field,[2] and has a subset of *algebraic integers* denoted by $\mathbb{Z}[\sqrt{-5}]$ whose elements are of the form $a + b\sqrt{-5}$ with $a, b \in \mathbb{Z}$. One can prove that $\mathbb{Z}[\sqrt{-5}]$ is a ring and is the analogue of $\mathbb{Z}$ in $\mathbb{Q}$. One can show that 3, 7, $1 \pm 2\sqrt{-5}$ are irreducible in $\mathbb{Z}[\sqrt{-5}]$,[3] but since

$$21 = 3 \times 7 = \left(1 + 2\sqrt{-5}\right)\left(1 - 2\sqrt{-5}\right),$$

we see that $\mathbb{Z}[\sqrt{-5}]$ is not a UFD, while $\mathbb{Z}$ is.

This problem lies at the heart of algebraic number theory. Lamé thought that, if x and y are chosen in $\mathbb{Z}$ so that $x + y$ and $x + \zeta_n^k y$ have no common factors in $\mathbb{Z}[\zeta_n]$ for any $0 < k \leqslant n - 1$, then the Fermat equation has a solution only if there are $z_k \in \mathbb{Z}$ such that $x + \zeta_n^k y = z_k^n$ for all $0 \leqslant k \leqslant n - 1$. This was actually his argument when he addressed a meeting of the *Académie des Sciences* on March 1, 1847 and where he announced he had solved Fermat's last theorem.[4] Liouville said that the assumption that allows equation $x + \zeta_n^k y = z_k^n$ to follow in the argument is that $\mathbb{Z}[\zeta_n]$ is a UFD for all n. But on April 28, 1847, Kummer proved that this is not the case in general, and later Cauchy showed that the first counterexample occurs for $n = 23$. This led Kummer to invent what he called *ideal numbers* to restore unique factorization.[5]

7.2   Algebraic Numbers

7.2.1   Some Group-Theoretic Results

In this chapter, we will essentially use the next result from Group Theory. Their proofs can easily be found in any textbook on this subject.

Theorem 7.1 (Finite Groups) *Let G be a finite group of order $|G|$.*

▷ (Lagrange). *Let H be a subgroup of G. Then H is a finite group and we have*

$$|G| = (G : H) \times |H|,$$

where $(G : H)$ is the index of H in G. In particular, $|H|$ divides $|G|$.
▷ (Cauchy). *Let p be a prime divisor of $|G|$. Then G has an element of order p.*

---

[2]Called an *imaginary quadratic field*.

[3]See Exercise 129.

[4]The FLT states that the Fermat equation has no solution in positive integers x, y, z as soon as $n \geqslant 3$. This was finally proved by Wiles in 1995.

[5]For an interesting account of the history of the birth of the ideal theory in the late nineteenth century, the reader is referred to [63, 92].

Proposition 7.1 (Free Abelian Groups) *Let G be a free abelian group of rank n.*

▷ *Let $\{e_1, \ldots, e_n\}$ be a basis for G and set $f_i = \sum_{j=1}^{n} a_{ij} e_j$ with $a_{ij} \in \mathbb{Z}$. Then $\{f_1, \ldots, f_n\}$ is a basis for G if and only if $\left| \det(a_{ij}) \right| = 1$.*

▷ *Let H be a subgroup of G. The group G/H is finite if and only if rank $G = $ rank H. In this case, if $\{e_1, \ldots, e_n\}$ is a basis for G and $\{f_1, \ldots, f_n\}$ is a basis for H such that $f_i = \sum_{j=1}^{n} a_{ij} e_j$ with $a_{ij} \in \mathbb{Z}$, then*

$$(G : H) = \left| \det(a_{ij}) \right|.$$

7.2.2   Polynomials

7.2.2.1   Irreducibility

Polynomials with coefficients in $\mathbb{Q}$ play an important role in this chapter. The main problem we shall have to deal with is to determine whether a given polynomial $P \in \mathbb{Q}[X]$ is irreducible over $\mathbb{Q}$ or not. After defining this concept, we will provide some useful irreducibility criteria.[6]

Definition 7.1

▷ A polynomial $P \in \mathbb{Z}[X]$ is *irreducible*, or *irreducible over* $\mathbb{Z}$, provided that $P \neq \pm 1$ and whenever $P = QR$ with $Q, R \in \mathbb{Z}[X]$, either $Q = \pm 1$ or $R = \pm 1$.

A polynomial not irreducible over $\mathbb{Z}$ and not $0,\ 1$ or -1 is called *reducible over* $\mathbb{Z}$.

The polynomials $0,\ 1$ and -1 are considered neither irreducible nor reducible over $\mathbb{Z}$.

▷ A polynomial $P \in \mathbb{Q}[X]$ is *irreducible*, or *irreducible over* $\mathbb{Q}$, provided that P is not constant and whenever $P = QR$ with $Q, R \in \mathbb{Q}[X]$, either $\deg Q = 0$ or $\deg R = 0$.

A non-constant polynomial not irreducible over $\mathbb{Q}$ is called *reducible over* $\mathbb{Q}$.

Constant polynomials are considered neither irreducible nor reducible over $\mathbb{Q}$.

For instance, the polynomial $X^2 + 1$ is irreducible over $\mathbb{Z}$ and $\mathbb{Q}$ and the polynomial $5X + 5$ is reducible over $\mathbb{Z}$ and irreducible over $\mathbb{Q}$. These examples suggest an important connection between irreducibilities over $\mathbb{Z}$ and $\mathbb{Q}$. Gauss's lemma implies the following answer.

Lemma 7.1 (Gauss's Lemma)   *If $P \in \mathbb{Z}[X]$ is irreducible over $\mathbb{Z}$, then it is irreducible over $\mathbb{Q}$. Furthermore, if $P \in \mathbb{Z}[X]$ is irreducible over $\mathbb{Q}$ and if the gcd of its coefficients is equal to 1, then P is irreducible over $\mathbb{Z}$.*

---

[6] We sometimes make use of the equality $P = P(X)$, where the right-hand side is the composition of the polynomial P with the polynomial X.

It follows that if $P \in \mathbb{Z}[X]$ is a reducible polynomial over $\mathbb{Q}$, then there exist $Q, R \in \mathbb{Z}[X]$ such that $P = QR$ with $\deg Q > 0$ and $\deg R > 0$. This is sometimes used to prove the irreducibility over $\mathbb{Q}$ of a polynomial $P \in \mathbb{Z}[X]$.

The following remarks, coming readily from the definition, can sometimes be of some help.

▷ Let $P \in \mathbb{Z}[X]$ and $a \in \mathbb{Z}$. If $P(X + a)$ is irreducible, then $P(X)$ is irreducible.
▷ If $P \in \mathbb{Z}[X]$ is an irreducible polynomial over $\mathbb{Z}$, then P has no multiple roots. Indeed, if there exists $\alpha \in \mathbb{C}$ such that $(X - \alpha)^2$ divides P, then α is a common root of P and its formal derivative P', so that $\gcd(P, P')$ is a non-constant polynomial of degree $< \deg P$ dividing[7] P.

7.2.2.2   Criteria for Irreducibility

The first useful irreducibility criterion, often referred to as Eisenstein's criterion, was first proved by Schönemann and shortly afterwards by Eisenstein.

Proposition 7.2 (Schönemann–Eisenstein) *Let* $P = a_n X^n + \cdots + a_1 X + a_0 \in \mathbb{Z}[X]$ *with* $n \in \mathbb{Z}_{\geqslant 1}$. *Suppose that there exists a prime number* p *such that* $p \nmid a_n$, $p \mid a_k$ *for all* $k < n$ *and* $p^2 \nmid a_0$. *Then* P *is irreducible over* $\mathbb{Q}$.

Proof Suppose that $P = QR$ with $Q, R \in \mathbb{Z}[X]$, $r = \deg Q > 0$ and $s = \deg R > 0$. Since $\mathbb{F}_p$ is a field, the ring $\mathbb{F}_p[X]$ is a UFD. Now we have

$$QR \equiv P \equiv a_n X^n \pmod{p}$$

and since $p \nmid a_n$, the leading coefficients of Q and R are not multiples of p, so that there exist $b, c \in \mathbb{Z}$ such that

$$Q \equiv b X^r \pmod{p} \quad \text{and} \quad R \equiv c X^s \pmod{p}.$$

Since $r, s > 0$, we get that p divides the constant terms of Q and R. This contradicts that $p^2 \nmid a_0$, which concludes the proof.   □

Note that the condition "over $\mathbb{Q}$" of the proposition cannot be removed. Indeed, by using Eisenstein's criterion with $p = 5$, we see that the polynomial $P = 3X^7 + 15X^3 + 15$ is irreducible over $\mathbb{Q}$, but reducible over $\mathbb{Z}$.

A polynomial is said to be *monic* if its leading coefficient is equal to 1. With this definition, one may slightly simplify Proposition 7.2. Indeed, in this case, the condition $p \nmid a_n$ is trivially true and the gcd of all the coefficients of the polynomial is equal to 1. One may deduce the following consequence.

---

[7]More generally, one can prove that if $\mathbb{K}$ is a field such that char $\mathbb{K} = 0$, then $P \in \mathbb{K}[X]$ can be written in the form $P = Q^2 R$ if and only if P and P' have a common factor of degree > 0.

Corollary 7.1 *Let* $P = X^n + a_{n-1}X^{n-1} + \cdots + a_1 X + a_0 \in \mathbb{Z}[X]$ *be a monic polynomial with* $n \in \mathbb{Z}_{\geqslant 1}$. *Suppose that there exists a prime number* p *such that* $p \mid a_k$ *for all* $0 \leqslant k \leqslant n - 1$ *and* $p^2 \nmid a_0$. *Then* P *is irreducible over* $\mathbb{Z}$.

Example 7.1

▷ Let p be a prime and m be an integer such that $p \nmid m$. Then $X^n - mp$ is irreducible over $\mathbb{Z}$.

▷ Let p be a prime. Then the *cyclotomic polynomial* $P = X^{p-1} + X^{p-2} + \cdots + X + 1$ is irreducible over $\mathbb{Z}$.

Indeed, it suffices to apply Corollary 7.1 to the polynomial

$$P(X + 1) = X^{p-1} + \binom{p}{1}X^{p-2} + \binom{p}{2}X^{p-3} + \cdots + \binom{p}{p-1}.$$

The next criterion shows how the reduction modulo a prime of a polynomial may be helpful. If p is a prime number and $P \in \mathbb{Z}[X]$, we write $\overline{P} \in \mathbb{F}_p[X]$, its reduction mod p.

Proposition 7.3 (Reduction Modulo p) *Let* $P = a_n X^n + \cdots + a_1 X + a_0 \in \mathbb{Z}[X]$ *and* p *be a prime number such that* $p \nmid a_n$. *If* $\overline{P}$ *is irreducible over* $\mathbb{F}_p$, *then* P *is irreducible over* $\mathbb{Q}$.

Proof Suppose that $P = QR$ with $Q, R \in \mathbb{Z}[X]$, $q = \deg Q > 0$ and $r = \deg R > 0$ and write b_q and c_r, the leading coefficients of Q and R. Since $\overline{P} = \overline{Q} \times \overline{R}$, we have $a_n \equiv b_q c_r \pmod{p}$. Since $\mathbb{F}_p$ is a field, the hypothesis $p \nmid a_n$ implies that $b_q \not\equiv 0 \pmod{p}$ and $c_r \not\equiv 0 \pmod{p}$. Since $\overline{P}$ is irreducible over $\mathbb{F}_p$, at least one of the two polynomials $\overline{Q}$ or $\overline{R}$ has degree 0. Suppose that $\deg \overline{Q} = 0$. Therefore $q = 0$, and P is irreducible over $\mathbb{Q}$.   □

Once again, if $P \in \mathbb{Z}[X]$ is monic, one can slightly simplify this result.

Corollary 7.2 *Let* $P = X^n + a_{n-1}X^{n-1} + \cdots + a_1 X + a_0 \in \mathbb{Z}[X]$ *be a monic polynomial and* p *be a prime number. If* $\overline{P}$ *is irreducible over* $\mathbb{F}_p$, *then* P *is irreducible over* $\mathbb{Z}$.

For instance, the reduction mod 2 of the polynomial $P = X^3 + 46246X^2 - 9987X + 258\,963$ is given by $P \equiv X^3 + X + 1 \pmod{2}$. Since $X^3 + X + \overline{1}$ has no roots in $\mathbb{F}_2$, it is irreducible over $\mathbb{F}_2$, and hence P is irreducible over $\mathbb{Z}$.

One may be careful that the converse of this result is generally untrue. There even exist irreducible polynomials over $\mathbb{Q}$ which are reducible over $\mathbb{F}_p$ for all primes p, as can be shown in the next result.

Lemma 7.2 *Let* $a, b \in \mathbb{Z}$. *Then the polynomial* $P = X^4 + aX^2 + b^2$ *is reducible over* $\mathbb{F}_p$ *for all primes* p.

Proof If $p = 2$, there are only four polynomials of the form indicated, all reducible. Suppose $p > 2$ is a prime number. One can choose an integer c such that $a \equiv 2c \pmod{p}$ which readily gives

$$P \equiv (X^2 + c)^2 - (c^2 - b^2)$$

$$\equiv (X^2 + b)^2 - (2b - 2c)X^2$$

$$\equiv (X^2 - b)^2 - (-2b - 2c)X^2 \pmod{p}.$$

Hence the result follows from the fact that one of the numbers $c^2 - b^2$, $2b - 2c$ or $-2b - 2c$ is a quadratic residue modulo p. By Proposition 3.5, if x is not a quadratic residue modulo p, then $x^{(p-1)/2} \equiv -1 \pmod{p}$, and therefore if two integers are non-quadratic residues modulo p, then their product is a quadratic residue modulo p. Now if $2b - 2c$ and $-2b - 2c$ are non-quadratic residues modulo p, then $(2b - 2c)(-2b - 2c) = 4(c^2 - b^2)$ is a quadratic residue modulo p and so is $c^2 - b^2$, which concludes the proof. $\square$

For instance, the polynomial $P = X^4 + 1$ is reducible over $\mathbb{F}_p$ for all primes p, but it is irreducible over $\mathbb{Z}$. Indeed, the only non-trivial real factors of P are the polynomials $X^2 \pm X\sqrt{2} + 1$ whose roots are $\frac{1}{2}(1 \pm i)$ and $\frac{1}{2}(-1 \pm i)$, but these polynomials do not belong to $\mathbb{Z}[X]$.

An irreducible polynomial P over $\mathbb{Q}$ such that $\deg P \geqslant 2$ has no roots in $\mathbb{Q}$, but the converse is untrue, since the polynomial $(X^2 + 1)(X^2 + 2)$ has no roots in $\mathbb{Q}$ but is reducible. On the other hand, if $\deg P = 2$ or 3, then P is irreducible over $\mathbb{Q}$ *if and only if* it has no roots in $\mathbb{Q}$, since any non-trivial factorization of P uses polynomials of degree 1 which have a root in $\mathbb{Q}$. These examples show that a link between roots and irreducibility must exist. The next result is another example that illustrates this subject.

Proposition 7.4 *Let* $P \in \mathbb{Z}[X]$ *such that* $P \neq \pm 1$, *and let* $m \in \mathbb{Z}$ *such that* $|P(m)|$ *is* 1 *or a prime number and such that* P *has no roots in the disc* $\{z \in \mathbb{C} : |z - m| \leqslant 1\}$. *Then* P *is irreducible over* $\mathbb{Z}$.

Proof Suppose that $P = QR$ with $Q, R \in \mathbb{Z}[X]$, $Q \neq \pm 1$ and $R \neq \pm 1$. By assumption, we have $|Q(m)| = 1$ or $|R(m)| = 1$. Without loss of generality, suppose that $|Q(m)| = 1$. Since $Q \neq \pm 1$, Q is not constant and we may write $Q = a \prod_{i=1}^{r}(X - \alpha_i)$ with $a \in \mathbb{Z} \setminus \{0\}$, $r \in \mathbb{Z}_{\geqslant 1}$ and $\alpha_1, \ldots, \alpha_r$ are the roots of Q counted to their multiplicity. Hence we have

$$1 = |Q(m)| = |a| \prod_{i=1}^{r} |m - \alpha_i|.$$

Since a is a non-zero integer, we have $|a| \geqslant 1$, so that there exists $j \in \{1, \ldots, r\}$ such that $|m - \alpha_j| \leqslant 1$ and $P(\alpha_j) = Q(\alpha_j)R(\alpha_j) = 0$, which is impossible since P has no roots in the disc $\{z \in \mathbb{C} : |z - m| \leqslant 1\}$. The proof is complete. $\square$

This result supposes that we have at our disposal some tools to locate the roots of polynomials. The following lemma, due to Eneström and Kakeya (see [4] for instance), is a useful tool to do the job.

Lemma 7.3 Let $P = a_n X^n + \cdots + a_1 X + a_0 \in \mathbb{R}[X]$ such that $a_0, \ldots, a_n > 0$. Then the roots of P are contained in the annulus

$$\min_{0 \leqslant i < n} \left(\frac{a_i}{a_{i+1}} \right) \leqslant |z| \leqslant \max_{0 \leqslant i < n} \left(\frac{a_i}{a_{i+1}} \right).$$

Example 7.2 Let $P = X^{p-1} + 2X^{p-2} + 3X^{p-3} + \cdots + (p-1)X + p$, where p is a prime number. By Lemma 7.3, all the roots of P are in the annulus $1 + 1/(p-1) \leqslant |z| \leqslant 2$. In particular, P has no roots in the disc $\{z \in \mathbb{C} : |z| \leqslant 1\}$ and since $|P(0)| = p$ is prime, we infer that P is irreducible over $\mathbb{Z}$ by Proposition 7.4 applied with $m = 0$.

7.2.2.3 Discriminant of a Polynomial

Definition 7.2 Let R be an integral domain, $P \in R|X]$, $\Bbbk$ be the quotient field of R, $r \geqslant 2$ be an integer, and let $P = a \prod_{i=1}^{r} (X - \alpha_i)$ be the factorization of P in an algebraic closure $\bar{\Bbbk}$ of $\Bbbk$. The *discriminant* of P is defined by[8]

$$\mathrm{disc}(P) = a^{2r-2} \prod_{k=1}^{r} \prod_{\ell=k+1}^{r} (\alpha_k - \alpha_\ell)^2.$$

The following examples are important in practice.

Example 7.3 Let p be an odd prime number, $a, b, c \in \mathbb{Z}$ and $m \geqslant 2$ be an integer.

▷ $\mathrm{disc} \left(aX^3 + bX^2 + cX + d \right) = -4b^3 d + (bc)^2 + 18abcd - 4ac^3 - 27(ad)^2$.
▷ $\mathrm{disc} \left(X^4 - 2(a+b)X^2 + (a-b)^2 \right) = \{64ab(a-b)\}^2$.
▷ $\mathrm{disc} \left(X^4 - 2bX^2 + b^2 - ac^2 \right) = \left(16ac^2 \right)^2 \left(b^2 - ac^2 \right)$.
▷ $\mathrm{disc} \left(X^m + aX + b \right) = (-1)^{m(m-1)/2} \{ (-1)^{m-1}(m-1)^{m-1}a^m + m^m b^{m-1} \}$.
▷ $\mathrm{disc} \left(X^{p-1} + \cdots + X + 1 \right) = (-1)^{(p-1)/2} p^{p-2}$.

---

[8]This definition comes from the theory of *resultants*. The word "discriminant" indicates that $\mathrm{disc}(P)$ does not vanish if all the roots α_i are distinct so that $\mathrm{disc}(P)$ *discriminates* the roots of P.

7.2.3   Algebraic Numbers

7.2.3.1   Field Extensions

Let $\mathbb{K}/\mathbb{k}$ be a field extension and $\alpha \in \mathbb{K}$. Define the homomorphism

$$F_\alpha : \mathbb{k}[X] \longrightarrow \mathbb{K}$$
$$P \longmapsto P(\alpha).$$

Since $\mathbb{K}$ is a field, the image $\mathrm{Im}\, F_\alpha$, often denoted by $\mathbb{k}[\alpha]$, is a subring of $\mathbb{K}$ and is then an integral domain. We also have $\mathbb{k}[X]/\ker F_\alpha \simeq \mathbb{k}[\alpha]$ and since $\mathbb{k}[\alpha]$ is an integral domain, we infer that $\ker F_\alpha$ is a prime ideal by Lemma 7.10. There are two cases.

▷ $\ker F_\alpha = (0)$.
▷ $\ker F_\alpha \neq (0)$. In this case, since $\mathbb{k}$ is a field, the ring $\mathbb{k}[X]$ is a PID, so that we deduce that $\ker F_\alpha$ is a maximal ideal[9] and is then of the shape $\ker F_\alpha = (P_\alpha)$, where P_α is an irreducible polynomial over $\mathbb{k}$.

Definition 7.3

▷ The number α is said to be *algebraic* over $\mathbb{k}$ if $\ker F_\alpha \neq (0)$, otherwise α is called *transcendental* over $\mathbb{k}$. In other words, α is algebraic over $\mathbb{k}$ if there exists a non-zero irreducible polynomial P_α over $\mathbb{k}$ such that α is a root of P_α. One may choose this polynomial to be monic, in which case it is unique, and is called the *minimal polynomial* of α over $\mathbb{K}$, denoted by μ_α, and we have

$$\mathbb{k}(\alpha) \simeq \mathbb{k}[X]/(\mu_\alpha) \quad \text{and} \quad \mathbb{k}(\alpha) = \mathbb{k}[\alpha] = \{P(\alpha) : P \in \mathbb{k}[X]\}.$$

▷ The degree of μ_α is called the *degree* of α over $\mathbb{K}$, written $\deg \alpha$.
▷ The extension $\mathbb{K}/\mathbb{k}$ is an *algebraic extension* if every element of $\mathbb{K}$ is algebraic over $\mathbb{k}$, otherwise it is called a *transcendental extension*.
▷ The field $\mathbb{k}$ is said to be *algebraically closed* if every non-constant polynomial of $\mathbb{k}[X]$ has a root in $\mathbb{k}$. An *algebraic closure* of a field $\mathbb{k}$ is an algebraic extension $\overline{\mathbb{k}}/\mathbb{k}$ such that the field $\overline{\mathbb{k}}$ is algebraically closed.
▷ $\theta \in \mathbb{C}$ is called an *algebraic number* if θ is algebraic over $\mathbb{Q}$. The set of algebraic numbers is $\overline{\mathbb{Q}}$. If θ is not an algebraic number, then it is called a *transcendental number*.

---

[9]In a PID, every non-zero prime ideal is maximal. See [33, Théorème X.3.1] for instance.

Note that $\overline{\mathbb{Q}}$ is *countable*. Indeed, it is the set of the roots of $P \in \mathbb{Z}[X]$. We define the following map:

$$\Phi : \qquad \mathbb{Z}[X] \qquad \longrightarrow \qquad\qquad \mathbb{Z}_{\geqslant 1}$$

$$P = \textstyle\sum_{i=0}^{n} a_i X^i \longmapsto N = \deg P + \textstyle\sum_{i=0}^{n} |a_i|.$$

Given $N \in \mathbb{Z}_{\geqslant 1}$, there are only finitely many polynomials P such that $\Phi(P) = N$, so that $\mathbb{Z}[X]$ is countable as a countable union of finite sets, and hence $\overline{\mathbb{Q}}$ is countable. The set of transcendental numbers is then uncountable and has no structure since the sum or product of two transcendental numbers could be algebraic. For instance, e is shown to be transcendental[10] over $\mathbb{Q}$ but $e - e = 0$.

Let $\mathbb{K}/\mathbb{k}$ be an extension and $\alpha \in \mathbb{K}$. It can easily be seen that, if α is algebraic over $\mathbb{k}$ and if its minimal polynomial is of degree d, then $[\mathbb{k}(\alpha) : \mathbb{k}] = d$ and the set $\{1, \alpha, \ldots, \alpha^{d-1}\}$ is a $\mathbb{k}$-basis of $\mathbb{k}(\alpha)$.

Example 7.4 The numbers i, $\sqrt{d}$ with $d \in \mathbb{Z} \setminus \{0, 1\}$ squarefree, $\sqrt[3]{2}$, $\rho = \zeta_3$, $\theta = \sqrt[4]{5} + \sqrt{5}$ are algebraic over $\mathbb{Q}$. Indeed, the minimal polynomials are, respectively, $X^2 + 1$, $X^2 - d$, $X^3 - 2$, $X^2 + X + 1$ since they are monic and have no roots in $\mathbb{Q}$ so that they are irreducible over $\mathbb{Z}$, and the minimal polynomial of θ is $X^4 - 10X^2 - 20X + 20$ since it is monic and irreducible over $\mathbb{Z}$ via Eisenstein's criterion applied with $p = 5$. We deduce that

$$[\mathbb{Q}(i) : \mathbb{Q}] = 2, \quad \left[\mathbb{Q}(\sqrt{d}) : \mathbb{Q}\right] = 2, \quad \left[\mathbb{Q}(\sqrt[3]{2}) : \mathbb{Q}\right] = 3$$

$$[\mathbb{Q}(\rho) : \mathbb{Q}] = 2 \quad \text{and} \quad \left[\mathbb{Q}(\sqrt[4]{5} + \sqrt{5}) : \mathbb{Q}\right] = 4.$$

The following result is an easy consequence of Bézout's theorem in $\mathbb{Q}[X]$.

Lemma 7.4 *Let α be an algebraic number and μ_α be the minimal polynomial of α. If P satisfies $P(\alpha) = 0$, then μ_α divides P.*

Proof If μ_α does not divide P, then since μ_α is irreducible, we have $\gcd(\mu_\alpha, P) = 1$, so that there exist $U, V \in \mathbb{Q}[X]$ such that $U(X)\mu_\alpha(X) + V(X)P(X) = 1$. However, evaluating this identity at $X = \alpha$ gives a contradiction.   □

The next result relates finite and algebraic extensions.

Lemma 7.5 *Let $\mathbb{K}/\mathbb{k}$ be a field extension and $\theta \in \mathbb{K}$. Then $\mathbb{k}(\theta)$ is a finite-dimensional $\mathbb{k}$-vector space if and only if θ is algebraic over $\mathbb{k}$. In particular, every finite extension is algebraic.*

---

[10]Hermite, 1873. In 1882, using essentially the same ideas, Lindemann proved that π is also transcendental over $\mathbb{Q}$ and hence showed that squaring the circle is impossible.

Proof $\Bbbk(\theta)$ is a sub-$\Bbbk$-vector space of $\Bbb{K}$ generated by $1, \theta, \theta^2, \ldots, \theta^n, \ldots$. If $\Bbbk(\theta)$ is a finite-dimensional $\Bbbk$-vector space, then $\dim \Bbbk(\theta) = \deg \mu_\theta$ by above. Conversely, if $\Bbbk(\theta)$ is a finite-dimensional $\Bbbk$-vector space with dimension n, then the vectors $1, \theta, \ldots, \theta^n$ are $\Bbbk$-linearly dependent so that there exist $(a_0, \ldots, a_n) \in \Bbbk^{n+1}$ such that $a_n \theta^n + a_{n-1} \theta^{n-1} + \cdots + a_0 = 0$, and hence θ is a root of the non-zero polynomial $a_n X^n + a_{n-1} X^{n-1} + \cdots + a_0$, and therefore θ is algebraic over $\Bbbk$. $\square$

7.2.3.2   Number Fields

Lemma 7.5 leads to the following definition.

Definition 7.4 An *algebraic number field* is a finite extension of $\Bbb{Q}$, written $\Bbb{K}/\Bbb{Q}$. The dimension of $\Bbb{K}$ as a $\Bbb{Q}$-vector space is called the *degree* of $\Bbb{K}/\Bbb{Q}$ and is denoted by $[\Bbb{K} : \Bbb{Q}]$.

Hence if $\Bbb{K}$ is an algebraic number field, then $\Bbb{K} = \Bbb{Q}(\alpha_1, \ldots, \alpha_n)$ for finitely many algebraic numbers $\alpha_1, \ldots, \alpha_n$, as for instance, a $\Bbb{Q}$-basis of the $\Bbb{Q}$-vector space $\Bbb{K}$. The following result improves on this observation.

Lemma 7.6 *If α and β are algebraic numbers, then there exists an algebraic number θ such that*

$$\Bbb{Q}(\alpha, \beta) = \Bbb{Q}(\theta).$$

Proof Let μ_α and μ_β be the minimal polynomials of α and β. We want to show that we can find $q \in \Bbb{Q}$ such that $\theta = \alpha + q\beta$ and $\Bbb{Q}(\alpha, \beta) = \Bbb{Q}(\theta)$. If such a q exists, then clearly $\Bbb{Q}(\theta) \subseteq \Bbb{Q}(\alpha, \beta)$. Set $f(X) = \mu_\alpha(\theta - qX) \in \Bbb{Q}(\theta)[X]$. Since

$$f(\beta) = \mu_\alpha(\theta - q\beta) = \mu_\alpha(\alpha) = 0$$

we infer that β is a root of f. Now we choose q such that β is the only common root of f and μ_β. This can be done since only a finite number of choices of q are thus ruled out. Therefore $\gcd(f, \mu_\beta) = a(X - \beta)$ with $a \in \Bbb{C} \setminus \{0\}$. Then $a(X - \beta) \in \Bbb{Q}(\theta)[X]$ which implies that $a, a\beta \in \Bbb{Q}(\theta)$ and so $\beta \in \Bbb{Q}(\theta)$. Now $\theta = \alpha + q\beta \in \Bbb{Q}(\theta)$ which implies that $\alpha \in \Bbb{Q}(\theta)$ and hence $\Bbb{Q}(\alpha, \beta) \subseteq \Bbb{Q}(\theta)$, so that $\Bbb{Q}(\alpha, \beta) = \Bbb{Q}(\theta)$ as required. $\square$

This result may be generalized quite easily by induction to show that for a set $\alpha_1, \ldots, \alpha_n$ of algebraic numbers, there exists an algebraic number θ such that

$$\Bbb{Q}(\alpha_1, \ldots, \alpha_n) = \Bbb{Q}(\theta).$$

We deduce that any algebraic number field $\Bbb{K}$ can be written as[11]

$$\Bbb{K} = \Bbb{Q}(\theta) \tag{7.1}$$

---

[11] This result is sometimes called the *theorem of the primitive element*.

for some algebraic number θ.

Definition 7.5 Let $\mathbb{K} = \mathbb{Q}(\theta)$ be an algebraic number field and μ_θ be the minimal polynomial of the algebraic number θ, abbreviated to μ for convenience.

▷ The polynomial μ is called a *defining polynomial* of $\mathbb{K}$.
▷ The roots of μ, which are all distinct since μ is irreducible, are called the *conjugates* of θ. Hence if $\deg \mu = n$, then θ has n conjugates, including itself, sometimes denoted by $\theta = \theta_1, \theta_2, \ldots, \theta_n$.
▷ For $i \in \{1, \ldots, n\}$, the field $\mathbb{K}_i = \mathbb{Q}(\theta_i)$ is called a *conjugate field* of $\mathbb{K}$.
▷ The homomorphism $\sigma_i : \mathbb{K} \to \mathbb{C}$ defined by $\sigma_i(\theta) = \theta_i$ is injective and called an *embedding* of $\mathbb{K}$ into $\mathbb{C}$.
It has been seen above that the set $\{1, \theta, \ldots, \theta^{n-1}\}$ is a $\mathbb{Q}$-basis for $\mathbb{K} = \mathbb{Q}(\theta)$. Therefore if $\alpha = a_0 + a_1\theta + \cdots + a_{n-1}\theta^{n-1} \in \mathbb{K}$ with $a_i \in \mathbb{Q}$, then for all $i \in \{1, \ldots, n\}$, we get

$$\sigma_i(\alpha) = a_0 + a_1\theta_i + \cdots + a_{n-1}\theta_i^{n-1}.$$

▷ If $\sigma_i(\mathbb{K}) \subseteq \mathbb{R}$, which happens if and only if $\sigma_i(\theta) \in \mathbb{R}$, we say that σ_i is *real*, otherwise σ_i is *complex*. Since complex conjugation is an automorphism of $\mathbb{C}$, it follows that $\overline{\sigma_i}$ is an embedding σ_j of $\mathbb{K}$ for some j. Hence σ_i is real if and only if $\overline{\sigma_i} = \sigma_i$ and since $\overline{\overline{\sigma_i}} = \sigma_i$, the complex embeddings come in conjugate pairs. We may enumerate in such a way that the system of all embeddings is

$$\underbrace{\sigma_1, \ldots, \sigma_{r_1}}_{\text{real}}, \underbrace{\sigma_{r_1+1}, \overline{\sigma_{r_1+1}}, \ldots, \sigma_{r_1+r_2}, \overline{\sigma_{r_1+r_2}}}_{\text{complex}}.$$

Then $r_1 + 2r_2 = [\mathbb{K} : \mathbb{Q}]$, and the pair (r_1, r_2) is called the *signature* of $\mathbb{K}$. One may note that r_1 and $2r_2$ are also, respectively, the number of real and complex roots of μ.
▷ The algebraic number field $\mathbb{K}$ is called a *Galois number field* if, for all $\alpha \in \mathbb{K}$, the minimal polynomial μ_α has all its roots in $\mathbb{K}$. This is equivalent saying that all the conjugate fields of $\mathbb{K}$ are identical to $\mathbb{K}$.
▷ The set of all embeddings of a Galois number field $\mathbb{K}$ is a group called the *Galois group* of $\mathbb{K}$ and denoted by $\mathrm{Gal}\,(\mathbb{K}/\mathbb{Q})$.
▷ A Galois number field $\mathbb{K}$ is said to be *abelian* if $\mathrm{Gal}\,(\mathbb{K}/\mathbb{Q})$ is abelian and *cyclic* if $\mathrm{Gal}\,(\mathbb{K}/\mathbb{Q})$ is cyclic.

Remark 7.1 From Galois theory, we know that, if $\mathbb{K}/\mathbb{Q}$ is Galois, then

$$|\mathrm{Gal}\,(\mathbb{K}/\mathbb{Q})| = [\mathbb{K} : \mathbb{Q}] = n,$$

and the signature of $\mathbb{K}$ must be of the form $(n, 0)$ or $\left(0, \frac{1}{2}n\right)$.

For instance, if $d \in \mathbb{Z} \setminus \{0, 1\}$ is squarefree, then the number field $\mathbb{K} = \mathbb{Q}(\sqrt{d})$ is a cyclic number field with Galois group $\mathrm{Gal}\,(\mathbb{K}/\mathbb{Q}) = \{\mathrm{id}, \sigma\}$, where id is the identity and $\sigma(a + b\sqrt{d}) = a - b\sqrt{d}$. On the other hand, the algebraic number field $\mathbb{K} = \mathbb{Q}\left(\sqrt[3]{2}\right)$ is not Galois since the two conjugate fields $\mathbb{Q}\left(\rho\sqrt[3]{2}\right)$ and $\mathbb{Q}\left(\rho^2\sqrt[3]{2}\right)$ are distinct from $\mathbb{K}$.

When $\mathbb{K} = \mathbb{Q}(\theta)$ is not Galois, one may define the *Galois closure* $\mathbb{K}^s$ of $\mathbb{K}$ to be the intersection of all subfields of $\overline{\mathbb{Q}}$, which are Galois and contain $\mathbb{K}$. This is also the splitting field of μ_θ, i.e. the field obtained by adjoining to $\mathbb{Q}$ all the roots of μ_θ. For instance, if $\mathbb{K} = \mathbb{Q}\left(\sqrt[3]{2}\right)$, then $\mathbb{K}^s = \mathbb{Q}(\sqrt[3]{2}, \rho)$.

7.2.4 The Ring of Integers

7.2.4.1 Algebraic Integers

Definition 7.6 A number $\alpha \in \mathbb{C}$ is an *algebraic integer* if there is a *monic* polynomial $P \in \mathbb{Z}[X]$ such that $P(\alpha) = 0$.

Example 7.5

▷ $\sqrt{-2}$ is an algebraic integer since it is a root of $X^2 + 2$.
▷ $\frac{1}{3}$ is not an algebraic integer.
▷ The number $\frac{1}{3}\sqrt{2}$ is an algebraic number since it has a root of $9X^2 - 2$, but it is not an algebraic integer. Indeed, assume the contrary. Then there exists a monic polynomial $P = X^n + \sum_{i=0}^{n-1} a_i X^i$ with $a_i \in \mathbb{Z}$ such that $P\left(\frac{1}{3}\sqrt{2}\right) = 0$. Clearing out the denominators we get

$$(\sqrt{2})^n + a_{n-1} \times 3 \times (\sqrt{2})^{n-1} + \cdots + a_0 \times 3^n = 0,$$

which implies that $3 \mid 2^{n/2}$ if n is even and $3 \mid 2^{(n-1)/2}$ if n is odd, which is false in either case. Hence $\frac{1}{3}\sqrt{2}$ is not an algebraic integer.

If α is an algebraic number of degree n, then the set $\{1, \alpha, \ldots, \alpha^{n-1}\}$ is a $\mathbb{Q}$-basis for $\mathbb{Q}(\alpha)$. Thus one may think of the structure of $\mathbb{Z}[\alpha] = \{P(\alpha) : P \in \mathbb{Z}[X]\}$ to try to determine whether α is an algebraic integer. This is the idea underlined in the following criterion.

Proposition 7.5 *Let $\alpha \in \mathbb{C}$. The following assertions are equivalent:*

▷ α *is an algebraic integer.*
▷ *The monic minimal polynomial μ_α lies in $\mathbb{Z}[X]$.*
▷ $\mathbb{Z}[\alpha]$ *is a finitely generated $\mathbb{Z}$-module.*
▷ *There exists a non-zero finitely generated $\mathbb{Z}$-module M such that $\alpha M \subseteq M$.*

Proof

▷ Let $P \in \mathbb{Z}[X]$ be a monic polynomial such that $P(\alpha) = 0$. By Lemma 7.4, we have $P = \mu_\alpha Q$ for some $Q \in \mathbb{Q}[X]$. We write $\mu_\alpha = (a/b)\mu^*$ with $a, b \in \mathbb{Z}$

and $\mu^* \in \mathbb{Z}[x]$ whose coefficients are coprime, and similarly $Q = (c/d)Q^*$ with $c, d \in \mathbb{Z}$ and $Q^* \in \mathbb{Z}[x]$ whose coefficients are also coprime. We infer that $bd\,P = ac\,\mu^*Q^*$, and Gauss's Lemma 7.1 implies that $bd = \pm ac$, so that $P = \pm\mu^*Q^*$. Therefore the leading coefficient of both μ^* and Q^* is ± 1 and since $\mu^*(\alpha) = 0$, we finally get $\mu_\alpha = \pm\mu^* \in \mathbb{Z}[X]$.

▷ Let $\mu_\alpha = X^n + a_{n-1}X^{n-1} + \cdots + a_0 \in \mathbb{Z}[X]$. It is sufficient to show that $\{1, \alpha, \ldots, \alpha^{n-1}\}$ generates $\mathbb{Z}[\alpha]$ as a $\mathbb{Z}$-module, i.e. for all $m \in \mathbb{Z}_{\geqslant 1}$, α^m is a linear combination of $\{1, \alpha, \ldots, \alpha^{n-1}\}$. The result is clear for $m < n$, so assume $m \geqslant n$ and suppose this holds for α^j with $j < m$. Then we have

$$\alpha^m = \alpha^{m-n}\alpha^n = \alpha^{m-n}\left(-a_0 - a_1\alpha - \cdots - a_{n-1}\alpha^{n-1}\right) = \sum_{j=0}^{n-1} \alpha^j \left(-\alpha^{m-n}a_j\right),$$

and the result follows by induction hypothesis.

▷ Obvious by choosing $M = \mathbb{Z}[\alpha]$.

▷ Let $m_1, \ldots, m_r$ be generators of M. Since $\alpha M \subseteq M$, we obtain the existence of $a_{ij} \in \mathbb{Z}$, with $(i, j) \in \{1, \ldots, r\}^2$, such that, for all $i \in \{1, \ldots, r\}$

$$\alpha m_i = \sum_{j=1}^{r} a_{ij}m_j$$

holds. Set $B \in M_r(\mathbb{Z})$ the matrix with entries $b_{ij} = a_{ij} - \alpha\delta_{ij}$, where $\delta_{ij} = 1$ if $i = j$, and 0 otherwise. Since M is non-zero, not all m_i can vanish and we infer that $\det B = 0$. Expanding this determinant yields an equation of the form $P(\alpha) = 0$ with $P \in \mathbb{Z}[X]$ monic. $\qquad\square$

7.2.4.2 The Ring $O_\mathbb{K}$

Corollary 7.3 (The Ring $O_\mathbb{K}$) *Let $\mathbb{K}$ be an algebraic number field and define $O_\mathbb{K}$ to be the set of all algebraic integers in $\mathbb{K}$. Then $O_\mathbb{K}$ is a ring called the ring of integers of $\mathbb{K}$.*

Proof By Proposition 7.5 we know that for $\alpha, \beta \in O_\mathbb{K}$, one can choose non-zero finitely generated $\mathbb{Z}$-modules M, N satisfying $\alpha M \subseteq M$ and $\beta N \subseteq N$. The $\mathbb{Z}$-module MN is finitely generated and non-zero, and we have

$$(\alpha \pm \beta)\,MN \subseteq MN \quad \text{and} \quad (\alpha\beta)\,MN \subseteq MN$$

so that $\alpha \pm \beta \in O_\mathbb{K}$ and $\alpha\beta \in O_\mathbb{K}$ by Proposition 7.5. $\qquad\square$

The next result is a refinement of (7.1).

Corollary 7.4 *Let $\mathbb{K}$ be an algebraic number field. Then there exists $\theta \in O_{\mathbb{K}}$ such that*

$$\mathbb{K} = \mathbb{Q}(\theta).$$

Proof By (7.1) there exists an algebraic number α such that $\mathbb{K} = \mathbb{Q}(\alpha)$. Let

$$\mu_\alpha = X^n + a_{n-1} X^{n-1} + \cdots + a_0 \in \mathbb{Q}[X]$$

and choose $d \in \mathbb{Z} \backslash \{0\}$ such that $da_i \in \mathbb{Z}$ for all $i \in \{0, \ldots, n-1\}$. Then the number $d\alpha$ is a root of a monic polynomial of $\mathbb{Z}[X]$ and hence $d\alpha \in O_{\mathbb{K}}$ by Proposition 7.5. Now it is clear that $\mathbb{Q}(d\alpha) = \mathbb{Q}(\alpha)$, which completes the proof by choosing $\theta = d\alpha$. □

7.2.4.3 The Index of $O_{\mathbb{K}}$

The ring $O_{\mathbb{K}}$ is the most important object of the theory, for it contains all the arithmetic information of $\mathbb{K}$. However, one must be careful with the following problem. Let $\mathbb{K} = \mathbb{Q}(\theta)$ with $\theta \in O_{\mathbb{K}}$. Then $\mathbb{Z}[\theta] \subseteq O_{\mathbb{K}}$, but it may be possible that $\mathbb{Z}[\theta] \neq O_{\mathbb{K}}$. For instance, $\mathbb{K} = \mathbb{Q}(\sqrt{-3})$ is an algebraic number field and $\sqrt{-3}$ is an algebraic integer. But $\rho = \zeta_3 \in \mathbb{K}$ and since it is a root of $X^2 + X + 1$, we have $\rho \in O_{\mathbb{K}}$, but $\rho \notin \mathbb{Z}[\sqrt{-3}]$.

   This leads to the following definition.

Definition 7.7 (Index) Let $\mathbb{K} = \mathbb{Q}(\theta)$ be an algebraic number field with $\theta \in O_{\mathbb{K}}$. The *index* of θ in $O_{\mathbb{K}}$ is the number $f = [O_{\mathbb{K}} : \mathbb{Z}[\theta]]$.

The arithmetic structure of f plays an important role in certain results of the theory. Dedekind provided the following criterion to determine whether a prime number p is not a divisor of f. See [13] for instance.

Proposition 7.6 (Dedekind) *Let $\mathbb{K} = \mathbb{Q}(\theta)$ be an algebraic number field with $\theta \in O_{\mathbb{K}}$, p be a prime number and $f = [O_{\mathbb{K}} : \mathbb{Z}[\theta]]$. Suppose that the decomposition of $\mu_\theta \in \mathbb{Z}[X]$ in $\mathbb{F}_p[X]$ is of the form*

$$\overline{\mu_\theta} = \prod_{i=1}^{g} \overline{P}_i^{e_i},$$

*where $g, e_i \geqslant 1$ are integers and $\overline{P}_i$ is irreducible in $\mathbb{F}_p[X]$ for all $i \in \{1, \ldots, g\}$.
Set*

$$T(X) = \frac{1}{p} \left\{ \mu_\theta(X) - \prod_{i=1}^{g} P_i^{e_i}(X) \right\}.$$

Then we have

$$p \nmid f \iff e_i = 1 \text{ or } \overline{P}_i \nmid \overline{T} \text{ in } \mathbb{F}_p[X] \quad (1 \leqslant i \leqslant g).$$

Diaz y Diaz [14] restated Dedekind's criterion as follows.

Proposition 7.7 (Diaz y Diaz) *With the notation of* Proposition 7.6, *for all* $i \in \{1, \ldots, g\}$, *let* R_i *be the remainder of the Euclidean division of* μ_θ *by* P_i. *Then we have*

$$p \nmid f \iff R_i \notin p^2 \mathbb{Z}[X] \quad \text{for all } i \in \{1, \ldots, g\} \text{ such that } e_i \geqslant 2.$$

We will prove in Exercise 137 the following third criterion when the minimal polynomial of θ is a p-Eisenstein, i.e. μ_θ satisfies the conditions of Corollary 7.1.

Proposition 7.8 *Let* $\mathbb{K} = \mathbb{Q}(\theta)$ *be an algebraic number field of degree n with $\theta \in O_{\mathbb{K}}$ and write*

$$\mu_\theta = X^n + a_{n-1} X^{n-1} + \cdots + a_0 \in \mathbb{Z}[X].$$

Let p be a prime number such that $p \mid a_i$ for all $0 \leqslant i \leqslant n - 1$ and $p^2 \nmid a_0$. Then $p \nmid f$.

7.2.5 Integral Bases

Let $\mathbb{K} = \mathbb{Q}(\theta)$ be an algebraic number field of degree n with $\theta \in O_{\mathbb{K}}$. We denote by $\theta_1, \ldots, \theta_n$ the conjugates of θ and $\sigma_1, \ldots, \sigma_n$ the embeddings of $\mathbb{K}$ in $\mathbb{C}$. It is a matter of fact that all the θ_i have the same minimal polynomial μ_θ.

7.2.5.1 Norm and Trace

We first define the following two numbers.

Definition 7.8 (Norm and Trace) Let $\alpha \in \mathbb{K}$. The *norm* and *trace* of α are defined by

$$N_{\mathbb{K}/\mathbb{Q}}(\alpha) = \prod_{i=1}^{n} \sigma_i(\alpha) \quad \text{and} \quad \text{Tr}_{\mathbb{K}/\mathbb{Q}}(\alpha) = \sum_{i=1}^{n} \sigma_i(\alpha).$$

Since σ_i are homomorphisms, we clearly have the following rules. Let $\alpha, \beta \in \mathbb{K}$.

▷ $N_{\mathbb{K}/\mathbb{Q}}(\alpha\beta) = N_{\mathbb{K}/\mathbb{Q}}(\alpha) N_{\mathbb{K}/\mathbb{Q}}(\beta)$ and $\text{Tr}_{\mathbb{K}/\mathbb{Q}}(\alpha + \beta) = \text{Tr}_{\mathbb{K}/\mathbb{Q}}(\alpha) + \text{Tr}_{\mathbb{K}/\mathbb{Q}}(\beta)$.
▷ $N_{\mathbb{K}/\mathbb{Q}}(1) = 1$ and $\text{Tr}_{\mathbb{K}/\mathbb{Q}}(1) = n$.
▷ For all $q \in \mathbb{Q}$, we have $N_{\mathbb{K}/\mathbb{Q}}(q\alpha) = q^n N_{\mathbb{K}/\mathbb{Q}}(\alpha)$ and $\text{Tr}_{\mathbb{K}/\mathbb{Q}}(q\alpha) = q\,\text{Tr}_{\mathbb{K}/\mathbb{Q}}(\alpha)$.

The next result shows that the norm and trace are rational numbers.

Proposition 7.9 *Let $\alpha \in \mathbb{K}$. Then $N_{\mathbb{K}/\mathbb{Q}}(\alpha) \in \mathbb{Q}$ and $\text{Tr}_{\mathbb{K}/\mathbb{Q}}(\alpha) \in \mathbb{Q}$. Furthermore, if $\alpha \in O_{\mathbb{K}}$, then $N_{\mathbb{K}/\mathbb{Q}}(\alpha) \in \mathbb{Z}$ and $\text{Tr}_{\mathbb{K}/\mathbb{Q}}(\alpha) \in \mathbb{Z}$.*

Proof We define the so-called *characteristic polynomial* of α

$$C_\alpha = \prod_{i=1}^{n} (X - \sigma_i(\alpha)) = X^n + b_{n-1}X^{n-1} + \cdots + b_0 \qquad (7.2)$$

so that $N_{\mathbb{K}/\mathbb{Q}}(\alpha) = (-1)^n b_0$ and $\text{Tr}_{\mathbb{K}/\mathbb{Q}}(\alpha) = -b_{n-1}$. Hence it suffices to show that $C_\alpha \in \mathbb{Q}[X]$. We first note that, if $\alpha = Q(\theta)$ for some $Q \in \mathbb{Q}[X]$, then $\sigma_i(\alpha) = Q(\theta_i)$ for all $i \in \{1, \ldots, n\}$. Then we may write

$$C_\alpha = \prod_{i} (X - Q(\theta_i)),$$

where θ_i runs through all the roots of the minimal polynomial μ_θ of θ whose coefficients are in $\mathbb{Z}$. Expanding the product we see that the coefficients of C_α are of the form $R(\theta_1, \ldots, \theta_n)$, where $R \in \mathbb{Q}[X_1, \ldots, X_n]$ is a symmetric polynomial. Therefore we have $C_\alpha \in \mathbb{Q}[X]$. The second part of the proposition follows from the following assertion

$$C_\alpha \text{ is a power of } \mu_\alpha. \qquad (7.3)$$

Indeed, if (7.3) is true and if $\alpha \in O_{\mathbb{K}}$, then $\mu_\alpha \in \mathbb{Z}[X]$ and $C_\alpha \in \mathbb{Z}[X]$ by Lemma 7.1. The rest of the text is devoted to the proof of (7.3). Since μ_α is irreducible, by factorizing C_α into irreducibles, we have $C_\alpha = \mu_\alpha^r P$ with μ_α and P are coprime and both monic. If P is not constant, then there exists $i \in \{1, \ldots, n\}$ such that $\sigma_i(\alpha)$ is a root of P, and hence θ_i is a root of the polynomial $P \circ Q$. By Lemma 7.4, we infer that $\mu_\theta \mid P \circ Q$, which implies in particular that $P \circ Q(\theta) = 0$, and thus

$$P(\alpha) = P \circ Q(\theta) = 0$$

so that $\mu_\alpha \mid P$, which is impossible since μ_α and P are coprime. We deduce that P is constant and monic, so that $P = 1$ and then $C_\alpha = \mu_\alpha^r$. □

Example 7.6 The norm and trace can sometimes be useful to determine the ring of integers and the index of some algebraic number fields. For instance, let $\mathbb{K} = \mathbb{Q}(\sqrt{-5})$. This is an algebraic number field of degree 2 by Example 7.4 with embeddings $\{\mathrm{id}, \sigma\}$, where $\sigma(a + b\sqrt{-5}) = a - b\sqrt{-5}$ for all $a, b \in \mathbb{Q}$. Hence if $\alpha = a + b\sqrt{-5} \in \mathbb{K}$, then

$$N_{\mathbb{K}/\mathbb{Q}}(\alpha) = a^2 + 5b^2 \quad \text{and} \quad \mathrm{Tr}_{\mathbb{K}/\mathbb{Q}}(\alpha) = 2a.$$

Assume now that $\alpha \in O_\mathbb{K}$. By Proposition 7.9, we have $2a \in \mathbb{Z}$ and $a^2 + 5b^2 \in \mathbb{Z}$, which implies that the denominator of a, and hence also of b, is at most 2. Writing $a = c/2$ and $b = d/2$, we must have $\frac{1}{4}(c^2 + 5d^2) \in \mathbb{Z}$, or equivalently $c^2 + 5d^2 \equiv 0 \pmod 4$. Since all squares are $\equiv 0$ or $1 \pmod 4$, we infer that c and d are even, and hence $a, b \in \mathbb{Z}$. Therefore $O_\mathbb{K} = \mathbb{Z}[\sqrt{-5}]$ and $f = 1$.

7.2.5.2 Integral Bases

It has been seen above that we can choose a $\mathbb{Q}$-basis $\{\alpha_1, \ldots, \alpha_n\}$ of $\mathbb{K}$ as a vector space over $\mathbb{Q}$. Now since $O_\mathbb{K}$ is a $\mathbb{Z}$-module, we may ask for a $\mathbb{Z}$-basis for $O_\mathbb{K}$.

Definition 7.9 (Integral Bases) Let $\mathbb{K}$ be an algebraic number field of degree n and $O_\mathbb{K}$ be its ring of integers. Then a $\mathbb{Z}$-basis of $O_\mathbb{K}$ is called an *integral basis* for $\mathbb{K}$. In other words, $\{\alpha_1, \ldots, \alpha_m\}$ is an integral basis if and only if $\alpha_i \in O_\mathbb{K}$ and every element of $O_\mathbb{K}$ can be uniquely expressed in the form

$$\sum_{i=1}^{m} a_i \alpha_i \quad \text{with} \quad a_i \in \mathbb{Z}.$$

By Corollary 7.4, it follows that an integral basis is a $\mathbb{Q}$-basis for $\mathbb{K}$, so that we have $m = n$.

7.2.5.3 Discriminants

Now one may wonder if such bases exist. If $\mathbb{K} = \mathbb{Q}(\theta)$, then a natural candidate is $\{1, \theta, \ldots, \theta^{n-1}\}$, which is a $\mathbb{Q}$-basis for $\mathbb{K}$ with $\theta \in O_\mathbb{K}$ and we may indeed take this basis if $O_\mathbb{K} = \mathbb{Z}[\theta]$, but we have seen above that this equality may be false. Thus we need more work to establish the existence of integral bases for *all* algebraic number fields. One possible proof[12] uses the following very important invariant of $\mathbb{K}$.

---

[12] See also [24] for another proof using the properties of sub-$\mathbb{Z}$-modules of finitely generated $\mathbb{Z}$-modules.

Definition 7.10 (Discriminants) Let $\mathbb{K}$ be an algebraic number field of degree n and $\alpha_1, \ldots, \alpha_n \in \mathbb{K}$.

▷ The *discriminant* of $\alpha_1, \ldots, \alpha_n$ is the number defined by

$$\Delta_{\mathbb{K}/\mathbb{Q}}(\alpha_1, \ldots, \alpha_n) = \left(\det(\sigma_i(\alpha_j))\right)^2,$$

where $\left(\sigma_i(\alpha_j)\right)_{i,j}$ is the matrix with entries $\sigma_i(\alpha_j)$ in the ith row and jth column.

▷ If $\{\alpha_1, \ldots, \alpha_n\}$ is an *integral basis* for $\mathbb{K}$, then $\Delta_{\mathbb{K}/\mathbb{Q}}(\alpha_1, \ldots, \alpha_n)$ is independent of the choice of that basis. It is called the *discriminant* of $\mathbb{K}$ and is denoted by $d_{\mathbb{K}}$.

When $(\alpha_1, \ldots, \alpha_n) = \left(1, \theta, \ldots, \theta^{n-1}\right)$, one may check that we have

$$\Delta_{\mathbb{K}/\mathbb{Q}}\left(1, \theta, \ldots, \theta^{n-1}\right) = \prod_{i=1}^{n} \prod_{j=i+1}^{n} \left(\sigma_j(\theta) - \sigma_i(\theta)\right)^2 = \mathrm{disc}(\mu_\theta) \qquad (7.4)$$

by Definition 7.2, which explains the word "discriminant". The next result provides the first basic properties of the discriminants.

Proposition 7.10 *Let* $\mathbb{K} = \mathbb{Q}(\theta)$ *be an algebraic number field of degree n with* $\theta \in O_{\mathbb{K}}$.

▷ *Let* $\alpha_1, \ldots, \alpha_n \in \mathbb{K}$. *Then we have*

$$\Delta_{\mathbb{K}/\mathbb{Q}}(\alpha_1, \ldots, \alpha_n) = \det\left(\mathrm{Tr}_{\mathbb{K}/\mathbb{Q}}\left(\alpha_i \alpha_j\right)\right).$$

▷ *If* $\{\alpha_1, \ldots, \alpha_n\}$ *is a $\mathbb{Q}$-basis for $\mathbb{K}$, then* $\Delta_{\mathbb{K}/\mathbb{Q}}(\alpha_1, \ldots, \alpha_n) \in \mathbb{Q}$. *Furthermore, if* $\alpha_i \in O_{\mathbb{K}}$, *then* $\Delta_{\mathbb{K}/\mathbb{Q}}(\alpha_1, \ldots, \alpha_n) \in \mathbb{Z}$.
In particular, we have $d_{\mathbb{K}} \in \mathbb{Z}$.

▷ *Let* $\alpha_1, \ldots, \alpha_n \in \mathbb{K}$. *Then we have*

$$\Delta_{\mathbb{K}/\mathbb{Q}}(\alpha_1, \ldots, \alpha_n) = 0 \iff \text{the } \alpha_i \text{ are } \mathbb{Q}\text{-linearly dependent}.$$

In particular, the discriminant of any $\mathbb{Q}$-basis for $\mathbb{K}$ is a non-zero rational number.

▷ *Let* $\{\alpha_1, \ldots, \alpha_n\}$ *and* $\{\beta_1, \ldots, \beta_n\}$ *be two $\mathbb{Q}$-basis for $\mathbb{K}$ such that* $\beta_i = \sum_{j=1}^{n} a_{ij} \alpha_j$ *with* $a_{ij} \in \mathbb{Q}$. *Then we have*

$$\Delta_{\mathbb{K}/\mathbb{Q}}(\beta_1, \ldots, \beta_n) = (\det(a_{ij}))^2 \Delta_{\mathbb{K}/\mathbb{Q}}(\alpha_1, \ldots, \alpha_n).$$

Proof

▷ Consider the matrix $M = (\sigma_i(\alpha_j)) \in \mathcal{M}_n(\mathbb{C})$. Then we have $M^T M = (m_{ij})$ with

$$m_{ij} = \sum_{k=1}^{n} \sigma_k(\alpha_i) \sigma_k(\alpha_j) = \mathrm{Tr}_{\mathbb{K}/\mathbb{Q}}\left(\alpha_i \alpha_j\right),$$

and we conclude by using the facts that $\det M^T = \det M$ and $\det(AB) = \det A \det B$.

▷ This follows readily from above and Proposition 7.9.

▷ If α_i are $\mathbb{Q}$-linearly dependent, then so are the columns of the matrix M defined above since $\mathbb{Q}$ is invariant by σ_i. Conversely, assume that $\Delta_{\mathbb{K}/\mathbb{Q}}(\alpha_1, \ldots, \alpha_n) = 0$. This implies that $\ker M^T M \neq \{0\}$ and since $M^T M$ has entries in $\mathbb{Q}$, there exists $q_i \in \mathbb{Q}$ such that, for all j, $\mathrm{Tr}_{\mathbb{K}/\mathbb{Q}}(x\alpha_j) = 0$ with $x = \sum_{i=1}^{n} q_i \alpha_i \neq 0$. If α_i are $\mathbb{Q}$-linearly independent, they generate $\mathbb{K}$ as a $\mathbb{Q}$-vector space and we have $\mathrm{Tr}_{\mathbb{K}/\mathbb{Q}}(xy) = 0$ for all $y \in \mathbb{K}$ with $x \neq 0$. But taking $y = x^{-1}$ gives $0 = \mathrm{Tr}_{\mathbb{K}/\mathbb{Q}}(1) = n$, which is impossible. Then α_i are $\mathbb{Q}$-linearly dependent.

▷ Setting $M = (\sigma_i(\alpha_j))$, $N = (\sigma_i(\beta_j))$ and $A = (a_{ij})$, we infer that

$$\Delta_{\mathbb{K}/\mathbb{Q}}(\beta_1, \ldots, \beta_n) = (\det N)^2 = (\det MA^T)^2$$

$$= (\det A)^2 (\det M)^2$$

$$= (\det(a_{ij}))^2 \Delta_{\mathbb{K}/\mathbb{Q}}(\alpha_1, \ldots, \alpha_n)$$

as asserted.

□

We are now in a position to answer the question of the existence of an integral basis in any algebraic number field.

Corollary 7.5 *Let* $\mathbb{K} = \mathbb{Q}(\theta)$ *be an algebraic number field of degree n, and let* $O_{\mathbb{K}}$ *be its ring of integers. Then* $O_{\mathbb{K}}$ *has an integral basis. Furthermore,* $O_{\mathbb{K}}$ *is a free* $\mathbb{Z}$*-module of rank n.*

Proof By Corollary 7.4, there exists a $\mathbb{Q}$-basis $\{\alpha_1, \ldots, \alpha_n\}$ for $\mathbb{K}$ with $\alpha_i \in O_{\mathbb{K}}$. It remains to show that there exists such a basis, which is a $\mathbb{Z}$-basis for $O_{\mathbb{K}}$. By Proposition 7.10, the discriminants of such bases are in $\mathbb{Z} \setminus \{0\}$. Thus we may choose a basis $\{\alpha_1, \ldots, \alpha_n\}$ with $\alpha_i \in O_{\mathbb{K}}$ and discriminant, abbreviated here in $\Delta_{\mathbb{K}/\mathbb{Q}}(\alpha)$, such that $|\Delta_{\mathbb{K}/\mathbb{Q}}(\alpha)|$ is minimal. Suppose that this basis is not a $\mathbb{Z}$-basis for $O_{\mathbb{K}}$. Then there exists $\beta \in O_{\mathbb{K}}$ such that $\beta = \sum_{i=1}^{n} q_i \alpha_i$ with $q_i \in \mathbb{Q}$ and at least one of these q_i is not an integer. Without loss of generality, assume that $q_1 \notin \mathbb{Z}$ and write $q_1 = \lfloor q_1 \rfloor + \{q_1\}$ with $0 < \{q_1\} < 1$. The matrix

$$\begin{pmatrix} \{q_1\} & q_2 & \cdots & q_n \\ 0 & 1 & \cdots & 0 \\ \vdots & \vdots & \vdots & \vdots \\ 0 & 0 & \cdots & 1 \end{pmatrix}$$

is non-singular since $\det A = \{q_1\} \neq 0$ so that $\{\gamma, \alpha_2, \ldots, \alpha_n\}$, with $\gamma = \beta - \lfloor q_1 \rfloor \alpha_1$, is a $\mathbb{Q}$-basis for $\mathbb{K}$ by Proposition 7.10, and also

$$\left| \Delta_{\mathbb{K}/\mathbb{Q}}(\gamma, \alpha_2, \ldots, \alpha_n) \right| = \{q_1\}^2 \left| \Delta_{\mathbb{K}/\mathbb{Q}}(\alpha) \right| < \left| \Delta_{\mathbb{K}/\mathbb{Q}}(\alpha) \right|,$$

which contradicts the minimality of $\left|\Delta_{\mathbb{K}/\mathbb{Q}}(\alpha)\right|$. Hence as a $\mathbb{Z}$-module, we get

$$O_{\mathbb{K}} = \mathbb{Z}\alpha_1 \oplus \mathbb{Z}\alpha_2 \oplus \cdots \oplus \mathbb{Z}\alpha_n,$$

which concludes the proof.   □

7.2.6   Theorems for $O_{\mathbb{K}}$

The determination of an explicit integral basis and of the discriminant of $O_{\mathbb{K}}$ is not an easy task. Our aim here is to collect some tools which can sometimes be helpful.

In what follows, $\mathbb{K} = \mathbb{Q}(\theta)$ is an algebraic number field of degree n with $\theta \in O_{\mathbb{K}}$, discriminant $d_{\mathbb{K}}$, signature (r_1, r_2). Let $f = [O_{\mathbb{K}} : \mathbb{Z}[\theta]]$ be the index of θ in $O_{\mathbb{K}}$ and $\mu_\theta \in \mathbb{Z}[X]$ be the minimal polynomial of $\mathbb{K}$. For convenience, we denote by $\Delta_{\mathbb{K}/\mathbb{Q}}(\theta)$ the discriminant of the $\mathbb{Q}$-basis $\{1, \theta, \ldots, \theta^{n-1}\}$.

7.2.6.1   Relations Between Discriminants

Theorem 7.2 *Let* $\{\beta_1, \ldots, \beta_n\}$ *be a* $\mathbb{Q}$-*basis for* $\mathbb{K}$ *such that* $\beta_i \in O_{\mathbb{K}}$. *Then*

$$\Delta_{\mathbb{K}/\mathbb{Q}}(\beta_1, \ldots, \beta_n) = [O_{\mathbb{K}} : N]^2 \times d_{\mathbb{K}},$$

where $N = \mathbb{Z}\beta_1 + \cdots + \mathbb{Z}\beta_n$. *In particular, we have*

$$\mathrm{disc}(\mu_\theta) = f^2 \times d_{\mathbb{K}}.$$

Proof Let $\{\alpha_1, \ldots, \alpha_n\}$ be an integral basis for $\mathbb{K}$. As a sub-$\mathbb{Z}$-module of the free $\mathbb{Z}$-module $O_{\mathbb{K}}$, $\mathbb{Z}[\theta]$ is a free $\mathbb{Z}$-module and thus has a basis $\{\gamma_1, \ldots, \gamma_n\}$ such that $\gamma_i = \sum_{j=1}^{n} m_{ij}\alpha_j$ with $m_{ij} \in \mathbb{Z}$. By Propositions 7.10 and 7.1 applied with $G = O_{\mathbb{K}}$ and $H = N$, we have

$$\Delta_{\mathbb{K}/\mathbb{Q}}(\gamma_1, \ldots, \gamma_n) = (\det(m_{ij}))^2 \times d_{\mathbb{K}} = [O_{\mathbb{K}} : N]^2 \times d_{\mathbb{K}}.$$

Now by Proposition 7.10, the discriminants $\Delta_{\mathbb{K}/\mathbb{Q}}(\beta_1, \ldots, \beta_n)$ and $\Delta_{\mathbb{K}/\mathbb{Q}}(\gamma_1, \ldots, \gamma_n)$ are integers and, with Proposition 7.10 again, we infer that $\Delta_{\mathbb{K}/\mathbb{Q}}(\beta_1, \ldots, \beta_n)$ divides $\Delta_{\mathbb{K}/\mathbb{Q}}(\gamma_1, \ldots, \gamma_n)$ and $\Delta_{\mathbb{K}/\mathbb{Q}}(\gamma_1, \ldots, \gamma_n)$ divides $\Delta_{\mathbb{K}/\mathbb{Q}}(\beta_1, \ldots, \beta_n)$ and their signs are equal, so that

$$\Delta_{\mathbb{K}/\mathbb{Q}}(\gamma_1, \ldots, \gamma_n) = \Delta_{\mathbb{K}/\mathbb{Q}}(\beta_1, \ldots, \beta_n),$$

which gives the asserted result.

The assertion $\mathrm{disc}\,(\mu_\theta) = f^2 \times d_{\mathbb{K}}$ follows by applying this result to $\{1, \theta, \ldots, \theta^{n-1}\}$ and using (7.4). $\square$

7.2.6.2   Stickelberger's and Kronecker's Theorems

Theorem 7.3

▷ (Stickelberger). *Let* $\alpha_1, \ldots, \alpha_n \in O_{\mathbb{K}}$. *Then* $\Delta_{\mathbb{K}/\mathbb{Q}}\,(\alpha_1, \ldots, \alpha_n) \equiv 0$ *or* $1 \pmod{4}$.
▷ (Kronecker). *The sign of* $d_{\mathbb{K}}$ *is* $(-1)^{r_2}$.

Proof

▷ Expanding the determinant $\det(\sigma_i(\alpha_j))$ and using the $n!$ terms enables us to write

$$\det(\sigma_i(\alpha_j)) = P - N,$$

where P is the contribution of the terms corresponding to permutations of even signature and N is the contribution of the terms corresponding to odd permutations. Hence

$$\Delta_{\mathbb{K}/\mathbb{Q}}\,(\alpha_1, \ldots, \alpha_n) = (P - N)^2 = (P + N)^2 - 4PN.$$

Now since $\sigma_i(P + N) = P + N$ and $\sigma_i(PN) = PN$ we have $P + N, PN \in \mathbb{Q}$ by Galois theory, and we have in fact $P + N, PN \in \mathbb{Z}$ since $\alpha_i \in O_{\mathbb{K}}$. The result follows from the fact that a square is always congruent to 0 or 1 modulo 4.

▷ By (7.4) and Theorem 7.2 we have

$$d_{\mathbb{K}} = f^{-2} \prod_{i<j} (\sigma_j(\theta) - \sigma_i(\theta))^2 .$$

Now a case-by-case examination shows that when conjugate terms are paired, all the factors become positive except for

$$\prod_{r_1 < i \leqslant r_1 + r_2} (\sigma_{i+r_2}(\theta) - \sigma_i(\theta))^2$$

whose sign is $(-1)^{r_2}$ since $\sigma_{i+r_2}(\theta) - \sigma_i(\theta)$ is purely imaginary.

$\square$

7.2.6.3   A Formula with the Minimal Polynomial

Proposition 7.11   *We have*

$$\Delta_{\mathbb{K}/\mathbb{Q}}(\theta) = (-1)^{n(n-1)/2} N_{\mathbb{K}/\mathbb{Q}}\left(\mu_\theta'(\theta)\right) .$$

Proof Writing $\mu_\theta = \prod_{j=1}^n (X - \sigma_j(\theta))$ we get

$$\mu_\theta'(x) = \sum_{j=1}^n \frac{\mu_\theta(x)}{x - \sigma_j(\theta)}$$

and thus

$$\mu_\theta'(\sigma_i(\theta)) = \prod_{j \neq i} (\sigma_i(\theta) - \sigma_j(\theta)).$$

We deduce that

$$N_{\mathbb{K}/\mathbb{Q}}\left(\mu_\theta'(\theta)\right) = \prod_{i=1}^n \mu_\theta'(\sigma_i(\theta)) = \prod_{i=1}^n \prod_{j \neq i} (\sigma_i(\theta) - \sigma_j(\theta))$$

$$= \prod_{i=1}^n \prod_{j=i+1}^n \left(-(\sigma_i(\theta) - \sigma_j(\theta))^2 \right)$$

$$= (-1)^{n(n-1)/2} \Delta_{\mathbb{K}/\mathbb{Q}}(\theta)$$

as asserted.   □

7.2.6.4   Some Criteria for Integral Bases

Theorem 7.4

▷ Let $\{\beta_1, \ldots, \beta_n\}$ be a $\mathbb{Q}$-basis for $\mathbb{K}$ such that $\beta_i \in O_{\mathbb{K}}$. If $\Delta_{\mathbb{K}/\mathbb{Q}}(\beta_1, \ldots, \beta_n)$ is squarefree, then $\{\beta_1, \ldots, \beta_n\}$ is an integral basis for $\mathbb{K}$.

▷ If $\mathrm{disc}(\mu_\theta)$ is squarefree or if $\mathrm{disc}(\mu_\theta) = 4D$ with D squarefree and $D \not\equiv 1$ (mod 4), then $\left\{1, \theta, \ldots, \theta^{n-1}\right\}$ is an integral basis for $\mathbb{K}$ and $d_{\mathbb{K}} = \mathrm{disc}(\mu_\theta)$.

▷ Suppose that, for any prime number p such that $p^2 \mid \Delta_{\mathbb{K}/\mathbb{Q}}(\theta)$, the polynomial μ_θ is a p-Eisenstein.[13] Then we have

$$O_{\mathbb{K}} = \mathbb{Z}[\theta].$$

▷ Suppose that $f > 1$. Then there exists $\alpha \in O_{\mathbb{K}}$ of the form

$$\alpha = p^{-1}\left(a_0 + a_1\theta + \cdots + a_{n-1}\theta^{n-1} \right)$$

where p is a prime number such that $p^2 \mid \Delta_{\mathbb{K}/\mathbb{Q}}(\theta)$ and $a_i \in \mathbb{Z}$ such that $0 \leqslant a_i < p$ for all $i \in \{0, \ldots, n-1\}$.

---

[13] See Proposition 7.8.

Proof

▷ Let $\{\alpha_1, \ldots, \alpha_n\}$ be an integral basis for $\mathbb{K}$ so that there exist $a_{ij} \in \mathbb{Z}$ satisfying $\beta_i = \sum_{j=1}^{n} a_{ij}\alpha_j$. By Proposition 7.10, we have

$$\Delta_{\mathbb{K}/\mathbb{Q}}(\beta_1, \ldots, \beta_n) = (\det(a_{ij}))^2 \Delta_{\mathbb{K}/\mathbb{Q}}(\alpha_1, \ldots, \alpha_n)$$

and we infer that $|\det(a_{ij})| = 1$ since $\Delta_{\mathbb{K}/\mathbb{Q}}(\beta_1, \ldots, \beta_n)$ is squarefree. We conclude by using Proposition 7.1.

▷ If $\mathrm{disc}(\mu_\theta)$ is squarefree, then we apply the above result. If $\mathrm{disc}(\mu_\theta) = 4D$ with D as stated in the proposition, we first note that $f^2 \mid 4D$ by Theorem 7.3 and (7.4), and hence $f = 1$ or $f = 2$ since D squarefree. If $f = 2$, then we have $4D = 4d_{\mathbb{K}}$ so that $D = d_{\mathbb{K}}$ and hence $d_{\mathbb{K}} \not\equiv 1 \pmod 4$, which contradicts Theorem 7.3. Hence $f = 1$ and then $\mathrm{disc}(\mu_\theta) = d_{\mathbb{K}}$.

▷ By Proposition 7.8, we infer that f is not divisible by p for any prime p such that $p^2 \mid \Delta_{\mathbb{K}/\mathbb{Q}}(\theta)$. Using Theorem 7.3, we derive $f = 1$ as asserted.

▷ Since $f > 1$, there exist $p \mid f$ and $\overline{\beta} \in O_{\mathbb{K}}/\mathbb{Z}[\theta]$ of order p by Cauchy's theorem,[14] and thus $p\beta \in \mathbb{Z}[\theta]$. Since $\{1, \theta, \theta^2, \ldots, \theta^{n-1}\}$ is a $\mathbb{Z}$-basis for $\mathbb{Z}[\theta]$, we get

$$p\beta = b_0 + b_1\theta + \cdots + b_{n-1}\theta^{n-1}$$

with $b_i \in \mathbb{Z}$. Now writing $b_i = pq_i + a_i$, with $0 \leqslant a_i < p$ and $q_i \in \mathbb{Z}$, implies that the number $\alpha = \beta - \sum_{i=0}^{n-1} q_i\theta^i \in O_{\mathbb{K}}$ satisfies the conditions of the proposition. Furthermore, we have $p \mid f$ so that $p^2 \mid \Delta_{\mathbb{K}/\mathbb{Q}}(\theta)$ by the first result above.

□

7.2.6.5 Examples

Example 7.7 Let $\mathbb{K} = \mathbb{Q}(6^{1/3})$. The polynomial $P = X^3 - 6$ satisfies Eisenstein's criterion with respect to 2 and 3 and hence is irreducible over $\mathbb{Z}$, so that $P = \mu_\theta$. We have $\mathrm{disc}(P) = -2^2 \times 3^5$ and the use of Theorem 7.4 implies that $O_{\mathbb{K}} = \mathbb{Z}[\theta]$. Therefore $\mathbb{K}$ is an algebraic number field of degree 3, called a *pure cubic field*, signature $(1, 1)$ so that $\mathbb{K}$ is not Galois, discriminant $d_{\mathbb{K}} = -2^2 \times 3^5$ and $\{1, \theta, \theta^2\}$ is an integral basis for $\mathbb{K}$.

□

Example 7.8 Let θ be a root of the polynomial $P = X^4 + X + 1$. P is irreducible over $\mathbb{Z}$ using Proposition 7.56, since $|P(m)|$ is 1 or prime for $m \in \{-6, -3, 0, 1, 2, 5, 6, 9, 11\}$. Let $\mathbb{K} = \mathbb{Q}(\theta)$ be the corresponding algebraic number field. Using Example 7.3, we obtain $\mathrm{disc}(P) = 229$. Since 229 is prime, we have $O_{\mathbb{K}} = \mathbb{Z}[\theta]$ and $\{1, \theta, \theta^2, \theta^3\}$ is an integral basis for $\mathbb{K}$ by Theorem 7.4. Hence $\mathbb{K}$ is

---

[14] See Theorem 7.1.

an algebraic number field of degree 4, signature $(0, 2)$, discriminant $d_{\mathbb{K}} = 229$ and $\mathcal{O}_{\mathbb{K}} = \mathbb{Z}[\theta]$.   □

Example 7.9 Let $\mathbb{K} = \mathbb{Q}(\sqrt{8 + 3\sqrt{7}})$. The minimal polynomial is $\mu_\theta = X^4 - 16X^2 + 1$, whose discriminant is equal to

$$\text{disc}(P) = (16 \times 7 \times 3^2)^2 (8^2 - 7 \times 3^2) = 2^8 \times 3^4 \times 7^2$$

by Example 7.3. Hence the only prime factors of f are 2, 3 or 7. Proposition 7.7 gives $2 \nmid f$, $3 \mid f$ and $7 \nmid f$ so that $f = 3$ or $f = 9$. In particular $f > 1$, so that using Theorem 7.4, we infer that there exists $\alpha \in \mathcal{O}_{\mathbb{K}}$ of the form

$$\alpha = \tfrac{1}{3}\left(a + b\theta + c\theta^2 + d\theta^3\right)$$

with integers $0 \leqslant a, b, c, d \leqslant 2$. Since $\text{Tr}_{\mathbb{K}/\mathbb{Q}}(\alpha) = \tfrac{4}{3}(a + 8c) \in \mathbb{Z}$, this implies that $3 \mid a + 8c$ so that $(a, c) = (0, 0)$, $(1, 1)$ or $(2, 2)$. In the first case we have

$$N_{\mathbb{K}/\mathbb{Q}}(\alpha) = \tfrac{1}{81}\left(b^4 + 32b^3 d + 258(bd)^2 + 32bd^3 + d^4\right) \in \mathbb{Z},$$

which implies that $(b, d) = (1, 1)$. We verify that $\tfrac{1}{3}\left(\theta + \theta^3\right) \in \mathcal{O}_{\mathbb{K}}$. If $(a, c) = (1, 1)$, then

$$N_{\mathbb{K}/\mathbb{Q}}(\alpha) = \tfrac{1}{81}\left(b^4 + 32b^3 d + b^2(258d^2 - 36) + b(32d^3 - 576d) \right.$$
$$\left. + d^4 - 4572d^2 + 324\right)$$

and since $N_{\mathbb{K}/\mathbb{Q}}(\alpha) \in \mathbb{Z}$, we get $(b, d) = (0, 0)$, $(1, 1)$ or $(2, 2)$. We check that $\tfrac{1}{3}\left(1 + \theta^2\right) \in \mathcal{O}_{\mathbb{K}}$, and let Δ be the discriminant of $\left\{1, \theta, \tfrac{1}{3}\left(1 + \theta^2\right), \tfrac{1}{3}\left(\theta + \theta^3\right)\right\}$. Straightforward computations give $\Delta = 2^8 \times 7^2$. Hence $|\Delta|$ is minimal, so that $f = 9$ and $d_{\mathbb{K}} = \Delta = 2^8 \times 7^2$ by Theorem 7.2. Furthermore,

$$\left\{1, \theta, \tfrac{1}{3}\left(1 + \theta^2\right), \tfrac{1}{3}\left(\theta + \theta^3\right)\right\}$$

is an integral basis for $\mathbb{K}$.   □

The algebraic number field $\mathbb{K}$ is said to be *monogenic*, or to have a *power basis*, if there exists $\alpha \in \mathcal{O}_{\mathbb{K}}$ such that $\mathcal{O}_{\mathbb{K}} = \mathbb{Z}[\alpha]$. The following example, due to Dedekind, shows that there exist some algebraic number fields which cannot be monogenic.

Example 7.10 (Dedekind) Let $\mathbb{K} = \mathbb{Q}(\theta)$ be an algebraic number field, where θ is a root of the polynomial $P = X^3 - X^2 - 2X - 8$. One may check that $\text{disc}(P) = -4 \times 503$ and that P is irreducible over $\mathbb{Q}$ by Exercise 41 for instance. Let $\beta = \tfrac{1}{2}(\theta^2 + \theta)$. A simple calculation shows that $\beta^3 - 3\beta^2 - 10\beta - 9 = 0$ so that $\beta \in \mathcal{O}_{\mathbb{K}}$ and, using

Definition 7.10, we have

$$\Delta_{\mathbb{K}/\mathbb{Q}}\left(1, \theta, \theta^2\right) = -4 \times 503 \quad \text{and} \quad \Delta_{\mathbb{K}/\mathbb{Q}}\left(1, \theta, \beta\right) = \tfrac{1}{4}\Delta_{\mathbb{K}/\mathbb{Q}}\left(1, \theta, \theta^2\right) = -503.$$

Since $\Delta_{\mathbb{K}/\mathbb{Q}}\left(1, \theta, \beta\right)$ is squarefree, we deduce that $\{1, \theta, \beta\}$ is an integral basis for $\mathbb{K}$ by Theorem 7.4. Now let $\alpha \in O_{\mathbb{K}}$ and set $\alpha = a + b\theta + c\beta$ with $a, b, c \in \mathbb{Z}$. We get

$$\alpha^2 = (a^2 + 6c^2 + 8bc) + (2c^2 - b^2 + 2ab)\theta + (2b^2 + 3c^2 + 2ac + 4bc)\beta$$

so that

$$\Delta_{\mathbb{K}/\mathbb{Q}}(1, \alpha, \alpha^2) \equiv (bc)^2(3c + b)^2 \pmod{2},$$

which is an even number in all cases. Hence this discriminant cannot be equal to -503, which proves that $O_{\mathbb{K}}$ has no integral basis of the form $\mathbb{Z}[\alpha]$. □

7.2.7 Usual Number Fields

In this section, we intend to study some examples of algebraic number fields often used in the literature. The calculation of an integral basis for these fields requires both basic principles seen in the former sections and a touch of some arithmetic technicality.

7.2.7.1 Quadratic Fields

Proposition 7.12 Let $d \in \mathbb{Z} \setminus \{0, 1\}$ be a squarefree integer and $\mathbb{K} = \mathbb{Q}(\sqrt{d})$ be a quadratic number field. We set

$$\omega = \begin{cases} \tfrac{1}{2}\left(1 + \sqrt{d}\right), & \text{if } d \equiv 1 \pmod{4} \\ \\ \sqrt{d}, & \text{otherwise.} \end{cases}$$

Then

d	Integral basis	Discriminant	Polynomial
$d \equiv 1 \pmod{4}$	$(1, \omega)$	d	$X^2 - X + \tfrac{1}{4}(1 - d)$
$d \not\equiv 1 \pmod{4}$	$(1, \omega)$	$4d$	$X^2 - d$

Proof We first look at the algebraic integers in $\mathbb{K}$ and let $\alpha \in \mathbb{K}$. Then there exist $a, b, c \in \mathbb{Z}$ with $(a, b, c) = 1$ and $c > 0$ such that

$$\alpha = \frac{1}{c}\left(a + b\sqrt{d}\right).$$

If $\alpha \in O_{\mathbb{K}}$, then $\mathrm{Tr}_{\mathbb{K}/\mathbb{Q}}(\alpha) = 2a/c \in \mathbb{Z}$ and $N_{\mathbb{K}/\mathbb{Q}}(\alpha) = (a^2 - db^2)/c^2 \in \mathbb{Z}$. If there exists a prime p dividing both a and c, then from the norm we infer that $p \mid b$ since d is squarefree, which contradicts the fact that $(a, b, c) = 1$. Hence $(a, c) = 1$ and the condition on the trace implies that $c \mid 2$.

▷ Suppose that $d \not\equiv 1 \pmod 4$. If $c = 2$, then from the norm we deduce that a and b have to be both odd and $a^2 - db^2 \equiv 0 \pmod 4$. This implies that $d \equiv 1 \pmod 4$, giving a contradiction. Hence $c = 1$ and since $\Delta_{\mathbb{K}/\mathbb{Q}}(1, \sqrt{d}) = 4d$ with d squarefree, we infer that $O_{\mathbb{K}} = \mathbb{Z}[\sqrt{d}]$ in this case by Theorem 7.4.

▷ Suppose that $d \equiv 1 \pmod 4$ and set $f = \left[O_{\mathbb{K}} : \mathbb{Z}[\sqrt{d}]\right]$. If $P = X^2 - d$, then by assumption on d we get $P \equiv (X + 1)^2 \pmod 2$, and hence $2 \mid f$ by Proposition 7.7. Since $f^2 \mid \mathrm{disc}(P) = 4d$, we also have $f^2 \mid 4$ since d is squarefree. Thus $f = 2$. Note that $\omega = \frac{1}{2}\left(1 + \sqrt{d}\right)$ is a root of the polynomial $X^2 - X - \frac{1}{4}(1 - d) \in \mathbb{Z}[X]$ so that $\omega \in O_{\mathbb{K}}$. Now $\Delta_{\mathbb{K}/\mathbb{Q}}(1, \omega) = d$ so that $O_{\mathbb{K}} = \mathbb{Z}\left[\frac{1}{2}\left(1 + \sqrt{d}\right)\right]$ by Theorem 7.4.

$\square$

7.2.7.2   Cyclotomic Fields

Let $\zeta_n = e_n(1) = e^{2i\pi/n}$. Then the n numbers $1, \zeta_n, \zeta_n^2, \ldots, \zeta_n^{n-1}$ are the nth roots of unity, forming a regular n-gon in the complex plane. If $(k, n) > 1$, then ζ_n^k is a root of unity or order $n/(n, k) < n$, whereas if $(n, k) = 1$, then ζ_n^k is not a root of lower order and is called a *primitive* nth root of unity. The number of these roots is then equal to $\varphi(n)$. We define the nth *cyclotomic polynomial* to be the monic polynomial Φ_n whose roots are the primitive nth root of unity so that

$$\Phi_n = \prod_{\substack{i=1 \\ (i,n)=1}}^{n} (X - \zeta_n^i).$$

We have

$$X^n - 1 = \prod_{k=1}^{n}(X - \zeta_n^k) = \prod_{d \mid n} \prod_{\substack{k=1 \\ (k,n)=n/d}}^{n} (X - \zeta_n^k)$$

$$= \prod_{\substack{d \mid n}} \prod_{\substack{i=1 \\ (i,d)=1}}^{d} (X - \zeta_n^{in/d}) = \prod_{d \mid n} \Phi_d$$

so that we derive by the Möbius inversion formula

$$\Phi_n = \prod_{d \mid n} (X^d - 1)^{\mu(n/d)}, \tag{7.5}$$

and hence $\Phi_n \in \mathbb{Z}[X]$ for all $n \geqslant 1$. In particular, if $n = p$ is a prime number, then

$$\Phi_p = X^{p-1} + X^{p-2} + \cdots + X + 1.$$

The irreducibility over $\mathbb{Z}$ of Φ_p has been proved in Example 7.1. We now intend to show that Φ_n is irreducible over $\mathbb{Z}$ for all positive integers n. A possible proof uses the following lemma established by Schönemann.

Lemma 7.7 *Let $A = (X - a_1) \cdots (X - a_r) \in \mathbb{Z}[X]$ be a monic polynomial, and let p be a prime number. Set $A_p = (X - a_1^p) \cdots (X - a_r^p)$. Then $\overline{A_p} = \overline{A}$ in $\mathbb{F}_p[X]$.*

We are now in a position to prove the irreducibility of Φ_n.

Lemma 7.8 *Let n be a positive integer. Then the polynomial Φ_n is irreducible over $\mathbb{Z}$.*

Proof We clearly have $\Phi_n \neq 0, \pm 1$ if $n \geqslant 1$. Suppose that $\Phi_n = PQ$ with $P, Q \in \mathbb{Q}[X]$ such that $\deg P > 0$. Lemma 7.1 implies in fact that $P, Q \in \mathbb{Z}[X]$. Let p be a prime number such that $p \nmid n$ and we define the polynomial R by

$$R = \prod_{\zeta \in Z_P} (X - \zeta^p)$$

where Z_P is the set of the roots of P, so that $\overline{R} = \overline{P}$ in $\mathbb{F}_p[X]$ by Lemma 7.7. Note that if $F = X^n - 1$ and since $nF(x) - xF'(x) = -n$, we have $(\overline{F}, \overline{F'}) = 1$ if $p \nmid n$, and therefore $\overline{F}$ is squarefree. Hence $\overline{\Phi_n}$ is squarefree by (7.5). Now we have $(\overline{Q}, \overline{R}) \mid \overline{R} = \overline{P}$ and $(\overline{Q}, \overline{R}) \mid \overline{Q}$ so that $(\overline{Q}, \overline{R})^2 \mid \overline{PQ} = \overline{\Phi_n}$, and therefore $(\overline{Q}, \overline{R}) = 1$ since $\overline{\Phi_n}$ is squarefree. Hence $(Q, R) = 1$ and since $R \mid \Phi_n$, we infer that $R \mid P$. Since these polynomials are monic and have the same degree, we get $R = P$. Now let ζ be a root of P. Then there exists a positive integer k such that $(k, n) = 1$ and ζ^k is a root of Φ_n. Write $k = p_1 \cdots p_r$, where all the non-necessarily distinct prime numbers p_i satisfy $p_i \nmid n$. By the argument above applied r times, we see that ζ^k is also a root of P, and applying this with all the roots of Φ_n, we deduce that $P = \Phi_n$. Hence Φ_n is irreducible over $\mathbb{Z}$.   □

Lemma 7.8 leads to the following definition. The algebraic number field $\mathbb{K} = \mathbb{Q}(\zeta_n)$ is called a *cyclotomic field*. Hence we have $[\mathbb{K} : \mathbb{Q}] = \varphi(n)$. It can be shown that $\mathbb{K}$

is monogenic [98], so that $O_{\mathbb{K}} = \mathbb{Z}[\zeta_n]$ for all $n \geq 1$. In what follows, the aim is to prove this result when $n = p$ is a prime number.

Proposition 7.13 *Let p be a prime number. Then $\mathbb{Q}(\zeta_p)$ is monogenic.*

Proof First note that since $\zeta_2 = -1$, we have $\mathbb{Q}(\zeta_2) = \mathbb{Q}$, and since $\mathbb{Q}(\zeta_3)$ is a quadratic field, we may suppose that $p \geq 5$. Write ζ instead of ζ_p for convenience. The number $\lambda = \zeta - 1$ plays an important part in the proof. Using Proposition 7.11, we derive

$$\Delta_{\mathbb{K}/\mathbb{Q}}(\lambda) = \Delta_{\mathbb{K}/\mathbb{Q}}(\zeta) = (-1)^{(p-1)/2} p^{p-2}$$

as also stated in Example 7.3. Now μ_λ is a p-Eisenstein as seen in Example 7.1, so that $p \nmid f$ by Proposition 7.8. Since $f^2 \times d_{\mathbb{K}} = (-1)^{(p-1)/2} p^{p-2}$, we infer that $f = 1$, and hence $\mathbb{K}$ is monogenic.   □

The proof above shows that, if $\mathbb{K} = \mathbb{Q}(\zeta_p)$, then $d_{\mathbb{K}} = (-1)^{(p-1)/2} p^{p-2}$. More generally, one can prove [98] that, if $\mathbb{K} = \mathbb{Q}(\zeta_n)$, then

$$d_{\mathbb{K}} = (-1)^{\varphi(n)/2} n^{\varphi(n)} \prod_{p|n} p^{-\varphi(n)/(p-1)}. \tag{7.6}$$

The next result shows that a cyclotomic field is an abelian algebraic number field.

Proposition 7.14 *The algebraic number field $\mathbb{K} = \mathbb{Q}(\zeta_n)$ is abelian with Galois group*

$$\mathrm{Gal}\,(\mathbb{Q}(\zeta_n)/\mathbb{Q}) \simeq (\mathbb{Z}/n\mathbb{Z})^\times .$$

Proof Set $\mathbb{K} = \mathbb{Q}(\zeta_n)$. Since ζ_n is a primitive nth root of unity and since every nth root of unity is a power of ζ_n, we deduce that the extension $\mathbb{K}/\mathbb{Q}$ is normal. Since Φ_n is irreducible over $\mathbb{Q}$ and char $\mathbb{Q} = 0$, we infer that the extension is separable, so that $\mathbb{K}/\mathbb{Q}$ is Galois. Now let $\sigma \in \mathrm{Gal}\,(\mathbb{Q}(\zeta_n)/\mathbb{Q})$. There exists an integer $a = a(\sigma)$ coprime to n such that $\sigma(\zeta_n) = \zeta_n^{a(\sigma)}$. Since the residue class $\bar{a}$ of $a(\sigma)$ modulo n is uniquely determined, one may define the map

$$\mathrm{Gal}\,(\mathbb{Q}(\zeta_n)/\mathbb{Q}) \longrightarrow (\mathbb{Z}/n\mathbb{Z})^\times$$
$$\sigma \longmapsto \bar{a}.$$

This map does not depend on the particular choice of the primitive nth root ζ_n. Furthermore, one may check that it is a group homomorphism. If $\bar{a} = \bar{1}$, then $\sigma(\zeta_n) = \zeta_n$, so that $\sigma = \mathrm{Id}$ and the map is injective. We infer that $\mathrm{Gal}\,(\mathbb{Q}(\zeta_n)/\mathbb{Q})$ is isomorphic to a subgroup of $(\mathbb{Z}/n\mathbb{Z})^\times$ and hence is abelian. Since $\mathrm{Gal}\,(\mathbb{Q}(\zeta_n)/\mathbb{Q})$ and $(\mathbb{Z}/n\mathbb{Z})^\times$ have the same order, the result follows.   □

7.2.7.3   Pure Cubic Fields

These are the fields of the form $\mathbb{K} = \mathbb{Q}(m^{1/3})$, where m is 3-free and written as $m = ab^2$ with $(a, b) = 1$, a, b squarefree and we assume that if $3 \mid m$, then $3 \mid a$ and $3 \nmid b$. Let $\theta = m^{1/3}$, which is a root of the polynomial $P = X^3 - m$. We have

$$\mathrm{disc}(P) = -3^3 \times m^2 = -3^3 \times a^2 \times b^4 = f^2 \times d_{\mathbb{K}}$$

so that writing $d_{\mathbb{K}} = -3^n \times a^\alpha \times b^\beta$, we get $f = 3^{(3-n)/2} \times a^{(2-\alpha)/2} \times b^{(4-\beta)/2}$. Hence $n = 1$ or 3, $\alpha = 0$ or 2 and $\beta = 0$, 2 or 4. Since P is a p-Eisenstein for any prime factor p of a, we have $p \mid a \Rightarrow p \nmid f$ by Proposition 7.8. In particular, if $3 \mid m$, then $3 \nmid f$ and $27a^2 \mid d_{\mathbb{K}}$. If $3 \nmid a$, then $3a^2 \mid d_{\mathbb{K}}$ so that $\alpha \geqslant 2$ in all cases. Hence $\alpha = 2$.

Now let $\lambda = \widetilde{m}^{1/3}$ with $\widetilde{m} = a^2 b$, which is a root of the polynomial $Q = X^3 - a^2 b$. Since

$$\begin{pmatrix} 1 \\ \lambda \\ \lambda^2 \end{pmatrix} = A \begin{pmatrix} 1 \\ \theta \\ \theta^2 \end{pmatrix}$$

with $\det A = -a \neq 0$, we deduce that $\mathrm{disc}(Q) = f^2 \times d_{\mathbb{K}}$. Now

$$\mathrm{disc}(Q) = -3^3 \times a^4 \times b^2,$$

which implies that $\beta \neq 4$. Since Q is a p-Eisenstein for any prime factor p of b, we have $p \mid b \Rightarrow p \nmid f$ and hence $b^2 \mid d_{\mathbb{K}}$ so that $\beta \geqslant 2$. Therefore $\beta = 2$. We then get

$$d_{\mathbb{K}} = \begin{cases} -27(ab)^2, & \text{if } 3 \mid m \\ \\ -3(ab)^2 \text{ or } -27(ab)^2, & \text{otherwise.} \end{cases}$$

We consider the following three cases.

▷ $m \not\equiv \pm 1 \pmod 9$. Thus $9 \nmid (m^3 - m)$ so that the polynomial $S = (X + m)^3 - m$ is a 3-Eisenstein. A root of S is $\theta - m$ and $\Delta_{\mathbb{K}/\mathbb{Q}}(\theta - m) = \Delta_{\mathbb{K}/\mathbb{Q}}(\theta) = -3^3 \times m^2$. Then $3 \nmid f$ so that $f = b$. Since $1, \theta, \lambda \in \mathcal{O}_{\mathbb{K}}$ and $\Delta_{\mathbb{K}/\mathbb{Q}}(1, \theta, \lambda) = b^{-2}\Delta_{\mathbb{K}/\mathbb{Q}}(\theta) = d_{\mathbb{K}}$, we infer that

$$\left\{ 1, \theta, \tfrac{1}{b}\theta^2 \right\}$$

is an integral basis for $\mathbb{K}$.

▷ $\widetilde{m} \equiv 1 \pmod 9$. Let $\nu = \frac{1}{3b}(b + ab\theta + \theta^2)$. One can observe that ν is a root of the polynomial

$$R = X^3 - X^2 - \tfrac{1}{3}(\widetilde{m} - 1)X - \tfrac{1}{27}(\widetilde{m} - 1)^2 \in \mathbb{Z}[X]$$

so that $\nu \in O_\mathbb{K}$. Since $3b\,\nu \in \mathbb{Z}[\theta]$, we infer that $O_\mathbb{K}/\mathbb{Z}[\theta]$ has an element of order $3b$, so that $3b \mid f$ by Theorem 7.1 and then $f = 3b$ and $d_\mathbb{K} = -3(ab)^2$. Now $\Delta_{\mathbb{K}/\mathbb{Q}}(1, \theta, \nu) = (3b)^{-2}\Delta_{\mathbb{K}/\mathbb{Q}}(\theta) = d_\mathbb{K}$ so that

$$\left\{ 1, \theta, \tfrac{1}{3b}\left(b + ab\theta + \theta^2 \right) \right\}$$

is an integral basis for $\mathbb{K}$.

▷ $\widetilde{m} \equiv -1 \pmod 9$. This case can be treated as above, except that we make use of $\varrho = \frac{1}{3b}(b + ab\theta - \theta^2)$, which is a root of the polynomial

$$S = X^3 - X^2 + \tfrac{1}{3}(\widetilde{m} + 1)X - \tfrac{1}{27}(\widetilde{m} + 1)^2 \in \mathbb{Z}[X]$$

instead of ν.

Also note that $\widetilde{m} \equiv \pm 1 \pmod 9 \iff m \equiv \pm 1 \pmod 9$. We may sum up the discussion in the following result.

Proposition 7.15 *Let* $\mathbb{K} = \mathbb{Q}(m^{1/3})$ *be a pure cubic field, where* $m = ab^2$ *is 3-free with* $(a, b) = 1$ *and* a, b *squarefree. Set* $\theta = m^{1/3} \in O_\mathbb{K}$ *and* $\widetilde{m} = a^2 b$.

▷ *If* $m \not\equiv \pm 1 \pmod 9$, *then* $d_\mathbb{K} = -27(ab)^2$ *and* $\left\{ 1, \theta, \tfrac{1}{b}\theta^2 \right\}$ *is an integral basis for* $\mathbb{K}$.

▷ *If* $\widetilde{m} \equiv \pm 1 \pmod 9$, *then* $d_\mathbb{K} = -3(ab)^2$ *and* $\left\{ 1, \theta, \tfrac{1}{3b}\left(b + ab\theta \pm \theta^2 \right) \right\}$ *is an integral basis for* $\mathbb{K}$.

7.2.7.4   Voronoï's Method for Cubic Fields

We investigate algebraic number fields $\mathbb{K} = \mathbb{Q}(\theta)$, where θ is a root of the polynomial $P = X^3 - aX + b \in \mathbb{Z}[X]$ with a 2-free or b 3-free. Using Example 7.3, we obtain $\mathrm{disc}(P) = 4a^3 - 27b^2$. In [97], Voronoï devised a method to compute an integral basis for $\mathbb{K}$. A new proof of this method was given in [2, 1] and rests on the evaluation of $d_\mathbb{K}$, which can be found in [59].

Proposition 7.16 (Voronoï) *Let* $\mathbb{K} = \mathbb{Q}(\theta)$ *be an algebraic number field, where* θ *is a root of the polynomial* $P = X^3 - aX + b \in \mathbb{Z}[X]$ *with* a *2-free or* b *3-free.*

▷ *Suppose that* $a \not\equiv 3 \pmod 9$ *or* $b^2 \not\equiv a + 1 \pmod{27}$ *and let* n^2 *be the largest square dividing* $\mathrm{disc}(P)$ *for which the system of congruences*

$$\begin{cases} x^3 - ax + b \equiv 0 \pmod{n^2} \\ \quad\; 3x^2 - a \equiv 0 \pmod{n} \end{cases}$$

is solvable for x. Then an integral basis for $\mathbb{K}$ *is given by*

$$\left\{ 1, \theta, \frac{1}{n} \left(x^2 - a + x\theta + \theta^2 \right) \right\}.$$

▷ *Suppose that* $a \equiv 3 \pmod 9$ *and* $b^2 \equiv a + 1 \pmod{27}$ *and let* n^2 *be the largest square dividing* $\frac{1}{729} \operatorname{disc}(P)$ *for which the system of congruences*

$$\begin{cases} x^3 - ax + b \equiv 0 \pmod{27n^2} \\ \quad\; 3x^2 - a \equiv 0 \pmod{9n} \end{cases}$$

is solvable for x. Then an integral basis for $\mathbb{K}$ *is given by*

$$\left\{ 1, \frac{1}{3}(\theta - x), \frac{1}{9n} \left(x^2 - a + x\theta + \theta^2 \right) \right\}.$$

Examples

▷ Let $\mathbb{K} = \mathbb{Q}(\theta)$, where θ is a root of $P = X^3 - 8X - 57$. Using Example 7.3, we get $\operatorname{disc}(P) = -5^2 \times 23 \times 149$ so that $n = 5$, $x = 4$, and thus $\left\{ 1, \theta, \frac{1}{5} \left(8 + 4\theta + \theta^2 \right) \right\}$ is an integral basis for $\mathbb{K}$.

▷ Let $\mathbb{K} = \mathbb{Q}(\theta)$, where θ is a root of $P = X^3 - 12X + 65$. Here, $729^{-1} \operatorname{disc}(P) = -3 \times 7^2$ so that $n = 7$, $x = -5$ so that $\left\{ 1, \frac{1}{3}(\theta + 5), \frac{1}{63} \left(13 - 5\theta + \theta^2 \right) \right\}$ is an integral basis for $\mathbb{K}$.

7.2.7.5   Pure Number Fields

A pure number field is an extension $\mathbb{K} = \mathbb{Q}\left(\sqrt[n]{a} \right)$, where the integer a is such that for each prime p dividing n, either $p \nmid a$ or the highest power of p dividing a is coprime to p. In [46], the authors established a general formula for the discriminant $d_{\mathbb{K}}$ of $\mathbb{K}$ by means of the prime powers decompositions of n and a. More precisely, assume $n = p_1^{e_1} \cdots p_r^{e_r}$, $|a| = q_1^{f_1} \cdots q_s^{f_s}$, and set $m_k := (n, f_k)$, $n_k := n/p_k^{e_k}$, $v_k := v_{p_k}\left(a^{p_k-1} - 1 \right) - 1$ and

$$c_k := \begin{cases} ne_k - 2n_k \sum_{j=1}^{\min(v_k, e_k)} p_k^{e_k - j}, & \text{if } v_k > 0 \\[2ex] ne_k, & \text{otherwise.} \end{cases}$$

With these parameters, we can state the following result.

Proposition 7.17 *Let* $\mathbb{K} = \mathbb{Q}\left(\sqrt[n]{a}\right)$ *be a pure number field. Then*

$$d_{\mathbb{K}} = \text{sgn}\left(a^{n-1}\right)(-1)^{\frac{(n-1)(n-2)}{2}}\prod_{k=1}^{r}p_k^{c_k}\prod_{k=1}^{s}q_k^{n-m_k}.$$

Furthermore, if $\theta = \sqrt[n]{a}$, *then* $\left\{1, \theta, \ldots, \theta^{n-1}\right\}$ *is an integral basis for* $\mathbb{K}$ *if and only if* a *is squarefree and, for each prime* p *dividing* n, $p^2 \nmid a^{p-1} - 1$.

7.2.7.6 Monogenic Number Fields

It is a long-standing problem in algebraic number theory to determine whether an algebraic number field $\mathbb{K}$ is monogenic. Indeed, the existence of an element $\theta \in O_{\mathbb{K}}$ such that $O_{\mathbb{K}} = \mathbb{Z}[\theta]$ makes the study of arithmetic in $\mathbb{K}$ considerably easier. From above we know that the quadratic fields and the cyclotomic fields are monogenic, but, in the general case, this property is relatively rare.

Let $\mathbb{K} = \mathbb{Q}(\theta)$ be an algebraic number field of degree n and $\{1, \alpha_2, \ldots, \alpha_n\}$ be an integral basis for $\mathbb{K}$. There exists a form $I(X_2, \ldots, X_n) \in \mathbb{Z}[X_2, \ldots, X_n]$, called the *index form*, of degree $\frac{1}{2}n(n-1)$ in $n-1$ variables $X_2, \ldots, X_n$ such that

$$\Delta_{\mathbb{K}/\mathbb{Q}}\left(\alpha_2 X_2 + \cdots + \alpha_n X_n\right) = I(X_2, \ldots, X_n)^2 d_{\mathbb{K}}.$$

Hence $\mathbb{K}$ is monogenic if and only if the *index form equation*

$$I(x_2, \ldots, x_n) = \pm 1$$

has a solution in $\mathbb{Z}^{n-1}$. The first effective upper bounds for the solutions of this equation were derived by Győry [36] by using Baker's lower bounds for logarithmic forms. As a consequence, it follows that up to translation by elements of $\mathbb{Z}$, there exist only finitely many generators of power integral bases in an algebraic number field.

In what follows, we provide some examples of monogenic algebraic number fields of low degree.

Cyclic Cubic Field

A *cyclic cubic field* is a number field $\mathbb{K}$ of degree 3, which is Galois such that $\text{Gal}(\mathbb{K}/\mathbb{Q}) \simeq C_3 \simeq \mathbb{Z}/3\mathbb{Z}$. A profound result[15] states that such a field is contained

---

[15] See Sect. 7.5.7.

in a cyclotomic field $\mathbb{Q}(\zeta_{f_{\mathbb{K}}})$ with $f_{\mathbb{K}}$ minimal, called the *conductor* of $\mathbb{K}$. In the case of cyclic cubic fields, it can be shown that one can always write

$$f_{\mathbb{K}} = \tfrac{1}{4}\left(a^2 + 27b^2\right)$$

with $b > 0$ and $a \equiv 1 \pmod 3$ if $f_{\mathbb{K}} \equiv 1 \pmod 3$, $a = 3a'$ with $a' \equiv 1 \pmod 3$ otherwise. In [32], the following necessary and sufficient condition is proved.

Proposition 7.18 *With the notation above, $\mathbb{K}$ is monogenic if and only if the equation $bu(u^2 - 9v^2) + av(u^2 - v^2) = 1$ has solutions $u, v \in \mathbb{Z}$.*

Biquadratic Field

A *biquadratic field* is a number field $\mathbb{K} = \mathbb{Q}(\sqrt{dm}, \sqrt{dn})$ with d, m, n squarefree, pairwise coprime such that $dm, dn, mn \neq 1$, $dm \equiv dn \pmod 4$, $d > 0$, $m > n$ and if $dm \equiv dn \equiv 1 \pmod 4$, then $d < \min(|m|, |n|)$. Such fields are Galois with Galois group $\mathrm{Gal}(\mathbb{K}/\mathbb{Q}) \simeq (\mathbb{Z}/2\mathbb{Z})^2$. Define $\delta \in \{0, 1\}$ such that $mn \equiv (-1)^\delta \pmod 4$. Then we have the following proposition [34].

Proposition 7.19 *Let $\mathbb{K} = \mathbb{Q}(\sqrt{dm}, \sqrt{dn})$ be a biquadratic field with d, m, n as above.*

▷ *If $dm \equiv dn \equiv 1 \pmod 4$, then $\mathbb{K}$ is not monogenic.*
▷ *If $dm \equiv dn \not\equiv 1 \pmod 4$, then $\mathbb{K}$ is monogenic if and only if the following two conditions are fulfilled.*

  (a) *$m - n = 2^{2-2\delta}d$.*
  (b) *The equation $2^\delta m(u^2 - v^2)^2 - 2^\delta n(u^2 + v^2)^2 = 4s$, where $s = \pm 1$, has solutions $u, v \in \mathbb{Z}$.*

Miscellaneous

In a series of papers [21, 57, 85], the authors consider some families of monic polynomials giving birth to infinitely many algebraic number fields of degrees 5 and 6 whose ring of integers are monogenic. The proofs rest on the following scheme:

▷ Prove that the polynomials are irreducible over $\mathbb{Z}$.
▷ Determine the Galois groups.
▷ Consider a subfamily of these polynomials, for instance, those that have a squarefree discriminant, use results from elementary number theory to show that this subfamily contains infinitely many polynomials.
▷ Prove that the algebraic number fields defined by these polynomials are monogenic.

Proposition 7.20 *Let $m \in \mathbb{Z} \setminus \{1\}$.*

▷ *Let $P_m = X^5 - 2X^4 + (m + 2)X^3 - (2m + 1)X^2 + mX + 1$ such that the number $4m^3 + 28m^2 + 24m + 47$ is squarefree. Then P_m is irreducible over $\mathbb{Z}$,*

$\mathrm{Gal}(P_m/\mathbb{Q}) \simeq D_5$ *and the algebraic number fields defined by* P_m *are distinct and monogenic.*

▷ *Let* $P_m = X^6 + (2m+2)X^4 + (2m-1)X^2 - 1$ *and assume that* $4m^2 + 2m + 7$ *is squarefree. Then* P_m *is irreducible over* $\mathbb{Z}$, $\mathrm{Gal}(P_m/\mathbb{Q}) \simeq \mathcal{A}_4$ *and the algebraic number fields defined by* P_m *are distinct and monogenic.*

▷ *Let* $P_m = X^6 - 4X^5 + 2X^4 - 3mX^3 + X^2 + 2X + 1$ *and suppose that the number* $729m^3 + 522m^2 + 1788m + 2648$ *is squarefree. Then* P_m *is irreducible over* $\mathbb{Z}$, $\mathrm{Gal}(P_m/\mathbb{Q}) \simeq \mathrm{PSL}(2,5) \simeq \mathcal{A}_5$ *and the algebraic number fields defined by* P_m *are distinct and monogenic.*

Proof of the Third Case Suppose that P_m is not irreducible. Since $P_m \neq 0, \pm 1$ and

$$P_m \equiv (X-1)(X^5 - X^3 - X^2 - 1) \pmod{3}$$

then we have $P_m = Q_m R_m$, where $Q_m, R_m \in \mathbb{Z}[X]$ are monic polynomials such that $\deg Q_m = 1$ and $\deg R_m = 5$. The condition $P_m(0) = 0$ implies that $Q_m = X \pm 1$. If $Q_m = X - 1$, then we have $0 = P_m(1) = 3 - 3m$ contradicting the fact that $m \neq 1$. If $Q_m = X + 1$, then similarly we have $0 = P_m(-1) = 3m + 7$ contradicting the fact that m is an integer. Hence P_m is irreducible over $\mathbb{Z}$. Furthermore, the factorization of P_m in $\mathbb{F}_3[X]$ above shows that $\mathrm{Gal}(P_m/\mathbb{Q})$ contains a 5-cycle so that 5 divides the order of $\mathrm{Gal}(P_m/\mathbb{Q})$. Since

$$\mathrm{disc}(P_m) = \left(729m^3 + 522m^2 + 1788m + 2648\right)^2$$

we have $\mathrm{Gal}(P_m/\mathbb{Q}) \subseteq \mathcal{A}_6$ by Lemma 7.20. Among the sixteen transitive subgroups of $\mathcal{S}_6$, which may be Galois groups of an irreducible polynomial of degree 6, the only groups having order divisible by 5 and contained in $\mathcal{A}_6$ are $\mathrm{PSL}(2,5)$ and $\mathcal{A}_6$. The final step is given by using a result of [37], where a factorization of a certain polynomial of degree 15,[16] along with the fact that $\mathrm{disc}(P_m)$ is a square, ensures that $\mathrm{Gal}(P_m/\mathbb{Q}) \simeq \mathrm{PSL}(2,5)$.

A theorem by Erdős [23] implies that there are infinitely many integers m, which are odd, such that $\delta(m) = 729m^3 + 522m^2 + 1788m + 2648$ is squarefree. Let p be an odd prime factor of $\mathrm{disc}(P_m) = \delta(m)^2$. Using Corollary 7.7, we infer that $p \mid d_{\mathbb{K}}$. Since $\delta(m)$ is odd and squarefree, we get $p > 2$ and $v_p(\mathrm{disc}(P_m)) = 2$. If $p \mid f$, then the relation $\mathrm{disc}(P_m) = f^2 \times d_{\mathbb{K}}$ and the fact that $p \mid d_{\mathbb{K}}$ give $p^3 \mid \mathrm{disc}(P_m)$, giving a contradiction with $v_p(\mathrm{disc}(P_m)) = 2$. Hence $f = 1$ and $\mathcal{O}_{\mathbb{K}} = \mathbb{Z}[\theta]$. □

---

[16]See Theorem 7.38.

7.2.8 Units and Regulators

7.2.8.1 Dirichlet's Unit Theorem

This section investigates the multiplicative group $O_{\mathbb{K}}^{\times}$ of the units in the ring $O_{\mathbb{K}}$. The structure of this group has been entirely determined by Dirichlet who proved the following important theorem sometimes called *Dirichlet's unit theorem*.

> **Theorem 7.5 (Dirichlet)** *Let* $\mathbb{K} = \mathbb{Q}(\theta)$ *be an algebraic number field with signature* (r_1, r_2), *and let* $\theta \in O_{\mathbb{K}}$. *We denote by* $W_{\mathbb{K}}$ *the subgroup of* $O_{\mathbb{K}}^{\times}$ *consisting of roots of unity in* $O_{\mathbb{K}}$. *Then* $O_{\mathbb{K}}^{\times}$ *is a finitely generated abelian group with*
>
> $$\operatorname{rank} O_{\mathbb{K}}^{\times} = r_1 + r_2 - 1 := r$$
>
> *and torsion subgroup equal to* $W_{\mathbb{K}}$. *More precisely, there exist units* $\varepsilon_1, \ldots, \varepsilon_r$, *called a* system of fundamental units, *such that every unit* $\varepsilon \in O_{\mathbb{K}}$ *can be written uniquely in the form*
>
> $$\varepsilon = \zeta \varepsilon_1^{n_1} \cdots \varepsilon_r^{n_r}$$
>
> *with* $\zeta \in W_{\mathbb{K}}$ *and* $n_1, \ldots, n_r \in \mathbb{Z}$. *The number* r *is called the* Dirichlet rank *of* $O_{\mathbb{K}}^{\times}$. *In other words, we have*
>
> $$O_{\mathbb{K}}^{\times} \simeq W_{\mathbb{K}} \times \mathbb{Z}^{r_1 + r_2 - 1}.$$

Example 7.11

▷ *Imaginary quadratic fields.* Let $\mathbb{K} = \mathbb{Q}(\sqrt{-d})$ be an imaginary quadratic field with $d > 0$ squarefree. Hence $(r_1, r_2) = (0, 1)$ so that $O_{\mathbb{K}}^{\times} \simeq W_{\mathbb{K}}$. One may check that

$$O_{\mathbb{K}}^{\times} = \begin{cases} \{\pm 1, \pm i\}, & \text{if } d = 1 \\ \{\pm 1, \pm \rho, \pm \rho^2\}, & \text{if } d = 3 \\ \{\pm 1\}, & \text{otherwise.} \end{cases}$$

▷ *Real quadratic fields.* Let $\mathbb{K} = \mathbb{Q}(\sqrt{d})$ be a real quadratic field with $d > 0$ squarefree and $d \neq 1$. Let $\sigma : a + b\sqrt{d} \mapsto a - b\sqrt{d}$ be the non-trivial embedding of $\mathbb{K}$. Since $(r_1, r_2) = (2, 0)$, there exists a fundamental unit, denoted by γ_d, such that

$$O_{\mathbb{K}}^{\times} = \left\{ \pm \gamma_d^k : k \in \mathbb{Z} \right\}.$$

The calculation of γ_d requires the theory of continued fractions in the following way.[17] Let ω be the number stated in Proposition 7.12, and assume that $-\sigma(\omega)$ has the continued fraction development $-\sigma(\omega) = [a_0, \overline{a_1, \ldots, a_t}]$. If p_{t-1}/q_{t-1} is the $(t-1)$th convergent, then

$$\gamma_d = p_{t-1} + \omega q_{t-1},$$

and we also have $N_{\mathbb{K}/\mathbb{Q}}(\gamma_d) = (-1)^t$.

We give below a sketch of proof of Dirichlet's unit theorem. To this end, we first derive some useful facts about units and roots of unity of $O_\mathbb{K}$ and define the so-called *regulator* of $\mathbb{K}$. The reader interested in this subject may refer to [24, 28, 47, 66, 78, 92].

A *root of unity* in $\mathbb{K}$ is a number $\zeta \in \mathbb{K}$ such that there exists a positive integer m such that $\zeta^m = 1$. Hence $\zeta \in O_\mathbb{K}$. The following lemma will be useful.

Lemma 7.9 *Let $A > 0$. The set $S = \{\alpha \in O_\mathbb{K} : |\sigma_1(\alpha)| \leqslant A, \ldots, |\sigma_n(\alpha)| \leqslant A\}$ is finite.*

Proof The characteristic polynomial (7.2) of $\alpha \in S$ belongs to $\mathbb{Z}[X]$ by (7.3), and its coefficients are all symmetric functions in $\sigma_i(\alpha)$ and hence are bounded. We infer that there are only a finite number of possibilities for this polynomial, e.g. $(2A+2)^{n^2}$ is such a bound. □

Proposition 7.21 *Let $\mathbb{K}$ be an algebraic number field of degree n, signature (r_1, r_2), and $O_\mathbb{K}$ is the ring of integers of $\mathbb{K}$. We set $W_\mathbb{K}$ as in Theorem 7.5.*

▷ *ε is a unit in $O_\mathbb{K}$ if and only if $|N_{\mathbb{K}/\mathbb{Q}}(\varepsilon)| = 1$.*
▷ *There are only finitely many roots of unity in $\mathbb{K}$.*
▷ *$W_\mathbb{K}$ is a cyclic group of even order. Furthermore, if $r_1 > 0$, then $|W_\mathbb{K}| = 2$.*
▷ *Let $\alpha \in O_\mathbb{K}$ and μ_α be its minimal polynomial. If $\mu_\alpha(m) = \pm 1$ for some $m \in \mathbb{Z}$, then $\alpha - m$ is a unit in $O_\mathbb{K}$.*

Proof

▷ If ε is a unit, then there exists $\lambda \in O_\mathbb{K}$ such that $\lambda\varepsilon = 1$ so that $N_{\mathbb{K}/\mathbb{Q}}(\varepsilon)N_{\mathbb{K}/\mathbb{Q}}(\lambda) = 1$ and these norms are integers. Conversely, if $\varepsilon \in O_\mathbb{K}$ satisfies $N_{\mathbb{K}/\mathbb{Q}}(\varepsilon) = \pm 1$, then one can write

$$\varepsilon \times \left(\pm \prod_{i=2}^{n} \sigma_i(\varepsilon) \right) = 1$$

so that ε is a unit in $O_\mathbb{K}$ as all terms are algebraic integers.
▷ Suppose that $\alpha^m = 1$. Then $|\alpha|^m = 1$ and hence $|\alpha| = 1$ and similarly $|\sigma_i(\alpha)| = 1$ for all $i \in \{1, \ldots, n\}$. The result follows from Lemma 7.9.

---

[17] See [78] for instance.

▷ Let $\alpha_1, \ldots, \alpha_k$ be the roots of unity in $\mathbb{K}$. Since $\alpha_i^{m_i} = 1$ for some $m_i \geqslant 1$, we get $\alpha_i = e_{m_i}(n_i)$ for some $0 \leqslant n_i \leqslant m_i - 1$. If we set $m = m_1 \cdots m_k$, then each α_i belongs to the cyclic group generated by $e_m(1)$, and therefore $W_{\mathbb{K}}$ is a subgroup of this group and hence is cyclic. Furthermore, since $\{-1, 1\} \subseteq W_{\mathbb{K}}$, then 2 divides $|W_{\mathbb{K}}|$ by Theorem 7.1. Finally, if $r_1 > 0$, then $W_{\mathbb{K}} = \{\pm 1\}$ since all other roots of unity are non-real.

▷ If $P(X) = \mu_\alpha(X + m)$, then $P \in \mathbb{Z}[X]$ is monic, satisfies $P(0) = \mu_\alpha(m) = \pm 1$ and $P(\alpha - m) = \mu_\alpha(\alpha) = 0$, so that the minimal polynomial of $\alpha - m \in O_{\mathbb{K}}$ divides P by Lemma 7.4, and thus has constant term ± 1. We infer that $N_{\mathbb{K}/\mathbb{Q}}(\alpha - m) = \pm 1$ and then $\alpha - m \in O_{\mathbb{K}}^\times$ by above.

$\square$

7.2.8.2 Sketch of Proof of Theorem 7.5

Dirichlet's unit theorem can be proved by embedding the unit group in a *logarithmic space*. More precisely, if $\mathbb{K}$ is an algebraic number field of degree n, signature (r_1, r_2) and embeddings σ_i, we may define a map

$$\phi : \mathbb{K} \longrightarrow \mathbb{R}^{r_1} \times \mathbb{C}^{r_2}$$
$$\alpha \longmapsto \phi(\alpha)$$

with $\phi(\alpha) = \big(\sigma_1(\alpha) \ldots, \sigma_{r_1}(\alpha), \sigma_{r_1+1}(\alpha), \ldots, \sigma_{r_1+r_2}(\alpha)\big)$. The map ϕ is an injective ring homomorphism and one can prove that, if $\{\alpha_1, \ldots, \alpha_n\}$ is a $\mathbb{Q}$-basis for $\mathbb{K}$, then the vectors $\phi(\alpha_1), \ldots, \phi(\alpha_n)$ are $\mathbb{R}$-linearly independent. This map enables us to get a geometric representation of algebraic numbers.

Define the following map

$$L : \mathbb{K} \setminus \{0\} \longrightarrow \mathbb{R}^{r_1+r_2}$$
$$\alpha \longmapsto (\delta_i \log |\sigma_i(\alpha)|) \,,$$

where $\delta_i = 1$ if $i \in \{1, \ldots, r_1\}$ and $\delta_i = 2$ if $i \in \{r_1 + 1, \ldots, r_1 + r_2\}$. This map is called the *logarithmic representation* of $\mathbb{K} \setminus \{0\}$ and $\mathbb{R}^{r_1+r_2}$ is the *logarithmic space*. It is easy to check that $L(\alpha\beta) = L(\alpha) + L(\beta)$ for all $\alpha, \beta \in \mathbb{K} \setminus \{0\}$.

Now consider the restriction $L : O_{\mathbb{K}}^\times \longrightarrow \mathbb{R}^{r_1+r_2}$. This abelian group homomorphism is not injective, but one may determine its kernel and image. Whereas the first one is easily described as we will see below, the determination of $\operatorname{Im} L$ uses the so-called *geometry of numbers* and more precisely a theorem by Minkowski applied to certain discrete additive subgroups of $\mathbb{R}^{r_1+r_2}$. A discrete[18] additive subgroup of $(\mathbb{R}^{r_1+r_2}, +)$ generated by the m-linearly independent vectors $e_1, \ldots, e_m$ is called a

---

[18]A subset of $\mathbb{R}^{r_1+r_2}$ is *discrete* if and only if it intersects every closed ball of centre O in a finite set.

lattice of dimension m. Hence we need to pass from the multiplicative group $O_{\mathbb{K}}^{\times}$ to an additive subgroup of $\mathbb{R}^{r_1+r_2}$, which is the reason why the logarithms are used.

We may sketch the proof of Theorem 7.5. To this end, let $\varepsilon \in O_{\mathbb{K}}^{\times}$.

From the definition of L above, we immediately see that $W_{\mathbb{K}} \subseteq \ker L$. Conversely, since $L(\varepsilon) = 0 \Leftrightarrow |\sigma_i(\varepsilon)| = 1$ for all i, we infer that $\ker L$ is a finite subgroup of $\mathbb{K} \setminus \{0\}$ by Lemma 7.9. Hence $\ker L$ is cyclic and must consist entirely of roots of unity. Therefore

$$\ker L \simeq W_{\mathbb{K}},$$

a result proved by Kronecker. Now set $L(\alpha) = \big(L_1(\alpha), \ldots, L_{r_1+r_2}(\alpha)\big)$. Then

$$\sum_{i=1}^{r_1+r_2} L_i(\alpha) = \log |N_{\mathbb{K}/\mathbb{Q}}(\alpha)|,$$

and Proposition 7.21 implies

$$\sum_{i=1}^{r_1+r_2} L_i(\varepsilon) = 0$$

so that $\operatorname{Im} L \subseteq \mathcal{H}$, where $\mathcal{H}$ is the hyperplane of $\mathbb{R}^{r_1+r_2}$ of equation $x_1 + \cdots + x_{r_1+r_2} = 0$. We have $\dim \mathcal{H} = r_1 + r_2 - 1$. Now let $\eta > 0$, and let $\| \ \|_2$ be the Euclidean norm on $\mathbb{R}^{r_1+r_2}$. Suppose that $\|L(\varepsilon)\|_2 < \eta$. This readily implies that $|\sigma_i(\varepsilon)| < e^{\eta}$ for all $i \in \{1, \ldots, r_1\}$ and $|\sigma_i(\varepsilon)| < e^{\eta/2}$ for all $i \in \{r_1 + 1, \ldots, r_1 + r_2\}$, so that the set of points $\sigma(\varepsilon)$ such that $\|L(\varepsilon)\|_2 < \eta$ is finite by Lemma 7.9, and hence $\operatorname{Im} L$ is a lattice of dimension $\leqslant r_1 + r_2 - 1$.

Since $W_{\mathbb{K}}$ is finite and $O_{\mathbb{K}}^{\times}/W_{\mathbb{K}} \simeq \operatorname{Im} L$, and since $\operatorname{Im} L$ is a lattice, and then a free abelian group, we infer that $O_{\mathbb{K}}^{\times}$ is a finitely generated abelian group of rank $\leqslant r_1 + r_2 - 1$. Using topological tools,[19] one may prove that the rank is actually equal to $r = r_1 + r_2 - 1$.

7.2.8.3 Regulator

The volume of the lattice $\operatorname{Im} L$ plays an important role in the theory and is called the *regulator* of $\mathbb{K}$. More precisely, we have the following definition.

Definition 7.11 (Regulator) Let $\mathbb{K}/\mathbb{Q}$ be an algebraic number field of signature (r_1, r_2) and $r = r_1 + r_2 - 1$ be the Dirichlet rank of $O_{\mathbb{K}}^{\times}$. Let $\varepsilon_1, \ldots, \varepsilon_r$ be a system of fundamental units of $O_{\mathbb{K}}^{\times}$, and define the matrix $U \in \mathcal{M}_{r,r+1}(\mathbb{R})$ whose ith row is

---

[19]See [28, Lemma 4.7, page 171].

the row-vector $L(\varepsilon_i)$ for $1 \leqslant i \leqslant r$. Let U_r be any $r \times r$ submatrix extracted from U. The *regulator* of $\mathbb{K}$ is the number $\mathcal{R}_{\mathbb{K}}$ defined by

$$\mathcal{R}_{\mathbb{K}} = |\det U_r| \,.$$

Note that the regulator is well-defined since changing a $\mathbb{Z}$-basis of $\mathbb{Z}^r$ into another one involves multiplication by a matrix of determinant ± 1 by Proposition 7.1.

Example 7.12   Let θ be a root of the polynomial $P = X^3 - X^2 - 11X - 1$. If P has a rational root, then it must be ± 1, but $P(\pm 1) \neq 0$, and hence P is irreducible over $\mathbb{Z}$. Define $\mathbb{K} = \mathbb{Q}(\theta)$ and hence $\theta \in O_{\mathbb{K}}$. The signature is $(r_1, r_2) = (3, 0)$, and therefore the Dirichlet rank of $O_{\mathbb{K}}^{\times}$ is $r = 2$. Since $P(0) = -1$, we have in fact $\theta \in O_{\mathbb{K}}^{\times}$. Now

$$(3 + \theta)^3 = \theta^3 + 9\theta^2 + 27\theta + 27 = 10\theta^2 + 38\theta + 28 = 2(\theta + 1)(5\theta + 14),$$

and we have $N_{\mathbb{K}/\mathbb{Q}}(1 + \theta) = -8$ and $N_{\mathbb{K}/\mathbb{Q}}(3 + \theta) = 4$ so that

$$64 = N_{\mathbb{K}/\mathbb{Q}}\left((3 + \theta)^3\right) = N_{\mathbb{K}/\mathbb{Q}}(2(\theta + 1)(5\theta + 14)) = -64 \times N_{\mathbb{K}/\mathbb{Q}}(5\theta + 14)\,.$$

Therefore $N_{\mathbb{K}/\mathbb{Q}}(5\theta + 14) = -1$ and hence $5\theta + 14 \in O_{\mathbb{K}}^{\times}$ by Proposition 7.21. If θ_2 and θ_3 are the conjugates of θ, we then have

$$U = \begin{pmatrix} \log |\theta| & \log |\theta_2| & \log |\theta_3| \\ \log |5\theta + 14| & \log |5\theta_2 + 14| & \log |5\theta_3 + 14| \end{pmatrix}$$

so that

$$\mathcal{R}_{\mathbb{K}} = \left| \det \begin{pmatrix} \log |\theta| & \log |\theta_2| \\ \log |5\theta + 14| & \log |5\theta_2 + 14| \end{pmatrix} \right| \approx 11.9265379\ldots$$

7.3   Ideal Theory

7.3.1   Arithmetic Properties

7.3.1.1   Prime Ideals

Let R be a ring. A *maximal ideal* of R is an ideal $\mathfrak{a} \neq R$ such that there are no ideals of R strictly between $\mathfrak{a}$ and R, i.e. such that every ideal containing strictly $\mathfrak{a}$ is equal to R. A *prime ideal* of R is an ideal $\mathfrak{p} \neq R$ such that, for all $x, y \in R$

satisfying $xy \in \mathfrak{p}$, then either $x \in \mathfrak{p}$ or $y \in \mathfrak{p}$; in other words, $\mathfrak{p}$ is a prime ideal of R if, whenever $\mathfrak{a}$ and $\mathfrak{b}$ are ideals of R such that $\mathfrak{ab} \subseteq \mathfrak{p}$, then either $\mathfrak{a} \subseteq \mathfrak{p}$ or $\mathfrak{b} \subseteq \mathfrak{p}$.

The role of maximal ideals in a commutative ring is illustrated in *Krull's lemma* stating that, in a non-zero commutative ring R, every ideal $\mathfrak{a} \neq R$ is contained in a maximal ideal of R.

Since $R/(0) \simeq R$, we deduce that, if R is an integral domain, then the zero ideal is prime by Lemma 7.10, so that a principal ideal (p) is prime if and only if p is a prime number or zero. It may seem surprising that 0 is excluded from the list of prime numbers, whereas (0) is a prime ideal.

The next result is a characterization of prime and maximal ideals.

Lemma 7.10 *Let R be a ring and $\mathfrak{a}$ be an ideal of R.*

▷ $\mathfrak{a}$ *is prime if and only if $R/\mathfrak{a}$ is an integral domain.*
▷ $\mathfrak{a}$ *is maximal if and only if $R/\mathfrak{a}$ is a field.*

Proof

▷ Assume that $\mathfrak{a}$ is prime. Then $\mathfrak{a} \neq R$ and hence $R/\mathfrak{a} \neq \{\bar{0}\}$. Let $\bar{x}, \bar{y} \in R/\mathfrak{a}$ such that $\bar{x}\,\bar{y} = \bar{0}$. We get $\overline{xy} = \bar{0}$ so that $xy \in \mathfrak{a}$. By definition of a prime ideal, we have $x \in \mathfrak{a}$ or $y \in \mathfrak{a}$ so that $\bar{x} = \bar{0}$ or $\bar{y} = \bar{0}$ and hence $R/\mathfrak{a}$ is an integral domain. Conversely, we have $\mathfrak{a} \neq R$ and if we suppose that $xy \in \mathfrak{a}$ with $x, y \in R$, then $\overline{xy} = \bar{0}$ and then $\bar{x}\,\bar{y} = \bar{0}$. Since $R/\mathfrak{a}$ is an integral domain, this implies that $\bar{x} = \bar{0}$ or $\bar{y} = \bar{0}$, and therefore $x \in \mathfrak{a}$ or $y \in \mathfrak{a}$ so that $\mathfrak{a}$ is prime.

▷ Suppose that $\mathfrak{a}$ is maximal. Then $\mathfrak{a} \neq R$ and hence $R/\mathfrak{a} \neq \{\bar{0}\}$. Let $\bar{x} \neq \bar{0} \in R/\mathfrak{a}$. We have $x \notin \mathfrak{a}$ and the ideal $\mathfrak{a} + Rx$ is such that $\mathfrak{a} + Rx \supsetneq \mathfrak{a}$, and hence $\mathfrak{a} + Rx = R$ since $\mathfrak{a}$ is maximal. One may then write $1 = y + rx$ with $y \in \mathfrak{a}$ and $r \in R$, so that $\bar{1} = \overline{rx} = \bar{r}\,\bar{x}$ which means that $\bar{x}$ is invertible and hence $R/\mathfrak{a}$ is a field. Conversely, we have $\mathfrak{a} \neq R$. Let $\mathfrak{b}$ be an ideal of R such that $\mathfrak{a} \subset \mathfrak{b} \subseteq R$, and let $b \in \mathfrak{b} \setminus \mathfrak{a}$. We then have $\bar{b} \neq \bar{0}$ and thus $\bar{b}$ is invertible, so that there exists $r \in R$ such that $\bar{r}\,\bar{b} = \bar{1}$ and hence $rb = 1 + c$ with $c \in \mathfrak{a}$, so that $1 = rb - c \in \mathfrak{b}$ which means that $\mathfrak{b} = R$, and therefore $\mathfrak{a}$ is maximal.

$\square$

The following lemma is also needed in the sequel.

Lemma 7.11 *Let $\mathbb{K}/\mathbb{Q}$ be an algebraic number field of degree n and $\mathfrak{a}$ be a non-zero ideal in $O_{\mathbb{K}}$. Then $O_{\mathbb{K}}/\mathfrak{a}$ is finite.*

Proof By Corollary 7.5, $O_{\mathbb{K}}$ is a free $\mathbb{Z}$-module of rank n. Since the ideal $\mathfrak{a}$ is non-zero and $\mathbb{Z}$ is a PID, $\mathfrak{a}$ is a free $\mathbb{Z}$-module of rank $r \leqslant n$. It remains to show that we actually have $r = n$. Let $\alpha \neq 0 \in \mathfrak{a}$. If $\mu_{\alpha} = X^r + a_{r-1}X^{r-1} + \cdots + a_0 \in \mathbb{Z}[X]$, then we have

$$a_0 = \alpha \left(-a_1 - \cdots - a_{r-1}\alpha^{r-2} - \alpha^{r-1} \right)$$

and hence $a_0 \in \mathfrak{a}$ since $-a_1 - \cdots - a_{r-1}\alpha^{r-2} - \alpha^{r-1} \in O_{\mathbb{K}}$. Hence we get $a_0 \neq 0 \in \mathfrak{a}$ and $a_0 \in \mathbb{Z}$. Now let $\{\theta_1, \ldots, \theta_n\}$ be an integral basis for $\mathbb{K}$. Since $a_0\theta_i \in \mathfrak{a}$ for all

$i = 1, \ldots, n$, we deduce that $\mathfrak{a}$ contains n $\mathbb{Q}$-linearly independent elements, so that $\mathfrak{a}$ is a free sub-$\mathbb{Z}$-module of rank n of $O_{\mathbb{K}}$ and since $(a_0) \subseteq \mathfrak{a} \subseteq O_{\mathbb{K}}$, the index $(O_{\mathbb{K}} : \mathfrak{a})$ divides $(O_{\mathbb{K}} : (a_0)) = |a_0|^n$ and hence $O_{\mathbb{K}}/\mathfrak{a}$ is finite. $\qquad \Box$

7.3.1.2 Dedekind Domains

We deduce readily from Lemma 7.10 that every maximal ideal is a prime ideal, but the converse is untrue in general, as it may be shown with the ring $R = \mathbb{R}[x, y]$ and the prime ideal (x), which is not maximal since $R/(x) \simeq \mathbb{R}[y]$ is not a field. This leads to the following definition.

Definition 7.12 (Dedekind Domain) Let R be an integral domain. R is a *Dedekind domain* if R is nœtherian, integrally closed[20] and such that every non-zero prime ideal is maximal.

The next result shows that the ring $O_{\mathbb{K}}$ is the right tool to deal with.

Proposition 7.22 *Let $\mathbb{K}/\mathbb{Q}$ be an algebraic number field. Then $O_{\mathbb{K}}$ is a Dedekind domain.*

Proof Let $\mathbb{A}$ be the set of all algebraic integers.

▷ The ring $O_{\mathbb{K}}$ is an integral domain since $\mathbb{A}$ is an integral domain and $O_{\mathbb{K}} = \mathbb{A} \cap \mathbb{K}$.
▷ The quotient field of $O_{\mathbb{K}}$ is $\mathbb{K}$. Indeed, if $\mathbb{F}$ is this quotient field, then it is clear that $\mathbb{F} \subseteq \mathbb{K}$. Conversely, if $\alpha = a/b \in \mathbb{K}$, where a is an algebraic integer and $b \in \mathbb{Z} \setminus \{0\}$, then $a = \alpha b \in \mathbb{K}$ and hence $a \in O_{\mathbb{K}}$, so that $\alpha = a/b \in \mathbb{F}$.
▷ $O_{\mathbb{K}}$ is a finitely generated $\mathbb{Z}$-module and hence $O_{\mathbb{K}}$ is nœtherian.
▷ Let $\alpha \in \mathbb{K}$ be a root of a monic polynomial $P = X^n + a_{n-1}X^{n-1} + \cdots + a_0 \in O_{\mathbb{K}}[X]$. As a $\mathbb{Z}$-module, $\mathbb{Z}[a_0, \ldots, a_{n-1}]$ is finitely generated since $a_i \in O_{\mathbb{K}}$. Set $M = \mathbb{Z}[a_0, \ldots, a_{n-1}, \alpha]$. Since $P(\alpha) = 0$, we can write α^n as a $\mathbb{Z}[a_0, \ldots, a_{n-1}]$-linear combination of α^i for all $0 \leqslant i < n$, so that as a $\mathbb{Z}$-module, M is also finitely generated. Since $\alpha M \subseteq M$, we infer that α is integral over $\mathbb{Z}$ by Proposition 7.5, and hence $\alpha \in O_{\mathbb{K}}$. Therefore $O_{\mathbb{K}}$ is integrally closed.
▷ Let $\mathfrak{p}$ be a non-zero prime ideal of $O_{\mathbb{K}}$. By Lemma 7.11, $O_{\mathbb{K}}/\mathfrak{p}$ is finite. Since $\mathfrak{p}$ is prime, then $O_{\mathbb{K}}/\mathfrak{p}$ is an integral domain by Lemma 7.10. Since a finite integral domain is a field, we infer that $O_{\mathbb{K}}/\mathfrak{p}$ is a field, and therefore the ideal $\mathfrak{p}$ is maximal by Lemma 7.10.

$\qquad \Box$

It follows in particular that $O_{\mathbb{K}}$ is a nœtherian ring so that every element of $O_{\mathbb{K}}$ can be written as a product of irreducible elements. We know that such decompositions may not be unique, but this problem may be overcome by considering ideals instead

---

[20]A domain R is *integrally closed* if α/β is a root of a monic polynomial lying in $R[X]$ with $\alpha, \beta \in R$, $\beta \neq 0$, then $\alpha/\beta \in R$.

of elements of $O_{\mathbb{K}}$. It remains to define more specific ideals which play the part of rational numbers.

7.3.2   Fractional Ideals

The study of the uniqueness of the factorization of ideals requires the behaviour of these ideals under multiplication to be known. It has been seen that this operation has nearly all the properties needed, but inverses need not exist. One may overcome this difficulty if we remember that ideals in $O_{\mathbb{K}}$ are also sub-$O_{\mathbb{K}}$-modules of $O_{\mathbb{K}}$. This gives birth to the concept of *fractional ideal*. Note that the following definition may be generalized in Dedekind domains.

Definition 7.13 Let $\mathbb{K}/\mathbb{Q}$ be an algebraic number field. A *fractional ideal* of $O_{\mathbb{K}}$ is a non-zero sub-$O_{\mathbb{K}}$-module $\mathfrak{a}$ of $\mathbb{K}$ such that there exists $\alpha \in O_{\mathbb{K}}$, $\alpha \neq 0$, satisfying

$$\alpha\mathfrak{a} \subseteq O_{\mathbb{K}}.$$

The collection of fractional ideals of $O_{\mathbb{K}}$ is denoted by $\mathcal{I}(\mathbb{K})$. The collection of principal fractional ideals of $O_{\mathbb{K}}$ is denoted[21] by $\mathcal{P}(\mathbb{K})$. Hence

$$\mathfrak{a} \in \mathcal{P}(\mathbb{K}) \Longleftrightarrow \mathfrak{a} = (a) = aO_{\mathbb{K}} \quad (a \in \mathbb{K}).$$

Obviously, any non-zero ideal of $O_{\mathbb{K}}$ is a fractional ideal by taking $\alpha = 1$. In order to underscore that the fractional ideal $\mathfrak{a}$ is actually contained in $O_{\mathbb{K}}$, we will say that $\mathfrak{a}$ is an *integral ideal*.

We may define the multiplication of fractional ideals in the same way as we did for ideals. This product is associative and commutative with identity element $(1) = O_{\mathbb{K}}$. It remains to define the inverse of a fractional ideal of $\mathbb{K}$ and to show that the set $\mathcal{I}(\mathbb{K})$ is a multiplicative abelian group.

We first define the notion of divisibility for integral ideals. Let $\mathfrak{a}$ and $\mathfrak{b}$ be integral ideals. We shall say that $\mathfrak{a}$ divides $\mathfrak{b}$, written $\mathfrak{a} \mid \mathfrak{b}$ if and only if $\mathfrak{a} \supseteq \mathfrak{b}$ so that

$$\mathfrak{a} \mid \mathfrak{b} \Longleftrightarrow \mathfrak{b} \subseteq \mathfrak{a}.$$

---

[21] In fact, we should rather denote these sets, respectively, by $\mathcal{I}(O_{\mathbb{K}})$ and $\mathcal{P}(O_{\mathbb{K}})$, but we have followed here the usual practice. Nevertheless, if R is any *number ring*, i.e. a domain for which the quotient field $\mathbb{K}$ is an algebraic number field, then the collection of all fractional ideals, resp. principal ideals, of R is denoted by $\mathcal{I}(R)$, resp. $\mathcal{P}(R)$. Similarly, the *class group* or *Picard group* of R, which is an invariant of R, is denoted by $\mathrm{Cl}(R)$ or $\mathrm{Pic}(R)$ and is defined as in Definition 7.18 by $\mathrm{Cl}(R) = \mathcal{I}(R)/\mathcal{P}(R)$. When R is the ring of integers of an algebraic number field $\mathbb{K}$, then $\mathrm{Cl}(O_{\mathbb{K}})$ depends only on $\mathbb{K}$ and is usually denoted by $C\ell(\mathbb{K})$. Hence the class group in the usual sense may be viewed as an invariant of $\mathbb{K}$.

Furthermore, the following observation, coming directly from the definition of a maximal ideal, will often be used.

Let $\mathfrak{p}$ be a maximal ideal of $O_{\mathbb{K}}$ and $\mathfrak{a}$ be an integral ideal such that $(1) \mid \mathfrak{a} \mid \mathfrak{p}$. Then

$$\mathfrak{a} = \mathfrak{p} \quad or \quad \mathfrak{a} = (1). \tag{7.7}$$

The following first lemma will prove useful.

Lemma 7.12 *Let $\mathfrak{a}$ be an integral ideal. Then there exist prime ideals $\mathfrak{p}_1, \ldots, \mathfrak{p}_r$ such that*

$$\mathfrak{a} \mid \mathfrak{p}_1 \cdots \mathfrak{p}_r.$$

Proof Let S be the set of integral ideals that do not satisfy the result of the lemma. If $S \neq \varnothing$ and since $O_{\mathbb{K}}$ is a nœtherian ring, there exists an ideal $\mathfrak{b} \in S$, which is maximal as an element of S. By assumption, $\mathfrak{b}$ is not prime so that there exist $\alpha_1, \alpha_2 \in O_{\mathbb{K}}$ such that $\alpha_1 \alpha_2 \in \mathfrak{b}$ and $\alpha_1 \notin \mathfrak{b}$ and $\alpha_2 \notin \mathfrak{b}$. Now set $\mathfrak{d}_i = \mathfrak{b} + (\alpha_i)$ for $i \in \{1, 2\}$. Since $\mathfrak{b}$ is maximal as an element of S, we have $\mathfrak{d}_i \notin S$, and therefore both ideals contain a product of prime ideals. We deduce that

$$\mathfrak{d}_1 \mathfrak{d}_2 = \mathfrak{b}^2 + \mathfrak{b}(\alpha_1) + \mathfrak{b}(\alpha_2) + (\alpha_1 \alpha_2) \subseteq \mathfrak{b},$$

and hence $\mathfrak{b} \notin S$, giving a contradiction.   □

The next lemma is a technical tool.

Lemma 7.13 *Let $\mathfrak{a}$ be an integral ideal. If $\alpha \in \mathbb{K}$ satisfies $\mathfrak{a} \mid \alpha \mathfrak{a}$, then $\alpha \in O_{\mathbb{K}}$.*

Proof As a finitely generated sub-$\mathbb{Z}$-module of $O_{\mathbb{K}}$ with rank n, let $\beta_1, \ldots, \beta_n$ be a $\mathbb{Z}$-basis of $\mathfrak{a}$. Since $\alpha \mathfrak{a} \subseteq \mathfrak{a}$, there exists an integer matrix $A = (a_{ij}) \in M_n(\mathbb{Z})$ such that, for all $i \in \{1, \ldots, n\}$, we have

$$\alpha \beta_i = \sum_{j=1}^{n} a_{ij} \beta_j,$$

which may be written as

$$(A - \alpha I_n) \begin{pmatrix} \beta_1 \\ \vdots \\ \beta_n \end{pmatrix} = 0$$

so that α is an eigenvalue of A and hence is an algebraic integer.   □

The natural candidate for the inverse of a fractional ideal $\mathfrak{a}$ will turn out to be

$$\mathfrak{a}^{-1} = \{\alpha \in \mathbb{K} : \alpha \mathfrak{a} \subseteq O_{\mathbb{K}}\}.$$

We first check that $\mathfrak{a}^{-1} \in \mathcal{I}(\mathbb{K})$. It is obviously a non-zero sub-$O_{\mathbb{K}}$-module and if $x \in \mathfrak{a}$ with $x \neq 0$, then $x\mathfrak{a}^{-1} \subseteq O_{\mathbb{K}}$ as required. Also note that, if $\mathfrak{a} \mid \mathfrak{b}$, then $\mathfrak{b}^{-1} \mid \mathfrak{a}^{-1}$. In particular, if $\mathfrak{a} \neq 0$ is an integral ideal, then

$$\mathfrak{a}^{-1} \mid (1). \tag{7.8}$$

We now are in a position to show the main result of this section.

Proposition 7.23 $\mathcal{I}(\mathbb{K})$ *is an abelian multiplicative group, called the* ideal group.

Proof It remains to show that, for all $\mathfrak{a} \in \mathcal{I}(\mathbb{K})$, we have

$$\mathfrak{a}\mathfrak{a}^{-1} = (1). \tag{7.9}$$

Let $\mathfrak{p}$ be a prime ideal. By definition of $\mathfrak{p}^{-1}$, we have $(1) \mid \mathfrak{p}^{-1}\mathfrak{p}$ and, using (7.8), we get $\mathfrak{p}^{-1}\mathfrak{p} \mid \mathfrak{p}$ and therefore $(1) \mid \mathfrak{p}^{-1}\mathfrak{p} \mid \mathfrak{p}$, so that by (7.7) either $\mathfrak{p}^{-1}\mathfrak{p} = \mathfrak{p}$ or $\mathfrak{p}^{-1}\mathfrak{p} = (1)$. In the latter case, (7.9) holds for prime ideals, so suppose that $\mathfrak{p} = \mathfrak{p}^{-1}\mathfrak{p}$ and let $\alpha \in \mathfrak{p}$, $\alpha \neq 0$. By Lemma 7.12, there exist prime ideals $\mathfrak{p}_1, \ldots, \mathfrak{p}_r$ such that

$$\mathfrak{p} \mid (\alpha) \mid \mathfrak{p}_1 \cdots \mathfrak{p}_r$$

with r minimal. If $\mathfrak{p} \nmid \mathfrak{p}_i$ for all i, then we can choose for each i an element $c_i \in \mathfrak{p}_i$ such that $c_i \notin \mathfrak{p}$. But then the product of c_i is in $\mathfrak{p}$, which is impossible since $\mathfrak{p}$ is a prime ideal. Therefore there exists a prime ideal $\mathfrak{p}_i$ such that $\mathfrak{p} \mid \mathfrak{p}_i$. Without loss of generality, assume that $\mathfrak{p} \mid \mathfrak{p}_1$. Since $O_{\mathbb{K}}$ is a Dedekind domain, $\mathfrak{p}_1$ is maximal, which implies that $\mathfrak{p} = \mathfrak{p}_1$. Because r is minimal, $(\alpha) \nmid \mathfrak{p}_2 \cdots \mathfrak{p}_r$ so that there exists $\beta \in \mathfrak{p}_2 \cdots \mathfrak{p}_r$ such that $\beta \notin (\alpha)$. Then

$$(\alpha) \mid \mathfrak{p}\mathfrak{p}_2 \cdots \mathfrak{p}_r \mid \mathfrak{p}(\beta),$$

which implies that $a = \beta/\alpha \in \mathfrak{p}^{-1}$. Since $\beta \notin (\alpha)$, we get $a \notin O_{\mathbb{K}}$. Now using $\mathfrak{p} = \mathfrak{p}^{-1}\mathfrak{p}$, we get $\mathfrak{p} \mid a\mathfrak{p}$ and Lemma 7.13 implies that $a \in O_{\mathbb{K}}$, giving a contradiction. Therefore (7.9) is true for prime ideals.

We now prove that (7.9) also holds for integral ideals. Suppose the contrary, there exists a non-zero integral ideal $\mathfrak{a}$, taken maximal among all non-zero integral ideals, which does not have an inverse. Let $\mathfrak{p}$ be a prime ideal such that $\mathfrak{p} \mid \mathfrak{a}$. Multiplying both sides by $\mathfrak{p}^{-1}$ implies that $(1) \mid \mathfrak{p}^{-1}\mathfrak{a} \mid \mathfrak{a}$. Arguing as above, $\mathfrak{a} = \mathfrak{p}^{-1}\mathfrak{a}$ gives a contradiction. By the maximality of $\mathfrak{a}$, we infer that $\mathfrak{p}^{-1}\mathfrak{a}$ has an inverse $\mathfrak{b}$ and hence $\mathfrak{p}^{-1}\mathfrak{b}$ is an inverse of $\mathfrak{a}$, leading to a contradiction. We infer that (7.9) is also true for integral ideals.

Since every fractional ideal $\mathfrak{a}$ can be written as $\mathfrak{a} = d\mathfrak{b}$ for some $d \in \mathbb{K}$ and integral ideal $\mathfrak{b}$, we deduce from above that (7.9) is true for fractional ideals.   □

It is noteworthy that the divisibility relation between integral ideals may be now viewed as the divisibility relation in $\mathbb{Z}$.

Let $\mathfrak{a}$ and $\mathfrak{b}$ be two non-zero integral ideals. Then $\mathfrak{a} \mid \mathfrak{b}$ if and only if there exists a suitable non-zero integral ideal $\mathfrak{c}$ such that $\mathfrak{b} = \mathfrak{a}\mathfrak{c}$.

Indeed, if $\mathfrak{b} = \mathfrak{a}\mathfrak{c}$, then $\mathfrak{b} \subseteq \mathfrak{a}$. Conversely, if $\mathfrak{b} \subseteq \mathfrak{a}$, then we get $\mathfrak{b}\mathfrak{a}^{-1} \subseteq \mathfrak{a}\mathfrak{a}^{-1} = O_{\mathbb{K}}$ so that $\mathfrak{c} = \mathfrak{b}\mathfrak{a}^{-1}$ is an integral ideal satisfying $\mathfrak{b} = \mathfrak{a}\mathfrak{c}$.

7.3.3   The Fundamental Theorem

We are now in a position to solve the problem of uniqueness in algebraic number fields. The following result shows that Kummer's favourite *ideal numbers* are the right tools to generalize the arithmetic in $\mathbb{Z}$.

Theorem 7.6 *Let $\mathfrak{a}$ be a non-zero integral ideal. Then $\mathfrak{a}$ can be written as a product*

$$\mathfrak{a} = \mathfrak{p}_1 \cdots \mathfrak{p}_r$$

of prime ideals of $O_{\mathbb{K}}$, and this decomposition is unique up to order.[22]

Proof Let S be the set of non-zero integral ideals that cannot be written as a product of non-zero prime ideals in $O_{\mathbb{K}}$. If $S \neq \varnothing$ and since $O_{\mathbb{K}}$ is a nœtherian ring, there exists an ideal $\mathfrak{b} \in S$, which is maximal as an element of S. Now let $\mathfrak{p}$ be a non-zero prime ideal such that $\mathfrak{p} \mid \mathfrak{b}$. If $\mathfrak{b}\mathfrak{p}^{-1} = \mathfrak{b}$, then we get $\mathfrak{p}^{-1} = (1)$ by Proposition 7.23, which is false. We deduce that $\mathfrak{b}$ is strictly contained in $\mathfrak{b}\mathfrak{p}^{-1}$ and hence $\mathfrak{b}\mathfrak{p}^{-1} \notin S$. We infer that there exist non-zero prime ideals $\mathfrak{p}_1, \ldots, \mathfrak{p}_r$ such that $\mathfrak{b}\mathfrak{p}^{-1} = \mathfrak{p}_1 \cdots \mathfrak{p}_r$, and therefore we get $\mathfrak{b} = \mathfrak{p}_1 \cdots \mathfrak{p}_r \mathfrak{p}$ so that $\mathfrak{b} \notin S$, leading to a contradiction. Hence $S = \varnothing$ and every non-zero integral ideal can be written as a product of prime ideals. Now let $\mathfrak{a}$ be a non-zero integral ideal and suppose that there exist non-zero prime ideals $\mathfrak{p}_1, \ldots, \mathfrak{p}_r$ and $\mathfrak{q}_1, \ldots, \mathfrak{q}_s$ such that

$$\mathfrak{a} = \mathfrak{p}_1 \cdots \mathfrak{p}_r = \mathfrak{q}_1 \cdots \mathfrak{q}_s.$$

If $\mathfrak{p}_1 \nmid \mathfrak{q}_i$ for all $i \in \{1, \ldots, s\}$, then there exist $c_i \in \mathfrak{q}_i$ such that $c_i \notin \mathfrak{p}_1$. By assumption, the product of c_i is in $\mathfrak{p}_1 \cdots \mathfrak{p}_r$, which is a subset of $\mathfrak{p}_1$, and hence $\mathfrak{p}_1$ is not a prime ideal, giving a contradiction. Thus there exists $i \in \{1, \ldots, s\}$ such that

---

[22]The zero ideal can also be written as a product of prime ideals, but the decomposition is not unique. Hence the case of the zero ideal is almost always excluded in this book.

$\mathfrak{p}_1 \mid \mathfrak{q}_i$, and without loss of generality, suppose that $\mathfrak{p}_1 \mid \mathfrak{q}_1$. Since $\mathfrak{q}_1$ is maximal, we get $\mathfrak{p}_1 = \mathfrak{q}_1$. By Proposition 7.23, we obtain $\mathfrak{p}_2 \cdots \mathfrak{p}_r = \mathfrak{q}_2 \cdots \mathfrak{q}_s$, and arguing as above we show that $\mathfrak{p}_2 = \mathfrak{q}_2$. Repeating the argument enables us to prove that $r = s$ and $\mathfrak{p}_i = \mathfrak{q}_i$ for all $i \in \{1, \ldots, r\}$ as required.   □

Remark 7.2

▷ As a consequence of Theorem 7.6, one can readily see that fractional ideals can also be written uniquely in the following form.

Let $\mathfrak{a} \in \mathcal{I}(\mathbb{K})$. Then there exist prime ideals $\mathfrak{p}_1, \ldots, \mathfrak{p}_r$ and $\mathfrak{q}_1, \ldots, \mathfrak{q}_s$ such that

$$\mathfrak{a} = (\mathfrak{p}_1 \cdots \mathfrak{p}_r)(\mathfrak{q}_1 \cdots \mathfrak{q}_s)^{-1},$$

the decomposition being unique up to order.

▷ As in the rational case, the fractional ideals can be uniquely written in the form

$$\mathfrak{a} = \mathfrak{p}_1^{e_1} \cdots \mathfrak{p}_g^{e_g},$$

where $\mathfrak{p}_i$ are *distinct* prime ideals and $e_i \in \mathbb{Z} \setminus \{0\}$ for all $i \in \{1, \ldots, g\}$. The ideal $(1) = \mathcal{O}_{\mathbb{K}}$ is regarded as the unique empty product of prime ideals. The integers e_i are the $\mathfrak{p}_i$-*adic valuations* of $\mathfrak{a}$, sometimes denoted by $v_{\mathfrak{p}_i}(\mathfrak{a})$. These valuations have the same properties as the ordinary valuations of rational numbers. In particular, if $\mathfrak{p}$ is a prime ideal and $\mathfrak{a}$ and $\mathfrak{b}$ are non-zero fractional ideals of $\mathbb{K}$, we have

$$v_{\mathfrak{p}}(\mathfrak{a}\mathfrak{b}) = v_{\mathfrak{p}}(\mathfrak{a}) + v_{\mathfrak{p}}(\mathfrak{b})$$
$$v_{\mathfrak{p}}(\mathfrak{a} + \mathfrak{b}) = \min\left(v_{\mathfrak{p}}(\mathfrak{a}), v_{\mathfrak{p}}(\mathfrak{b})\right)$$
$$v_{\mathfrak{p}}(\mathfrak{a} \cap \mathfrak{b}) = \max\left(v_{\mathfrak{p}}(\mathfrak{a}), v_{\mathfrak{p}}(\mathfrak{b})\right).$$

Example 7.13 Let $\mathbb{K} = \mathbb{Q}\left(\sqrt{-5}\right)$, which we know is not a UFD since

$$6 = 2 \times 3 = (1 + \sqrt{-5})(1 - \sqrt{-5}).$$

Theorem 7.6 shows how to restore unique factorization. Set

$$\mathfrak{p}_2 = (2, 1 + \sqrt{-5}) = 2\mathcal{O}_{\mathbb{K}} + (1 + \sqrt{-5})\mathcal{O}_{\mathbb{K}}$$

$\mathfrak{p}_3 = (3, 1 + \sqrt{-5})$ and $\mathfrak{p}_3' = (3, 1 - \sqrt{-5})$. One can check that these ideals are prime and

$$(6) = \mathfrak{p}_3\mathfrak{p}_3'\mathfrak{p}_2^2.$$

7.3.4   Applications

Let $\mathbb{K}/\mathbb{Q}$ be an algebraic number field of degree n and $O_{\mathbb{K}}$ be its ring of integers.

7.3.4.1   Generators of Fractional Ideals

As a $\mathbb{Z}$-module of rank n, any integral ideal can be generated by n elements of $O_{\mathbb{K}}$. The purpose of this section is to prove that any fractional ideal can be generated as an $O_{\mathbb{K}}$-module by at most two elements. We start with an application of the Chinese remainder theorem.

Lemma 7.14 *Let $\mathfrak{a}$ and $\mathfrak{b}$ be non-zero integral ideals. Then there exists $\alpha \in \mathfrak{a}$ such that the ideals $(\alpha)\mathfrak{a}^{-1}$ and $\mathfrak{b}$ are coprime.*

Proof Using Theorem 7.6, we can write $\mathfrak{b} = \mathfrak{p}_1^{f_1} \cdots \mathfrak{p}_g^{f_g}$, where $\mathfrak{p}_i$ are distinct prime ideals and f_i are non-negative integers. We set $\mathfrak{a} = \mathfrak{p}_1^{e_1} \cdots \mathfrak{p}_g^{e_g} \mathfrak{c}$ with $\mathfrak{b} + \mathfrak{c} = (1)$ and e_i are non-negative integers. For all $i \in \{1, \ldots, g\}$, we pick up an element $a_i \in \mathfrak{p}_i^{e_i} \setminus \mathfrak{p}_i^{e_i+1}$. Applying the Chinese remainder theorem, we infer that there exists $\alpha \in O_{\mathbb{K}}$ such that

$$\alpha \equiv a_i \pmod{\mathfrak{p}_i^{e_i+1}} \quad \text{and} \quad \alpha \equiv 0 \pmod{\mathfrak{c}}.$$

Since $a_i \in \mathfrak{p}_i^{e_i} \setminus \mathfrak{p}_i^{e_i+1}$, we have $\alpha \equiv 0 \pmod{\mathfrak{p}_i^{e_i}}$ and $\alpha \not\equiv 0 \pmod{\mathfrak{p}_i^{e_i+1}}$ so that we get $(\alpha) = \mathfrak{a}\mathfrak{d}$ with $\mathfrak{b} + \mathfrak{d} = (1)$, which concludes the proof. $\square$

Proposition 7.24 *Let $\mathfrak{a} \in \mathcal{I}(\mathbb{K})$. Then there exist $\alpha, \beta \in \mathbb{K}$ such that*

$$\mathfrak{a} = (\alpha, \beta) = \alpha O_{\mathbb{K}} + \beta O_{\mathbb{K}}.$$

Proof The zero ideal is generated by one element and if $\mathfrak{a} \in \mathcal{I}(\mathbb{K})$ is not an integral ideal, then there exists $d \in \mathbb{K}$ such that $d\mathfrak{a} \in O_{\mathbb{K}}$, and since $\mathfrak{a}$ and $d\mathfrak{a}$ have the same number of generators, we may suppose that $\mathfrak{a}$ is a non-zero integral ideal. Let $\beta \neq 0 \in \mathfrak{a}$ and set $\mathfrak{b} = (\beta)$. By Lemma 7.14, there exists $\alpha \in \mathfrak{a}$ such that the ideals $(\alpha)\mathfrak{a}^{-1}$ and $\mathfrak{b}$ are coprime. Since $\alpha, \beta \in \mathfrak{a}$, we have $\mathfrak{a} \mid (\alpha, \beta)$. Let $\mathfrak{p}$ be a prime divisor of (α, β) and e be its $\mathfrak{p}$-adic valuation. Since $(\alpha)\mathfrak{a}^{-1}$ and $\mathfrak{b}$ are coprime, we infer that $\mathfrak{p}^e \nmid (\alpha)\mathfrak{a}^{-1}$, but since $\mathfrak{p}^e \mid (\alpha) = (\alpha)\mathfrak{a}^{-1}\mathfrak{a}$, we deduce that $\mathfrak{p}^e \mid \mathfrak{a}$. Therefore we have $(\alpha, \beta) \mid \mathfrak{a}$ and hence $\mathfrak{a} = (\alpha, \beta)$ as asserted. $\square$

7.3.4.2   Unique Factorization

It is a well-known fact that if a ring R is a PID, then it is a UFD. We will prove that if R is a ring of integers of an algebraic number field, then the converse is true.

Theorem 7.7 *Let* $\mathbb{K}/\mathbb{Q}$ *be an algebraic number field of degree n. Then* $\mathcal{O}_{\mathbb{K}}$ *is a* UFD *if and only if it is a* PID.

Proof The factorization of elements of $\mathcal{O}_{\mathbb{K}}$ into irreducibles exists since $\mathcal{O}_{\mathbb{K}}$ is nœtherian.[23] Suppose that this factorization is unique. By Theorem 7.6, it suffices to show that every prime ideal is principal. Let $\mathfrak{p}$ be a prime ideal and $\alpha \neq 0 \in \mathfrak{p}$. Let $\alpha = \pi_1 \cdots \pi_r$ be the decomposition of α into irreducibles in $\mathcal{O}_{\mathbb{K}}$. Then $(\alpha) = (\pi_1) \cdots (\pi_r)$. Now every (π) is a prime ideal for if $ab \in (\pi)$, then $\pi \mid ab$ and by assumption of unique factorization we get $\pi \mid a$ or $\pi \mid b$ and hence $a \in (\pi)$ or $b \in (\pi)$. Therefore $\mathfrak{p}$ divides a product of principal prime ideals and, by Theorem 7.6, we infer that $\mathfrak{p}$ is itself one of these principal prime ideals.   □

7.3.5   Norm of an Ideal

7.3.5.1   Definition

Let $\mathfrak{a}$ be a non-zero integral ideal. By Lemma 7.11, the order of $\mathcal{O}_{\mathbb{K}}/\mathfrak{a}$ is finite. This leads to the following definition.

Definition 7.14 Let $\mathbb{K}/\mathbb{Q}$ be an algebraic number field and $\mathfrak{a}$ be a non-zero integral ideal. The *norm* of $\mathfrak{a}$ is the integer defined by

$$\mathcal{N}_{\mathbb{K}/\mathbb{Q}}(\mathfrak{a}) = |\mathcal{O}_{\mathbb{K}}/\mathfrak{a}| .$$

Example 7.14 Let $\mathbb{K} = \mathbb{Q}(\sqrt{-6})$. By Proposition 7.12, we get $\mathcal{O}_{\mathbb{K}} = \mathbb{Z}[\sqrt{-6}]$. Define the ideals $\mathfrak{a} = (1 + \sqrt{-6})$, $\mathfrak{b} = (2, \sqrt{-6})$ and $\mathfrak{c} = (3, \sqrt{-6})$.

▷ Let $\alpha = a + b\sqrt{-6}$, $\beta = c + d\sqrt{-6} \in \mathbb{Z}[\sqrt{-6}]$. We have $\alpha \equiv \beta \pmod{\mathfrak{a}}$ if and only if

$$\frac{a - c + (b - d)\sqrt{-6}}{1 + \sqrt{-6}} \in \mathbb{Z}[\sqrt{-6}].$$

Since

$$\frac{a - c + (b - d)\sqrt{-6}}{1 + \sqrt{-6}} = \frac{a - c + 6(b - d)}{7} + \frac{b - d - a + c}{7}\sqrt{-6}$$

we infer that $\alpha \equiv \beta \pmod{\mathfrak{a}} \iff a - c \equiv b - d \pmod 7$ so that $\mathcal{N}_{\mathbb{K}/\mathbb{Q}}(\mathfrak{a}) = 7$.
▷ One may check that $\mathfrak{b} = \{2a + b\sqrt{-6} : a, b \in \mathbb{Z}\}$ so that $a + b\sqrt{-6} \equiv c + d\sqrt{-6} \pmod{\mathfrak{b}}$ if and only if $a \equiv c \pmod 2$ and hence $\mathcal{N}_{\mathbb{K}/\mathbb{Q}}(\mathfrak{b}) = 2$.

---

[23]In any nœtherian ring, every element can be written as a product of irreducible elements.

▷ Similarly, we have $\mathfrak{c} = \{3a + b\sqrt{-6} : a, b \in \mathbb{Z}\}$ so that $a + b\sqrt{-6} \equiv c + d\sqrt{-6} \pmod{\mathfrak{c}}$ if and only if $a \equiv c \pmod 3$ and hence $\mathcal{N}_{\mathbb{K}/\mathbb{Q}}(\mathfrak{c}) = 3$.

7.3.5.2 Basic Properties

The norm of an integral ideal will in a way generalize the role of the integers. The next result summarizes the first properties of this operator.

Theorem 7.8 *Let $\mathbb{K}/\mathbb{Q}$ be an algebraic number field of degree n with discriminant $d_{\mathbb{K}}$.*

▷ *Let $\mathfrak{a}$ be a non-zero integral ideal and $\{\alpha_1, \ldots, \alpha_n\}$ be a $\mathbb{Z}$-basis for $\mathfrak{a}$. Then*

$$\Delta_{\mathbb{K}/\mathbb{Q}} (\alpha_1, \ldots, \alpha_n) = \left(\mathcal{N}_{\mathbb{K}/\mathbb{Q}}(\mathfrak{a})\right)^2 d_{\mathbb{K}}.$$

▷ *Let $\alpha \neq 0 \in O_{\mathbb{K}}$. Then*

$$\mathcal{N}_{\mathbb{K}/\mathbb{Q}} ((\alpha)) = \left|\mathcal{N}_{\mathbb{K}/\mathbb{Q}}(\alpha)\right|.$$

Proof Let $\{\theta_1, \ldots, \theta_n\}$ be an integral basis for $\mathbb{K}$

▷ Assume that

$$\alpha_i = \sum_{j=1}^n a_{ij}\theta_j$$

with $a_{ij} \in \mathbb{Z}$ for all $i, j \in \{1, \ldots, n\}$. By Proposition 7.1, we have

$$\mathcal{N}_{\mathbb{K}/\mathbb{Q}}(\mathfrak{a}) = |O_{\mathbb{K}}/\mathfrak{a}| = |\det(a_{ij})|$$

and we conclude using Proposition 7.10.

▷ $\{\alpha\theta_1, \ldots, \alpha\theta_n\}$ is a $\mathbb{Z}$-basis for (α), and writing $\alpha\theta_i = \sum_{j=1}^n a_{ij}\theta_j$ with $a_{ij} \in \mathbb{Z}$, we have $\mathcal{N}_{\mathbb{K}/\mathbb{Q}}(\alpha) = \det(a_{ij})$, so that by Proposition 7.10, we get

$$\Delta_{\mathbb{K}/\mathbb{Q}}(\alpha\theta_1, \ldots, \alpha\theta_n) = \mathcal{N}_{\mathbb{K}/\mathbb{Q}}(\alpha)^2 d_{\mathbb{K}}.$$

We conclude by using the previous result, which implies that

$$\Delta_{\mathbb{K}/\mathbb{Q}}(\alpha\theta_1, \ldots, \alpha\theta_n) = \mathcal{N}_{\mathbb{K}/\mathbb{Q}} ((\alpha))^2 d_{\mathbb{K}}$$

and the fact that $\mathcal{N}_{\mathbb{K}/\mathbb{Q}} ((\alpha))$ is a positive integer.

$\square$

7.3.5.3 Multiplicativity

The following result is fundamental.

> **Theorem 7.9** *Let $\mathbb{K}/\mathbb{Q}$ be an algebraic number field and $\mathfrak{a}$ and $\mathfrak{b}$ be non-zero integral ideals. Then*
>
> $$\mathcal{N}_{\mathbb{K}/\mathbb{Q}}(\mathfrak{ab}) = \mathcal{N}_{\mathbb{K}/\mathbb{Q}}(\mathfrak{a})\,\mathcal{N}_{\mathbb{K}/\mathbb{Q}}(\mathfrak{b}).$$

Proof By Theorem 7.6 and induction on the number of factors, it is sufficient to prove

$$\mathcal{N}_{\mathbb{K}/\mathbb{Q}}(\mathfrak{ap}) = \mathcal{N}_{\mathbb{K}/\mathbb{Q}}(\mathfrak{a})\,\mathcal{N}_{\mathbb{K}/\mathbb{Q}}(\mathfrak{p}),$$

where $\mathfrak{p}$ is a prime ideal. First note that, since $\mathfrak{a} \mid \mathfrak{ap}$, using an isomorphism theorem[24] we get

$$\mathcal{N}_{\mathbb{K}/\mathbb{Q}}(\mathfrak{ap}) = \mathcal{N}_{\mathbb{K}/\mathbb{Q}}(\mathfrak{a}) \times |\mathfrak{a}/\mathfrak{ap}|$$

so that it is sufficient to prove that

$$|\mathfrak{a}/\mathfrak{ap}| = \mathcal{N}_{\mathbb{K}/\mathbb{Q}}(\mathfrak{p}). \tag{7.10}$$

First notice that from Theorem 7.6 we deduce that $\mathfrak{ap}$ is strictly contained in $\mathfrak{a}$. Now we will prove that there is no ideal $\mathfrak{b}$ strictly contained between $\mathfrak{a}$ and $\mathfrak{ap}$. Suppose the contrary, we then have $\mathfrak{a} \mid \mathfrak{b} \mid \mathfrak{ap}$ and multiplying by $\mathfrak{a}^{-1}$ gives $(1) \mid \mathfrak{a}^{-1}\mathfrak{b} \mid \mathfrak{p}$, and hence $\mathfrak{a}^{-1}\mathfrak{b} = (1)$ or $\mathfrak{a}^{-1}\mathfrak{b} = \mathfrak{p}$ by (7.7), so that $\mathfrak{b} = \mathfrak{a}$ or $\mathfrak{b} = \mathfrak{ap}$ as required. Now choose $a \in \mathfrak{a} \setminus \mathfrak{ap}$ and define the map

$$\begin{aligned} F_a : \mathcal{O}_{\mathbb{K}} &\longrightarrow \mathfrak{a}/\mathfrak{ap} \\ \alpha &\longmapsto a\alpha + \mathfrak{ap} \end{aligned}$$

Since there are no ideals between $\mathfrak{a}$ and $\mathfrak{ap}$, this implies that $\mathfrak{a}$ is generated by a and $\mathfrak{ap}$, so that F_a is surjective. Furthermore, if $\alpha \in \mathfrak{p}$, then $a\alpha \in \mathfrak{ap}$ and hence $F_a(\alpha) = 0 + \mathfrak{ap}$ so that $\mathfrak{p} \subseteq \ker F_a$. Since $\mathfrak{p}$ is maximal, we get $\ker F_a$ is either (1) or $\mathfrak{p}$. But $F_a(1) = a + \mathfrak{ap} \neq 0 + \mathfrak{ap}$ since $a \notin \mathfrak{a}\,\mathfrak{p}$, so $\ker F_a = \mathfrak{p}$. By an isomorphism theorem,[25] we deduce that $\mathcal{O}_{\mathbb{K}}/\mathfrak{p} \simeq \mathfrak{a}/\mathfrak{ap}$ proving (7.10).   □

---

[24] If $\mathfrak{a}_1 \subset \mathfrak{a}_2$ are two ideals of a commutative ring R, then $R/\mathfrak{a}_2 \simeq (R/\mathfrak{a}_1)\,/\,(\mathfrak{a}_2/\mathfrak{a}_1)$.

[25] Let R and L be commutative rings and $f : R \to L$ be a ring homomorphism. Then $R/\ker f \simeq \operatorname{Im} f$.

Thus the norm map is completely multiplicative, which is certainly its most important property. Note also that, for each non-zero integral ideal $\mathfrak{a}$, we have

$$N_{\mathbb{K}/\mathbb{Q}}(\mathfrak{a}) \in \mathfrak{a} \qquad (7.11)$$

by applying Lagrange's theorem seen in Theorem 7.1 to the additive group $O_{\mathbb{K}}/\mathfrak{a}$. Furthermore, if $\sigma : \mathbb{K} \to \mathbb{K}$ is any field homomorphism, then

$$N_{\mathbb{K}/\mathbb{Q}}(\sigma \mathfrak{a}) = N_{\mathbb{K}/\mathbb{Q}}(\mathfrak{a}). \qquad (7.12)$$

For instance, let $\mathbb{K} = \mathbb{Q}(\sqrt{-6})$ and take $\mathfrak{b} = (2, \sqrt{-6})$ of Example 7.14. If σ is the non-trivial embedding of $\mathbb{K}$, we have $\sigma \mathfrak{b} = (2, -\sqrt{-6})$ and hence by (7.12) and Theorem 7.9, we get

$$N_{\mathbb{K}/\mathbb{Q}}(\mathfrak{b})^2 = N_{\mathbb{K}/\mathbb{Q}}(\mathfrak{b}\sigma\mathfrak{b}) = N_{\mathbb{K}/\mathbb{Q}}(4, 2\sqrt{-6}, -2\sqrt{-6}, -6)$$
$$= N_{\mathbb{K}/\mathbb{Q}}((2)) = |N_{\mathbb{K}/\mathbb{Q}}(2)| = 4.$$

The next result provides some useful consequences of Theorem 7.9.

Corollary 7.6 *Let $\mathbb{K}/\mathbb{Q}$ be an algebraic number field of degree n.*

▷ *Let $\mathfrak{p}$ be a non-zero integral ideal. If $N_{\mathbb{K}/\mathbb{Q}}(\mathfrak{p})$ is a prime number, then $\mathfrak{p}$ is a prime ideal.*
▷ *There are only finitely many integral ideals with a given norm.*

Proof

▷ This follows from the complete multiplicativity of the norm.
▷ If $N_{\mathbb{K}/\mathbb{Q}}(\mathfrak{a}) = m$ for some positive integer m, then $\mathfrak{a} \mid (m)$ by (7.11). By uniqueness of factorization, m has only finitely many factors.

□

Example 7.15

▷ Let $\mathbb{K} = \mathbb{Q}(\sqrt{-10})$. We have[26]

$$(14) = \mathfrak{p}_1 \mathfrak{p}_2 \mathfrak{p}_3^2,$$

where $\mathfrak{p}_1 = (7, 2 + \sqrt{-10})$, $\mathfrak{p}_2 = \sigma \mathfrak{p}_1 = (7, 2 - \sqrt{-10})$ and $\mathfrak{p}_3 = (2, \sqrt{-10})$. Suppose that 14 lies in some integral ideal $\mathfrak{a}$. Then $\mathfrak{a} \mid (14)$ so that by Theorem 7.6 we get

$$\mathfrak{a} = \mathfrak{p}_1^{e_1} \mathfrak{p}_2^{e_2} \mathfrak{p}_3^{e_3}$$

---

[26]For instance, using the PARI/GP system.

with $e_1, e_2 \in \{0, 1\}$ and $e_3 \in \{0, 1, 2\}$ and hence 14 belongs only to a finite number of ideals. Also by Theorem 7.8, we get $N_{\mathbb{K}/\mathbb{Q}}((14)) = |N_{\mathbb{K}/\mathbb{Q}}(14)| = 14^2$. Now how many ideals $\mathfrak{a}$ have norm 14? This can only happen when $\mathfrak{a} \mid (14)$ by (7.11), so that by above we get

$$14 = N_{\mathbb{K}/\mathbb{Q}}(\mathfrak{a}) = 7^{e_1 + e_2} \times 2^{e_3},$$

which implies that $e_1 + e_2 = e_3 = 1$, giving two integral ideals. Note also that by a similar argument we have only one ideal with norm 2 and two ideals with norm 7. Let $r_{\mathbb{K}}(m)$ be the number of integral ideals with norm m. This example shows that $r_{\mathbb{K}}(2 \times 7) = r_{\mathbb{K}}(2) r_{\mathbb{K}}(7)$. We shall return to this function in Sect. 7.4.

▷ Let $\mathbb{K} = \mathbb{Q}(\zeta_8)$ with defining polynomial $\Phi_8 = X^4 + 1$. We have $d_{\mathbb{K}} = 256$ and Theorem 7.11 below gives

$$(2) = \mathfrak{p}^4$$

with $\mathfrak{p} = (2, 1 + \zeta_8)$. A similar argument shows that $\mathfrak{p}$ is the only prime ideal with norm 2. Let us prove that $\mathfrak{p}$ is principal. It is first clear that $\mathfrak{p} \subseteq (1 + \zeta_8)$. Conversely, since

$$2 = (1 + \zeta_8)^2 (\zeta_8 - \zeta_8^2)$$

we infer that $(1 + \zeta_8) \subseteq \mathfrak{p}$. Therefore $\mathfrak{p} = (1 + \zeta_8)$ and then $\mathfrak{p}$ is principal. We will show later that $O_{\mathbb{K}}$ is actually a PID.

7.3.6   Factorization of (p)

7.3.6.1   Primes Above Primes

By definition, if $\mathfrak{p}$ is a prime ideal in $O_{\mathbb{K}}$, then $\mathfrak{p} \cap \mathbb{Z}$ is a prime ideal in $\mathbb{Z}$ and is therefore of the form (p) for some prime number p. This leads to the following definition.

Definition 7.15 Let $\mathbb{K}/\mathbb{Q}$ be an algebraic number field, p be a prime number and $\mathfrak{p}$ be a prime ideal of $O_{\mathbb{K}}$. We shall say that $\mathfrak{p}$ *lies above* p, written $\mathfrak{p} \mid p$, if $\mathfrak{p} \cap \mathbb{Z} = p\mathbb{Z}$.

When $\mathfrak{p}$ lies above p, then $p \in \mathfrak{p}$ so that $\mathfrak{p} \mid (p)$ as ideals of $O_{\mathbb{K}}$, which explains the notation. We also notice that, for any prime ideal $\mathfrak{p}$ of $O_{\mathbb{K}}$, there exists a unique prime number p such that $\mathfrak{p}$ lies above p. Indeed, suppose that there is another prime number $q \neq p$ such that $\mathfrak{p} \mid q$. We have $\mathfrak{p} \mid (p) + (q) = (1)$, which is impossible since $\mathfrak{p}$ is maximal.

To find prime ideals of $O_{\mathbb{K}}$, we need to factorize ideals generated by prime numbers. Pay careful attention to the fact that the ideal (p) is not in general a prime ideal of $O_{\mathbb{K}}$. Theorem 7.11 below, due to Kummer for a particular case and extended by Dedekind, relates the prime ideal factorization of (p) to the decomposition of $\overline{\mu}_\theta$ into irreducible polynomials in $\mathbb{F}_p[X]$. Kummer–Dedekind's theorem holds for any algebraic number field $\mathbb{K} = \mathbb{Q}(\theta)$ such that p does not divide the index f of θ. We do not have a similar result in the case where $p \mid f$, but algorithms do exist which perform the factorization of (p) in this case. However, recall that in practice $p^2 \nmid \Delta_{\mathbb{K}/\mathbb{Q}}(1, \theta, \ldots, \theta^{n-1})$ is sufficient to ensure that $p \nmid f$, and hence all but finitely many prime numbers p are covered by this result. We start with the following general situation.

Lemma 7.15 *Let $\mathbb{K}/\mathbb{Q}$ be an algebraic number field of degree n, p be a prime number and*

$$(p) = \mathfrak{p}_1^{e_1} \cdots \mathfrak{p}_g^{e_g} \tag{7.13}$$

be the factorization of (p) into prime ideals.

▷ *There exist positive integers f_i such that*

$$\mathcal{N}_{\mathbb{K}/\mathbb{Q}}(\mathfrak{p}_i) = p^{f_i} \quad and \quad \sum_{i=1}^{g} e_i f_i = n.$$

▷ *The index $[O_\mathbb{K}/\mathfrak{p}_i : \mathbb{Z}/p\mathbb{Z}]$ is finite and $f_i = [O_\mathbb{K}/\mathfrak{p}_i : \mathbb{Z}/p\mathbb{Z}]$.*

Proof

▷ First, since $\mathfrak{p}_i \mid (p)$, we get $\mathcal{N}_{\mathbb{K}/\mathbb{Q}}(\mathfrak{p}_i) \mid \mathcal{N}_{\mathbb{K}/\mathbb{Q}}((p)) = p^n$, so that there exist integers $f_i \in \{1, \ldots, n\}$ such that $\mathcal{N}_{\mathbb{K}/\mathbb{Q}}(\mathfrak{p}_i) = p^{f_i}$. Furthermore, since $\mathfrak{p}_i^{e_i} + \mathfrak{p}_j^{e_j} = (1)$ for all $i \neq j$, the Chinese remainder theorem implies that

$$O_\mathbb{K}/(p) \simeq O_\mathbb{K}/\mathfrak{p}_1^{e_1} \oplus \cdots \oplus O_\mathbb{K}/\mathfrak{p}_g^{e_g}$$

and hence, by Theorem 7.9, we get

$$p^n = \mathcal{N}_{\mathbb{K}/\mathbb{Q}}((p)) = \mathcal{N}_{\mathbb{K}/\mathbb{Q}}(\mathfrak{p}_1^{e_1}) \cdots \mathcal{N}_{\mathbb{K}/\mathbb{Q}}(\mathfrak{p}_g^{e_g}) = \mathcal{N}_{\mathbb{K}/\mathbb{Q}}(\mathfrak{p}_1)^{e_1} \cdots \mathcal{N}_{\mathbb{K}/\mathbb{Q}}(\mathfrak{p}_g)^{e_g},$$

which implies the asserted result.
▷ The map

$$\mathbb{Z}/p\mathbb{Z} \longrightarrow O_\mathbb{K}/\mathfrak{p}_i$$
$$\overline{a} \longmapsto a + \mathfrak{p}_i$$

is well-defined and one may check that it is a homomorphism. Now $\mathbb{Z}/p\mathbb{Z} \simeq \mathbb{F}_p$ is a finite field, and the same is true for $O_\mathbb{K}/\mathfrak{p}_i$ by Lemmas 7.10 and 7.11 and the fact that $\mathfrak{p}_i$ is maximal. Hence the map is injective and $[O_{\mathbb{K}}/\mathfrak{p}_i : \mathbb{Z}/p\mathbb{Z}]$ is the dimension of $O_{\mathbb{K}}/\mathfrak{p}_i$ considered as a $\mathbb{Z}/p\mathbb{Z}$-vector space. Furthermore, we have

$$\mathcal{N}_{\mathbb{K}/\mathbb{Q}}(\mathfrak{p}_i) = |O_{\mathbb{K}}/\mathfrak{p}_i| = p^{[O_{\mathbb{K}}/\mathfrak{p}_i:\mathbb{Z}/p\mathbb{Z}]}$$

and we conclude by using the previous result.

$\square$

7.3.6.2 Ramification

Definition 7.16 Let $\mathbb{K}/\mathbb{Q}$ be an algebraic number field of degree n and p be a prime number and consider the factorization (7.13) of (p) into prime ideals.

▷ The integer e_i is the *ramification index* of $\mathfrak{p}_i$ over $\mathbb{Z}$, denoted by $e(\mathfrak{p}_i \mid p)$.
▷ The integer f_i is the *inertial degree*, or *residue class degree*, of $\mathfrak{p}_i$ over $\mathbb{Z}$, denoted by $f(\mathfrak{p}_i \mid p)$.
▷ The integer g is the *decomposition number* of p over $\mathbb{Z}$, denoted by g_p.
▷ The prime number p is said to be *ramified* in $O_{\mathbb{K}}$ if and only if there exists some $i \in \{1, \ldots, g\}$ such that $e_i \geqslant 2$. Similarly, the prime ideals $\mathfrak{p}_i$ such that $e_i \geqslant 2$ are called *ramified prime ideals*.
▷ The prime number p is said to be *unramified* in $O_{\mathbb{K}}$ if $e_1 = \cdots = e_g = 1$. Two cases are then possible.

$$p \text{ is } inert \quad \begin{cases} g = 1 \\ e_1 = 1, \ f_1 = n \end{cases} \quad (p) = \mathfrak{p}$$

$$p \text{ } splits \text{ } completely \quad \begin{cases} g = n \\ e_i = f_i = 1 \end{cases} \quad (p) = \mathfrak{p}_1 \cdots \mathfrak{p}_n$$

Note that there are intermediate cases which do not deserve a special name. The ramified primes are characterized by the following result.

Theorem 7.10 *Let $\mathbb{K}/\mathbb{Q}$ be an algebraic number field with discriminant $d_{\mathbb{K}}$ and p be a prime number. Then p is ramified in $\mathbb{K}$ if and only if $p \mid d_{\mathbb{K}}$. In particular, there are $\omega(|d_{\mathbb{K}}|)$ ramified primes.*

We do not supply here a proof of this theorem but, as often in algebraic number theory, the result is more important than the proof itself. As a corollary, we provide the following criterion.[27]

---

[27] See also [1, 85].

Corollary 7.7 Let $\mathbb{K} = \mathbb{Q}(\theta)$ be an algebraic number field of degree n, discriminant $d_{\mathbb{K}}$, $\theta \in O_{\mathbb{K}}$ and p be a prime number. Let $\mu_\theta = X^n + a_{n-1}X^{n-1} + \cdots + a_0 \in \mathbb{Z}[X]$ be the minimal polynomial of θ. Suppose that $p \parallel a_0$ and $p \mid a_1$. Then $p \mid d_{\mathbb{K}}$.

Proof Suppose that p is not ramified. Then there exist distinct prime ideals $\mathfrak{p}_1, \ldots, \mathfrak{p}_g$ such that

$$(p) = \mathfrak{p}_1 \cdots \mathfrak{p}_g.$$

Since $p \parallel a_0$, we get $(a_0) = \mathfrak{p}_1 \cdots \mathfrak{p}_g(b)$ for some $b \in \mathbb{Z}$ with $p \nmid b$ and hence $\mathfrak{p}_i \nmid (b)$ for all $i \in \{1, \ldots, g\}$. Now $N_{\mathbb{K}/\mathbb{Q}}(\theta) = (-1)^n a_0 \equiv 0 \pmod p$ so that there exists a prime ideal $\mathfrak{p} \in \{\mathfrak{p}_1, \ldots, \mathfrak{p}_g\}$ such that $\mathfrak{p} \mid (\theta)$. As $\mathfrak{p} \mid (a_1)$, we get

$$a_0 = a_0 - \mu_\theta(\theta) = -a_1\theta - \cdots - a_{n-1}\theta^{n-1} - \theta^n$$

so that $\mathfrak{p}^2 \mid (a_0)$ contradicting the assumption $p \parallel a_0$. Hence p is ramified and we conclude by using Theorem 7.10.   □

7.3.6.3   Galois Extensions

In the case of Galois extensions, the result is simpler.

Proposition 7.25 Let $\mathbb{K}/\mathbb{Q}$ be a Galois extension of degree n and p be a prime number. Then the ramification indexes are all equal, say to e_p, and the inertial degrees are all equal, say to f_p. Hence we have

$$(p) = \left(\prod_{i=1}^{g_p} \mathfrak{p}_i\right)^{e_p}$$

with $e_p f_p g_p = n$. Furthermore, $\mathrm{Gal}(\mathbb{K}/\mathbb{Q})$ operates transitively on the prime ideal above p, i.e. for all prime ideals $\mathfrak{p}_i$, $\mathfrak{p}_j$ above p, there exists $\sigma \in \mathrm{Gal}(\mathbb{K}/\mathbb{Q})$ such that $\sigma(\mathfrak{p}_i) = \mathfrak{p}_j$.

7.3.6.4   The Theorem of Kummer–Dedekind

We now state the main result of this section. The idea is to prove and use the fact that, if $p \nmid f$, then

$$O_{\mathbb{K}}/(p) \simeq \mathbb{F}_p[X]/(\overline{\mu_\theta}).$$

Theorem 7.11 (Kummer–Dedekind) *Let* $\mathbb{K} = \mathbb{Q}(\theta)$ *be an algebraic number field of degree* n *with* $\theta \in O_{\mathbb{K}}$, *and* p *be a prime number such that* $p \nmid f$. *Let* $\mu \in \mathbb{Z}[X]$ *be the minimal polynomial of* θ *and suppose that* μ *factorizes over* $\mathbb{F}_p[X]$ *as*

$$\overline{\mu} = \prod_{i=1}^{g} \overline{P_i}^{e_i},$$

where $\overline{P_i}$ *are distinct monic irreducible polynomials in* $\mathbb{F}_p[X]$. *Then we have*

$$(p) = \prod_{i=1}^{g} \mathfrak{p}_i^{e_i},$$

where $\mathfrak{p}_i s$ *are pairwise distinct and given by* $\mathfrak{p}_i = (p, P_i(\theta)) = pO_{\mathbb{K}} + P_i(\theta)O_{\mathbb{K}}$, *where* P_i *is any lifting of* $\overline{P_i}$ *in* $\mathbb{Z}[X]$. *We also have*

$$\mathcal{N}_{\mathbb{K}/\mathbb{Q}}(\mathfrak{p}_i) = p^{\deg \overline{P_i}}.$$

Proof Let p be a prime number such that $p \nmid f$.

▷ Step 1. We will prove that

$$O_{\mathbb{K}}/(p) \simeq \mathbb{F}_p[X]/(\overline{\mu}). \tag{7.14}$$

Let $\{\alpha_1, \ldots, \alpha_n\}$ be an integral basis for $\mathbb{K}$. Then $\{1, \theta, \ldots, \theta^{n-1}\}$ is a $\mathbb{Z}$-basis for $\mathbb{Z}[\theta]$ for which there exists a matrix $A = (a_{ij}) \in \mathcal{M}_n(\mathbb{Z})$ such that $\theta^{i-1} = \sum_{j=1}^{n} a_{ij}\alpha_j$ for all $i \in \{1, \ldots, n\}$. Since $\det(a_{ij}) = \pm f$ and $p \nmid f$, we infer that the reduction matrix $\overline{A} = (\overline{a_{ij}}) \in \mathcal{M}_n(\mathbb{F}_p)$ is invertible and therefore induces an isomorphism

$$O_{\mathbb{K}}/(p) \simeq \mathbb{Z}[\theta]/p\mathbb{Z}[\theta].$$

Now since μ is irreducible over $\mathbb{Z}$, we have $\mathbb{Z}[X]/(\mu) \simeq \mathbb{Z}[\theta]$, which implies that

$$\mathbb{Z}[\theta]/p\mathbb{Z}[\theta] \simeq \mathbb{Z}[X]/(\mu, p) \simeq \mathbb{F}_p[X]/(\overline{\mu}),$$

and hence (7.14) is proved.

▷ Step 2. Let $(p) = \mathfrak{q}_1^{k_1} \cdots \mathfrak{q}_r^{k_r}$ given by Theorem 7.6 for some pairwise distinct prime ideals $\mathfrak{q}_i$ and positive integers r and k_i.

◦ The maximal ideals of $O_{\mathbb{K}}/(p)$ are of the form $\mathfrak{q}/(p)$, where $\mathfrak{q}$ is a prime ideal dividing (p). We infer that $\mathfrak{q} \in \{\mathfrak{q}_1, \ldots, \mathfrak{q}_r\}$ and r is thus the number of maximal ideals in $O_{\mathbb{K}}/(p)$, which are then of the form $\mathfrak{q}_i/(p)$ for all $i \in \{1, \ldots, r\}$.

◦ Similarly, we infer that the maximal ideals of $\mathbb{F}_p[X]/(\overline{\mu})$ are the ideals $(\overline{P_i})/(\overline{\mu})$, and hence g counts the number of maximal ideals in $\mathbb{F}_p[X]/(\overline{\mu})$.

Hence we have proved that r is the number of maximal ideals in $O_{\mathbb{K}}/(p)$ and g is that of maximal ideals in $\mathbb{F}_p[X]/(\overline{\mu})$. By (7.14), we get $r = g$.

▷ Step 3. For all $i \in \{1, \ldots, g\}$, let P_i be any lifting of $\overline{P_i}$ in $\mathbb{Z}[X]$ and set $\mathfrak{p}_i = (p, P_i(\theta))$. Similarly as for (7.14), one can prove the following isomorphisms for all $i \in \{1, \ldots, g\}$

$$O_{\mathbb{K}}/\mathfrak{p}_i \simeq \mathbb{Z}[\theta]/(p, P_i(\theta))\mathbb{Z}[\theta] \simeq \mathbb{Z}[X]/(p, P_i(X))\mathbb{Z}[X] \simeq \mathbb{F}_p[X]/(\overline{P_i}).$$
(7.15)

Since the polynomials $\overline{P_i}$ are irreducible over $\mathbb{F}_p[X]$, the ideals $(\overline{P_i})$ are maximal. Therefore (7.15) implies that the ideals $\mathfrak{p}_i$ are maximal and hence are prime ideals of $O_{\mathbb{K}}$. We also infer that $N_{\mathbb{K}/\mathbb{Q}}(\mathfrak{p}_i) = p^{\deg \overline{P_i}}$. Furthermore, we have

$$\mathfrak{p}_1^{e_1} \cdots \mathfrak{p}_g^{e_g} = (p, P_1(\theta))^{e_1} \cdots (p, P_g(\theta))^{e_g} \subseteq (p, P_1(\theta)^{e_1} \cdots P_g(\theta)^{e_g})$$

and since $\mu \equiv P_1^{e_1} \cdots P_g^{e_g} \pmod{p}$ and $\mu(\theta) = 0$, we get $P_1(\theta)^{e_1} \cdots P_g(\theta)^{e_g} \in p\mathbb{Z}[\theta]$ and hence

$$\mathfrak{p}_1^{e_1} \cdots \mathfrak{p}_g^{e_g} \subseteq (p).$$

On the other hand, since $\mathfrak{p}_i = (p, P_i(\theta)) \supseteq (p)$, we get $\mathfrak{p}_i \mid (p)$ for all $i \in \{1, \ldots, g\}$. This implies that

$$(p) = \mathfrak{p}_1^{a_1} \cdots \mathfrak{p}_g^{a_g}$$

for some non-negative integers a_i satisfying $a_i \leqslant e_i$ for all $i \in \{1, \ldots, g\}$. Comparing the norms as in Lemma 7.15, we obtain

$$n = a_1 \deg \overline{P_1} + \cdots + a_g \deg \overline{P_g}$$

and by the Chinese remainder theorem we also have

$$\mathbb{F}_p[X]/(\overline{\mu}) \simeq \bigoplus_{i=1}^{g} \mathbb{F}_p[X]/(\overline{P_i})^{e_i}$$

implying that

$$n = e_1 \deg \overline{P_1} + \cdots + e_g \deg \overline{P_g}.$$

Setting $d_i = \deg \overline{P_i}$ and summarizing the above results, we have finally obtained $a_i \leqslant e_i$ and $a_1 d_1 + \cdots + a_g d_g = e_1 d_1 + \cdots + e_g d_g$, which implies that $a_i = e_i$ for all $i \in \{1, \ldots, g\}$. The proof is complete.

$\square$

Remark 7.3 The fact that $\mathfrak{p}_i = (p, P_i(\theta))$ is not really useful in practice. What is more important yet is to know the *nature* of the prime ideals above p, e.g. whether they are ramified or not.

7.3.7 *Quadratic Fields*

7.3.7.1 The Legendre–Jacobi–Kronecker Symbol

Definition 7.17 Let p be an odd prime number and n be a positive integer. The *Legendre symbol* (n/p) is defined by $(n/p) = 0$ if $p \mid n$, and otherwise

$$\left(\frac{n}{p}\right) = \begin{cases} 1, & \text{if } n \text{ is residue quadratic } \bmod p \\ -1, & \text{if } n \text{ is not residue quadratic } \bmod p. \end{cases}$$

One of the main properties of this symbol is the *quadratic reciprocity law* stating that, if p and q are distinct odd primes

$$\left(\frac{p}{q}\right)\left(\frac{q}{p}\right) = (-1)^{(p-1)(q-1)/4}.$$

We also have the two *complementary laws*

$$\left(\frac{-1}{p}\right) = (-1)^{(p-1)/2} \quad \text{and} \quad \left(\frac{2}{p}\right) = (-1)^{(p^2-1)/8}.$$

There are a lot of proofs in the literature. Nevertheless, it is noteworthy that the Legendre symbol (n/p) is a real primitive Dirichlet character. In fact, we have the following useful result.[28]

---

[28] See [15].

Lemma 7.16 *Let $q \geq 2$ be an integer and χ be a real primitive Dirichlet character modulo q. If p is an odd prime, then*

$$\chi(p) = \left(\frac{\chi(-1)q}{p} \right).$$

As for all Dirichlet characters, the Legendre symbol may be extended to all integers by complete multiplicativity. More precisely, if $m, n \in \mathbb{Z}$, the *Jacobi–Kronecker symbol*, still denoted by (n/m), is defined in the following way:

▷ $(n/1) = 1$, $(n/-1) = \mathrm{sgn}(n)$, $(-1/2) = 1$, $(-1/n) = (-1)^{(n-1)/2}$ if $n \geq 1$ is odd, and

$$\left(\frac{n}{2} \right) = \begin{cases} (-1)^{(n^2-1)/8}, & \text{if } 2 \nmid n \\ \\ 0, & \text{otherwise.} \end{cases}$$

▷ If $m = \pm p_1^{e_1} \cdots p_r^{e_r}$, then

$$\left(\frac{n}{m} \right) = \left(\frac{n}{\pm 1} \right) \prod_{i=1}^{r} \left(\frac{n}{p_i} \right)^{e_i},$$

where (n/p_i) are Legendre symbols. A generalized quadratic reciprocity law still exists with this symbol. *Let $m, n \in \mathbb{Z} \setminus \{0\}$ and set $n = 2^e n_1$ and $m = 2^f m_1$ with n_1, m_1 odd. Then*

$$\left(\frac{n}{m} \right) = (-1)^{((m_1-1)(n_1-1)+(\mathrm{sgn}(m)-1)(\mathrm{sgn}(n)-1))/4} \left(\frac{m}{n} \right).$$

In particular, if D_1 and D_2 are non-zero integers congruent to 0 or 1 modulo 4, then

$$\left(\frac{D_2}{D_1} \right) = (-1)^{((\mathrm{sgn}(D_1)-1)(\mathrm{sgn}(D_2)-1))/4} \left(\frac{D_1}{D_2} \right).$$

Lemma 7.16 was generalized by Dirichlet for Kronecker symbols.[29]

Lemma 7.17 *Let χ be a real primitive Dirichlet character modulo q. Then $D = \chi(-1)q$ is a discriminant of a quadratic field and*

$$\chi(n) = \left(\frac{D}{n} \right).$$

---

[29] See [15, 77].

*Conversely, if D is a discriminant of a quadratic field, then the Kronecker symbol
defines a real primitive Dirichlet character modulo $q = |D|$.*

7.3.7.2   Prime Ideal Decomposition of (p)

Let $d \in \mathbb{Z} \setminus \{0, 1\}$ squarefree and $\mathbb{K} = \mathbb{Q}(\sqrt{d})$ be a quadratic field with discriminant
$d_{\mathbb{K}}$. By Proposition 7.12, $\{1, \omega\}$ is an integral basis for $\mathbb{K}$, where $\omega = \frac{1}{2}\left(d_{\mathbb{K}} + \sqrt{d_{\mathbb{K}}}\right)$
in every case. Applying Corollary 7.7 and Theorem 7.11, we derive the following
result.

Proposition 7.26 *Let p be a prime number and $(d_{\mathbb{K}}/\cdot)$ be a Kronecker symbol.
Then we have*

$(d_{\mathbb{K}}/p)$	*Factorization of* (p)	*Nature*	*Remarks*
-1	$(p) = \mathfrak{p}$	*Inert*	
1	$(p) = \mathfrak{p}_1 \mathfrak{p}_2$	*Completely split*	$\mathfrak{p}_i = \left(p, \omega - \frac{d_{\mathbb{K}} \pm b}{2}\right)$ *with b satisfying* $b^2 \equiv d_{\mathbb{K}} \pmod{4p}$
0	$(p) = \mathfrak{p}^2$	*Totally ramified*	$\mathfrak{p} = (p, \omega)$ *except when $p = 2$ and $d_{\mathbb{K}} = 16$, where $\mathfrak{p} = (p, 1 + \omega)$*

7.3.7.3   The Conductor of a Quadratic Field

Proposition 7.27 *Let $d \in \mathbb{Z} \setminus \{0, 1\}$ squarefree and $\mathbb{K} = \mathbb{Q}(\sqrt{d})$ be a quadratic
field with discriminant $d_{\mathbb{K}}$. Then $\mathbb{K}$ is a subfield of $\mathbb{Q}(\zeta_{|d_{\mathbb{K}}|})$.*

Proof

 ▷ If $d_{\mathbb{K}}$ is odd, we have $d_{\mathbb{K}} = d \equiv 1 \pmod 4$. Write $d = \pm \prod_i p_i$, where p_i
   are pairwise distinct odd prime numbers. Define $p_i^* = (-1)^{(p_i-1)/2} p_i$ so that

$p_i^* \equiv 1 \pmod 4$ and $d_{\mathbb{K}} = \prod_i p_i^*$. Using quadratic Gauss sums, one can show[30] that for all odd primes p

$$\sum_{j \pmod p} \left(\frac{j}{p}\right) \zeta_p^j = \sqrt{p^*},$$

which implies that $\mathbb{Q}(\sqrt{p_i^*}) \subset \mathbb{Q}(\zeta_{p_i})$. We infer that $\mathbb{Q}(\sqrt{p_i^*}) \subset \mathbb{Q}(\zeta_{|d_{\mathbb{K}}|})$, and hence each p_i^* is a square in $\mathbb{Q}(\zeta_{|d_{\mathbb{K}}|})$, so that $d_{\mathbb{K}}$ is also a square in $\mathbb{Q}(\zeta_{|d_{\mathbb{K}}|})$. Therefore $\mathbb{K}$ is a subfield of $\mathbb{Q}(\zeta_{|d_{\mathbb{K}}|})$ in this case.

▷ If $d_{\mathbb{K}}$ is even, we have $d_{\mathbb{K}} = 4d$ with $d \equiv 2, 3 \pmod 4$. We may suppose that $d \neq -1$ since $\mathbb{Q}(\sqrt{-1}) = \mathbb{Q}(\zeta_4)$. Set $d = \pm 2^e \prod_i p_i$ with $e \in \{0, 1\}$ and p_i are odd prime numbers and hence $d_{\mathbb{K}} = \pm 2^e \prod_i p_i^*$. As above, p_i^* is a square in $\mathbb{Q}(\zeta_{|d_{\mathbb{K}}|})$ and since $4 \mid d_{\mathbb{K}}$ and $\mathbb{Q}(\sqrt{-1}) = \mathbb{Q}(\zeta_4)$, -1 is also a square in $\mathbb{Q}(\zeta_{|d_{\mathbb{K}}|})$. Therefore $d_{\mathbb{K}}$ is a square in $\mathbb{Q}(\zeta_{|d_{\mathbb{K}}|})$ if $e = 0$. If $e = 1$, then $8 \mid d_{\mathbb{K}}$ and then $\mathbb{Q}(\zeta_8) \subset \mathbb{Q}(\zeta_{|d_{\mathbb{K}}|})$, so that 2 is a square in $\mathbb{Q}(\zeta_{|d_{\mathbb{K}}|})$, and hence $d_{\mathbb{K}}$ is a square in $\mathbb{Q}(\zeta_{|d_{\mathbb{K}}|})$ in this case, as required.

□

We say that $|d_{\mathbb{K}}|$ is the *conductor* of $\mathbb{Q}(\sqrt{d})$. Proposition 7.27 is a particular case of the Kronecker–Weber theorem[31] stating that every *abelian* extension $\mathbb{K}$ of $\mathbb{Q}$ is contained in a cyclotomic field $\mathbb{Q}(\zeta_f)$.

7.3.8 The Class Group

7.3.8.1 Definition

Let $\mathbb{K}/\mathbb{Q}$ be an algebraic number field of degree n, ring of integers $O_{\mathbb{K}}$, signature (r_1, r_2) and embeddings $\sigma_1, \ldots, \sigma_n$. Recall that $I(\mathbb{K})$ is the abelian group of fractional ideals of $O_{\mathbb{K}}$ and $\mathcal{P}(\mathbb{K})$ is the subset of $I(\mathbb{K})$ of principal ideals. In view of $(a)(b)^{-1} = (ab^{-1})$, we see that $\mathcal{P}(\mathbb{K})$ is a subgroup of $I(\mathbb{K})$, and since $I(\mathbb{K})$ is abelian, we infer that $\mathcal{P}(\mathbb{K})$ is a normal subgroup of $I(\mathbb{K})$, so that the quotient group $I(\mathbb{K})/\mathcal{P}(\mathbb{K})$ is well-defined and abelian.

Definition 7.18 (Class Group) The abelian quotient group $I(\mathbb{K})/\mathcal{P}(\mathbb{K})$ is called the *ideal class group* of $O_{\mathbb{K}}$ and is denoted[32] by $C\ell(\mathbb{K})$.

For $\mathfrak{a} \in I(\mathbb{K})$, we write $\bar{\mathfrak{a}}$ the corresponding class in $C\ell(\mathbb{K})$. We have

$$\bar{\mathfrak{a}} = \overline{(1)} \iff \mathfrak{a} = (a) = aO_{\mathbb{K}}.$$

---

[30] See [43, Proposition 4.1] for instance.

[31] See Theorem 7.30.

[32] See footnote 21 for the notation.

For instance, in $\mathbb{K} = \mathbb{Q}(\sqrt{6})$, set $\mathfrak{p} = (2, \sqrt{6})$. We have $\mathcal{N}_{\mathbb{K}/\mathbb{Q}}(\mathfrak{p}) = 2$ so that $\mathfrak{p}$ is not principal. Since $(2) = \mathfrak{p}^2$ by Theorem 7.11, we have $\overline{\mathfrak{p}}^2 = \overline{(1)}$ and hence $\overline{\mathfrak{p}}$ has order 2 in $C\ell(\mathbb{K})$.

7.3.8.2   Finiteness of the Class Group

The main result is the following fundamental finiteness theorem of $C\ell(\mathbb{K})$, which is one of the three important finiteness theorems in algebraic number theory, namely

▷ $C\ell(\mathbb{K})$ is finite.
▷ $O_{\mathbb{K}}^*$ is a finitely generated abelian group.
▷ Given $D > 0$, the set of algebraic number fields $\mathbb{K}$ such that $|d_{\mathbb{K}}| \leqslant D$ is finite.

> **Theorem 7.12** *Let $\mathbb{K}/\mathbb{Q}$ be an algebraic number field of degree n. The ideal class group $C\ell(\mathbb{K})$ of $\mathbb{K}$ is a finite abelian group. Furthermore, $C\ell(\mathbb{K})$ is generated by the classes of prime ideals $\mathfrak{p}$ in $O_{\mathbb{K}}$ such that $\mathcal{N}_{\mathbb{K}/\mathbb{Q}}(\mathfrak{p}) \leqslant c$ for some $c > 0$.*

This result leads to the following crucial definition.

Definition 7.19 (Class Number)   The order of $C\ell(\mathbb{K})$ is called the *class number* of $\mathbb{K}$ and is denoted by $h_{\mathbb{K}}$.

The proof of Theorem 7.12 requires the following lemma.

Lemma 7.18   *There exists a constant $c_{\mathbb{K}} > 0$, called the* Hurwitz constant *and depending only on $\mathbb{K}$, such that every ideal class of $\mathbb{K}$ contains an integral ideal $\mathfrak{a}$ such that $\mathcal{N}_{\mathbb{K}/\mathbb{Q}}(\mathfrak{a}) \leqslant c_{\mathbb{K}}$.*

Proof   We first prove that there exists a constant $c_{\mathbb{K}} > 0$, called the Hurwitz constant and depending only on $\mathbb{K}$, such that in every non-zero integral ideal $\mathfrak{a}$ there is a non-zero β such that

$$|N_{\mathbb{K}/\mathbb{Q}}(\beta)| \leqslant c_{\mathbb{K}} \, \mathcal{N}_{\mathbb{K}/\mathbb{Q}}(\mathfrak{a}).$$

To this end, let $\{\alpha_1, \ldots, \alpha_n\}$ be an integral basis for $\mathbb{K}$ and set

$$c_{\mathbb{K}} = \prod_{j=1}^{n} \left(\sum_{i=1}^{n} |\sigma_j(\alpha_i)| \right)$$

and $r = \lfloor \mathcal{N}_{\mathbb{K}/\mathbb{Q}}(\mathfrak{a})^{1/n} \rfloor$. The set

$$S = \{a_1\alpha_1 + \cdots + a_n\alpha_n : a_i \in \mathbb{Z}, \, 0 \leqslant a_i \leqslant r\}$$

has $(r + 1)^n$ elements and so $|S| > N_{\mathbb{K}/\mathbb{Q}}(\mathfrak{a}) = |O_{\mathbb{K}}/\mathfrak{a}|$. The elements of S cannot lie in distinct cosets of $\mathfrak{a}$ in $O_{\mathbb{K}}$, and therefore there are $\beta_1, \beta_2 \in S$ such that $\beta_1 \neq \beta_2$ and $\beta_1 \equiv \beta_2 \pmod{\mathfrak{a}}$ by the Dirichlet's pigeon-hole principle. Set $\beta = \beta_1 - \beta_2$. Then $\beta \in \mathfrak{a} \setminus \{0\}$ and is such that $\beta = \sum_{i=1}^{n} b_i \alpha_i$ with $|b_i| \leqslant r$. Now we have

$$|N_{\mathbb{K}/\mathbb{Q}}(\beta)| = \prod_{j=1}^{n} |\sigma_j(\beta)| \leqslant \prod_{j=1}^{n} \sum_{i=1}^{n} |b_i| |\sigma_j(\alpha_i)| \leqslant c_{\mathbb{K}} r^n \leqslant c_{\mathbb{K}} N_{\mathbb{K}/\mathbb{Q}}(\mathfrak{a})$$

as asserted. Now let $\mathfrak{b}$ be a fractional ideal in the given ideal class. Without loss of generality, one may assume that $\mathfrak{b}^{-1}$ is an integral ideal. Choose an element $\beta \in \mathfrak{b}^{-1}$ such that $|N_{\mathbb{K}/\mathbb{Q}}(\beta)| \leqslant c_{\mathbb{K}} N_{\mathbb{K}/\mathbb{Q}}(\mathfrak{b}^{-1})$ and set $\mathfrak{a} = \beta \mathfrak{b}$.

▷ Since $\beta \in \mathfrak{b}^{-1}$ and $\mathfrak{b}\mathfrak{b}^{-1} = O_{\mathbb{K}}$, we deduce that $\mathfrak{a}$ is an integral ideal.
▷ We have $\mathfrak{a}\mathfrak{b}^{-1} = (\beta)$ so that

$$N_{\mathbb{K}/\mathbb{Q}}(\mathfrak{a}) \, N_{\mathbb{K}/\mathbb{Q}}(\mathfrak{b}^{-1}) = |N_{\mathbb{K}/\mathbb{Q}}(\beta)| \leqslant c_{\mathbb{K}} N_{\mathbb{K}/\mathbb{Q}}(\mathfrak{b}^{-1}),$$

which implies the desired result.

$\square$

Now we are in a position to prove Theorem 7.12.

Proof of Theorem 7.12 Let $\overline{\mathfrak{b}}$ be an ideal class in $C\ell(\mathbb{K})$. By Lemma 7.18, $\overline{\mathfrak{b}}$ contains a non-zero integral ideal $\mathfrak{a}$ such that $N_{\mathbb{K}/\mathbb{Q}}(\mathfrak{a}) \leqslant c_{\mathbb{K}}$. From Corollary 7.6, there are only finitely many integral ideals with a given norm so that there are only finitely many choices for $\mathfrak{a}$. Since $\overline{\mathfrak{b}} = \overline{\mathfrak{a}}$, we infer that there are only finitely many ideal classes $\overline{\mathfrak{b}}$.

$\square$

7.3.8.3 The Minkowski Bound

Note that the Hurwitz constant can be very large, and hence useless to help compute $h_{\mathbb{K}}$. One can improve this constant by using the *geometry of numbers*, which enables us to derive the following bound, called *Minkowski bound*.

Theorem 7.13 (Minkowski Bound) *Let $\mathbb{K}/\mathbb{Q}$ be an algebraic number field of degree n, signature (r_1, r_2) and discriminant $d_{\mathbb{K}}$. Then every ideal class of $\mathbb{K}$ contains an integral ideal $\mathfrak{a}$ such that*

$$N_{\mathbb{K}/\mathbb{Q}}(\mathfrak{a}) \leqslant M_{\mathbb{K}} |d_{\mathbb{K}}|^{1/2}, \tag{7.16}$$

(continued)

Theorem 7.13 (continued)
where

$$M_{\mathbb{K}} := \left(\frac{4}{\pi}\right)^{r_2} \frac{n!}{n^n}. \tag{7.17}$$

$M_{\mathbb{K}}$ *is called the* Minkowski constant *attached to* $\mathbb{K}$.

We will see later how analytic methods can supersede the Minkowski constant.

Remark 7.4

▷ The fractional ideals $\mathfrak{a}$ such that $O_{\mathbb{K}} \subset \mathfrak{a}$ and $N_{\mathbb{K}/\mathbb{Q}}(\mathfrak{a}) \leqslant c_{\mathbb{K}}$ represent all ideal classes, but that does not imply that the number of ideal classes, i.e. $h_{\mathbb{K}}$, is bounded above by $c_{\mathbb{K}}$ since several fractional ideals containing $O_{\mathbb{K}}$ may have the same norm.

▷ To get upper and lower bounds for $h_{\mathbb{K}}$ is one of the highlights in algebraic number theory, mainly because the class number is a sort of measure of the default of principality of an algebraic number field. Indeed, if $h_{\mathbb{K}} = 1$, then $\mathcal{I}(\mathbb{K}) = \mathcal{P}(\mathbb{K})$ and $O_{\mathbb{K}}$ is a PID, and hence a UFD. Conversely, if $O_{\mathbb{K}}$ is a UFD, it is also a PID by Theorem 7.7, so that $\mathcal{I}(\mathbb{K}) = \mathcal{P}(\mathbb{K})$ and thus $h_{\mathbb{K}} = 1$. We will provide some upper bounds using multiplicative methods in the next section.

▷ The discussion above allows us to construct an algorithm to compute $h_{\mathbb{K}}$. Indeed, to determine the representatives of the ideal classes, we only need to look at the integral ideals with norms upper bounded by a fixed constant c. We infer that every prime ideal dividing these integral ideals has a norm bounded by this constant, and since the norm of a prime ideal is a prime power, it is then sufficient to look at the prime numbers bounded by the constant. Indeed, forming all possible products of the prime ideals lying above $p \leqslant c$ will yield all ideals of norm $\leqslant c$. This yields the following algorithm:

⋄ Given an algebraic number field $\mathbb{K} = \mathbb{Q}(\theta)$, determine its degree, signature, discriminant and compute $b_{\mathbb{K}} = M_{\mathbb{K}}|d_{\mathbb{K}}|^{1/2}$.

⋄ Determine the prime ideal factorization of (p) for all primes $p \leqslant b_{\mathbb{K}}$.

⋄ Find *all* dependence relations between the integral ideals having norm $\leqslant b_{\mathbb{K}}$. A useful method is to compute $N_{\mathbb{K}/\mathbb{Q}}(\theta + m)$ for some $m \in \mathbb{Z}$ and use those values that only involve the primes $p \leqslant b_{\mathbb{K}}$.

It should also be noticed that, if every prime number $p \leqslant b_{\mathbb{K}}$ factorizes into a product of prime ideals, each of which is principal, then $h_{\mathbb{K}} = 1$.

Example 7.16 Let $\mathbb{K} = \mathbb{Q}(\sqrt{229})$. A defining polynomial of $\mathbb{K}$ is $P = X^2 - X - 57$, and we have $n = 2$, $(r_1, r_2) = (2, 0)$, $\{1, \omega\}$ is an integral basis for $\mathbb{K}$ with $\omega =$

$\frac{1}{2}\left(1 + \sqrt{229}\right)$ and $d_{\mathbb{K}} = 229$. We also have $M_{\mathbb{K}}|d_{\mathbb{K}}|^{1/2} \approx 5.351$. We factorize (2), (3) and (5) using Theorem 7.11 or Proposition 7.26, which gives

$$(2) = \mathfrak{p}_2, \quad (3) = \mathfrak{p}_3\mathfrak{p}_3' \quad \text{and} \quad (5) = \mathfrak{p}_5\mathfrak{p}_5'.$$

Furthermore, we have

$$(7 - \omega) = \mathfrak{p}_3\mathfrak{p}_5 \quad \text{and} \quad (6 + \omega) = \mathfrak{p}_3'\mathfrak{p}_5'.$$

We infer that $\overline{\mathfrak{p}_2} = \overline{(1)}$, and

$$\left. \begin{array}{l} \overline{\mathfrak{p}_3} \cdot \overline{\mathfrak{p}_5} = \overline{(1)} \\ \overline{\mathfrak{p}_3} \cdot \overline{\mathfrak{p}_3'} = \overline{(1)} \end{array} \right\} \implies \overline{\mathfrak{p}_5} = \overline{\mathfrak{p}_3'} = \left(\overline{\mathfrak{p}_3}\right)^{-1}$$

and

$$\left. \begin{array}{l} \overline{\mathfrak{p}_3'} \cdot \overline{\mathfrak{p}_5'} = \overline{(1)} \\ \overline{\mathfrak{p}_3} \cdot \overline{\mathfrak{p}_3'} = \overline{(1)} \end{array} \right\} \implies \overline{\mathfrak{p}_5'} = \overline{\mathfrak{p}_3},$$

and thus all the ideal classes lie among the following three distinct classes $\overline{(1)} = \overline{\mathfrak{p}_2}$, $\overline{\mathfrak{p}_3}$ and $\left(\overline{\mathfrak{p}_3}\right)^{-1}$, so that $h_{\mathbb{K}} = 3$.

Example 7.17 Let θ be a root of the polynomial $P = X^3 - X^2 - 11X - 1$. We have $n = 3$, $(r_1, r_2) = (3, 0)$, $d_{\mathbb{K}} = 1304$ and $M_{\mathbb{K}}|d_{\mathbb{K}}|^{1/2} \approx 8.025$. It was seen in Example 7.12 that θ is a unit in $\mathbb{K}$ so that $(\theta) = (1)$. Furthermore, using Theorem 7.11, we get

$$(2) = \mathfrak{p}_2^2\mathfrak{p}_2', \quad (3) = \mathfrak{p}_3\mathfrak{p}_3', \quad (5) = \mathfrak{p}_5 \quad \text{and} \quad (7) = \mathfrak{p}_7.$$

We also have

$$(1 - \theta) = \mathfrak{p}_2\mathfrak{p}_2'\mathfrak{p}_3, \quad (1 + \theta) = \mathfrak{p}_2\mathfrak{p}_2'^2, \quad (2 + \theta) = \mathfrak{p}_3^2 \quad \text{and} \quad (3 + \theta) = \mathfrak{p}_2\mathfrak{p}_2'.$$

Using $(1 - \theta)$, $(3 + \theta)$ and (3) we get

$$\left. \begin{array}{l} \overline{\mathfrak{p}_2} \cdot \overline{\mathfrak{p}_2'} \cdot \overline{\mathfrak{p}_3} = \overline{(1)} \\ \overline{\mathfrak{p}_2} \cdot \overline{\mathfrak{p}_2'} = \overline{(1)} \end{array} \right\} \implies \overline{\mathfrak{p}_3} = \overline{\mathfrak{p}_3'} = \overline{(1)}$$

and using $(1 + \theta)$, $(3 + \theta)$ and (2) we get

$$\left. \begin{array}{l} \overline{\mathfrak{p}_2} \cdot \overline{\mathfrak{p}_2'}^2 = \overline{(1)} \\ \overline{\mathfrak{p}_2} \cdot \overline{\mathfrak{p}_2'} = \overline{(1)} \end{array} \right\} \implies \overline{\mathfrak{p}_2} = \overline{\mathfrak{p}_2'} = \overline{(1)}$$

and since $\overline{\mathfrak{p}_5} = \overline{\mathfrak{p}_7} = \overline{(1)}$, we get $h_{\mathbb{K}} = 1$ and hence $O_{\mathbb{K}}$ is a PID.

The following result is of great use in the theory of Diophantine equations.

Proposition 7.28 *Let $\mathbb{K}$ be an algebraic number field with class number $h_{\mathbb{K}}$ and $\mathfrak{a}$ be an integral ideal of $\mathbb{K}$. Let p be a prime number such that $p \nmid h_{\mathbb{K}}$. If $\mathfrak{a}^p$ is principal, then $\mathfrak{a}$ is principal.*

Proof Since $p \nmid h_{\mathbb{K}}$, we have $uh_{\mathbb{K}} + pv = 1$ for some $u, v \in \mathbb{Z}$. By assumption, $\overline{\mathfrak{a}}^p = \overline{(1)}$ and by Lagrange's Theorem 7.1, $\overline{\mathfrak{a}}^{h_{\mathbb{K}}} = \overline{(1)}$, so that

$$\overline{\mathfrak{a}} = \overline{\mathfrak{a}}^{uh_{\mathbb{K}}+pv} = \left(\overline{\mathfrak{a}}^{h_{\mathbb{K}}}\right)^u \left(\overline{\mathfrak{a}}^p\right)^v = \overline{(1)},$$

which implies the asserted result.   □

Example 7.18 Let us have a look at Bachet's Diophantine equation $x^2 + 5 = y^3$. Let $(x, y) \in \mathbb{Z}^2$ be a solution to this equation. One may assume that y is odd, otherwise x is odd and we have $x^2 \equiv 3 \pmod 4$, which is impossible. One may also suppose that x and y are coprime. Consider the quadratic field $\mathbb{K} = \mathbb{Q}(\sqrt{-5})$. By Proposition 7.12, we have $O_{\mathbb{K}} = \mathbb{Z}[\sqrt{-5}]$. In this ring, we have

$$(x + \sqrt{-5})(x - \sqrt{-5}) = y^3.$$

Since y is odd and x and y are coprime, we infer that the ideals $(x + \sqrt{-5})$ and $(x - \sqrt{-5})$ are coprime, and hence there exist two coprime integral ideals $\mathfrak{a}$ and $\mathfrak{b}$ such that $(x + \sqrt{-5}) = \mathfrak{a}^3$ and $(x - \sqrt{-5}) = \mathfrak{b}^3$. Therefore $\mathfrak{a}^3$ is principal. Using the method of Example 7.16 or 7.17, we derive $h_{\mathbb{K}} = 2$, so that $\mathfrak{a}$ is principal by Proposition 7.28. By Example 7.11, the units in $\mathbb{K}$ are ± 1, which are cubes, and hence we deduce that there exist $a, b \in \mathbb{Z}$ such that

$$x + \sqrt{-5} = (a + b\sqrt{-5})^3.$$

Equating imaginary parts implies that $3a^2b - 5b^3 = 1$ and hence $b \mid 1$, so that $b = \pm 1$. This gives $3a^2 - 5 = \pm 1$, which is impossible in both cases. Hence Bachet's equation has no solution in $\mathbb{Z}^2$.

7.4   The Dedekind Zeta-Function

In what follows, $\mathbb{K} = \mathbb{Q}(\theta)$ is an algebraic number field of degree n, and let $O_{\mathbb{K}}$ be its ring of integers. The purpose of this section is to use tools from analytic number theory to get an answer to certain questions of algebraic number theory.

7.4.1   The Function $r_{\mathbb{K}}$

Definition 7.20 The function $r_{\mathbb{K}}$ is defined by $r_{\mathbb{K}}(1) = 1$ and, for all integers $m \geqslant 2$

$$r_{\mathbb{K}}(m) = \sum_{\mathcal{N}_{\mathbb{K}/\mathbb{Q}}(\mathfrak{a})=m} 1.$$

Hence $r_{\mathbb{K}}(m)$ is the number of non-zero integral ideals with norm m.

The following result shows that $r_{\mathbb{K}}$ is a multiplicative function. Besides, we shall see later that this function is not really far from the usual Piltz–Dirichlet divisor function τ_n. Hence we will be able to apply to $r_{\mathbb{K}}$ the fundamental results from Chap. 4.

Proposition 7.29 *The function $r_{\mathbb{K}}$ is multiplicative, and, for all prime powers p^α*

$$r_{\mathbb{K}}\left(p^\alpha\right) = \mathcal{D}_{(f_1,\dots,f_g)}(\alpha),$$

where $\mathcal{D}_{(f_1,\dots,f_g)}(\alpha)$ is the denumerant[33] of α with respect to g and to the set $\{f_1, \dots, f_g\}$.

Proof

▷ Let $\mathfrak{a}$ be an integral ideal of norm $\mathcal{N}_{\mathbb{K}/\mathbb{Q}}(\mathfrak{a}) = bc$ with $(b, c) = 1$ and set $\mathfrak{a} = \prod_{\mathfrak{p}} \mathfrak{p}^{a_{\mathfrak{p}}}$ and

$$\mathfrak{b} = \prod_{\substack{\mathfrak{p} \\ (\mathcal{N}_{\mathbb{K}/\mathbb{Q}}(\mathfrak{p}),b)>1}} \mathfrak{p}^{a_{\mathfrak{p}}} \quad \text{and} \quad \mathfrak{c} = \prod_{\substack{\mathfrak{p} \\ (\mathcal{N}_{\mathbb{K}/\mathbb{Q}}(\mathfrak{p}),c)>1}} \mathfrak{p}^{a_{\mathfrak{p}}}.$$

Then $\mathfrak{a} = \mathfrak{bc}$ with $\mathcal{N}_{\mathbb{K}/\mathbb{Q}}(\mathfrak{b}) = b$ and $\mathcal{N}_{\mathbb{K}/\mathbb{Q}}(\mathfrak{c}) = c$. Hence, if $(b, c) = 1$, any ideal of norm bc can be uniquely written as the product of integral ideals of norm b and of norm c, respectively. This establishes the multiplicativity of $r_{\mathbb{K}}$.

▷ Let p be a prime number. By Lemma 7.15, we have

$$(p) = \prod_{i=1}^{g} \mathfrak{p}_i^{e_i}$$

with $1 \leqslant g \leqslant n$ and $e_1 f_1 + \cdots + e_g f_g = n$. If $\mathfrak{a}$ is a non-zero integral ideal satisfying $\mathcal{N}_{\mathbb{K}/\mathbb{Q}}(\mathfrak{a}) = p^\alpha$ for some $\alpha \in \mathbb{Z}_{\geqslant 1}$, then $\mathfrak{a} \mid (p^\alpha)$ by (7.11) so that $\mathfrak{a} = \mathfrak{p}_1^{a_1} \cdots \mathfrak{p}_g^{a_g}$ for some $(a_1, \dots, a_g) \in \left(\mathbb{Z}_{\geqslant 0}\right)^g$, and comparing the norms we get $f_1 a_1 + \cdots + f_g a_g = \alpha$. Hence $\mathfrak{a}$ induces a solution $(a_1, \dots, a_g) \in \left(\mathbb{Z}_{\geqslant 0}\right)^g$ of the

---

[33] See Definition 2.3.

Diophantine equation $f_1 x_1 + \cdots + f_g x_g = \alpha$, and if $\mathfrak{b}$ is another non-zero integral ideal such that $N_{\mathbb{K}/\mathbb{Q}}(\mathfrak{b}) = p^\alpha$ and inducing the same solution, then proceeding similarly we infer that $\mathfrak{b} = \mathfrak{p}_1^{b_1} \cdots \mathfrak{p}_g^{b_g}$ for some $(b_1, \ldots, b_g) \in \left(\mathbb{Z}_{\geqslant 0}\right)^g$ satisfying $f_1 b_1 + \cdots + f_g b_g = \alpha$, and therefore $b_i = a_i$ for all $i \in \{1, \ldots, g\}$ so that $\mathfrak{b} = \mathfrak{a}$. Conversely, if $(a_1, \ldots, a_g) \in \left(\mathbb{Z}_{\geqslant 0}\right)^g$ is a solution of the equation $f_1 x_1 + \cdots + f_g x_g = \alpha$, then $\mathfrak{a} = \mathfrak{p}_1^{a_1} \cdots \mathfrak{p}_g^{a_g}$ is a non-zero integral ideal satisfying $N_{\mathbb{K}/\mathbb{Q}}(\mathfrak{a}) = p^\alpha$ by Theorem 7.9. We have thus defined a one-to-one correspondence between the set of non-zero integral ideals having a norm equal to p^α and the set of the solutions in $\left(\mathbb{Z}_{\geqslant 0}\right)^g$ of the equation $f_1 x_1 + \cdots + f_g x_g = \alpha$, completing the proof since these two sets are finite.

$\square$

The following particular cases are easy.

Corollary 7.8 *With the notation of Proposition 7.29, we have the following results.*

▷ *If $\mathbb{K}$ is Galois over $\mathbb{Q}$, then*

$$r_{\mathbb{K}}\left(p^\alpha\right) = \begin{cases} \tau_g \left(p^{\alpha/f_p}\right), & \text{if } f_p \mid \alpha \\ 0, & \text{otherwise.} \end{cases}$$

▷ *If all the prime ideals above p are of degree 1, then $r_{\mathbb{K}}\left(p^\alpha\right) = \tau_g\left(p^\alpha\right)$.*
▷ *In every case, we have the inequality $r_{\mathbb{K}}(m) \leqslant \tau_n(m)$.*

Proof

▷ If $\mathbb{K}$ is Galois over $\mathbb{Q}$, then $f_1 = \cdots = f_g = f_p$ by Proposition 7.25, so that

$$r_{\mathbb{K}}\left(p^\alpha\right) = \mathcal{D}_{(1,\ldots,1)}\left(\frac{\alpha}{f_p}\right) = \begin{cases} \tau_g \left(p^{\alpha/f_p}\right), & \text{if } f_p \mid \alpha \\ 0, & \text{otherwise} \end{cases}$$

by (4.26).
▷ If $f_i = 1$ for all $i \in \{1, \ldots, g\}$, then $r_{\mathbb{K}}\left(p^\alpha\right) = \mathcal{D}_{(1,\ldots,1)}(\alpha) = \tau_g\left(p^\alpha\right)$ by (4.26).
▷ The two functions are multiplicative, so that it is sufficient to prove the inequality for prime powers. Now by (4.26) we get for all prime powers p^α

$$r_{\mathbb{K}}\left(p^\alpha\right) \leqslant \mathcal{D}_{(1,\ldots,1)}(\alpha) = \tau_g\left(p^\alpha\right) \leqslant \tau_n(p^\alpha)$$

as required.

$\square$

Remark 7.5

▷ It should be noticed that from Corollary 7.8 we infer that $r_{\mathbb{K}}(p)$ counts the number of prime ideals above p with inertial degree 1.
▷ The following special cases may be useful in practice. When $g = 1$, and then $ef = n$, we have

$$r_{\mathbb{K}}(p) = \begin{cases} 1, & \text{if } f = 1 \\ 0, & \text{otherwise} \end{cases} \quad \text{and} \quad r_{\mathbb{K}}(p^\alpha) = \begin{cases} 1, & \text{if } n \mid \alpha \times e \\ 0, & \text{otherwise.} \end{cases}$$

Hence if p is inert, so that $e = g = 1$, we get

$$r_{\mathbb{K}}(p^\alpha) = \begin{cases} 1, & \text{if } n \mid \alpha \\ 0, & \text{otherwise.} \end{cases}$$

7.4.2 The Function $\zeta_{\mathbb{K}}$

As in the rational case, one may attach to an algebraic number field $\mathbb{K}$ a generating function, which contains all the arithmetic information of $\mathbb{K}$.

Definition 7.21 Let $\mathbb{K}/\mathbb{Q}$ be an algebraic number field. The *Dedekind zeta-function* $\zeta_{\mathbb{K}}$ of $\mathbb{K}$ is the Dirichlet series of the multiplicative function $r_{\mathbb{K}}$, so that for all $s = \sigma + it \in \mathbb{C}$ such that $\sigma > 1$, we have

$$\zeta_{\mathbb{K}}(s) = \sum_{m=1}^{\infty} \frac{r_{\mathbb{K}}(m)}{m^s} = \sum_{\mathfrak{a}} \frac{1}{\mathcal{N}_{\mathbb{K}/\mathbb{Q}}(\mathfrak{a})^s} = \prod_{\mathfrak{p}} \left(1 - \frac{1}{\mathcal{N}_{\mathbb{K}/\mathbb{Q}}(\mathfrak{p})^s} \right)^{-1},$$

where the second sum is taken over non-zero integral ideals, the product runs through all prime ideals of $\mathcal{O}_{\mathbb{K}}$ and the last equality comes from Corollary 4.5. The absolute convergence in the half-plane $\sigma > 1$ follows from the estimate

$$\sum_{\mathcal{N}_{\mathbb{K}/\mathbb{Q}}(\mathfrak{p}) \leqslant x} \left| \frac{1}{\mathcal{N}_{\mathbb{K}/\mathbb{Q}}(\mathfrak{p})^s} \right| \leqslant n \sum_{p \leqslant x} \frac{1}{p^\sigma} \quad (\sigma > 1).$$

It should be noticed that $\zeta_{\mathbb{K}}(s)$ has no zero in the region $\sigma > 1$ since none of the factors of the Euler product have any zeros therein.

Remark 7.6

▷ We readily get $\zeta_{\mathbb{Q}}(s) = \zeta(s)$, so that the Dedekind zeta-function generalizes the ordinary Riemann zeta-function.

▷ If $\mathbb{K}$ is Galois over $\mathbb{Q}$, then the Euler product simplifies

$$\zeta_{\mathbb{K}}(s) = \prod_{p} \left(1 - \frac{1}{p^{f_p s}}\right)^{-g_p}.$$

▷ Using Corollary 7.8, we infer that for all real numbers $\sigma > 1$, we have

$$\zeta_{\mathbb{K}}(\sigma) \leqslant \zeta(\sigma)^n. \tag{7.18}$$

We shall see below that the function $\zeta_{\mathbb{K}}(s)$ has an analytic continuation to a meromorphic function in the whole complex plane with a simple pole at $s = 1$. Therefore this inequality does not reflect the arithmetic nature of the Dedekind zeta-function, since the right-hand side has at $s = 1$ a multiple pole of order n.

▷ It may be shown that if two algebraic number fields have the same Dedekind zeta-function, then they have the same degree, the same signature and the same discriminant.[34]

▷ One may define the *Möbius* function for $\mathbb{K}$ in the same way as in the rational case. Let $\mu_{\mathbb{K}}$ be the $O_{\mathbb{K}}$-arithmetic function defined by $\mu_{\mathbb{K}}(O_{\mathbb{K}}) = 1$ and, for all non-zero integral ideals $\mathfrak{a}$

$$\mu_{\mathbb{K}}(\mathfrak{a}) = \begin{cases} (-1)^r, & \text{if } \mathfrak{a} = \mathfrak{p}_1 \cdots \mathfrak{p}_r \\ \\ 0, & \text{otherwise.} \end{cases}$$

One can prove that the Dirichlet series of $\mu_{\mathbb{K}}$ is $\zeta_{\mathbb{K}}(s)^{-1}$ and we also have for all $s \in \mathbb{C}$ such that $\sigma > 1$

$$\left|\zeta_{\mathbb{K}}(s)^{-1}\right| \leqslant \zeta_{\mathbb{K}}(\sigma).$$

7.4.3 Functional Equation

Let $\mathbb{K}/\mathbb{Q}$ be an algebraic number field of degree n, signature (r_1, r_2), discriminant $d_{\mathbb{K}}$, class number $h_{\mathbb{K}}$, regulator $\mathcal{R}_{\mathbb{K}}$, and let $w_{\mathbb{K}}$ be the number of roots of unity contained in $\mathbb{K}$.[35] According to the usual practice, we set

$$\Gamma_{\mathbb{K}}(s) = \Gamma(s/2)^{r_1} \Gamma(s)^{r_2} \quad \text{and} \quad A_{\mathbb{K}} = 2^{-r_2} \pi^{-n/2} |d_{\mathbb{K}}|^{1/2},$$

---

[34] See [13].
[35] See Proposition 7.21.

where $\Gamma(s)$ is the usual Gamma-function. The function $\Gamma_{\mathbb{K}}$ is sometimes called the *Gamma-function of* $\mathbb{K}$.[36]

The main properties of the Dedekind zeta-function generalize those of the Riemann zeta-function. In particular, $\zeta_{\mathbb{K}}(s)$ satisfies a functional equation, as seen in (4.35), and an approximate functional equation. The former was discovered by Hecke using n-dimensional analytic methods analogous to the case of ζ. Using adelic Chevalley's language, Tate gave another proof of the functional equation in his thesis.[37]

Theorem 7.14 (Functional Equation)

▷ *The function* $\zeta_{\mathbb{K}}(s)$ *can be extended analytically in the whole complex plane to a meromorphic function having a simple pole at* $s = 1$ *with residue* $\kappa_{\mathbb{K}}$ *equal to*

$$\kappa_{\mathbb{K}} = \frac{2^{r_1}(2\pi)^{r_2} h_{\mathbb{K}} \mathcal{R}_{\mathbb{K}}}{w_{\mathbb{K}} |d_{\mathbb{K}}|^{1/2}}. \tag{7.19}$$

This identity is called the analytic class number formula.

▷ *For all* $s \in \mathbb{C} \setminus \{1\}$, *the function* $\xi_{\mathbb{K}}(s) = A_{\mathbb{K}}^s \Gamma_{\mathbb{K}}(s)\zeta_{\mathbb{K}}(s)$ *satisfies the following functional equation*[38]

$$\xi_{\mathbb{K}}(s) = \xi_{\mathbb{K}}(1 - s).$$

Furthermore, for all real numbers $\sigma > 1$, *we also have*

$$\kappa_{\mathbb{K}} \leqslant (2\pi)^{r_2}\sigma(\sigma - 1)\xi_{\mathbb{K}}(\sigma)|d_{\mathbb{K}}|^{-1/2}. \tag{7.20}$$

Note that the functional equation of $\zeta_{\mathbb{K}}(s)$ may take the following alternative form [53, Satz 156].

$$\zeta_{\mathbb{K}}(1 - s) = \zeta_{\mathbb{K}}(s) \, d_{\mathbb{K}}^{s-1/2} \left(\cos \tfrac{\pi s}{2}\right)^{r_1+r_2} \left(\sin \tfrac{\pi s}{2}\right)^{r_2} \left(\pi^{-s} 2^{1-s} \Gamma(s)\right)^n.$$

---

[36] The factor $A_{\mathbb{K}}^s \Gamma_{\mathbb{K}}(s)$ is sometimes called the *Euler factor* of $\zeta_{\mathbb{K}}(s)$.

[37] See [55, 65, 66] for an exhaustive account of Hecke's method and Tate's work.

[38] Some authors [13, 66] consider the "weighted" function $2^{r_2}\xi_{\mathbb{K}}(s)$ instead.

7.4.4   Explicit Convexity Bound

The Phragmén–Lindelöf principle is the starting point of the convexity bound for $\zeta_{\mathbb{K}}$, for which we use the following specific notation. We set for any $t \in \mathbb{R}$

$$\mathcal{L}_{\mathbb{K}}(t) := \log\left(|d_{\mathbb{K}}|^{1/n}\, \frac{e^{2\gamma}(|t|+3)}{2\pi}\right). \tag{7.21}$$

We now may state the main result of this section.

Theorem 7.15 Let $a \geqslant 1$ and assume that $|d_{\mathbb{K}}| \geqslant \left(\frac{2\pi}{3}\right)^n e^{n(2a-2\gamma)}$. Then, for any $t \in \mathbb{R}$

$$\left|\zeta_{\mathbb{K}}\left(\tfrac{1}{2}+it\right)\right| \leqslant 3\sqrt{3}\left(\frac{e^{2a}}{2\pi a^4}\right)^{n/4}(|t|+3)^{n/4}\,|d_{\mathbb{K}}|^{1/4}\,\mathcal{L}_{\mathbb{K}}(t)^n,$$

where $\mathcal{L}_{\mathbb{K}}(t)$ is given in (7.21). If $|t| \geqslant 3$, then $3\sqrt{3}$ may be replaced by $\frac{45}{37}\sqrt{3}$.

The proof uses in a crucial way the following lemma, due to Rademacher [71], which can be seen as an extension of the Phragmén–Lindelöf principle seen in Lemma 3.14.

Lemma 7.19 For all $0 < \delta \leqslant \frac{1}{2}$ and $s = \sigma + it \in \mathbb{C}$ such that $-\delta \leqslant \sigma \leqslant 1 + \delta$

$$|(1-s)\zeta_{\mathbb{K}}(s)| \leqslant \left(\frac{1+\delta}{1-\delta}\right)^{\frac{1+\delta-\sigma}{1+2\delta}}|1+s|\left(|d_{\mathbb{K}}|^{1/n}\,\frac{|1+s|}{2\pi}\right)^{n(1+\delta-\sigma)/2}\zeta(1+\delta)^n.$$

Proof of Theorem 7.15 Lemma 7.19, applied with $s = \frac{1}{2}+it$, yields for any $0 < \delta \leqslant \frac{1}{2}$

$$\left|\zeta_{\mathbb{K}}\left(\tfrac{1}{2}+it\right)\right| \leqslant \left(\frac{1+\delta}{1-\delta}\right)^{1/2}\left|\frac{\frac{3}{2}+it}{\frac{1}{2}-it}\right||d_{\mathbb{K}}|^{1/4+\delta/2}\left(\frac{\left|\frac{3}{2}+it\right|}{2\pi}\right)^{n(1/4+\delta/2)}\zeta(1+\delta)^n.$$

The first two terms are bounded by $3\sqrt{3}$ if $t \in \mathbb{R}$, or $\frac{45}{37}\sqrt{3}$ if $|t| \geqslant 3$, and

$$\zeta(1+\delta)^n \leqslant e^{\gamma n\delta}\delta^{-n}$$

so that

$$\left| \zeta_{\mathbb{K}}(\tfrac{1}{2} + it) \right| \leqslant 3\sqrt{3} \, |d_{\mathbb{K}}|^{1/4} \left(\frac{|t| + 3}{2\pi} \right)^{n/4} \left(\frac{e^{2\gamma}(|t| + 3) \, |d_{\mathbb{K}}|^{1/n}}{2\pi} \right)^{n\delta/2} \delta^{-n},$$

and choosing $\delta = a \mathcal{L}_{\mathbb{K}}(t)^{-1}$ yields the asserted bound. Note that the inequality $|d_{\mathbb{K}}| \geqslant \left(\frac{2\pi}{3} \right)^n e^{n(2a - 2\gamma)}$ allows us to ensure that $0 < \delta \leqslant \frac{1}{2}$ for all $t \in \mathbb{R}$. $\qquad\square$

Remark 7.7 A similar argument can be used to derive the following bound. Assuming as above that $|d_{\mathbb{K}}| \geqslant 3^{-n} \left(2\pi e^{2a - 2\gamma} \right)^n$ for some $a \geqslant 1$, then, for all $0 \leqslant \sigma \leqslant 1$ and $t \in \mathbb{R}$, we have

$$|\zeta_{\mathbb{K}}(\sigma + it)| < \mathcal{K}_{\sigma,t} \left(\frac{e^a}{a^2} \right)^{n/2} \left(d_K^{1/n} \frac{|t| + 3}{2\pi} \right)^{n(1-\sigma)/2} \mathcal{L}_K(t)^n, \qquad (7.22)$$

where $\mathcal{L}_K(t)$ is given in (7.21) and

$$\mathcal{K}_{\sigma,t} := \begin{cases} \dfrac{6}{1 - \sigma}, & \text{if } 0 \leqslant \sigma < 1 \text{ and } t \in \mathbb{R} \\ 3\left(1 + \sqrt{2} \right), & \text{if } 0 \leqslant \sigma \leqslant 1 \text{ and } |t| \geqslant 1. \end{cases}$$

7.4.5 Subconvexity Bound

The exponent $\frac{n}{4}$ in Theorem 7.15 may be considered as a trivial bound for the Dedekind zeta-function on the critical line. As for the Riemann zeta-function, any improvement of this exponent is called a "subconvexity bound", but generalizing the ideas in the case $\mathbb{K} = \mathbb{Q}$ to the general case is not an easy task.

In [39], Heath-Brown used a multidimensional version of the Van der Corput's method for estimating exponential sums seen in Chap. 6 to prove the following estimate, which is still currently the best to date.

Theorem 7.16 *Let $\mathbb{K}/\mathbb{Q}$ be any algebraic number field of degree n. Then, for any $t \geqslant 1$ and any fixed $\varepsilon \in (0, 1)$*

$$\zeta_{\mathbb{K}} \left(\tfrac{1}{2} + it \right) \ll_{\varepsilon, \mathbb{K}} t^{n/6 + \varepsilon}.$$

7.4.6   Zero-Free Region

The next step is to get a zero-free region for the Dedekind zeta-function. As for the Dirichlet L-functions, a Deuring–Heilbronn phenomenon appears for $\zeta_{\mathbb{K}}(s)$. More precisely, improving on a previous result by Stark [87], Kadiri [48] proved the following result.

Theorem 7.17   *If $|d_{\mathbb{K}}|$ is sufficiently large, $\zeta_{\mathbb{K}}(s)$ has at most one zero in the region*

$$\sigma \geqslant 1 - \frac{1}{2 \log |d_{\mathbb{K}}|}, \quad |t| \leqslant \frac{1}{2 \log |d_{\mathbb{K}}|}.$$

If it exists, this zero is simple and is real.

The next theorem summarizes the main results dealing with zero-free regions of $\zeta_{\mathbb{K}}(s)$ and also provides the analogue of (3.44) for $\zeta_{\mathbb{K}}(s)$, which is used to get these zero-free regions.

Theorem 7.18 (Zero-Free Regions)   *Let $s = \sigma + it \in \mathbb{C}$ and the constants $c_1, c_3 > 0$ and $c_2 > 1$ be absolute.*

▷ *In the region $1 - 1/(n+1) \leqslant \sigma \leqslant 1$ and $t \geqslant e$, we have*

$$|\zeta_{\mathbb{K}}(s)| \leqslant e^{c_1 n^8 |d_{\mathbb{K}}|^2} t^{600 n^2 \{n(1-\sigma)\}^{3/2}} (\log t)^{2/3}.$$

▷ *The function $\zeta_{\mathbb{K}}(s)$ has no zero in the region*

$$\sigma \geqslant 1 - \frac{1}{c_2 n^{11} |d_{\mathbb{K}}|^3 (\log t)^{2/3} (\log \log t)^{1/3}}, \quad t \geqslant 4.$$

The proofs of these estimates can be found in [8, 89]. The second result is a refinement of classical theorems by Landau [51] and Sokolovskiĭ [83]. In [48], Kadiri derived the following explicit zero-free region for $\zeta_{\mathbb{K}}(s)$.

Theorem 7.19   *If $|d_{\mathbb{K}}|$ is sufficiently large, $\zeta_{\mathbb{K}}(s)$ has no zero in the region*

$$\sigma \geqslant 1 - \frac{1}{12.55 \log |d_{\mathbb{K}}| + 9.69 n \log |t| + 3.03 n + 58.63}, \quad |t| \geqslant 1.$$

Remark 7.8   As in the rational case,[39] an approximate functional equation for $\zeta_{\mathbb{K}}(s)$ has been stated in [12], where it is proved that, for some $H > 0$, $x, y > H$ satisfying

---

[39] See Theorem 3.11.

$xy = |d_{\mathbb{K}}| \left(\frac{|t|}{2\pi}\right)^n$ and $c_1 < x/y < c_2$, we have

$$\zeta_{\mathbb{K}}(s) = \sum_{\mathcal{N}_{\mathbb{K}/\mathbb{Q}}(\mathfrak{a}) \leqslant x} \frac{1}{\mathcal{N}_{\mathbb{K}/\mathbb{Q}}(\mathfrak{a})^s}$$

$$+ A_{\mathbb{K}}^{1-2s} \frac{\Gamma_{\mathbb{K}}(1-s)}{\Gamma_{\mathbb{K}}(s)} \sum_{\mathcal{N}_{\mathbb{K}/\mathbb{Q}}(\mathfrak{a}) \leqslant y} \frac{1}{\mathcal{N}_{\mathbb{K}/\mathbb{Q}}(\mathfrak{a})^{1-s}} + O\left(x^{1-\sigma-1/n} \log x\right).$$

The functional equation shows that $\zeta_{\mathbb{K}}(s)$ may have trivial zeros at negative integers $-m$ with $m \in \mathbb{Z}_{\geqslant 1}$. In fact, it can be shown that if m is even, then the order of the possible zero is equal to $r_1 + r_2$, while for m odd the order equals r_2. The Euler product implies that all other zeros $\rho = \beta + i\gamma$ satisfy $0 < \beta < 1$ and are called *non-trivial zeros*. Furthermore, we deduce that the only fields for which some of the values of $\zeta_{\mathbb{K}}(-m)$ can be non-zeros are totally real fields. The next result shows that these values are nevertheless rational numbers.

Theorem 7.20 (Siegel–Klingen) *Let $\mathbb{K}/\mathbb{Q}$ be a totally real algebraic number field, i.e. $r_2 = 0$. For all positive integers m, we have $\zeta_{\mathbb{K}}(1 - 2m) \in \mathbb{Q}$.*

Combining this result with the functional equation and assuming $\mathbb{K}$ to be totally real of degree n, we get for all $m \in \mathbb{N}$

$$\zeta_{\mathbb{K}}(2m) = q_m \pi^{2mn} d_{\mathbb{K}}^{-1/2}$$

for some $q_m \in \mathbb{Q} \setminus \{0\}$.

7.4.7   Application to the Class Number

The problem of getting upper and lower bounds for the class number $h_{\mathbb{K}}$ of an algebraic number field $\mathbb{K}$ has a long history. In particular, the search for quadratic fields with class number one became in the early days one of the most important questions in algebraic number theory. For imaginary quadratic fields, the answer came in 1966. For real quadratic fields, Gauss conjectured that there are infinitely many such fields having class number one, but the question remains open, as will be seen in Sect. 7.5.2 below.

One of the first ideas to get an upper bound for $h_{\mathbb{K}}$ is the use of the functional equation, in particular the inequality (7.20). Using (7.18) and choosing σ near 1 enable us to get very good upper estimates for $h_{\mathbb{K}}\mathcal{R}_{\mathbb{K}}$. In particular, this approach enables Louboutin [60] to get the following estimate,[40] valid for all totally real

_____
[40] A proof is supplied in Exercise 142.

algebraic number fields of degree $n \geqslant 2$

$$h_{\mathbb{K}} \mathcal{R}_{\mathbb{K}} \leqslant d_{\mathbb{K}}^{1/2} \left(\frac{e \log d_{\mathbb{K}}}{4n - 4} \right)^{n-1} . \tag{7.23}$$

Another way to estimate $h_{\mathbb{K}}$ lies in Theorem 7.12 and the proof of Lemma 7.18. Let $b_{\mathbb{K}}$ be any positive real number such that every ideal class contains a non-zero integral ideal $\mathfrak{a}$ such that

$$\mathcal{N}_{\mathbb{K}/\mathbb{Q}}(\mathfrak{a}) \leqslant b_{\mathbb{K}}.$$

From Remark 7.4, we know that $b_{\mathbb{K}}$ could be the Minkowski bound, and we infer that

$$h_{\mathbb{K}} \leqslant \sum_{\mathcal{N}_{\mathbb{K}/\mathbb{Q}}(\mathfrak{a}) \leqslant b_{\mathbb{K}}} 1 = \sum_{m \leqslant b_{\mathbb{K}}} r_{\mathbb{K}}(m).$$

It is therefore important to have at our disposal upper bounds for the average order of $r_{\mathbb{K}}$, which *do not depend on* the invariants of $\mathbb{K}$. By Theorem 7.28 below, we derive

$$\sum_{m \leqslant x} r_{\mathbb{K}}(m) = \kappa_{\mathbb{K}} x + o_{\mathbb{K}}(x),$$

but $\kappa_{\mathbb{K}}$ contains $h_{\mathbb{K}}$. Another interesting result is the estimate[41]

$$\sum_{m \leqslant x} r_{\mathbb{K}}(m)^2 \ll_{\mathbb{K}} x (\log x)^{n-1},$$

but the implied constant depends on the usual invariants of $\mathbb{K}$. Using Corollary 7.8 and Exercise 78, we get the following result.

Proposition 7.30 *Let $\mathbb{K}/\mathbb{Q}$ be an algebraic number field of degree n and class number $h_{\mathbb{K}}$. Let $b_{\mathbb{K}}$ such that every ideal class contains a non-zero integral ideal $\mathfrak{a}$ such that $\mathcal{N}_{\mathbb{K}/\mathbb{Q}}(\mathfrak{a}) \leqslant b_{\mathbb{K}}$. Then*

$$h_{\mathbb{K}} \leqslant b_{\mathbb{K}} \sum_{j=0}^{n-1} \binom{n-1}{j} \frac{(\log b_{\mathbb{K}})^j}{j!} \leqslant \frac{b_{\mathbb{K}} (\log b_{\mathbb{K}} + n - 1)^{n-1}}{(n-1)!}.$$

In [11], the function τ_n has been studied more carefully. This leads to the following improvement.

---

[41] See [12].

Proposition 7.31 *Let $\mathbb{K}/\mathbb{Q}$ be an algebraic number field of degree $n \geqslant 2$ and class number $h_{\mathbb{K}}$. Let $b_{\mathbb{K}} \geqslant 6$ such that every ideal class contains a non-zero integral ideal $\mathfrak{a}$ such that $N_{\mathbb{K}/\mathbb{Q}}(\mathfrak{a}) \leqslant b_{\mathbb{K}}$. Then*

$$h_{\mathbb{K}} \leqslant 2\, b_{\mathbb{K}} (\log b_{\mathbb{K}})^{n-1}.$$

Applied with $b_{\mathbb{K}} = M_{\mathbb{K}}|d_{\mathbb{K}}|^{1/2}$, where $M_{\mathbb{K}}$ is the Minkowski constant (7.17), we get

$$h_{\mathbb{K}} \leqslant 2^{2-n} M_{\mathbb{K}} |d_{\mathbb{K}}|^{1/2} \left(\log M_{\mathbb{K}}^2 |d_{\mathbb{K}}| \right)^{n-1}$$

as soon as $|d_{\mathbb{K}}| \geqslant 36 M_{\mathbb{K}}^{-2}$.

7.4.8 Lower Bounds for $|d_{\mathbb{K}}|$

In 1881, Kronecker asked whether $|d_{\mathbb{K}}| > 1$ holds for all algebraic number fields $\mathbb{K}$ of degree $n \geqslant 2$. This question remained open until 1890 when Minkowski created the geometry of numbers and discovered the Minkowski bound (7.16). This lower bound follows readily by using the fact that $N_{\mathbb{K}/\mathbb{Q}}(\mathfrak{a}) \geqslant 1$ and the easy bounds $r_2 \leqslant \frac{1}{2} n$ and $n^n/n! \geqslant 2^{n-1}$. By Theorem 7.10, we infer that every algebraic number field $\mathbb{K} \neq \mathbb{Q}$ has at least a ramified prime number, which plays an important role in the proof of the Kronecker–Weber theorem. Furthermore, Minkowski observed that his method provides lower bounds tending to ∞ with the degree of $\mathbb{K}$. More precisely, using Stirling's bounds[42] we see that, for all $n \geqslant 3$, we have

$$\log |d_{\mathbb{K}}| \geqslant \left(2 + \log \tfrac{\pi}{4}\right) n + r_1 \log \tfrac{4}{\pi} - \log 2\pi n - \tfrac{1}{6n} > n - 1. \qquad (7.24)$$

In the late 1960s and early 1970s, Stark [88] used the functional equation and the Hadamard's factorization theorem to get a lower bound which eventually supersedes the geometric methods. This is the purpose of the next result, in which the digamma function $\Psi(\sigma) = \Gamma'(\sigma)/\Gamma(\sigma)$ appears.

Proposition 7.32 *Let $\mathbb{K}/\mathbb{Q}$ be an algebraic number field of degree n, discriminant $d_{\mathbb{K}}$ and signature (r_1, r_2). For all real numbers $\sigma > 1$, we have*

$$\log |d_{\mathbb{K}}| \geqslant r_1 \left(\log \pi - \Psi\left(\frac{\sigma}{2}\right) \right) + 2r_2 \left(\log 2\pi - \Psi(\sigma) \right) - \frac{2}{\sigma} - \frac{2}{\sigma - 1}.$$

---

[42] For all $n \in \mathbb{Z}_{\geqslant 1}$, $\left(\frac{n}{e}\right)^n \sqrt{2\pi n} \leqslant n! \leqslant \left(\frac{n}{e}\right)^n \sqrt{2\pi n} \times e^{1/(12n)}$.

Proof We proceed as in Proposition 3.10. Define

$$F_{\mathbb{K}}(s) = s(s-1)\xi_{\mathbb{K}}(s) = s(s-1)A_{\mathbb{K}}^{s}\Gamma_{\mathbb{K}}(s)\zeta_{\mathbb{K}}(s).$$

$F_{\mathbb{K}}(s)$ is an entire function of order 1, so that by the Hadamard's factorization theorem, there exist suitable constants a, b such that

$$F_{\mathbb{K}}(s) = e^{a+bs}\prod_{\rho}\left(1 - \frac{s}{\rho}\right)e^{s/\rho},$$

where the product runs through all zeros $\rho = \beta + i\gamma$ of $F_{\mathbb{K}}(s)$, which are exactly the non-trivial zeros of $\zeta_{\mathbb{K}}(s)$. The logarithmic differentiation provides

$$\frac{F_{\mathbb{K}}'}{F_{\mathbb{K}}}(s) = b + \sum_{\rho}\left(\frac{1}{s-\rho} + \frac{1}{\rho}\right), \tag{7.25}$$

where the sum is absolutely convergent. Taking $s = 0$ gives $b = F'(0)/F(0)$, and using the functional equation of $\zeta_{\mathbb{K}}(s)$ gives $F_{\mathbb{K}}(s) = F_{\mathbb{K}}(1-s)$, so that

$$b = \frac{F_{\mathbb{K}}'(0)}{F_{\mathbb{K}}(0)} = -\frac{F_{\mathbb{K}}'(1)}{F_{\mathbb{K}}(1)} = -b - \sum_{\rho}\left(\frac{1}{\rho} + \frac{1}{1-\rho}\right).$$

Now if ρ is a zero of $F_{\mathbb{K}}(s)$, so is $\overline{\rho}$ and since $F_{\mathbb{K}}(s) = F_{\mathbb{K}}(1-s)$, we infer that $1-\rho$ is also a zero of $F_{\mathbb{K}}$. This gives

$$b = -\frac{1}{2}\sum_{\rho}\left(\frac{1}{\rho} + \frac{1}{\overline{\rho}}\right)$$

and (7.25) becomes

$$\frac{F_{\mathbb{K}}'}{F_{\mathbb{K}}}(s) = \frac{1}{2}\sum_{\rho}\left(\frac{1}{s-\rho} + \frac{1}{s-\overline{\rho}}\right)$$

so that, by the definition of $F_{\mathbb{K}}(s)$, we derive

$$\log|d_{\mathbb{K}}| = r_1\left(\log\pi - \Psi\left(\frac{s}{2}\right)\right) + 2r_2\left(\log 2\pi - \Psi(s)\right) - \frac{2}{s} - \frac{2}{s-1}$$
$$+ \sum_{\rho}\left(\frac{1}{s-\rho} + \frac{1}{s-\overline{\rho}}\right) - \frac{2\zeta_{\mathbb{K}}'}{\zeta_{\mathbb{K}}}(s).$$

Now we have

$$\sum_{\rho}\left(\frac{1}{\sigma - \rho} + \frac{1}{\sigma - \bar{\rho}}\right) = 2\sum_{\rho}\frac{\sigma - \beta}{|\sigma - \rho|^2} > 0,$$

and we have from the Euler product of Definition 7.21

$$-\frac{\zeta'_{\mathbb{K}}}{\zeta_{\mathbb{K}}}(\sigma) = \sum_{\mathfrak{p}}\frac{\log \mathcal{N}_{\mathbb{K}/\mathbb{Q}}(\mathfrak{p})}{\mathcal{N}_{\mathbb{K}/\mathbb{Q}}(\mathfrak{p})^{\sigma} - 1} > 0,$$

which concludes the proof.   □

Now applying Gautschi's inequality, we get the following lower bounds for $|d_{\mathbb{K}}|$.

Corollary 7.9 *Let $\mathbb{K}/\mathbb{Q}$ be an algebraic number field of degree $n \geqslant 2$ and discriminant $d_{\mathbb{K}}$. Then we have*

$$|d_{\mathbb{K}}| > \max(e^{n-1}, e^{-8}(2\pi)^n).$$

Furthermore, if $n \geqslant 50$, then $|d_{\mathbb{K}}| > (2\pi)^n$.

Proof If $\mathbb{K}$ is a quadratic field, then $|d_{\mathbb{K}}| \geqslant 3 > e$, so that we may suppose that $n \geqslant 3$. The inequality $|d_{\mathbb{K}}| > e^{n-1}$ has been seen in (7.24). By Gautschi's inequality [20, Corollary 3], we have for all $h > 0$

$$\Psi(1 + h) < \log(h + e^{-\gamma}) \quad \text{and} \quad \Psi\left(\frac{1+h}{2}\right) < \log\left(\frac{h}{2} + e^{-\gamma - \log 4}\right).$$

Using these inequalities in Proposition 7.32 with $\sigma = 1 + h$ ($h > 0$), we get

$$\log|d_{\mathbb{K}}| \geqslant r_1\left(\log \pi - \Psi\left(\frac{1+h}{2}\right)\right) + 2r_2\left(\log 2\pi - \Psi(1+h)\right) - \frac{2}{1+h} - \frac{2}{h}$$

$$> \log\left\{\left(\frac{4\pi}{2h + e^{-\gamma}}\right)^{r_1}\left(\frac{2\pi}{h + e^{-\gamma}}\right)^{2r_2}\right\} - \frac{2}{1+h} - \frac{2}{h}$$

$$= \log\left\{\left(\frac{2\pi}{h + e^{-\gamma}}\right)^{n}\left(\frac{2h + 2e^{-\gamma}}{2h + e^{-\gamma}}\right)^{r_1}\right\} - \frac{2}{1+h} - \frac{2}{h}$$

$$\geqslant n \log\left(\frac{2\pi}{h + e^{-\gamma}}\right) - \frac{2}{1+h} - \frac{2}{h}.$$

Choosing $h = \sqrt{2e^{-\gamma}/n}$ gives

$$\log|d_{\mathbb{K}}| > n \log 2\pi + f(n),$$

where

$$f(n) = \gamma n - 2\sqrt{2e^\gamma n}\left(1 + \frac{1}{\sqrt{2e^\gamma n} + 2}\right) > -8$$

for all $n \geqslant 2$ implying that $|d_{\mathbb{K}}| > e^{-8}(2\pi)^n$. If $n \geqslant 50$, the trivial bound

$$\frac{1}{\sqrt{2e^\gamma n} + 2} < \frac{1}{\sqrt{2e^\gamma n}}$$

implies

$$\log|d_{\mathbb{K}}| > n\log 2\pi + \gamma n - 2\sqrt{2e^\gamma n} - 2 > n\log 2\pi$$

as asserted.   □

Let us notice that the proof of Proposition 7.32 rests on the fact that, for all $\sigma > 1$, we have

$$-\frac{\zeta'_{\mathbb{K}}(\sigma)}{\zeta_{\mathbb{K}}(\sigma)} + \frac{1}{2}\sum_\rho\left(\frac{1}{\sigma - \rho} + \frac{1}{\sigma - \overline{\rho}}\right) > 0.$$

In [67], Odlyzko showed that this quantity is in fact quite large and used this to get substantial improvements on the estimates of Corollary 7.9. Later Serre, Odlyzko and Poitou used Guinand's and Weil's explicit formulae to bound $|d_{\mathbb{K}}|$ in a somewhat much more elegant and efficient way than with Proposition 7.32, both under ERH[43] and unconditionally. For instance, one may prove [69] without ERH that

$$\log|d_{\mathbb{K}}| \geqslant n(\gamma + \log 4\pi) + r_1 - 8.6n^{1/3}.$$

7.4.9 *Quadratic Fields*

Let $d \in \mathbb{Z}\setminus\{0, 1\}$ squarefree and $\mathbb{K} = \mathbb{Q}(\sqrt{d})$ be a quadratic field with discriminant $d_{\mathbb{K}}$. According to Lemma 7.17, the Kronecker symbol $(d_{\mathbb{K}}/\cdot)$ is a real primitive Dirichlet character. We define

$$L_{d_{\mathbb{K}}}(s) = \sum_{m=1}^{\infty} \frac{(d_{\mathbb{K}}/m)}{m^s} \qquad\qquad (7.26)$$

---

43 See Footnote 6 in Chap. 3.

its associated Dirichlet L-series, which is absolutely convergent in the half-plane $\sigma > 0$. Our aim is to prove the following factorization of $\zeta_{\mathbb{K}}$, which may be viewed as a sort of analytic translation of the quadratic reciprocity law.

Proposition 7.33 *For all real numbers $\sigma > 1$, we have $\zeta_{\mathbb{K}}(\sigma) = \zeta(\sigma)\, L_{d_{\mathbb{K}}}(\sigma)$.*

Proof First note that using the Euler product of L-functions, we get

$$
L_{d_{\mathbb{K}}}(\sigma) = \prod_{p}\left(1 - \frac{(d_{\mathbb{K}}/p)}{p^{\sigma}}\right)^{-1} = \prod_{(d_{\mathbb{K}}/p)=1}\left(1 - \frac{1}{p^{\sigma}}\right)^{-1}\prod_{(d_{\mathbb{K}}/p)=-1}\left(1 + \frac{1}{p^{\sigma}}\right)^{-1}.
$$

Now using Remark 7.6 and Proposition 7.26, we get for $\sigma > 1$

$$
\zeta_{\mathbb{K}}(\sigma) = \prod_{(d_{\mathbb{K}}/p)=1}\left(1 - \frac{1}{p^{\sigma}}\right)^{-2}\prod_{(d_{\mathbb{K}}/p)=-1}\left(1 - \frac{1}{p^{2\sigma}}\right)^{-1}\prod_{p\mid d_{\mathbb{K}}}\left(1 - \frac{1}{p^{\sigma}}\right)^{-1}
$$

$$
= L_{d_{\mathbb{K}}}(\sigma)\prod_{(d_{\mathbb{K}}/p)=1}\left(1 - \frac{1}{p^{\sigma}}\right)^{-1}\prod_{(d_{\mathbb{K}}/p)=-1}\left(1 - \frac{1}{p^{\sigma}}\right)^{-1}\prod_{p\mid d_{\mathbb{K}}}\left(1 - \frac{1}{p^{\sigma}}\right)^{-1}
$$

$$
= L_{d_{\mathbb{K}}}(\sigma)\prod_{p}\left(1 - \frac{1}{p^{\sigma}}\right)^{-1} = \zeta(\sigma)\, L_{d_{\mathbb{K}}}(\sigma)
$$

as required.   □

Now letting $\sigma \to 1$ in Proposition 7.33 and using the analytic class number formula give $L_{d_{\mathbb{K}}}(1) = \kappa_{\mathbb{K}}$, which implies that

$$
h_{\mathbb{K}} = \begin{cases} \dfrac{L_{d_{\mathbb{K}}}(1)\, d_{\mathbb{K}}^{1/2}}{2\log \varepsilon_{\mathbb{K}}}, & \text{if } d > 0 \\[2em] \dfrac{L_{d_{\mathbb{K}}}(1)\, |d_{\mathbb{K}}|^{1/2} w_{\mathbb{K}}}{2\pi}, & \text{if } d < 0 \end{cases}, \tag{7.27}
$$

where $\varepsilon_{\mathbb{K}}$ is the fundamental unit of $\mathbb{K}$ in the case of $d > 0$ and $w_{\mathbb{K}} = 2, 4, 6$ according to whether $d_{\mathbb{K}} < -4$ or $d_{\mathbb{K}} = -4, -3$. This is the *Dirichlet class number formula for quadratic fields*.

7.5 Selected Problems in Algebraic Number Theory

7.5.1 Computations of Galois Groups

7.5.1.1 Fundamental Tools

The purpose of this section is to supply some usual tools from Galois theory to help compute some Galois groups of Galois extensions. We refer the reader to [54, 70, 82] for more information and some proofs. We use group-theoretic notation for the usual transitive subgroups of S_n, the *symmetric group* on $\{1, \ldots, n\}$. For instance $\mathcal{A}_n$ is the *alternating group* on $\{1, \ldots, n\}$, C_n is the *cyclic group* of order n, D_n is the *dihedral group* of order $2n$, which is the group of symmetries of a regular n-gon, and so on. For any positive integer n and prime number p, $\mathrm{PSL}(n, p)$ means $\mathrm{PSL}_n(\mathbb{F}_p)$.

Let $\mathbb{K}/\mathbb{Q}$ be a Galois extension of $\mathbb{Q}$ and $P \in \mathbb{Z}[X]$ be a defining polynomial of $\mathbb{K}$. Since P is irreducible, $\mathrm{Gal}(\mathbb{K}/\mathbb{Q}) \simeq \mathrm{Gal}(P/\mathbb{Q})$ is transitive considered as a subgroup of S_n, i.e. for all $i, j \in \{1, \ldots, n\}$, there exists $\sigma \in \mathrm{Gal}(P/\mathbb{Q})$ such that $\sigma(i) = j$. The notation $a \in \mathbb{Q}^2$ means that a is a square in $\mathbb{Q}$.

The first criterion enables us to check whether $\mathrm{Gal}(P/\mathbb{Q}) \subseteq \mathcal{A}_n$ or not.

Lemma 7.20 $\mathrm{Gal}(P/\mathbb{Q}) \subseteq \mathcal{A}_n$ *if and only if* $\mathrm{disc}(P) \in \mathbb{Q}^2$.

Proof Since P is irreducible over $\mathbb{Q}$, the roots $\theta_1, \ldots, \theta_n$ of P are all distinct. By Definition 7.2, we have $\mathrm{disc}(P) = d^2$ with $d \neq 0$ and

$$d = \prod_{1 \leqslant i < j \leqslant n} \left(\theta_i - \theta_j \right).$$

If d is an algebraic integer, then, for all $\sigma \in \mathrm{Gal}(P/\mathbb{Q})$, we have $\sigma(d) = \epsilon(d)d$, where $\epsilon(d)$ is the signature of σ.

▷ If $\sigma \in \mathcal{A}_n$, then $\epsilon(d) = 1$ and therefore $\sigma(d) = d$ and we infer $d \in \mathbb{Z}$ from Galois theory.
▷ If $d \in \mathbb{Z}$, we have $\sigma(d) = d$ because σ fixes $\mathbb{Q}$. Since $d \neq 0$, we get $\epsilon(d) = 1$ and thus $\sigma \in \mathcal{A}_n$.

$\square$

The next tool, due to Dedekind, relies on the factorization of $\overline{P}$ in $\mathbb{F}_p[X]$ for some suitable prime number p.

Proposition 7.34 (Dedekind) *Let* $P \in \mathbb{Z}[X]$ *be a monic and irreducible polynomial and* $p \nmid \mathrm{disc}(P)$ *be a prime number. Assume that in* $\mathbb{F}_p[X]$ *we have the factorization*

$$\overline{P} = \prod_{i=1}^{g} \overline{P_i},$$

where $\overline{P_i}$ are irreducible polynomials over $\mathbb{F}_p$. Then $\mathrm{Gal}(P/\mathbb{Q})$ contains a permutation, which is the product of distinct cycles σ_i of length $\deg \overline{P_i}$.

For Galois groups over $\mathbb{F}_p$, the following proposition, which is a particular case of a result due to Frobenius, relies on the ramification of p in the corresponding Galois extension.

Proposition 7.35 (Frobenius) *Let $\mathbb{K}/\mathbb{Q}$ be a Galois extension of $\mathbb{Q}$ of degree n and $P \in \mathbb{Z}[X]$ be a defining monic irreducible polynomial of $\mathbb{K}$. Let $p \nmid \mathrm{disc}(P)$ be a prime number with inertial degree f_p and assume that $\overline{P}$ is squarefree in $\mathbb{F}_p[X]$. Then*

$$\mathrm{Gal}(\overline{P}/\mathbb{F}_p) \simeq C_{f_p}$$

The following tools are often useful in the determination of certain Galois groups.

Lemma 7.21

▷ *Let H be a transitive subgroup of S_n. If H contains a transposition and a $(n-1)$-cycle, then $H = S_n$.*
▷ *Let p be a prime number and H be a subgroup of S_p. If H contains a transposition and an element of order p, then $H = S_p$.*
▷ *Let $P \in \mathbb{Q}[X]$ such that $\deg P = p$ is a prime number. If P has exactly two non-real roots, then $\mathrm{Gal}(P/\mathbb{Q}) \simeq S_p$.*

One may also notice the next lemma due to Jordan.

Lemma 7.22 (Jordan) *Let H be a transitive subgroup of S_n. If H contains a p-cycle for some prime number p satisfying $\frac{1}{2}n < p < n-2$, then H contains $\mathcal{A}_n$.*

When the degree of P is small, it is easy to compute $\mathrm{Gal}(P/\mathbb{Q})$.

Lemma 7.23 *Let $P \in \mathbb{Q}[X]$ be irreducible.*

▷ *If $P = X^2 + pX + q$, then $\mathrm{Gal}(P/\mathbb{Q}) \simeq C_2$.*
▷ *If $P = X^3 + pX + q$, then $\mathrm{Gal}(P/\mathbb{Q}) \simeq \begin{cases} \mathcal{A}_3 \simeq C_3, & \text{if } \mathrm{disc}(P) \in \mathbb{Q}^2 \\ S_3, & \text{otherwise.} \end{cases}$*
▷ *If $P = X^4 - 2bX^2 + b^2 - ac^2$ with $a, b, c \in \mathbb{Q}$ such that $a \notin \mathbb{Q}^2$ and $a(b^2 - ac^2) \in \mathbb{Q}^2$, then*

$$\mathrm{Gal}(P/\mathbb{Q}) \simeq C_4.$$

▷ *If $P = X^4 - 2(a+b)X^2 + (a-b)^2$ with $a, b \in \mathbb{Q} \setminus \{0\}$ such that $a, b, ab^{-1} \notin \mathbb{Q}^2$, then*

$$\mathrm{Gal}(P/\mathbb{Q}) \simeq C_2 \times C_2 = \mathcal{V}_4.$$

$\mathcal{V}_4$ *is called the Klein 4-group.*

7.5.1.2   Resolvents

Let $F \in \mathbb{Z}[X_1, \ldots, X_n]$, G be a subgroup of $\mathcal{S}_n$ and $P \in \mathbb{Z}[X]$ with roots $\alpha_1, \ldots, \alpha_n$. The *stabilizer* H of F in G is the group

$$H = \left\{ \sigma \in G : F\left(X_{\sigma(1)}, \ldots, X_{\sigma(n)}\right) = F(X_1, \ldots, X_n) \right\}.$$

The *resolvent* $\mathrm{Res}_G(F, P)$ associated with F and P is the polynomial defined by

$$\mathrm{Res}_G(F, P) = \prod_{\sigma \in G/H} \left(X - F\left(\alpha_{\sigma(1)}, \ldots, \alpha_{\sigma(n)}\right) \right),$$

where the product ranges over $|G/H|$ cosets representative of G/H. When $G = \mathcal{S}_n$, we omit the subscript in the notation $\mathrm{Res}_G(F, P)$. By the fundamental theorem on symmetric polynomials, the coefficients of $\mathrm{Res}_G(F, P)$ can be expressed as polynomials over $\mathbb{Q}$ in the coefficients of P. If P is monic, then these coefficients are algebraic integers, and hence rational integers. We also notice that $\mathrm{Res}_G(F, P)$ is independent of the ordering of the roots of P. An important special case is the *linear resolvent polynomial* defined with

$$F(X_1, \ldots, X_n) = \sum_{i=1}^{n} a_i X_i \quad (a_i \in \mathbb{Q}).$$

The resolvents are often used in computational algebra in order to construct algorithms to compute Galois groups of polynomials. Most of these algorithms rest on the following result [13].

Proposition 7.36 *If* $\mathrm{Res}_G(F, P)$ *has a simple root in* $\mathbb{Z}$*, then* $\mathrm{Gal}(P/\mathbb{Q})$ *is conjugate under* G *to a subgroup of* H*.*

In practice, one does not need to compute explicitly the resolvent. It suffices to compute numerical approximations of the roots of P and determine numerically $F\left(\alpha_{\sigma(1)}, \ldots, \alpha_{\sigma(n)}\right)$. These approximations are in general accurate enough to guarantee that we can correctly recognize when $\mathrm{Res}_G(F, P)$ has a simple root in $\mathbb{Z}$. Algorithms up to degree 7 are detailed in [13], where the choices of polynomials F and corresponding systems of representatives of G/H are also given. It should be noticed that, for polynomials of degree 7, one can use the following simple resolvent

$$R = \prod_{1 \leqslant i < j < k \leqslant 7} \left(X - \alpha_i - \alpha_j - \alpha_k \right),$$

which is a polynomial of degree 35. It is an exercise in Galois theory to show that, if $R = R_1 R_2$ with R_i irreducibles such that $\deg R_1 = 7$ and $\deg R_2 = 28$, then

$$\mathrm{Gal}(P/\mathbb{Q}) \simeq \mathrm{PSL}(3, 2) \simeq \mathrm{PSL}(2, 7),$$

which is the unique simple group of order 168. For instance, this is the case for the polynomial $P = X^7 - 7X^3 + 14X^2 - 7X + 1$.

7.5.1.3   Usual Examples

▷ deg $P = 4$. There are up to conjugacy five transitive subgroups of $\mathcal{S}_4$, i.e.

$$\mathcal{S}_4, \ \mathcal{A}_4, \ D_4, \ \mathcal{V}_4, \ C_4.$$

With $F = X_1 X_2 + X_3 X_4$ and $G = \mathcal{S}_4$, a system of representatives of G/H is given by

$$G/H = \{\text{Id}, (12), (14)\}$$

and, if $P = X^4 + aX^3 + bX^2 + cX + d$, we have

$$\text{Res}_G(F, P) = X^3 - bX^2 - (ac - 4d)X - (a^2 d + 4bd + c^2).$$

If $\text{Res}_G(F, P)$ has no root in $\mathbb{Z}$, then $\text{Gal}(P/\mathbb{Q}) \simeq \mathcal{S}_4$ or $\mathcal{A}_4$ by Proposition 7.36, and one can use Lemma 7.20 to determine precisely $\text{Gal}(P/\mathbb{Q})$. If $\text{Res}_G(F, P)$ has a root in $\mathbb{Z}$, then $\text{Gal}(P/\mathbb{Q}) \simeq C_4$, $\mathcal{V}_4$ or D_4. We have $\mathcal{V}_4 \subset D_4 \cap \mathcal{A}_4$ and hence if $\text{disc}(P) \in \mathbb{Q}^2$, then $\text{Gal}(P/\mathbb{Q}) \simeq \mathcal{V}_4$, otherwise $\text{Gal}(P/\mathbb{Q}) \simeq C_4$ or D_4. Another resolvent is used to distinguish between the two.

When $P = X^4 + aX^2 + b$, we have the following more precise result.

Proposition 7.37 Let $P = X^4 + aX^2 + b$ be irreducible over $\mathbb{Q}$ with roots $\pm\alpha$ and $\pm\beta$. We have

$$\text{Gal}(P/\mathbb{Q}) \simeq \begin{cases} C_4, & \text{if } \alpha\beta^{-1} - \alpha^{-1}\beta \in \mathbb{Q} \\ \mathcal{V}_4, & \text{if } \alpha\beta \in \mathbb{Q} \text{ or } \alpha^2 - \beta^2 \in \mathbb{Q} \\ D_4, & \text{otherwise.} \end{cases}$$

▷ deg $P = 5$. There are up to conjugacy five transitive subgroups of $\mathcal{S}_5$, i.e.

$$\mathcal{S}_5, \ \mathcal{A}_5, \ D_5, \ M_{20} = \langle(12345), (2354)\rangle, \ C_5.$$

We have the inclusions $C_5 \subset D_5 \subset \mathcal{A}_5 \cap M_{20}$. One uses a first resolvent with $G = \mathcal{S}_5$ and $H = M_{20}$ so that

$$G/H = \{\text{Id}, (12), (13), (14), (15), (25)\}$$

and if $\text{Res}_G(F, P)$ has no root in $\mathbb{Z}$, then $\text{Gal}(P/\mathbb{Q}) \simeq \mathcal{S}_5$ or $\mathcal{A}_5$, and use again Lemma 7.20 to determine precisely $\text{Gal}(P/\mathbb{Q})$. If $\text{Res}_G(F, P)$ has a root in $\mathbb{Z}$ and

$\mathrm{disc}(P) \notin \mathbb{Q}^2$, then $\mathrm{Gal}(P/\mathbb{Q}) \simeq M_{20}$, otherwise $\mathrm{Gal}(P/\mathbb{Q}) \simeq C_5$ or D_5. One uses again another resolvent to finalize the determination.

▷ $\deg P = 6$. There are up to conjugacy sixteen transitive subgroups of S_6. This case has been completely solved in [37], where the author uses specializations of three resolvents of degrees 2, 10 and 15 denoted, respectively, by f_2, f_{10} and f_{15}. Now f_2 is just the polynomial $f_2 = X^2 - \mathrm{disc}(P)$, and the coefficients of f_{10} and f_{15} are given in [37] and hence are known. Furthermore, there are twelve solvable subgroups of S_6, i.e. the groups $C_3^2 \rtimes D_4$ and $S_4 \times C_2$ and their proper subgroups. As an example, let us give one of the results of this article.

Proposition 7.38 *Let* $P \in \mathbb{Z}[X]$ *be an irreducible polynomial of degree* 6 *with discriminant* D *and set* $G = \mathrm{Gal}(P/\mathbb{Q})$.

▷ *G is solvable if and only if one of the following statements holds:*

  ◇ f_{10} *has a rational root and then* $G \subseteq C_3^2 \rtimes D_4$.
  ◇ f_{15} *has a rational root with multiplicity* $\neq 5$.
  ◇ f_{15} *has a rational root with multiplicity* 5, *and* f_{10} *is a product of quartic and sextic irreducible polynomials.*

▷ *Assume that* $G \simeq C_3^2 \rtimes D_4$, $C_3^2 \rtimes C_4$, $D_3 \times D_3$ *or* $C_3 \times D_3$. *Then*

$$G \simeq C_3^2 \rtimes C_4 \iff D \in \mathbb{Q}^2.$$

▷ *If G is not solvable, then*

  ◇ $G \simeq S_6$ *if and only if* f_{15} *is irreducible over* $\mathbb{Q}$ *and* $D \notin \mathbb{Q}^2$.
  ◇ $G \simeq \mathcal{A}_6$ *if and only if* f_{15} *is irreducible over* $\mathbb{Q}$ *and* $D \in \mathbb{Q}^2$.
  ◇ $G \simeq S_5 \simeq \mathrm{PGL}(2, 5)$ *if and only if* f_{15} *is reducible over* $\mathbb{Q}$ *and* $D \notin \mathbb{Q}^2$.
  ◇ $G \simeq \mathcal{A}_5 \simeq \mathrm{PSL}(2, 5)$ *if and only if* f_{15} *is reducible over* $\mathbb{Q}$ *and* $D \in \mathbb{Q}^2$.

7.5.2   Gauss's Class Number Problems

7.5.2.1   Euler's Polynomials

Let $P = X^2 + X + 41$. One may readily check that $P(n)$ is prime for $n \in \{0, \ldots, 39\}$ but $P(40) = 41^2$ is not a prime number. This polynomial, discovered by Euler in 1772, was one of the first polynomials, which can provide a finite subset of prime numbers as long as n lies in a subset of non-negative integers. One may ask for a polynomial giving *all* the prime numbers, but it is a folklore in prime number theory that such a single-variable polynomial cannot exist. However, one may formulate the problem in the following way: are there polynomials of the form $P_q = X^2 + X + q$, with q prime, such that $P_q(n)$ is a prime number for all $n \in \{0, \ldots, q - 2\}$? The answer is given by the following result.

Theorem 7.21 *Let q be a prime number and $P_q = X^2 + X + q$. Then, $P_q(n)$ is a prime number for all $n \in \{0, \ldots, q-2\}$ if and only if $q \in \{2, 3, 5, 11, 17, 41\}$.*

7.5.2.2 Class Number One Problems

The above result is the consequence of two profound theorems. The first one relates the values of q to the class number of the imaginary quadratic field $\mathbb{K} = \mathbb{Q}(\sqrt{1-4q})$. More precisely, we have the following theorem[44], which goes back to Rabinowitch.

Proposition 7.39 *Let q be a prime number, $P_q = X^2 + X + q$ and $\mathbb{K} = \mathbb{Q}(\sqrt{1-4q})$. Then $P_q(n)$ is a prime number for all $n \in \{0, \ldots, q-2\}$ if and only if $h_{\mathbb{K}} = 1$.*

In 1966/7, with two different methods, Baker [7] and Stark [86] discovered *all* the imaginary quadratic fields with class number one, proving the following result, which easily implies Theorem 7.21.

> **Theorem 7.22** *Let $d < 0$ squarefree and $\mathbb{K} = \mathbb{Q}(\sqrt{d})$. Then $h_{\mathbb{K}} = 1$ if and only if*
>
> $$d \in \{-1, -2, -3, -7, -11, -19, -43, -67, -163\}.$$

In Articles 303 and 304 of his *Disquisitiones Arithmeticæ* [30], Gauss used the language of binary quadratic forms to formulate several conjectures, which remain open nowadays. Translated into the language of modern algebraic number theory, i.e. Dedekind's language, the two particular conjectures below can be stated as follows, where $\mathbb{K} = \mathbb{Q}(\sqrt{d})$ is a quadratic field with class number denoted here by $h(d)$.

In Article 303, Gauss conjectured that $h(d) \to \infty$ as $d \to -\infty$. He also included the following table:

$h(d)$	1	2	3	4	5
Number of fields	9	18	16	54	25
Largest $|d|$	163	427	907	1555	2683

---

[44]See [31, 75] and the references therein.

translated into a table of quadratic fields with small class numbers, and he surmised that this table is complete. Given any $h \in \mathbb{Z}_{\geqslant 1}$, the problem of finding all imaginary quadratic fields of class number h is called *Gauss's class number h-problem for imaginary quadratic fields*.

For real quadratic fields, translated into modern language, Gauss surmised in Article 304 that there are infinitely many real quadratic fields with class number one. This conjecture still remains open today, and we do not know if there are infinitely many number fields *of arbitrary degree* having class number one, or even just bounded.

7.5.2.3   Background of Gauss' Conjecture

The conjecture $h(d) \to \infty$ as $d \to -\infty$ has a curious and interesting story [31]. In 1918, Landau published the following result which he attributed to a lecture given by Hecke.

Proposition 7.40 *Let $d < 0$ and χ be an odd, real and primitive Dirichlet character modulo d. If $L(\sigma, \chi) \neq 0$ for all real numbers $\sigma > 1 - c_1 / \log|d|$, then we have*

$$h(d) > \frac{c_2 |d|^{1/2}}{\log|d|}.$$

We infer from this result that the GRH[45] implies Gauss's conjecture.

But in the 1930s, Deuring, Mordell and Heilbronn proved that the *falsity* of GRH *also* implies Gauss's conjecture. Hence this conjecture is true and was the first result to be proved by assuming the truth and the falsity of GRH.

The flaw of this method is that the result is *not effective*, since if the GRH is false, all constants depend on the Siegel's zero[46] of $L(s, \chi)$ located off the line $\sigma = \frac{1}{2}$. Refining this proof, in order to work with Gauss's class number one problem for imaginary quadratic fields, Heilbronn and Linfoot proved that there are at most ten imaginary quadratic fields with class number one, i.e. the nine fields of Theorem 7.22 plus possibly an unknown field. The existence of this tenth imaginary UFD quadratic field reflects the ineffectivity of the Deuring–Heilbronn phenomenon and would also contradict the truth of the GRH. Then one can imagine that this problem led to intense research.

The solution of Gauss's class number one problem was found by Baker [7] and Stark [86], with completely different methods. Baker used an idea of Gelfond and Linnik who proved that this problem could be solved if one had linear independence of three logarithms, while Stark showed that a tenth imaginary quadratic field

---

[45] The *Generalized Riemann Hypothesis*, or GRH for short, asserts that the Dirichlet L-functions have *all* their zeros lying on the line $\sigma = \frac{1}{2}$.

[46] See Theorems 3.31 and 3.32.

cannot exist. At this point, it should be mentioned that 15 years earlier, Heegner, a High School teacher, announced he had solved the class number one problem [41]. Unfortunately, his work contained some "gaps" in his proof and his paper was dismissed at the time. As pointed out by Goldfeld [31], Heegner died before anyone really understood his discoveries.

7.5.2.4   Elliptic Curves

The general solution to Gauss's class number problem for imaginary quadratic fields comes from another area in number theory. Goldfeld showed that if an elliptic curve over $\mathbb{Q}$ having certain properties could exist, it would provide a lower bound of $h(d)$ sufficiently accurate to be effective and then solve the conjecture. Gross and Zagier discovered such an elliptic curve, and the method, called today the *Goldfeld–Gross–Zagier* theorem, provides a solution to this long-standing problem.

Let us be more precise. To each elliptic curve E with minimal Weierstrass equation $y^2 = 4x^3 - g_2 x - g_3$ and discriminant $\mathrm{disc}(E) = g_2^3 - 27g_3^2 \neq 0$, one can define a positive integer N, called the *conductor* of E, having the same prime factors as $\mathrm{disc}(E)$ and dividing it, and one can also define an L-function, called the *Hasse–Weil L*-function, by the Euler product

$$L(E, s) = \prod_{p \nmid \mathrm{disc}(E)} \left(1 - \frac{a_p}{p^s} + \frac{1}{p^{2s-1}}\right)^{-1} \prod_{p \mid \mathrm{disc}(E)} \left(1 - \frac{a_p}{p^s}\right)^{-1},$$

where $a_p = p + 1 - |E(\mathbb{F}_p)|$ and $|E(\mathbb{F}_p)|$ is the number of points of E over $\mathbb{F}_p$, including the point at infinity. It has been shown by Hasse that $|a_p| \leqslant 2\sqrt{p}$, so that the Euler product is absolutely convergent in the half-plane $\sigma > \frac{3}{2}$. One of the most important results on elliptic curves over $\mathbb{Q}$ is a theorem proved by Wiles and other mathematicians[47] stating that the function $L(E, s)$ has an analytic continuation in the whole complex plane into a holomorphic function and satisfies a functional equation. Another fundamental result is Mordell's theorem stating that the group $E(\mathbb{Q})$ of rational points of E is a finitely generated abelian group. The rank of this group is called the *algebraic rank* of E, in comparison to the *analytic rank* of E, which is defined as the order of vanishing of $L(E, s)$ at $s = 1$. These two ranks are related in one of the most famous conjectures in number theory.

Conjecture 7.1 (Birch and Swinnerton-Dyer)   For all elliptic curves over $\mathbb{Q}$, the algebraic rank and the analytic rank are equal.

In recent decades, some progress has been made toward this conjecture. In particular, it can be proved that, *if the analytic rank is equal to 0, respectively, 1,*

---

[47]Implying in particular, along with a result of Ribet, Fermat's last theorem.

then the algebraic rank is equal to 0, *respectively,* 1. The Goldfeld–Gross–Zagier method enables us to prove the following result [45, Theorem 23.2].

Proposition 7.41 *There exists an absolute, effectively computable constant $c > 0$ such that for all imaginary quadratic fields $\mathbb{K} = \mathbb{Q}(\sqrt{d})$ with $d < 0$, we have*

$$h(d) > c \prod_{p|d} \left(1 + \frac{1}{p}\right)^{-3} \left(1 + \frac{2p^{1/2}}{p+1}\right)^{-1} \log|d|.$$

7.5.3 The Brauer–Siegel Theorem

The functional equation of $\zeta_{\mathbb{K}}(s)$ and the inequality (7.20) imply that, for all $\sigma > 1$, we have

$$\kappa_{\mathbb{K}} \leqslant \sigma(\sigma - 1) \, 2^{r_2(1-\sigma)} \pi^{r_2 - n\sigma/2} |d_{\mathbb{K}}|^{(\sigma-1)/2} \Gamma_{\mathbb{K}}(\sigma) \zeta_{\mathbb{K}}(\sigma).$$

Alzer's inequality [3] states that

$$\Gamma(x) < \begin{cases} x^{x-1-\gamma}, & \text{if } x > 1 \\[2mm] x^{\delta(x-1)-\gamma}, & \text{if } 0 < x < 1 \end{cases},$$

where $\gamma \approx 0.5772\ldots$ is the Euler constant and $\delta = \frac{1}{2}(\zeta(2) - \gamma) \approx 0.5338\ldots$, so that for all $1 < \sigma < 2$ we get

$$\Gamma_{\mathbb{K}}(\sigma) < (\sigma/2)^{r_1\{\delta(\sigma/2-1)-\gamma\}} \, \sigma^{r_2(\sigma-1-\gamma)}$$

$$= \sigma^{r_1\{\delta(\sigma/2-1)-\gamma\}+r_2(\sigma-1-\gamma)} \, 2^{-r_1\{\delta(\sigma/2-1)-\gamma\}}$$

$$= \sigma^{(\sigma-1)(r_1\delta+r_2)-\gamma(n-r_2)-r_1\delta\sigma/2} \, 2^{-r_1\{\delta(\sigma/2-1)-\gamma\}}$$

and using inequality (7.18) along with $\zeta(\sigma) \leqslant \sigma/(\sigma - 1)$ and following the ideas of [61], we may write

$$\kappa_{\mathbb{K}} \leqslant \frac{|d_{\mathbb{K}}|^{(\sigma-1)/2}}{(\sigma - 1)^{n-1}} F(\sigma)$$

with

$$F(\sigma) = \sigma^{n+1+(\sigma-1)(r_1\delta+r_2)-\gamma(n-r_2)-r_1\delta\sigma/2} \, 2^{r_2(1-\sigma)-r_1\{\delta(\sigma/2-1)-\gamma\}} \, \pi^{r_2-n\sigma/2}.$$

Now we have for all $\sigma \geq 1$ and $n \geq 4$

$$2\sigma^2 \left(\frac{F'}{F}\right)' (\sigma) = 2n(\gamma - 1) + r_1\delta(\sigma + 2) + 2r_2(\sigma + 1 - \gamma) - 2$$

$$= n\{\delta\sigma + 2(\delta + \gamma - 1)\} + 2r_2\{\sigma(1 - \delta) + 1 - 2\delta - \gamma\} - 2$$

$$\geq n(3\delta + 2\gamma - 2) + 2r_2(2 - 3\delta - \gamma) - 2$$

$$\geq n\gamma - 2 > 0.$$

Since F is positive, this implies that F is convex on $[1, +\infty)$, so that for $\sigma \in [1, 2]$ we derive

$$F(\sigma) \leq \max (F(1), F(2))$$

with

$$F(1) = 2^{(n-2r_2)(\gamma+\delta/2)} \pi^{r_2-n/2} \leq \left(\frac{2^{\gamma+\delta/2}}{\sqrt{\pi}}\right)^n < 2^{(n-1)/2}$$

$$F(2) = 2^{n+1-r_2\gamma} \pi^{r_2-n} \leq 2 \left(\frac{2^{1-\gamma/2}}{\sqrt{\pi}}\right)^n < 2^{(n-1)/2}$$

for all $n \geq 3$. Now choose

$$\sigma = 1 + \frac{2(n-1)}{\log |d_\mathbb{K}|}.$$

By Corollary 7.9, we have $1 < \sigma < 2$ as soon as $n \geq 3$, and hence using the inequality above, we get

$$\kappa_\mathbb{K} < \left(\frac{e}{\sqrt{2}} \frac{\log |d_\mathbb{K}|}{n-1}\right)^{n-1}$$

for $n \geq 4$. Using the class number formula (7.19), we infer the following result.

Theorem 7.23 *Let $\mathbb{K}/\mathbb{Q}$ be an algebraic number field of degree $n \geq 4$, discriminant $d_\mathbb{K}$, class number $h_\mathbb{K}$, regulator $\mathcal{R}_\mathbb{K}$ and let $w_\mathbb{K}$ be the number of roots of unity in $\mathbb{K}$. Then we have*

$$h_\mathbb{K}\mathcal{R}_\mathbb{K} < \frac{w_\mathbb{K}}{2} \left(\frac{2}{\pi}\right)^{r_2} \left(\frac{e}{2\sqrt{2}} \frac{\log |d_\mathbb{K}|}{n-1}\right)^{n-1} |d_\mathbb{K}|^{1/2}.$$

It is natural to ask for lower bounds of $\kappa_{\mathbb{K}}$. Hecke's integral representation of $\zeta_{\mathbb{K}}(s)$ enables us to prove that, for $0 < \beta < 1$ satisfying $\zeta_{\mathbb{K}}(\beta) \leqslant 0$, then [55, 65]

$$\kappa_{\mathbb{K}} \geqslant \beta(1 - \beta)2^{-n}e^{-4\pi n}|d_{\mathbb{K}}|^{(\beta-1)/2}. \tag{7.28}$$

Using this inequality, one can show that, for all $\varepsilon > 0$, there exists $c_\varepsilon > 0$ such that, for all algebraic number fields $\mathbb{K}$ *Galois* over $\mathbb{Q}$

$$\kappa_{\mathbb{K}} \geqslant c_\varepsilon |d_{\mathbb{K}}|^{-\varepsilon}. \tag{7.29}$$

This implies that there exists a constant $c_1 > 0$ such that, for all algebraic number fields $\mathbb{K}$ *Galois* over $\mathbb{Q}$, we have

$$|\log h_{\mathbb{K}}\mathcal{R}_{\mathbb{K}}| \geqslant c_1 \log |d_{\mathbb{K}}|^{1/2}.$$

Using this lower bound with Proposition 7.23, we get the *Brauer–Siegel theorem*.

Theorem 7.24 (Brauer–Siegel) *If $\mathbb{K}$ ranges over a sequence of algebraic number fields of degree n Galois over $\mathbb{Q}$ for which $n/\log|d_{\mathbb{K}}|$ tends to 0, then*

$$\log h_{\mathbb{K}}\mathcal{R}_{\mathbb{K}} \sim \log |d_{\mathbb{K}}|^{1/2}.$$

However, the lower bound (7.29) for all fields Galois over $\mathbb{Q}$ is *ineffective*. As in Chap. 3, any attempt at improving it effectively or at providing a value of c_ε for a sufficiently small $\varepsilon > 0$ has been unsuccessful. We have seen above that a possible exceptional zero of $\zeta_{\mathbb{K}}(s)$ causes trouble in the effectiveness of this constant. Stark [87] also observed that algebraic number fields of small degrees, especially quadratic fields, are a real obstacle to any improvement.

We end this section by pointing out that one may improve on (7.28) in some special cases. For instance, if $\mathbb{K}/\mathbb{Q}$ is a totally imaginary algebraic number field of degree $n \geqslant 4$ and discriminant $d_{\mathbb{K}}$ satisfying $|d_{\mathbb{K}}| \geqslant 2683^n$, then Louboutin [62] showed that

$$\kappa_{\mathbb{K}} \geqslant (1 - \beta)|d_{\mathbb{K}}|^{(\beta-1)/2}$$

if $\zeta_{\mathbb{K}}(\beta) \leqslant 0$ for some $1 - 2/\log|d_{\mathbb{K}}| \leqslant \beta < 1$.

7.5.4   The Class Number Formula

The Dirichlet class number formula (7.27) for quadratic fields may be generalized to any finite abelian extension of $\mathbb{Q}$ in the following way. Recall that, if p is a prime, then f_p and g_p are, respectively, its inertial degree and decomposition number in $\mathbb{K}$.

Theorem 7.25 (Class Number Formula) *Let $\mathbb{K}/\mathbb{Q}$ be an abelian number field with conductor $f_{\mathbb{K}}$ and character group $X_{f_{\mathbb{K}}}(\mathbb{K})$ denoted by X for convenience. Then, for all $\sigma > 1$*

$$\zeta_{\mathbb{K}}(\sigma) = \zeta(\sigma) \prod_{\substack{\chi \in X \\ \chi \neq \chi_0}} L(\sigma, \chi^{\star}),$$

where $\chi^{\star}$ is the primitive Dirichlet character that induces χ. In particular

$$h_{\mathbb{K}} \mathcal{R}_{\mathbb{K}} = \frac{|d_{\mathbb{K}}|^{1/2} w_{\mathbb{K}}}{2^{r_1} (2\pi)^{r_2}} \prod_{\substack{\chi \in X \\ \chi \neq \chi_0}} L(1, \chi^{\star}).$$

Proof The starting point is the following identity sometimes called the *product formula for characters* [43, Lemme IV.4.3].

Lemma 7.24 *Let G be a finite abelian group and $\widehat{G}$ be the group of characters of G. If $a \in G$ is an element of order r, then we have for all z*

$$\prod_{\chi \in \widehat{G}} (1 - \chi(a)z) = \left(1 - z^r\right)^{|G|/r}.$$

Assume first that $p \nmid f_{\mathbb{K}}$ so that p is unramified in $\mathbb{Q}\left(\zeta_{f_{\mathbb{K}}}\right)$. Applying Lemma 7.24 with $z = p^{-\sigma}$, where $\sigma > 1$, $G = \mathrm{Gal}\,(\mathbb{K}/\mathbb{Q})$, $a = p$ so that $r = f_p$,[48] we get

$$\prod_{\chi \in X} \left(1 - \frac{\chi(p)}{p^{\sigma}}\right)^{-1} = \left(1 - \frac{1}{p^{f_p \sigma}}\right)^{-g_p}.$$

---

[48] See Theorems 7.31 and 7.33 below.

Now if $p \mid f_{\mathbb{K}}$, then for all $\sigma > 1$, we still have [65, page 416]

$$\prod_{\chi \in X} \left(1 - \frac{\chi^{\star}(p)}{p^{\sigma}}\right) = \left(1 - \frac{1}{p^{f_p \sigma}}\right)^{g_p},$$

where $\chi^{\star}$ is the primitive Dirichlet character that induces χ. Multiplying out through all primes p and using (3.34), we get

$$\zeta_{\mathbb{K}}(\sigma) = \prod_{p} \left(1 - \frac{1}{p^{f_p \sigma}}\right)^{-g_p} = \prod_{p \mid f_k} \left(1 - \frac{1}{p^{f_p \sigma}}\right)^{-g_p} \prod_{\chi \in X} L(\sigma, \chi)$$

$$= \prod_{\chi \in X} L(\sigma, \chi^{\star}) \prod_{p \mid f_k} \left(1 - \frac{1}{p^{f_p \sigma}}\right)^{-g_p} \prod_{\chi \in X} \left(1 - \frac{\chi^{\star}(p)}{p^{\sigma}}\right)$$

$$= \prod_{\chi \in X} L(\sigma, \chi^{\star}) = \zeta(\sigma) \prod_{\substack{\chi \in X \\ \chi \neq \chi_0}} L(\sigma, \chi^{\star})$$

showing the first part of the theorem. By Corollary 3.8, the function $L(s, \chi)$ is analytic in the half-plane $\sigma > 0$ as long as $\chi \neq \chi_0$. Hence letting $\sigma \to 1^{+}$ in the previous identity and taking the residues of $\zeta(s)$ and $\zeta_{\mathbb{K}}(s)$ at $s = 1$ into account give the second asserted result.   □

See Theorem 7.43 below for an extension of this result.

7.5.5 The Prime Ideal Theorem

Let $\mathbb{K}/\mathbb{Q}$ be an algebraic number field of degree n and $\pi_{\mathbb{K}}(x)$ be the number of prime ideals $\mathfrak{p}$ in $O_{\mathbb{K}}$ such that $N_{\mathbb{K}/\mathbb{Q}}(\mathfrak{p}) \leqslant x$, i.e.

$$\pi_{\mathbb{K}}(x) = \sum_{N_{\mathbb{K}/\mathbb{Q}}(\mathfrak{p}) \leqslant x} 1.$$

Note that $\pi_{\mathbb{Q}}(x) = \pi(x)$ is the usual prime counting function.[49] The so-called *Prime Ideal Theorem*, or PIT, is the evaluation of $\pi_{\mathbb{K}}(x)$. It was first proved by Landau in [51], and, subsequently, the error term was gradually improved to an order of magnitude similar to that of (3.46) with a Vinogradov–Korobov type zero-free region as it was seen in Theorem 7.18. This gives the following result.

---

[49] See Definition 3.3.

Theorem 7.26 (Prime Ideal Theorem) *Let $\mathbb{K}$ be an algebraic number field. There exists a constant $c_{\mathbb{K}} > 0$, depending on the usual invariants of $\mathbb{K}$, such that, for all x sufficiently large*

$$\pi_{\mathbb{K}}(x) = \mathrm{Li}(x) + O_{\mathbb{K}}\left\{x \exp\left(-c_{\mathbb{K}}(\log x)^{3/5}(\log\log x)^{-1/5}\right)\right\}.$$

Effective versions of the PIT have been stated by many authors. For instance, we quote the following estimate.

Theorem 7.27 (Effective Prime Ideal Theorem) *For all $x \geq e^{c_0 n \log^2 \sqrt{|d_{\mathbb{K}}|}}$, we have*

$$\left|\pi_{\mathbb{K}}(x) - \mathrm{Li}(x)\right| \leq \mathrm{Li}\left(x^{\beta}\right) + c_1 x \exp\left(-c_2\sqrt{\frac{\log x}{n}}\right),$$

where the constants c_i are absolute and the term containing β is present if and only if $\zeta_{\mathbb{K}}(s)$ has an exceptional zero β in the region

$$1 - \frac{1}{4\log|d_{\mathbb{K}}|} \leq \sigma < 1.$$

The analogue of Corollaries 3.4 and 3.5 in algebraic number fields does also exist [58].

Proposition 7.42 (Mertens Theorems in Number Fields) *There exist absolute constants $c_0, \ldots, c_4 > 0$ satisfying the following assertions.*

▷ *For all $x \geq \max\left(e^{c_0 n \log^2 \sqrt{|d_{\mathbb{K}}|}}, e^{1024 c_1^{-2} n \log^2(\sqrt{n}/c_2)}\right)$, we have*

$$\sum_{\mathcal{N}_{\mathbb{K}/\mathbb{Q}}(\mathfrak{p}) \leq x} \frac{1}{\mathcal{N}_{\mathbb{K}/\mathbb{Q}}(\mathfrak{p})} = \log\log x + B_{\mathbb{K}} + R_{\mathbb{K}}(x)$$

with

$$B_{\mathbb{K}} = \log\kappa_{\mathbb{K}} + \gamma + \sum_{\mathfrak{p}}\left\{\log\left(1 - \frac{1}{\mathcal{N}_{\mathbb{K}/\mathbb{Q}}(\mathfrak{p})}\right) + \frac{1}{\mathcal{N}_{\mathbb{K}/\mathbb{Q}}(\mathfrak{p})}\right\}$$

and

$$|R_{\mathbb{K}}(x)| \leq \frac{c_3}{\beta\log x}\left(\frac{2-\beta}{1-\beta}\right) + \frac{2c_4}{\log x},$$

and the term containing β is present if and only if $\zeta_{\mathbb{K}}(s)$ has an exceptional zero β in the region

$$1 - \frac{1}{4 \log |d_{\mathbb{K}}|} \leqslant \sigma < 1.$$

▷ *For x sufficiently large, we have*

$$\prod_{\mathcal{N}_{\mathbb{K}/\mathbb{Q}}(\mathfrak{p}) \leqslant x} \left(1 - \frac{1}{\mathcal{N}_{\mathbb{K}/\mathbb{Q}}(\mathfrak{p})}\right)^{-1} = \kappa_{\mathbb{K}} e^{\gamma} \log x \left\{1 + O_{\mathbb{K}}\left(\frac{1}{\log x}\right)\right\}.$$

7.5.6 The Ideal Theorem

7.5.6.1 Background

Let $\mathbb{K}$ be an algebraic number field of fixed degree $n \geqslant 2$. The *Ideal Theorem* deals with the investigation of the values of $\theta_{\mathbb{K}}$ in the error term of the asymptotic formula

$$\sum_{m \leqslant x} r_{\mathbb{K}}(m) = \kappa_{\mathbb{K}} x + O\left(x^{\theta_{\mathbb{K}} + \varepsilon}\right).$$

Since $L(s, r_{\mathbb{K}}) = \zeta_{\mathbb{K}}(s)$, this problem is similar to the Dirichlet–Piltz divisor problem seen in Theorem 4.62, except that it is somewhat more difficult since the values of $r_{\mathbb{K}}(p^{\alpha})$ strongly depend on the nature of $\mathbb{K}$. When the field $\mathbb{K}$ is abelian, then $\zeta_{\mathbb{K}}(s)$ is essentially a product of Dirichlet L-functions as seen in Theorem 7.25, and the problem may be treated as in the Dirichlet–Piltz divisor problem.[50]

The first non-trivial result in the general case is attributed to Weber [99] who showed circa 1895 that

$$\theta_{\mathbb{K}} \leqslant 1 - \frac{1}{n}.$$

With the dazzling progress of complex analysis and methods of contour integration, as seen in Corollary 4.8, Weber's result was quickly superseded by Landau who first noticed [52, pp. 106–107] that the Schnee–Landau theorem implies

$$\theta_{\mathbb{K}} \leqslant 1 - \frac{2}{n+2},$$

---

[50] See Theorem 7.29 below.

and next using a more precise contour and some averaging arguments he was able to prove [53, Satz 210] that

$$\theta_{\mathbb{K}} \leqslant 1 - \frac{2}{n+1}.$$

In the opposite direction, Landau [53, Satz 211] also proved that

$$\theta_{\mathbb{K}} > \frac{1}{2} - \frac{1}{2n}.$$

As usual, it is surmised that this estimate is the right order of magnitude, but this is still an open problem.

7.5.6.2   Main Results

It is well-known that, if the degree of $\mathbb{K}$ is quite small, the Perron summation formula does not necessarily give the best error term, especially when one does not have bounds for the power mean values of $\zeta_{\mathbb{K}}(s)$ at our disposal.

In the following table, we summarize the latest known values for $\theta_{\mathbb{K}}$. For a proof of these results, see [44, 64, 56]. For a proof of the case $n \geqslant 4$, see Exercise 146.

> **Theorem 7.28** *Let $\mathbb{K}$ be an algebraic number field of degree $n \geqslant 2$. Then, for any $\varepsilon > 0$*
>
> $$\sum_{m \leqslant x} r_{\mathbb{K}}(m) = \kappa_{\mathbb{K}} x + O_{\varepsilon, \mathbb{K}} \left(x^{\theta_{\mathbb{K}} + \varepsilon} \right),$$
>
> *where $\kappa_{\mathbb{K}}$ is given in (7.19) and*
>
> | n | 2 | 3 | $n \geqslant 4$ |
> |---|---|---|---|
> | $\theta_{\mathbb{K}}$ | $\frac{23}{73}$ | $\frac{43}{96}$ | $1 - \frac{3}{n+6}$ |

It should be pointed out that these values can be improved in many special cases [25], in particular when $\zeta_{\mathbb{K}}(s)$ can be factorized in a product of Dirichlet's or Artin's L-functions for which power mean value results are known.

Theorem 7.29 *Let $\mathbb{K}$ be a number fields belonging to one of the following classes:*

▷ $\mathbb{K}$ is abelian;
▷ $\mathbb{K}$ is the Galois closure of a pure cubic field, i.e. a Galois field with Galois group S_3;
▷ $\mathbb{K}$ is one of the following fields:

$$\mathbb{K} = \mathbb{Q}\left(\sqrt[4]{m}\right), \ \mathbb{K} = \mathbb{Q}\left(\sqrt{-1}, \sqrt[4]{m}\right), \ \mathbb{K} = \mathbb{Q}\left(\sqrt[4]{\varepsilon_m}\right) \ or \ \mathbb{K} = \mathbb{Q}\left(\sqrt{-1}, \sqrt[4]{\varepsilon_m}\right),$$

where m is a squarefree number and ε_m is the fundamental unit of the quadratic field $\mathbb{Q}\left(\sqrt{m}\right)$.

Then, for any $\varepsilon > 0$

$$\sum_{m \leqslant x} r_{\mathbb{K}}(m) = \kappa_{\mathbb{K}} x + O_{\varepsilon, \mathbb{K}}\left(x^{1 - \frac{3}{n+2} + \varepsilon}\right).$$

7.5.7   The Kronecker–Weber Theorem

The study of cyclotomic fields arose naturally in the early days of algebraic number theory owing to Fermat's equation. In order to generalize, one may ask which are the possible abelian extensions of $\mathbb{Q}$, i.e. Galois extensions $\mathbb{K}/\mathbb{Q}$ such that $\mathrm{Gal}(\mathbb{K}/\mathbb{Q})$ is abelian. Stated in this way, that question can certainly not be solved. But a great achievement of the nineteenth century is in the next important result, which generalizes Proposition 7.27.

> **Theorem 7.30 (Kronecker–Weber)** *If $\mathbb{K}/\mathbb{Q}$ is a finite abelian extension, then there exists a positive integer f such that $\mathbb{K} \subseteq \mathbb{Q}(\zeta_f)$.*

This result was first stated by Kronecker in 1853 who provided an incomplete proof, which reveals some difficulties with extensions of degree 2^α for some α. In 1886, Weber gave the first proof but also had an error[51] at 2. Both authors used the theory of Lagrange resolvents. Later, Hilbert, and then Speiser, used ramification theory to give a proof, which is now often considered as the classic one. It is noteworthy that Hilbert's strategy works partly because $\mathbb{Q}$ does not have any proper unramified abelian extension. Nowadays, many proofs do exist in the literature, mostly based upon Hilbert's method [35], localization methods [98] or on the fact that they are a simple consequence of results belonging to class field theory.

---

[51] This error remained unnoticed for about 90 years.

We shall not prove this result and we shall refer the reader to [63, 76] for proofs using ramification theory. Nevertheless, it seems interesting to have a look at a particular case of Theorem 7.30 using the following lemma.

Lemma 7.25 (Reduction Principle) *Assume that* Theorem 7.30 *is true for abelian number fields with prime power degrees. Then it is true for all abelian number fields.*

Proof Let $\mathbb{K}/\mathbb{Q}$ be an abelian number field with Galois group $G = \text{Gal}(\mathbb{K}/\mathbb{Q})$. Since G is abelian and setting $|G| = p_1^{\alpha_1} \cdots p_r^{\alpha_r}$, we have by (3.1)

$$G \simeq H_1 \oplus \cdots \oplus H_r,$$

where the H_i are the p_i-Sylow subgroups of G and hence $|H_i| = p_i^{\alpha_i}$. Set

$$E_i = \bigoplus_{j \neq i} H_j.$$

The E_i are subgroups of G and let $\mathbb{K}_i$ be the fixed field of E_i. From Galois theory, we have

$$[\mathbb{K}_i : \mathbb{Q}] = (G : E_i) = |H_i| = p_i^{\alpha_i}$$

and any automorphism that fixes the compositum $\mathbb{K}_1 \cdots \mathbb{K}_r$ fixes all of the $\mathbb{K}_i$ so that

$$\text{Gal}(\mathbb{K}/\mathbb{K}_1 \cdots \mathbb{K}_r) \subseteq \bigcap_{i=1}^{r} E_i = \{0\},$$

which implies that $\mathbb{K} = \mathbb{K}_1 \cdots \mathbb{K}_r$. Now by assumption there exists a primitive root of unity ζ_i of order s_i such that $\mathbb{K}_i \subseteq \mathbb{Q}(\zeta_i)$ and hence

$$\mathbb{K} = \mathbb{K}_1 \cdots \mathbb{K}_r \subseteq \mathbb{Q}(\zeta_1, \ldots, \zeta_r) \subseteq \mathbb{Q}\left(\zeta_{[s_1, \ldots, s_r]}\right),$$

which concludes the proof.   □

Now suppose that $\mathbb{K}/\mathbb{Q}$ is a finite abelian extension satisfying

$$[\mathbb{K} : \mathbb{Q}] = p^{\alpha} \quad \text{and} \quad d_{\mathbb{K}} = p^{\beta}$$

for some odd prime number p and positive integers α and β. Define $\mathbb{L} = \mathbb{Q}\left(\zeta_{p^{\alpha+1}}\right)$ with degree $[\mathbb{L} : \mathbb{Q}] = \varphi\left(p^{\alpha+1}\right) = p^{\alpha}(p-1)$ and

$$d_{\mathbb{L}} = (-1)^{p^{\alpha}(p-1)/2} p^{p^{\alpha}(p\alpha-1)}.$$

Fig. 7.1 The compositum
$\mathbb{K}\mathbb{F}$

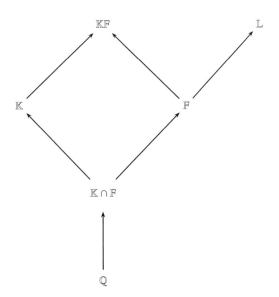

Let $\mathbb{F}$ be the fixed field of the subgroup H of $\mathrm{Gal}\,(\mathbb{L}/\mathbb{Q})$ of order $p-1$ so that $[\mathbb{F}:\mathbb{Q}]=p^{\alpha}$. Since p is odd, $\mathrm{Gal}\,(\mathbb{L}/\mathbb{Q})$ is cyclic, then G/H is also cyclic and hence $\mathbb{F}/\mathbb{Q}$ is a cyclic number field. Furthermore, if q is a prime dividing $d_{\mathbb{F}}$, then q ramifies in $\mathbb{F}$ and then ramifies in $\mathbb{L}$, so that $q=p$ and $d_{\mathbb{F}}$ is a power of p. We consider the compositum $\mathbb{K}\mathbb{F}$ (Fig. 7.1).

We have

$$[\mathbb{K}\mathbb{F}:\mathbb{Q}]=[\mathbb{K}\mathbb{F}:\mathbb{F}]\,[\mathbb{F}:\mathbb{Q}]=[\mathbb{K}:\mathbb{K}\cap\mathbb{F}]\,[\mathbb{F}:\mathbb{Q}]$$

so that $[\mathbb{K}\mathbb{F}:\mathbb{Q}]$ is a power of p. As above, let q be a prime dividing $d_{\mathbb{K}\mathbb{F}}$, and hence q is ramified in $\mathbb{K}\mathbb{F}$, which implies that q is ramified in $\mathbb{K}$ or q is ramified in $\mathbb{F}$, and therefore $q=p$ and $d_{\mathbb{K}\mathbb{F}}$ is a power of p. Next, it can be shown [76] that $\mathbb{K}\mathbb{F}$ is a cyclic number field. We infer that the subgroup $\mathrm{Gal}\,(\mathbb{K}\mathbb{F}/\mathbb{K}\cap\mathbb{F})$ of $\mathrm{Gal}\,(\mathbb{K}\mathbb{F}/\mathbb{Q})$ is cyclic. Now we have

$$\mathrm{Gal}\,(\mathbb{K}\mathbb{F}/\mathbb{K}\cap\mathbb{F})\simeq\mathrm{Gal}\,(\mathbb{K}/\mathbb{K}\cap\mathbb{F})\oplus\mathrm{Gal}\,(\mathbb{F}/\mathbb{K}\cap\mathbb{F})$$

and using [76, Lemma 2 p. 277] we infer that one of the groups on the right-hand side is trivial, so that either $\mathbb{K}=\mathbb{K}\cap\mathbb{F}$ or $\mathbb{F}=\mathbb{K}\cap\mathbb{F}$. In the first case, we deduce that $\mathbb{K}\subseteq\mathbb{F}$, whereas the second case gives $\mathbb{F}\subseteq\mathbb{K}$, but since they have the same degree, we get $\mathbb{F}=\mathbb{K}$ in both cases and hence

$$\mathbb{K}=\mathbb{F}\subseteq\mathbb{L}=\mathbb{Q}\left(\zeta_{p^{\alpha+1}}\right).$$

Using Proposition 7.27 and Lemma 7.25, we get the following result.

Lemma 7.26 *Let* $\mathbb{K} = \mathbb{K}_1 \cdots \mathbb{K}_r$ *be an abelian number field such that either* $[\mathbb{K}_i : \mathbb{Q}] = 2$ *or* $[\mathbb{K}_i : \mathbb{Q}] = p_i^{\alpha_i}$ *and* $d_{\mathbb{K}_i} = p_i^{\beta_i}$ *for some odd primes* p_i. *Then* $\mathbb{K}$ *is contained in a cyclotomic field.*

7.5.8   Class Field Theory Over $\mathbb{Q}$

7.5.8.1   Background

From the discussion above, it turns out that describing all abelian extensions of an algebraic number field is a long-standing problem. When examining Abel's work, Kronecker observed that certain abelian extensions of imaginary quadratic fields may be generated by adjoining certain values of *automorphic functions* arising from elliptic curves.

Definition 7.22 Let $\mathcal{H} = \{s \in \mathbb{C} : \mathrm{Im}\, s > 0\}$ be the Poincaré half-plane and Γ be a subgroup of $\mathrm{SL}(2, \mathbb{Z})$. An *automorphic function for* Γ is a meromorphic function f defined on $\mathcal{H}$ having a Fourier series expansion of the form $f(z) = \sum_{n=N}^{\infty} a_n e(nz)$ for some $N \in \mathbb{Z}$, and satisfying

$$f\left(\frac{az + b}{cz + d}\right) = f(z)$$

for all $M = \begin{pmatrix} a & b \\ c & d \end{pmatrix} \in \Gamma$.

Example 7.19 The j-function defined on $\mathcal{H}$ by

$$j(z) = \frac{1728\, g_2(z)^3}{g_2(z)^3 - 27 g_3(z)^2} \tag{7.30}$$

with

$$g_2(z) = \sum_{m,n} \frac{60}{(m + nz)^4} \quad \text{and} \quad g_3(z) = \sum_{m,n} \frac{140}{(m + nz)^6},$$

and where the summations are over all ordered pairs of integers $(m, n) \neq (0, 0)$ is an automorphic function.

It can be shown that if $\mathbb{K} = \mathbb{Q}(\sqrt{-d})$ with $d > 0$ squarefree such that $d \not\equiv 3 \pmod 4$, then $\mathbb{K}\left(j\left(\sqrt{-d}\right)\right)$ is an abelian extension[52] of $\mathbb{K}$.

---

[52]This is in fact the *Hilbert class field* of $\mathbb{K}$. See Theorem 7.36.

Kronecker wondered whether *all* finite abelian extensions of imaginary quadratic fields could be obtained in this way.[53] The generalization of this problem was later addressed by Hilbert in 1900 when he presented a series of 23 problems for the new century at the International Congress of Mathematicians in Paris, and the generalization of Kronecker's conjecture is Hilbert's 12th problem.[54]

Historically, the class field theory began when Gauss tried to decide when the congruence equation $x^2 - a \equiv 0 \pmod{p}$ has a solution, where $p \nmid a$ is a prime. The answer is given by his quadratic reciprocity law stating that, *if p and q are odd primes not dividing a and $p \equiv q \pmod{4a}$, then $x^2 - a \equiv 0 \pmod{p}$ has a solution if and only if $x^2 - a \equiv 0 \pmod{q}$ has a solution.* Hence, whether or not the congruence $x^2 - a \equiv 0 \pmod{p}$ has a solution depends only on the arithmetic progression mod $4a$ to which p belongs.

Proposition 7.26 also shows that an odd prime p splits completely in a quadratic field $\mathbb{Q}(\sqrt{d})$ if and only if $p \nmid d$ and the congruence $x^2 - d \equiv 0 \pmod{p}$ has a solution.

Let $\mathbb{K} = \mathbb{Q}(\zeta_m)$, where $m \in \mathbb{Z}_{\geqslant 1}$. It can be proved [98, Theorem 2.13] that, if $p \neq m$ is a prime number and f is the smallest positive integer such that $p^f \equiv 1 \pmod{m}$, then $(p) = \mathfrak{p}_1 \cdots \mathfrak{p}_g$ with $g = \varphi(m)/f$. In particular,

$$p \text{ splits completely in } \mathbb{Q}(\zeta_m) \text{ if and only if } p \equiv 1 \pmod{m}.$$

All these examples above show that we have a decomposition theory in terms of congruence conditions. Therefore, the purposes of class field theory over $\mathbb{Q}$ are threefold.

▷ To describe every finite abelian extension of $\mathbb{Q}$ in terms of the arithmetic of $\mathbb{Q}$.
▷ To canonically realize the abelian group Gal($\mathbb{K}/\mathbb{Q}$) in terms of the arithmetic of $\mathbb{Q}$.
▷ To describe the decomposition of a prime in terms of the arithmetic of $\mathbb{Q}$, i.e. by giving congruence conditions.

We now intend to describe the class field theory over $\mathbb{Q}$ following [29].

7.5.8.2   Artin Map

Let p be a prime number and $\mathbb{F}_p$ be the field with p elements. By Theorem 2.3, the Frobenius map $x \in \mathbb{F}_p \mapsto x^p$ is nothing but the identity. We generalize this concept to abelian number fields in the following way.

Proposition 7.43 *Let $\mathbb{K}/\mathbb{Q}$ be an abelian extension with a defining monic polynomial $P \in \mathbb{Z}[X]$, and let p be a prime number such that $p \nmid \mathrm{disc}(P)$. Then there exists an element $\phi_p \in \mathrm{Gal}(\mathbb{K}/\mathbb{Q})$ such that the Frobenius map in $\mathbb{F}_p$ is the reduction of*

---

[53]This is now referred to as *Kronecker Jurgendtraum*, i.e. Kronecker's dream of his youth. This was completely proved by Takagi in 1920.

[54]Among the 23 problems introduced by Hilbert, some of them have been solved and others are still open. Note that Hilbert's 8th problem is the Riemann hypothesis.

ϕ_p *modulo* p. *This means that, if* $\mathbb{K} = \mathbb{Q}(\theta)$, *where* θ *is a root of the polynomial* P, *there exist rational numbers* $r_0, \dots, r_{n-1}$, *the denominators of which are not divisible by* p, *such that*

$$\theta^p = \phi_p(\theta) + p \sum_{k=0}^{n-1} r_k \theta^k.$$

This element ϕ_p *is called the* Artin symbol *and is denoted by* $\left(\frac{\mathbb{K}/\mathbb{Q}}{p} \right)$.

Remark 7.9 In the case $n = 2$, the Artin symbol is virtually identical to the Legendre symbol $\left(\frac{\text{disc } P}{p} \right)$, explaining the notation. We will extend this symbol in Definition 7.25 below.

Example 7.20 Let $\mathbb{K} = \mathbb{Q}(\theta)$, where θ is a root of the polynomial $P = X^3 - 3X - 1$. It can be easily checked that P is irreducible over $\mathbb{Q}$ since P has no integer root. Since $\text{disc}(P) = 3^4 \in \mathbb{Q}^2$, we have Gal $(\mathbb{K}/\mathbb{Q}) \simeq \mathcal{A}_3 \simeq C_3$ by Lemma 7.20, so that $\mathbb{K}/\mathbb{Q}$ is cyclic, and let σ be a generator of Gal $(\mathbb{K}/\mathbb{Q})$. Since $-\theta^2 + 2$ and $\theta^2 - \theta - 2$ are the two other roots of P, set $\sigma(\theta) = -\theta^2 + 2$ so that $\sigma^2(\theta) = \theta^2 - \theta - 2$ and Gal $(\mathbb{K}/\mathbb{Q}) = \{\text{id}, \sigma, \sigma^2\}$. Now we have

$$\theta^2 \equiv -\theta^2 + 2 \equiv \sigma(\theta) \pmod{2}$$

and hence

$$\left(\frac{\mathbb{K}/\mathbb{Q}}{2} \right) = \sigma.$$

Similarly, since

$$\theta^5 = \theta^3 \theta^2 = (1 + 3\theta)\theta^2 = \theta^2 + 3\theta^3 = \theta^2 + 9\theta + 3 \equiv \theta^2 - \theta - 2 \equiv \sigma^2(\theta) \pmod{5}$$

we have

$$\left(\frac{\mathbb{K}/\mathbb{Q}}{5} \right) = \sigma^2.$$

One may continue in this way to determine other values for primes not dividing 3. However, it seems more appropriate to look for a rule which can dictate the Artin symbol for all primes $p \neq 3$. This is the purpose of the *Artin reciprocity law*.

Let $\mathbb{K}/\mathbb{Q}$ be an abelian number field. By Theorem 7.30, there exists a positive integer f such that $\mathbb{K} \subseteq \mathbb{Q}(\zeta_f)$. Such an integer is called an *admissible modulus* or *cycle* of

$\mathbb{K}$. Now let $a = \prod_{p|a} p^{v_p(a)} \in \mathbb{Z}_{\geqslant 1}$ such that $(a, f) = 1$, where f is an admissible modulus of $\mathbb{K}$. The *Artin map* is defined by

$$\mathrm{Art}_{\mathbb{K}/\mathbb{Q}} : (\mathbb{Z}/f\mathbb{Z})^{\times} \longrightarrow \quad \mathrm{Gal}\,(\mathbb{K}/\mathbb{Q})$$

$$\bar{a} \quad \longmapsto \quad \prod_{p|a} \left(\frac{\mathbb{K}/\mathbb{Q}}{p} \right)^{v_p(a)}. \tag{7.31}$$

Example 7.21 Let m be a positive integer, $\mathbb{K} = \mathbb{Q}(\zeta_m)$, and let $p \nmid m$ be a prime number. By Proposition 7.43, we have $\left(\frac{\mathbb{Q}(\zeta_m)/\mathbb{Q}}{p} \right) = \sigma_p$ where, for all $a \in \mathbb{Z}_{\geqslant 1}$ such that $(a, m) = 1$, σ_a is defined by $\sigma_a(\zeta_m) = \zeta_m^a$. Now if $a = \prod_p p^{v_p(a)} \in \mathbb{Z}_{\geqslant 1}$ such that $(a, m) = 1$, then

$$\left(\frac{\mathbb{Q}(\zeta_m)/\mathbb{Q}}{a} \right) = \prod_p \left(\frac{\mathbb{Q}(\zeta_m)/\mathbb{Q}}{p} \right)^{v_p(a)} = \prod_p \sigma_p^{v_p(a)} = \sigma_a.$$

Furthermore, we also have

$$\left(\frac{\mathbb{Q}(\zeta_m)/\mathbb{Q}}{a} \right)(\zeta_m) = \zeta_m \iff \sigma_a(\zeta_m) = \zeta_m \iff \zeta_m^a = \zeta_m \iff a \equiv 1 \pmod{m},$$

which is called the *cyclotomic reciprocity law*.

7.5.8.3   Main Theorems

It can be proved that the Artin map is *surjective*. One of the main purposes of the theory is then the study of the kernel denoted by $H_{\mathbb{K}, f}$ and called the *Artin group*. Using $\mathrm{Gal}\left(\mathbb{Q}(\zeta_f)/\mathbb{Q}\right) \simeq (\mathbb{Z}/f\mathbb{Z})^*$, we have $H_{\mathbb{K}, f} \simeq \mathrm{Gal}\left(\mathbb{Q}(\zeta_f)/\mathbb{K}\right)$ and from Galois theory we have the correspondence

$$
\begin{array}{ccccc}
\mathbb{Q}(\zeta_f) & \supseteq & \mathbb{K} & \supseteq & \mathbb{Q} \\
\updownarrow & & \updownarrow & & \updownarrow \\
\mathrm{Gal}\left(\mathbb{Q}(\zeta_f)/\mathbb{Q}(\zeta_f)\right) & \subseteq & \mathrm{Gal}\left(\mathbb{Q}(\zeta_f)/\mathbb{K}\right) & \subseteq & \mathrm{Gal}\left(\mathbb{Q}(\zeta_f)/\mathbb{Q}\right) \\
\updownarrow & & \updownarrow & & \updownarrow \\
\{1\} & \subseteq & H_{\mathbb{K}, f} & \subseteq & (\mathbb{Z}/f\mathbb{Z})^*
\end{array}
$$

so that $\mathbb{K}$ is the fixed field of $H_{\mathbb{K}, f}$. This proves the first result of class field theory over $\mathbb{Q}$.

Theorem 7.31 (Artin Reciprocity) *Let $\mathbb{K}/\mathbb{Q}$ be an abelian number field with an admissible modulus f. Then the following sequence is exact*

$$1 \longrightarrow H_{\mathbb{K},f} \hookrightarrow (\mathbb{Z}/f\mathbb{Z})^{\times} \longrightarrow \mathrm{Gal}\,(\mathbb{K}/\mathbb{Q}) \longrightarrow 1$$

so that

$$\mathrm{Gal}\,(\mathbb{K}/\mathbb{Q}) \simeq (\mathbb{Z}/f\mathbb{Z})^{\times}/H_{\mathbb{K},f}.$$

In other words, Theorem 7.31 realizes canonically $\mathrm{Gal}(\mathbb{K}/\mathbb{Q})$ in terms of arithmetic of $\mathbb{Q}$, and each abelian extension of $\mathbb{Q}$ is described in terms of arithmetic of $\mathbb{Q}$.

The minimal admissible modulus of $\mathbb{K}$ is called the *conductor* of $\mathbb{K}$ and is denoted by $f_{\mathbb{K}}$. Since $\mathbb{K} \subseteq \mathbb{Q}(\zeta_f) \cap \mathbb{Q}(\zeta_{f_{\mathbb{K}}}) = \mathbb{Q}\left(\zeta_{(f,f_{\mathbb{K}})}\right)$, we infer that $f_{\mathbb{K}} \mid f$. For instance, the conductor of a cyclic extension is described in the next result[29].

Lemma 7.27 *Let $\mathbb{K}/\mathbb{Q}$ be a cyclic extension with odd prime degree p. Then*

$$f_{\mathbb{K}} = p^{\theta} p_1 \cdots p_r$$

for some $r \in \mathbb{Z}_{\geq 1}$, with $\theta \in \{0, 2\}$, and p_i are primes satisfying $p_i \equiv 1 \pmod{p}$.

We have the following result on ramification.

Theorem 7.32 (Conductor-Ramification Theorem) *Let $\mathbb{K}/\mathbb{Q}$ be an abelian number field with conductor $f_{\mathbb{K}}$ and p be a prime number. Then p is ramified in $\mathbb{K}$ if and only if $p \mid f_{\mathbb{K}}$.*

This implies in particular that, if p is a prime number such that $p \nmid f$, then p is unramified in $\mathbb{Q}(\zeta_f)$. This may be generalized to any finite abelian extension of $\mathbb{Q}$.

Theorem 7.33 (Decomposition Theorem) *Let $\mathbb{K}/\mathbb{Q}$ be an abelian number field with an admissible modulus f, and let $p \nmid f$. Then the order of $\overline{p}$ in $(\mathbb{Z}/f\mathbb{Z})^*/H_{\mathbb{K},f}$ is its inertial degree f_p in $\mathbb{K}$.*

It is customary to denote by $\mathrm{Spl}\,(\mathbb{K}/\mathbb{Q})$ the set of prime numbers which split completely in $\mathbb{K}$. Using $e_p f_p g_p = [\mathbb{K} : \mathbb{Q}]$, we have

$$p \in \mathrm{Spl}\,(\mathbb{K}/\mathbb{Q}) \iff g_p = [\mathbb{K} : \mathbb{Q}] \iff e_p = f_p = 1$$

and applying Theorems 7.32 and 7.33 with $f = f_{\mathbb{K}}$, we get

$$p \in \mathrm{Spl}\,(\mathbb{K}/\mathbb{Q}) \Longleftrightarrow p \nmid f_{\mathbb{K}} \text{ and } p \in H_{\mathbb{K},f_{\mathbb{K}}}. \tag{7.32}$$

We infer that

$$p \in \mathrm{Spl}\,(\mathbb{K}/\mathbb{Q}) \Longleftrightarrow \exists \bar{a} \in H_{\mathbb{K},f_{\mathbb{K}}} : p \equiv a \pmod{f_{\mathbb{K}}}.$$

This clearly accomplishes the third goal of class field theory.

7.5.8.4 Characters

Definition 7.23 Let $\mathbb{K}/\mathbb{Q}$ be an abelian extension with an admissible modulus f. We define the *character group* of $\mathbb{K}$ as the group $X_f(\mathbb{K})$ of characters χ of $(\mathbb{Z}/f\mathbb{Z})^{\times}$ such that $\chi(a) = 1$ for all $a \in H_{\mathbb{K},f}$. We first extend χ in a Dirichlet character modulo f, denoted again by χ, by setting $\chi(a) = 0$ for all a such that $(a, f) > 1$, and then to a primitive Dirichlet character. If $\chi \in X_f(\mathbb{K})$, we denote by f_χ its conductor.

The character group $X_f(\mathbb{K})$ satisfies the following properties [65, Proposition 8.4].

▷ $X_f(\mathbb{K}) \simeq \mathrm{Gal}\,(\mathbb{K}/\mathbb{Q})$ and hence $\left| X_f(\mathbb{K}) \right| = [\mathbb{K} : \mathbb{Q}]$.
▷ $X_f(\mathbb{K}) \simeq (\mathbb{Z}/f\mathbb{Z})^{\times} \Longleftrightarrow \mathbb{K} = \mathbb{Q}(\zeta_f)$ and $X_f(\mathbb{K}) = \{\chi_0\} \Longleftrightarrow \mathbb{K} = \mathbb{Q}$.
▷ If $r_2 = 0$, then every character of $X_f(\mathbb{K})$ is even. If $r_1 = 0$, then $X_f(\mathbb{K})$ contains the same number of even and odd characters, and the set of even characters is the subgroup of $X_f(\mathbb{K})$ equal to $X_f(\mathbb{K}^+)$, where $\mathbb{K}^+$ is the maximal real subfield of $\mathbb{K}$.

The next result is another highlight of class field theory.

Theorem 7.34 (Conductor-Discriminant Formula) *Let $\mathbb{K}/\mathbb{Q}$ be an abelian extension with admissible modulus f and character group $X_f(\mathbb{K})$. Then*

$$f_{\mathbb{K}} = \mathrm{lcm}\left\{ f_\chi : \chi \in X_f(\mathbb{K}) \right\}$$

and

$$|d_{\mathbb{K}}| = \prod_{\chi \in X_f(\mathbb{K})} f_\chi.$$

(continued)

Theorem 7.34 (continued)
In particular, we have $f_{\mathbb{K}} \mid d_{\mathbb{K}}$ and hence we always have the following tower

$$\mathbb{Q} \subseteq \mathbb{K} \subseteq \mathbb{Q}(\zeta_{f_{\mathbb{K}}}) \subseteq \mathbb{Q}(\zeta_{|d_{\mathbb{K}}|}).$$

7.5.9 Primes of the Form $x^2 + ny^2$

7.5.9.1 Places

The class field theory over $\mathbb{Q}$ may be generalized to any finite abelian extension $\mathbb{L}/\mathbb{K}$, but since the Kronecker–Weber theorem is only valid with the ground field $\mathbb{Q}$, the concept of admissible modulus must be rewritten.

The starting point is a generalization of the notion of prime ideals of an algebraic number field in the following way. Let $\mathbb{K}$ be an algebraic number field of degree n. An *absolute value* of $\mathbb{K}$ is a map $|\cdot| : \mathbb{K} \to \mathbb{R}$ satisfying

▷ $|x| \geqslant 0$ and $|x| = 0 \Leftrightarrow x = 0$.
▷ $|xy| = |x||y|$.
▷ $|x + y| \leqslant |x| + |y|$.

If we replace the third condition by the stronger condition

▷ $|x + y| \leqslant \max(|x|, |y|)$,

then the absolute value is said to be *non-Archimedean*, and *Archimedean* otherwise. If $|\cdot|_1$ and $|\cdot|_2$ are two absolute values such that there exist constants $0 < c_0 \leqslant c_1$ such that $c_0|x|_1 \leqslant |x|_2 \leqslant c_1|x|_1$ for all $x \in \mathbb{K}$, then the absolute values are equivalent, and the set of equivalence classes is called the *places* of $\mathbb{K}$. There are two types of places.

▷ **The finite places.** If $\mathfrak{p}$ is a prime ideal of $O_{\mathbb{K}}$ and p is the prime below $\mathfrak{p}$, then we define for all $x \in \mathbb{K}$

$$|x|_{\mathfrak{p}} = p^{-v_{\mathfrak{p}}(x)/e_{\mathfrak{p}}},$$

where $e_{\mathfrak{p}}$ is the ramification index of $\mathfrak{p}$ in $\mathbb{K}$ and $v_{\mathfrak{p}}(x) = v_{\mathfrak{p}}(xO_{\mathbb{K}})$. This absolute value is non-Archimedean and if $\mathfrak{p}_1 \neq \mathfrak{p}_2$, then $|\cdot|_{\mathfrak{p}_1}$ and $|\cdot|_{\mathfrak{p}_2}$ are not equivalent. Furthermore, any non-Archimedean absolute value is equivalent to one of these.
▷ **The infinite places.** Let σ be one of the embeddings of $\mathbb{K}$. Then we define for all $x \in \mathbb{K}$

$$|x|_{\sigma} = |\sigma(x)|.$$

If $\sigma(\mathbb{K}) \subseteq \mathbb{R}$, then the place is called real, and complex otherwise. For a complex place σ, the conjugate $\overline{\sigma}$ defines the same place. We infer that the number of infinite places of $\mathbb{K}$ is given by $r_1 + r_2$. For instance, if $\mathbb{K} = \mathbb{Q}$, there is only one infinite place, often denoted by ∞, given by $|x|_{\mathrm{id}} = |x|$, where the right-hand side is the ordinary absolute value on $\mathbb{Q}$.

It can be proved that there is no other place for a number field $\mathbb{K}$, and therefore there are infinitely many finite places and finitely many infinite places in $\mathbb{K}$.

Let $\mathbb{L}/\mathbb{K}$ be a finite extension *unramified at all places*, which is then equivalent to the two following assertions. If $P = X^n + a_{n-1}X^{n-1} + \cdots + a_0 \in O_{\mathbb{K}}[X]$ is a defining polynomial of $\mathbb{L}/\mathbb{K}$, then

▷ Every prime ideal of $\mathbb{K}$ is unramified in $\mathbb{L}$, or equivalently the relative discriminant $\mathfrak{D}_{\mathbb{L}/\mathbb{K}}$ is equal to $O_{\mathbb{K}}$.
▷ Let σ_i be the embeddings of $\mathbb{K}$ in $\mathbb{L}$. Then either $\sigma_i(\mathbb{K}) \not\subseteq \mathbb{R}$ or $\sigma_i(\mathbb{K}) \subseteq \mathbb{R}$, and P^{σ_i} has no non-real roots, where $P^{\sigma_i} = X^n + \sigma_i(a_{n-1})X^{n-1} + \cdots + \sigma_i(a_0)$.

Note that if $r_1 = 0$, then no infinite place ramifies in $\mathbb{L}$, so that $\mathbb{L}/\mathbb{K}$ unramified at all places is *equivalent* to the first point above. On the other hand, in the case $r_1 \geqslant 1$, we say that $\mathbb{L}/\mathbb{K}$ is unramified *outside* ∞ if and only if the sole first point above is satisfied.

Example 7.22 The extension $\mathbb{Q}(\zeta_{12})/\mathbb{Q}(\sqrt{3})$ is unramified outside ∞ since

$$\mathfrak{D}_{\mathbb{Q}(\zeta_{12})/\mathbb{Q}(\sqrt{3})} = \mathbb{Z}[\sqrt{3}]$$

but is not unramified since the two infinite places ramify in $\mathbb{Q}(\zeta_{12})$. Indeed, we have

$$P = X^2 - X\sqrt{3} + 1,$$

and the two infinite places of $\mathbb{Q}(\sqrt{3})$ are $\{\mathrm{id}, \sigma\}$, where $\sigma(a + b\sqrt{3}) = a - b\sqrt{3}$. We have $\sigma_i\left(\mathbb{Q}(\sqrt{3})\right) \subseteq \mathbb{R}$ and the polynomials P^{σ_i} have non-real roots.

7.5.9.2   Hilbert Class Field

We have previously seen that Hilbert's proof of the Kronecker–Weber theorem works in part, thanks to the fact that $\mathbb{Q}$ has no unramified abelian extension larger than $\mathbb{Q}$. It is certainly the reason why Hilbert focused on unramified abelian extensions. However, it should be mentioned that, at that time, the word "unramified" meant at finite places only, so that Hilbert's class fields were studied in the *narrow sense*, i.e. the Galois group is isomorphic to the narrow ideal class group, where fractional ideals are identified if and only if their ratio is a principal ideal having a totally positive generator.[55]

The next result, conjectured by Hilbert, is an important tool in class field theory.

Theorem 7.35 (Hilbert Class Field) *Let $\mathbb{K}$ be a number field. Then there exists a unique maximal unramified abelian extension of $\mathbb{K}$ denoted by $\mathbb{K}(1)/\mathbb{K}$, in the sense that each unramified abelian extension of $\mathbb{K}$ is isomorphic to a sub-extension of $\mathbb{K}(1)$. The extension $\mathbb{K}(1)/\mathbb{K}$ is called the* Hilbert class field *of $\mathbb{K}$ and satisfies the following properties.*

  ▷ $\mathrm{Gal}\,(\mathbb{K}(1)/\mathbb{K}) \simeq C\ell(\mathbb{K})$ *and therefore $h_{\mathbb{K}} = [\mathbb{K}(1) : \mathbb{K}]$.*
  ▷ *Let $\mathfrak{p}$ be a prime ideal in $\mathbb{K}$. Then $\mathfrak{p}$ splits completely in $\mathbb{K}(1)$ if and only if $\mathfrak{p}$ is principal in $\mathbb{K}$.*
  ▷ *Every integral ideal of $\mathbb{K}$ becomes principal as an integral ideal of $\mathbb{K}(1)$.*

Example

The following extensions are examples of Hilbert's class fields over their ground fields.

$$\mathbb{Q}\left(\sqrt{2}, \sqrt{5}\right) \qquad \mathbb{Q}\left(\sqrt{-23}, \alpha\right) \qquad \mathbb{Q}\left(\sqrt{-14}, \sqrt{2\sqrt{2}-1}\right) \qquad \mathbb{Q}\left(\theta, \sqrt{u(5-\theta^2)}\right)$$

$$\Big|\,2 \qquad\qquad \Big|\,3 \qquad\qquad\qquad \Big|\,4 \qquad\qquad\qquad\qquad \Big|\,2$$

$$\mathbb{Q}\left(\sqrt{10}\right) \qquad \mathbb{Q}\left(\sqrt{-23}\right) \qquad \mathbb{Q}\left(\sqrt{-14}\right) \qquad\qquad \mathbb{Q}(\theta)$$

Here we have $\alpha^3 = \alpha - 1$, $\theta^3 = 11$ and $u = 89 + 40\theta + 18\theta^2$. The last example is picked up from [47].

Hilbert proved the existence of the Hilbert class field of $\mathbb{K}$ when $\mathbb{K}$ is a quadratic field with class number 2. Hilbert's student Furtwängler proved the first point in 1907 and the third one, called the *capitulation property*, in 1930.

The second point above may be generalized as follows.

Proposition 7.44 *If $\mathfrak{p}$ is a prime ideal of $\mathbb{K}$ and if f is the smallest power of $\mathfrak{p}$ such that $\mathfrak{p}^f$ is a principal ideal of $\mathbb{K}$, then $\mathfrak{p}$ splits into $h_{\mathbb{K}}/f$ distinct prime ideals of $\mathbb{K}(1)$ with degree f.*

The next result studies the normality of the extension $\mathbb{K}(1)/\mathbb{Q}$.

---

[55]For a definition of totally positive numbers in algebraic number fields, see [65, page 44]. See also Sect. 7.5.11.1 below.

Lemma 7.28 *If* $\mathbb{K}/\mathbb{Q}$ *is Galois, then* $\mathbb{K}(1)/\mathbb{Q}$ *is Galois.*

Proof Considering $\mathbb{K}$ and $\mathbb{K}(1)$ in a fixed algebraic closure $\overline{\mathbb{Q}}$ of $\mathbb{Q}$, let $\sigma \in$ $\mathrm{Gal}\left(\overline{\mathbb{Q}}/\mathbb{Q}\right)$. Since $\mathbb{K}$ is Galois over $\mathbb{Q}$, we have $\sigma(\mathbb{K}) = \mathbb{K}$. But $\sigma(\mathbb{K}(1))$ is the Hilbert class field of $\sigma(\mathbb{K}) = \mathbb{K}$, so that using the maximality of $\mathbb{K}(1)$, we infer that $\sigma(\mathbb{K}(1)) = \mathbb{K}(1)$ and hence $\mathbb{K}(1)$ is Galois over $\mathbb{Q}$. $\square$

7.5.9.3   A Particular Case

One of the most beautiful applications of the Hilbert class field lies in the solution to the following problem. Let $n \in \mathbb{Z}_{\geqslant 1}$ satisfying

$$n \text{ is squarefree and } n \not\equiv 3 \ (\mathrm{mod}\, 4). \tag{7.33}$$

We ask for necessary and sufficient conditions for a prime number p to be expressed in the form $p = x^2 + ny^2$ for some integers x, y. We use the imaginary quadratic field $\mathbb{K} = \mathbb{Q}\left(\sqrt{-n}\right)$ and notice that (7.33) and Proposition 7.12 imply that $O_{\mathbb{K}} = \mathbb{Z}\left[\sqrt{-n}\right]$. The theoretic solution to this problem is given by the following result.

Theorem 7.36 *Let* $n \in \mathbb{Z}_{\geqslant 1}$ *satisfying* (7.33), *and let* $\mathbb{K} = \mathbb{Q}\left(\sqrt{-n}\right)$ *and* $\mathbb{K}(1)$ *be the Hilbert class field of* $\mathbb{K}$. *If* p *is an odd prime such that* $p \nmid n$, *then*

$$p = x^2 + ny^2 \Longleftrightarrow p \in \mathrm{Spl}\left(\mathbb{K}(1)/\mathbb{Q}\right).$$

Proof First note that $d_{\mathbb{K}} = -4n$ by (7.33) and Proposition 7.12, and hence since $p \nmid n$ is odd, we infer that $p \nmid d_{\mathbb{K}}$ and therefore p is unramified in $\mathbb{K}$.

▷ Assume that $p = x^2 + ny^2$. Then $(p) = \left(x + y\sqrt{-n}\right)\left(x - y\sqrt{-n}\right) = \mathfrak{p}\overline{\mathfrak{p}}$, say. Since p is unramified, we get $\mathfrak{p} \neq \overline{\mathfrak{p}}$. Conversely, if $(p) = \mathfrak{p}\overline{\mathfrak{p}}$ with $\mathfrak{p}$ principal, then $\mathfrak{p} = \left(x + y\sqrt{-n}\right)$ for some integers x, y since $O_{\mathbb{K}} = \mathbb{Z}\left[\sqrt{-n}\right]$. This implies that $(p) = (x^2 + ny^2)$, and then $p = x^2 + ny^2$. At this step, we have then proved

$$p = x^2 + ny^2 \Longleftrightarrow (p) = \mathfrak{p}\overline{\mathfrak{p}}, \ \mathfrak{p} \neq \overline{\mathfrak{p}} \text{ and } \mathfrak{p} \text{ principal}.$$

▷ By Theorem 7.35, we derive

$$p = x^2 + ny^2 \Longleftrightarrow (p) = \mathfrak{p}\overline{\mathfrak{p}}, \ \mathfrak{p} \neq \overline{\mathfrak{p}} \text{ and } \mathfrak{p} \in \mathrm{Spl}\left(\mathbb{K}(1)/\mathbb{K}\right),$$

which is in turn equivalent to the fact that p splits completely in $\mathbb{K}$ and that a prime ideal of $\mathbb{K}$ dividing p splits completely in $\mathbb{K}(1)$. By Lemma 7.28, we deduce that $\mathbb{K}(1)$ is Galois over $\mathbb{Q}$, which implies that the previous assertion is equivalent to p splitting completely in $\mathbb{K}(1)$.

$\square$

Therefore, the problem is twofold. We first have to determine the Hilbert class field of an imaginary quadratic field. The next elegant theorem gives an answer by using the elliptic modular function j defined in (7.30).

Proposition 7.45 *Let $n \in \mathbb{Z}_{\geq 1}$ satisfying (7.33), and let $\mathbb{K} = \mathbb{Q}\left(\sqrt{-n}\right)$ and $\mathbb{K}(1)$ be the Hilbert class field of $\mathbb{K}$. Then*

$$\mathbb{K}(1) = \mathbb{K}\left(j\left(\sqrt{-n}\right)\right).$$

However, it should be mentioned that in practice it is a very difficult matter to compute these j-invariants, so that many authors provided results in certain particular cases. For instance, Herz [42] proved a somewhat general theorem giving any unramified cyclic cubic extension of an imaginary quadratic field. Also for real quadratic fields $\mathbb{K}$ with class number 2, it is known [16] that there exists a proper divisor δ of $d_\mathbb{K}$ satisfying $\delta \equiv 0, 1 \pmod 4$ such that $\mathbb{K}(1) = \mathbb{K}\left(\sqrt{\delta}\right)$.

The second point is to determine which primes split completely in $\mathbb{K}(1)/\mathbb{Q}$. We might use (7.32), but the computation of $H_{\mathbb{K}(1),f}$ is very often uneasy. The following alternative tool provides a useful criterion in this direction [17, Proposition 5.29].

Proposition 7.46 *Let $\mathbb{K}$ be an imaginary quadratic field and $\mathbb{L}$ be a finite extension of $\mathbb{K}$, which is Galois over $\mathbb{Q}$. Then there exists an algebraic integer θ such that $\mathbb{L} = \mathbb{K}(\theta)$ and if $P \in \mathbb{Z}[X]$ is the monic minimal polynomial of θ and p is a prime such that $p \nmid \mathrm{disc}(P)$, then*

$$p \in \mathrm{Spl}\,(\mathbb{L}/\mathbb{Q}) \iff \left(\frac{d_\mathbb{K}}{p}\right) = 1 \text{ and } P(x) \equiv 0 \pmod p \text{ has an integer solution.}$$

Example 7.23 Let $p \neq 5$ be an odd prime number. We will prove that

$$p = x^2 + 5y^2 \iff p \equiv 1 \text{ or } 9 \pmod{20}.$$

Proof Let $\mathbb{K} = \mathbb{Q}(\sqrt{-5})$. It is known that $j\left(\sqrt{-5}\right) = 282880\sqrt{5} + 632000$ so that $\mathbb{K}(1) = \mathbb{K}\left(\sqrt{5}\right)$ by Proposition 7.45. Using Theorem 7.36 and Proposition 7.46, we derive

$$p = x^2 + 5y^2 \iff \left(\frac{-20}{p}\right) = 1 \text{ and } x^2 - 5 \equiv 0 \pmod p \text{ has an integer solution}$$

$$\iff \left(\frac{5}{p}\right) = \left(\frac{-5}{p}\right) = 1,$$

where we used $(-20/p) = (4/p) \times (-5/p) = (-5/p)$. Now by the quadratic reciprocity law and since p is odd, we infer that

$$\left(\frac{5}{p}\right) = 1 \iff \left(\frac{p}{5}\right) = 1 \iff p \equiv \pm 1 \ (\mathrm{mod}\, 5).$$

Using $(-5/p) = (-1/p) \times (5/p) = (-1)^{(p-1)/2}(5/p)$, we get

$$\left(\frac{-5}{p}\right) = 1 \iff p \equiv 1,\, 3,\, 7,\, 9 \ (\mathrm{mod}\, 20).$$

Hence

$$p = x^2 + 5y^2 \iff p \equiv \pm 1 \ (\mathrm{mod}\, 5) \text{ and } p \equiv 1,\, 3,\, 7,\, 9 \ (\mathrm{mod}\, 20)$$

implying the stated result.   □

7.5.9.4   The General Case

If the integer n does not satisfy (7.33), then $\mathbb{Z}\left[\sqrt{-n}\right]$ need not be the maximal order of $\mathbb{Q}\left(\sqrt{-n}\right)$. We then look for a generalization of the Hilbert class field, which will be given by class field theory. Let O be any order of an imaginary quadratic field of conductor f and class group $C\ell(O)$. Then the existence theorem from class field theory implies that there exists an abelian extension $\mathbb{L}/\mathbb{K}$, called a *ring class field* of O, such that all primes of $\mathbb{K}$ ramified in $\mathbb{L}$ divide $fO_{\mathbb{K}}$ and the Artin map induces the following isomorphism

$$\mathrm{Gal}\,(\mathbb{L}/\mathbb{K}) \simeq C\ell(O).$$

We first provide some information about orders in quadratic fields. In what follows, $\mathbb{K}$ is a quadratic field.

▷ An *order* in $\mathbb{K}$ is a free $\mathbb{Z}$-module of rank 2 of $\mathbb{K}$ containing 1.
▷ An order O has always a finite index f in $O_{\mathbb{K}}$, called the *conductor* of O. Furthermore, if d_O is the discriminant of O, then

$$d_O = f^2 d_{\mathbb{K}}$$

and if $O = \mathbb{Z}\left[\sqrt{-n}\right]$, then $d_O = -4n$.
▷ If $I(O)$ is the set of invertible fractional ideals of the order O and $P(O)$ is the set of principal ideals of O, then $C\ell(O) = I(O)/P(O)$ is the *ideal class group*, or *Picard group*, of O and $h_O = |C\ell(O)|$ is the *class number* of O.
▷ If $O \subsetneq O_{\mathbb{K}}$, then O *is not* in general a Dedekind ring.

We summarize the main properties of orders of an imaginary quadratic field in the next lemma.

Lemma 7.29 *Let $n \in \mathbb{Z}_{\geqslant 1}$ squarefree and $\mathbb{K} = \mathbb{Q}\left(\sqrt{-n}\right)$ with discriminant $d_{\mathbb{K}}$ and class number $h_{\mathbb{K}}$. Let O be any order of $\mathbb{K}$ with discriminant d_O, class number h_O and conductor f.*

▷ *We have*

$$h_O = \frac{h_{\mathbb{K}} \times f}{\left[(O_{\mathbb{K}})^\times : O^\times\right]} \prod_{p|f} \left(1 - \left(\frac{d_{\mathbb{K}}}{p}\right)\frac{1}{p}\right),$$

where the index $\left[(O_{\mathbb{K}})^\times : O^\times\right]$ is given by

$$\left[(O_{\mathbb{K}})^\times : O^\times\right] = \begin{cases} 2, & \text{if } n = 1 \\ 3, & \text{if } n = 3 \\ 1, & \text{otherwise.} \end{cases}$$

▷ *A representative of a class of invertible fractional ideals of O is given by*

$$\left(a, \tfrac{1}{2}\left(-b + \sqrt{d_O}\right)\right)$$

with $0 < a \leqslant \sqrt{-\tfrac{1}{3}d_O}$, $|b| \leqslant a \leqslant c$, where $c = \frac{1}{4a}\left(b^2 - d_O\right) \in \mathbb{Z}_{\geqslant 1}$ is such that $(a, b, c) = 1$. Furthermore, if $b \geqslant 0$, then either $|b| = a$ or $a = c$.

The elliptic modular invariant j seen in (7.30) may be defined on orders of $\mathbb{K}$ in the following way. For all $\tau \in \mathcal{H}$ and orders $(1, \tau)$, we set

$$j\left((1, \tau)\right) = j(\tau).$$

It can be shown that, if O is any order in $\mathbb{K}$, then $j(O) \in \mathbb{R}$. Furthermore, if the class of $\mathfrak{a}$ has an order $\leqslant 2$ in $C\ell(O)$, then $j(\mathfrak{a}) \in \mathbb{R}$.[56] The next result generalizes Theorem 7.36 and Proposition 7.45.

Theorem 7.37 *Let n be a positive integer, $\mathbb{K} = \mathbb{Q}\left(\sqrt{-n}\right)$ and $O = \mathbb{Z}\left[\sqrt{-n}\right]$ with class number h_O. Let $\mathfrak{a}$ be any representative of a class of invertible fractional ideals of O. Then the extension*

$$\mathbb{L} = \mathbb{K}\left(j\left(\mathfrak{a}\right)\right)$$

---

[56]See [17, Exercise 11.1].

is the ring class field of O. Furthermore, if $\mathfrak{a}_1, \ldots, \mathfrak{a}_{h_O}$ *are representatives of the classes of invertible fractional ideals of O, then the polynomial*

$$H_O = \prod_{k=1}^{h_O} (X - j(\mathfrak{a}_k)) \in \mathbb{Z}[X]$$

is a defining polynomial of $\mathbb{L}$.

That result contains Proposition 7.45 since, if n is squarefree and satisfies (7.33), then $\mathbb{Z}\left[\sqrt{-n}\right] = O_\mathbb{K}$ so that $f = 1, d_O = d_\mathbb{K}, h_O = h_\mathbb{K}$ and

$$\mathbb{K}\left(j(\mathfrak{a})\right) = \mathbb{K}\left(j(O_\mathbb{K})\right) = \mathbb{K}\left(j\left(\sqrt{-n}\right)\right) = \mathbb{K}(1).$$

Let $n \in \mathbb{Z}_{\geq 1}$. Proceeding similarly as in Theorem 7.36,[57] one may prove that, if $p \nmid n$ is an odd prime, then

$$p = x^2 + ny^2 \Longleftrightarrow p \in \mathrm{Spl}\,(\mathbb{L}/\mathbb{Q})$$

where $\mathbb{L}$ is the ring class field of the order $\mathbb{Z}[\sqrt{-n}]$ of $\mathbb{Q}(\sqrt{-n})$. Now applying Proposition 7.46 and Theorem 7.37, we obtain the general solution to the problem of detecting primes of the form $x^2 + ny^2$.

Corollary 7.10 *Let n be a positive integer and* $p \nmid n$ *be an odd prime. With the notation of* Theorem 7.37, *we have*

$$p = x^2 + ny^2 \Longleftrightarrow \left(\frac{-n}{p}\right) = 1 \text{ and } H_O(x) \equiv 0 \pmod{p} \text{ has an integer solution.}$$

Example 7.24 We wish to determine which primes p can be expressed in the form $p = x^2 + 11y^2$. We work in $\mathbb{K} = \mathbb{Q}\left(\sqrt{-11}\right)$ with $d_\mathbb{K} = -11$ and $h_\mathbb{K} = 1$. The order $O = \mathbb{Z}\left[\sqrt{-11}\right]$ has discriminant $d_O = -44$ and conductor $f = \sqrt{44/11} = 2$. Using Lemma 7.29, we derive

$$h_O = \frac{1 \times 2}{1} \prod_{p | 2} \left(1 - \left(\frac{-11}{p}\right) \frac{1}{p}\right) = 3.$$

By Lemma 7.29, representatives of the three classes of invertible fractional ideals of O are given by

$$\mathfrak{a}_1 = \left(1, \sqrt{-11}\right), \quad \mathfrak{a}_2 = \left(3, -1 + \sqrt{-11}\right) \quad \text{and} \quad \mathfrak{a}_3 = \left(3, 1 + \sqrt{-11}\right).$$

---

[57] See [17, Theorem 9.4].

Using PARI/GP, we get

$$H_O = X^3 - aX^2 + bX - \left(2^4 \times 11 \times 17 \times 29\right)^3, \tag{7.34}$$

where $a = 2^4 \times 1709 \times 41057$ and $b = 2^8 \times 3 \times 11^4 \times 24049$. As in Example 7.23, we have

$$\left(\frac{-11}{p}\right) = \left(\frac{-1}{p}\right)\left(\frac{11}{p}\right) = (-1)^{(p-1)/2}\left(\frac{11}{p}\right) = \begin{cases} (11/p), & \text{if } p \equiv 1 \pmod 4 \\ -(11/p), & \text{if } p \equiv 3 \pmod 4. \end{cases}$$

Now using the quadratic reciprocity law, if p and q are primes such that $q \equiv 3 \pmod 4$, then

$$\left(\frac{q}{p}\right) = 1 \iff p \equiv \pm\alpha^2 \pmod{4q}$$

for some odd integer α such that $q \nmid \alpha$. This implies that

$$\left(\frac{11}{p}\right) = 1 \iff p \equiv \pm 1, \pm 5, \pm 9, \pm 25, \pm 37 \pmod{44}$$

so that

$$\left(\frac{-11}{p}\right) = 1 \iff p \equiv 1, 3, 5, 9, 15, 23, 25, 27, 31, 37 \pmod{44}.$$

This gives the following result.

Corollary 7.11 *Let $p \neq 11$ be an odd prime number. Then $p = x^2 + 11y^2$ if and only if*

▷ *$p \equiv 1, 3, 5, 9, 15, 23, 25, 27, 31, 37 \pmod{44}$,*
▷ *the equation $x^3 - ax^2 + bx - \left(2^4 \times 11 \times 17 \times 29\right)^3 \equiv 0 \pmod p$ has an integer solution, where $a = 2^4 \times 1709 \times 41057$ and $b = 2^8 \times 3 \times 11^4 \times 24049$.*

Remark 7.10 The theory predicts that the constant term of H_O is indeed a perfect cube, which happens if and only if $3 \nmid d_O$. In this case, Gross and Zagier, improving on an earlier work of Deuring, showed that, if p is a prime factor of the constant term of H_O, then $(d_O/p) \neq 1$ and either $p = 3$ or $p \equiv 2 \pmod 3$. Furthermore, we also have

$$p \leqslant \tfrac{3}{4}|d_O|,$$

which explains why the prime factors of the constant term of H_O are so small in (7.34).

7.5.10 The Chebotarëv's Density Theorem

7.5.10.1 Densities

There exist several densities in number theory. The densities in question here can be defined as follows.

Definition 7.24 Let $\mathbb{K}$ be a number field and S be a set of primes of $\mathbb{K}$.

▷ The *Dirichlet density*, or *analytic density*, of S is given by

$$d(S) := \lim_{\sigma \to 1^+} \frac{\sum_{\mathfrak{p} \in S} \mathcal{N}_{\mathbb{K}/\mathbb{Q}}(\mathfrak{p})^{-\sigma}}{\sum_{\mathfrak{p}} \mathcal{N}_{\mathbb{K}/\mathbb{Q}}(\mathfrak{p})^{-\sigma}} = \lim_{\sigma \to 1^+} \frac{\sum_{\mathfrak{p} \in S} \mathcal{N}_{\mathbb{K}/\mathbb{Q}}(\mathfrak{p})^{-\sigma}}{\log \frac{1}{\sigma - 1}}$$

if the limit exists.[58]

▷ The *natural density* of S is given by

$$\delta(S) := \lim_{x \to \infty} \frac{\left|\{\mathfrak{p} \in S : \mathcal{N}_{\mathbb{K}/\mathbb{Q}}(\mathfrak{p}) \leqslant x\}\right|}{\left|\{\mathfrak{p} : \mathcal{N}_{\mathbb{K}/\mathbb{Q}}(\mathfrak{p}) \leqslant x\}\right|}$$

if the limit exists.

Example 7.25 Let $K = \mathbb{Q}\left(\sqrt{-1}\right)$ and S be the set of primes of $\mathbb{K}$ that lie above primes congruent to 3 (mod 4). Such primes are inert in $\mathbb{K}$ by Proposition 7.26, so that $\sum_{\mathfrak{p} \in S} \mathcal{N}_{\mathbb{K}/\mathbb{Q}}(\mathfrak{p})^{-\sigma} = \sum_{p \equiv 3 \,(\text{mod } 4)} p^{-2\sigma} \ll 1$, and hence $d(S) = 0$.

7.5.10.2 Artin Symbol

Let $\mathbb{L}/\mathbb{K}$ be a relative Galois number field with Galois group G. We extend the definition of the Artin symbol seen in Proposition 7.14 to the non-abelian case as follows.

Definition 7.25 Let $\mathfrak{p} \subset \mathcal{O}_{\mathbb{K}}$ be an unramified prime. For any prime ideal $\mathfrak{P} \subset \mathcal{O}_{\mathbb{L}}$ lying above $\mathfrak{p}$, the *Artin symbol*, or *Frobenius element*, is the Frobenius map satisfying

$$\alpha \mapsto \alpha^{\mathcal{N}_{\mathbb{K}/\mathbb{Q}}(\mathfrak{p})} \pmod{\mathfrak{P}}$$

---

[58]The equality of the two expressions for $d(S)$ rests on the fact that $\zeta_{\mathbb{K}}(s)$ has a simple pole at $s = 1$, hence $\log \zeta_{\mathbb{K}}(\sigma) = \log \frac{1}{\sigma - 1} + O(1)$ as $\sigma \to 1^+$. On the other hand, by the Euler product of $\zeta_{\mathbb{K}}$, we derive

$$\log \zeta_{\mathbb{K}}(\sigma) = -\sum_{\mathfrak{p}} \log\left(1 - \mathcal{N}_{\mathbb{K}/\mathbb{Q}}(\mathfrak{p})^{-\sigma}\right) = \sum_{\mathfrak{p}} \mathcal{N}_{\mathbb{K}/\mathbb{Q}}(\mathfrak{p})^{-\sigma} + O(1)$$

as $\sigma \to 1^+$, concluding the proof of the claim.

for all $\alpha \in O_{\mathbb{L}}$. This map is usually denoted by $\left(\frac{\mathbb{L}/\mathbb{K}}{\mathfrak{P}}\right)$ and belongs to G.

The fact that $\mathfrak{p}$ is unramified ensures that such map is well-defined and is unique, up to the choice of the prime $\mathfrak{P}$ lying over $\mathfrak{p}$. Indeed, given another choice $\mathfrak{Q} \mid \mathfrak{p}$, then there exists $\sigma \in G$ such that $\mathfrak{Q} = \sigma(\mathfrak{P})$ and hence

$$\left(\frac{\mathbb{L}/\mathbb{K}}{\mathfrak{Q}}\right) = \sigma\left(\frac{\mathbb{L}/\mathbb{K}}{\mathfrak{P}}\right)\sigma^{-1}.$$

This is the reason why we must treat not just the elements themselves but the entire conjugacy class. It is therefore natural to consider

$$\left(\frac{\mathbb{L}/\mathbb{K}}{\mathfrak{p}}\right) = \left\{\left(\frac{\mathbb{L}/\mathbb{K}}{\mathfrak{P}}\right) : \mathfrak{P} \mid \mathfrak{p}\right\}.$$

Then $\left(\frac{\mathbb{L}/\mathbb{K}}{\mathfrak{p}}\right)$ is a conjugacy class in G, and we will refer to this as the *Artin symbol* of $\mathfrak{p}$, with no risk of ambiguity.

Remark 7.11 When G is abelian, we recover the Artin symbol given in Proposition 7.14, and we note that in this case the set $\left(\frac{\mathbb{L}/\mathbb{K}}{\mathfrak{p}}\right)$ contains only a single element—though formally the element and the set containing this element are two different objects, we may identify the two. We will often abuse notation and denote by $\left(\frac{\mathbb{L}/\mathbb{K}}{\mathfrak{p}}\right)$ any element of the conjugacy class. Thus, for an unramified prime $\mathfrak{p} \subset O_{\mathbb{K}}$, $\left(\frac{\mathbb{L}/\mathbb{K}}{\mathfrak{p}}\right)$ is either an element of G or a conjugacy class of G, depending on whether G is abelian or not.

7.5.10.3 Chebotarëv's Theorem

Roughly speaking, the Chebotarëv's density theorem [93] says that, for any relative Galois number field $\mathbb{L}/\mathbb{K}$ with Galois group G and any conjugacy class C in G, the number of primes $\mathfrak{p}$ of $O_{\mathbb{K}}$ is proportional to the size of C.

Theorem 7.38 (Chebotarëv) *Let $\mathbb{L}/\mathbb{K}$ be a relative Galois number field with Galois group G and C be any conjugacy class in G. Then the set of primes $\mathfrak{p}$, unramified in $\mathbb{L}$ and such that $\left(\frac{\mathbb{L}/\mathbb{K}}{\mathfrak{p}}\right) = C$ has Dirichlet and natural densities*

$$\frac{|C|}{|G|}$$

in the set of primes $\mathfrak{p}$ unramified in $\mathbb{L}$.

Define

$$\pi_C(x) := \sum_{\substack{N_{\mathbb{K}/\mathbb{Q}}(\mathfrak{p}) \leqslant x \\ \left(\frac{L/\mathbb{K}}{\mathfrak{p}}\right) = C \\ \mathfrak{p} \text{ unramified in } \mathbb{L}}} 1.$$

In modern language, the Chebotarëv's density theorem can be rephrased as

$$\pi_C(x) \sim \frac{|C|}{|G|} \operatorname{Li}(x) \quad (x \to \infty).$$

The proof of this important result predated the class field theory and also inspired Artin's original proof of class field theory. As the authors pointed out in [91], Chebotarëv was not very familiar with the class field theory at the time he proved his result, and although it can be proved using tools from class field theory, it can be seen that Chebotarëv's proof was more important for class field theory than class field theory was for his proof.

As usual in analytic number theory, further advances were made when one derived some effective versions of Chebotarëv's result. It was first done in [50], in which the authors established both conditional and unconditional statements with effective error terms. This was subsequently made completely explicit in [100], where the following estimate is shown.

Theorem 7.39 *For all* $x \geqslant \exp\left(8n_{\mathbb{L}} \log^2\left(150{,}867|d_{\mathbb{L}}|^{44/5}\right)\right)$, *we have*

$$\left|\pi_C(x) - \frac{|C|}{|G|} \operatorname{Li}(x)\right| \leqslant \frac{|C|}{|G|} \operatorname{Li}\left(x^\beta\right) + 7.84 \times 10^{14} x \exp\left(-\frac{1}{99}\sqrt{\frac{\log x}{n_{\mathbb{L}}}}\right),$$

where $n_{\mathbb{L}} = [\mathbb{L} : \mathbb{Q}]$, *and the term with* β *is present only if* $\zeta_{\mathbb{L}}$ *has an exceptional zero* β *satisfying* $1 - (4 \log |d_{\mathbb{L}}|)^{-1} \leqslant \beta < 1$.

We already talked about the "Siegel zero" β of $\zeta_{\mathbb{L}}$ in Theorem 7.17, and a little more is known about it. For instance, Stark [87, (27)] proved that there exists an effectively computable absolute constant $c > 0$ such that

$$1 - \beta > \frac{1}{c \, |d_{\mathbb{L}}|^{1/n_{\mathbb{L}}}}$$

and from [9] it is very likely that $c = \zeta(2)$.

7.5.10.4 Dirichlet's Theorem

While the Chebotarëv's density theorem is a statement about how prime ideals behave under Galois number fields, it is also a generalization of Dirichlet's Theorem 3.20 on primes in arithmetic progressions.

Theorem 7.40 (Dirichlet) *Let a and q be positive coprime integers. Then the Dirichlet and natural densities of the set of primes p such that $p \equiv a \pmod{q}$ is $\varphi(q)^{-1}$.*

Proof We apply Theorem 7.38 to $\mathbb{K} = \mathbb{Q}$ and $\mathbb{L} = \mathbb{Q}(\zeta_q)$, where ζ_q is primitive qth root of unity. By Proposition 7.14, $G \simeq (\mathbb{Z}/q\mathbb{Z})^\times$, where the isomorphism is given explicitly by identifying an element $a \in (\mathbb{Z}/q\mathbb{Z})^\times$ with the unique automorphism σ_a of $\mathbb{Q}(\zeta_q)$ such that $\sigma_a(\zeta_q) = \zeta_q^a$. Now let $p \nmid q$ be a prime number. By Example 7.21

$$\left(\frac{\mathbb{Q}(\zeta_q)/\mathbb{Q}}{p} \right) = \sigma_p$$

and therefore, for any $a \in (\mathbb{Z}/q\mathbb{Z})^\times$

$$\left(\frac{\mathbb{Q}(\zeta_q)/\mathbb{Q}}{p} \right) = \sigma_a \iff \zeta_q^p = \zeta_q^a \iff p \equiv a \pmod{q}.$$

By Theorem 7.38, we infer that the Dirichlet and natural densities of the set of primes p such that $p \equiv a \pmod{q}$ is

$$\frac{1}{\left| (\mathbb{Z}/q\mathbb{Z})^\times \right|} = \frac{1}{\varphi(q)}$$

as asserted.   □

7.5.10.5 An Application

We end this section with a theorem due to Frobenius [65], which is a particular case of Theorem 7.39.

Proposition 7.47 *Let $P \in \mathbb{Z}[X]$ irreducible. Then the Dirichlet density of the set of prime numbers satisfying $p \nmid \mathrm{disc}(P)$ for which $\overline{P}$ splits completely in $\mathbb{F}_p[X]$ is equal to $|\mathrm{Gal}(P/\mathbb{Q})|^{-1}$.*

Example 7.26 $P = X^5 + 4X^3 + 7X^2 + 2X + 9$. Since $\overline{P} = X^5 + X^2 + \overline{1}$ is irreducible in $\mathbb{F}_2[X]$, we infer that P is irreducible over $\mathbb{Z}$ by Corollary 7.2. Since $\mathrm{disc}(P) = 2503 \times 7759$, we derive $\mathrm{Gal}(P/\mathbb{Q}) \not\subseteq \mathcal{A}_5$ by Lemma 7.20. Furthermore, we have

$$P \equiv X(X+1)(X^3 + 2X^2 + 2X + 2) \pmod{3}$$

and

$$P \equiv (X + 1)(X^4 + 2X^3 + 3X^2 + 3X + 3) \pmod{5}$$

so that $\mathrm{Gal}(P/\mathbb{Q}) \simeq \mathcal{S}_5$ using Table 5C of [82].

Example 7.27 $P = X^6 + 2X^5 + 3X^4 + 4X^3 + 5X^2 + 6X + 7$. By Example 7.2, P is irreducible over $\mathbb{Z}$. We also have

$$P \equiv (X + 1)(X^5 + 4X^4 + 11X^3 + 5X + 3) \pmod{13}$$

so that $5 \mid \mathrm{Gal}(P/\mathbb{Q})$ by Proposition 7.34. Hence $\mathrm{Gal}(P/\mathbb{Q})$ lies among the following four groups:

$$\mathrm{PSL}(2, 5), \ \mathrm{PGL}(2, 5), \ \mathcal{A}_6, \ \mathcal{S}_6,$$

which are the only transitive subgroups of $\mathcal{S}_6$ having an order divisible by 5. Since

$$\mathrm{disc}(P) = -2^{16} \times 7^4$$

we have $\mathrm{Gal}(P/\mathbb{Q}) \not\subseteq \mathcal{A}_6$ by Lemma 7.20 and thus $\mathrm{Gal}(P/\mathbb{Q}) \not\simeq \mathrm{PSL}(2, 5), \mathcal{A}_6$. Since $\mathrm{PGL}(2, 5)$ and $\mathcal{S}_6$ are non-solvable groups, we infer that $\mathrm{Gal}(P/\mathbb{Q})$ is not solvable. One may compute the polynomial f_{15} of Theorem 7.38 and prove that it is reducible. Since $\mathrm{disc}(P)$ is not a square, we deduce from Theorem 7.38 that

$$\mathrm{Gal}(P/\mathbb{Q}) \simeq \mathrm{PGL}(2, 5) \simeq \mathcal{S}_5.$$

One may check that, of the first 10,000 prime numbers distinct from 2 and 7, P splits completely relative to 78 of them, which gives a proportion of $1/128$. Proposition 7.47 says that this proportion tends to $1/120$.

7.5.11   Artin L-Functions

The question of generalizing Dirichlet characters and L-functions to algebraic number fields quite rapidly and naturally arose in the late 19th century. The first attempt was made by Kummer and Dedekind by introducing the Dedekind zeta-function $\zeta_{\mathbb{K}}$ of an algebraic number field $\mathbb{K}$, generalizing the usual Riemann zeta-function.

As for the Dirichlet character χ, the naive approach would be to generalize $\mathbb{Z}/q\mathbb{Z}$ by $\mathcal{O}_{\mathbb{K}}/\mathfrak{a}$ for some non-zero integral ideal $\mathfrak{a}$ of $\mathcal{O}_{\mathbb{K}}$. Unfortunately, this idea cannot work, for there is no natural lift from $\mathcal{O}_{\mathbb{K}}/\mathfrak{a}$ to $\mathcal{O}_{\mathbb{K}}$, since the latter is not a UFD. On the other hand, Dedekind showed that we do have unique prime factorization of ideals, so Weber's idea is to lift to characters of groups of ideal classes.

7.5.11.1   An Overview of Class Field Theory

▷ A *modulus*, or *arithmetic cycle*, is a formal product $\mathfrak{m} = \mathfrak{m}_0\mathfrak{m}_\infty$, where $\mathfrak{m}_0$ is a non-zero integral ideal of $O_{\mathbb{K}}$ and $\mathfrak{m}_\infty$ is a set of *real* embeddings of $\mathbb{K}$ in $\mathbb{C}$. Thus, $\mathfrak{m}$ can be uniquely written as a formal infinite product with only finitely many non-trivial terms.

$$\mathfrak{m} = \mathfrak{m}_0\mathfrak{m}_\infty = \prod_{\mathfrak{p}} \mathfrak{p}^{n_{\mathfrak{p}}(\mathfrak{m})} \prod_{v} v^{n_v(\mathfrak{m})},$$

where $\mathfrak{p}$ runs through the set of non-zero prime ideals of $O_{\mathbb{K}}$ and $n_{\mathfrak{p}}(\mathfrak{m}) \in \mathbb{Z}_{\geq 0}$ with $n_{\mathfrak{p}}(\mathfrak{m}) \geq 1 \Leftrightarrow \mathfrak{p} \mid \mathfrak{m}_0$. Also, v runs through the set of real places of $\mathbb{K}$ with $n_v(\mathfrak{m}) \in \{0, 1\}$.

▷ Let $\alpha \in \mathbb{K}^*$ and $\mathfrak{m} = \mathfrak{m}_0\mathfrak{m}_\infty$ be a modulus. We write $\alpha \equiv 1 \ (\mathrm{mod}^\star \mathfrak{m})$ to mean that, for all $\mathfrak{p} \mid \mathfrak{m}_0$, $v_{\mathfrak{p}}(\alpha - 1) \geq v_{\mathfrak{p}}(\mathfrak{m}_0)$ and, for all $\sigma \in \mathfrak{m}_\infty$, $\sigma(\alpha) > 0$.

▷ We define the generalized Euler totient Φ in the following way. If $\mathfrak{m} = \mathfrak{m}_0\mathfrak{m}_\infty$ is a modulus, then

$$\Phi(\mathfrak{m}) := \left|(O_{\mathbb{K}}/\mathfrak{m})^\times\right| = 2^{|\mathfrak{m}_\infty|} N_{\mathbb{K}/\mathbb{Q}}(\mathfrak{m}_0) \prod_{\mathfrak{p}\mid\mathfrak{m}_0} \left(1 - \frac{1}{N_{\mathbb{K}/\mathbb{Q}}(\mathfrak{p})}\right).$$

Proposition 7.48 (Generalized Euler–Fermat Theorem)   *For all $x \in O_{\mathbb{K}}$ satisfying $(x) + \mathfrak{m}_0 = O_{\mathbb{K}}$*

$$x^{\varphi(\mathfrak{m}_0)} \equiv 1 \ (\mathrm{mod}\ \mathfrak{m}_0).$$

▷ Let $\mathfrak{m} = \mathfrak{m}_0\mathfrak{m}_\infty$ be a modulus. The set $I_{\mathfrak{m}} = I_{\mathfrak{m}}(\mathbb{K})$ is given by

$$I_{\mathfrak{m}} = \{\mathfrak{a}/\mathfrak{b} : \mathfrak{a}, \mathfrak{b} \subset O_{\mathbb{K}}, \ \mathfrak{a} + \mathfrak{m}_0 = \mathfrak{b} + \mathfrak{m}_0 = O_{\mathbb{K}}\}.$$

We also set

$$\mathbb{K}_{\mathfrak{m}}^* = \{\alpha \in \mathbb{K}^* : \alpha \equiv 1 \ (\mathrm{mod}^\star \mathfrak{m})\}$$

and define $O_{\mathfrak{m}}^\times = O_{\mathbb{K}}^\times \cap \mathbb{K}_{\mathfrak{m}}^*$. Finally, the *ray group* modulo $\mathfrak{m}$ is the set

$$P_{\mathfrak{m}} = P_{\mathfrak{m}}(\mathbb{K}) = \{\alpha O_{\mathbb{K}} : \alpha \equiv 1 \ (\mathrm{mod}^\star \mathfrak{m})\}.$$

In other words, $P_{\mathfrak{m}}$ is the subgroup of principal ideals of the shape (a/b) such that $a, b \in O_{\mathbb{K}}$ are coprime to $\mathfrak{m}$, $a \equiv b \pmod{\mathfrak{m}}$ and a/b is *totally positive*, that is, all its real conjugates are positive.

Definition 7.26 Let $\mathfrak{m} = \mathfrak{m}_0\mathfrak{m}_\infty$ be a modulus. The *narrow ray class group* modulo $\mathfrak{m}$, or more simply the *ray class group* modulo $\mathfrak{m}$, is the quotient $C\ell_\mathfrak{m} = C\ell_\mathfrak{m}(\mathbb{K})$ given by

$$C\ell_\mathfrak{m}(\mathbb{K}) = I_\mathfrak{m}(\mathbb{K})/P_\mathfrak{m}(\mathbb{K}).$$

We usually write $[\mathfrak{a}]_\mathfrak{m}$ for the image in $C\ell_\mathfrak{m}$ of $\mathfrak{a} \in I_\mathfrak{m}$.

Example 7.28 When $\mathbb{K} = \mathbb{Q}$ and $\mathfrak{m} = (m)\infty$, then there is an isomorphism from $(\mathbb{Z}/m\mathbb{Z})^\times$ to $C\ell_\mathfrak{m}(\mathbb{Q})$ given by $\bar{a} \mapsto [a]_\mathfrak{m}$, where a is any *positive* integer prime to m.

Proposition 7.49 *Let $\mathfrak{m} = \mathfrak{m}_0\mathfrak{m}_\infty$ be a modulus. The ray class group modulo $\mathfrak{m}$ is a finite abelian group and its order, denoted by $h_\mathfrak{m}$, is given by*

$$h_\mathfrak{m} = h_\mathbb{K} \times \frac{\Phi(\mathfrak{m})}{\left(O_\mathbb{K}^\times : O_\mathfrak{m}^\times\right)}.$$

Example 7.29 Let $\mathbb{K} = \mathbb{Q}\left(\sqrt{2}\right)$ so that $h_\mathbb{K} = 1$ and take $\mathfrak{m} := \mathfrak{p} = \left(7 + 2\sqrt{2}\right)$ with $N_{\mathbb{K}/\mathbb{Q}}(\mathfrak{p}) = 41$, and hence $\Phi(\mathfrak{p}) = 40$. By Theorem 7.5, $O_\mathbb{K}^\times = \left\{\pm(1 + \sqrt{2})^k : k \in \mathbb{Z}\right\}$. Now the order $(\mathrm{mod}\,\mathfrak{p})$ of -1 is 2 and, by Proposition 7.48, the order $(\mathrm{mod}\,\mathfrak{p})$ of $1 + \sqrt{2}$ divides 40. Writing $\bar{x}$ instead of x $(\mathrm{mod}\,\mathfrak{p})$ and using $\overline{41} = \overline{0}$, we derive successively $\overline{(\sqrt{2})} = \overline{(-7/2)} = \overline{((41 - 7)/2)} = \overline{17}$, so that $\overline{1 + \sqrt{2}} = \overline{18}$, and implying that the order $(\mathrm{mod}\,\mathfrak{p})$ of $1 + \sqrt{2}$ is 5. Hence $(O_\mathbb{K}^\times : O_\mathfrak{p}^\times) = [2, 5] = 10$, and therefore $h_\mathfrak{p} = 4$ by Proposition 7.49.

Now let $\mathbb{L}/\mathbb{K}$ be an *abelian* extension, and let $\mathfrak{p}$ be a prime ideal of $O_\mathbb{K}$. An element $x \in \mathbb{K}^*$ is a *local norm* modulo $\mathfrak{p}$ if, for all $n \in \mathbb{Z}_{\geqslant 0}$, there exists $y_n \in \mathbb{L}^*$ such that

$$\frac{x}{N_{\mathbb{L}/\mathbb{K}}(y_n)} \equiv 1 \;(\mathrm{mod}^\star \mathfrak{p}^n).$$

Define

$$k_\mathfrak{p} = \inf_{n \in \mathbb{Z}_{\geqslant 0}} \left\{x \equiv 1 \;(\mathrm{mod}^\star \mathfrak{p}^k) : x \text{ is a local norm } (\mathrm{mod}\,\mathfrak{p})\right\}.$$

Note that $k_\mathfrak{p} = 0$ if and only if $\mathfrak{p}$ is unramified in $\mathbb{L}/\mathbb{K}$.

Definition 7.27 (Conductor) Let $\mathbb{L}/\mathbb{K}$ be an abelian extension. The *conductor* of $\mathbb{L}/\mathbb{K}$ is the arithmetic cycle $\mathfrak{f} = \mathfrak{f}_0\mathfrak{f}_\infty$, where $\mathfrak{f}_0 = \prod_\mathfrak{p} \mathfrak{p}^{k_\mathfrak{p}}$ and $\mathfrak{f}_\infty$ is the set of real places of $\mathbb{K}$ ramified in $\mathbb{L}$.

We are now in a position to define one of the main objects of the class field theory.

Definition 7.28 (Ray Class Field) Let $m = m_0 m_\infty$ be a modulus and $\mathbb{K}$ be a number field. The *ray class field* of modulus m of $\mathbb{K}$ is the greatest abelian extension of $\mathbb{K}$, denoted by $\mathbb{K}(m)/\mathbb{K}$, with a minimal conductor dividing m.

Note that, when $m = (1)$, $\mathbb{K}(1)$ is the *Hilbert class field* of $\mathbb{K}$ seen in Theorem 7.35. While not immediately apparent from the definition, ray class fields are uniquely determined by m, so it makes sense to speak of the ray class field for the modulus m. The existence and unicity of this field is one of the highlights of class field theory proved by Takagi. In fact, much more is known, as we can see in the following result.

Theorem 7.41 *Let $\mathbb{L}/\mathbb{K}$ be an abelian extension of conductor $\mathfrak{f}$ and m be a modulus such that $\mathfrak{f} \mid m$.*[59] *Let H_m be the Artin group of $\mathbb{L}/\mathbb{K}$, i.e. the kernel of the Artin map $\mathrm{Art}_{\mathbb{L}/\mathbb{K}}$ given by*

$$\mathrm{Art}_{\mathbb{L}/\mathbb{K}} : I_m \longrightarrow \mathrm{Gal}\,(\mathbb{L}/\mathbb{K})$$

$$\mathfrak{a} \longmapsto \prod_{\mathfrak{p}\mid\mathfrak{a}} \left(\frac{\mathbb{L}/\mathbb{K}}{\mathfrak{p}}\right)^{v_\mathfrak{p}(\mathfrak{a})}$$

and generalizing the map (7.31). Then $\mathbb{L} \subset \mathbb{K}(m)$ and we have the following exact sequence

$$1 \to H_m \to C\ell_m \to \mathrm{Gal}(\mathbb{L}/\mathbb{K}) \to 1$$

and in particular

$$\mathrm{Gal}\,(\mathbb{L}/\mathbb{K}) \simeq C\ell_m/H_m, \quad \mathrm{Gal}\,(\mathbb{K}(m)/\mathbb{K}) \simeq C\ell_m \quad and \quad \mathrm{Gal}\,(\mathbb{K}(m)/\mathbb{L}) \simeq H_m.$$

Example 7.30

▷ When $\mathbb{K} = \mathbb{Q}$ and $m = (m)\infty$, $\mathbb{K}(m) = \mathbb{Q}(\zeta_m)$.
▷ When $\mathbb{K} = \mathbb{Q}(\sqrt{-1})$ and $m = (2)$, $\mathbb{K}(m) = \mathbb{K}$ since, using Proposition 7.49, we derive

$$[\mathbb{K}(2) : \mathbb{K}] = h_{(2)} = h_\mathbb{K} \times \frac{\Phi((2))}{\left(O_\mathbb{K}^\times : O_{(2)}^\times\right)} = \frac{1 \times 2}{2} = 1.$$

---

[59] This means that $m_0 \subset \mathfrak{f}_0$ and $\mathfrak{f}_\infty \subset m_\infty$.

7.5.11.2   Ray Class Characters

Definition 7.29 Let $\mathbb{K}$ be a number field and $\mathfrak{m} = \mathfrak{m}_0 \mathfrak{m}_\infty$ be a modulus. A *ray class character*, or *Weber character*, of modulus $\mathfrak{m}$ is a homomorphism $\chi : I_\mathfrak{m} \to S^1$[60] such that $P_\mathfrak{m} \subset \ker \chi$. In other words, χ is a character of $C\ell_\mathfrak{m}$. The *Weber L-function* attached to a ray class character χ is the Dirichlet series of χ, i.e.

$$L_\mathfrak{m}(s, \chi) = \sum_{\substack{\mathfrak{a} \subset \mathcal{O}_\mathbb{K} \\ \mathfrak{a} \in I_\mathfrak{m}}} \frac{\chi(\mathfrak{a})}{N_{\mathbb{K}/\mathbb{Q}}(\mathfrak{a})^s} = \prod_{\mathfrak{p} \nmid \mathfrak{m}_0} \left(1 - \frac{\chi(\mathfrak{p})}{N_{\mathbb{K}/\mathbb{Q}}(\mathfrak{p})^s} \right)^{-1} \quad (\sigma > 1).$$

By standard comparisons, the above sum and Euler product converge absolutely to the same limit on the half-plane $\sigma > 1$.

Example 7.31 Assume that $\mathfrak{m}$ is trivial, i.e. $\mathfrak{m}_0 = \mathcal{O}_\mathbb{K}$ and $\mathfrak{m}_\infty = \varnothing$, so that $C\ell_\mathfrak{m} = C\ell(\mathbb{K})$ is the class group. If χ_0 is the principal character of $C\ell_\mathfrak{m}$, then

$$L_{(1)}(s, \chi_0) = \sum_{\mathfrak{a} \subset \mathcal{O}_\mathbb{K} \setminus \{(0)\}} \frac{1}{N_{\mathbb{K}/\mathbb{Q}}(\mathfrak{a})^s} = \zeta_\mathbb{K}(s).$$

Example 7.32 Take $\mathbb{K} = \mathbb{Q}$ and $\mathfrak{m} = (m)\infty$ for some $m \in \mathbb{Z}_{\geq 1}$. By Example 7.28, we have $C\ell_\mathfrak{m}(\mathbb{Q}) \simeq (\mathbb{Z}/m\mathbb{Z})^\times$. Hence a character χ of $C\ell_\mathfrak{m}$ coincides with a Dirichlet character modulo $\mathfrak{m}$, and $L_\mathfrak{m}(s, \chi)$ is then just the ordinary associated Dirichlet L-function.

The concept of "primitive" character can be defined in a similar way as usual primitive Dirichlet characters. In the following lines, we adopt the presentation given in [84].

Let $\mathfrak{m}$ and $\mathfrak{n}$ be two modulus such that $\mathfrak{n} \mid \mathfrak{m}$. Then there is a surjective homomorphism

$$\pi_{\mathfrak{m},\mathfrak{n}} : C\ell_\mathfrak{m} \longrightarrow C\ell_\mathfrak{n}$$
$$[\mathfrak{a}]_\mathfrak{m} \longmapsto [\mathfrak{a}]_\mathfrak{n},$$

and it can be shown [84] that, if ℓ is another modulus that divides $\mathfrak{m}$, then

$$\ker\left(\pi_{\mathfrak{m},\mathfrak{n}}\right) \ker\left(\pi_{\mathfrak{m},\ell}\right) = \ker\left(\pi_{\mathfrak{m},(\mathfrak{n},\ell)}\right)$$

so that there exists a unique minimal[61] modulus $\mathfrak{f}_\chi$ such that there exists a character $\chi^\star$ of $C\ell_{\mathfrak{f}_\chi}$ such that $\chi = \chi^\star \circ \ker\left(\pi_{\mathfrak{m},\mathfrak{f}_\chi}\right)$.

---

[60] $S^1 = \{z \in \mathbb{C} : |z| = 1\}$.
[61] With respect to divisibility.

Definition 7.30 The modulus $\mathfrak{f}_\chi$ defined above is the *conductor* of χ, and $\chi^\star$ is the *primitive* character that induces χ. When a character χ is primitive, i.e. $\mathfrak{m} = \mathfrak{f}_\chi$, we write $L(s, \chi)$ for the attached Weber L-function $L_{\mathfrak{f}_\chi}(s, \chi)$. Note that, when χ is imprimitive, then $L_\mathfrak{m}(s, \chi)$ and $L(s, \chi^\star)$ differ at most by finitely many Euler factors, since, as in (3.34)

$$L_\mathfrak{m}(s, \chi) = L(s, \chi^\star) \prod_{\substack{\mathfrak{p} \mid \mathfrak{m} \\ \mathfrak{p} \nmid \mathfrak{f}_\chi}} \left(1 - \frac{\chi^\star(\mathfrak{p})}{N_{\mathbb{K}/\mathbb{Q}}(\mathfrak{p})^s} \right).$$

Weber L-functions satisfy a functional equation generalizing that of Dirichlet L-functions seen in Theorem 3.24.

Theorem 7.42 Let $\mathbb{K}$ be an algebraic number field and χ be a primitive *ray class character of conductor* $\mathfrak{f}_\chi = \mathfrak{f}_0\mathfrak{f}_\infty$ *of* $C\ell_{\mathfrak{f}_\chi}(\mathbb{K})$. *Then* $L(s, \chi)$ *is analytic at s except possibly when* $s = 1$, *where it has a simple pole if and only if* $\chi = \chi_0$. *Furthermore, if* $a_1(\chi)$ *and* $a_2(\chi)$ *denote, respectively, the number of real places v such that* $v \nmid \mathfrak{f}_\infty$ *and* $v \mid \mathfrak{f}_\infty$ *so that* $a_1(\chi) + a_2(\chi) = r_1$, *then the completed L-function*

$$\Lambda(s, \chi) := \left(|d_\mathbb{K}| N_{\mathbb{K}/\mathbb{Q}}(\mathfrak{f}_0) \right)^{s/2} 2^{r_2(1-s)} \pi^{-\frac{1}{2}(ns+a_2(\chi))}$$

$$\times \Gamma \left(\tfrac{s}{2} \right)^{a_1(\chi)} \Gamma \left(\tfrac{1+s}{2} \right)^{a_2(\chi)} \Gamma(s)^{r_2} L(s, \chi)$$

satisfies the functional equation

$$\Lambda(1 - s, \chi) = \frac{i^{a_2(\chi)} \left(N_{\mathbb{K}/\mathbb{Q}}(\mathfrak{f}_0) \right)^{1/2}}{\tau(\overline{\chi})} \Lambda(s, \overline{\chi}),$$

where $\tau(\chi)$ is the Gauss sum. Note also that $|\tau(\chi)|^2 = N_{\mathbb{K}/\mathbb{Q}}(\mathfrak{f}_0)$.

Let us finally state the following result, proved by Hecke, and generalize Theorem 7.25.

Theorem 7.43 *Let $\mathbb{K}$ be an algebraic number field and $\mathfrak{m}$ be a modulus for $\mathbb{K}$. If $\mathbb{L}/\mathbb{K}$ is a sub-extension of $\mathbb{K}(\mathfrak{m})/\mathbb{K}$, then*

$$\zeta_{\mathbb{L}}(s) = \zeta_{\mathbb{K}}(s) \prod_{\chi \neq \chi_0} L(s, \chi^\star) \quad (\sigma > 1).$$

In particular, the function $\zeta_{\mathbb{L}}/\zeta_{\mathbb{K}}$ is entire.

7.5.11.3 Hecke *L*-Functions

To account for infinite primes of $\mathbb{K}$, Hecke went further into the generalization of Dirichlet characters by introducing his *Grössencharakteren*[62] and associated *L*-functions, for which we briefly describe the notions.

When generalizing Dirichlet characters, one of the major obstacles was to get functions of ideals, and not elements. As pointed out in [81, p. 33], even in $\mathbb{Z}$ this is not completely obvious. If χ is an *odd* Dirichlet character,[63] then $(-n) = (n)$ but $\chi(-n) \neq \chi(n)$, so χ is not well-defined in this case. We can easily overcome this problem by taking the sign into account. Indeed, if $\chi(-1) = (-1)^e$ for $e \in \{0, 1\}$, then we have the function on ideals by setting

$$\chi((n)) = \chi(n) \left(\frac{n}{|n|} \right)^e. \tag{7.35}$$

Let $\mathbb{K}$ be a number field of signature (r_1, r_2) and set

$$\mathbb{K}^\infty = \left(\mathbb{R}^\times \right)^{r_1} \times \left(\mathbb{C}^\times \right)^{r_2}.$$

Since $O_{\mathbb{K}}^\times \hookrightarrow \mathbb{K}^\infty$, any character of the units will come from a character of $\mathbb{K}^\infty$.

Definition 7.31 Let $\mathfrak{m}$ be a modulus for $\mathbb{K}$ and assume that $C\ell(\mathbb{K}) = \{a_1, \ldots, a_h\}$. A function $\chi : I_{\mathfrak{m}}(\mathbb{K}) \to S^1$ is a *Hecke character* modulo $\mathfrak{m}$ if it can be written in the shape

$$\chi(a_i(a)) = \chi_{cl}(a_i) \chi_f(a) \chi_\infty(a),$$

---

[62]Called today *Hecke characters*.
[63]That is, $\chi(-1) = -1$.

where χ_{cl} is a character of $C\ell(\mathbb{K})$, χ_f is a character of $(O_\mathbb{K}/\mathfrak{m})^\times$ and χ_∞ is a character of $\mathbb{K}^\infty$. The attached *Hecke L-function* is defined for $\sigma > 1$ by

$$L(s, \chi) = \sum_{\mathfrak{a} \in I_\mathfrak{m}(\mathbb{K})} \frac{\chi(\mathfrak{a})}{N_{\mathbb{K}/\mathbb{Q}}(\mathfrak{a})^s} = \prod_\mathfrak{p} \left(1 - \frac{\chi(\mathfrak{p})}{N_{\mathbb{K}/\mathbb{Q}}(\mathfrak{p})^s}\right)^{-1}.$$

The next proposition provides an explicit form of the character χ_∞ generalizing the above discussion leading to (7.35). For a proof, see [66, Proposition VII.6.7] or [81, Proposition 1.4.18].

Proposition 7.50 *If χ_∞ is a character of $\mathbb{K}^\infty$, then*

$$\chi_\infty (x_1, \ldots, x_{r_1}, z_1, \ldots, z_{r_2}) = \prod_{k=1}^{r_1} \left(\frac{x_k}{|x_k|}\right)^{e_k} |x_k|^{is_k} \prod_{k=1}^{r_2} \left(\frac{z_k}{|z_k|}\right)^{f_k} |z_k|^{it_k},$$

where $e_k \in \{0, 1\}$, $f_k \in \mathbb{Z}$ and $s_k, t_k \in \mathbb{R}$.

The notion of *conductor* of a Hecke character can be defined as above.

Definition 7.32 Let χ be a Hecke character modulo $\mathfrak{m}$ with finite part χ_f. χ_f need not be a primitive character of $(O_\mathbb{K}/\mathfrak{m})^\times$. If it were not primitive, then it would factor through $(O_\mathbb{K}/\mathfrak{n})^\times$ for some $\mathfrak{n} \mid \mathfrak{m}$. If it factors through $(O_\mathbb{K}/\mathfrak{n}_1)^\times$ and $(O_\mathbb{K}/\mathfrak{n}_2)^\times$, then it will have to factor through $(O_\mathbb{K}/(\mathfrak{n}_1, \mathfrak{n}_2))^\times$. The *conductor* $\mathfrak{f}_\chi$ of χ is then the smallest of these $\mathfrak{n}$. A Hecke character is said to be *primitive* if its modulus is equal to its conductor.

Any Hecke character modulo $\mathfrak{m}$ corresponds to a unique Hecke character modulo $\mathfrak{f}_\chi$. However as in (3.34), the Euler factorization of their L-functions will differ by finitely many terms. To overcome this problem, we slightly change the definition of the L-function for an imprimitive Hecke character.

Definition 7.33 The *Hecke L-function* of an imprimitive Hecke character χ is the Hecke L-function of the primitive character $\chi^\star$ that induces χ. We usually denote the finite part of $\mathfrak{f}_\chi$ by $\mathfrak{f}_0$.

Hecke proved the following result extending Theorem 7.42.

Theorem 7.44 *Let χ be a Hecke character modulo $\mathfrak{m}$ for a number field $\mathbb{K}$, and define the Gamma-factor of χ by*

$$L_\infty(s, \chi) = \prod_{k=1}^{r_1} \pi^{-\frac{1}{2}(s+e_k-is_k)} \Gamma\left(\frac{s + e_k - is_k}{2}\right)$$

(continued)

Theorem 7.44 (continued)

$$\times \prod_{k=1}^{r_2} 2(2\pi)^{-(s+f_k-it_k)} \Gamma \left(s + f_k - it_k \right),$$

where the numbers e_k, f_k, s_k, t_k are given in Proposition 7.50. Then $L(s, \chi)$ has a meromorphic continuation to the whole complex plane, which is holomorphic everywhere, except for two simple poles in the case when χ_f is trivial, χ_{cl} is trivial, and all the es and fs are 0. Furthermore, the completed Hecke L-function

$$\Lambda(s, \chi) := \left(|d_{\mathbb{K}}| N_{\mathbb{K}/\mathbb{Q}}(\mathfrak{f}_0) \right)^{s/2} L_\infty(s, \chi) L(s, \chi)$$

satisfies the functional equation

$$\Lambda(1 - s, \chi) = w(\chi) \Lambda \left(s, \overline{\chi} \right),$$

where $w(\chi)$ is the so-called "root number" and satisfies $|w(\chi)| = 1$.

Hecke's proof of this result is a generalization of Theorem 3.7. Hecke used much more complicated θ-functions and showed that their Mellin transforms were Hecke L-functions times a factor involving the Gamma-function. See [40, pp. 178–197] or [66, Chapter VII, §7].

7.5.11.4   Representation Theory of Finite Groups

Artin's intellectual passion was class field theory and representation theory of finite groups. Recognizing the utility of studying groups in terms of their matrix representations, he focused attention on homomorphisms $\rho : \text{Gal}(\mathbb{K}/\mathbb{Q}) \to \text{GL}(m, \mathbb{C})$. This leads to our first definition.

Definition 7.34 Let G be a finite group. A *representation* of G over $\mathbb{C}$ is a pair (V, ρ), where V is a finite dimensional $\mathbb{C}$-vector space and ρ is a group homomorphism $\rho : G \to \text{GL}(V)$. The *degree* of the representation is $\dim V$ and it is denoted by $\deg \rho$. For any $g \in G$, the action of $\rho(g) := \rho_g$ on a vector $v \in V$ is denoted by $\rho_g \cdot v$.

Example 7.33 The representation $(\mathbb{C}, \rho_0)$ of a group G given by $\rho_0 : G \to \mathbb{C}^*$ such that $\rho_0(g) = 1$ is called the *trivial* representation. The representation $(\mathbb{C}, \rho)$ of the group $\mathbb{Z}/n\mathbb{Z}$ given by $\rho(\overline{m}) = e_n(m)$ is also of degree one.

Let (V, ρ) be a representation of a finite group G and W be a subspace of V. W is said to be *G-subspace* if, for all $g \in G$, $\rho_g(W) \subseteq W$. This leads to the following important definition.

Definition 7.35 A non-zero representation (V, ρ) of a finite group G is said to be *irreducible* if the only G-subspaces of V are $\{0\}$ and V.

Example 7.34 We give here some quite important facts about irreducible representations. The proofs can be found in [90].

▷ Any degree one representation is irreducible, since $\mathbb{C}$ has no proper non-zero subspaces.
▷ If (V, ρ) is an irreducible representation of a finite group G, then $\deg \rho$ divides $|G|$.
▷ Any irreducible representation of a finite *abelian* group has degree one.

Definition 7.36 Let (V, ρ) be a representation of a finite group G and W be a G-subspace of V. Then $(W, \rho|_W)$ given by $\rho|_W : G \to \mathrm{GL}(W)$, $g \mapsto \rho(g)|_W$ is a representation of G, called a *subrepresentation*.

The next notion of a *character* of a representation will be important in the genesis of Artin L-functions.

Definition 7.37 Let (V, ρ) be a representation of a finite group G. The *character* $\chi_\rho : G \to \mathbb{C}$ is defined by

$$\chi_\rho(g) = \mathrm{Tr}\left(\rho_g\right).$$

The character of an irreducible representation is called an *irreducible character*.

Example 7.35 Let (V, ρ) be a representation of a finite group G.

▷ If $\deg \rho = 1$, then $\chi_\rho = \rho$, and it is customary to equate a degree 1 representation with its character.
▷ $\chi_\rho(1) = \deg \rho$. Indeed, $\mathrm{Tr}(\rho_1) = \mathrm{Tr}(\mathrm{id}) = \dim V = \deg \rho$.
▷ For all $g, h \in G$, $\chi_\rho\left(hgh^{-1}\right) = \chi_\rho(g)$. Indeed, the left-hand side is equal to

$$\mathrm{Tr}\left(\rho_{hgh^{-1}}\right) = \mathrm{Tr}\left(\rho_h \rho_g \rho_h^{-1}\right) = \mathrm{Tr}\left(\rho_h^{-1} \rho_h \rho_g\right) = \mathrm{Tr}\left(\rho_g\right) = \chi_\rho(g).$$

Example 7.36 (Regular Representation) Let G be a finite group. Define the *regular representation* of G to be the pair $\left(\mathbb{C}G, \rho_{\mathrm{reg}}\right)$ by $\rho_{\mathrm{reg}} : G \to \mathrm{GL}\left(\mathbb{C}G\right)$ such that, for all $g \in G$ and all $v = \sum_{h \in G} c_h h \in \mathbb{C}G$

$$\rho_{\mathrm{reg}}(g) \cdot v = \sum_{h \in G} c_h gh.$$

If χ_{reg} is the character of the regular representation, then, for all $g \in G$

$$\chi_{\text{reg}}(g) = \begin{cases} |G|, & \text{if } g = 1 \\ 0, & \text{otherwise.} \end{cases}$$

Furthermore, it can be proved [90, Theorem 4.4.4] that χ_{reg} is a linear combination of irreducible characters $\chi_1, \ldots, \chi_r$ with multiplicities $\chi_1(1), \ldots, \chi_r(1)$, i.e.

$$\chi_{\text{reg}} = \sum_{k=1}^{r} \chi_k(1)\chi_k. \tag{7.36}$$

Note that (7.36) applied at $1 = 1_G$ yields

$$\sum_{\rho}(\deg \rho)^2 = |G|, \tag{7.37}$$

where the sum runs over the irreducible representations of G.

Definition 7.38 A function $f : G \to \mathbb{C}$ is a *class function*, or *central function*, if it is constant on conjugacy classes of G, i.e. for all $g, h \in G$

$$f\left(hgh^{-1}\right) = f(g).$$

By the example above, characters of representation are class functions. It can be proved [90, Propositions 4.3.7 and 4.3.8] that the set of class functions is a finite-dimensional subspace of the vector space $\mathbb{C}[G]$ of all function $G \to \mathbb{C}$ with dimension equal to $|C\ell(G)|$, where $C\ell(G)$ is the set of conjugacy classes of G. $\mathbb{C}[G]$ can be equipped with the inner product

$$\langle f_1, f_2 \rangle = \frac{1}{|G|} \sum_{g \in G} f_1(g)\overline{f_2(g)}.$$

This allows the characters to have some *orthogonality relations* [90, Theorem 4.3.9].

Proposition 7.51 *Let (V_1, ρ_1) and (V_2, ρ_2) be irreducible representations of G. Then*

$$\langle \chi_{\rho_1}, \chi_{\rho_2} \rangle = \begin{cases} 1, & \text{if } \rho_1 \sim \rho_2 \\ 0, & \text{otherwise,} \end{cases}$$

where $\rho_1 \sim \rho_2$ means that there exists an isomorphism $T : V_1 \to V_2$ such that, for all $g \in G$, $\rho_1(g)T = T\rho_2(g)$.

Table 7.1 Irreducible
characters of S_3

	Id	(12)	(123)	Degree
χ_0	1	1	1	1
χ_φ	1	-1	1	1
χ_ρ	2	0	-1	2

This result implies that the number of equivalence classes of irreducible representations of G does not exceed $|C\ell(G)|$. In fact, it can be proved [90, Corollary 4.4.8] that the number of equivalence classes of irreducible representations of G is equal to $h_G := |C\ell(G)|$, the *class number* of G.

Let us treat the following example dealing with the irreducible characters of $G = S_3 = \{(1), (123), (321), (12), (23), (31)\}$, the group of permutations of three elements.

Example 7.37 Since S_3 has three conjugacy classes $C_1 = \{(1)\}$, $C_2 = \{(123), (321)\}$ and $C_3 = \{(12), (23), (31)\}$, there are three inequivalent irreducible representations of S_3. First, we have the principal character $\chi_0 : S_3 \to \mathbb{C}^*$ given by $\chi_0(\sigma) = 1$. Define the representation $\rho : S_3 \to GL(2, \mathbb{C})$ specified on the generators (12) and (123) by

$$\rho_{(12)} = \begin{pmatrix} -1 & -1 \\ 0 & 1 \end{pmatrix} \quad \text{and} \quad \rho_{(123)} = \begin{pmatrix} -1 & -1 \\ 1 & 0 \end{pmatrix}.$$

From [90, Example 3.1.14], ρ is an irreducible representation of S_3 of degree 2. From (7.37), the last irreducible representation φ of S_3 has degree one, and the associated character[64] can be defined [90, Example 4.3.17] by setting

$$\chi_\varphi(\sigma) = \begin{cases} 1, & \text{if } \sigma \text{ is even}; \\ -1, & \text{if } \sigma \text{ is odd.} \end{cases}$$

Now if $\rho : G \to GL(m, \mathbb{C})$ is given by $\rho_g = \big(\rho_{ij}(g)\big)$, then $\chi_\rho(g) = \rho_{1,1}(g) + \cdots + \rho_{m,m}(g)$, so that, for the representation ρ defined above, we have $\chi_\rho(\text{id}) = 2$, $\chi_\rho((12)) = 0$ and $\chi_\rho((123)) = -1$, yielding the following Table 7.1.

7.5.11.5 Ramification

Now let us return to the problem of building L-functions for non-abelian Galois extensions $\mathbb{L}/\mathbb{K}$. We first need the following definition from ramification theory.

Definition 7.39 Let $\mathbb{L}/\mathbb{K}$ be a Galois extension of number fields with Galois group $G = \text{Gal}\,(\mathbb{L}/\mathbb{K})$, $\mathfrak{p}$ be a prime ideal of $O_\mathbb{K}$ and $\mathfrak{P} \mid \mathfrak{p}$.

---

[64]Determined by the subgroup $C_1 \cup C_2$.

▷ The *decomposition group* $\mathcal{D}_{\mathfrak{P}}$ is the subgroup of G defined by

$$\mathcal{D}_{\mathfrak{P}} = \{\sigma \in G : \sigma(\mathfrak{P}) = \mathfrak{P}\}.$$

▷ The *inertia group* $I_{\mathfrak{P}}$ is the subgroup of G defined by

$$I_{\mathfrak{P}} = \{\sigma \in G : \forall x \in O_{\mathbb{L}}, \ \sigma(x) \equiv x \pmod{\mathfrak{P}}\}.$$

The main properties of these groups are synthesized in the next result [66, Proposition I.9.6].

Proposition 7.52 *Let* $\mathbb{L}/\mathbb{K}$ *be a Galois extension of number fields with Galois group* $G = \mathrm{Gal}(\mathbb{L}/\mathbb{K})$, $\mathfrak{p}$ *be a prime ideal of* $O_{\mathbb{K}}$ *with ramification index* e *and decomposition number* g *and* $\mathfrak{P} \mid \mathfrak{p}$. *Assume that the residue field extension* $(O_{\mathbb{L}}/\mathfrak{P})/(O_{\mathbb{K}}/\mathfrak{p})$ *is separable. Then we have the exact sequence*

$$1 \to I_{\mathfrak{P}} \to \mathcal{D}_{\mathfrak{P}} \to \mathrm{Gal}((O_{\mathbb{L}}/\mathfrak{P})/(O_{\mathbb{K}}/\mathfrak{p})) \to 1,$$

and hence $\mathcal{D}_{\mathfrak{P}}/I_{\mathfrak{P}} \simeq \mathrm{Gal}((O_{\mathbb{L}}/\mathfrak{P})/(O_{\mathbb{K}}/\mathfrak{p}))$ *so that* $\mathcal{D}_{\mathfrak{P}}/I_{\mathfrak{P}}$ *is cyclic.*[65] *Also*

$$(G : \mathcal{D}_{\mathfrak{P}}) = g \quad and \quad |I_{\mathfrak{P}}| = e.$$

Hence when $\mathfrak{p}$ is unramified in $O_{\mathbb{L}}$, i.e. $e = 1$, then $I_{\mathfrak{P}}$ is trivial and the decomposition group $\mathcal{D}_{\mathfrak{P}}$ is cyclic. A generator is the Frobenius element, given by the Artin symbol seen in Definition 7.25 and that we denote here by $\phi_{\mathfrak{P}}$ [66, Exercise 2, p. 58].

7.5.11.6 Artin *L*-Functions

Born in 1898 in Vienna and died in 1962, Emil Artin wanted to extend the Hecke–Weber factorization, describing Dedekind zeta-functions in abelian extensions, to a non-abelian formulation. The piece that really seems to have piqued his interest was the factorization of zeta-functions and *L*-functions as embodied in Theorem 7.43. To make things clearer about Artin's ideas, consider the Galois extension $\mathbb{K}/\mathbb{Q}$ with Galois group G. With the representation $\rho : G \to \mathrm{GL}(m, \mathbb{C})$, one may transfer the problem of analyzing conjugacy classes in G to the analogous problem in $\mathrm{GL}(m, \mathbb{C})$, where the corresponding classes are completely determined by their characteristic polynomials

$$\det\left(\mathrm{id} - \rho(\phi_p)p^{-s}\right),$$

---

[65] Since $(O_{\mathbb{L}}/\mathfrak{P})/(O_{\mathbb{K}}/\mathfrak{p})$ is a Galois extension of finite fields.

and it is then natural to set

$$L(s, \rho) := \prod_p \det\left(\mathrm{id} - \rho(\phi_p)p^{-s}\right)^{-1}.$$

When $\mathbb{K}$ is abelian, then $\deg \rho = 1$, and hence $\rho = \chi_\rho$ so that we may write

$$L(s, \rho) = L(s, \chi_\rho) = \prod_p \det\left(\mathrm{id} - \chi_\rho(\phi_p)p^{-s}\right)^{-1}.$$

Furthermore, Artin shows that $L(s, \rho)$ is nothing more than a Dirichlet L-function, eventually allowing him to prove Theorem 7.31.

If one wishes to transfer the above discussion to the general case of a Galois extension $\mathbb{L}/\mathbb{K}$ with Galois group G, it should be pointed out that the change from the abelian case to the non-abelian case brings up the two following problems:

▷ When $\mathbb{L}/\mathbb{K}$ is not abelian, there are multiple decomposition groups $\mathcal{D}_{\mathfrak{P}}$.
▷ If $\mathfrak{p}$ ramifies in $O_{\mathbb{L}}$, then the Frobenius element $\phi_{\mathfrak{P}}$ to a given decomposition group is a coset, not just an element.

The first hurdle can be overcome by recalling that the action of G on the set of primes in $O_{\mathbb{L}}$ above $\mathfrak{p}$ is transitive, so that for any pair of primes $\mathfrak{P}_1$ and $\mathfrak{P}_2$ lying above $\mathfrak{p}$, there exists an automorphism in G, which simultaneously conjugates $\mathcal{D}_{\mathfrak{P}_1}$ into $\mathcal{D}_{\mathfrak{P}_2}$, $I_{\mathfrak{P}_1}$ into $I_{\mathfrak{P}_2}$ and $\phi_{\mathfrak{P}_1}$ into $\phi_{\mathfrak{P}_2}$. In particular, we can write $\phi_{\mathfrak{p}}$ in place of $\phi_{\mathfrak{P}}$ without loss of information. As a first step, we may then define the Artin L-function as follows.

Definition 7.40 Let $\mathbb{L}/\mathbb{K}$ be a Galois extension of number field with Galois group G and relative discriminant $\mathcal{D}_{\mathbb{L}/\mathbb{K}}$, and let (V, ρ) be a representation of G with character χ_ρ. The *unramified* Artin L-function is given by

$$L_{\mathrm{unr}}(s, \rho, \mathbb{L}/\mathbb{K}) = L_{\mathrm{unr}}(s, \chi_\rho, \mathbb{L}/\mathbb{K}) = \prod_{\mathfrak{p} \nmid \mathcal{D}_{\mathbb{L}/\mathbb{K}}} \det\left(\mathrm{id} - \rho(\phi_{\mathfrak{p}}) N_{\mathbb{K}/\mathbb{Q}}(\mathfrak{p})^{-s}\right)^{-1}.$$

Since conjugating $\rho(\phi_{\mathfrak{p}})$ by another matrix in the above definition does not change the value of the L-function because the determinant does not depend on choice of basis, the above definition only depends on V up to isomorphism. Also note that the Euler product is absolutely convergent in the half-plane $\sigma > 1$ since it is bounded by $\zeta_{\mathbb{K}}^{\deg \rho}$.

Remark 7.12 The second notation $L_{\mathrm{unr}}(s, \chi_\rho, \mathbb{L}/\mathbb{K})$ may be explained as follows. For any $\mathfrak{p}$ unramified in $O_{\mathbb{L}}$, set for $\sigma > 1$

$$L_{\mathfrak{p}}(s, \rho) = \det\left(\mathrm{id} - \rho(\phi_{\mathfrak{p}}) N_{\mathbb{K}/\mathbb{Q}}(\mathfrak{p})^{-s}\right)^{-1}.$$

Taking logarithm on both sides yields

$$\log L_{\mathfrak{p}}(s, \rho) = -\log \det \left(\mathrm{id} - \rho(\phi_{\mathfrak{p}}) \mathcal{N}_{\mathbb{K}/\mathbb{Q}}(\mathfrak{p})^{-s} \right),$$

and diagonalizing $\rho(\phi_{\mathfrak{p}})$ with eigenvalues $\lambda_1, \ldots, \lambda_m$ gives

$$\log L_{\mathfrak{p}}(s, \rho) = -\log \prod_{k=1}^{m} \left(1 - \lambda_k \, \mathcal{N}_{\mathbb{K}/\mathbb{Q}}(\mathfrak{p})^{-s} \right) = \sum_{k=1}^{m} \sum_{\ell=1}^{\infty} \frac{\lambda_k^{\ell}}{\ell} \, \mathcal{N}_{\mathbb{K}/\mathbb{Q}}(\mathfrak{p})^{-s\ell}$$

$$= \sum_{\ell=1}^{\infty} \frac{\chi_\rho \left(\phi_{\mathfrak{p}}^{\ell} \right)}{\ell} \, \mathcal{N}_{\mathbb{K}/\mathbb{Q}}(\mathfrak{p})^{-s\ell}.$$

To obtain an L-function which has good analytic properties, such as a functional equation, it is necessary to also define Euler factors at the primes $\mathfrak{p}$ which are ramified in $O_{\mathbb{L}}$ and also at infinite primes of $\mathbb{K}$. But it is certainly not an easy task. Indeed, 7 years elapsed between Artin's first article [5] and the second one [6], in which he took the ramified primes and the infinite places into account.

At the ramified primes, the Frobenius element $\phi_{\mathfrak{p}}$ is not well-defined because of the existence of a non-trivial inertia group. Let us have a more precise glance at this situation. Let $\mathfrak{p}$ be a ramified prime in $O_{\mathbb{L}}$, and let $\mathfrak{P} \mid \mathfrak{p}$. Also take a representation (V, ρ) of $G = \mathrm{Gal}\,(\mathbb{L}/\mathbb{K})$. Firstly, Artin begun by observing that, prime-by-prime, the Euler factor seen in Remark 7.12 only depends on $\rho|_{\mathcal{D}_{\mathfrak{P}}}$, the restriction of ρ to the decomposition group at $\mathfrak{P}$, still denoted by ρ by convenience. The idea is to restrict the action of this representation to a subspace of V on which $\mathcal{I}_{\mathfrak{P}}$ acts trivially. More precisely, set

$$V^{\mathcal{I}_{\mathfrak{P}}} = \left\{ v \in V : \forall \sigma \in \mathcal{I}_{\mathfrak{P}}, \, \rho_\sigma \cdot v = v \right\}.$$

Let $\sigma \in \mathcal{I}_{\mathfrak{P}}$. Since $\mathcal{I}_{\mathfrak{P}}$ is normal in G, $\phi_{\mathfrak{P}} \sigma \phi_{\mathfrak{P}}^{-1} \in \mathcal{I}_{\mathfrak{P}}$. Now let $v \in V^{\mathcal{I}_{\mathfrak{P}}}$. Then

$$\rho_{\phi_{\mathfrak{P}} \sigma \phi_{\mathfrak{P}}^{-1}} \cdot \left(\rho_{\phi_{\mathfrak{P}}} \cdot v \right) = \left(\rho_{\phi_{\mathfrak{P}}} \rho_\sigma \rho_{\phi_{\mathfrak{P}}}^{-1} \right) \cdot \left(\rho_{\phi_{\mathfrak{P}}} \cdot v \right) = \rho_{\phi_{\mathfrak{P}}} \cdot \left(\rho_\sigma \cdot v \right) = \rho_{\phi_{\mathfrak{P}}} \cdot v$$

so that $\rho_{\phi_{\mathfrak{P}}} \cdot v \in V^{\mathcal{I}_{\mathfrak{P}}}$, and therefore ρ restricts to a subrepresentation $\rho|_{V^{\mathcal{I}_{\mathfrak{P}}}}$ on $V^{\mathcal{I}_{\mathfrak{P}}}$. Furthermore, by the conjugacy argument made above, we see that any $\phi_{\mathfrak{P}}$ will have the same characteristic polynomials on $V^{\mathcal{I}_{\mathfrak{P}}}$. This finally leads to the following definition.

Definition 7.41 Let $\mathbb{L}/\mathbb{K}$ be a Galois extension of number field with Galois group G, and let (V, ρ) be a representation of G with character χ_ρ. The *Artin L-function* is given by

$$L(s, \chi_\rho, \mathbb{L}/\mathbb{K}) = \prod_\mathfrak{p} \det \left(\mathrm{id} - \rho(\phi_\mathfrak{p})\big|_{V^{I_\mathfrak{P}}} \, \mathcal{N}_{\mathbb{K}/\mathbb{Q}}(\mathfrak{p})^{-s} \right)^{-1}.$$

Remark 7.13 If ρ_0 is the trivial representation of G, then

$$\det \left(\mathrm{id} - \rho_0(\phi_\mathfrak{p}) \, \mathcal{N}_{\mathbb{K}/\mathbb{Q}}(\mathfrak{p})^{-s} \right)^{-1} = \left(1 - \mathcal{N}_{\mathbb{K}/\mathbb{Q}}(\mathfrak{p})^{-s} \right)^{-1}$$

so that, if χ_0 is the principal character of $\mathrm{Gal}(\mathbb{L}/\mathbb{K})$, then

$$L(s, \chi_0, \mathbb{L}/\mathbb{K}) = \zeta_\mathbb{K}(s) \quad (\sigma > 1).$$

7.5.11.7   Main Properties of Artin *L*-Functions

First note that, as in Remark 7.12, we still have for any prime $\mathfrak{p}$

$$L_\mathfrak{p}(s, \chi_\rho, \mathbb{L}/\mathbb{K}) := \det \left(\mathrm{id} - \rho(\phi_\mathfrak{p})\big|_{V^{I_\mathfrak{P}}} \, \mathcal{N}_{\mathbb{K}/\mathbb{Q}}(\mathfrak{p})^{-s} \right)^{-1}$$

$$= \exp \left(\sum_{\ell=1}^{\infty} \frac{\chi_\rho\left(\phi_\mathfrak{p}^\ell\right)}{\ell} \, \mathcal{N}_{\mathbb{K}/\mathbb{Q}}(\mathfrak{p})^{-s\ell} \right).$$

This immediately implies that

$$L_\mathfrak{p}(s, \chi_1 + \chi_2, \mathbb{L}/\mathbb{K}) = L_\mathfrak{p}(s, \chi_1, \mathbb{L}/\mathbb{K}) L_\mathfrak{p}(s, \chi_2, \mathbb{L}/\mathbb{K}),$$

from which we derive the first following result.

Proposition 7.53 (Additivity) *For any characters χ_1 and χ_2 of* $\mathrm{Gal}(\mathbb{L}/\mathbb{K})$, *we have*

$$L(s, \chi_1 + \chi_2, \mathbb{L}/\mathbb{K}) = L(s, \chi_1, \mathbb{L}/\mathbb{K}) L(s, \chi_2, \mathbb{L}/\mathbb{K}).$$

The second main result on Artin *L*-function is called *inductivity*. It relies on *induced representation theory*,[66] for which we refer the reader to the book [90, Chapter 8]. We only describe here the main objects we wish to deal with.

Definition 7.42 Let G be a finite group, H be a subgroup of G and (V, ρ) be a representation of H with character χ_ρ. The *induced character* of χ_ρ on G is the

---

[66]For our purpose, it suffices to know that *induction* is a procedure which enables us to construct a representation of a finite group G from a representation of a subgroup H of G.

character, denoted by $\mathrm{Ind}_H^G \chi_\rho$, defined for all $g \in G$ by

$$\mathrm{Ind}_H^G \chi_\rho(g) = \frac{1}{|H|} \sum_{\substack{x \in G \\ xgx^{-1} \in H}} \chi_\rho\left(xgx^{-1}\right).$$

Example 7.38 By Example 7.36, we easily derive that, if χ_0 is the character of the trivial representation ρ_0, then

$$\mathrm{Ind}_{\{1\}}^G \chi_0 = \chi_{\mathrm{reg}}.$$

Now assume we have a tower of number fields $\mathbb{K} \subset \mathbb{M} \subset \mathbb{L}$ all Galois over $\mathbb{K}$, and set $G = \mathrm{Gal}(\mathbb{L}/\mathbb{K})$ and $H = \mathrm{Gal}(\mathbb{L}/\mathbb{M})$, being identified with a subgroup of G. By direct computations,[67] Artin shows the following identity.

Theorem 7.45 (Inductivity) *With the above notation, we have for all characters χ_ρ of H and all $\sigma > 1$*

$$L\left(s, \chi_\rho, \mathbb{L}/\mathbb{M}\right) = L\left(s, \mathrm{Ind}_H^G \chi_\rho, \mathbb{L}/\mathbb{K}\right).$$

This result may be summarized in the following diagram, with $d = (\mathbb{M} : \mathbb{K})$.

$$\begin{array}{ccc} 1 \longrightarrow H = \mathrm{Gal}(\mathbb{L}/\mathbb{M}) & \longrightarrow & G = \mathrm{Gal}(\mathbb{L}/\mathbb{K}) \\ \downarrow{\rho} & & \downarrow{\mathrm{Ind}_H^G \rho} \\ GL(m, \mathbb{C}) & & GL(md, \mathbb{C}) \end{array}$$

As a corollary, let us apply this result with $\mathbb{M} = \mathbb{L}$ and $\chi_\rho = \chi_0$. On the one hand, by Example 7.38 and Remark 7.13, Theorem 7.45 yields

$$L\left(s, \chi_{\mathrm{reg}}, \mathbb{L}/\mathbb{K}\right) = \zeta_\mathbb{L}(s).$$

On the other hand, by (7.36) and Proposition 7.53, we derive

$$L\left(s, \chi_{\mathrm{reg}}, \mathbb{L}/\mathbb{K}\right) = \prod_{k=1}^{r} L(s, \chi_k, \mathbb{L}/\mathbb{K})^{\chi_k(1)}.$$

Since among the characters $\chi_1, \ldots, \chi_r$, there is the principal character χ_0 of G, we get the so-called *Artin factorization* generalizing Theorems 7.25 and 7.43.

---

[67]One can also use a result from representation theory, namely Mackey's theorem. See [90, Theorem 8.3.5].

Theorem 7.46 (Artin) *Let* $\mathbb{L}/\mathbb{K}$ *be a Galois extension of number fields. Then*

$$\zeta_\mathbb{L}(s) = \zeta_\mathbb{K}(s) \prod_{\chi_\rho \neq \chi_0} L(s, \chi_\rho, \mathbb{L}/\mathbb{K})^{\chi_\rho(1)},$$

where the product runs over non-principal irreducible characters of $\mathrm{Gal}(\mathbb{L}/\mathbb{K})$.

Example 7.39 Let $\mathbb{K}_3$ be a cubic field with discriminant $d_\mathbb{K} < 0$ and signature $(1, 1)$. Hence $\mathbb{K}_3$ is not Galois, and let $\mathbb{K}_6$ be a normal closure of $\mathbb{K}_3$ over $\mathbb{Q}$. Then $\mathbb{K}_6$ is a non-abelian extension of degree 6 of $\mathbb{Q}$ with $\mathrm{Gal}(\mathbb{K}_6/\mathbb{Q}) \simeq S_3$. By Theorem 7.46 and Example 7.37, we derive

$$\zeta_{\mathbb{K}_6}(s) = \zeta(s) L\left(s, \chi_\varphi, \mathbb{K}_6/\mathbb{Q}\right) L\left(s, \chi_\rho, \mathbb{K}_6/\mathbb{Q}\right)^2,$$

where χ_φ and χ_ρ are given in Table 7.1.

As the other L-functions defined above, the Artin L-function satisfies a functional equation. It involves several ingredients, the first of which being the *Artin conductor* f_χ, which is too complicated to define here and for which we refer to [66, Chapter VII, §11]. Next, we need an expression for the Euler factor at the Archimedean places of $\mathbb{K}$, which can be defined in the following shape. Set

$$\gamma\left(s, \chi_\rho, \mathbb{L}/\mathbb{K}\right) = \prod_v L_v\left(s, \chi_\rho, \mathbb{L}/\mathbb{K}\right),$$

where the product runs over all infinite places of $\mathbb{K}$, and the local Euler factor $L_v\left(s, \chi_\rho, \mathbb{L}/\mathbb{K}\right)$ is given by

$$L_v\left(s, \chi_\rho, \mathbb{L}/\mathbb{K}\right) = \begin{cases} \left\{2(2\pi)^{-s}\Gamma(s)\right\}^{\chi_\rho(1)}, & \text{if } v \text{ is complex}; \\ \left\{\pi^{-s/2}\Gamma\left(\frac{s}{2}\right)\right\}^a \left\{\pi^{-(s+1)/2}\Gamma\left(\frac{s+1}{2}\right)\right\}^b, & \text{if } v \text{ is real}; \end{cases}$$

where a and b satisfy $a + b = \chi_\rho(1)$.[68] Finally, set

$$A_\chi = |d_\mathbb{K}|^{\chi_\rho(1)} \mathcal{N}_{\mathbb{K}/\mathbb{Q}}\left(f_\chi\right).$$

---

[68]If w is an infinite place of $\mathbb{L}$ lying above v, and if ϕ_w is a generator of $G_w := \{g \in G : \rho_g \cdot w = w\}$, then $a = \frac{1}{2}\left(\chi_\rho(1) + \chi_\rho\left(\phi_w\right)\right)$ and $b = \frac{1}{2}\left(\chi_\rho(1) - \chi_\rho\left(\phi_w\right)\right)$. See [66, p. 535].

Theorem 7.47 (Functional Equation) *With the notation above, the com-*
pleted *Artin L-function*

$$\Lambda\left(s, \chi_\rho, \mathbb{L}/\mathbb{K}\right) = A_\chi^{s/2} \gamma\left(s, \chi_\rho, \mathbb{L}/\mathbb{K}\right) L\left(s, \chi_\rho, \mathbb{L}/\mathbb{K}\right)$$

admits a meromorphic continuation to $\mathbb{C}$, *analytic except for poles at* $s = 0$
and $s = 1$, *and satisfies the functional equation*

$$\Lambda\left(1 - s, \chi_\rho, \mathbb{L}/\mathbb{K}\right) = w\left(\chi_\rho\right) \Lambda\left(s, \overline{\chi_\rho}, \mathbb{L}/\mathbb{K}\right),$$

where $w\left(\chi_\rho\right)$ *is the "root number" satisfying* $\left|w\left(\chi_\rho\right)\right| = 1$.

7.5.11.8   The Great Conjectures

We start this section with the *Dedekind conjecture.*

Conjecture 7.2 (Dedekind) Let $\mathbb{L}/\mathbb{K}$ be a finite extension of number fields. Then
the quotient $\zeta_\mathbb{L}(s)/\zeta_\mathbb{K}(s)$ is entire.

This is still an open problem, but some particular cases are known.

Theorem 7.48 (Aramata–Brauer) *If the extension* $\mathbb{L}/\mathbb{K}$ *is Galois, then the*
Dedekind conjecture is true.

In the non-Galois case, we have the following result due to Uchida [94] and van der
Waall [96].

Theorem 7.49 (Uchida–van der Waall) *If* $\mathbb{L}$ *is contained in a* solvable *extension*
of $\mathbb{K}$, *then the Dedekind conjecture is true.*

We now turn to a conjecture generalizing Dedekind's, namely Artin's conjecture,
which may be stated as follows.

Conjecture 7.3 (Artin) For each non-principal irreducible character χ_ρ, the function $L\left(s, \chi_\rho, \mathbb{L}/\mathbb{K}\right)$ is entire.

It can be proved [66, Proposition VII.12.5] that, if $\deg \rho = 1$, then the corresponding
Artin *L*-function coincides with a Hecke *L*-function. It follows that Conjecture 7.3
is true for representations of degree one.

By definition, the function $\Lambda\left(s, \chi_\rho, \mathbb{L}/\mathbb{K}\right)$ has neither zero nor poles in the
half-plane $\sigma > 1$. From the functional equation of Theorem 7.47, we see that
$\Lambda\left(s, \chi_\rho, \mathbb{L}/\mathbb{K}\right)$ does not vanish in $\sigma < 0$. The poles in the half-plane $\sigma < 0$ of the
analytic continuation of the Gamma-factor must be offset by the zeros of the Artin
L-function. Therefore the function $L\left(s, \chi_\rho, \mathbb{L}/\mathbb{K}\right)$ has *trivial zeros* in $s = -2m$ or
$s = -2m - 1$ for $m \in \mathbb{Z}_{\geq 0}$. Also, the only poles and zeros that the completed
Artin *L*-function can have are in the strip $0 \leq \sigma \leq 1$, and since it has only a

possible pole at $s = 0$ in this range, it follows that the functions $\Lambda\left(s, \chi_\rho, \mathbb{L}/\mathbb{K}\right)$ and $L\left(s, \chi_\rho, \mathbb{L}/\mathbb{K}\right)$ have the same poles and zeros in the "critical" strip $0 < \sigma < 1$. As for the Riemann zeta-function, this leads to the following conjecture, named *Grand Riemann Hypothesis* and abbreviated in GRH.

Conjecture 7.4 (GRH) The non-trivial zeros of the Artin L-functions are on the "critical" line $\sigma = \frac{1}{2}$.

7.5.11.9 Application

There is quite a lot of work to deal with Conjectures 7.3 and 7.4, so that we systematically assume these two conjectures in this section. As an application, we go back to the Chebotarëv's density theorem, for which assuming the above conjectures enables us to considerably improve the error term.

Let $\mathbb{L}/\mathbb{Q}$ be a finite Galois extension, $G = \mathrm{Gal}\,(\mathbb{L}/\mathbb{Q})$ and let M be the product of the primes dividing $d_\mathbb{L}$. For any class function f on G, set

$$\pi(x, f) = \sum_{\substack{p \leqslant x \\ p \nmid M}} f\left(\phi_p\right),$$

where, as usual, ϕ_p is the conjugacy class of the Frobenius element associated with p in G. Also define

$$m(f) = \frac{1}{|G|} \sum_{g \in G} f(g)$$

the mean value of f over G. Let $\widehat{G}$ be the set of equivalence classes of irreducible complex representations of G, and let $\widehat{f} : \widehat{G} \to \mathbb{C}$ be the *discrete Fourier transform* of f given by

$$\widehat{f}(\rho) = \langle f, \chi_\rho \rangle = \frac{1}{|G|} \sum_{g \in G} f(g)\,\overline{\chi}_\rho(g).$$

Definition 7.43 (Littlewood Complexity) With the notation above, the number

$$\lambda(f) := \sum_{\rho \in \widehat{G}} \left|\widehat{f}(\rho)\right| \deg \rho$$

is called the *Littlewood norm*. When $f = 1_D$ is the characteristic function of a subset $D \subseteq G$, it is denoted by $\lambda(D)$ and is named the *Littlewood complexity* of D.

It is no easy task to give good estimates of $\lambda(f)$ in general. The trivial bound is $|\lambda(f)| \leqslant |G| \|f\|_1$.[69] This bound can be significantly improved by using the Cauchy–Schwarz inequality, yielding

$$|\lambda(f)| \leqslant |G|^{1/2} \|f\|_2.$$

Indeed, by Cauchy–Schwarz, Parseval identity and (7.37)

$$|\lambda(f)| \leqslant \left(\sum_{\rho \in \widehat{G}} |\widehat{f}(\rho)|^2\right)^{1/2} \left(\sum_{\rho \in \widehat{G}} (\deg \rho)^2\right)^{1/2} = |G|^{1/2} \|f\|_2$$

as claimed. When $f = \mathbf{1}_D$, this, respectively, gives $|\lambda(D)| \leqslant |D|$ and $|\lambda(D)| \leqslant |D|^{1/2}$. Note that this Littlewood complexity may sometimes be evaluated. For instance, if $G = S_n$, then for any $D \subset S_n$, one has $\lambda(D) = n^{-1}2^{n-1}$ [10, proposition 10].

In [10], the author provides the following effective version of the Chebotarëv's density theorem, assuming Conjectures 7.4 and 7.3, the latter being adapted to the representations ρ of G such that $\widehat{f}(\rho) \neq 0$.

Theorem 7.50 *Let* $\mathbb{L}/\mathbb{Q}$ *be a finite Galois extension,* $G = \mathrm{Gal}\,(\mathbb{L}/\mathbb{Q})$, M *be the product of the primes dividing* $d_{\mathbb{L}}$ *and* f *be a class function on* G. *Assuming Conjectures* 7.3 *and* 7.4, *there exists an absolute constant* $c_0 > 0$ *such that, for all* $x \geqslant 3$

$$|\pi(x, f) - m(f)\pi(x)| < c_0\, x^{1/2}\lambda(f) \log\,(xM|G|).$$

Let us state two nice applications of this result given by Bellaïche in [10].

Corollary 7.12 *Let* $P \in \mathbb{Z}[X]$ *be a monic, irreducible polynomial of degree* n, *and let* M *be the product of the primes dividing* $\mathrm{disc}(P)$. *Assume Conjectures* 7.3 *and* 7.4, *then*

▷ *there exists a prime* $p \ll n^2 (\log M + n \log n)^2$, $p \nmid M$, *such that* P *has at least one root modulo* p;
▷ *there exists a prime* $p \ll n^4 (\log M + n \log n)^2$ *such that* P *has at least two roots modulo* p;
▷ *there exists a prime* $p \ll n^4 (\log M + n \log n)^2$ *such that* P *has no root modulo* p.

---

[69]The usual norms on G are $\|f\|_k := \left(|G|^{-1} \sum_{g \in G} |f(g)|^k\right)^{1/k}$ for $k \in \{1, 2\}$.

Note that the bounds above are polynomials in n, whereas they would be in the shape $n(n!)^a \log n$ with $a \in \{1, 2\}$, i.e. super-exponential bounds, if we used previous results. As the author points out in [10, p. 42], some work by Weinberger and Adleman-Odlyzko have already obtained polynomial bounds for the first case of the corollary, using Theorem 7.26 in the field $\mathbb{Q}[X]/(P)$, which is of degree n, instead of a splitting field of P which can be of degree $n!$, as is the case in the proof of this corollary. Notice that the former field need not be Galois, while the latter is always a normal extension of $\mathbb{Q}$.

As a second application, let us turn to the field of elliptic curves, already seen in Sect. 7.5.2.4 from which we take the notation.

Certain public key cryptographic systems, based on intractability of the discrete logarithm problem, can be implemented using the group of points on an elliptic curve E defined over $\mathbb{F}_p$. In that case one wants the cyclic subgroup generated by certain points to have order divisible by a large prime. In [49], Koblitz fixed an elliptic curve E and tried to choose a prime finite field $\mathbb{F}_p$ such that $|E(\mathbb{F}_p)|$ is prime. This leads to the following natural definition. Let E be a non-CM-elliptic curve over a field $\mathbb{K}$ and set

$$\pi_E(x) := \sum_{\substack{p \leqslant x \\ p \nmid \mathrm{disc}(E) \\ |E(\mathbb{F}_p)| \text{ is prime}}} 1.$$

Let us specify two points.

▷ An *isogeny* between two elliptic curves E_1 and E_2 is a holomorphic $\mathbb{Z}$-linear map F of algebraic curves from E_1 to E_2 sending the zero element of E_1 to the zero element of E_2. If $\mathbb{K}$ is algebraically closed, then F is surjective and preserves the law of the elliptic curves.

▷ An elliptic curve E has *complex multiplication* if E has an endomorphism ring[70] strictly larger than $\mathbb{Z}$. If char $\mathbb{K} = 0$, the CM-elliptic curves have an endomorphism ring, which is isomorphic to an order in an imaginary quadratic field.

In [49, Conjecture A], Koblitz then proposed the following conjecture.

Conjecture 7.5 (Koblitz) Let E be an elliptic curve of discriminant disc(E) defined over $\mathbb{Z}$, which is not $\mathbb{Q}$-isogenous to a curve with non-trivial $\mathbb{Q}$-torsion and which does not have complex multiplication. Then there exists a constant $C_E > 0$ such that, as $x \to \infty$

$$\pi_E(x) \sim C_E \frac{x}{(\log x)^2}.$$

---

[70]This is the ring of all endomorphisms of E, including those defined over extensions of the base field of E.

As a consequence of Theorem 7.50 and some large sieve techniques, Bellaïche [10] proved the next estimate.

Corollary 7.13 *With the same hypotheses as in Conjecture 7.5 and assuming Conjectures 7.3 and 7.4, we have as $x \to \infty$*

$$\pi_E(x) < C_E \ (8 + o(1)) \ \frac{x}{(\log x)^2}.$$

7.5.12   Analytic Methods

7.5.12.1   Ideal Classes

Let $\mathbb{K}/\mathbb{Q}$ be an algebraic number field of degree n, signature (r_1, r_2), discriminant $d_{\mathbb{K}}$, and let $\zeta_{\mathbb{K}}(s)$ be its Dedekind zeta-function. Recall that

$$\Gamma_{\mathbb{K}}(s) = \Gamma\left(\tfrac{1}{2}s\right)^{r_1} \Gamma(s)^{r_2} \quad \text{and} \quad A_{\mathbb{K}} = 2^{-r_2} \pi^{-n/2} |d_{\mathbb{K}}|^{1/2},$$

where $\Gamma(s)$ is the usual Gamma-function. We have previously seen that bounds for class numbers may require the computation of a minimal positive real number $b_{\mathbb{K}}$ such that every ideal class contains a non-zero integral ideal $\mathfrak{a}$ satisfying $N_{\mathbb{K}/\mathbb{Q}}(\mathfrak{a}) \leqslant b_{\mathbb{K}}$. From the work of Minkowski who created and used deep results from the geometry of numbers, we know that $b_{\mathbb{K}} = M_{\mathbb{K}} |d_{\mathbb{K}}|^{1/2}$ is admissible, where the Minkowski constant $M_{\mathbb{K}}$ is defined in (7.17). In the 1950s, Rogers, and later Mulholland, improved on Minkowski's bound using essentially the same ideas.

Stark's work as previously seen in Proposition 7.32 was the starting point of the resurgence of analytic methods in algebraic number theory. In the late 1970s, using the functional equation, given below, of the partial Dedekind zeta-function associated with a given ideal class and its conjugate class, Zimmert [101] succeeded in establishing an inequality which supersedes the previous results. The *conjugate class* $C^\star$ of an ideal class C is defined by $CC^\star = \mathcal{D}$, where $\mathcal{D}$ is the class of the different[71] of $O_{\mathbb{K}}$.

---

[71] The *different* of $O_{\mathbb{K}}$ is the integral ideal $\mathcal{D}_{\mathbb{K}/\mathbb{Q}}$ defined as the inverse of the fractional ideal

$$\{y \in \mathbb{K} : \mathrm{Tr}_{\mathbb{K}/\mathbb{Q}}(xy) \in \mathbb{Z} \text{ for all } x \in O_{\mathbb{K}}\}$$

called the *codifferent*. It can be proved that the ramified prime ideals of $O_{\mathbb{K}}$ are the prime ideals dividing $\mathcal{D}_{\mathbb{K}/\mathbb{Q}}$ and $N_{\mathbb{K}/\mathbb{Q}}\left(\mathcal{D}_{\mathbb{K}/\mathbb{Q}}\right) = |d_{\mathbb{K}}|$. In fact, much more is true. Indeed, let p be a prime number and $\mathfrak{p}$ be a prime ideal above p. If e is the ramification index $e(\mathfrak{p} \mid p)$, then $\mathfrak{p}^{e-1} \mid \mathcal{D}_{\mathbb{K}/\mathbb{Q}}$.

Let $\zeta(s, C)$ be the *partial Dedekind zeta-function* associated with the ideal class C defined by

$$\zeta(s, C) = \sum_{\mathfrak{a} \in C} \frac{1}{N_{K/\mathbb{Q}}(\mathfrak{a})^s}.$$

As for $\zeta_K(s)$, this function has a functional equation of a similar form.[72]

Proposition 7.54 *Define* $\xi_K(s, C) = A_K^s \Gamma_K(s) \zeta(s, C)$. *Then this function is analytic except for simple poles at $s = 0$, $s = 1$, and*

$$\xi_K(s, C) = \xi_K(1 - s, C^\star).$$

Zimmert's ideas rest on the next general result dealing with Dirichlet series with non-negative coefficients.

Proposition 7.55 (Zimmert) *Let* $F(s) = L(s, f)$ *and* $G(s) = L(s, g)$ *be two Dirichlet series with finite abscissas of convergence and such that $f(n), g(n) \geqslant 0$. Also assume that*

▷ *F and G can be extended analytically in the whole complex plane to meromorphic functions having a simple pole at $s = 1$.*
▷ *They satisfy the functional equation $\xi_F(s) = \xi_G(1 - s)$, where*

$$\xi_F(s) = \left(2^{-b} A\right)^s \Gamma\left(\tfrac{1}{2}s\right)^{a-b} \Gamma(s)^b F(s)$$

and similarly for $\xi_G(s)$, for some positive real number A and integers $a \geqslant b \geqslant 0$.
▷ *The function $s(s - 1)\xi_F(s)$ is entire of order 1.*
▷ $\operatorname*{Res}_{s=1}(F(s)) = \operatorname*{Res}_{s=1}(G(s)) = \kappa.$

Then, for all real numbers $x > 0$, $\sigma > 1$ and $0 \leqslant \alpha < \sigma - 1$

$$G(\sigma)\left\{\log x - 2\frac{\xi_{G'}}{\xi_G}(\sigma) + \frac{G'}{G}(\sigma) - \frac{2}{\sigma - \alpha - 1}\right\}$$

$$\leqslant 2^{1-b} \pi^{(a-b)/2} A\kappa \left\{x^\sigma \frac{G(2\sigma)}{\xi_G(2\sigma)} t(\sigma, \alpha) - x^{\sigma-1} \frac{G(2\sigma - 1)}{\xi_G(2\sigma - 1)} t(\sigma - 1, \alpha - 1)\right\},$$

$$(7.38)$$

where

$$t(\sigma, \alpha) = \frac{\alpha + 1}{\sigma^2(2\sigma - \alpha - 1)}.$$

---

[72] See [55, 66].

Sketch of the Proof The starting point is the inequality

$$\frac{1}{2\pi i}\int_{2-i\infty}^{2+i\infty}\frac{\xi_F(z)F(z+2\sigma-1)R_{\alpha,\sigma}(z)}{\xi_F(z+2\sigma-1)}x^z dz \geq 0 \qquad (7.39)$$

valid for all $x > 0$, $\sigma > 1$, and where $R_{\alpha,\sigma}(z)$ is a certain rational function of z depending on the parameters α and σ, and satisfying $R_{\alpha,\sigma}(x) > 0$ for all $x > 0$. By the four hypotheses Proposition 7.55, one may shift the line of integration of (7.39) from $\mathrm{Re}\, z = 2$ to $\mathrm{Re}\, z = 1 - \sigma$, and picking up the residues at $z = 0$, $z = 1$ and at the poles of $R_{\alpha,\sigma}(z)$, and using the functional equation above, Theorem 3.3 and estimates for $\zeta_{\mathbb{K}}(s, C)$ in the spirit of Theorem 7.14, the inequality (7.38) follows after some tedious computations.   □

We may apply (7.38) with

$$F(s) = \zeta\left(s, C^\star\right) \quad \text{and} \quad G(s) = \zeta\left(s, C\right)$$

so that $a = r_1 + r_2$, $b = r_2$ and $A = 2^b A_{\mathbb{K}}$, and choosing

$$x = \frac{G(2\sigma-1)\xi_G(2\sigma)}{\xi_G(2\sigma-1)G(2\sigma)}\times\frac{t(\sigma-1,\alpha-1)}{t(\sigma,\alpha)} = A_{\mathbb{K}}\times\frac{\Gamma_{\mathbb{K}}(2\sigma)}{\Gamma_{\mathbb{K}}(2\sigma-1)}\times\frac{t(\sigma-1,\alpha-1)}{t(\sigma,\alpha)}$$

we derive

$$\log A_{\mathbb{K}}+\frac{G'}{G}(\sigma) \leq 2\frac{\xi_{G'}}{\xi_G}(\sigma)-\log\frac{\Gamma_{\mathbb{K}}(2\sigma)}{\Gamma_{\mathbb{K}}(2\sigma-1)}-\log\frac{t(\sigma-1,\alpha-1)}{t(\sigma,\alpha)}+\frac{2}{\sigma-\alpha-1}.$$

Note that

$$\frac{\xi_{G'}}{\xi_G}(\sigma) = \log A_{\mathbb{K}} + \frac{\Gamma'_{\mathbb{K}}}{\Gamma_{\mathbb{K}}}(\sigma) + \frac{G'}{G}(\sigma),$$

and hence we obtain

$$-2\frac{\Gamma'_{\mathbb{K}}}{\Gamma_{\mathbb{K}}}(\sigma)+\log\frac{\Gamma_{\mathbb{K}}(2\sigma)}{\Gamma_{\mathbb{K}}(2\sigma-1)}+\log\frac{t(\sigma-1,\alpha-1)}{t(\sigma,\alpha)}-\frac{2}{\sigma-\alpha-1} \leq \log A_{\mathbb{K}}+\frac{G'}{G}(\sigma).$$

Now let $\mathfrak{a}_0$ be an ideal in C with the smallest norm. Then for all $\sigma > 1$ we have

$$G'(\sigma) = -\sum_{\mathfrak{a}\in C}\frac{\log \mathcal{N}_{\mathbb{K}/\mathbb{Q}}(\mathfrak{a})}{\mathcal{N}_{\mathbb{K}/\mathbb{Q}}(\mathfrak{a})^\sigma} \leq -G(\sigma)\log \mathcal{N}_{\mathbb{K}/\mathbb{Q}}(\mathfrak{a}_0)$$

and therefore

$$
\log N_{K/\mathbb{Q}}(\mathfrak{a}_0) \leqslant \log A_K + 2 \frac{\Gamma'_K}{\Gamma_K}(\sigma) - \log \frac{\Gamma_K(2\sigma)}{\Gamma_K(2\sigma - 1)}
$$
$$
- \log \frac{t(\sigma - 1, \alpha - 1)}{t(\sigma, \alpha)} + \frac{2}{\sigma - \alpha - 1}.
$$

Choosing α optimally implies Zimmert's theorem in the following shape.

Theorem 7.51 (Zimmert) *Let* $\sigma > 1$ *be a real number. Then every ideal class contains a non-zero integral ideal* $\mathfrak{a}$ *such that*

$$
N_{K/\mathbb{Q}}(\mathfrak{a}) \leqslant |d_K|^{1/2} \exp \left\{ 2 \frac{\Gamma'_K}{\Gamma_K}(\sigma) - \log \frac{\Gamma_K(2\sigma)}{\Gamma_K(2\sigma - 1)} - \log \left(2^{r_2} \pi^{n/2} \right) + r(\sigma) \right\},
$$

where

$$
r(\sigma) = \frac{2s(\sigma)}{\sigma(\sigma - 1)} + 2\log \left(1 - \frac{1}{\sigma} \right) + \log \left(\frac{(s(\sigma) + \sigma)(s(\sigma) - \sigma + 1)}{(s(\sigma) - \sigma)(s(\sigma) + \sigma - 1)} \right)
$$

and $s(\sigma) = \sqrt{3\sigma^2 - 3\sigma + 1}$.

Example 7.40 The following table supplies numbers $C_K > 1$ such that every ideal class contains a non-zero integral ideal $\mathfrak{a}$ such that

$$
N_{K/\mathbb{Q}}(\mathfrak{a}) \leqslant C_K^{-1} |d_K|^{1/2}.
$$

The signatures $(2, 0)$ and $(0, 1)$ are due to Gauss and one has to suppose $d_K \geqslant 8$ for the real quadratic field case.

(r_1, r_2)	$(2, 0)$	$(0, 1)$	$(0, 2)$	$(1, 1)$	$(3, 0)$	$(2, 1)$	$(3, 1)$	$(4, 0)$
C_K	$\sqrt{8}$	$\sqrt{3}$	6.792	3.355	4.636	9.749	32.12	14.45

Choosing $\sigma = 1 + 2n^{-1/2}$, Theorem 7.51 implies that every ideal class contains a non-zero integral ideal $\mathfrak{a}$ such that

$$
N_{K/\mathbb{Q}}(\mathfrak{a}) \leqslant |d_K|^{1/2} e^{-r_1 (\gamma + \log 4) - 2r_2 \left(\gamma + \log \sqrt{2\pi} \right) + O(\sqrt{n})}
$$

as $n \to \infty$.

It can be checked [26, (4.1)] that the inequality (7.39) still holds if $R_{\alpha,\sigma}(z)$ belongs to the class of rational functions defined by

$$R(z) = \left(\frac{z}{z+2\sigma-1}\right)^a \left(\frac{z+1}{z+2\sigma}\right)^b \sum_{k=0}^m a_k \prod_{j=0}^{r_k} \frac{1}{z+b_{jk}}$$

with $\sigma > 1$, $m \in \mathbb{Z}_{\geqslant 0}$, $a_k, b_{jk} \geqslant 0$ and $r_k \in \mathbb{Z}_{\geqslant 1}$. In [19], the author refined Zimmert's results for small degrees by taking another rational function in this class.

7.5.12.2 Regulator

As can be seen in (7.23) or Proposition 7.23, a lower bound for $\mathcal{R}_{\mathbb{K}}$ is needed to get a good estimate for the class number. In the 1930s, Remak [73] provided such lower bounds depending on the degree n of $\mathbb{K}$ when this field is totally real, i.e. $r_1 = n$. Using the geometry of numbers, he proved that there exist constants $c_0 > 0$ and $c_1 > 1$ such that $\mathcal{R}_{\mathbb{K}} \geqslant c_0 c_1^n$. This was later improved by Pohst [68] who showed that, if $\mathbb{K}$ is totally real, then

$$\mathcal{R}_{\mathbb{K}} > n^{-1/2} \Gamma\left(\tfrac{1}{2}(n+3)\right)^{-1} \left(\sqrt{\tfrac{1}{2}n\pi} \log \Phi\right)^{n-1},$$

where $\Phi = \tfrac{1}{2}\left(1 + \sqrt{5}\right)$ is the golden ratio.

In view of (7.23), one may ask for some bounds depending on the discriminant. This was first achieved in the early 1950s by Remak [74] who proved that if $\mathbb{K}$ *is not a totally complex quadratic extension of a totally real field,*[73] then there exists a constant $c_n > 0$ such that

$$\mathcal{R}_{\mathbb{K}} \geqslant c_n \log \frac{|d_{\mathbb{K}}|}{n^n}$$

and for algebraic number fields without proper subfields, he showed that there exists a constant $C_n > 0$ such that

$$\mathcal{R}_{\mathbb{K}} \geqslant C_n \left(\log \frac{|d_{\mathbb{K}}|}{n^n}\right)^r,$$

where $r = r_1 + r_2 - 1$ is the Dirichlet rank of $O_{\mathbb{K}}^*$. This result was generalized independently by Silverman [80], Friedman [26] and Uchida [95], still using the geometry of numbers. Their main theorems are summarized in the next result.

---

[73] Such fields are usually called CM-fields.

Theorem 7.52 *Let $\mathbb{K}$ be an algebraic number field of degree $n \geqslant 2$, signature (r_1, r_2), discriminant $d_{\mathbb{K}}$, regulator $\mathcal{R}_{\mathbb{K}}$, and let $r = r_1 + r_2 - 1$ be the Dirichlet rank of $O_{\mathbb{K}}^*$.*

▷ *(Silverman–Friedman). Let ρ be the maximal Dirichlet rank of the unit groups of all proper subfields of $\mathbb{K}$. Then there exists an effectively computable absolute constant $c > 0$ such that*

$$\mathcal{R}_{\mathbb{K}} > \frac{c}{n^{2n}} \left(\log \frac{|d_{\mathbb{K}}|}{n^n} \right)^{r-\rho}.$$

▷ *(Uchida). Let $\alpha > 1$ and $\mathbb{F}$ be a maximal subfield of $\mathbb{K}$ such that $|d_{\mathbb{F}}| < |d_{\mathbb{K}}|^{m^{-\alpha}}$, where $m = [\mathbb{K} : \mathbb{F}]$. If λ is the Dirichlet rank of $O_{\mathbb{F}}^*$, then there exist constants $c_n, d_n > 0$ depending only on n such that*

$$\mathcal{R}_{\mathbb{K}} > c_n (\log d_n |d_{\mathbb{K}}|)^{r-\lambda}.$$

Sharpening the constants in the above results is a very difficult problem as n grows. However, for small degrees, there exist some optimal results. For $n = 2$, we have the following easy lemma.

Lemma 7.30 *Let $d \geqslant 5$ be a squarefree number and $\mathbb{K} = \mathbb{Q}\left(\sqrt{d}\right)$ be a real quadratic field. Then*

$$\mathcal{R}_{\mathbb{K}} \geqslant \log\left(\tfrac{1}{2} \left(\sqrt{d-4} + \sqrt{d} \right) \right).$$

The equality occurs when $d = 5$.

Proof If $\varepsilon_d = \tfrac{1}{2}\left(u_1 + v_1\sqrt{d} \right)$ is the fundamental unit of $\mathbb{K}$, then ε_d is equal to the fundamental solution of the equation $u^2 - dv^2 = \pm 4$, and hence

$$\varepsilon_d = \tfrac{1}{2}\left(u_1 + v_1\sqrt{d} \right) = \tfrac{1}{2}\left(\sqrt{dv_1^2 - 4} + v_1\sqrt{d} \right) \geqslant \tfrac{1}{2}\left(\sqrt{d-4} + \sqrt{d} \right)$$

implying the asserted result since $\mathcal{R}_{\mathbb{K}} = \log \varepsilon_d$. $\qquad\square$

For cubic and quartic fields, Cusick [18] showed the following lower bounds:

$[\mathbb{K} : \mathbb{Q}]$	(r_1, r_2)	Lower bound $\mathcal{R}_\mathbb{K} \geqslant \dots$
3	$(3, 0)$	$\frac{1}{16}\left(\log \frac{d_\mathbb{K}}{4}\right)^2$
3	$(1, 1)$	$\frac{1}{3}\log \frac{|d_\mathbb{K}|}{27}$
4	$(4, 0)$	$\frac{1}{80\sqrt{10}}\left(\log \frac{d_\mathbb{K}}{16}\right)^3$
4	$(0, 2)$	$\frac{1}{4}\log \frac{d_\mathbb{K}}{256}$

Finally, for a quintic cyclic number field $\mathbb{K}$, Schoof and Washington [79] proved that

$$\mathcal{R}_\mathbb{K} \geqslant \frac{1}{6400}\left(\log \frac{d_\mathbb{K}}{16}\right)^4.$$

The analytic tools can also be used to derive lower bounds for the regulator. Using (7.38) applied with $\alpha = 0$ and G the partial Dedekind zeta-function associated with the trivial class, Zimmert [101] was able to prove that, for each algebraic number field $\mathbb{K}$ with signature (r_1, r_2) and number of roots of unity $w_\mathbb{K}$, we have for all $\sigma > 1$

$$\frac{\mathcal{R}_\mathbb{K}}{w_\mathbb{K}} \geqslant 2^{-(r_1+1)}\sigma\,(2\sigma - 1)\,\Gamma_\mathbb{K}(2\sigma)\exp\left(-2\sigma\frac{\Gamma'_\mathbb{K}}{\Gamma_\mathbb{K}}(\sigma) - \frac{3\sigma - 1}{\sigma - 1}\right),$$

and he chose $\sigma = 2$ to get

$$\frac{\mathcal{R}_\mathbb{K}}{w_\mathbb{K}} \geqslant 0.02\, e^{0.46 r_1 + 0.2 r_2}.$$

This implies in particular that the smallest regulator of an algebraic number field is $\geqslant 0.056$ and that the smallest regulator of a totally real algebraic number field is $\log \Phi$, i.e. the regulator of $\mathbb{Q}\left(\sqrt{5}\right)$.

Zimmert's ideas were later extended by Friedman and Skoruppa. The former showed in [26] that the ratio $\mathcal{R}_\mathbb{K}/w_\mathbb{K}$ can be expressed as a sum of rapidly convergent series

$$\frac{\mathcal{R}_\mathbb{K}}{w_\mathbb{K}} = \sum_\mathfrak{a} f\left(\frac{N_{\mathbb{K}/\mathbb{Q}}(\mathfrak{a})^2}{|d_\mathbb{K}|}\right) + \sum_\mathfrak{b} f\left(\frac{N_{\mathbb{K}/\mathbb{Q}}(\mathfrak{b})^2}{|d_\mathbb{K}|}\right), \tag{7.40}$$

where $\mathfrak{a}$ runs through the principal ideals of $O_\mathbb{K}$, $\mathfrak{b}$ runs through the integral ideals in the ideal class of the different of $O_\mathbb{K}$ and $f : (0, \infty) \to \mathbb{R}$ is defined by

$$f(x) = \frac{2^{-(r_1+1)}}{2\pi i} \int_{2-i\infty}^{2+i\infty} \xi_\mathbb{K}(s)\zeta_\mathbb{K}(s)^{-1} (x|d_\mathbb{K}|)^{-s/2} (2s-1)\, ds.$$

This function is C^∞, takes both positive and negative values, and, for all $r_1 + r_2 \geqslant 2$, we have

$$\lim_{x \to 0^+} f(x) = -\infty.$$

However, studying more carefully the behaviour of f, Friedman [27] proved that $f(x)$ has a single simple zero and that this conclusion still holds for all derivatives $f^{(k)}$ of f. This implies that

$$\frac{\mathcal{R}_\mathbb{K}}{w_\mathbb{K}} \geqslant f\left(\frac{1}{|d_\mathbb{K}|}\right).$$

Indeed, assume that $f\left(|d_\mathbb{K}|^{-1}\right) > 0$, otherwise the inequality is trivial, then by the previous observations we infer that $f(x) > 0$ as long as $x \geqslant |d_\mathbb{K}|^{-1}$, and the result follows by dropping out all terms in (7.40) except the one corresponding to the ideal $(1) = O_\mathbb{K}$.

Friedman pointed out that this bound is useful only when the discriminant lies in a certain range, otherwise Zimmert's bound gives better results. For instance, let $\mathbb{K}$ be the totally complex number field of degree 36 with discriminant $d_\mathbb{K} = 3^{18} \times 4057^9$ taken from [14, § 12.2.2]. Friedman's result provides $\mathcal{R}_\mathbb{K} w_\mathbb{K}^{-1} > 27839$, whereas Zimmert's bound only gives[74] $\mathcal{R}_\mathbb{K} w_\mathbb{K}^{-1} > 121$, but we also have the useless bound $f\left(10^{-17}d_\mathbb{K}^{-1}\right) \approx -3.6 \times 10^7$.

---

[74] PARI/GP provides $\mathcal{R}_\mathbb{K} w_\mathbb{K}^{-1} \approx 172495$.

Exercises

129 Show that 3, 7, $1 \pm 2\sqrt{-5}$ are irreducibles in $\mathbb{Z}[\sqrt{-5}]$.

130 Show that $\theta = \sqrt[3]{2} + \sqrt[5]{2}$ is algebraic over $\mathbb{Q}$.

131 Let $P = X^3 - X + 1$ and α be a root of P. Set $\beta = \left(2\alpha^2 - 3\alpha + 2\right)^{-1}$. Determine the minimal polynomial of β.

132 Show that $\mathbb{Q}\left(\sqrt{5}, \sqrt[4]{2}\right) = \mathbb{Q}\left(\sqrt{5} + \sqrt[4]{2}\right)$. Deduce $\left[\mathbb{Q}\left(\sqrt{5}, \sqrt[4]{2}\right) : \mathbb{Q}\right]$.

133 (Ore's irreducibility criterion). In this exercise, we intend to show the following criterion due to Ore.

Proposition 7.56 (Ore) *Let $P \in \mathbb{Z}[X]$ be of degree n. If there exist at least $n + 5$ integers m such that $|P(m)|$ is 1 or a prime number, then P is irreducible over $\mathbb{Z}$.*

(a) Show that, if $Q \in \mathbb{Z}[X]$ is of degree $d \geqslant 1$, then there exist at most $d + 2$ integers m such that $Q(m) = \pm 1$.

(b) Proceeding as in Exercise 41, finalize the proof of Proposition 7.56.

(c) Is the polynomial $P = X^6 + 8X^5 + 22X^4 + 22X^3 + 5X^2 + 6X + 1$ irreducible over $\mathbb{Z}$?

134 Let $P \in \mathbb{Z}[X]$ be a *monic* polynomial for which $P(0) \neq 0$. Suppose further that P has exactly one root α with multiplicity 1 such that $|\alpha| \geqslant 1$. Prove that P is irreducible over $\mathbb{Z}$.

135 With the help of *Rouché's theorem*,[75] prove that

$$P = X^{16} - 8X^{15} - 4X^{14} - 2X^{13} - \sum_{i=0}^{12} X^i$$

is irreducible over $\mathbb{Z}$.

136 (Schur). Let $n \in \mathbb{Z}_{\geqslant 0}$ and set

$$P_n = \frac{X^n}{n!} + \frac{X^{n-1}}{(n-1)!} + \cdots + X + 1.$$

The aim of this exercise is to show the following result due to Schur.

Proposition 7.57 (Schur) *P_n is irreducible over $\mathbb{Q}$.*

---

[75]**Lemma 7.31 (Rouché)** *Let $P, Q \in \mathbb{C}[X]$ such that, for all $z \in \mathbb{C}$ such that $|z| = 1$, the strict inequality $|P(z) + Q(z)| < |P(z)| + |Q(z)|$ holds. Then P and Q have the same total number of roots, counting multiplicity, in the open disc $\{z \in \mathbb{C} : |z| < 1\}$.*

The proof will make use of the following generalization of Bertrand's postulate seen in Corollary 3.1.[76]

Lemma 7.32 *Let* $k \leqslant \ell$ *be positive integers. At least one of the numbers* $\ell + 1, \ldots, \ell + k$ *is divisible by a prime number* $p > k$.

Suppose that $F_n = n! P_n$ is reducible over $\mathbb{Q}$. Since F_n is monic, it must have an irreducible monic factor $A_m \in \mathbb{Z}[X]$ of degree $m \leqslant \frac{1}{2}n$ defined by

$$A_m = X^m + a_{m-1} X^{m-1} + \cdots + a_0.$$

1. Prove that each prime divisor of $n(n-1) \cdots (n - m + 1)$ divides a_0.
2. Let p be a prime factor of a_0 and θ be a root of A_m. Define $\mathbb{K} = \mathbb{Q}(\theta)$ so that $[\mathbb{K} : \mathbb{Q}] = m$.

   a. Prove that there is some prime ideal $\mathfrak{p}$ lying above p and dividing (θ).
   b. Write $(\theta) = \mathfrak{p}^\alpha \mathfrak{a}$ and $(p) = \mathfrak{p}^e \mathfrak{b}$ for some $\alpha, e \in \mathbb{N}$ and integral ideals $\mathfrak{a}$ and $\mathfrak{b}$ not divisible by $\mathfrak{p}$. Using the fact that $F_n(\theta) = 0$, prove that there exists an integer $k \in \{1, \ldots, n\}$ such that $v_\mathfrak{p}\left(n! \theta^k / k!\right) \leqslant v_\mathfrak{p}(n!)$, and using the identity $v_\mathfrak{p}(r!) = e v_p(r!)$, deduce that $k\alpha \leqslant e v_p(k!)$.
   c. Using Exercise 36, deduce that $p \leqslant m$.

3. Conclude the proof of Schur's theorem with the help of Lemma 7.32.

137 │   This exercise provides a proof of Proposition 7.8 from which we take the notation. We set $M = \mathbb{Z} + \mathbb{Z}\theta + \cdots + \mathbb{Z}\theta^{n-1}$.

(a) Show that $\theta^n / p \in \mathcal{O}_\mathbb{K}$ and that $N_{\mathbb{K}/\mathbb{Q}}(\theta) \not\equiv 0 \pmod{p^2}$. In what follows, we suppose that $p \mid f$.
(b) Explain why there exist $\alpha \in \mathcal{O}_\mathbb{K} \setminus M$ and $b_0, \ldots, b_{n-1} \in \mathbb{Z}$ not all divisible by p and such that $p\alpha = b_0 + b_1\theta + \cdots + b_{n-1}\theta^{n-1}$.
(c) Let j be the smallest index such that $p \nmid b_j$ and set

$$\beta = \alpha - \frac{1}{p}\left(b_0 + b_1\theta + \cdots + b_{j-1}\theta^{j-1}\right).$$

Show that $\beta \in \mathcal{O}_\mathbb{K}$ and deduce that $\frac{1}{p}(b_j\theta^{n-1}) \in \mathcal{O}_\mathbb{K}$. Prove the contradiction by taking the norm of this number.

138 │   Let $P = X^3 - 189X + 756$ and θ be a root of P.

(a) Show that P is irreducible over $\mathbb{Z}$, and let $\mathbb{K} = \mathbb{Q}(\theta)$ be the corresponding algebraic number field.
(b) Determine an integral basis for $\mathbb{K}$.

---

[76] See [22] for instance.

(c) Compute the Galois group $\mathrm{Gal}(\mathbb{K}/\mathbb{Q})$ and the regulator $\mathcal{R}_\mathbb{K}$ of $\mathbb{K}$.

139 Factorize (3) and the ideal $\mathfrak{a} = (1 + 2\sqrt{-2})$ in $\mathbb{K} = \mathbb{Q}(\sqrt{-2})$, and deduce that there exists no non-negative integer n such that $(1 + 2\sqrt{-2})^n = 3^n$ and that $\frac{1}{\pi} \arccos\left(\frac{1}{3}\right) \notin \mathbb{Q}$.

140 Let a and b be two squarefree numbers such that $a \equiv 1 \pmod 3$ and $b \equiv 1 \pmod 3$. Prove that the biquadratic field $\mathbb{K} = \mathbb{Q}\left(\sqrt{a}, \sqrt{b}\right)$ is not monogenic.

141 Let $\mathbb{K}$ be a cubic field such that $d_\mathbb{K} < 0$ and p be any prime number. It can be shown [38] that the factorization of (p) in $\mathcal{O}_\mathbb{K}$ can be divided into the following five different types:

Type	I	II	III	IV	V
(p)	$\mathfrak{p}\mathfrak{p}'\mathfrak{p}''$	$\mathfrak{p}$	$\mathfrak{p}\mathfrak{q}$	$\mathfrak{p}^2\mathfrak{q}$	$\mathfrak{p}^3$
Degree	1	3	1 and 2	1	1

Let $\alpha \in \mathbb{Z}_{\geq 1}$. Compute $r_\mathbb{K}(p^\alpha)$ in each case.

142 The purpose of this exercise is to provide a proof of the inequality (7.23) following Louboutin's ideas [60]. Let $\mathbb{K}/\mathbb{Q}$ be a totally real algebraic number field of degree $n \geq 2$, discriminant $d_\mathbb{K} > 1$ and class number $h_\mathbb{K}$.

(a) Set $\sigma_0 = 1 + \dfrac{2n - 2}{\log d_\mathbb{K}}$. Prove that $1 < \sigma_0 < 2$.

(b) Let $\sigma > 1$. Using the bounds $\zeta_\mathbb{K}(\sigma) \leq \zeta(\sigma)^n \leq \left(\frac{\sigma}{\sigma-1}\right)^n$ and writing the inequality (7.20) in the form

$$\kappa_\mathbb{K} \leq \frac{d_\mathbb{K}^{(\sigma-1)/2}}{(\sigma - 1)^{n-1}} f_n(\sigma)$$

show that the function $g_n : \sigma \mapsto \dfrac{2\sigma f_n'(\sigma)}{n f_n(\sigma)}$ is convex on $(0, +\infty)$.

(c) Deduce that, for all $\sigma \in [1, 2]$, we have $g_n(\sigma) \leq 0$.

(d) Finalize the proof of (7.23).

143 In [72], Ramaré proved that, if χ is a primitive even Dirichlet character of conductor f_χ, then

$$|L(1, \chi)| \leq \tfrac{1}{2} \log f_\chi. \tag{7.41}$$

Let $\mathbb{K}/\mathbb{Q}$ be an abelian real number field of degree n, discriminant $d_{\mathbb{K}}$, class number $h_{\mathbb{K}}$ and regulator $\mathcal{R}_{\mathbb{K}}$. With the help of (7.41) and the class number formula, show that[77]

$$h_{\mathbb{K}}\mathcal{R}_{\mathbb{K}} \leqslant d_{\mathbb{K}}^{1/2}\left(\frac{\log d_{\mathbb{K}}}{4n-4}\right)^{n-1}.$$

144 Determine the prime numbers $p \geqslant 7$, which can be expressed in the form

$$p = x^2 + 15y^2.$$

145 With the help of Theorem 7.16 and Lemma 3.14, prove that, for all $\varepsilon > 0$, $\sigma \in [0, 1]$ and $t \geqslant e$

$$\zeta_{\mathbb{K}}(\sigma + it) \ll_{\mathbb{K},\varepsilon} \begin{cases} t^{n(1-\sigma)/3+\varepsilon}, & \text{if } \frac{1}{2} \leqslant \sigma \leqslant 1 \\[2mm] t^{n(3-4\sigma)/6+\varepsilon}, & \text{if } 0 \leqslant \sigma \leqslant \frac{1}{2}. \end{cases}$$

146 Let $\mathbb{K}$ be an algebraic number field of degree n. Using Theorems 3.25 and 7.16, along with the estimates of the previous exercise, prove the following result given in [56], namely, for all $x \geqslant |d_K|^{1/2}$ and $\varepsilon > 0$

$$\sum_{m\leqslant x} r_{\mathbb{K}}(m) = \kappa_{\mathbb{K}}x + O_{\varepsilon,\mathbb{K}}\left(x^{1-\frac{3}{n+6}+\varepsilon}\right).$$

References

1. Alaca, S., Williams, K.S.: Introductory Algebraic Number Theory. Cambridge University Press, Cambridge (2004)
2. Alaca, S., Williams, K.S.: On Voronoï's method for finding an integral basis of a cubic field. Util Math. **65**, 163–166 (2004)
3. Alzer, H.: Inequalities for the Gamma function. Proc. Amer. Math. Soc. **128**, 141–147 (2000)
4. Anderson, N., Saff, E.B., Varga, R.S.: On the Eneström–Kakeya theorem and its sharpness. Linear Alg. Appl. **28**, 5–16 (1979)
5. Artin, E.: Über eine neue art von L-Reihen. Abh. Math. Sem. Hamburg **3**, 89–108 (1923)
6. Artin, E.: Zur Theorie der L-Reihen mit allgemeinen Gruppencharakteren. Abh. Math. Sem. Hamburg **8**, 292–206 (1930)
7. Baker, A.: Linear forms in the logarithms of algebraic numbers. Mathematika **13**, 204–216 (1966)
8. Bartz, K.M.: On a theorem of Sokolovskii. Acta Arith. **34**, 113–126 (1978)
9. Bateman, P.T., Grosswald, E.: Imaginary quadratic fields with unique factorization. Illinois J. Math. **6**, 187–192 (1962)

---

[77] This improves the bound (7.23) by a factor e^{n-1}.

10. Bellaïche, J.: Théorème de Chebotarev et complexité de Littlewood. Ann. Sci Éc Norm Supér **49**, 579–632 (2016)
11. Bordellès, O.: Explicit upper bounds for the average order of $d_n(m)$ and application to class number. J. Inequal Pure Appl. Math. **3**, Art. 38 (2002)
12. Chadrasekharan, K., Narasimhan, R.: The approximate functional equation for a class of zeta-functions. Math. Annal **152**, 30–64 (1963)
13. Cohen, H.: A Course in Computational Algebraic Number Theory. Graduate Texts in Mathematics, vol. 138. Springer, Berlin (1993)
14. Cohen, H.: Advanced Topics in Computational Algebraic Number Theory. Graduate Texts in Mathematics, vol. 193. Springer, Berlin (2000)
15. Cohen, H.: Number Theory Volume I: Tools and Diophantine Equations. Graduate Texts in Mathematics, vol. 239. Springer, Berlin (2007)
16. Cohen, H., Roblot, X.R.: Computing the Hilbert class field of real quadratic fields. Math. Comp. **69**, 1229–1244 (2000)
17. Cox, D.A.: Primes of the Form $x^2 + ny^2$. Wiley, New York (1989)
18. Cusick, T.W.: Lower bounds for regulators. In: Number Theory, Proceedings of the Journées Arithmétiques held at Noordwijkerhout/The Netherlands 1983. Lecture Notes in Mathematics, vol. 1068, pp. 63–73 (1984)
19. de la Maza, A.C.: Bounds for the smallest norm in an ideal class. Math. Comp. **71**, 1745–1758 (2001)
20. Elezović, N., Giordano, C., Pečarić, J.: The best bounds in Gautschi's inequality. Math. Ineq. Appl. **3**, 239–252 (2000)
21. Eloff, D., Spearman, B.K., Williams, K.S.: A_4-sextic fields with a power basis. Missouri J. Math. Sci. **19**, 188–194 (2007)
22. Erdős, P.: A theorem of Sylvester and Schur. J. London Math. Soc. **9**, 282–288 (1934)
23. Erdős, P.: Arithmetic properties of polynomials. J. London Math. Soc. **28**, 416–425 (1953)
24. Esmonde, J., Murty, M.R.: Problems in Algebraic Number Theory. Graduate Texts in Mathematics, vol. 190. Springer, Berlin (1999)
25. Fomenko, O.M.: On the Dedekind zeta function, II. J. Math. Sci. **207**, 923–933 (2015)
26. Friedman, E.: Analytic formulas for the regulator of a number field. Invent Math. **98**, 599–622 (1989)
27. Friedman, E.: Regulators and total positivity. Publ. Mat. (Proceedings of the Primeras Jornadas de Teoría de Números), 119–130 (2007)
28. Fröhlich, A., Taylor, M.J.: Algebraic Number Theory. Cambridge Studies in Advanced Mathematics, vol. 27. Cambridge University Press, Cambridge (1991)
29. Garbanati, D.: Class field theory summarized. Rocky Mountain J. Math. **11**, 195–225 (1981)
30. Gauss, C.F.: Disquisitiones Arithmeticæ. Springer, Berlin (1986), Reprint of the Yale University Press, New Haven, 1966 edition
31. Goldfeld, D.: Gauss' class number problem for imaginary quadratic fields. Bull. Amer. Math. Soc. **13**, 23–37 (1985)
32. Gras, M.N.: Nombre de classes, unités et bases d'entiers des extensions cubiques cycliques de $\mathbb{Q}$. Mem. SMF **37**, 101–106 (1974)
33. Gras, G., Gras, M.N.: Algèbre Fondamentale, Arithmétique. Ellipses (2004)
34. Gras, M.N., Tanoé, F.: Corps biquadratiques monogènes. Manuscripta Math. **86**, 63–79 (1995)
35. Greenberg, M.: An elementary proof of the Kronecker–Weber theorem. Amer. Math. Monthly **81**, 601–607 (1974); Correction **81**, 803 (1975)
36. Győry, K.: Sur les polynômes à coefficients entiers et de discriminant donné, III. Publ. Math. Debrecen **23**, 141–165 (1976)
37. Hagedorn, T.R.: General formulas for solving solvable sextic equations. J. Algebra **233**, 704–757 (2000)
38. Hasse, H.: Arithmetischen theorie der kubischen Zahlkörper auf klassenkörper-theoretischer Grundlage. Math. Z **31**, 565–582 (1930)
39. Heath-Brown, D.R.: The growth rate of the Dedekind zeta-function on the critical line. Acta Arith. **49**, 323–339 (1988)

40. Hecke, E.: Mathematische Werke. Vandernhoeck and Ruprecht, Göttingen (1959)
41. Heegner, K.: Diophantische analysis und Modulfunktionen. Math. Z **56**, 227–253 (1952)
42. Herz, C.S.: Construction of class fields. In: Seminar on Complex Multiplication. Lecture Notes in Mathematics, no. 21, pp. VII–1–VII–21. Springer, Berlin (1966)
43. Hindry, M.: Arithmétique. Calvage & Mounet, Montrouge (2008)
44. Huxley, M.N., Watt, N.: The number of ideals in a quadratic field, II. Israel J. Math. **120**, 125–153 (2000)
45. Iwaniec, H., Kowalski, E.: Analytic Number Theory. Colloquium Publications, vol. 53. American Mathematical Society, Providence (2004)
46. Jakhar, A., Khanduja, S.K., Sangwan, N.: On the discriminant of pure number fields. Preprint 10 pp (2020). https://arxiv.org/abs/2005.01300
47. Janusz, G.: Algebraic Number Fields. Graduate Studies in Mathematics, vol. 7, 2nd edn. American Mathematical Society, Providence (1996)
48. Kadiri, H.: Explicit zero-free regions for Dedekind zeta functions. Int. J. Number Theory **8**, 125–147 (2012)
49. Koblitz, N.: Primality of the number of points on an elliptic curve over a finite field. Pac. J. Math. **131**, 157–165 (1988)
50. Lagarias, J.C., Odlyzdo, A.M.: Algebraic Number Fields, L-Functions and Galois Properties.Effective versions of the Chebotarev density theorem, pp. 409–464. Academic, New York (1977)
51. Landau, E.: Neuer Beweis des primzahlsatzes und beweis des primidealsatzes. Math. Ann. **56**, 645–670 (1903)
52. Landau, E.: Über die heckesche funktionalgleichung. Nachrichten von der Königlichen Gesellschaft der Wissenschaften zu Göttingen, pp. 102–111 (1917)
53. Landau, E.: Einführung in die Elementare und Analytische Theorie der Algebraischen Zahlen und der Ideale. AMS Chelsea Publishing, New York (2005), Reprint of the 1949 Edition
54. Lang, S.: Algebra, 3rd edn. Addison Wesley, Boston (1993)
55. Lang, S.: Algebraic Number Theory. Graduate Texts in Mathematics, vol. 110, 2nd edn. Springer, Berlin (1994)
56. Lao, H.: On the distribution of integral ideals and Hecke grössencharacters. Chinese Ann. Math. Ser. B **31**, 385–392 (2010)
57. Lavallee, M.J., Spearman, B.K., Williams, K.S., Yang, Q.: Dihedral quintic fields with a power basis. Math. J. Okayama Univ. **47**, 75–79 (2005)
58. Lebacque, P.: Generalized mertens and Brauer–Siegel theorems. Acta Arith. **130**, 333–350 (2007)
59. Llorente, P., Nart, E.: Effective determination of the decomposition of the rational primes in a cubic field. Proc. Amer. Math. Soc. **87**, 579–582 (1983)
60. Louboutin, S.: Majorations explicites du résidu au point 1 des fonctions zêta de certains corps de nombres. J. Math. Soc. Japan **50**, 57–69 (1998)
61. Louboutin, S.: Explicit bounds for residues of Dedekind zeta functions, values of L-functions at $s = 1$, and relative class number. J. Num. Theory **85**, 263–282 (2000)
62. Louboutin, S.: Explicit lower bounds for residues of Dedekind zeta functions at $s = 1$ and relative class number of CM-fields. Trans. Amer. Math. Soc. **355**, 3079–3098 (2003)
63. Mollin, R.A.: Algebraic Number Theory. Chapman and Hall/CRC, London (1999)
64. Müller, W.: On the distribution of ideals in cubic number fields. Monat. Math. **106**, 211–219 (1988)
65. Narkiewicz, W.: Elementary and Analytic Theory of Algebraic Numbers. Springer Monographs in Mathematics, 3rd edn. Springer, Berlin (2004)
66. Neukirch, J.: Algebraic Number Theory. A Series of Comprehensive Studies in Mathematics, vol. 322. Springer, Berlin (2010)
67. Odlyzko, A.M.: Lower bounds for discriminants of number fields. Acta Arith. **29**, 275–297 (1976)
68. Pohst, M.: Regulatorabschätzungen für total reelle algebraische Zahlkörper. J. Num. Theory **9**, 459–492 (1977)

69. Poitou, G.: Minorations de discriminants (d'après A. M. Odlyzko). Séminaire Bourbaki, vol 1975/76, 28ème année, pp. 136–153. Springer, Berlin (1977)
70. Prasolov, V.V.: Polynomials. ACM, vol. 11. Springer, Berlin (2004)
71. Rademacher, H.: On the phragmén–lindelöf theorem and some applications. Math. Z **72**, 192–204 (1959)
72. Ramaré, O.: Approximate formulæ for $L(1, \chi)$. Acta Arith. **100**, 245–266 (2001)
73. Remak, R.: Über die Abchätzung des absoluten Betrages des Regulator eines algebraisches Zahlkörpers nach unten. J. Reine Angew. Math. **167**, 360–378 (1932)
74. Remak, R.: Über Grössenbeziehungen zwischen Diskriminante und Regulator eines algebraischen Zahlkörpers. Compositio Math. **10**, 245–285 (1952)
75. Ribenboim, P.: Euler's famous prime generating polynomial and the class number of imaginary quadratic fields. L'Ens Math. **34**, 23–42 (1988)
76. Ribenboim, P.: Classical Theory of Algebraic Numbers. Universitext. Springer, Berlin (2001)
77. Rose, H.E.: A Course in Number Theory. Oxford Science Publications, Oxford (1994)
78. Samuel, P.: Théorie Algébrique des Nombres. Collection Méthodes, Hermann (1971)
79. Schoof, R., Washington, L.C.: Quintic polynomials and real cyclotomic fields with large class number. Math. Comp. **50**, 543–556 (1988)
80. Silverman, J.: An inequality relating the regulator and the discriminant of a number field. J. Num. Theory **19**, 437–442 (1984)
81. Snyder, N.: Artin's L-functions: A Historical Approach. PhD Thesis, Harvard University, Cambridge, Massachusetts (2002)
82. Soicher, L.: The computation of Galois groups. PhD Thesis, Concordia University, Montréal (1981)
83. Sokolovskiĭ, A.V.: A theorem on the zeros of the Dedekind zeta function and the distance between neighbouring prime ideals (Russian). Acta Arith. **13**, 321–334 (1968)
84. Solomon, D.: Equivariant L-functions at $s = 0$ and $s = 1$. Publications Mathématiques de Besançon, Actes de la Conférence 'Fonctions L et arithmétique', pp. 129–156 (2010)
85. Spearman, B.K., Watanabe, A., Williams, K.S.: PSL$(2, 5)$ sextic fields with a power basis. Kodai Math. J. **29**, 5–12 (2006)
86. Stark, H.M.: A complete determination of the complex quadratic fields of class-number one. Mich. Math. J. **14**, 1–27 (1967)
87. Stark, H.M.: Some effective cases of the Brauer–Siegel theorem. Inventiones Math. **23**, 123–152 (1974)
88. Stark, H.M.: The analytic theory of algebraic numbers. Bull. Amer. Math. Soc. **81**, 961–972 (1975)
89. Stas, W.: On the order of the Dedekind zeta-function near the line $\sigma = 1$. Acta Arith. **35**, 195–202 (1979)
90. Steinberg, B.: Representation Theory of Finite Groups. Universitext. Springer, Berlin (2012)
91. Stevenhagen, P., Lenstra, H.W.: Chebotarëv and his density theorem. Math. Intell. **18**, 26–37 (1996)
92. Stewart, I., Tall, D.: Algebraic Number Theory and Fermat's Last Theorem, 3rd edn. A K Peters, Natick (2002)
93. Tschebotareff, N.: Die bestimmung der dichtigkeit einer menge von primzahlen, welche zu einer gegebenen substitutionsklasse gehren. Math. Ann. **95**, 191–228 (1926)
94. Uchida, K.: On Artin L-functions. Tohuku Math J. **27**, 75–81 (1975)
95. Uchida, K.: On Silverman's estimate of regulators. Tôhoku Math. J. **46**, 141–145 (1994)
96. van der Waall, R.: On a conjecture of Dedekind on zeta functions. Nederl akad Wetensch Proc. Ser. A **37**, 86–86 (1975)
97. Voronoï, G.: Concerning algebraic integers derivable from a root of an equation of the third degree. Master's Thesis, St. Petersburg (1894)
98. Washington, L.C.: Introduction to Cyclotomic Fields. GTM 83, Springer (1982), 2nd edition (1997)
99. Weber, H.: Lehrbuch der Algebra, vol II. Vieweg und Sohn, Braunschweig (1896)

100. Winckler, B.: Théorème de Chebotarev effectif. Preprint 27 pp (2013) . https://arxiv.org/abs/1311.5715v1
101. Zimmert, R.: Ideale kleiner Norm in Idealklassen und eine Regulatorabschätzung. Invent Math. **62**, 367–380 (1981)

Hints and Answers to Exercises

Chapter 1

1 If $1 \leqslant q < a$, let b_q be the quotient of the Euclidean division of a by q. We have

$$b \in S_q \Longleftrightarrow \left\lfloor \frac{a}{b} \right\rfloor = q \Longleftrightarrow \frac{a}{b} - 1 < q \leqslant \frac{a}{b} \Longleftrightarrow \frac{a}{q+1} < b \leqslant \frac{a}{q}$$

and since $b \in \mathbb{Z}_{\geqslant 1}$, this is equivalent to $b_{q+1} < b \leqslant b_q$. Hence we have

$$\sum_{q=1}^{a} |S_q| = \sum_{q=1}^{a-1} |S_q| + 1 = \sum_{q=1}^{a-1} (b_q - b_{q+1}) + 1 = b_1 - b_a + 1 = a - 1 + 1 = a.$$

2 For (a), the Euclidean division of n by m gives $n = qm + r$ with $0 \leqslant r \leqslant m - 1$ so that

$$\frac{n+1}{m} - 1 = q + \frac{r+1}{m} - 1 \leqslant q + \frac{m-1+1}{m} - 1 = q = \left\lfloor \frac{n}{m} \right\rfloor.$$

For (b), the proof is similar except that we use $m \nmid n \implies 1 \leqslant r \leqslant m - 1$. The identity (c) is obvious if $m \mid n$. Otherwise, we have by (a) and (b)

$$0 \leqslant \left\lfloor \frac{n}{m} \right\rfloor - \left\lfloor \frac{n-1}{m} \right\rfloor \leqslant 1 - \frac{1}{m} < 1$$

and we conclude the proof by noticing that the difference above is an integer.

© Springer Nature Switzerland AG 2020
O. Bordellès, *Arithmetic Tales*, Universitext,
https://doi.org/10.1007/978-3-030-54946-6

3 The first identity follows from Theorem 1.3 and letting $x \to \infty$. The second identity follows from

$$\sum_{n>x} f(n)g(n) = \sum_{n=1}^{\infty} f(n)g(n) - \sum_{n \leqslant x} f(n)g(n).$$

4 Using Theorem 1.3 we get for all integers $N > M$

$$\sum_{n=1}^{N} \frac{a_n}{n} = \frac{1}{N} \sum_{n=1}^{N} a_n + \int_1^M \frac{1}{t^2} \left(\sum_{n \leqslant t} a_n \right) dt + \int_M^N \frac{1}{t^2} \left(\sum_{n \leqslant t} a_n \right) dt.$$

By assumption, the first term on the right-hand side tends to 0 as $N \to \infty$ and the second integral converges since $\left| \sum_{n \leqslant t} a_n \right| \leqslant M$. Hence we obtain

$$\left| \sum_{n=1}^{\infty} \frac{a_n}{n} \right| \leqslant \int_1^M \frac{1}{t^2} \left| \sum_{n \leqslant t} a_n \right| dt + \int_M^\infty \frac{1}{t^2} \left| \sum_{n \leqslant t} a_n \right| dt$$

$$\leqslant \int_1^M \frac{dt}{t} + M \int_M^\infty \frac{dt}{t^2} = \log M + 1$$

as asserted.

5 One may assume that $N \geqslant 2$. By Proposition 1.4, we derive

$$\sum_{k=1}^{n-1} \frac{k}{N} (a_{k+1} - a_k) + \sum_{k=n}^{N-1} \frac{k-N}{N} (a_{k+1} - a_k)$$

$$= \sum_{k=1}^{N-1} \frac{k}{N} (a_{k+1} - a_k) - \sum_{k=n}^{N-1} (a_{k+1} - a_k)$$

$$= \frac{1}{N} \left\{ (N-1) \sum_{k=1}^{N-1} (a_{k+1} - a_k) - \sum_{k=1}^{N-2} \sum_{j=1}^{k} (a_{j+1} - a_j) \right\} - a_N + a_n$$

$$= \left(1 - \frac{1}{N} \right) (a_N - a_1) - \frac{1}{N} \sum_{k=1}^{N-2} (a_{k+1} - a_1) - a_N + a_n$$

$$= a_n - \frac{a_1 + a_N}{N} - \frac{1}{N} \sum_{k=2}^{N-1} a_k = a_n - \frac{1}{N} \sum_{k=1}^{N} a_k.$$

so that

$$a_n = \frac{1}{N} \sum_{k=1}^{N} a_k + \sum_{k=1}^{N-1} \Delta_N(n, k) (a_{k+1} - a_k),$$

where

$$\Delta_N(n, k) = \frac{1}{N} \times \begin{cases} k, & \text{if } 1 \leqslant k \leqslant n - 1 \\ k - N, & \text{if } n \leqslant k \leqslant N \end{cases}$$

and the result follows from the trivial estimate $|\Delta_N(n, k)| \leqslant 1$.

6 We use Theorem 1.4 with $k = 2$, $f(t) = (x - t)^n$, $a = 0, b = x$ so that $f(0) = x^n$, $f'(0) = -nx^{n-1}$ and $f(x) = f'(x) = 0$. Noticing that $-\frac{1}{12} \leqslant B_2(\{t\}) \leqslant \frac{1}{6}$ for any $t \in \mathbb{R}$, we get

$$\sum_{0 \leqslant j \leqslant x} (x - j)^n = x^n + \sum_{0 < j \leqslant x} (x - j)^n$$

$$= x^n + \int_0^x (x - t)^n \, dt + \psi(0) f(0) - \tfrac{1}{2} B_2(0) f'(0)$$

$$- \tfrac{1}{2} n(n - 1) \int_0^x B_2(\{t\}) (x - t)^{n-2} \, dt$$

$$= \tfrac{1}{n+1} x^{n+1} + \tfrac{1}{2} x^n + \tfrac{1}{12} nx^{n-1} + R_n$$

and the inequalities $-\frac{1}{12} \leqslant B_2(\{t\}) \leqslant \frac{1}{6}$ imply that $\frac{-1}{12} nx^{n-1} \leqslant R_n \leqslant \frac{1}{24} nx^{n-1}$. The second inequality follows easily from Newton's formula yielding

$$\tfrac{1}{n+1} \left(x + \tfrac{1}{2}\right)^{n+1} = \tfrac{1}{n+1} x^{n+1} + \tfrac{1}{2} x^n + \tfrac{1}{8} nx^{n-1} + \text{positive terms.}$$

7 This is an immediate consequence of Theorem 1.3 using

$$\sum_{p \leqslant x} f(p) = \sum_{p \leqslant x} \left(\frac{f(p)}{\log p} \times \log p\right).$$

8

- The identity (1.5) follows easily from an integration by parts.
- By (1.5), it suffices to show that, for all $x \geqslant 2$

$$\int_2^x \frac{dt}{(\log t)^2} \leqslant \frac{x}{\log x}. \tag{7.42}$$

This inequality is easily seen to be true for $x \in [2, e^6)$, hence assume $x \geqslant e^6$. Let $f(x) = \dfrac{1}{(\log x)^2}$ and F be the function defined on $[e^6, +\infty)$ by

$$F(x) = \frac{x}{\log x} - \int_2^x \frac{dt}{(\log t)^2}.$$

Since $F'(x) = \dfrac{\log x - 2}{(\log x)^2}$, we infer that F is increasing on $[e^6, +\infty)$, and using a Hermite–Hadamard type inequality [8, Corollary 1.1], we derive for all $x \geqslant e^6$

$$F(x) \geqslant F\left(e^6\right) = \frac{e^6}{6} - \int_2^{e^6} f(t)\, dt \geqslant \frac{e^6}{6} - \frac{e^6 - 2}{2} \inf_{0 \leqslant t \leqslant 1} L(t) > 0,$$

where $L(t) := f\left(e^6 t + 2(1 - t)\right) + t f(2) + (1 - t) f\left(e^6\right)$, completing the proof of (7.42).

- We start by checking numerically the lower bound for all $x \in [e^2, 16]$, and then assume $x > 16$. Using the Hermite–Hadamard inequality [8, (1.1)], we derive, for all $x \geqslant 16$

$$\int_2^x \frac{dt}{(\log t)^2} \geqslant \frac{x - 2}{\log^2(x/2 + 1)} \geqslant \frac{7}{2(\log 3)^3} > \frac{2}{\log 2}$$

and the result then follows then by using (1.5).

- Studying the function $x \mapsto \dfrac{2 \log x}{\log(2x \log x)}$, we derive

$$\frac{2 \log x}{\log(2x \log x)} \geqslant 1 \quad (x \geqslant 2).$$

Now by the previous lower bound, we get for $x \geqslant e^2$

$$\mathrm{Li}(2x \log x) \geqslant \frac{2x \log x}{\log(2x \log x)} \geqslant x$$

which is equivalent to the desired upper bound for Li^{-1} since Li is strictly increasing. The lower bound is proved in a similar way using

$$\mathrm{Li}\left(\tfrac{1}{2} x \log x\right) \leqslant \frac{x \log x}{\log\left(\tfrac{1}{2} x \log x\right)} \leqslant x \quad \left(x \geqslant e^2\right).$$

9 Using Exercise 7 with $f(x) = 1$ and (1.5), we get

$$\pi(x) = \frac{\theta(x)}{\log x} + \int_2^x \frac{\theta(t)}{t(\log t)^2} \, dt$$

$$= \frac{x}{\log x} + \int_2^x \frac{dt}{(\log t)^2} + \frac{\theta(x) - x}{\log x} + \int_2^x \frac{\theta(t) - t}{t(\log t)^2} \, dt$$

$$= \mathrm{Li}(x) + \frac{2}{\log 2} + \frac{\theta(x) - x}{\log x} + \int_2^x \frac{\theta(t) - t}{t(\log t)^2} \, dt$$

and hence

$$|\pi(x) - \mathrm{Li}(x)| \leqslant \frac{2}{\log 2} + \frac{R(x)}{\log x} + \int_2^x \frac{R(t)}{t(\log t)^2} \, dt$$

with

$$\int_2^x \frac{R(t)}{t(\log t)^2} \, dt = \left(\int_2^{\sqrt{x}} + \int_{\sqrt{x}}^x \right) \frac{R(t)}{t(\log t)^2} \, dt$$

$$< \frac{\sqrt{x}}{(\log 2)^2} + R(x) \int_{\sqrt{x}}^x \frac{dt}{t(\log t)^2}$$

$$= \frac{\sqrt{x}}{(\log 2)^2} + \frac{R(x)}{\log x} \leqslant \frac{R(x)}{\log x} \left(1 + \frac{1}{(\log 2)^2} \right)$$

and therefore

$$|\pi(x) - \mathrm{Li}(x)| < \frac{R(x)}{\log x} \left(2 + \frac{1}{(\log 2)^2} \right) + \frac{2}{\log 2} < \frac{5R(x)}{\log x}.$$

10 Note that the inequalities $p_n \geqslant a \geqslant 13$ imply $n \geqslant 6$. Since $\pi(p_n) = n$, we derive

$$|n - \mathrm{Li}(p_n)| \leqslant R(p_n) \quad (p_n \geqslant a)$$

and hence $p_n = \mathrm{Li}^{-1}(n + S(n))$ where $|S(n)| \leqslant R(p_n) \leqslant R(2n \log n)$ by Corollary 3.3. The mean value theorem yields the existence of a real number c_n between n and $n + S(n)$ such that

$$p_n - \mathrm{Li}^{-1}(n) = S(n) \times \left(\mathrm{Li}^{-1} \right)'(c_n).$$

Since $n, n + S(n) \in [n - |S(n)|, n + |S(n)|]$, we get

$$\left| p_n - \mathrm{Li}^{-1}(n) \right| \leqslant |S(n)| \sup_{n-|S(n)| \leqslant x \leqslant n+|S(n)|} \left| \left(\mathrm{Li}^{-1} \right)'(x) \right|$$

$$\leqslant R(2n \log n) \log \left\{ \mathrm{Li}^{-1} \left(n + |S(n)| \right) \right\}$$

and using Exercise 8 and the trivial bound $|S(n)| \leqslant R(2n \log n) < 2n \log n$, we derive for all $n \geqslant 6$

$$\left| p_n - \mathrm{Li}^{-1}(n) \right| \leqslant R(2n \log n) \log \{ (2n + 4n \log n) \log(n + 2n \log n) \}$$

$$\leqslant 3 \log n \, R(2n \log n)$$

as required.

11 Using Corollary 1.3, we derive for all $x \geqslant 1$

$$\log (\lfloor x \rfloor !) = \sum_{n \leqslant x} \log n$$

$$= \int_1^x \log t \, dt - \tfrac{1}{8} - \psi(x) \log x + \tfrac{\psi(x)^2}{2x} + \tfrac{1}{2} \int_1^x \left(\tfrac{\psi(t)}{t} \right)^2 dt$$

and the result follows at once from the bound $|\psi(t)| \leqslant \tfrac{1}{2}$.

Chapter 2

12 Let $S = \{0, \ldots, \lfloor \sqrt{p} \rfloor\}$ and define $f : S^2 \to \{0, \ldots, p-1\}$ by $f(u, v) \equiv au - v \pmod{p}$. We first prove that f is not injective. Indeed, we have $|S^2| = (\lfloor \sqrt{p} \rfloor + 1)^2 > p = |\{0, \ldots, p-1\}|$ so that f is not injective by the Dirichlet pigeon-hole principle. Now let (u_1, v_1) and (u_2, v_2) be two pairs such that $(u_1, v_1) \neq (u_2, v_2)$ and $f(u_1, v_1) = f(u_2, v_2)$, and set $u = u_1 - u_2$ and $v = v_1 - v_2$. We have

$$f(u_1, v_1) = f(u_2, v_2) \iff au_1 - v_1 \equiv au_2 - v_2 \pmod{p} \iff au \equiv v \pmod{p}.$$

Furthermore, $|u| = |u_1 - u_2| \leqslant \lfloor \sqrt{p} \rfloor < \sqrt{p}$ and similarly $|v| < \sqrt{p}$. If $u = 0$, then we have $v \equiv 0 \pmod{p}$ and hence $v = 0$ since $|v| < \sqrt{p}$. This is impossible in view of the condition $(u_1, v_1) \neq (u_2, v_2)$. If $v = 0$, then we have $au \equiv 0 \pmod{p}$ and since $p \nmid a$ and p is prime, then we get $u = 0$, which is also impossible.

13 Using Theorem 2.7, we derive

$$\mathcal{D}_{(1,1,2)}(n) = \tfrac{1}{8}\left\{ 1 + 2\binom{n+1}{n} + 4\binom{n+2}{n} + (-1)^n \right\}$$

$$= \tfrac{1}{8}\left(2n^2 + 8n + 7 + (-1)^n \right)$$

$$= \tfrac{1}{4}(n+2)^2 + \begin{cases} 0, & \text{if } n \text{ is even}; \\[2mm] -\tfrac{1}{4}, & \text{if } n \text{ is odd}; \end{cases}$$

$$= \left\lfloor \tfrac{1}{4}(n+2)^2 \right\rfloor.$$

14 Using Theorem 2.7, we first derive

$$\mathcal{D}_{(1,2,4)}(n) = \tfrac{9}{32} + \tfrac{5}{32}(-1)^n + \tfrac{1}{16}\binom{n+1}{n}(4 + (-1)^n)$$

$$+ \tfrac{1}{8}\binom{n+2}{n} + \tfrac{1}{8}\left(\cos\tfrac{n\pi}{2} + \sin\tfrac{n\pi}{2} \right)$$

$$= \tfrac{1}{32}\left(2n^2 + 14n + 21 + (-1)^n(2n+7) + 4(-1)^{\lfloor n/2 \rfloor} \right)$$

so that the equation has exactly 225 solutions when $n = 56$ or $n = 57$.

15 First, it can easily be proved by induction that u_n is odd, and thus $(u_n, u_{n-1}) = 1$ since $u_{n-1}^2 - u_n = 2$ and u_n and u_{n-1} are odd. Next, by induction, it can be shown that, for any integer $n \geqslant 3$, we have

$$u_n - 2 = u_{n-2}^2 u_{n-3}^2 \cdots u_1^2(u_2 - 2).$$

Therefore, for all $r \in \{2, \ldots, n-1\}$, setting $d_r = (u_n, u_{n-r})$, we have $d_r \mid u_n$ and $d_r \mid u_{n-r}$, so that d_r divides

$$u_n - u_{n-r}^2 u_{n-2}^2 \cdots u_{n-r+1}^2 u_{n-r-1}^2 \cdots u_1^2(u_2 - 2) = 2$$

and we conclude the proof by using the fact that both u_n and u_{n-r} are odd.

16 *The solution is due to T. Andreescu & D. Andrica* [1]. Assume first that $a = 0$. Then, for all $m \in \mathbb{Z}$, there exists $y \in \mathbb{Z}$ such that $m = by$, so that $b = \pm 1$, and then $cx = n \pm dm$, hence c divides $n \pm dm$ for all $m, n \in \mathbb{Z}$, implying that $c = \pm 1$, and then $ad - bc = -bc = \pm 1$ as asserted. We may then assume $abcd \neq 0$. Define $\delta := ad - bc$. If $\delta = 0$, then $\frac{c}{a} = \frac{d}{b} := k$, and hence $n = cx + dy = k(ax + by) = km$ for all $m, n \in \mathbb{Z}$, which is absurd. Thus, $\delta \neq 0$, and solving the system $ax + by = m$ and $cx + dy = n$ gives $x = \frac{1}{\delta}(bm - dn)$ and $y = \frac{1}{\delta}(an - cm)$. Therefore, for all $m, n \in \mathbb{Z}$, we have the expressions of the integers x, y. In particular, for $(m, n) = (1, 0)$, $x_1 = \delta^{-1}d$ and $y_1 = -\delta^{-1}c$ are integers and, for $(m, n) = (0, 1)$, $x_2 = -\delta^{-1}b$ and $y_2 = \delta^{-1}a$ are also integers. We infer that

$$x_1 y_2 - x_2 y_1 = \frac{ad - bc}{\delta^2} = \frac{1}{\delta} \in \mathbb{Z}$$

and hence $\delta = \pm 1$.

17 Using Theorem 2.7 with $k = 3$ immediately yields

$$\mathcal{D}_{(a,b,c)}(n) = c_1 + c_2 \binom{n+1}{n} + c_3 \binom{n+2}{n}$$

$$+ \sum_{h=1}^{a-1} A_a(h)\zeta_a^{-nh} + \sum_{h=1}^{b-1} A_b(h)\zeta_b^{-nh} + \sum_{h=1}^{c-1} A_c(h)\zeta_c^{-nh} \qquad (7.43)$$

with $c_2 = \frac{a+b+c-3}{2abc}$, $c_3 = (abc)^{-1}$,

$$A_a(h) = a^{-1}\left(1 - \zeta_a^{bh}\right)^{-1}\left(1 - \zeta_a^{ch}\right)^{-1}, \quad A_b(h) = b^{-1}\left(1 - \zeta_b^{ch}\right)^{-1}\left(1 - \zeta_b^{ah}\right)^{-1}$$

and $A_c(h) = c^{-1}\left(1 - \zeta_c^{ah}\right)^{-1}\left(1 - \zeta_c^{bh}\right)^{-1}$. Now substituting $z = 0$ in (2.8) gives

$$1 = c_1 + c_2 + c_3 + \sum_{h=1}^{a-1} A_a(h) + \sum_{h=1}^{b-1} A_b(h) + \sum_{h=1}^{c-1} A_c(h)$$

and subtracting this identity from (7.43) yields the asserted result.

Remark It can be proved that

$$S_a = \frac{c_a b_a(b_a - 1)}{2} + a - a \sum_{h=1}^{b_a - 1} \left\lfloor \frac{hc_a}{a} \right\rfloor - \frac{b_a(a + 1)}{2},$$

where b_a and c_a are given by $b b_a \equiv -n \pmod{a}$ and $c c_a \equiv -n \pmod{a}$, and similarly for S_b and S_c.

18 Let $u \in (0, b)$ be a solution of $au \equiv c \pmod{b}$. Hence $c = au + bv$ for some $v \in \mathbb{Z}$. If $v > 0$, then we are done. Otherwise, set $v = -y < 0$ and write

$$c = au + bv = ab - ab + au - by = ab - (b - u)a - by$$

with $x := b - u > 0$ and $y = -v > 0$. Furthermore, c is not simultaneously representable in both forms. Indeed, if $ab - ax_1 - by_1 = ax_2 + by_2$ with $x_1, y_1, x_2, y_2 > 0$, then $ab = a(x_1 + x_2) + b(y_1 + y_2)$ and hence $a \mid y_1 + y_2$, which is impossible since $y_1 + y_2 < a$.

19 If $c \notin \langle a, b \rangle$, then $a \nmid c$ or $b \nmid c$, so that $c = ab - ax - by$ for some $x, y > 0$ by the previous exercise. Conversely, assume $c = ab - ax - by$ for some $x, y > 0$. We first see that, since $c > 0$, then $\max(ax, by) < ab$ and hence $x < b$ and $y < a$. If $a \mid c$, then $ab - ax - by = aq$ for some $q \in \mathbb{Z}_{\geq 1}$, and then $a(b - x - q) = by$, and since $(a, b) = 1$, this implies $a \mid y$ by Gauss lemma, which is impossible since $a > y$. Therefore $a \nmid c$, and a similar argument shows that $b \nmid c$. By the previous exercise, this implies that $c \notin \langle a, b \rangle$.

20 First note that $g(a, b, c) \leq g(a, b)$.

- Assume $c \in \langle a, b \rangle$. We will show that $g(a, b, c) = g(a, b)$. Assume the contrary, so that $g(a, b, c) < g(a, b)$. Since $g(a, b, c)$ is the largest positive integer not representable as a non-negative linear combination of a, b, c, we deduce that there exist $u, v, w \in \mathbb{Z}_{\geq 0}$ such that $g(a, b) = au + bv + cw$. Since $c = ax + by$ for some $x, y \geq 0$, we derive

$$g(a, b) = au + bv + w(ax + by) = a(xw + u) + b(yw + v)$$

which is impossible since $g(a, b)$ is the largest positive integer not representable as a non-negative linear combination of a, b. Thus, $g(a, b, c) = g(a, b)$.
- Conversely, assume $c \notin \langle a, b \rangle$. By the previous exercise, $c = ab - ax - by$ for some $x, y > 0$, and then, using Proposition 2.4

$$g(a, b) = ab - a - b = (ab - ax - by) + ax - a + by - b = c + a(x - 1) + b(y - 1)$$

with $x - 1 \in \mathbb{Z}_{\geq 0}$ and $y - 1 \in \mathbb{Z}_{\geq 0}$, so that $g(a, b)$ is representable as a non-negative linear combination of a, b, c, and then $g(a, b) \neq g(a, b, c)$.

21 Note that $g(a, b, c) \notin \langle a, b, c \rangle$ and since $\langle a, b \rangle \subset \langle a, b, c \rangle$, we infer $g(a, b, c) \notin \langle a, b \rangle$. By Exercise 19, there exist $x, y > 0$ such that $g(a, b, c) = ab - ax - by$. Hence

$$g(a, b, c) = g(a, b) + a + b - ax - by = g(a, b) - a(x - 1) - b(y - 1)$$

with $x - 1 \in \mathbb{Z}_{\geq 0}$ and $y - 1 \in \mathbb{Z}_{\geq 0}$, as required.

22 If $c \in \langle a, b \rangle$, then $234 = g(a, b, c) = g(a, b) = (a - 1)(b - 1) - 1$ by Proposition 2.4 and Theorem 2.1. Hence $a - 1$ and $b - 1$ divide 235, so that $a = 6$ and $b = 48$, which is impossible since $(6, 48) \neq 1$. Therefore $c \notin \langle a, b \rangle$, and there exist $x, y > 0$ such that $c = ab - ax - by$ and $g(a, b, c) = g(a, b) - ax$ or $g(a, b, c) = g(a, b) - ay$. Assume the first case with $x \leq \frac{1}{2}b < 25$.[1] Note that (2.3) yields

$$234 \leq \tfrac{b}{2}(a + 2) - a < \tfrac{50}{2}(a + 2) - a$$

---

[1] Recall that $y \leq \frac{1}{2}a$ and $ax < by$ entail $x \leq \frac{1}{2}b$.

and hence $a > 7$, and similarly $234 < \frac{b}{2}(b+2) - a$, so that $b > 21$. Also $234 = g(a, b) - ax$ entails

$$234 > a^2 - a(1+x) - b > a^2 - 26a - 50$$

which gives $a \leqslant 34$. Subtracting the equation $c = ab - ax - by$ from the equation $234 = ab - a - b - ax$ entails

$$b(y-1) = 234 + a - c.$$

Furthermore, since $x = a^{-1}(ab - a - b - 234) = b - a^{-1}(by + c)$, we infer that $a^{-1}(by + c)$ is an integer. Now the Python script

```
 1 def gcd(a,b):
 2     if a%b==0:
 3         return b
 4     return gcd(b,a%b)
 5
 6 def coprime(a,b):
 7     return(gcd(a,b)==1)
 8
 9 for b in range(22,50):
10     for a in range(8,min(35,b)):
11         if coprime(a,b):
12             for c in range(51,150):
13                 for y in range(0,a//2+1):
14                     if b*(y-1)==234+a-c and ((b*y+c)/a)%1==0:
15                         print(a,b,c)
```

yields 16 solutions, each of which having $a > 10$. If $a = 11$, then the equation $234 = g(a, b) - ax$ and Proposition 2.4 yield $10b - 11x = 245$, and hence $b = -245 + 11k$ and $x = -245 + 10k$ for some $k \in \mathbb{Z}$. Since $b \in (21, 50)$, this gives $b = 30$ and $x = 5$, and hence $c = 125$ by the script above. Conversely, Proposition 2.5 with $\kappa = 4, \rho = 6, q = 2, r = 1, \lambda = 0$, yields

$$g(11, 30, 125) = 30(11 - 6 - 1) + 125(2 - 0 - 1) - 11 = 234.$$

23 | Using Theorem 2.1 we get

$$r = 0 \iff \left\lfloor \frac{n}{ab} \right\rfloor = \frac{n}{ab} - 1 + \frac{aa' + bb'}{ab} \iff \left\{ \frac{n}{ab} \right\} = 1 - \frac{aa' + bb'}{ab}$$

which is equivalent to

$$0 \leqslant 1 - \frac{aa' + bb'}{ab} < 1$$

giving the asserted result. The case $r = 1$ is similar.

24

(a) The first identity is trivial if $a = 1$. If $a \geqslant 2$, it follows from the logarithmic derivative of (2.11) and taking $x = 1$. For the second identity, we have

$$\prod_{\substack{j=1 \\ j \neq k}}^{a} \frac{1}{e_a(k) - e_a(j)} = e_a(-k(a-1)) \prod_{\substack{j=1 \\ j \neq k}}^{a} \frac{1}{1 - e_a(j - k)}$$

$$= e_a(k) \prod_{h=1}^{a-1} \frac{1}{1 - e_a(h)} = \frac{e_a(k)}{a},$$

where we used (2.11) with $x = 1$.

(b) By Proposition 2.11, we infer that $F(z) = z^{n+1} f(z)$ is the generating function of $\mathcal{D}_2(n)$. Therefore

$$f(z) = \frac{1}{z^{n+1}} \sum_{k=0}^{\infty} \mathcal{D}_2(k) z^k = \frac{\mathcal{D}_2(n)}{z} + \sum_{\substack{k=0 \\ k \neq n}}^{\infty} \mathcal{D}_2(k) z^{k-n-1}$$

so that

$$\operatorname*{Res}_{z=0} f(z) = \mathcal{D}_2(n).$$

The non-zero poles of f are 1 of order 2, $e_a(k)$ of order 1 for all $1 \leqslant k \leqslant a-1$, and $e_b(k)$ of order 1 for all $1 \leqslant k \leqslant b-1$. Since

$$z^a - 1 = \prod_{j=1}^{a} (z - e_a(j)) \quad \text{and} \quad z^b - 1 = \prod_{j=1}^{b} (z - e_b(j))$$

we get

$$\operatorname*{Res}_{z=1} f(z) = G'(1),$$

where

$$G(z) = (z-1)^2 f(z) = z^{-n-1} \prod_{k=1}^{a-1} (z - e_a(k))^{-1} \prod_{k=1}^{b-1} (z - e_b(k))^{-1}$$

so that

$$\frac{G'}{G}(z) = -\frac{n+1}{z} - \sum_{k=1}^{a-1} \frac{1}{z - e_a(k)} - \sum_{k=1}^{b-1} \frac{1}{z - e_b(k)}.$$

Now (2.11) with $x = 1$ implies that $G(1) = (ab)^{-1}$ and then we get using the previous question

$$\operatorname*{Res}_{z=1} f(z) = \frac{1}{ab} \left\{ -n - 1 - \sum_{k=1}^{a-1} \frac{1}{1 - e_a(k)} - \sum_{k=1}^{b-1} \frac{1}{1 - e_b(k)} \right\}$$

$$= \frac{1}{ab} \left\{ -n - 1 - \frac{a-1}{2} - \frac{b-1}{2} \right\}$$

$$= -\frac{a+b+2n}{2ab}.$$

Finally, for all $k \in \{1, \ldots, a-1\}$, we have using the previous question

$$\operatorname*{Res}_{z=e_a(k)} f(z) = \frac{1}{e_a(kb) - 1} \times \frac{1}{e_a(k(n+1))} \times \prod_{\substack{j=1 \\ j \neq k}}^{a} \frac{1}{e_a(k) - e_a(j)}$$

$$= \frac{1}{a\, e_a(kn)(e_a(kb) - 1)}$$

and similarly

$$\operatorname*{Res}_{z=e_b(k)} f(z) = \frac{1}{b\, e_b(kn)(e_b(ka) - 1)}$$

for all $k \in \{1, \ldots, b-1\}$.

(c) Since $\lim_{|z| \to \infty} zf(z) = 0$, Jordan's first lemma implies that

$$\lim_{R \to \infty} \frac{1}{2\pi i} \int_{|z|=R} f(z)\, dz = 0.$$

and Cauchy's residue theorem then gives

$$D_2(n) = \frac{a+b+2n}{2ab} + \frac{1}{a}\sum_{k=1}^{a-1}\frac{1}{e_a(kn)(1-e_a(kb))}$$

$$+ \frac{1}{b}\sum_{k=1}^{b-1}\frac{1}{e_b(kn)(1-e_b(ka))} \qquad (7.44)$$

which is similar to (2.4).

(d) If $b = 1$, then the equation is $ax + y = n$ so that $\frac{n}{a} \geq x$ and then

$$D_2(n) = \left|[0, \tfrac{n}{a}] \cap \mathbb{Z}\right| = \left[\tfrac{n}{a}\right] + 1 = \tfrac{n}{a} - \left\{\tfrac{n}{a}\right\} + 1.$$

Replacing in (7.44) gives

$$\frac{n}{a} - \left\{\frac{n}{a}\right\} + 1 = \frac{a+1+2n}{2a} + \frac{1}{a}\sum_{k=1}^{a-1}\frac{1}{e_a(kn)(1-e_a(k))}$$

so that

$$\frac{1}{a}\sum_{k=1}^{a-1}\frac{1}{e_a(kn)(1-e_a(k))} = \frac{a-1}{2a} - \left\{\frac{n}{a}\right\}.$$

(e) By above we get

$$\frac{1}{a}\sum_{k=1}^{a-1}\frac{1}{e_a(kn)(1-e_a(kb))} = \frac{1}{a}\sum_{k=1}^{a-1}\frac{1}{e_a(k\bar{b}n)(1-e_a(k))} = \frac{a-1}{2a} - \left\{\frac{n\bar{b}}{a}\right\}$$

and similarly

$$\frac{1}{b}\sum_{k=1}^{b-1}\frac{1}{e_b(kn)(1-e_b(ka))} = \frac{b-1}{2b} - \left\{\frac{n\bar{a}}{b}\right\}$$

and therefore

$$D_2(n) = \frac{a+b+2n}{2ab} + \frac{a-1}{2a} - \left\{\frac{n\bar{b}}{a}\right\} + \frac{b-1}{2b} - \left\{\frac{n\bar{a}}{b}\right\}$$

$$= \frac{n}{ab} + 1 - \left\{\frac{n\bar{b}}{a}\right\} - \left\{\frac{n\bar{a}}{b}\right\}.$$

Chapter 3

25 Let q be a prime divisor of $2^p - 1$ and $\delta = \mathrm{ord}_q(2)$. Thus we have $2^p \equiv 1 \pmod{q}$ and hence $\delta \mid p$. Since $\delta \neq 1$, we get $\delta = p$, and then using Proposition 3.2 we have $p \mid q - 1$. Thus there exists $h \in \mathbb{Z}_{\geqslant 1}$ such that $q = 1 + ph$. Furthermore, q is odd since $2^p - 1$ is odd, and hence ph is even, and therefore h is even since p is odd.

26 The inequality may be numerically checked for all $n \in \{33, \ldots, 65\}$ so that we suppose that $n \geqslant 66$. Among the integers $\{2, \ldots, n\}$, we remove the $\left\lfloor \frac{1}{2}n \right\rfloor - 1$ even integers $\neq 2$ and the $\left\lfloor \frac{1}{3}n \right\rfloor - 1$ integers $\neq 3$ multiples of 3. However, the $\left\lfloor \frac{1}{6}n \right\rfloor$ integers multiples of 6 have been removed twice, and removing the numbers 25, 35, 55 and 65, we derive

$$\pi(n) \leqslant n - \left(\left\lfloor \tfrac{1}{2}n \right\rfloor - 1 \right) - \left(\left\lfloor \tfrac{1}{3}n \right\rfloor - 1 \right) + \left\lfloor \tfrac{1}{6}n \right\rfloor - 4$$

and the inequalities $x - 1 < \lfloor x \rfloor \leqslant x$ imply the asserted result.

27 The formula is easily checked for $n \in \{3, 4\}$, hence assume $n \geqslant 5$. Recall the notation $e_q(x) := e^{2i\pi x/q}$ and set

$$f(k) := \frac{e_k((k-1)!) - 1}{e_k(-1) - 1} \qquad (k \geqslant 5).$$

If k is prime, then by Wilson's theorem $e_k((k-1)!) = e_k(-1)$, so that $f(k) = 1$. If $k \geqslant 5$ is composite, then

- either $k = ab$ with $1 \leqslant a \neq b \leqslant k - 1$, and hence $a \mid (k-1)!$ and $b \mid (k-1)!$ so that $k \mid (k-1)!$;
- or $k = p^2$ for some prime $p > 2$ and using Exercise 35, we derive

$$v_p((k-1)!) = \left\lfloor p - \tfrac{1}{p} \right\rfloor \geqslant 2$$

so that $k = p^2 \mid (k-1)!$;

and therefore, $f(k) = 0$ if k is composite, completing the proof.

28 This sequence has $6k - 5$ integers. Furthermore, if $n \geqslant 6$ is even, then $n^2 + 2$ is also even and then is composite. Similarly, if $n \equiv \pm 1 \pmod 6$, then $n^2 + 2 \equiv 3 \pmod 6$ and $n^2 + 2$ is odd and composite. Therefore

$$\pi_k \leqslant 6k - 5 - (3k - 2) - 2(k - 1) = k - 1 < k.$$

29 Let $N \geqslant 2$ be an integer. Using Theorem 1.3 and a weak version of Corollary 3.4, we get

$$\sum_{p \leqslant N} \frac{1}{p \log p} = \frac{1}{\log N} \sum_{p \leqslant N} \frac{1}{p} + \int_2^N \frac{1}{t (\log t)^2} \left(\sum_{p \leqslant t} \frac{1}{p} \right) dt$$

$$= \frac{\log \log N + O(1)}{\log N} + \int_2^N \frac{\log \log t + O(1)}{t (\log t)^2} \, dt = O(1)$$

implying the asserted result.

30 The sum is clearly convergent. By Exercise 7 and Corollary 3.12, we have[2]

$$\sum_p \frac{\log p}{p^2} = \left(\sum_{p \leqslant 100} + \sum_{p > 100} \right) \frac{\log p}{p^2}$$

$$= \sum_{p \leqslant 100} \frac{\log p}{p^2} - \frac{\theta(100)}{10^4} + 2 \int_{100}^\infty \frac{\theta(t)}{t^3} \, dt$$

$$< 0.484 - 0.0075 + 2.000162 \int_{100}^\infty t^{-2} \, dt < \tfrac{1}{2}$$

as required.

31 *This solution is due to Claude Quitté.* Wolstenholme's theorem [13, Theorem 115] asserts that, if $p > 3$ is prime, then the numerator of H_{p-1} is divisible by p^2. Write $H_{p-1} = \frac{\alpha}{\beta}$ with $p \nmid \beta$ and $\alpha = kp^2$ for some $k \in \mathbb{Z}_{\geqslant 1}$. We get

$$H_p = H_{p-1} + \frac{1}{p} = \frac{\alpha p + \beta}{\beta p}.$$

---

[2]Exercise 7 is used in the following form. If $f \in C^1[2, +\infty)$ such that $f(x)\theta(x)/\log x \to 0$ as $x \to \infty$, then

$$\sum_{p > x} f(p) = -\frac{f(x)\theta(x)}{\log x} - \int_x^\infty \theta(t) \frac{d}{dt} \left(\frac{f(t)}{\log t} \right) dt.$$

Since $(a, b) = 1$, we derive $a = \alpha p + \beta$ and $b = \beta p$, so that

$$ap = \alpha p^2 + \beta p = \alpha p^2 + b = kp^4 + b \equiv b \pmod{p^4}.$$

32 Assume that $\frac{n}{p(n)}$ is composite, and let q be a prime divisor. Then $q \leqslant \sqrt{\frac{n}{p(n)}} < n^{1/3}$, which contradicts the assumption on q.

33

(a) If $n \equiv 1 \pmod{p}$, then $0 \equiv n^2 + n + 1 \equiv 3 \pmod{p}$ which is impossible since $p \geqslant 5$. If $n^2 \equiv 1 \pmod{p}$, then $n \equiv \pm 1 \pmod{p}$. Since $n \not\equiv 1 \pmod{p}$, this implies that $n \equiv -1 \pmod{p}$ and then $0 \equiv n^2 + n + 1 \equiv 1 \pmod{p}$ giving a contradiction again. Since $n^3 - 1 = (n - 1)(n^2 + n + 1)$, we derive $n^3 \equiv 1 \pmod{p}$ and hence $\mathrm{ord}_p(n) = 3$. We infer that $3 \mid p - 1$ by Proposition 3.2. If $p = 1 + 3k$ for some even integer k, then $p \equiv 1 \pmod 6$. If $p = 1 + 3(1 + 2h) = 2(2 + 3h)$ for some $h \in \mathbb{Z}_{\geqslant 1}$, then p is composite, giving a contradiction. Hence $p \equiv 1 \pmod 6$.

(b) Assume that the set of primes of the form $p \equiv 1 \pmod 6$ is finite and write all these primes as

$$p_1 = 7 < p_2 < \cdots < p_m.$$

Set $M = (p_1 \cdots p_m)^2 + p_1 \cdots p_m + 1$ supposed to be composite without loss of generality. Using the previous question, the prime factors of M are all congruent to 1 modulo 6 and then there exists an index $i \in \{1, \ldots, m\}$ such that $p_i \mid M$, giving a contradiction since we also have $p_i \mid (M - 1)$.

34 Assume first that $(a, b) = 1$. Then obviously $(b, a^n - b^n) = 1$, and hence b has an inverse $\overline{b}$ modulo $a^n - b^n$ and

$$a^n \equiv b^n \pmod{a^n - b^n} \iff (a\overline{b})^n \equiv 1 \pmod{a^n - b^n}.$$

If there exists $m < n$ such that $(a\overline{b})^m \equiv 1 \pmod{a^n - b^n}$, then $a^m \equiv b^m \pmod{a^n - b^n}$, and therefore $(a^n - b^n) \mid (a^m - b^m)$, so that $n \leqslant m$, leading to a contradiction. Hence $\mathrm{ord}_{a^n - b^n}(a\overline{b}) = n$, so that $n \mid \varphi(a^n - b^n)$. Now suppose that $(a, b) = d > 1$. Write $a = da_1, b = db_1$ with $(a_1, b_1) = 1$ and set $\Delta := (d^n, a_1^n - b_1^n)$. Then

$$\varphi(a^n - b^n) = \varphi(d^n)\varphi(a_1^n - b_1^n)\frac{\Delta}{\varphi(\Delta)}$$

$$= d^n \varphi(a_1^n - b_1^n) \prod_{\substack{p \mid d \\ p \nmid \Delta}} \left(1 - \frac{1}{p}\right)$$

and by the previous case, we derive $n \mid \varphi(a_1^n - b_1^n)$, concluding the proof.

35 Since the p-adic valuation of a product is the sum of the p-adic valuations, we derive

$$v_p(n!) = \sum_{m=1}^{n} v_p(m) = \sum_{m=1}^{n} \sum_{k=1}^{v_p(m)} 1 = \sum_{k=1}^{\infty} \sum_{\substack{m=1 \\ v_p(m) \geq k}}^{n} 1$$

and we conclude the proof by noticing that the inner sum counts the number of multiples of p^k not greater than n which is precisely equal to $\lfloor n/p^k \rfloor$.

36 Let $p \leq n$ be a prime number and set $N := \lfloor \log n / \log p \rfloor$. By the previous exercise and the inequalities $x - 1 < \lfloor x \rfloor \leq x$, we get

$$n \sum_{k=1}^{N} \frac{1}{p^k} - N < v_p(n!) \leq n \sum_{k=1}^{N} \frac{1}{p^k}$$

so that

$$\frac{n}{p-1}\left(1 - \frac{1}{p^N}\right) - N < v_p(n!) \leq \frac{n}{p-1}\left(1 - \frac{1}{p^N}\right)$$

and using $\log n / \log p - 1 < N \leq \log n / \log p$ gives

$$\frac{n}{p-1}\left(1 - \frac{p}{n}\right) - \frac{\log n}{\log p} < v_p(n!) \leq \frac{n}{p-1}\left(1 - \frac{1}{n}\right).$$

This may be written in the form

$$\frac{1}{(p-1)\log n} \leq \frac{\dfrac{n}{p-1} - v_p(n!)}{\log n} < \frac{1}{\log p} + \frac{p}{(p-1)\log n}$$

and since $p \geq 2$, we get

$$0 < \frac{\dfrac{n}{p-1} - v_p(n!)}{\log n} < \frac{1}{\log 2} + \frac{2}{\log n} = O(1)$$

as required.

Remark Used with $p = 5$, this asymptotic formula shows that, if n is sufficiently large, the decimal expansion of $n!$ ends up with approximately $\frac{1}{4}n$ zeros.

37 From Exercise 35, we derive

$$r \leqslant v_p\left(\binom{2n}{n}\right) = v_p((2n)!) - 2v_p(n!)$$

$$= \sum_{k=1}^{\lfloor \log 2n/\log p \rfloor} \left\lfloor \frac{2n}{p^k} \right\rfloor - 2 \sum_{k=1}^{\lfloor \log n/\log p \rfloor} \left\lfloor \frac{n}{p^k} \right\rfloor$$

$$= \sum_{k=1}^{\lfloor \log 2n/\log p \rfloor} \left(\left\lfloor \frac{2n}{p^k} \right\rfloor - 2 \left\lfloor \frac{n}{p^k} \right\rfloor \right) \leqslant \sum_{k=1}^{\lfloor \log 2n/\log p \rfloor} 1 \leqslant \frac{\log 2n}{\log p}$$

as required.

38 Since a square is congruent to 0 or 1 modulo 4, we see that $a^2 - b^2$ cannot be congruent to 2 modulo 4. Conversely, let $n \not\equiv 2 \pmod 4$. Then either n is odd or it is a multiple of 4. If n is odd, then

$$n = \left(\tfrac{1}{2}(n+1)\right)^2 - \left(\tfrac{1}{2}(n-1)\right)^2$$

is a difference of two squares. If $4 \mid n$, then

$$n = \left(\tfrac{1}{4}n + 1\right)^2 - \left(\tfrac{1}{4}n - 1\right)^2$$

is also a difference of two squares. Finally, if $n \geqslant 4$, then $n! \equiv 0 \pmod 4$ and hence $n!$ can be expressed as a difference of two squares in this case. For instance, we have $13! = 78{,}912^2 - 288^2 = 112{,}296^2 - 79{,}896^2$. Furthermore, $2! \equiv 2 \pmod 4$ and $3! \equiv 2 \pmod 4$ so that neither $2!$ nor $3!$ can be expressed as a difference of two squares.

39 Let $x \geqslant 2$ be a real number and p_N be the greatest prime satisfying $\leqslant x$. By the Prime Number Theorem, we get as $x \to \infty$

$$\sup_{p_n \leqslant x} d_n \geqslant \frac{1}{\pi(x)} \sum_{p_n \leqslant x} d_n \geqslant \frac{p_{N+1} - 2}{\pi(x)} \geqslant \frac{x-1}{\pi(x)} \sim \log x.$$

40 By assumption, $pn^2 \geqslant m^2 + 1$, and since $p \equiv 3 \pmod 4$, -1 is not a quadratic résidue modulo p. Therefore $pn^2 \geqslant m^2 + 2$, whence

$$n\sqrt{p} \geqslant \sqrt{m^2 + 2} = m\sqrt{1 + \frac{2}{m^2}} \geqslant m\left(1 + \frac{2}{3m^2}\right) = m + \frac{2}{3m}$$

which is equivalent to the asserted result, where we used the inequality

$$\sqrt{1 + x} \geqslant 1 + \tfrac{1}{3}x \quad (0 \leqslant x \leqslant 3).$$

41 Suppose that P is not irreducible over $\mathbb{Z}$. Then $P = QR$ for some $Q, R \in \mathbb{Z}[X]$ such that $Q, R \neq \pm 1$. Set $d = \deg Q$ and $\delta = \deg R$ so that $n = d + \delta$. Since $Q \neq \pm 1$, each polynomial $Q \pm 1$ has at most d roots. Therefore, there are at most d integers m such that $Q(m) = 1$ and at most d integers m such that $Q(m) = -1$, so that there are at most $2d$ integers m such that $Q(m) = \pm 1$. Similarly, there are at most 2δ integers m such that $R(m) = \pm 1$. Now if $|P(m)| = |Q(m)| \times |R(m)|$ is prime, then either $Q(m) = \pm 1$ or $R(m) = \pm 1$. We infer that there are at most $2d + 2\delta = 2n$ integers m such that $|P(m)|$ is prime, as required.

The polynomials P_1 and P_2 are both irreducible over $\mathbb{Z}$ by applying this criterion, respectively, with

$$m \in \{2, 3, 4, 5, 6, 9, 11, 12, 15\} \quad \text{and} \quad m \in \{3, 5, 8, 9, 12, 14, 15, 17, 21\}.$$

42 Since $(7, 15) = 1$, the sequence $7 - 15k$ contains infinitely many primes by Theorem 3.20. Now $7 - 15k + 2 = 3(3 + 5k)$ and $7 - 15k - 2 = 5(1 + 3k)$ and hence these two numbers are composite. We deduce that the primes contained in the sequence $7 - 15k$ cannot lie in a pair of twin primes.

43 The sequence (d_j) of the positive divisors of n is strictly increasing so that $d_j \geqslant j$ for all j. Note also that $d_j d_{k+1-j} = n$ for all $j \in \{1, \ldots, k\}$ so that

$$d_j = \frac{n}{d_{k+1-j}} \leqslant \frac{n}{k+1-j} \tag{7.45}$$

and hence

$$\sum_{j=2}^{k} d_{j-1} d_j \leqslant \sum_{j=2}^{k} \frac{n^2}{(k+2-j)(k+1-j)}$$

$$= n^2 \sum_{j=2}^{k} \left(\frac{1}{k+1-j} - \frac{1}{k+2-j}\right)$$

$$= n^2 \left(1 - \frac{1}{k}\right) < n^2.$$

44 *The following proof is picked up from* [17].

We first check numerically the assertion for $2 \leqslant n \leqslant 30$, and then assume that there exists an integer $n \geqslant 31$ which cannot be written as the sum of two squarefree numbers. Set

- $Q(x)$ be the counting function of the squarefree numbers.
- $m_1 = 1, m_2 = 2, \ldots$ be the sequence of the squarefree numbers.

Let $k \in \mathbb{Z}_{\geqslant 1}$ such that $m_1 < m_2 < \cdots < m_k < n \leqslant m_{k+1}$ so that $k \geqslant Q(n) - 1$. For all $i, j \in \{1, \ldots, k\}$, we have $1 \leqslant n - m_i < n$ and $n - m_i \neq m_j$. Since

$$\{m_1, m_2, \ldots, m_k, n - m_1, n - m_2, \ldots, n - m_k\} \subset \{1, \ldots, n - 1\}$$

we derive

$$n - 1 \geqslant 2k \geqslant 2Q(n) - 2$$

and using Proposition 4.21, we get

$$n \geqslant 2Q(n) - 1 \geqslant \frac{2n}{\zeta(2)} - \sqrt{n} - 1$$

so that $\sqrt{n} < 5.48$, and hence $n < 31$, implying a contradiction.

45 *This proof comes from* [14].

1. For $x \geqslant 2$, define $G(x) = x^{-1} \pi(x) \log x$.

- For each $r \in \mathbb{Z}_{\geqslant 1}$ satisfying $\left[\alpha^r, \alpha^{r+1}\right] \cap \mathbb{P} = \varnothing$, we have

$$G(\alpha^r) = r \alpha^{-r} \pi(\alpha^r) \log \alpha \quad \text{and} \quad G(\alpha^{r+1}) = (r + 1) \alpha^{-r-1} \pi(\alpha^r) \log \alpha$$

so that

$$\frac{G(\alpha^{r+1})}{G(\alpha^r)} = \frac{r + 1}{\alpha r} := \alpha^{\varepsilon(r)}$$

and hence

$$\log_\alpha G(\alpha^{r+1}) - \log_\alpha G(\alpha^r) = \varepsilon(r) - 1.$$

- By the PNT, $G(x) \xrightarrow[x \to \infty]{} 1$, so that $\log_\alpha G(x) \xrightarrow[x \to \infty]{} 0,$[3] and then there exists $h_1(\alpha) \in \mathbb{Z}_{\geq 1}$ such that

$$r > h_1(\alpha) \implies \log_\alpha G(\alpha^{r+1}) - \log_\alpha G(\alpha^r) > -\tfrac{1}{2}.$$

On the other hand, since $\varepsilon(r) \xrightarrow[r \to \infty]{} 0$, there exists $h_2(\alpha) \in \mathbb{Z}_{\geq 1}$ such that

$$r > h_2(\alpha) \implies \varepsilon(r) < \tfrac{1}{2}.$$

We then set $m_\alpha := \max(h_1(\alpha), h_2(\alpha))$. From the above, for all integers $r > m_\alpha$ such that $[\alpha^r, \alpha^{r+1}] \cap \mathbb{P} = \varnothing$, we have

$$-\tfrac{1}{2} < \log_\alpha G(\alpha^{r+1}) - \log_\alpha G(\alpha^r) = \varepsilon(r) - 1 < -\tfrac{1}{2}$$

which is impossible and completes the proof.

2. It suffices to show that, for all $c > \varepsilon > 0$, we have $[c - \varepsilon, c + \varepsilon] \cap F(\mathbb{P}) \neq \varnothing$, or equivalently

$$[(c - \varepsilon)q, (c + \varepsilon)q] \cap \mathbb{P} \neq \varnothing$$

for some prime number q. Let $c > \varepsilon > 0$ and choose $\alpha \in \mathbb{R}$ satisfying

$$1 < \alpha^2 < \frac{c + \varepsilon}{c - \varepsilon}.$$

Take the integer m_α from the previous question and choose a prime number q satisfying

$$q > \frac{\alpha^{m_\alpha}}{c - \varepsilon}$$

which is always possible. Then

$$\log_\alpha((c + \varepsilon)q) - \log_\alpha((c - \varepsilon)q) = \log_\alpha\left(\frac{c + \varepsilon}{c - \varepsilon}\right) > \log_\alpha(\alpha^2) = 2$$

so that the interval $\left[\log_\alpha((c - \varepsilon)q), \log_\alpha((c + \varepsilon)q)\right]$ contains at least two consecutive integers n and $n + 1$, and therefore

$$\left[\alpha^n, \alpha^{n+1}\right] \subseteq [(c - \varepsilon)q, (c + \varepsilon)q].$$

---

[3] Here we set $\log_\alpha x := \frac{\log x}{\log \alpha}$.

Since $\alpha^{m_\alpha} < (c - \varepsilon)q < \alpha^n$, we derive $n > m_\alpha$ and we may apply the previous question implying that $[\alpha^n, \alpha^{n+1}] \cap \mathbb{P} \neq \varnothing$ and hence

$$[(c - \varepsilon)q, (c + \varepsilon)q] \cap \mathbb{P} \neq \varnothing$$

as required.

46 If $|T(x, y)| \geqslant 1$, then $f(x, y) = 2$ is prime. If $|T(x, y)| < 1$, then $T(x, y) = 0$ so that $y! + 1 = x(y + 1)$ and hence $y! \equiv -1 \pmod{y + 1}$, so that $y + 1$ is prime by Wilson's theorem. Now let $p \geqslant 3$ be a prime number. Thus, $f(x, y) = p$ implies that $y = p - 1$, and since $T(x, y) = 0$, we get

$$x = \frac{y! + 1}{y + 1} = \frac{(p - 1)! + 1}{p} \in \mathbb{Z}_{\geqslant 0}$$

by Wilson's theorem. Therefore, p has the unique pre-image

$$(x, y) = \left(\frac{1}{p} \left((p - 1)! + 1 \right) , \ p - 1 \right).$$

47 The series converges by the Leibniz test. Let S be its sum and let $n \in \mathbb{Z}_{\geqslant 2}$. By (3.4), we get

$$\log(n!) = \sum_{j=1}^{n} \log j = n \log n - n + \frac{1}{2} \log(2\pi n) + O\left(\frac{1}{n}\right). \tag{7.46}$$

On the other hand, using Theorem 1.3, we also have

$$\log(n!) = n \log n - \int_{1}^{n} \frac{\lfloor t \rfloor}{t} \, dt$$

$$= n \log n - \sum_{j=1}^{n-1} \int_{j}^{j+1} \frac{j}{t} \, dt$$

$$= n \log n - \sum_{j=1}^{n-1} j \log \left(1 + \frac{1}{j}\right)$$

$$= n \log n - \sum_{j=1}^{n-1} j \sum_{k=1}^{\infty} \frac{(-1)^{k-1}}{k j^k}$$

$$= n \log n - \sum_{k=1}^{\infty} \frac{(-1)^{k-1}}{k} \sum_{j=1}^{n-1} \frac{1}{j^{k-1}},$$

where the interchange of summations is justified by absolute convergence. Therefore

$$\log(n!) = n\log n - n + 1 + \frac{1}{2}\sum_{j=1}^{n}\frac{1}{j} - \frac{1}{2n} - \sum_{k=3}^{\infty}\frac{(-1)^{k-1}}{k}\sum_{j=1}^{n-1}\frac{1}{j^{k-1}}$$

$$= n\log n - n + 1 + \frac{1}{2}\left(\log n + \gamma + O\left(\frac{1}{n}\right)\right) - \sum_{k=3}^{\infty}\frac{(-1)^{k-1}}{k}\left(\zeta(k-1) - \sum_{j\geq n}\frac{1}{j^{k-1}}\right)$$

$$= n\log n - n + 1 + \frac{1}{2}\left(\log n + \gamma + O\left(\frac{1}{n}\right)\right) - S + \sum_{k=3}^{\infty}\frac{(-1)^{k-1}}{k}\sum_{j\geq n}\frac{1}{j^{k-1}}.$$

By Corollary 1.2, we see that the last sum is $\ll n^{-1}$, so that

$$\log(n!) = n\log n - n + 1 + \frac{1}{2}(\log n + \gamma) - S + O\left(\frac{1}{n}\right) \qquad (7.47)$$

and the result follows by comparing (7.46) with (7.47) and letting $n \to \infty$.

48 Let $\sigma > 1$. We have

$$N^{\sigma-1}\sum_{n=N}^{\infty}\frac{1}{n^{\sigma}} = N^{\sigma-1}\sum_{j=1}^{\infty}\sum_{n=jN}^{(j+1)N-1}\frac{1}{n^{\sigma}}$$

$$= \frac{1}{N}\sum_{j=1}^{\infty}\sum_{n=jN}^{(j+1)N-1}\left(\frac{N}{n}\right)^{\sigma}$$

$$\leq \frac{1}{N}\sum_{j=1}^{\infty}\left(\frac{N}{jN}\right)^{\sigma}\sum_{n=jN}^{(j+1)N-1}1 = \zeta(\sigma).$$

The second inequality is similar.

49 Assume on the contrary that there exists $(x, y, z) \in (\mathbb{Z}_{\geq 1})^3$ such that $x^p + y^p = z^p$ and $z \geq \min(x^2, y^2)$. Then, using the Geometric–Arithmetic Mean inequality, we get

$$x^p = (z - y)\sum_{k=0}^{p-1}y^k z^{p-1-k}$$

$$\geq p(z - y)\left(\prod_{k=0}^{p-1}y^k z^{p-1-k}\right)^{1/p} = p(z - y)(yz)^{(p-1)/2} > (yz)^{(p-1)/2}$$

and similarly $y^p > (xz)^{(p-1)/2}$. Since $p \geqslant 3$, this implies that $x^3 > yz$ and $y^3 > xz$, and hence

$$x^9 > y^3 z^3 > xz^4 \quad \text{and} \quad y^9 > x^3 z^3 > yz^4$$

so that $z < \max\left(x^2, y^2\right)$, yielding a contradiction.

50 Assume $\sigma \in \left[0, \frac{1}{2}\right]$. From (3.18), $\zeta(it) \ll t^{1/2} \log t$ so that, using Lemma 3.14, we derive

$$\zeta(\sigma + it) \ll_{\varepsilon} \left(t^{\frac{1}{2}\left(\frac{1}{2}-\sigma\right)} t^{\alpha\sigma}\right)^2 t^{\varepsilon} \ll t^{\sigma(2\alpha-1)+\frac{1}{2}+\varepsilon} \quad \left(0 \leqslant \sigma \leqslant \frac{1}{2}, \ t \geqslant e\right).$$

The case $\sigma \in \left[\frac{1}{2}, 1\right]$ is similar, (3.18) being replaced by (3.17).

Chapter 4

51

(a) To each divisor d of n corresponds a unique divisor d' such that $dd' = n$. Hence either d or d' must be $\leqslant \sqrt{n}$ so that

$$\tau(n) \leqslant 2 \sum_{\substack{d \mid n \\ d \leqslant \sqrt{n}}} 1 \leqslant 2\sqrt{n}.$$

(b) Since n is composite, it has a prime factor q such that $q \leqslant \sqrt{n}$, and hence

$$\varphi(n) = n \prod_{p \mid n} \left(1 - \frac{1}{p}\right) \leqslant n\left(1 - \frac{1}{q}\right) \leqslant n\left(1 - \frac{1}{\sqrt{n}}\right)$$

as required.

Remark A more precise estimate than that of (a) may be easily seen, namely

$$\tau(n) = 2 \sum_{\substack{d \mid n \\ d \leqslant \sqrt{n}}} 1 - \mathbf{1}_{\square}(n), \tag{7.48}$$

where $\mathbf{1}_{\square}(n) = 1$ if n is a perfect square, 0 otherwise.

52 Using (7.48), we derive

$$\tau_3(n) = \sum_{d|n} \tau(d) = 2 \sum_{d|n} \sum_{\substack{e|d \\ e \leqslant \sqrt{d}}} 1 - \sum_{d^2|n} 1$$

$$= 2 \sum_{\substack{e|n \\ e \leqslant \sqrt{n}}} \sum_{\substack{f|(n/e) \\ f \geqslant e}} 1 - \sum_{d^2|n} 1 = 2 \sum_{\substack{e|n \\ e \leqslant \sqrt{n}}} \tau^+(n/e; e) - \sum_{d^2|n} 1.$$

53 Let $S(x)$ be the sum on the left-hand side. If $n > 1$ is neither a prime number nor a square, then $n \geqslant 6$ and $\tau(n) \geqslant 4$ is even, and if $n = m^2$ is composite, then $\tau(n) \geqslant 3$ and $m \geqslant 2$, whence

$$S(x) = 2 + \pi(x) + \sum_{\substack{n \leqslant x \\ n \text{ composite} \\ n \text{ not a square}}} f(n) + \sum_{\substack{n \leqslant x \\ n \text{ composite} \\ n \text{ square}}} f(n)$$

$$\leqslant 2 + \pi(x) + \sum_{n=6}^{\infty} \left(\frac{1}{2}\right)^n + \sum_{m=2}^{\infty} \left(\frac{2}{3}\right)^{m^2}$$

$$< \pi(x) + 2.25636 < \pi(x) + \tfrac{61}{27}.$$

54 The following ideas are due to Pólya (see [13] for instance). Fix a small real number $\varepsilon > 0$. By assumption, the product

$$\prod_{\substack{p^\alpha \\ |f(p^\alpha)| \geqslant 1}} f(p^\alpha)$$

is finite and set M its value. Furthermore, except for finitely many integers, each integer n has at least a prime power p^A such that

$$\left| f\left(p^A\right) \right| < \frac{\varepsilon}{|M|}.$$

Thus

$$|f(n)| = \left| f\left(p_1^{\alpha_1} \cdots p^A \cdots\right) \right| = \left| f\left(p_1^{\alpha_1}\right) \right| \left| f\left(p_2^{\alpha_2}\right) \right| \cdots \left| f\left(p^A\right) \right| \cdots < |M| \times \frac{\varepsilon}{|M|} = \varepsilon.$$

This implies the asserted result since ε may be as small as we want. We apply now this result to the positive multiplicative function $f(n) = \tau(n)n^{-\varepsilon}$. In view of the inequality

$$\frac{\alpha + 1}{p^{\varepsilon\alpha}} \leqslant \frac{2(1 + \log p^{\alpha})}{p^{\varepsilon\alpha}} \xrightarrow[p^{\alpha} \to \infty]{} 0$$

we derive

$$\lim_{p^{\alpha} \to \infty} f\left(p^{\alpha}\right) = \lim_{p^{\alpha} \to \infty} \frac{\alpha + 1}{p^{\varepsilon\alpha}} = 0$$

and hence $\tau(n) = O(n^{\varepsilon})$.

55 Since $\binom{k+\alpha-1}{\alpha} \leqslant (\alpha+1)^{k-1}$, we immediately derive $\tau_k(p^{\alpha}) \leqslant (\tau(p^{\alpha}))^{k-1}$, and hence $\tau_k(n) \leqslant \tau(n)^{k-1}$ by multiplicativity. The result then follows from Theorem 4.30.

56 *The method is similar for the four identities, this is why we only give the details for the first one.* We have to show that $\tau^3 \star \mathbf{1} = (\tau \star \mathbf{1})^2$. As seen in Sect. 4.1.3.4, we need to verify this identity only for prime powers. Using the well-known identity

$$\sum_{j=0}^{N} j^3 = \left(\sum_{j=0}^{N} j\right)^2$$

we get

$$\left(\tau^3 \star \mathbf{1}\right)\left(p^{\alpha}\right) = \sum_{j=0}^{\alpha} (j+1)^3 = \left(\sum_{j=0}^{\alpha} (j+1)\right)^2 = (\tau \star \mathbf{1})^2 \left(p^{\alpha}\right).$$

57

(a) We make use of the convolution identity $\Lambda_j = \log^j \star \mu$ and, by the Möbius inversion formula, we also have $\log^j = \Lambda_j \star \mathbf{1}$ so that

$$\Lambda_k(n) = \sum_{d\mid n} \mu(d) \log^k(n/d) = \sum_{d\mid n} \mu(d) \log^{k-1}(n/d) \log(n/d)$$

$$= \log n \sum_{d\mid n} \mu(d) \log^{k-1}(n/d) - \sum_{d\mid n} \mu(d) \log d \log^{k-1}(n/d)$$

$$= \Lambda_{k-1}(n) \log n - \left(\log^{k-1} \star \mu \log\right)(n)$$

and the identity $\mu \star \mathbf{1} = e_1$ together with the use of Lemma 4.2 implies that

$$\log^{k-1} \star \mu \log = \left(\log^{k-1} \star \mu\right) \star (\mathbf{1} \star \mu \log) = -(\Lambda_{k-1} \star \Lambda)$$

giving the asserted identity.

(b) We proceed as in Theorem 4.1. Since $(m, n) = 1$ and using Newton's formula, we get

$$\Lambda_k(mn) = \sum_{a|m} \sum_{b|n} \mu(ab) \log^k \left(\frac{mn}{ab}\right)$$

$$= \sum_{a|m} \mu(a) \sum_{b|n} \mu(b) (\log(m/a) + \log(n/b))^k$$

$$= \sum_{a|m} \mu(a) \sum_{b|n} \mu(b) \sum_{j=0}^{k} \binom{k}{j} \log^j (m/a) \log^{k-j} (n/b)$$

$$= \sum_{j=0}^{k} \binom{k}{j} \left(\sum_{a|m} \mu(a) \log^j (m/a)\right) \left(\sum_{b|n} \mu(b) \log^{k-j} (n/b)\right)$$

$$= \sum_{j=0}^{k} \binom{k}{j} \Lambda_j(m) \Lambda_{k-j}(n).$$

58 Define the multiplicative function f by $f(1) = 1$ and, for all prime powers

$$f(p^\alpha) = -\frac{\binom{2\alpha}{\alpha}}{4^\alpha (2\alpha - 1)}.$$

Then $f \star f$ is multiplicative by Theorem 4.1 and hence $(f \star f)(1) = 1 = \mu(1)$. Furthermore, for all primes p, we have

$$(f \star f)(p) = 2p = -\frac{2}{4} \times \binom{2}{1} = -1 = \mu(p)$$

and for all prime powers p^α such that $\alpha \geqslant 2$, we have

$$(f \star f)(p^\alpha) = \sum_{j=0}^{\alpha} f\left(p^j\right) f\left(p^{\alpha-j}\right)$$

$$= \sum_{j=0}^{\alpha} \left(-\frac{\binom{2j}{j}}{4^j(2j-1)} \right) \left(-\frac{\binom{2\alpha-2j}{\alpha-j}}{4^{\alpha-j}(2\alpha-2j-1)} \right)$$

$$= 4^{-\alpha} \sum_{j=0}^{\alpha} \frac{\binom{2j}{j}\binom{2\alpha-2j}{\alpha-j}}{(2j-1)(2\alpha-2j-1)} = 0 = \mu\left(p^{\alpha}\right),$$

where we used [10, identity 3.93], which concludes the proof.

59 We proceed as in Theorem 4.1 or in Exercise 57 above.

$$(f \star g)(mn) = \sum_{a|m} \sum_{b|n} f(ab)g\left(\frac{mn}{ab}\right)$$

$$= \sum_{a|m} \sum_{b|n} (f(a) + f(b)) \, g\left(\frac{m}{a}\right) g\left(\frac{n}{b}\right)$$

$$= \sum_{a|m} f(a)g\left(\frac{m}{a}\right) \sum_{b|n} g\left(\frac{n}{b}\right) + \sum_{a|m} g\left(\frac{m}{a}\right) \sum_{b|n} f(b)g\left(\frac{n}{b}\right)$$

$$= (f \star g)(m)(g \star 1)(n) + (g \star 1)(m)(f \star g)(n)$$

as required. We apply this identity with $g = \mu$ and f any additive function. First, $(f \star \mu)(1) = f(1) = 0$. If $n = p_1^{\alpha_1} p_2^{\alpha_2}$, we get using (4.6)

$$(f \star \mu)(n) = (f \star \mu)\left(p_1^{\alpha_1}\right) e_1\left(p_2^{\alpha_2}\right) + (f \star \mu)\left(p_2^{\alpha_2}\right) e_1\left(p_1^{\alpha_1}\right) = 0$$

and by induction this result is still true for all $n = p_1^{\alpha_1} \cdots p_r^{\alpha_r}$ and $r \geqslant 2$. Note that if $f(p^{\alpha}) \neq f(p^{\alpha-1})$, then $(f \star \mu)(p^{\alpha}) \neq 0$ by (4.14).

60 This is [12, Lemma 61.1].

61 We first easily check that $\sum_{p|n} \dfrac{1}{p} < 1$ for $2 \leqslant n \leqslant 15$. It can also be numerically checked that the inequality

$$\sum_{p|n} \frac{1}{p} < \log\log\log n + 1$$

holds in the range $16 \leqslant n < 30{,}000$. Assume $n \geqslant 30{,}000$ and let $t \in (1, n]$ be a parameter at our disposal. Write

$$\sum_{p|n} \frac{1}{p} = \left(\sum_{\substack{p|n \\ p \leqslant t}} + \sum_{\substack{p|n \\ p > t}} \right) \frac{1}{p} \leqslant \sum_{p \leqslant t} \frac{1}{p} + \frac{\omega(n)}{t}.$$

From Proposition 4.20, we derive $\omega(n) < \frac{3}{5}\log n$ since $n \geqslant 30{,}000$. Using Corollary 3.13, we get

$$\sum_{p|n} \frac{1}{p} < \log\log t + 0.2615 + (\log t)^{-2} + \tfrac{3}{5}t^{-1}\log n.$$

Now let $a > 0$ and choose $t = a\log n$. We obtain

$$\sum_{p|n} \frac{1}{p} < \log\log\log n + \frac{\log a}{\log\log n} + 0.2615 + \frac{1}{(\log(a\log n))^2} + \frac{3}{5a}$$

$$\leqslant \log\log\log n + \frac{\log a}{\log\log 30{,}000} + 0.2615 + \frac{1}{(\log(a\log 30{,}000))^2} + \frac{3}{5a}$$

and the asserted bound follows from choosing $a = 1.74$.

62 We first numerically check that, for $3 \leqslant n \leqslant 600{,}000$ and $n \notin \{3, 4, 5, 6, 10, 12, 30\}$, the inequality

$$\sum_{p|n} \frac{\log p}{p} < \log\log n + 1 + \gamma - E$$

holds, where $E \approx 1.332582275\ldots$ is the constant appearing in Corollary 3.13, and for $n \in \{3, 4, 5, 6, 10, 12, 30\}$, we check that

$$\sum_{p|n} \frac{\log p}{p-1} < \log\log n + 1.$$

Since

$$\sum_{p|n} \frac{\log p}{p-1} = \sum_{p|n} \frac{\log p}{p} + \sum_{p|n} \frac{\log p}{p(p-1)} \leqslant \sum_{p|n} \frac{\log p}{p} + \sum_{p} \frac{\log p}{p(p-1)} = \sum_{p|n} \frac{\log p}{p} + E - \gamma$$

we infer that the inequality of this exercise holds for any $3 \leqslant n \leqslant 600000$. Now suppose $n > 600000$ and let $t \in (1, n]$ be a parameter at our disposal. Using Corollary 3.13 and (3.61), we derive

$$\sum_{p|n} \frac{\log p}{p-1} \leqslant \sum_{p|n} \frac{\log p}{p} + E - \gamma = \sum_{\substack{p|n \\ p \leqslant t}} \frac{\log p}{p} + \sum_{\substack{p|n \\ p > t}} \frac{\log p}{p} + E - \gamma$$

$$< \sum_{p \leqslant t} \frac{\log p}{p} + \frac{\log n}{t} + E - \gamma = \log t + \frac{\log n}{t} - \gamma + R(t),$$

where

$$
|R(t)| < \begin{cases} 2t^{-1/2}, & \text{if } 0 < t < 113; \\ (\log t)^{-1}, & \text{if } t \geqslant 32. \end{cases}
$$

- Assume first that $600000 < n < e^{\frac{113}{1.3}}$. Choose $t = 1.3 \log n$ so that $17 < t < 113$, and thus

$$
\sum_{p|n} \frac{\log p}{p-1} < \log\log n + \log 1.3 + \frac{1}{1.3} - \gamma + \frac{2}{\sqrt{1.3\log n}} < \log\log n + 0.9353.
$$

- Suppose that $n \geqslant e^{\frac{113}{1.3}}$. Choose $t = \log n$ so that $t > 86.923$, and hence

$$
\sum_{p|n} \frac{\log p}{p-1} < \log\log n - \gamma + 1 + \frac{1}{\log\log n} < \log\log n + 0.65.
$$

63 Using Exercise 61, we derive, for all $n \geqslant 16$

$$
\log\left(\frac{\varphi(n)}{n}\right) = \sum_{p|n} \log\left(1 - \frac{1}{p}\right) \geqslant -\sum_{p|n}\left(\frac{1}{p} + \frac{1}{p^2}\right) \geqslant -\log\log\log n - 1 - \log\zeta(2)
$$

implying the asserted first lower bound. The second one follows immediately from

$$
\sigma(n) = n \prod_{p^\alpha \| n} \frac{p - p^{-\alpha}}{p - 1} \leqslant n \prod_{p|n} \frac{p}{p-1} = \frac{n^2}{\varphi(n)}.
$$

Remark Better values of the constant exist in the literature. For instance, Ivić [16] proved that

$$
\sigma(n) \leqslant 2.59\, n \log\log n \quad (n \geqslant 7)
$$

while Robin [19] got the estimate

$$
\sigma(n) < e^\gamma n \log\log n + \frac{0.6483n}{\log\log n} \quad (n \geqslant 3)
$$

and more recently Axler [2] derived the better bound

$$
\sigma(n) < e^\gamma n \log\log n + \frac{0.1209n}{(\log\log n)^2} \quad (n \geqslant 2521).
$$

64

(a) One may assume that $n \geqslant 2$ is expressed in the form $n = p_1^{\alpha_1} \cdots p_r^{\alpha_r}$ with $\alpha_j \geqslant 1$. Then $n^n = p_1^{n\alpha_1} \cdots p_r^{n\alpha_r}$ and hence

$$\varphi(n)\sigma\left(n^n\right) = n \prod_{j=1}^{r}\left(1 - \frac{1}{p_j}\right) \prod_{j=1}^{r}\left(\frac{p_j^{n\alpha_j+1} - 1}{p_j - 1}\right)$$

$$= n \prod_{j=1}^{r} p^{n\alpha_j}\left(1 - \frac{1}{p_j^{n\alpha_j+1}}\right)$$

$$= n^{n+1} \prod_{j=1}^{r}\left(1 - \frac{1}{p_j^{n\alpha_j+1}}\right)$$

which implies the asserted upper bound by noticing that the product is $\leqslant 1$. For the lower bound, we use $\alpha_j \geqslant 1$ which provides

$$\varphi(n)\sigma\left(n^n\right) \geqslant n^{n+1} \prod_{j=1}^{r}\left(1 - \frac{1}{p_j^{n+1}}\right) \geqslant n^{n+1} \prod_{p}\left(1 - \frac{1}{p^{n+1}}\right) = \frac{n^{n+1}}{\zeta(n + 1)}$$

as asserted.

(b) By the previous question, we first have $f(n) \geqslant 0$ and

$$f(n) \leqslant \frac{n}{\varphi(n)} - \frac{n^{n+1}}{n^n \varphi(n)\zeta(n + 1)} = \frac{n(\zeta(n + 1) - 1)}{\varphi(n)\zeta(n + 1)}.$$

Now by Exercise 48, we derive

$$\zeta(n + 1) - 1 = \sum_{k=2}^{\infty} \frac{1}{k^{n+1}} \leqslant \frac{\zeta(n + 1)}{2^n}$$

and the estimate

$$\frac{n}{\varphi(n)} < e^{\gamma}\zeta(2) \log\log n$$

from Exercise 63, and valid for all $n \geqslant 16$, yields

$$0 \leqslant f(n) < 2^{-n}e^{\gamma}\zeta(2) \log\log n$$

for all $n \geqslant 16$, and hence the series $\sum_{n \geqslant 1} f(n)$ converges. Note that

$$\sum_{n=1}^{\infty} \left(\frac{n}{\varphi(n)} - \frac{\sigma(n^n)}{n^n} \right) \approx 0.298603\ldots$$

65 First note that it suffices to show the identity for n squarefree. Now if $\mu(n)^2 = 1$ and if $p \mid n$, then $\frac{\varphi(n)}{p-1} = \varphi\left(\frac{n}{p}\right)$, so that

$$-\frac{\varphi(n)}{n} \sum_{p|n} \frac{\log p}{p-1} = -\frac{1}{n} \sum_{p|n} \varphi\left(\frac{n}{p}\right) \log p = -\frac{1}{n} (\Lambda \star \varphi)(n)$$

$$= -\frac{1}{n} (-\mu \log \star \mathbf{1} \star \mu \star \mathrm{id})(n)$$

$$= \frac{1}{n} (\mu \log \star \mathrm{id})(n) = \sum_{d|n} \frac{\mu(d) \log d}{d}.$$

66 Taking the multiplicativity of f and the additivity of g into account, we derive

$$\sum_{d|n} \mu(d)^e f(d) g(d) = \sum_{d|n} \mu(d)^e f(d) \sum_{p|d} g(p) = \sum_{p|n} g(p) \sum_{\substack{h|(n/p) \\ p \nmid h}} \mu(ph)^e f(ph)$$

$$= (-1)^e \sum_{p|n} f(p) g(p) \sum_{\substack{h|(n/p) \\ p \nmid h}} \mu(h)^e f(h)$$

$$= (-1)^e \sum_{p|n} f(p) g(p) \prod_{\substack{q \text{ prime} \\ q|(n/p) \\ q \neq p}} \left(1 + (-1)^e f(q)\right)$$

and the product is equal to

$$\left(1 + (-1)^e f(p)\right)^{-1} \prod_{p|n} \left(1 + (-1)^e f(p)\right)$$

completing the proof.

Example Let $k, n \in \mathbb{Z}_{\geq 1}$.

$$\sum_{d|n} \frac{\mu(d) \log d}{d^k} = -\frac{J_k(n)}{n^k} \sum_{p|n} \frac{\log p}{p^k - 1}.$$

$$\sum_{d|n} \frac{\mu(d)^2 \log d}{d^k} = \frac{\Psi_k(n)}{n^k} \sum_{p|n} \frac{\log p}{p^k + 1}.$$

$$\sum_{d|n} \frac{\mu(d) \log d}{\tau(d)} = -2^{-\omega(n)} \log \gamma(n).$$

$$\sum_{d|n} \mu(d)^e \tau_k(d) \log d = \left(1 + (-1)^e k\right)^{\omega(n)} \times \frac{k \log \gamma(n)}{k + (-1)^e} \quad \left(k \in \mathbb{Z}_{\geq 2}\right).$$

The first example above generalizes Exercise 65.

67 For any $s \in \mathbb{R}_{\geq 0}$, define $\mathrm{id}_s(n) = n^s$ and

$$F(s) := (f \star \mathrm{id}_s)(n) = n^s \prod_{p^\alpha \| n} \sum_{j=0}^{\alpha} \frac{f(p^j)}{p^{js}}.$$

Then

$$\log\left(\frac{n}{t}\right) \sum_{\substack{d|n \\ d \leq t}} f(d) \leq \sum_{\substack{d|n \\ d \leq t}} f(d) \log\left(\frac{n}{d}\right) \leq (f \star \log)(n) = F'(0)$$

with $F(0) = (f \star 1)(n)$ and

$$\frac{F'}{F}(s) = \log n + \sum_{p^\alpha \| n} \frac{d}{ds} \log\left(\sum_{j=0}^{\alpha} \frac{f(p^j)}{p^{js}}\right)$$

completing the proof. *The idea of this proof is due to* [7].

Example Let $k \in \mathbb{Z}_{\geq 2}$.

$$\sum_{\substack{d|n \\ d \leq t}} \mu(d)^2 \leq \frac{2^{\omega(n)} \log n}{2 \log(n/t)}.$$

$$\sum_{\substack{d|n \\ d \leq t}} \tau(d) \leq \frac{\tau_3(n) \log n}{3 \log(n/t)}.$$

$$\sum_{\substack{d|n \\ d \leqslant t}} \varphi(d) \leqslant \frac{n}{\log(n/t)} \sum_{p|n} \frac{\log p}{p-1} \leqslant \frac{n(\log \log n + 1)}{\log(n/t)}.$$

$$\sum_{\substack{d|n \\ d \leqslant t}} J_k(d) \leqslant \frac{n^k}{\log(n/t)} \sum_{p|n} \frac{\log p}{p^k-1} \leqslant \frac{2n^k}{\log(n/t)}.$$

68 We easily check that

$$f_k(n) = \begin{cases} 1, & \text{if } n = m^k ; \\ 0, & \text{otherwise} \end{cases} \quad \text{and} \quad f_k^{-1}(n) = \begin{cases} \mu(m), & \text{if } n = m^k ; \\ 0, & \text{otherwise.} \end{cases}$$

69 By (4.14), the condition (4.84) implies that $g = f \star \mu \geqslant 0$. Hence we have by Proposition 4.3

$$\sum_{n \leqslant x} f(n) = \sum_{n \leqslant x} (g \star \mathbf{1})(n) = \sum_{d \leqslant x} g(d) \left\lfloor \frac{x}{d} \right\rfloor \leqslant x \sum_{d \leqslant x} \frac{g(d)}{d} \leqslant x \prod_{p \leqslant x} \left(1 + \sum_{\alpha=1}^{\infty} \frac{g(p^{\alpha})}{p^{\alpha}} \right)$$

and we conclude by using $g(p^{\alpha}) = f(p^{\alpha}) - f(p^{\alpha-1})$.

Example

- When $f = \tau_k$, this result yields

$$\sum_{n \leqslant x} \tau_k(n) \leqslant x \prod_{p \leqslant x} \left(1 - \frac{1}{p} \right)^{1-k}.$$

Now the explicit upper bound of the second Mertens theorem provided by Corollary 3.13 implies that

$$\sum_{n \leqslant x} \tau_k(n) < (2e^{\gamma})^{k-1} x (\log x)^{k-1}$$

for all $x \geqslant e$.
- When $f = k^{\omega}$, we derive

$$\sum_{n \leqslant x} k^{\omega(n)} \leqslant x \prod_{p \leqslant x} \left(1 + \frac{k-1}{p} \right) \leqslant x \exp \left((k-1) \sum_{p \leqslant x} \frac{1}{p} \right)$$

and Corollary 3.13 implies that

$$\sum_{n \leqslant x} k^{\omega(n)} \leqslant x \exp\{(k-1)(\log\log x + 1)\} = e^{k-1} x (\log x)^{k-1}$$

for all $x \geqslant 4$. A more precise estimate will be established in Exercise 81.

70 The identity is true for $k = 2$ by Theorem 4.28. Suppose it is true for some integer $k \geqslant 2$. By (5), (6) and induction hypothesis, we get

$$S_{k+1}(x) = x \sum_{d \leqslant x} \frac{\tau_k(d)}{d} + O\left(S_k(x)\right)$$

$$= x \int_1^x \frac{S_k(t)}{t^2} \, dt + O\left(x(\log x)^{k-1}\right)$$

$$= x \int_1^x \left\{ \frac{(\log t)^{k-1}}{(k-1)!} + O\left((\log t)^{k-2}\right) \right\} \frac{dt}{t} + O\left(x(\log x)^{k-1}\right)$$

$$= \frac{x(\log x)^k}{k!} + O\left(x(\log x)^{k-1}\right).$$

71 Let U satisfy $1 \leqslant U \leqslant x$. Using Theorem 4.4, we get

$$\sum_{n \leqslant x} (\chi \star \mathrm{id})(n) = \sum_{n \leqslant U} \chi(n) \sum_{k \leqslant x/n} k + \sum_{n \leqslant x/U} n \sum_{k \leqslant x/n} \chi(k) - \sum_{n \leqslant U} \chi(n) \sum_{n \leqslant x/U} n$$

$$= \frac{1}{2} \sum_{n \leqslant U} \chi(n) \left\lfloor \frac{x}{n} \right\rfloor \left(\left\lfloor \frac{x}{n} \right\rfloor + 1 \right) + O\left\{ \sum_{n \leqslant x/U} n \left| \sum_{k \leqslant x/n} \chi(k) \right| \right\}$$

$$+ O\left\{ \left(\sum_{n \leqslant x/U} n \right) \left| \sum_{n \leqslant U} \chi(n) \right| \right\}.$$

The use of the Pólya–Vinogradov inequality[4] and the estimate

$$\left\lfloor \frac{x}{n} \right\rfloor \left(\left\lfloor \frac{x}{n} \right\rfloor + 1 \right) = \frac{x^2}{n^2} - \frac{2x}{n} \psi\left(\frac{x}{n}\right) + O(1)$$

---

[4]See Theorem 6.10.

gives

$$\sum_{n \leqslant x} (\chi \star \mathrm{id})\,(n) = \frac{x^2}{2} \sum_{n \leqslant U} \frac{\chi(n)}{n^2} - x \sum_{n \leqslant U} \frac{\chi(n)}{n} \psi\left(\frac{x}{n}\right) + O\left(x^2 U^{-2} q^{1/2} \log q + U\right).$$

Now

$$\sum_{n \leqslant U} \frac{\chi(n)}{n^2} = L(2, \chi) - \sum_{n > U} \frac{\chi(n)}{n^2}$$

and, as in Corollary 3.8, we get by Abel's summation and the Pólya–Vinogradov inequality the estimate

$$\left|\sum_{n > U} \frac{\chi(n)}{n^2}\right| \leqslant \frac{2\sqrt{3q}\,\log q}{U^2}$$

so that the choice of $U = x^{2/3} q^{1/6}$ gives the asserted result.

Remark When χ is a quadratic character modulo q, and thus is the Dirichlet character attached to the quadratic field $\mathbb{K} = \mathbb{Q}(\sqrt{d})$ where $d = \chi(-1)q$ by Lemma 7.17, it can be proved that[5]

$$(\chi \star \mathrm{id})(n) = \sum_{i=1}^{n} r_{\mathbb{K}}((i, n))$$

so that this estimate is used in the problem of the composition of the gcd and the multiplicative function $r_{\mathbb{K}}$. See [4] for instance.

72

(a) The function $\mathrm{id} \times \varphi^{-1}$ satisfies the condition of Theorem 4.9 with $a = \log 16$ and $b = 6$, so that

$$\sum_{n \leqslant x} \frac{n}{\varphi(n)} \ll \frac{x}{\log x} \exp\left(\sum_{p \leqslant x} \frac{1}{p - 1}\right) \ll x$$

and the result follows by partial summation.

---

[5] See Definition 7.20 and Exercise 141 for the function r_K.

(b) It can easily be checked that $g(n) = \dfrac{\mu(n)^2}{\varphi(n)}$, and therefore

$$L(s, g) = \prod_p \left(1 + \frac{1}{p^s(p-1)}\right)$$

which is a Dirichlet series absolutely convergent in the half-plane $\sigma > 0$ by Corollary 4.5. Also note that, for any $s \in \mathbb{C}$ such that $\sigma > 0$

$$\frac{L'}{L}(s, g) = -\sum_p \frac{\log p}{p^s(p-1)+1}.$$

(c) Putting all together we derive

$$\sum_{n \leqslant x} \frac{1}{\varphi(n)} = \sum_{n \leqslant x} \frac{(g \star 1)(n)}{n} = \sum_{d \leqslant x} \frac{g(d)}{d} \sum_{k \leqslant x/d} \frac{1}{k}$$

$$= \sum_{d \leqslant x} \frac{g(d)}{d} \left(\log \frac{x}{d} + \gamma + O\left(\frac{d}{x}\right)\right)$$

$$= (\log x + \gamma) \sum_{d \leqslant x} \frac{g(d)}{d} - \sum_{d \leqslant x} \frac{g(d)\log d}{d} + O\left(\frac{1}{x} \sum_{d \leqslant x} g(d)\right).$$

Now from question (a) we have $\displaystyle\sum_{d \leqslant x} g(d) \ll \log x$, and hence by partial summation

$$\sum_{d > x} \frac{g(d)}{d} = -\frac{1}{x} \sum_{d \leqslant x} g(d) + \int_x^\infty \frac{1}{t^2} \left(\sum_{d \leqslant t} g(d)\right) dt \ll \frac{\log x}{x}$$

and similarly

$$\sum_{d > x} \frac{g(d)\log d}{d} \ll \frac{(\log x)^2}{x}$$

and thus

$$\sum_{d \leqslant x} \frac{g(d)}{d} = L(1, g) + O\left(\frac{\log x}{x}\right)$$

$$-\sum_{d \leqslant x} \frac{g(d)\log d}{d} = L'(1, g) + O\left(\frac{(\log x)^2}{x}\right) = L(1, g) \times \frac{L'}{L}(1, g) + O\left(\frac{(\log x)^2}{x}\right)$$

so that

$$\sum_{n \leqslant x} \frac{1}{\varphi(n)} = L(1, g) \left(\log x + \gamma + \frac{L'}{L}(1, g) \right) + O\left(\frac{(\log x)^2}{x} \right)$$

and we conclude the proof with

$$L(1, g) = \prod_p \left(1 + \frac{1}{p(p-1)} \right) = \frac{\zeta(2)\zeta(3)}{\zeta(6)}.$$

(d) Denote $\mathbf{1}_{(\cdot, q)=1}$ the characteristic function of the integers coprime to q and set

$$g_q := g \times \mathbf{1}_{(\cdot, q)=1}.$$

As above, we first derive

$$\sum_{\substack{n \leqslant x \\ (n,q)=1}} \frac{1}{\varphi(n)} = \sum_{d \leqslant x} \frac{g_q(d)}{d} \sum_{\substack{k \leqslant x/d \\ (k,q)=1}} \frac{1}{k}$$

$$= \sum_{d \leqslant x} \frac{g_q(d)}{d} \left\{ \frac{\varphi(q)}{q} \left(\log \frac{x}{d} + \gamma + E(q) \right) + O\left(\frac{2^{\omega(q)}d}{x} \right) \right\}$$

$$= \frac{\varphi(q)}{q} (\log x + \gamma + E(q)) \sum_{d \leqslant x} \frac{g_q(d)}{d} - \frac{\varphi(q)}{q} \sum_{d \leqslant x} \frac{g_q(d) \log d}{d}$$

$$+ O\left(\frac{2^{\omega(q)}}{x} \sum_{d \leqslant x} g_q(d) \right),$$

where we used Example 4.32 and where $E(q) := \sum_{p|q} \frac{\log p}{p-1}$. Now, for $\sigma > 0$

$$L(s, g_q) = \prod_{p|q} \left(1 + \frac{1}{p^s(p-1)} \right)^{-1} L(s, g);$$

$$\frac{L'}{L}(s, g_q) = \frac{L'}{L}(s, g) + \sum_{p|q} \frac{\log p}{p^s(p-1)+1}$$

so that the bounds above for g also hold for g_q, namely

$$\sum_{d \leqslant x} g_q(d) \ll \log x, \quad \sum_{d > x} \frac{g_q(d)}{d} \ll \frac{\log x}{x} \quad \text{and} \quad \sum_{d > x} \frac{g_q(d) \log d}{d} \ll \frac{(\log x)^2}{x}.$$

and hence

$$\sum_{d\leqslant x} \frac{g_q(d)}{d} = L(1, g_q) + O\left(\frac{\log x}{x}\right) = C_1 + O\left(\frac{\log x}{x}\right)$$

$$-\sum_{d\leqslant x} \frac{g_q(d)\log d}{d} = L(1, g_q) \times \frac{L'}{L}(1, g_q) + O\left(\frac{(\log x)^2}{x}\right)$$

with

$$L(1, g_q) \times \frac{L'}{L}(1, g_q) = C_1 \left(-\sum_p \frac{\log p}{p(p-1)+1} + \sum_{p|q} \frac{\log p}{p(p-1)+1}\right).$$

Putting all together yields

$$\sum_{\substack{n\leqslant x \\ (n,q)=1}} \frac{1}{\varphi(n)} = \frac{C_1\varphi(q)}{q}\log x + \frac{C_1\varphi(q)}{q}\left(\gamma + E(q) - \sum_p \frac{\log p}{p(p-1)+1} + \sum_{p|q} \frac{\log p}{p(p-1)+1}\right)$$

$$+ O\left\{\frac{\log x}{x}\left(\frac{\varphi(q)E(q)}{q} + 2^{\omega(q)} + \log x\right)\right\}$$

and we conclude the proof noting that, from Exercise 66, we have

$$\frac{\varphi(q)E(q)}{q} \leqslant \sum_{d|q} \frac{\mu(d)^2\log d}{d} \leqslant \sum_{d|q} \mu(d)^2 = 2^{\omega(q)}$$

and also

$$C_1\left(\gamma + E(q) - \sum_p \frac{\log p}{p(p-1)+1} + \sum_{p|q} \frac{\log p}{p(p-1)+1}\right) = C_2.$$

73 We have

$$\sum_{k=1}^n f((k,n)) = \sum_{d|n} f(d) \sum_{\substack{k\leqslant n/d \\ (k,n/d)=1}} 1 = \sum_{d|n} f(d)\varphi\left(\frac{n}{d}\right) = (f \star \varphi)(n).$$

74 If $S(n)$ is the sum on the left-hand side, then

$$S(n) = \sum_{j=1}^n \sum_{\frac{n}{j+1}<k\leqslant \frac{n}{j}} (k, j)$$

and since

$$\sum_{k\leqslant \frac{n}{j}}(k,j) = \sum_{d\mid j} d \sum_{\substack{k\leqslant n/j \\ (k,j)=d}} 1 = \sum_{d\mid j} d \sum_{\substack{h\leqslant n/(dj) \\ (h,j/d)=1}} 1 = \sum_{d\mid j} d\varphi\left(\frac{n}{dj}, \frac{j}{d}\right)$$

we derive

$$S(n) = \sum_{j=1}^{n} \sum_{d\mid j} d\left\{\varphi\left(\frac{n}{dj}, \frac{j}{d}\right) - \varphi\left(\frac{n}{d(j+1)}, \frac{j}{d}\right)\right\}$$

$$= \sum_{d\leqslant n} d \sum_{k\leqslant n/d} \left\{\varphi\left(\frac{n}{kd^2}, k\right) - \varphi\left(\frac{n}{d(kd+1)}, k\right)\right\}.$$

75   We first have

$$\sum_{T<n\leqslant x} f\left(\lfloor x/n\rfloor\right) = \sum_{d\leqslant x/T} f(d) \sum_{\substack{\frac{x}{d+1}<n\leqslant\frac{x}{d} \\ T<n\leqslant x}} 1.$$

Now note that when $d \leqslant \frac{x}{N} - 1$, the interval $\left(\frac{x}{d+1}, \frac{x}{d}\right]$ is indeed a subinterval of $(T, x]$ since $x/d \leqslant x$ and

$$\frac{x}{d+1} \geqslant \frac{x}{xT^{-1} - 1 + 1} = T.$$

Therefore

$$\sum_{T<n\leqslant x} f\left(\lfloor x/n\rfloor\right) = \sum_{d\leqslant x/T-1} f(d) \sum_{\frac{x}{d+1}<n\leqslant\frac{x}{d}} 1 + \sum_{x/T-1<d\leqslant x/T} f(d) \sum_{\substack{\frac{x}{d+1}<n\leqslant\frac{x}{d} \\ T<n\leqslant x}} 1$$

$$= \sum_{d\leqslant x/T} f(d) \left(\left\lfloor\frac{x}{d}\right\rfloor - \left\lfloor\frac{x}{d+1}\right\rfloor\right)$$

$$- \sum_{x/T-1<d\leqslant x/T} f(d) \left(\sum_{\frac{x}{d+1}<n\leqslant\frac{x}{d}} 1 - \sum_{\substack{\frac{x}{d+1}<n\leqslant\frac{x}{d} \\ T<n\leqslant x}} 1\right)$$

and since $\lfloor x/T \rfloor$ is the only integer in the interval $\left(\frac{x}{T} - 1, \frac{x}{T} \right]$, the last sum is

$$= f\left(\lfloor x/T \rfloor\right) \sum_{\substack{\lfloor x/T \rfloor + 1 < n \leqslant \lfloor x/T \rfloor \\ n \leqslant T}} 1 = f\left(\lfloor x/T \rfloor\right) \sum_{\lfloor x/T \rfloor + 1 < n \leqslant T} 1 \leqslant f\left(\lfloor x/T \rfloor\right)\left(1 + T^2 x^{-1}\right)$$

as required.

76

(a) By partial summation, we obtain

$$\sum_{n \leqslant z} n\tau(n) = z \sum_{n \leqslant z} \tau(n) - \int_1^z \left(\sum_{n \leqslant t} \tau(n) \right) dt$$

$$= z \left\{ z(\log z + 2\gamma - 1) + O\left(z^{\theta + \varepsilon}\right) \right\}$$

$$- \int_1^z \left\{ t(\log t + 2\gamma - 1) + O\left(t^{\theta + \varepsilon}\right) \right\} dt$$

$$= \tfrac{1}{2} z^2 \log z + z^2 \left(\gamma - \tfrac{1}{4} \right) + O\left(z^{1 + \theta + \varepsilon}\right)$$

as asserted.

(b) Cesáro's identity from Exercise 73 with $f = \mathrm{id}$ yields $\mathcal{P} = \varphi \star \mathrm{id}$, and (4.7) gives

$$\mathcal{P} = \mu \star \mathrm{id} \star \mathrm{id} = \mu \star \mathrm{id} \times (1 \star 1) = \mu \star (\mathrm{id} \times \tau),$$

where we used the complete multiplicativity of the function id.

(c) By above and Proposition 4.3, we have

$$\sum_{n \leqslant x} \mathcal{P}(n) = \sum_{d \leqslant x} \mu(d) \sum_{k \leqslant x/d} k\tau(k)$$

$$= \sum_{d \leqslant x} \mu(d) \left\{ \frac{x^2}{d^2} \left(\frac{1}{2} \log \frac{x}{d} + \gamma - \frac{1}{4} \right) + O\left(\left(\frac{x}{d}\right)^{1 + \theta + \varepsilon} \right) \right\}$$

$$= x^2 \left\{ \left(\frac{\log x}{2} + \gamma - \frac{1}{4} \right) \sum_{d \leqslant x} \frac{\mu(d)}{d^2} - \sum_{d \leqslant x} \frac{\mu(d) \log d}{2d^2} \right\}$$

$$+ O\left(x^{1 + \theta + \varepsilon} \sum_{d \leqslant x} \frac{1}{d^{1 + \theta + \varepsilon}} \right).$$

Now Exercise 48 yields

$$\sum_{d \leqslant x} \frac{\mu(d)}{d^2} = \sum_{d=1}^{\infty} \frac{\mu(d)}{d^2} + O\left(\sum_{d>x} \frac{1}{d^2}\right) = \frac{1}{\zeta(2)} + O\left(\frac{1}{x}\right)$$

and the use of Theorem 4.7 implies that

$$\sum_{d \leqslant x} \frac{\mu(d) \log d}{d^2} = \sum_{d=1}^{\infty} \frac{\mu(d) \log d}{d^2} + O\left(\frac{\log x}{x}\right) = \frac{\zeta'(2)}{\zeta(2)^2} + O\left(\frac{\log x}{x}\right)$$

so that

$$\sum_{n \leqslant x} P(n) = \frac{x^2}{\zeta(2)}\left(\frac{\log x}{2} + \gamma - \frac{1}{4}\right) - \frac{x^2 \zeta'(2)}{2\zeta(2)^2} + O\left(x^{1+\theta+\varepsilon}\right)$$

$$= \frac{x^2}{2\zeta(2)}\left(\log x + 2\gamma - \frac{1}{2} - \frac{\zeta'(2)}{\zeta(2)}\right) + O\left(x^{1+\theta+\varepsilon}\right).$$

The best value for θ to date is given by Bourgain and Watt in Corollary 6.12. We get for all $\varepsilon > 0$

$$\sum_{n \leqslant x} P(n) = \frac{x^2}{2\zeta(2)}\left(\log x + 2\gamma - \frac{1}{2} - \frac{\zeta'(2)}{\zeta(2)}\right) + O\left(x^{2175/1648+\varepsilon}\right).$$

77 Let $k \in \mathbb{Z}_{\geqslant 1}$.

(a) The right-hand side is equal to

$$\sum_{\substack{med \leqslant x \\ d|k \\ e|k}} \mu(d)\mu(e)\tau(m) = \sum_{\substack{n \leqslant x \, ed|n \\ d|k \\ e|k}} \sum \mu(d)\mu(e)\tau\left(\frac{n}{ed}\right) = \sum_{n \leqslant x} (\mu \mathbf{1}_k \star \mu \mathbf{1}_k \star \tau)(n),$$

where $\mathbf{1}_k(m) = 1$ if $m \mid k$ and 0 otherwise. Hence on the one hand

$$L\left(s, \mathbf{1}_{(\cdot,k)=1}\tau\right) = \zeta(s)^2 \prod_{p|k}\left(1 - \frac{1}{p^s}\right)^2$$

and on the other hand

$$L(s, \mu \mathbf{1}_k) = \prod_{p|k}\left(1 - \frac{1}{p^s}\right)$$

so that

$$L\left(s, \mu\mathbf{1}_k \star \mu\mathbf{1}_k \star \tau\right) = \zeta(s)^2 \prod_{p|k}\left(1 - \frac{1}{p^s}\right)^2 = L\left(s, \mathbf{1}_{(\cdot,k)=1}\tau\right).$$

(b) With the help of (4.85), we derive

$$\sum_{\substack{n \leqslant x \\ (n,k)=1}} \tau(n) = \sum_{d|k} \sum_{e|k} \mu(d)\mu(e)\left\{\frac{x}{ed}\log\frac{x}{ed} + \frac{x}{ed}(2\gamma - 1) + O\left(\left(\frac{x}{ed}\right)^{\theta+\varepsilon}\right)\right\}$$

$$= x\sum_{d|k}\frac{\mu(d)}{d}\sum_{e|k}\frac{\mu(e)}{e}\log\frac{x}{ed} + x(2\gamma - 1)\left(\sum_{d|k}\frac{\mu(d)}{d}\right)^2$$

$$+ O\left(x^{\theta+\varepsilon}\left(\sum_{d|k}\frac{\mu(d)^2}{d^\theta}\right)^2\right)$$

$$= x\,(\log x + 2\gamma - 1)\left(\sum_{d|k}\frac{\mu(d)}{d}\right)^2 - x\sum_{d|k}\frac{\mu(d)}{d}\sum_{e|k}\frac{\mu(e)}{e}\log(ed)$$

$$+ O\left(x^{\theta+\varepsilon}\left(\sum_{d|k}\frac{\mu(d)^2}{d^\theta}\right)^2\right)$$

and note that

$$\sum_{d|k}\frac{\mu(d)}{d} = \frac{\varphi(k)}{k}, \quad \sum_{d|k}\frac{\mu(d)^2}{d^\theta} = \prod_{p|k}\left(1 + \frac{1}{p^\theta}\right)$$

and

$$\sum_{d|k}\frac{\mu(d)}{d}\sum_{e|k}\frac{\mu(e)}{e}\log(ed) = 2\sum_{d|k}\frac{\mu(d)\log d}{d}\sum_{e|k}\frac{\mu(e)}{e}$$

$$= \frac{2\varphi(k)}{k}\sum_{d|k}\frac{\mu(d)\log d}{d}$$

$$= 2\left(\frac{\varphi(k)}{k}\right)^2\sum_{p|k}\frac{\log p}{p - 1}$$

as required.

78

(a) Using $\tau_{k+1} = \tau_k \star 1$, we get with Proposition 4.3

$$S_{k+1}(x) = \sum_{n \leqslant x} (\tau_k \star 1)(n) = \sum_{d \leqslant x} \tau_k(d) \left\lfloor \frac{x}{d} \right\rfloor \tag{5}$$

and the inequalities $x - 1 < \lfloor x \rfloor \leqslant x$ then give

$$x \sum_{d \leqslant x} \frac{\tau_k(d)}{d} - S_k(x) < S_{k+1}(x) \leqslant x \sum_{d \leqslant x} \frac{\tau_k(d)}{d}$$

and using Theorem 1.3 we get

$$\sum_{d \leqslant x} \frac{\tau_k(d)}{d} = \frac{S_k(x)}{x} + \int_1^x \frac{S_k(t)}{t^2} \, dt \tag{6}$$

which concludes the proof.

(b) The inequalities are true for $k = 1$. Assume they are true for some positive integer k. By the previous question and the induction hypothesis, we have

$$S_{k+1}(x) \leqslant x \sum_{j=0}^{k-1} \binom{k-1}{j} \frac{(\log x)^j}{j!} + x \int_1^x \frac{1}{t} \left(\sum_{j=0}^{k-1} \binom{k-1}{j} \frac{(\log t)^j}{j!} \right) dt$$

$$= x \sum_{j=0}^{k-1} \binom{k-1}{j} \frac{1}{j!} \left\{ (\log x)^j + \int_1^x \frac{(\log t)^j}{t} \, dt \right\}$$

$$= x \sum_{j=0}^{k-1} \binom{k-1}{j} \frac{1}{j!} \left\{ (\log x)^j + \frac{(\log x)^{j+1}}{j+1} \right\}$$

$$= x \left(1 + \frac{(\log x)^k}{k!} \right) + x \sum_{j=1}^{k-1} \left\{ \binom{k-1}{j} + \binom{k-1}{j-1} \right\} \frac{(\log x)^j}{j!}$$

$$= x \left(1 + \frac{(\log x)^k}{k!} \right) + x \sum_{j=1}^{k-1} \binom{k}{j} \frac{(\log x)^j}{j!} = x \sum_{j=0}^{k} \binom{k}{j} \frac{(\log x)^j}{j!}$$

and

$$S_{k+1}(x) > x \sum_{j=0}^{k-1} \frac{(-1)^{k+j+1}}{j!} \int_1^x \frac{(\log t)^j}{t}\, dt + (-1)^k x \int_1^x \frac{dt}{t^2}$$

$$= x \sum_{j=0}^{k-1} (-1)^{k+j+1} \frac{(\log x)^{j+1}}{(j+1)!} + (-1)^k (x-1)$$

$$= x \sum_{j=1}^{k} (-1)^{k+j+2} \frac{(\log x)^j}{j!} + (-1)^{k+2} x + (-1)^{k+1}$$

$$= x \sum_{j=0}^{k} (-1)^{k+j+2} \frac{(\log x)^j}{j!} + (-1)^{k+1}$$

completing the proof.

(c) The result follows from the upper bound of the previous question and the inequality

$$\frac{1}{j!} = \frac{1}{(k-1)!} \times \prod_{i=0}^{k-j-2} (i+j+1) \leqslant \frac{(k-j-2+j+1)^{k-j-1}}{(k-1)!} = \frac{(k-1)^{k-j-1}}{(k-1)!}$$

so that

$$S_k(x) \leqslant \frac{x}{(k-1)!} \sum_{j=0}^{k-1} \binom{k-1}{j} (\log x)^j (k-1)^{k-1-j} = \frac{x(\log x + k - 1)^{k-1}}{(k-1)!}.$$

Remark One may proceed slightly differently by using the arithmetic–geometric mean inequality which implies that, for all $k \geqslant 3$ and $0 \leqslant j \leqslant k-3$, we have

$$\prod_{i=0}^{k-j-2} (i+j+1) = (k-1) \prod_{i=0}^{k-j-3} (i+j+1)$$

$$\leqslant (k-1) \left(\frac{1}{k-j-2} \sum_{i=0}^{k-j-3} (i+j+1) \right)^{k-j-2}$$

$$= (k-1)\left(\frac{k+j-1}{2}\right)^{k-j-2} \leqslant (k-1)(k-2)^{k-j-2}$$

$$= \frac{k-1}{k-2}(k-2)^{k-j-1}$$

since $k \geqslant 3$, and this inequality remains clearly true if $j \in \{k-2, k-1\}$, so that for all $k \geqslant 3$, we get

$$S_k(x) \leqslant \frac{x(\log x + k - 2)^{k-1}}{(k-2)(k-2)!}.$$

It was proved in [3] that the denominator may be replaced by $(k-1)!$.

79 Define

$$S_k^\star(x) = \sum_{n \leqslant x} \tau_k^\star(n).$$

We prove the inequality by induction on k, the case $k = 1$ being clearly true via

$$S_1^\star(x) = \lfloor x \rfloor - 1 < x.$$

Assume the inequality is true for some $k \in \mathbb{Z}_{\geqslant 1}$. By induction hypothesis we have

$$S_{k+1}^\star(x) = \sum_{2 \leqslant n \leqslant x 2^{-k}} S_k^\star\left(\frac{x}{n}\right) \leqslant \frac{x}{(k-1)!} \sum_{2 \leqslant n \leqslant x 2^{-k}} \frac{1}{n}\left(\log\frac{x}{n}\right)^{k-1}.$$

Now when $x < 2^{k+1}$, we have $S_{k+1}^\star(x) = 0$ and otherwise

$$S_{k+1}^\star(x) \leqslant \frac{x}{(k-1)!} \int_1^{x 2^{-k}} \left(\log\frac{x}{t}\right)^{k-1}\frac{dt}{t} = \frac{x(\log x)^k}{k!} - \frac{x(k \log 2)^k}{k!} < \frac{x(\log x)^k}{k!}$$

as required.

80 By induction on k, the case $k = 1$ following from (4.38). Assume the inequality is true for some $k \geqslant 1$ and let $s_k(x)$ be the sum on the left-hand side. Using the convolution identity $(k+1)^\omega = \left(\mu^2 \times k^\omega\right) \star \mathbf{1}$, we derive

$$s_{k+1}(x) = \sum_{d \leqslant x} \frac{k^{\omega(d)}}{d} \sum_{h \leqslant x/d} \frac{\mu(hd)^2}{h} \leqslant \sum_{d \leqslant x} \frac{\mu(d)^2 k^{\omega(d)}}{d} \sum_{h \leqslant x/d} \frac{\mu(h)^2}{h}$$

$$\leqslant \frac{1}{\zeta(2)} \sum_{d \leqslant x} \frac{\mu(d)^2 k^{\omega(d)}}{d}\left(\log\frac{x}{d} + 2\right)$$

$$= \frac{1}{\zeta(2)} \left(\int_1^x \frac{s_k(t)}{t} \, dt + 2s_k(x) \right)$$

$$\leq \frac{1}{\zeta(2)^{k+1}} \left\{ \int_1^x \left(\frac{\log(t)^k}{k!t} + 2^k \sum_{j=0}^{k-1} \frac{1}{j!} \binom{k}{j} \frac{(\log t)^j}{2^j t} \right) dt \right.$$

$$\left. +2 \left(\frac{\log(x)^k}{k!} + 2^k \sum_{j=0}^{k-1} \frac{1}{j!} \binom{k}{j} \left(\frac{\log x}{2} \right)^j \right) \right\}$$

$$= \frac{1}{\zeta(2)^{k+1}} \left(\frac{(\log x)^{k+1}}{(k+1)!} + 2^k \sum_{j=0}^{k-1} \frac{1}{j!} \binom{k}{j} \frac{(\log x)^{j+1}}{2^j(j+1)} + \frac{2(\log x)^k}{k!} \right.$$

$$\left. +2^{k+1} \sum_{j=0}^{k-1} \frac{1}{j!} \binom{k}{j} \left(\frac{\log x}{2} \right)^j \right)$$

$$= \frac{1}{\zeta(2)^{k+1}} \left(\frac{(\log x)^{k+1}}{(k+1)!} + \frac{2(\log x)^k}{k!} + 2^{k+1} \sum_{j=1}^{k} \frac{1}{j!} \binom{k}{j-1} \left(\frac{\log x}{2} \right)^j \right.$$

$$\left. +2^{k+1} \sum_{j=0}^{k-1} \frac{1}{j!} \binom{k}{j} \left(\frac{\log x}{2} \right)^j \right)$$

$$= \frac{1}{\zeta(2)^{k+1}} \left(\frac{(\log x)^{k+1}}{(k+1)!} + \frac{(2k+2)(\log x)^k}{k!} \right.$$

$$\left. +2^{k+1} \sum_{j=1}^{k-1} \frac{1}{j!} \left(\binom{k}{j-1} + \binom{k}{j} \right) \left(\frac{\log x}{2} \right)^j + 2^{k+1} \right)$$

$$= \frac{1}{\zeta(2)^{k+1}} \left(\frac{(\log x)^{k+1}}{(k+1)!} + \frac{(2k+2)(\log x)^k}{k!} + 2^{k+1} \sum_{j=1}^{k-1} \frac{1}{j!} \binom{k+1}{j} \left(\frac{\log x}{2} \right)^j + 2^{k+1} \right)$$

$$= \frac{1}{\zeta(2)^{k+1}} \left(\frac{(\log x)^{k+1}}{(k+1)!} + 2^{k+1} \sum_{j=0}^{k} \frac{1}{j!} \binom{k+1}{j} \left(\frac{\log x}{2} \right)^j \right)$$

completing the proof.

81 The result is obvious if $k = 1$, so assume $k \geq 2$. Since $k^\omega = \left(\mu^2 \times (k-1)^\omega \right) \star 1$, we get

$$\sum_{n \leq x} k^{\omega(n)} = \sum_{d \leq x} \mu(d)^2 (k-1)^{\omega(d)} \left\lfloor \frac{x}{d} \right\rfloor \leq x \sum_{d \leq x} \frac{\mu(d)^2 (k-1)^{\omega(d)}}{d}$$

and using the inequality of the previous exercise, we derive

$$\sum_{n \leqslant x} k^{\omega(n)} \leqslant \frac{x}{\zeta(2)^{k-1}} \left(\frac{(\log x)^{k-1}}{(k-1)!} + 2^{k-1} \sum_{j=0}^{k-2} \frac{1}{j!} \binom{k-1}{j} \left(\frac{\log x}{2} \right)^j \right).$$

We conclude the proof noticing that

$$\sum_{j=0}^{k-2} \frac{1}{j!} \binom{k-1}{j} \left(\frac{\log x}{2} \right)^j \leqslant (k-1) \sum_{j=0}^{k-2} \frac{1}{j!} \binom{k-2}{j} \left(\frac{\log x}{2} \right)^j$$

$$\leqslant \frac{k-1}{(k-2)!} \sum_{j=0}^{k-2} \binom{k-2}{j} \left(\frac{\log x}{2} \right)^j (k-2)^{k-2-j}$$

$$= \frac{k-1}{(k-2)!} \left(\frac{\log x}{2} + k - 2 \right)^{k-2}$$

implying the asserted inequality.

Remark This implies in particular that, for any $k \in \mathbb{Z}_{\geqslant 1}$ and any $x \geqslant e^{2(k-1)^2}$

$$\sum_{n \leqslant x} k^{\omega(n)} \leqslant \frac{ex(\log x)^{k-1}}{(k-1)! \, \zeta(2)^{k-1}}.$$

This bound is far better than that of Exercise 69 for large x.

82 We adapt the proof of Theorem 4.25. Set $f_k := (-k)^\omega$. Usual computations show that, for any $s = \sigma + it \in \mathbb{C}$ such that $\sigma > 1$

$$L(s, f_k) = \zeta(s)^{-k} \zeta(2s)^{-\frac{1}{2}k(k+1)} G_k(s),$$

where $G_k(s)$ is a Dirichlet series absolutely convergent in the half-plane $\sigma > \frac{1}{3}$. Set $c := 57.54^{-1}$, $\kappa := 1 + \frac{1}{\log x}$,

$$T := e^{c(\log x)^{3/5}(\log \log x)^{-1/5}} \quad \text{and} \quad \alpha := \alpha(T) = c(\log T)^{-2/3}(\log \log T)^{-1/3}.$$

We use Theorem 3.26 with $H = \sqrt{T}$ yielding

$$\sum_{n \leqslant x} (-k)^{\omega(n)} = \frac{1}{2\pi i} \int_{\kappa - iT}^{\kappa + iT} \frac{G_k(s)}{\zeta(s)^k \zeta(2s)^{\frac{1}{2}k(k+1)}} \frac{x^s}{s} \, ds$$

$$+ O \left(\sum_{x - x/\sqrt{T} < n \leqslant x + x/\sqrt{T}} |f_k(n)| + \frac{x^\kappa}{\sqrt{T}} \sum_{n=1}^{\infty} \frac{|f_k(n)|}{n^\kappa} \right).$$

By (3.59), $\zeta(s)$ has no zero in the region $\sigma \geqslant 1 - c(\log |t|)^{-2/3}(\log \log |t|)^{-1/3}$ and $|t| \geqslant 3$, so that we may shift the line of integration to the left and apply Cauchy's theorem in the rectangle with vertices $\kappa \pm iT$, $1 - \alpha \pm iT$. In this region

$$\zeta(s)^{-1} \ll (\log(|t| + 2))^{2/3} (\log \log(|t| + 2))^{1/3}.$$

Therefore, the contribution of the horizontal sides does not exceed

$$\ll xT^{-1}(\log T)^{2k/3}(\log \log T)^{k/3}$$

and the contribution of the vertical side is bounded by

$$\ll x^{1-\alpha}(\log T)^{1+2k/3}(\log \log T)^{k/3}.$$

Since $T \ll x^{1-\varepsilon}$, Theorem 4.17 yields

$$\sum_{x-x/\sqrt{T}<n\leqslant x+x/\sqrt{T}} |f_k(n)| \leqslant \sum_{x-x/\sqrt{T}<n\leqslant x+x/\sqrt{T}} \tau_k(n) \ll \frac{x}{\sqrt{T}}(\log x)^{k-1}.$$

With the choice of κ, the 2nd error term does not exceed

$$\leqslant \frac{x^\kappa}{\sqrt{T}} \sum_{n=1}^{\infty} \frac{\tau_k(n)}{n^\kappa} = \frac{x^\kappa}{\sqrt{T}}\zeta(\kappa)^k \ll \frac{x}{\sqrt{T}}(\log x)^k.$$

Since the path of integration does not surround the origin, nor the poles of the integrand, Cauchy's theorem and the choice of T give the asserted estimate for any $c_0 \leqslant \frac{1}{4}c$ and any real number x satisfying $x \geqslant \exp(c_1 k^{10/3})$, say, where $c_1 \geqslant 1$ is absolute.

83 Let $a \in \mathbb{Z}_{\geqslant 2}$.

(a) Since

$$\lfloor ax \rfloor - a\lfloor x \rfloor = \begin{cases} 0, & \text{si } \{x\} < \frac{1}{a} \\[2mm] 1, & \text{si } \frac{1}{a} \leqslant \{x\} < \frac{2}{a} \\[2mm] \vdots & \vdots \\[2mm] a-1, & \text{si } \{x\} \geqslant \frac{a-1}{a} \end{cases}$$

the sum on the left-hand side is equal to

$$= \sum_{k=1}^{an} f(k) \left(\left\lfloor \frac{an}{k} \right\rfloor - a \left\lfloor \frac{n}{k} \right\rfloor \right) = \sum_{k=1}^{an} f(k) \left\lfloor \frac{an}{k} \right\rfloor - a \sum_{k=1}^{n} f(k) \left\lfloor \frac{n}{k} \right\rfloor = G(an) - aG(n)$$

as asserted.

(b) Follows from (4.85) and the previous question used with $f = 1$.

84 We check numerically the inequality for $x \in [1, 4000]$ by studying the difference

$$\frac{2}{25(k+1)} + \frac{1}{(k+1)^{3/2}} - \left| \frac{1}{\zeta(2)} - \sum_{n=1}^{k} \frac{\mu(n)}{n^2} \right| \qquad (k = 1, \ldots, 4001)$$

so that one may assume that $x > 4000$. By partial summation

$$\sum_{n>x} \frac{\mu(n)}{n^2} = -\frac{m(x)}{x} + \int_{x}^{\infty} \frac{m(t)}{t^2} \, dt,$$

where $m(t) := \sum_{n \leqslant t} \mu(n) n^{-1}$, and the bound [18, Theorem 1.2]

$$|m(t)| \leqslant \frac{1}{25 \log t} \qquad (t \geqslant 3470)$$

implies that

$$\left| \sum_{n>x} \frac{\mu(n)}{n^2} \right| \leqslant \frac{1}{25 \log x} \left(\frac{1}{x} + \int_{x}^{\infty} \frac{dt}{t^2} \right) = \frac{2}{25x \log x} < \frac{2}{25x}$$

since $x > 4000$.

85 Since each square-full number may be uniquely written in the shape $n = a^2 b^3$ with b squarefree, we obtain

$$\sum_{n \leqslant x} s_2(n) = \sum_{\substack{a^2 b^3 \leqslant x \\ \mu_2(b) = 1}} 1 = \sum_{b \leqslant x^{1/3}} \mu_2(b) \sum_{a \leqslant (x/b^3)^{1/2}} 1$$

$$\leqslant x^{1/2} \sum_{b \leqslant x^{1/3}} \frac{\mu_2(b)}{b^{3/2}} \leqslant \frac{\zeta(3/2) \, x^{1/2}}{\zeta(3)} < 3x^{1/2},$$

where we used Lemma 3.13 in the last inequality. The second estimate follows by partial summation as seen in Exercise 3, so that

$$\sum_{n>x} \frac{s_2(n)}{n} = -\frac{1}{x} \sum_{n \leqslant x} s_2(n) + \int_x^\infty \left(\sum_{n \leqslant t} s_2(n) \right) \frac{dt}{t^2} < 3 \int_x^\infty \frac{dt}{t^{3/2}} = \frac{6}{x^{1/2}}$$

as asserted.

86 Define $g = f \star \mu$. Then g is multiplicative by Theorem 4.1, $|g(p)| = |f(p) - 1| \leqslant p^{-1}$ and, for all prime powers p^α with $\alpha \geqslant 2$, we have $|g(p^\alpha)| \leqslant 1$, so that $|g(n)| \leqslant 1$ for all $n \in \mathbb{Z}_{\geqslant 1}$. Let $x \geqslant 2$ be a large real number. We have

$$\sum_{p \leqslant x} \sum_{\alpha=1}^\infty \frac{|g(p^\alpha)|}{p^\alpha} = \sum_{p \leqslant x} \frac{|f(p) - 1|}{p} + \sum_{p \leqslant x} \sum_{\alpha=2}^\infty \frac{|g(p^\alpha)|}{p^\alpha}$$

$$\leqslant \sum_{p \leqslant x} \frac{1}{p^2} + \sum_{p \leqslant x} \sum_{\alpha=2}^\infty \frac{1}{p^\alpha}$$

$$\leqslant 3 \sum_{p \leqslant x} \frac{1}{p^2} < \frac{3}{2}$$

by (4.23), so that the series $\sum_{d \geqslant 1} g(d)/d$ converges absolutely by Theorem 4.8 and we have

$$1 + \sum_{\alpha=1}^\infty \frac{g(p^\alpha)}{p^\alpha} = 1 + \sum_{\alpha=1}^\infty \frac{f(p^\alpha) - f(p^{\alpha-1})}{p^\alpha} = \left(1 - \frac{1}{p} \right) \left(1 + \sum_{\alpha=1}^\infty \frac{f(p^\alpha)}{p^\alpha} \right).$$

Using Theorem 4.2 and Proposition 4.3, we get

$$\sum_{n \leqslant x} f(n) = \sum_{n \leqslant x} (g \star \mathbf{1})(n) = \sum_{d \leqslant x} g(d) \left\lfloor \frac{x}{d} \right\rfloor$$

$$= x \sum_{d \leqslant x} \frac{g(d)}{d} + O \left(\sum_{d \leqslant x} |g(d)| \right)$$

$$= x \sum_{d=1}^\infty \frac{g(d)}{d} + O \left(\sum_{d \leqslant x} |g(d)| + x \sum_{d>x} \frac{|g(d)|}{d} \right)$$

and by partial summation, we obtain

$$\sum_{d>x} \frac{|g(d)|}{d} = -\frac{1}{x} \sum_{d \leqslant x} |g(d)| + \int_x^\infty \left(\sum_{d \leqslant t} |g(d)| \right) \frac{dt}{t^2}$$

and

$$\sum_{d=1}^\infty \frac{g(d)}{d} = \prod_p \left(1 + \sum_{\alpha=1}^\infty \frac{g(p^\alpha)}{p^\alpha} \right) = \prod_p \left(1 - \frac{1}{p} \right) \left(1 + \sum_{\alpha=1}^\infty \frac{f(p^\alpha)}{p^\alpha} \right)$$

so that

$$\sum_{n \leqslant x} f(n) = x \prod_p \left(1 - \frac{1}{p} \right) \left(1 + \sum_{\alpha=1}^\infty \frac{f(p^\alpha)}{p^\alpha} \right) + R(x)$$

with

$$|R(x)| \ll \sum_{d \leqslant x} |g(d)| + x \int_x^\infty \left(\sum_{d \leqslant t} |g(d)| \right) \frac{dt}{t^2}.$$

It remains to estimate the sum $\sum_{d \leqslant x} |g(d)|$. To do this we use the unique decomposition of each positive integer d in the form $d = ab$ with $\mu_2(a) = s_2(b) = 1$ and $(a, b) = 1$. Also note that, for all squarefree numbers a, we have $|g(a)| \leqslant a^{-1}$. Using Exercise 85, we obtain

$$\sum_{d \leqslant x} |g(d)| \leqslant \sum_{a \leqslant x} \mu_2(a) |g(a)| \sum_{b \leqslant x/a} s_2(b) |g(b)|$$

$$\leqslant \sum_{a \leqslant x} \frac{\mu_2(a)}{a} \sum_{b \leqslant x/a} s_2(b)$$

$$< 3x^{1/2} \sum_{a \leqslant x} \frac{\mu_2(a)}{a^{3/2}} \leqslant \frac{3\zeta(3/2)x^{1/2}}{\zeta(3)}.$$

Hence

$$|R(x)| \ll x^{1/2} + x \int_x^\infty \frac{dt}{t^{3/2}} \ll x^{1/2}$$

completing the proof.

87 This is a direct application of Exercise 86 since $\beta(p) = 1$ and $\beta \star \mu = s_2$.

88 Again a direct application of Exercise 86 with $f(n) = \varphi(n)\gamma_2(n)/n^2$ since, for all prime powers p^α, we have

$$f(p) = 1 - \frac{1}{p} \quad \text{and} \quad (f \star \mu)(p^\alpha) = \begin{cases} -1/p, & \text{if } \alpha = 1 \\ -p^{-\alpha}(p-1)^2, & \text{if } \alpha \geqslant 2. \end{cases}$$

89

(a) This follows from

$$D_k(mn) = \tau_k(mn)k^{-\omega(mn)} = \tau_k(mn)k^{-\omega(m)-\omega(n)+\omega((m,n))}$$

$$\leqslant \tau_k(m)k^{-\omega(m)}\tau_k(n)k^{-\omega(n)}k^{\omega((m,n))}$$

$$= D_k(m)D_k(n)k^{\omega((m,n))} \leqslant D_k(m)D_k(n)k^{\omega(\min(m,n))}.$$

(b) Since the functions $D_k \star \mu$ and g_k are multiplicative, it suffices to show that the equality holds on prime powers p^α. When $\alpha = 1$, we have $(D_k \star \mu)(p) = D_k(p) - 1 = 0 = g_k(p)$. Now assume $\alpha \geqslant 2$. Then, using (4.26)

$$(D_k \star \mu)(p^\alpha) = D_k(p^\alpha) - D_k\left(p^{\alpha-1}\right)$$

$$= \frac{1}{k}\left\{\binom{k+\alpha-1}{\alpha} - \binom{k+\alpha-2}{\alpha-1}\right\}$$

$$= \frac{1}{k}\binom{k+\alpha-2}{\alpha} = \frac{k-1}{k}D_{k-1}\left(p^\alpha\right)$$

as required.

(c) Use the same argument as in Exercise 86.

(d) Each square-full integer n may be written as $n = a^2b^3$ with $\mu(b)^2 = 1$, so that, for any $t \in \mathbb{R}_{\geqslant 1}$

$$\sum_{n \leqslant t} g_k(n) = \frac{k-1}{k} \sum_{b \leqslant t^{1/3}} \mu(b)^2 \sum_{a \leqslant \sqrt{t/b^3}} D_{k-1}\left(a^2b^3\right)$$

and using the above inequality we derive

$$\sum_{n \leqslant t} g_k(n) \leqslant \frac{k-1}{k} \sum_{b \leqslant t^{1/3}} \mu(b)^2(k-1)^{2\omega(b)} D_{k-1}(b)^3 \sum_{a \leqslant \sqrt{t/b^3}} D_{k-1}(a)^2(k-1)^{\omega(a)}.$$

Now Theorem 4.9 yields

$$\sum_{a \leqslant z} D_{k-1}(a)^2 (k-1)^{\omega(a)} \ll \frac{z}{\log z} \exp\left(\sum_{p \leqslant z} \frac{D_{k-1}(p)^2 (k-1)^{\omega(p)}}{p}\right) \ll z(\log z)^{k-2}$$

so that

$$\sum_{n \leqslant t} g_k(n) \ll t^{1/2} (\log t)^{k-2} \sum_{b \leqslant t^{1/3}} \frac{\mu(b)^2 (k-1)^{2\omega(b)} D_{k-1}(b)^3}{b^{3/2}} \ll t^{1/2} (\log t)^{k-2}$$

and hence

$$\sum_{n \leqslant x} D_k(n) = x \sum_{n=1}^{\infty} \frac{g_k(n)}{n} + O_k\left(x^{1/2} (\log x)^{k-2}\right).$$

It remains to show that the series is equal to the constant C_k given (4.87). We have

$$\sum_{n=1}^{\infty} \frac{g_k(n)}{n} = \prod_p \left(1 + \frac{k-1}{k} \sum_{\alpha=2}^{\infty} \frac{D_{k-1}(p^\alpha)}{p^\alpha}\right)$$

$$= \prod_p \left(1 + \frac{1}{k} \sum_{\alpha=2}^{\infty} \frac{1}{p^\alpha} \binom{k+\alpha-2}{\alpha}\right)$$

$$= \prod_p \left(1 + \frac{1}{kp} \sum_{\alpha=1}^{\infty} \frac{1}{p^\alpha} \binom{k+\alpha-1}{\alpha+1}\right)$$

$$= \prod_p \left(1 + \frac{k-1}{kp} \sum_{\alpha=1}^{\infty} \frac{1}{(\alpha+1)p^\alpha} \binom{k+\alpha-1}{\alpha}\right)$$

$$= \prod_p \left\{1 - \frac{1}{k} - \frac{k-1}{kp} + \frac{1}{k}\left(1 - \frac{1}{p}\right)^{1-k}\right\}$$

$$= \prod_p \left\{\left(1 - \frac{1}{p}\right)\left(1 - \frac{1}{k}\right) + \frac{1}{k}\left(1 - \frac{1}{p}\right)^{1-k}\right\}$$

$$= \prod_p \left(1 - \frac{1}{p}\right)\left(1 - \frac{1}{k} + \frac{1}{k}\left(1 - \frac{1}{p}\right)^{-k}\right) = C_k,$$

where we used the identity

$$x(k-1) \sum_{\alpha=0}^{\infty} \frac{x^\alpha}{\alpha+1} \binom{k+\alpha-1}{\alpha} = (1-x)^{1-k} - 1 \quad \left(k \in \mathbb{Z}_{\geqslant 2}, |x| < 1\right).$$

The first values of C_k are given in the following table.

k	2	3	4	5	6
C_k	1.43	2.22	3.8	7.11	14.45

90

(a) Using Corollary 4.7, we derive

$$\sum_{n \leqslant x} \frac{f(n)}{n^2} = C\sqrt{\log x} + O\left((\log x)^{-1/2}\right)$$

with

$$C := \frac{2}{\sqrt{\pi}} \prod_p \frac{p^{3/2}}{\sqrt{p-1}} \log\left(1 + \frac{1}{p}\right) \approx 1.42762$$

and the asserted result follows by partial summation.

(b) It is clear that, if $p \geqslant 3$ is prime, then $f(p) = \frac{1}{2}(p+1) \in \mathbb{Z}$. Since f is multiplicative, we infer that, if n is an odd squarefree integer, then $f(n)$ is an integer.

(c) Let $\alpha \geqslant 1$. We have

$$f(p^\alpha) = \frac{p^{\alpha+1}-1}{(\alpha+1)(p-1)}.$$

Now we prove that, if $N > 1$ is such that $P^-(N) \nmid p - 1$, then the congruence $p^N \equiv 1 \pmod{N(p-1)}$ has no solution. Indeed, assume there exists a solution $N > 1$ and suppose $q := P^-(N)$ satisfies $q \nmid p - 1$. Then $p^N \equiv 1 \pmod q$, implying that $\mathrm{ord}_q(p) \mid N$. But since $\mathrm{ord}_q(p) < q$ and that q is the smallest prime divisor of N, we get $\mathrm{ord}_q(p) = 1$ and hence $q \mid p - 1$, which is impossible. Thus, $f(p^\alpha)$ is not an integer as soon as $\alpha \geqslant 1$ whenever $P^-(\alpha+1) \nmid p - 1$.

91

(a) Let $g(n) = \mu(n)/\tau(n)$ and $G = g \star \mathbf{1}$. By Theorem 4.1, G is multiplicative and, for all prime powers p^α, we have

$$G(p^\alpha) = 1 + \sum_{j=1}^{\alpha} g(p^j) = 1 - \tau(p)^{-1} = \frac{1}{2}$$

so that, for all $n \in \mathbb{Z}_{\geqslant 1}$, we get

$$G(n) = 2^{-\omega(n)}$$

and hence using Proposition 4.3 we obtain

$$\sum_{n \leqslant x} G(n) = \sum_{n \leqslant x} (g \star 1)(n) = \sum_{d \leqslant x} g(d) \left\lfloor \frac{x}{d} \right\rfloor$$

$$= x \sum_{d \leqslant x} \frac{g(d)}{d} - \sum_{d \leqslant x} g(d) \left\{ \frac{x}{d} \right\}$$

$$= x F(x) - \sum_{d \leqslant x} g(d) \left\{ \frac{x}{d} \right\}$$

as required.

(b) Using the inequalities $0 \leqslant \{x\} < 1$ we get

$$\left| \sum_{n \leqslant x} \frac{\mu(n)}{\tau(n)} \left\{ \frac{x}{n} \right\} \right| < \sum_{n \leqslant x} \frac{\mu^2(n)}{\tau(n)} = \sum_{n \leqslant x} \frac{\mu^2(n)}{2^{\omega(n)}} \leqslant \sum_{n \leqslant x} \frac{1}{2^{\omega(n)}}$$

so that

$$\sum_{n \leqslant x} \frac{1}{2^{\omega(n)}} + \sum_{n \leqslant x} \frac{\mu(n)}{\tau(n)} \left\{ \frac{x}{n} \right\} > 0.$$

Furthermore, we also have

$$F(x) = |F(x)| \leqslant \frac{1}{x} \left(\sum_{n \leqslant x} \frac{1}{2^{\omega(n)}} + \left| \sum_{n \leqslant x} \frac{\mu(n)}{\tau(n)} \left\{ \frac{x}{n} \right\} \right| \right) < \frac{2}{x} \sum_{n \leqslant x} \frac{1}{2^{\omega(n)}}.$$

The function $2^{-\omega}$ satisfies the Wirsing conditions (4.24) with $\lambda_1 = \frac{1}{2}$ and $\lambda_2 = 1$ so that we may apply Theorem 4.9 with $(a, b) = \left(\log 2, \frac{3}{2} \right)$ by Lemma 4.5. This gives

$$\sum_{n \leqslant x} \frac{1}{2^{\omega(n)}} \leqslant e^{3/2} \left(\frac{5}{2} + \log 2 \right) \frac{x}{\log ex} \exp \left(\frac{1}{2} \sum_{p \leqslant x} \frac{1}{p} \right).$$

By Corollary 3.13, we have

$$\sum_{p \leqslant x} \frac{1}{p} < \log \log x + \tfrac{1}{2}$$

as soon as $x \geqslant 8$ implying that

$$\sum_{n \leqslant x} \frac{1}{2^{\omega(n)}} < \frac{19x}{(\log x)^{1/2}}$$

and hence we finally get for all $x \geqslant 8$

$$0 < F(x) < 38 \, (\log x)^{-1/2}.$$

92

(a) We have $f(p^{\alpha}) = t^{\alpha} - t^{\alpha-1}$ and using Theorem 4.2 and Proposition 4.3, we get

$$\sum_{n \leqslant x} t^{\Omega(n)} = \sum_{n \leqslant x} (f \star \mathbf{1})(n) = \sum_{d \leqslant x} f(d) \left\lfloor \frac{x}{d} \right\rfloor$$

$$\leqslant x \sum_{d \leqslant x} \frac{f(d)}{d} \leqslant x \prod_{p \leqslant x} \left(1 + \sum_{\alpha=1}^{\infty} \frac{t^{\alpha} - t^{\alpha-1}}{p^{\alpha}} \right)$$

$$= x \prod_{p \leqslant x} \left(1 + \frac{t-1}{p-t} \right).$$

We treat the cases $p = 2$ and $p \geqslant 3$ separately which gives

$$\sum_{n \leqslant x} t^{\Omega(n)} \leqslant \frac{x}{2-t} \prod_{3 \leqslant p \leqslant x} \left(1 + \frac{t-1}{p-t} \right)$$

$$\leqslant \frac{x}{2-t} \prod_{3 \leqslant p \leqslant x} \left(1 + \frac{1}{p-2} \right)$$

$$\leqslant \frac{x}{2-t} \exp \left(\sum_{3 \leqslant p \leqslant x} \frac{1}{p-2} \right) \ll \frac{x \log x}{2-t}$$

as required.

(b) We have

$$N_k(x) = \sum_{\substack{n \leqslant x \\ \Omega(n)=k}} t^{-\Omega(n)} t^{\Omega(n)} \leqslant t^{-k} \sum_{n \leqslant x} t^{\Omega(n)} \ll \frac{t^{-k} x \log x}{2-t}$$

and the choice of

$$t = \frac{2k}{k+1}$$

gives the asserted estimate.

93

1.(a) Since n is squarefree, a positive integer d is a divisor of n if and only if either
d divides n/p or $d \mid n$ is a multiple of p, so that

$$\sum_{d|n} f(d) = \sum_{d|(n/p)} f(d) + \sum_{\substack{d|n \\ p|d}} f(d) = \sum_{d|(n/p)} f(d) + \sum_{k|(n/p)} f(kp).$$

Since f is multiplicative and $k \mid (n/p)$ implies that $(k, p) = 1$, we infer that

$$\sum_{d|n} f(d) = \sum_{d|(n/p)} f(d) + f(p) \sum_{k|(n/p)} f(k) = (1 + f(p)) \sum_{k|(n/p)} f(k)$$

giving (4.91).
(b) First note that $f(d) \geqslant 0$ for all divisors d of n necessarily squarefree. Thus,
using (4.91), we get

$$\sum_{d|n} f(d) \log d = \sum_{d|n} f(d) \sum_{p|d} \log p = \sum_{p|n} \log p \sum_{k|(n/p)} f(kp)$$

$$= \sum_{p|n} f(p) \log p \sum_{k|(n/p)} f(k)$$

$$= \sum_{p|n} \left(\frac{f(p) \log p}{1 + f(p)} \sum_{d|n} f(d) \right)$$

$$= \left(\sum_{d|n} f(d) \right) \left(\sum_{p|n} \frac{f(p) \log p}{1 + f(p)} \right)$$

as required. Now the function $t \mapsto t/(1 + t)$ is increasing on $[0, \lambda]$ and since $f(d) \geqslant 0$ for all divisors d of n, we deduce that

$$\sum_{d|n} f(d) \log d \leqslant \frac{\lambda}{1 + \lambda} \left(\sum_{d|n} f(d) \right) \left(\sum_{p|n} \log p \right) = \frac{\lambda \log n}{1 + \lambda} \left(\sum_{d|n} f(d) \right).$$

2. First note that $\log(n^{1/a}/d) \leqslant \log(n^{1/a})$ and since $f(d) \geqslant 0$, we get

$$\sum_{\substack{d|n \\ d \leqslant n^{1/a}}} f(d) \frac{\log(n^{1/a}/d)}{\log(n^{1/a})} \leqslant \sum_{\substack{d|n \\ d \leqslant n^{1/a}}} f(d).$$

Note also that, if $d > n^{1/a}$, then $\log(n^{1/a}/d) < 0$ and hence

$$\sum_{d|n} f(d) \frac{\log(n^{1/a}/d)}{\log(n^{1/a})} \leqslant \sum_{\substack{d|n \\ d \leqslant n^{1/a}}} f(d) \frac{\log(n^{1/a}/d)}{\log(n^{1/a})}.$$

We infer that

$$\sum_{\substack{d|n \\ d \leqslant n^{1/a}}} f(d) \geqslant \sum_{d|n} f(d) \frac{\log(n^{1/a}/d)}{\log(n^{1/a})} = \sum_{d|n} f(d) - \frac{a}{\log n} \sum_{d|n} f(d) \log d$$

and using (4.92) gives

$$\sum_{\substack{d|n \\ d \leqslant n^{1/a}}} f(d) \geqslant \sum_{d|n} f(d) \left(1 - \frac{\lambda a}{1 + \lambda} \right).$$

implying the desired estimate since $\lambda < (a - 1)^{-1}$.

94 Let $S_n = (s_{ij})$ and $T_n = (t_{ij})$ be the matrices defined by

$$s_{ij} = \begin{cases} 1, & \text{if } i \mid j \\ 0, & \text{otherwise} \end{cases} \quad \text{and} \quad t_{ij} = \begin{cases} M(n/i), & \text{if } j = 1 \\ 1, & \text{if } i = j \geqslant 2 \\ 0, & \text{otherwise.} \end{cases}$$

We will prove the following result.

Lemma 7.33 *We have $R_n = S_n T_n$. In particular we have $\det R_n = M(n)$.*

Proof Set $S_n T_n = (x_{ij})$. If $j = 1$ we have

$$x_{i1} = \sum_{k=1}^{n} s_{ik} t_{k1} = \sum_{\substack{k \leqslant n \\ i|k}} M\left(\frac{n}{k}\right) = \sum_{d \leqslant n/i} M\left(\frac{n/i}{d}\right) = 1 = r_{i1}$$

by (4.15). If $j \geqslant 2$, then $t_{1j} = 0$ and thus

$$x_{ij} = \sum_{k=2}^{n} s_{ik} t_{kj} = s_{ij} = \begin{cases} 1, & \text{if } i \mid j \\ 0, & \text{otherwise} \end{cases} = r_{ij}$$

which is the desired result. The second assertion follows at once from

$$\det R_n = \det S_n \det T_n = \det T_n = M(n).$$

The proof is complete.   □

95

(a) We may suppose $n > 1$. Let p^α be a prime power. Using Bernoulli's inequality, we get

$$\left(t^\omega \star 1\right)\left(p^\alpha\right) = 1 + \alpha t \leqslant (1 + t)^\alpha = (1 + t)^{\Omega(p^\alpha)}$$

implying the first inequality by multiplicativity.

(b) Write $n = p_1^{\alpha_1} \cdots p_r^{\alpha_r}$ and, for each $j \in \{1, \ldots, k\}$, let d_j be a divisor of n with $\omega(d_j) = j \leqslant k$. The number of such divisors is at most equal to the number of integers which are the products of j prime powers from the list

$$p_1, p_1^2, \ldots, p_1^{\alpha_1}, p_2, p_2^2, \ldots, p_2^{\alpha_2}, \ldots, p_r, p_r^2, \ldots, p_r^{\alpha_r}.$$

Since this list contains $\Omega(n)$ elements, we infer that the number of divisors d_j is at most $\binom{\Omega(n)}{j}$ and hence

$$\sum_{\substack{d|n \\ \omega(d) \leqslant k}} t^{\omega(d)} \leqslant 1 + \sum_{j=1}^{k} \binom{\Omega(n)}{j} t^j = \sum_{j=0}^{k} \binom{\Omega(n)}{j} t^j$$

as asserted. It is easy to see that this inequality generalizes the previous one and, with a little more work, one can prove that

$$\sum_{\substack{d|n \\ \omega(d) \leqslant k}} t^{\omega(d)} = \sum_{j=0}^{k} t^j \sum_{1 \leqslant i_1 < i_2 < \cdots < i_j \leqslant \omega(n)} \alpha_{i_1} \cdots \alpha_{i_j}.$$

96 Let S_n be the sum on the left-hand side. We have

$$S_n = \sum_{d|n} \frac{1}{d} \sum_{\substack{h \leqslant n/d \\ (h,n/d)=1}} \frac{\mu(h)}{h} = \frac{1}{n} \sum_{d|n} d \sum_{\substack{h \leqslant d \\ (h,d)=1}} \frac{\mu(h)}{h}.$$

Now Corollary 4.1 implies that $|S_n| \leqslant \dfrac{\sigma(n)}{n}$, and we conclude using Exercise 63.

97

(a) First note that f is multiplicative and, if n is squarefree, then $f(f(n)) = f(1) = 1$. Hence it is sufficient to prove the inequality when n is square-full, otherwise, writing uniquely $n = ab$ with a squarefree, b square-full and $(a, b) = 1$, we obtain

$$f(f(n)) = f(f(ab)) = f(f(a)f(b)) = f(f(b)) \leqslant b \leqslant n.$$

Assume n square-full written in the shape $n = p_1^{a_1} \cdots p_k^{a_k}$ with $a_i \geqslant 2$ for $i \in \{1, \dots, k\}$. Write

$$a_i = q_1^{b_{i,1}} \dots q_\ell^{b_{i,\ell}}$$

with q_i primes, $\ell \in \mathbb{Z}_{\geqslant 1}$ and $b_{i,j} \in \mathbb{Z}_{\geqslant 0}$. Then

$$f(n) = \prod_{i=1}^k a_i^{p_i} = \prod_{i=1}^k \prod_{j=1}^\ell q_j^{p_i b_{i,j}} = \prod_{j=1}^\ell \prod_{i=1}^k q_j^{p_i b_{i,j}} = \prod_{j=1}^\ell q_j^{\sum_{i=1}^k p_i b_{i,j}}$$

whence

$$f(f(n)) = \prod_{j=1}^\ell \left(\sum_{i=1}^k p_i b_{i,j} \right)^{q_j}$$

since $\sum_{i=1}^k p_i b_{i,j} \neq 0$. Using the inequality $a_1 b_1 + a_2 b_2 + \cdots + a_k b_k \leqslant a_1^{b_1} a_2^{b_2} \cdots a_k^{b_k}$ for $a_i \geqslant 2$ and $b_i \geqslant 0$, we derive

$$f(f(n)) \leqslant \prod_{j=1}^\ell \prod_{i=1}^k p_i^{b_{i,j} q_j} = \prod_{i=1}^k \prod_{j=1}^\ell p_i^{b_{i,j} q_j} = \prod_{i=1}^k p_i^{\sum_{j=1}^\ell b_{i,j} q_j}$$

and finally

$$f(f(n)) \leqslant \prod_{i=1}^k p_i^{\prod_{j=1}^\ell q_j^{b_{i,j}}} = \prod_{i=1}^k p_i^{a_i} = n.$$

(b) Let n be square-full, written uniquely in the form $n = a^2 b^3$ with b squarefree. Thus

$$f\left(2^a \times 3^b\right) = a^2 b^3 = n.$$

98 Let S be the sum of the double series $\sum_{m,n=1}^{\infty} (mn\,[m,n])^{-1}$. We have

$$S = \sum_{m,n=1}^{\infty} \frac{1}{mn[m,n]} = \sum_{m,n=1}^{\infty} \frac{(m,n)}{(mn)^2} = \sum_{m=1}^{\infty} \frac{1}{m^2} \sum_{d|m} d \sum_{\substack{n=1 \\ (m,n)=d}}^{\infty} \frac{1}{n^2}$$

$$= \sum_{m=1}^{\infty} \frac{1}{m^2} \sum_{d|m} d \sum_{\substack{k=1 \\ (m/d,k)=1}}^{\infty} \frac{1}{k^2 d^2} = \sum_{m=1}^{\infty} \frac{1}{m^2} \sum_{d|m} \frac{1}{d} \sum_{\substack{k=1 \\ (m/d,k)=1}}^{\infty} \frac{1}{k^2},$$

where, for any $N \in \mathbb{Z}_{\geqslant 1}$

$$\sum_{\substack{k=1 \\ (N,k)=1}}^{\infty} \frac{1}{k^2} = \sum_{d|N} \mu(d) \sum_{m=1}^{\infty} \frac{1}{(md)^2} = \zeta(2) \sum_{d|N} \frac{\mu(d)}{d^2}$$

$$= \zeta(2) \prod_{p|N} \left(1 - \frac{1}{p^2}\right) = \frac{\zeta(2)\varphi(N)\Psi(N)}{N^2}$$

therefore

$$S = \zeta(2) \sum_{m=1}^{\infty} \frac{1}{m^2} \sum_{d|m} \frac{1}{d} \left(\frac{m}{d}\right)^{-2} \varphi\left(\frac{m}{d}\right) \Psi\left(\frac{m}{d}\right)$$

$$= \zeta(2) \sum_{m=1}^{\infty} \frac{1}{m^4} \sum_{d|m} d\varphi\left(\frac{m}{d}\right) \Psi\left(\frac{m}{d}\right) = \zeta(2) \sum_{m=1}^{\infty} \frac{(\mathrm{Id} \star \varphi\psi)(m)}{m^4}$$

$$= \zeta(2) L(4, \mathrm{Id} \star \varphi\psi) = \zeta(2)\zeta(3) L(4, \varphi\Psi)$$

and it is easy to check that $L(s, \varphi\Psi) = \dfrac{\zeta(s-2)}{\zeta(s)}$ (Re $s > 3$), and hence

$$S = \frac{\zeta(2)^2 \zeta(3)}{\zeta(4)} = \frac{5\zeta(3)}{2}$$

as required.

99

(a) We readily have

$$\sum_{dk\leqslant x} f(d,k) = \sum_{n\leqslant x}\sum_{dk=n} f(d,k) = \sum_{n\leqslant x}\sum_{d|n} f(d,n/d).$$

(b) By the previous question

$$\sum_{mn\leqslant x} \tau(m)\tau(n) = \sum_{n\leqslant x}\sum_{d|n}\tau(d)\tau(n/d) = \sum_{n\leqslant x}\tau_4(n)$$

and the result follows using Corollary 4.19 with $r = 4$.

100 Let $r \in \mathbb{Z}_{\geqslant 2}$.

(a) Note that the left-hand side is equal to

$$\sum_{n_1\leqslant x,\ldots,n_r\leqslant x}\ \sum_{n_1\cdots n_r|n}\mu(n_1\cdots n_r)$$

so that the asserted result follows from Example 4.9.

(b) Set $u(n_1,\ldots,n_r) := \mu(n_1\cdots n_r)\left\lfloor\dfrac{x}{n_1\cdots n_r}\right\rfloor$. Since u is symmetric with respect to the r variables $n_1,\ldots,n_r$, multiplying the left-hand side by $r!$ amounts to summing in the hypercube $[1,x]^r$, but we must take the diagonals into account. By a sieving argument, we get

$$\sum_{n_1,\ldots,n_r\leqslant x} u(n_1,\ldots,n_r) = r!\sum_{1\leqslant n_1<\cdots<n_r\leqslant x} u(n_1,\ldots,n_r)$$

$$+ \sum_{j=2}^{r}(-1)^j(j-1)\binom{r}{j}\sum_{n_j,\ldots,n_r\leqslant x} u\left(n_j,\ldots,n_j,n_{j+1},\ldots,n_r\right)$$

where the variable n_j appears j times in the inner sum. Now since

$$u\left(n_j,\ldots,n_j,n_{j+1},\ldots,n_r\right) = \mu\left(n_j^j n_{j+1}\cdots n_r\right)\left\lfloor\dfrac{x}{n_j^j n_{j+1}\cdots n_r}\right\rfloor$$

so that $u\left(n_j, \ldots, n_j, n_{j+1}, \ldots, n_r\right) = 0$ as soon as $n_j > 1$ since $j \geqslant 2$, we infer that the inner sum is

$$
= \begin{cases}
\displaystyle\sum_{n_{j+1},\ldots,n_r \leqslant x} \mu\left(n_{j+1}\cdots n_r\right)\left\lfloor \frac{x}{n_{j+1}\cdots n_r} \right\rfloor = \sum_{n \leqslant x}(1 - r + j)^{\omega(n)} & \text{if } 2 \leqslant j \leqslant r - 2 \\[2em]
\displaystyle\sum_{n_{r-1},n_r \leqslant x} \mu\left(n_{r-1}^{r-1}n_r\right)\left\lfloor \frac{x}{n_{r-1}^{r-1}n_r} \right\rfloor = 1 & \text{if } j = r - 1 \\[2em]
\displaystyle\sum_{n_r \leqslant x} \mu\left(n_r^r\right)\left\lfloor \frac{x}{n_r^r} \right\rfloor = \lfloor x \rfloor & \text{if } j = r
\end{cases}
$$

where we used the previous question when $2 \leqslant j \leqslant r - 2$, implying the first asserted result. Next, the asymptotic formula follows immediately from this identity, question (a) and Exercise 82.

(b) Similar to question (a) with the use of the second identity in Example 4.9.

101 An easy induction shows that, for any $r \in \mathbb{Z}_{\geqslant 2}$

$$
T_{r-1}(x) = \sum_{n_1 \leqslant x} \sum_{n_2 \leqslant x/n_1} \cdots \sum_{n_{r-2} \leqslant \frac{x}{n_1 \cdots n_{r-3}}} T_1\left(\frac{x}{n_1 \cdots n_{r-2}}\right).
$$

Now

$$
\begin{aligned}
S_r(x) &= r \sum_{n_1 \leqslant x} \sum_{n_2 \leqslant x/n_1} \cdots \sum_{n_{r-2} \leqslant \frac{x}{n_1 \cdots n_{r-3}}} \sum_{n_{r-1} \leqslant \frac{x}{n_1 \cdots n_{r-2}}} f\left(n_{r-1}\right) \sum_{n_r \leqslant \frac{x}{n_1 \cdots n_{r-1}}} \left\lfloor \frac{x}{n_1 \cdots n_r} \right\rfloor \\[1em]
&= r \sum_{n_1 \leqslant x} \sum_{n_2 \leqslant x/n_1} \cdots \sum_{n_{r-2} \leqslant \frac{x}{n_1 \cdots n_{r-3}}} \sum_{n_{r-1} \leqslant \frac{x}{n_1 \cdots n_{r-2}}} \left\lfloor \frac{x}{n_1 \cdots n_{r-1}} \right\rfloor \sum_{n_r \mid n_{r-1}} f\left(n_r\right) \\[1em]
&= r \sum_{n_1 \leqslant x} \sum_{n_2 \leqslant x/n_1} \cdots \sum_{n_{r-2} \leqslant \frac{x}{n_1 \cdots n_{r-3}}} \sum_{n_{r-1} \leqslant \frac{x}{n_1 \cdots n_{r-2}}} \left\lfloor \frac{x}{n_1 \cdots n_{r-1}} \right\rfloor (f \star \mathbf{1})\left(n_{r-1}\right) \\[1em]
&= r \sum_{n_1 \leqslant x} \sum_{n_2 \leqslant x/n_1} \cdots \sum_{n_{r-2} \leqslant \frac{x}{n_1 \cdots n_{r-3}}} T_1\left(\frac{x}{n_1 \cdots n_{r-2}}\right) \\[1em]
&= r\, T_{r-1}(x)
\end{aligned}
$$

by the above identity. When $f = \mu$, then $\mu \star 1 = e_1$ and hence $T_r(x) = \sum_{n \leqslant x} \tau_r(n)$ by induction. Therefore we derive

$$\sum_{n_1 \leqslant x, \ldots, n_r \leqslant x} (\mu(n_1) + \cdots + \mu(n_r)) \left\lfloor \frac{x}{n_1 \cdots n_r} \right\rfloor = r \sum_{n \leqslant x} \tau_{r-1}(n).$$

For instance, with $r = 3$ and using Corollary 6.12, we obtain for any $\varepsilon > 0$

$$\sum_{n_1, n_2, n_3 \leqslant x} (\mu(n_1) + \mu(n_2) + \mu(n_3)) \left\lfloor \frac{x}{n_1 n_2 n_3} \right\rfloor = 3x \log x + (6\gamma - 3)x + O_\varepsilon \left(x^{\frac{517}{1648} + \varepsilon} \right).$$

Chapter 5

102 We have $0 \leqslant \|x\| \leqslant \frac{1}{2}$, and the function ψ is odd and 1-periodic, implying that the function $\| \cdot \|$ is even and 1-periodic. For the first inequality, it suffices to suppose that $|x|, |y| \leqslant \frac{1}{2}$ giving in this case

$$\left| \|x\| - \|y\| \right| = \left| |x| - |y| \right| \leqslant |x - y|.$$

The second inequality may be proved similarly, noticing that for all $x \in \mathbb{R}$, there exists a unique $\theta_x \in \left] -\frac{1}{2}, \frac{1}{2} \right]$ such that $x = \lfloor x \rceil + \theta_x$, so that by periodicity we get $\|x + y\| = \|\theta_x + \theta_y\|$ showing that we may suppose that $|x|, |y| \leqslant \frac{1}{2}$, and we also may restrict ourselves to $0 \leqslant x, y \leqslant \frac{1}{2}$ since the function $\| \cdot \|$ is even. In this case, we finally have

$$\|x + y\| = \begin{cases} x + y = \|x\| + \|y\|, & \text{if } 0 \leqslant x + y \leqslant \frac{1}{2} \\ 1 - (x + y) \leqslant x + y = \|x\| + \|y\|, & \text{if } \frac{1}{2} \leqslant x + y \leqslant 1 \end{cases}$$

as required.

103 The functions $x \longmapsto |\sin(\pi x)|$ and $x \longmapsto \|x\|$ are both even and 1-periodic, so that it suffices to prove the asserted inequality for all $x \in \left[0, \frac{1}{2} \right]$. In this interval, the inequality takes the shape

$$2x \leqslant \sin(\pi x) \leqslant \pi x.$$

which is well-known. For instance, if $f(x) = \sin(\pi x) - 2x$, then $f''(x) = -\pi^2 \sin(\pi x) \leqslant 0$ for all $x \in \left[0, \frac{1}{2}\right]$, so that f is concave on this interval and therefore

$$f(x) \geqslant \min\left(f(0), f\left(\tfrac{1}{2}\right)\right) = 0$$

as required.

104 We have $\left\|\frac{m}{n}\right\| = \min\left(\frac{1}{2} - \psi\left(\frac{m}{n}\right), \frac{1}{2} - \psi\left(\frac{m}{n}\right)\right)$. Now from Exercise 2

- $\left\lfloor\frac{m}{n}\right\rfloor \geqslant \frac{m+1}{n} - 1 \Longleftrightarrow \frac{1}{2} - \psi\left(\frac{m}{n}\right) \geqslant \frac{1}{n}$;
- When $n \nmid m$, $\left\lfloor\frac{m}{n}\right\rfloor \leqslant \frac{m-1}{n} \Longleftrightarrow \frac{1}{2} + \psi\left(\frac{m}{n}\right) \geqslant \frac{1}{n}$.

Remark Obviously, $\left\|\frac{m}{n}\right\| = 0$ when $n \mid m$.

105 *This is [15, Lemma 3.1.1] but, for the sake of completeness, we give here a proof.* One can assume that f is non-decreasing, so that the length of interval $[f(N), f(2N)]$ satisfies

$$c_1 N \lambda_1 \leqslant M := f(2N) - f(N) \leqslant c_2 N \lambda_1$$

by the mean value theorem. Let $n \in \mathcal{E}(f, N, \delta)$. There exists $(m, d) \in \mathbb{Z} \times \left(0, \frac{1}{4}\right]$ such that

$$f(n) = m + d \quad \text{with} \quad |d| < \delta$$

and either $m \in [f(N), f(2N)]$, or $m \in [f(N) - \delta, f(N)) \cup (f(2N), f(2N) + \delta]$.

- Assume first that $m \in [f(N), f(2N)]$. Therefore there exists $y \in (f(N), f(2N))$ such that

$$n = f^{-1}(m + d) = f^{-1}(m) + d\left(f^{-1}\right)'(y)$$

and hence

$$f^{-1}(m) = n - d\left(f^{-1}\right)'(y) \quad \text{with} \quad \left|d\left(f^{-1}\right)'(y)\right| < \frac{\delta}{c_1 \lambda_1}$$

so that $m \in \mathcal{E}\left(f^{-1}, M, \frac{\delta}{c_1 \lambda_1}\right)$ with $c_1 N \lambda_1 \leqslant M \leqslant c_2 N \lambda_1$.
- Now suppose that $m \in (f(2N), f(2N) + \delta]$ so that $m = f(2N) + d'$ with $0 < d' \leqslant \delta$. Note that, since f is non-decreasing, $f(n) \leqslant f(2N) < m$, and thus

$$f(n) = m + d = f(2N) + d + d'$$

with $-\delta < d < 0$ and $0 < d' \leqslant \delta$. Since $|d' + d| = |d' - (-d)|$ with $0 < -d, d' < \delta$, there exists d_1 such that $f(n) = f(2N) + d_1$ with $-\delta < d_1 < 0$. Consequently there exists $y_1 \in (f(2N) - \delta, f(2N))$ such that

$$n = f^{-1}(f(2N) + d_1) = 2N + d_1 \left(f^{-1}\right)'(y_1).$$

The function f^{-1} is continuous and increasing, so that $f^{-1}(y_1) \in (N, 2N)$ and hence

$$|n - 2N| \leqslant \left|d_1 \left(f^{-1}\right)'(y_1)\right| < \frac{\delta}{c_1 \lambda_1}.$$

- By a similar argument, one can prove in the case $m \in [f(N) - \delta, f(N))$ that there exists $y_2 \in (f(N), f(N) + \delta)$ such that

$$|n - N| \leqslant \left|d_2 \left(f^{-1}\right)'(y_2)\right| < \frac{\delta}{c_1 \lambda_1}.$$

We complete the proof noticing that the number of integers n giving values of m outside the interval $[f(N), f(2N)]$ does not exceed

$$2 \left(\frac{\delta}{c_1 \lambda_1} + 1\right)$$

as required.

106

(a) If $N < x^{1/5}$, one may take $T = \mathcal{S}(f, N, \delta)$ since then $x^{-1/6} N^{5/6} < 1$. Now suppose that $N \geqslant x^{1/5}$ and let $a, b \in \mathbb{Z}_{\geqslant 1}$ and $n, n + a$ and $n + a + b$ be three consecutive elements of $\mathcal{S}(f, N, \delta)$ such that

$$1 \leqslant a, b \leqslant 2^{2/3} x^{-1/6} N^{5/6}.$$

As in Lemma 5.17, we will show that there are only two possibilities for the choice of b. The result will then follow by taking each 4th element of $\mathcal{S}(f, N, \delta)$.

There exist non-zero integers m_i and real numbers δ_i such that

$$f(n) = m_1 + \delta_1$$
$$f(n + a) = m_2 + \delta_2$$
$$f(n + a + b) = m_3 + \delta_3$$

with $|\delta_i| < \delta$ for $i \in \{1, 2, 3\}$. In fact, each integer m_i is positive since, for all $u \in [N, 2N]$, we have $f(u) \geqslant \sqrt{\frac{x}{2N}} \geqslant 2^{-1/2}$ and $\delta \leqslant c_0 N^{-1}$. Using the given polynomials P and Q, we get

$$f(n)P(n, a) - f(n + a)Q(n, a) = \left(\frac{x}{n}\right)^{1/2}(4n + a) - \left(\frac{x}{n + a}\right)^{1/2}(4n + 3a)$$

$$= \left(\frac{x}{n}\right)^{1/2}\frac{(n + a)^{1/2}(4n + a) - n^{1/2}(4n + 3a)}{(n + a)^{1/2}}$$

$$= \frac{x^{1/2}a^3}{n^{1/2}(n + a)^{1/2}D(n, a)},$$

where

$$D(n, a) = (n + a)^{1/2}(4n + a) + n^{1/2}(4n + 3a) \geqslant 8N^{3/2}$$

so that

$$0 < f(n)P(n, a) - f(n + a)Q(n, a) \leqslant \tfrac{1}{8}x^{1/2}a^3N^{-5/2} < \tfrac{1}{2}.$$

On the other hand, we have

$$f(n)P(n, a) - f(n + a)Q(n, a) = m_1 P(n, a) - m_2 Q(n, a) + \varepsilon$$

with

$$|\varepsilon| \leqslant 7\delta(n + a) \leqslant 14N\delta < \tfrac{1}{2}.$$

Hence by Lemma 5.16, we obtain

$$m_1 P(n, a) - m_2 Q(n, a) = 0. \tag{7}$$

Similarly we have

$$m_2 P(n + a, b) - m_3 Q(n + a, b) = 0$$
$$m_1 P(n, a + b) - m_3 Q(n, a + b) = 0$$

and eliminating m_3 we obtain

$$m_2 P(n + a, b)Q(n, a + b) - m_1 P(n, a + b)Q(n + a, b) = 0$$

implying that

$$3b^2(m_1 - m_2) + \kappa_1 b + 2\kappa_2 = 0, \tag{8}$$

where

$$\kappa_1 = a\,(7m_1 - 15m_2) - 16n(m_1 - m_2)$$

$$\kappa_2 = a^2\,(m_1 - 3m_2) - an\,(-20m_1 + 28m_2) - 16n^2(m_1 - m_2).$$

If $m_1 = m_2$, then by (7) we have $P(n, a) = Q(n, a)$ and then $4n+a = 4n+3a$, so that $a = 0$ which is impossible since $a \geqslant 1$. Therefore $m_1 \neq m_2$ and (8) is a quadratic equation in b, concluding the proof.

(b) Hence we deduce that

$$\mathcal{R}\left(\sqrt{\frac{x}{n}}, N, \delta\right) \ll |T| + 1 \ll \frac{N}{x^{-1/6}N^{5/6}} + 1 \ll (Nx)^{1/6}.$$

107 We follow the proof of Corollary 5.2 from which we borrow the notation. If $16 \leqslant y \leqslant x^{1/5}$, then obviously

$$\left| \sum_{x < n \leqslant x+y} \mu_2(n) - \frac{y}{\zeta(2)} \right| < 3y \leqslant 3\,x^{1/15}y^{2/3}.$$

Suppose that $x^{1/5} < y < \frac{1}{4}x^{1/2}$. We may write

$$\sum_{2\sqrt{y} < n \leqslant \sqrt{x}} \left(\left\lfloor \frac{x+y}{n^2} \right\rfloor - \left\lfloor \frac{x}{n^2} \right\rfloor \right) = \left(\sum_{2\sqrt{y} < n \leqslant c_0^{-1}y} + \sum_{c_0^{-1}y < n \leqslant \sqrt{x}} \right) \left(\left\lfloor \frac{x+y}{n^2} \right\rfloor - \left\lfloor \frac{x}{n^2} \right\rfloor \right)$$

$$= \Sigma_1 + \Sigma_2,$$

where c_0 is the constant appearing in (5.33). We use Theorem 5.7 with $k = 4$ for Σ_1 and Theorem 5.10 for Σ_2 which gives

$$\Sigma_1 \ll \left(x^{1/10}y^{2/5} + y^{2/3} + (xy)^{1/7} \right) \log x \ll x^{1/15}y^{2/3} \log x$$

and

$$\Sigma_2 \ll \max_{c_0^{-1}y < n \leqslant \sqrt{x}} \left(x^{1/5} + x^{1/15}yN^{-1/3} \right) \log x$$

$$\ll \left(x^{1/5} + x^{1/15}y^{2/3} \right) \log x \ll x^{1/15}y^{2/3} \log x$$

since $y > x^{1/5}$. This completes the proof since clearly $y^{1/2} \leqslant x^{1/15}y^{2/3}$.

108 The proof is exactly the same as that of Corollary 5.2 except that Theorem 5.10 is replaced by Theorems 5.11 and 5.7 is used with $k = 2r$ instead of $k = 4$. We omit the details.

109

(a) Let $L < T \leqslant (x + y)^{1/3}$ be any parameter at our disposal. We have

$$\sum_{L < b \leqslant (x+y)^{1/3}} \left(\left\lfloor \sqrt{\frac{x+y}{b^3}} \right\rfloor - \left\lfloor \sqrt{\frac{x}{b^3}} \right\rfloor \right)$$

$$= \left(\sum_{L < b \leqslant T} + \sum_{T < b \leqslant (x+y)^{1/3}} \right) \left(\left\lfloor \sqrt{\frac{x+y}{b^3}} \right\rfloor - \left\lfloor \sqrt{\frac{x}{b^3}} \right\rfloor \right)$$

$$= \sum_{L < b \leqslant T} \left(\left\lfloor \sqrt{\frac{x+y}{b^3}} \right\rfloor - \left\lfloor \sqrt{\frac{x}{b^3}} \right\rfloor \right) + \sum_{T < b \leqslant (x+y)^{1/3}} \sum_{x < a^2 b^3 \leqslant x+y} 1$$

$$= \sum_{L < b \leqslant T} \left(\left\lfloor \sqrt{\frac{x+y}{b^3}} \right\rfloor - \left\lfloor \sqrt{\frac{x}{b^3}} \right\rfloor \right) + \sum_{a \leqslant \sqrt{\frac{x+y}{T^3}}} \sum_{\left(\frac{x}{a^2}\right)^{1/3} < b \leqslant \left(\frac{x+y}{a^2}\right)^{1/3}} 1$$

$$= \sum_{L < b \leqslant T} \left(\left\lfloor \sqrt{\frac{x+y}{b^3}} \right\rfloor - \left\lfloor \sqrt{\frac{x}{b^3}} \right\rfloor \right) + \sum_{a \leqslant \sqrt{\frac{x+y}{T^3}}} \left(\left\lfloor \left(\frac{x+y}{a^2}\right)^{1/3} \right\rfloor - \left\lfloor \left(\frac{x}{a^2}\right)^{1/3} \right\rfloor \right)$$

$$= \sum_{L < b \leqslant T} \left(\left\lfloor \sqrt{\frac{x+y}{b^3}} \right\rfloor - \left\lfloor \sqrt{\frac{x}{b^3}} \right\rfloor \right)$$

$$+ \left(\sum_{a \leqslant L} + \sum_{L < a \leqslant \sqrt{\frac{x+y}{T^3}}} \right) \left(\left\lfloor \left(\frac{x+y}{a^2}\right)^{1/3} \right\rfloor - \left\lfloor \left(\frac{x}{a^2}\right)^{1/3} \right\rfloor \right)$$

$$\ll \sum_{L < b \leqslant T} \left(\left\lfloor \sqrt{\frac{x+y}{b^3}} \right\rfloor - \left\lfloor \sqrt{\frac{x}{b^3}} \right\rfloor \right) + \sum_{a \leqslant L} \left(\frac{y}{(ax)^{2/3}} + 1 \right)$$

$$+ \sum_{L < a \leqslant \sqrt{\frac{x+y}{T^3}}} \left(\left\lfloor \left(\frac{x+y}{a^2}\right)^{1/3} \right\rfloor - \left\lfloor \left(\frac{x}{a^2}\right)^{1/3} \right\rfloor \right)$$

$$\ll \sum_{L < b \leqslant T} \left(\left\lfloor \sqrt{\frac{x+y}{b^3}} \right\rfloor - \left\lfloor \sqrt{\frac{x}{b^3}} \right\rfloor \right) + y x^{-2/3} L^{1/3} + L$$

$$+ \sum_{L < a \leqslant \sqrt{\frac{2x}{T^3}}} \left(\left\lfloor \left(\frac{x+y}{a^2}\right)^{1/3} \right\rfloor - \left\lfloor \left(\frac{x}{a^2}\right)^{1/3} \right\rfloor \right).$$

Using (5.45), we get $yx^{-2/3}L^{1/3} < L$ and, for all $A, B > L$, we infer

$$\frac{y}{\sqrt{xB^3}} < \frac{y}{\sqrt{xL^3}} = x^{1/4}y^{-1/2}(\log x)^{3/4} \leqslant \tfrac{1}{4}$$

$$\frac{y}{(Ax)^{2/3}} < \frac{y}{(xL)^{2/3}} = \left(x^{-1}y \log x\right)^{1/3} \leqslant \tfrac{1}{4}$$

as soon as $x \geqslant 3$. Now the choice of $T = (2x)^{1/5}$ and the usual splitting argument provide the final result.

(b) Using the previous question, we have

$$\sum_{x<n\leqslant x+y} s_2(n) = \sum_{b\leqslant(x+y)^{1/3}} \mu_2(b) \sum_{\sqrt{\frac{x}{b^3}}<a\leqslant\sqrt{\frac{x+y}{b^3}}} 1$$

$$= \left(\sum_{b\leqslant L} + \sum_{L<b\leqslant(x+y)^{1/3}}\right) \mu_2(b) \left(\left\lfloor\sqrt{\frac{x+y}{b^3}}\right\rfloor - \left\lfloor\sqrt{\frac{x}{b^3}}\right\rfloor\right)$$

$$= \sum_{b\leqslant L} \mu_2(b) \left(\left\lfloor\sqrt{\frac{x+y}{b^3}}\right\rfloor - \left\lfloor\sqrt{\frac{x}{b^3}}\right\rfloor\right) + O\{(R_1 + R_2)\log x + L\}$$

and since

$$\sqrt{\frac{x+y}{b^3}} - \sqrt{\frac{x}{b^3}} = \frac{y}{2\sqrt{xb^3}} + O\left(\frac{y^2}{(bx)^{3/2}}\right)$$

we infer that

$$\sum_{b\leqslant L} \mu_2(b) \left(\left\lfloor\sqrt{\frac{x+y}{b^3}}\right\rfloor - \left\lfloor\sqrt{\frac{x}{b^3}}\right\rfloor\right) = \frac{y}{2x^{1/2}} \sum_{b\leqslant L} \frac{\mu_2(b)}{b^{3/2}} + O\left(\frac{y^2}{x^{3/2}} + L\right)$$

$$= \frac{y}{2x^{1/2}} \sum_{b=1}^{\infty} \frac{\mu_2(b)}{b^{3/2}} + O\left(L + \frac{y}{x^{1/2}} \sum_{b>L} \frac{1}{b^{3/2}}\right)$$

$$= \frac{\zeta(3/2)}{2\zeta(3)} \frac{y}{x^{1/2}} + O\left(L + \left(y^2x^{-1}\log x\right)^{1/4}\right)$$

$$= \frac{\zeta(3/2)}{2\zeta(3)} \frac{y}{x^{1/2}} + O(L),$$

where we used (5.45) which implies that L dominates all the other terms.

(c) To bound R_2, we split the sum into two subsums estimating trivially in the interval $(L, x^{2/15}]$ and using Theorem 5.6 in the interval $(x^{2/15}, (2x)^{1/5}]$ giving

$$\mathcal{R}\left(\left(\frac{x}{a^2}\right)^{1/3}, A, \frac{y}{(Ax)^{2/3}}\right) \ll (Ax)^{1/9} + y\left(Ax^{-2}\right)^{1/3} + \left(x^{-1}y^3 A^{-4}\right)^{1/6} + Ayx^{-1}$$

so that

$$\max_{x^{2/15}<A\leqslant(2x)^{1/5}} \mathcal{R}\left(\left(\frac{x}{a^2}\right)^{1/3}, A, \frac{y}{(Ax)^{2/3}}\right) \ll x^{2/15} + yx^{-3/5} + y^{1/2}x^{-23/90}.$$

Now (5.45) gives

$$R_2 \ll x^{2/15} + \frac{L}{\log x}.$$

We treat R_1 by using Theorem 5.7 with $k = 3$ giving

$$\mathcal{R}\left(\sqrt{\frac{x}{b^3}}, B, \frac{y}{\sqrt{xB^3}}\right) \ll \left(B^3 x\right)^{1/12} + \left(yB^{-1}\right)^{1/4} + y(Bx)^{-1/2} + B\left(yx^{-1}\right)^{1/2}$$

so that

$$\max_{L<B\leqslant(2x)^{1/5}} \mathcal{R}\left(\sqrt{\frac{x}{b^3}}, B, \frac{y}{\sqrt{xB^3}}\right) \ll x^{2/15} + y^{1/2}x^{-1/4}(\log x)^{1/4} + y^{1/2}x^{-3/10}$$

and using (5.45) implies also that

$$R_1 \ll x^{2/15} + \frac{L}{\log x}$$

concluding the proof.

(d) *Bounds for R_2.* In the range $(L, (64x)^{1/8}]$, we use Theorem 5.7 with $k = 3$ giving

$$\mathcal{R}\left(\left(\frac{x}{a^2}\right)^{1/3}, A, \frac{y}{(Ax)^{2/3}}\right) \ll \left(A^7 x\right)^{1/18} + \left(A^2 x^{-1}y^3\right)^{1/12} + y\left(x^{-2}A\right)^{1/3} + A\left(yx^{-1}\right)^{1/2}$$

so that

$$\max_{L<A\leqslant(64x)^{1/8}} \mathcal{R}\left(\left(\frac{x}{a^2}\right)^{1/3}, A, \frac{y}{(Ax)^{2/3}}\right) \ll x^{5/48} + y^{1/4}x^{-1/16} + yx^{-5/8} + y^{1/2}x^{-3/8}.$$

In the range $\left((64x)^{1/8}, (2x)^{1/5}\right]$, we use Theorem 5.8 implying that

$$\mathcal{R}\left(\left(\frac{x}{a^2}\right)^{1/3}, A, \frac{y}{(Ax)^{2/3}}\right) \ll \left\{(Ax)^{1/10} + \left(A^4x^3\right)^{1/15} + y\left(Ax^{-2}\right)^{1/3} + \left(A^{-2}xy^3\right)^{1/24}\right.$$

$$\left. + \left(A^8x^{-1}y^3\right)^{1/21} + \left(Ax^{-1}y^2\right)^{1/5} + A\left(yx^{-1}\right)^{1/2}\right\}(\log A)^{2/5}$$

so that

$$\max_{(64x)^{1/8} < A \leqslant (2x)^{1/5}} \mathcal{R}\left(\left(\frac{x}{a^2}\right)^{1/3}, A, \frac{y}{(Ax)^{2/3}}\right) \ll \left\{x^{3/25} + yx^{-3/5} + x^{1/32}y^{1/8} + x^{1/35}y^{1/7}\right.$$

$$\left. + y^{2/5}x^{-4/25} + y^{1/2}x^{-3/10}\right\}(\log x)^{2/5}$$

and the lower bound of (5.46) implies that

$$R_2 \ll x^{3/25}(\log x)^{2/5} + \frac{L}{\log x}.$$

Bounds for R_1. In the range $\left(L, (16x)^{1/6}\right]$, we use Theorem 5.7 with $k = 3$ giving

$$\mathcal{R}\left(\sqrt{\frac{x}{b^3}}, B, \frac{y}{\sqrt{xB^3}}\right) \ll \left(B^3x\right)^{1/12} + \left(yB^{-1}\right)^{1/4} + y(Bx)^{-1/2} + B\left(yx^{-1}\right)^{1/2}$$

so that

$$\max_{L < B \leqslant (16x)^{1/6}} \mathcal{R}\left(\sqrt{\frac{x}{b^3}}, B, \frac{y}{\sqrt{xB^3}}\right) \ll x^{1/8}(\log x)^{1/8} + y^{1/2}x^{-1/4}(\log x)^{1/4} + y^{1/2}x^{-3/8}.$$

In the range $\left((16x)^{1/6}, (2x)^{1/5}\right]$, we use Theorem 5.8 implying that

$$\mathcal{R}\left(\sqrt{\frac{x}{b^3}}, B, \frac{y}{\sqrt{xB^3}}\right) \ll \left\{\left(xB^{-1}\right)^{3/20} + (Bx)^{1/10} + y(Bx)^{-1/2} + \left(xy^2B^{-3}\right)^{1/16}\right.$$

$$\left. + (By)^{1/7} + \left(x^{-1}y^4B^{-3}\right)^{1/10} + B\left(yx^{-1}\right)^{1/2}\right\}(\log B)^{2/5}$$

so that

$$\max_{(16x)^{1/6} < B \leqslant (2x)^{1/5}} \mathcal{R}\left(\sqrt{\frac{x}{b^3}}, B, \frac{y}{\sqrt{xB^3}}\right) \ll \left\{x^{1/8} + yx^{-7/12} + x^{1/32}y^{1/8} + x^{1/35}y^{1/7}\right.$$

$$\left. + y^{2/5}x^{-3/20} + y^{1/2}x^{-3/10}\right\}(\log x)^{2/5}$$

and the lower bound of (5.46) implies that

$$R_1 \ll x^{1/8}(\log x)^{2/5} + \frac{L}{\log x}.$$

The proof is complete.

Remark Splitting R_1 into more parts and using Theorem 5.9 in the "critical" part, the author [20] proved that, under a more restricted range than (5.46), the exponent $1/8$ may be reduced to $\frac{19}{154} \approx 0.12333\ldots$ Note that the result obtained for R_2 above shows that we might expect to get the bound $\frac{3}{25} = 0.12$. However, this estimate for R_1 still remains open.

110 Let $r \geqslant 2$ be an integer and recall that the multiplicative function $\tau^{(r)}$ is given by

$$\tau^{(r)}(n) = \sum_{d^r \mid n} 1.$$

Clearly, we have $\tau^{(r)}(p^\alpha) = 1 + \lfloor \alpha/r \rfloor$ so that $\tau^{(r)}$ satisfies the hypotheses of Theorem 4.17. Furthermore, we have using Lemma 5.2

$$\mathcal{R}\left(\frac{x}{n^r}, N, \delta\right) \leqslant \sum_{N \leqslant d \leqslant 2N}\left(\left\lfloor \frac{x}{d^r} + \delta \right\rfloor - \left\lfloor \frac{x}{d^r} - \delta \right\rfloor\right) = \sum_{N \leqslant d \leqslant 2N} \sum_{x - d^r\delta < md^r \leqslant x + d^r\delta} 1$$

$$\leqslant \sum_{N \leqslant d \leqslant 2N} \sum_{x-(2N)^r\delta \leqslant md^r \leqslant x+(2N)^r\delta} 1 = \sum_{x-(2N)^r\delta \leqslant n \leqslant x+(2N)^r\delta} \sum_{\substack{d^r \mid n \\ N \leqslant d \leqslant 2N}} 1$$

$$\leqslant \sum_{x-(2N)^r\delta \leqslant n \leqslant x+(2N)^r\delta} \tau^{(r)}(n)$$

$$\ll \frac{2^{r+1}N^r\delta}{\log x} \exp\left(\sum_{p \leqslant x} \frac{\tau^{(r)}(p)}{p}\right) + x^\varepsilon \ll_{r,\varepsilon} N^r\delta,$$

where we used Theorem 4.17 with the fact that $r \geqslant 2$, and Corollary 3.4.

111

(a) Using respectively the inequalities $\delta^{-1} \geqslant c_0^{-1}N$ and $N^{-1} > \left(c_0^{-1}x\right)^{-1/3}$, we get

$$\frac{A}{R} = N^{-2/3}x^{1/6}\delta^{-1/6} \geqslant c_0^{-1/6}N^{-1/2} > c_0^{-1/6}\left(c_0^{-1}x\right)^{-1/6}x^{1/6} = 1$$

and similarly using $N \geqslant 2\,(x\delta)^{-1/2}$, we obtain

$$\frac{N}{2A} = 2^{-2/3}N^{1/3}(x\delta)^{1/6} \geqslant (x\delta)^{1/6}(x\delta)^{-1/6} = 1$$

so that

$$R < A \leqslant \tfrac{1}{2}N.$$

(b) Inserting this value of A in the proof of Theorem 5.10 we get

$$|T| \ll \left(N^2\delta x\right)^{1/6} + \left(x\delta^{-2}N^{-4}\right)^{1/3} + \left(x\delta^7 N^{11}\right)^{1/9}$$

implying the asserted result.

(c) For all $c_0^{-1}y \leqslant N \leqslant \left(c_0^{-1}x\right)^{1/3}$, we infer that

$$\mathcal{R}\left(\frac{x}{n^2}, N, \frac{y}{N^2}\right) \ll x^{1/5} + (xy)^{1/6} + \left(xy^{-2}\right)^{1/3} + \left(xy^7 N^{-3}\right)^{1/9}$$

so that

$$\max_{c_0^{-1}y \leqslant N \leqslant \left(c_0^{-1}x\right)^{1/3}} \mathcal{R}\left(\frac{x}{n^2}, N, \frac{y}{N^2}\right) \ll x^{1/5}+(xy)^{1/6}+\left(xy^{-2}\right)^{1/3}+\left(xy^4\right)^{1/9} \ll \left(xy^4\right)^{1/9}$$

since $x^{1/5} \leqslant y \leqslant x^{1/3}$. Furthermore, it has been proved in Corollary 5.2 that

$$\max_{2\sqrt{y} \leqslant N \leqslant c_0^{-1}y} \mathcal{R}\left(\frac{x}{n^2}, N, \frac{y}{N^2}\right) \ll x^{1/10}y^{2/5}+(xy)^{1/7}+y^{2/3}+\left(x^{-1}y^4\right)^{1/3} \ll \left(xy^4\right)^{1/9}$$

since $x^{1/5} \leqslant y \leqslant x^{1/3}$. Finally, by (5.29), we have

$$\max_{N \leqslant 2c_0^{2/3}x^{1/3}} \mathcal{R}\left(\sqrt{\frac{x}{n}}, N, \frac{y}{\sqrt{Nx}}\right) \ll x^{1/5}(\log x)^{2/5}$$

if c_0 is sufficiently small. Clearly $y^{1/2} \leqslant \left(xy^4\right)^{1/9}$, so that Lemma 5.3 with $A = 2\sqrt{y}$ and $B = \left(c_0^{-1}x\right)^{1/3}$ implies the asserted result. Note also that, since $y \geqslant x^{1/5}$, then

$$x^{1/15}y^{2/3} \geqslant \left(xy^4\right)^{1/9}.$$

112 Using Theorem 5.6 we get for all $4y < N \leqslant x$

$$\mathcal{R}\left(\frac{x}{n}, N, \frac{y}{N}\right) \ll x^{1/3} + y + (xy)^{1/2}N^{-1} + N\left(yx^{-1}\right)$$

so that

$$\max_{4y<N\leqslant x} \mathcal{R}\left(\frac{x}{n}, N, \frac{y}{N}\right) \ll x^{1/3} + y + \left(xy^{-1}\right)^{1/2}$$

and the second term dominates the others in view of $y \geqslant x^{1/3}$.

113 We use induction on k, the case $k = 2a$ coming from (5.26) and the fact that $k = 2a \geqslant 4$. Assume that the estimate is true for some $k \geqslant 2a$. By induction hypothesis and (5.26) used with $k+1$ instead of k, we get $\mathcal{R}(f, N, \delta) \ll \min(E, F)$ where

$$E = \max\left(T^{\frac{2}{(k+1)(k+2)}} N^{\frac{k}{k+2}}, \; N\delta^{\frac{2}{k(k+1)}}, \; N\left(\delta T^{-1}\right)^{\frac{1}{k+1}}\right) = \max(e_1, e_2, e_3)$$

and

$$F = \max\left(T^{\frac{2}{k(k+1)}} N^{\frac{k-1}{k+1}}, \; N\delta^{\frac{1}{a(2a-1)}}\right) = \max(f_1, f_2)$$

say. The result is proved except in the cases $\min(e_2, f_1)$ and $\min(e_3, f_1)$. As in Proposition 5.2, the following inequality

$$\min(x, y) \leqslant x^a y^{1-a}$$

with $0 \leqslant a \leqslant 1$, is used.

- *Case* $\min(e_2, f_1)$. We choose $a = \frac{k-2a}{(k-a)(k+2)} \in [0, 1]$ which gives

$$\mathcal{R}(f, N, \delta) \ll T^{\frac{2}{(k+1)(k+2)}} N^{\frac{k}{k+2}} \left(TN^{-a}\delta^{1-2a/k}\right)^{\frac{2}{(k-a)(k+1)(k+2)}} \ll T^{\frac{2}{(k+1)(k+2)}} N^{\frac{k}{k+2}}$$

by (5.50) and the fact that $k \geqslant 2a$ and $\delta < \frac{1}{4}$.

- *Case* $\min(e_3, f_1)$. We choose $a = \frac{2}{k+2}$ which gives

$$\mathcal{R}(f, N, \delta) \ll T^{\frac{2}{(k+1)(k+2)}} N^{\frac{k}{k+2}} \left(N\delta T^{-1}\right)^{\frac{2}{(k+1)(k+2)}} \ll T^{\frac{2}{(k+1)(k+2)}} N^{\frac{k}{k+2}}$$

by (5.50) again. This completes the proof.

Using this result with $a = 2$ we get

$$\mathcal{R}(f, N, \delta) \ll T^{\frac{2}{k(k+1)}} N^{\frac{k-1}{k+1}} + N\delta^{1/6}$$

if $N\delta \leqslant T \leqslant N^2$. This result is useful since the condition $T \leqslant N^2$ (i.e. $\lambda_2 \leqslant 1$) is often satisfied in the usual applications.

Chapter 6

114

(a) We have

$$|e(x) - e(y)| = |e(y)| \times |e(x - y) - 1| = |e(x - y) - 1| = 2|\sin \pi(x - y)|$$

and we conclude using Exercise 103.

(b) If $\alpha \in \mathbb{Z}$, then $e(\alpha n + \beta) = e(\beta)$ so that

$$\left| \sum_{n=M+1}^{N} e(\alpha n + \beta) \right| = \left| \sum_{n=M+1}^{N} 1 \right| = N - M.$$

Assume that $\alpha \notin \mathbb{Z}$. Using the previous inequality we derive

$$\left| \sum_{n=M+1}^{N} e(\alpha n + \beta) \right| = \left| \sum_{n=M+1}^{N} e(\alpha n) \right| = \frac{|e(N\alpha) - e(M\alpha)|}{|e(\alpha) - 1|} \leqslant \frac{2}{4\|\alpha\|} = \frac{1}{2\|\alpha\|}$$

as asserted.

115

(a) We have

$$\left| \sum_{N < n \leqslant N_1} e(\pm f(n)) \right| \leqslant \left| \sum_{\substack{N < n \leqslant N_1 \\ \|f'(n)\| < \delta}} e(\pm f(n)) \right| + \left| \sum_{\substack{N < n \leqslant N_1 \\ \|f'(n)\| \geqslant \delta}} e(\pm f(n)) \right|$$

$$\leqslant \mathcal{R}(f', N, \delta) + \left| \sum_{\substack{N < n \leqslant N_1 \\ \|f'(n)\| \geqslant \delta}} e(\pm f(n)) \right|.$$

Since f' is non-decreasing and $f'(N_1) - f'(N) \ll N\lambda_2$ by the mean value theorem, the interval $f'([N, N_1])$ has at most $\ll N\lambda_2 + 1$ integers. It follows that the set $\{x \in [N, N_1] : \|f'(x)\| \geqslant \delta\}$ can be partitioned in at most $\ll N\lambda_2 + 1$ subintervals, and the Kusmin–Landau inequality applied on each of these intervals implies the asserted estimate.

(b) Applying Theorem 5.3 we get

$$\sum_{N<n\leqslant N_1} e(f(n)) \ll N\lambda_2 + N\delta + \delta\lambda_2^{-1} + N\lambda_2\delta^{-1} + \delta^{-1} + 1$$

$$\ll N\lambda_2\delta^{-1} + N\delta + \delta\lambda_2^{-1} + \delta^{-1}$$

and choosing $\delta = \lambda_2^{1/2}$ gives

$$\sum_{N<n\leqslant N_1} e(f(n)) \ll N\lambda_2^{1/2} + \lambda_2^{-1/2}$$

as required.

116 Using Proposition 6.3, we derive

$$\sum_{(20x)^{1/2}<n\leqslant x^{3/4}} \psi\left(\frac{x}{n}\right) \ll \max_{(20x)^{1/2}<N\leqslant x^{3/4}} \left|\sum_{N<n\leqslant 2N} \psi\left(\frac{x}{n}\right)\right| \log x$$

$$\ll \max_{(20x)^{1/2}<N\leqslant x^{3/4}} \left(\frac{x}{N} + \frac{N^2}{x}\right) \log x \ll x^{1/2} \log x$$

as required.

117 It is an exercise of partial summation. From Corollary 6.1, we derive

$$\sum_{N<n\leqslant 2N} \psi\left(f(n) + u(n)\right) \ll \frac{N}{H} + \sum_{h\leqslant H} \frac{1}{h} \left|\sum_{N<n\leqslant 2N} e\left(hf(n)\right) e\left(hu(n)\right)\right|$$

and by Corollary 1.1 and Exercise 114, we get

$$\left|\sum_{N<n\leqslant 2N} e\left(hf(n)\right) e\left(hu(n)\right)\right|$$

$$\leqslant \left(1 + 2\pi h \sum_{N<n\leqslant 2N-1} |u(n+1) - u(n)|\right) \max_{N\leqslant N_1\leqslant 2N} \left|\sum_{N\leqslant n\leqslant N_1} e\left(hf(n)\right)\right|$$

$$\ll (1 + hU) \max_{N \leqslant N_1 \leqslant 2N} \left| \sum_{N \leqslant n \leqslant N_1} e\,(hf(n)) \right|.$$

118 Squaring out we get

$$\left| \sum_{h=0}^{H-1} e(ha) \right|^2 = \sum_{h_1=0}^{H-1} e\,(h_1 a) \sum_{h_2=0}^{H-1} e\,(-h_2 a) = \sum_{h_1=0}^{H-1} \sum_{h_2=0}^{H-1} e\,((h_1 - h_2)a)\,.$$

Now set $h_1 = h + k$ and $h_2 = k$ so that

$$\left\{ \begin{array}{l} 0 \leqslant h_1 \leqslant H - 1 \\ 0 \leqslant h_2 \leqslant H - 1 \end{array} \right. \Longleftrightarrow \left\{ \begin{array}{l} 0 \leqslant k \leqslant H - 1 \\ -h \leqslant k \leqslant H - 1 - h \end{array} \right.$$

$$\Longleftrightarrow \left\{ \begin{array}{l} |h| \leqslant H - 1 \\ 0 \leqslant k \leqslant H - 1 - |h| \end{array} \right.$$

and hence

$$\left| \sum_{h=0}^{H-1} e(ha) \right|^2 = \sum_{|h| \leqslant H-1} \sum_{k=0}^{H-1-|h|} e(ha) = \sum_{|h| \leqslant H-1} (H - |h|)\, e(ha)$$

as required.

119 We may obviously assume that $\mathcal{R}(f, N, \delta) \neq 0$, otherwise the inequality (6.53) is trivial.

(a) There exist $m \in \mathbb{Z}$ and $\delta_0 \in \mathbb{R}$ such that $f(n) = m + \delta_0$ with $|\delta_0| < \delta$, so that

$$hf(n) = hm + h\delta_0 \quad \text{with} \quad |h\delta_0| < (H - 1)\delta \leqslant (K - 1)\delta \leqslant \tfrac{1}{8}$$

and thus

$$\mathrm{Re}\,\{e(hf(n))\} = \cos(2\pi h f(n)) = \cos(2\pi hm + 2\pi h\delta_0) = \cos(2\pi h\delta_0) > \frac{\sqrt{2}}{2}$$

since $2\pi |h\delta_0| < \tfrac{1}{4}\pi$.

(b) Summing the previous inequality over n and h running, respectively, through the whole set $\mathcal{S}(f, N, \delta)$ and the integers $\{0, \ldots, H - 1\}$, we obtain

$$H\mathcal{R}(f, N, \delta) \leqslant \sqrt{2} \sum_{n \in \mathcal{S}(f,N,\delta)} \mathrm{Re}\left(\sum_{h=0}^{H-1} e(hf(n)) \right) \leqslant \sqrt{2} \sum_{n \in \mathcal{S}(f,N,\delta)} \left| \sum_{h=0}^{H-1} e(hf(n)) \right|.$$

Applying the Cauchy–Schwarz inequality gives

$$
\mathcal{R}(f, N, \delta) \leqslant \frac{\sqrt{2}}{H} \left(\sum_{n \in \mathcal{S}(f, N, \delta)} 1 \right)^{1/2} \left(\sum_{n \in \mathcal{S}(f, N, \delta)} \left| \sum_{h=0}^{H-1} e(h f(n)) \right|^2 \right)^{1/2}
$$

$$
= \frac{\sqrt{2}}{H} \mathcal{R}(f, N, \delta)^{1/2} \left(\sum_{n \in \mathcal{S}(f, N, \delta)} \left| \sum_{h=0}^{H-1} e(h f(n)) \right|^2 \right)^{1/2}
$$

so that squaring out we get

$$
\mathcal{R}(f, N, \delta) \leqslant \frac{2}{H^2} \sum_{n \in \mathcal{S}(f, N, \delta)} \left| \sum_{h=0}^{H-1} e(h f(n)) \right|^2 \leqslant \frac{2}{H^2} \sum_{N \leqslant n \leqslant 2N} \left| \sum_{h=0}^{H-1} e(h f(n)) \right|^2
$$

as asserted. Now using Exercise 118 we obtain

$$
\mathcal{R}(f, N, \delta) \leqslant \frac{2}{H^2} \sum_{N \leqslant n \leqslant 2N} \sum_{|h| \leqslant H-1} (H - |h|) \, e \, (h f(n))
$$

and treating the cases $h = 0$ and $h \neq 0$ separately we get

$$
\mathcal{R}(f, N, \delta) \leqslant \frac{2(N+1)}{H} + \frac{2}{H^2} \sum_{N \leqslant n \leqslant 2N} \sum_{\substack{|h| \leqslant H-1 \\ h \neq 0}} (H - |h|) \, e \, (h f(n))
$$

$$
\leqslant \frac{4N}{H} + \frac{2}{H} \sum_{N \leqslant n \leqslant 2N} \sum_{\substack{|h| \leqslant H-1 \\ h \neq 0}} \left(1 - \frac{|h|}{H} \right) e \, (h f(n))
$$

$$
= \frac{4N}{H} + \frac{2}{H} \sum_{h=1}^{H-1} \left(1 - \frac{h}{H} \right) \sum_{N \leqslant n \leqslant 2N} \{ e(h f(n)) + e(-h f(n)) \}
$$

$$
\leqslant \frac{4N}{H} + \frac{4}{H} \sum_{h=1}^{H-1} \mathrm{Re} \left(\sum_{N \leqslant n \leqslant 2N} e \, (h f(n)) \right)
$$

$$
\leqslant \frac{4N}{H} + \frac{4}{H} \sum_{h=1}^{H-1} \left| \sum_{N \leqslant n \leqslant 2N} e \, (h f(n)) \right|
$$

completing the proof of (6.53).

120 By Definition 6.2 and the inequality (6.53), we get for all integers $1 \leqslant H \ll \delta^{-1}$

$$\mathcal{R}(f, N, \delta) \ll NH^{-1} + H^{-1} \sum_{h \leqslant H} \left((hT)^k N^{l-k} + N(hT)^{-1} \right)$$

$$\ll NH^{-1} + (HT)^k N^{\ell-k} + NT^{-1} \ll NH^{-1} + (HT)^k N^{\ell-k}$$

since $N \leqslant 8T$. We conclude the proof by using Lemma 5.6.

121 *Squarefree problem.* Let x, y be real numbers satisfying (5.3). Using the exponent pair (6.17), we get for all $N \leqslant 2x^{1/3}$

$$\mathcal{R}\left(\sqrt{\frac{x}{n}}, N, \frac{y}{\sqrt{Nx}} \right) \ll \left(x^{97} N^{167} \right)^{1/696} + \left(x^{97} N^{-27} \right)^{1/502} + y \left(Nx^{-1} \right)^{1/2}$$

so that

$$\max_{N \leqslant 2x^{1/3}} \mathcal{R}\left(\sqrt{\frac{x}{n}}, N, \frac{y}{\sqrt{Nx}} \right) \ll x^{229/1044} + yx^{-1/3}.$$

Next we take the exponent pair

$$BA\left(\tfrac{13}{84} + \varepsilon, \ \tfrac{55}{84} + \varepsilon \right) = \left(\tfrac{55}{194} + \varepsilon, \ \tfrac{55}{97} + \varepsilon \right),$$

where the small real number $\varepsilon > 0$ need not have the same occurrence at each computation. This yields for all $2\sqrt{y} < N \leqslant 2x^{1/3}$ and all $\varepsilon > 0$

$$\mathcal{R}\left(\frac{x}{n^2}, N, \frac{y}{N^2} \right) \ll x^{55/249+\varepsilon} + \left(xN^{-1} \right)^{55/194+\varepsilon} + yN^{-1}$$

so that

$$\max_{x^{1/4} < N \leqslant 2x^{1/3}} \mathcal{R}\left(\frac{x}{n^2}, N, \frac{y}{N^2} \right) \ll x^{55/249+\varepsilon} + y^{1/2}.$$

The exponent pair $\left(\tfrac{1}{6}, \tfrac{2}{3} \right)$ provides the bound

$$\mathcal{R}\left(\frac{x}{n^2}, N, \frac{y}{N^2} \right) \ll \left(N^2 x \right)^{1/7} + (Nx)^{1/6} + yN^{-1}$$

for all $2\sqrt{y} < N \leqslant 2x^{1/3}$, so that

$$\max_{2\sqrt{y} < N \leqslant x^{1/4}} \mathcal{R}\left(\frac{x}{n^2}, N, \frac{y}{N^2} \right) \ll x^{3/14} + y^{1/2}.$$

Hence using Lemma 5.3 with $A = 2\sqrt{y}$ and $B = x^{1/3}$ we get for all x, y satisfying (5.3) and $\varepsilon > 0$

$$\sum_{x < n \leqslant x+y} \mu_2(n) = \frac{y}{\zeta(2)} + O_\varepsilon \left(x^{55/249+\varepsilon} + y^{1/2} \right).$$

Note that $\frac{55}{249} \approx 0.22088\ldots$

- *Square-full problem.* We take up the notation of Exercise 109 and let x, y be real numbers satisfying (5.45). Using the exponent pair (6.17), we get

$$\mathcal{R}\left(\left(\frac{x}{a^2} \right)^{1/3}, A, \frac{y}{(Ax)^{2/3}} \right) \ll \left(x^{97} A^{202} \right)^{1/1044} + \left(x^{97} A^{-89} \right)^{1/753} + y \left(Ax^{-2} \right)^{1/3}$$

so that

$$\max_{L < A \leqslant (2x)^{1/5}} \mathcal{R}\left(\left(\frac{x}{a^2} \right)^{1/3}, A, \frac{y}{(Ax)^{2/3}} \right) \ll x^{229/1740} + yx^{-3/5}.$$

Using the exponent pair

$$B A^2 \left(\tfrac{13}{84} + \varepsilon, \ \tfrac{55}{84} + \varepsilon \right) = \left(\tfrac{76}{207} + \varepsilon, \ \tfrac{110}{207} + \varepsilon \right)$$

we get

$$\mathcal{R}\left(\sqrt{\frac{x}{b^3}}, B, \frac{y}{\sqrt{x B^3}} \right) \ll \left(x^{38} B^{-4} \right)^{1/283} + \left(x^{38} B^{-80} \right)^{1/207} + \frac{y}{\sqrt{Bx}}$$

so that

$$\max_{x^{3359/17,557} < B \leqslant (2x)^{1/5}} \mathcal{R}\left(\sqrt{\frac{x}{b^3}}, B, \frac{y}{\sqrt{x B^3}} \right) \ll x^{2310/17,557} + yx^{-10,458/17,557}.$$

Taking the exponent pair $\left(\tfrac{55}{194} + \varepsilon, \tfrac{55}{97} + \varepsilon \right)$ again yields

$$\mathcal{R}\left(\sqrt{\frac{x}{b^3}}, B, \frac{y}{\sqrt{x B^3}} \right) \ll (Bx)^{55/498+\varepsilon} + \left(x B^{-1} \right)^{55/388+\varepsilon} + \frac{y}{\sqrt{Bx}}$$

valid for all $L < B \leqslant (2x)^{1/5}$, so that

$$\max_{x^{1/8} < B \leqslant x^{1671/8621}} \mathcal{R}\left(\sqrt{\frac{x}{b^3}}, B, \frac{y}{\sqrt{x B^3}} \right) \ll x^{2310/17,557} + yx^{-9/16}.$$

Completing the proof with the trivial bound for R_1 in the range $\left(L, x^{1/8}\right]$ we finally get for all x, y satisfying (5.45)

$$\sum_{x < n \leqslant x+y} s_2(n) = \frac{\zeta(3/2)}{2\zeta(3)} \frac{y}{x^{1/2}} + O\left(x^{2310/17557} \log x + L\right).$$

Note that $\frac{2310}{17557} \approx 0.13157\ldots$

122

(a) This is [11, Lemma 2.11].

(b) Let $1 \leqslant T < t$ and let (k, ℓ) be an exponent pair. We have

$$\left|\sum_{n \leqslant t} n^{-\sigma-it}\right| \leqslant \left|\sum_{n \leqslant T} n^{-\sigma-it}\right| + \left|\sum_{T < n \leqslant t} n^{-\sigma-it}\right|$$

$$\ll \sum_{n \leqslant T} n^{-\sigma} + \max_{T < N \leqslant t} \left|\sum_{N < n \leqslant 2N} n^{-\sigma-it}\right| \log t$$

$$\ll \sum_{n \leqslant T} n^{-\sigma} + \max_{T < N \leqslant t} N^{-\sigma} \max_{N \leqslant N_1 \leqslant 2N} \left|\sum_{N \leqslant n \leqslant N_1} n^{-it}\right| \log t$$

$$\ll T^{1-\sigma} + \max_{T < N \leqslant t} N^{-\sigma} \max_{N \leqslant N_1 \leqslant 2N} \left(t^k N^{\ell-k} + Nt^{-1}\right) \log t$$

$$\ll T^{1-\sigma} + \max_{T < N \leqslant t} \left(t^k N^{\ell-k-\sigma} + t^{-1} N^{1-\sigma}\right) \log t.$$

Now since $\ell - k \leqslant \frac{1}{2}$ and $\frac{1}{2} \leqslant \sigma \leqslant 1$, we deduce that $k + \sigma \geqslant \ell$. This implies that

$$\sum_{n \leqslant t} n^{-\sigma-it} \ll T^{1-\sigma} + t^k T^{\ell-k-\sigma} \log t$$

and the choice of $T = t^{\frac{k}{1+k-\ell}}$ gives

$$\sum_{n \leqslant t} n^{-\sigma-it} \ll t^{\frac{k(1-\sigma)}{1+k-\ell}} \log t.$$

Note that $0 \leqslant k \leqslant \frac{1}{2} \leqslant \ell \leqslant 1$ and $\ell - k \leqslant \frac{1}{2}$ imply that $\frac{1}{2} \leqslant 1 + k - \ell \leqslant 1$. Using the previous question, we derive

$$\zeta(\sigma + it) \ll \left(t^{\frac{k(1-\sigma)}{1+k-\ell}} + t^{1-2\sigma}\right) \log t$$

and the second term is clearly absorbed by the first one. With $\sigma = \frac{1}{2}$ this gives for all $t \geqslant 3$

$$\zeta\left(\tfrac{1}{2} + it\right) \ll t^{\frac{k}{2(1+k-\ell)}} \log t$$

and Bourgain and Watt's exponent pair $\left(\frac{13}{84} + \varepsilon, \frac{1}{2} + \frac{13}{84} + \varepsilon\right)$ provides the bound

$$\zeta\left(\tfrac{1}{2} + it\right) \ll t^{13/84+\varepsilon}$$

for all $t \geqslant 3$ and $\varepsilon > 0$, which is the best result up to now.

123 From Lemma 6.15 applied to the function $f : n \longmapsto (n + M - N)\alpha$, we derive using Theorem 5.3 and setting $K := \left\lfloor \frac{\log H}{\log 2} \right\rfloor$

$$\sum_{M < n \leqslant M+N} \min\left(H, \frac{1}{\|n\alpha\|}\right) = H \sum_{N < n \leqslant 2N} \min\left(1, \frac{1}{H\|(n+M-N)\alpha\|}\right)$$

$$< 24N + 2H \sum_{k=0}^{K-2} 2^{-k} \mathcal{R}\left((n + M - N)\alpha, N, 2^k H^{-1}\right)$$

$$< 24N + 2H \sum_{k=0}^{K-2} 2^{-k}\left(2N\alpha + 2^{k+2}NH^{-1} + 2^{k+1}(H\alpha)^{-1} + 1\right)$$

$$\leqslant 24N + 8HN\alpha + 16NK + 4\alpha^{-1}K + 4H$$

implying the asserted result since $K \geqslant 2 \Rightarrow 24N \leqslant 12NK$.

124

(a) For $\psi(x/n)$, this is exactly Corollary 6.7, and note that, for any $h \in \mathbb{Z}_{\geqslant 1}$

$$e\left(h\left(\frac{x}{n} + \frac{1}{2}\right)\right) = (-1)^h e\left(\frac{hx}{n}\right)$$

so that, by Corollary 6.1, we derive for any $H \in \mathbb{Z}_{\geqslant 1}$

$$\left|\sum_{N < n \leqslant 2N} \psi\left(\frac{x}{n} + \frac{1}{2}\right)\right| \ll \frac{N}{H} + \sum_{h \leqslant H} \frac{1}{h}\left|\sum_{N < n \leqslant 2N} e\left(h\left(\frac{x}{n} + \frac{1}{2}\right)\right)\right|$$

$$\ll \frac{N}{H} + \sum_{h \leqslant H} \frac{1}{h}\left|\sum_{N < n \leqslant 2N} e\left(\frac{hx}{n}\right)\right|$$

yielding the same estimate by proceeding as in Corollary 6.7.

(b) Since $\lfloor z \rfloor + \left\lfloor z + \frac{1}{2} \right\rfloor = \lfloor 2z \rfloor$, we first get

$$S(x) = x \sum_{k \leqslant K} \frac{\left\lfloor \frac{x}{k} + \frac{1}{2} \right\rfloor - \left\lfloor \frac{x}{k} \right\rfloor}{\left\lfloor \frac{x}{k} \right\rfloor \left(2 \left\lfloor \frac{x}{k} \right\rfloor + 1 \right)}$$

and using

$$\frac{1}{\left\lfloor \frac{x}{k} \right\rfloor \left(2 \left\lfloor \frac{x}{k} \right\rfloor + 1 \right)} = \frac{k^2}{2x^2} + O\left(\frac{k^3}{x^3} \right)$$

we derive

$$S(x) = x \sum_{k \leqslant K} \left\{ \frac{k^2}{2x^2} + O\left(\frac{k^3}{x^3} \right) \right\} \left\{ \frac{1}{2} + \psi\left(\frac{x}{k} \right) - \psi\left(\frac{x}{k} + \frac{1}{2} \right) \right\}$$

$$= \frac{1}{4x} \sum_{k \leqslant K} k^2 + \frac{1}{2x} \sum_{k \leqslant K} k^2 \left(\psi\left(\frac{x}{k} \right) - \psi\left(\frac{x}{k} + \frac{1}{2} \right) \right) + O\left(K^4 x^{-2} \right)$$

$$= \frac{K(K+1)(2K+1)}{24x} + \frac{S_1(x)}{2x} + O(1), \tag{7.49}$$

where, by Abel summation

$$|S_1(x)| := \left| \sum_{k \leqslant K} k^2 \left(\psi\left(\frac{x}{k} \right) - \psi\left(\frac{x}{k} + \frac{1}{2} \right) \right) \right|$$

$$\leqslant 2K^2 \max_{k \leqslant K} \left| \sum_{m \leqslant k} \left(\psi\left(\frac{x}{m} \right) - \psi\left(\frac{x}{m} + \frac{1}{2} \right) \right) \right|$$

$$\ll K^2 \max_{k \leqslant K} \max_{M \leqslant k} \left| \sum_{M < m \leqslant 2M} \left(\psi\left(\frac{x}{m} \right) - \psi\left(\frac{x}{m} + \frac{1}{2} \right) \right) \right| \log k$$

$$\ll K^2 \max_{k \leqslant K} \max_{M \leqslant k} \left(\left(x^k M^{\ell-k} \right)^{\frac{1}{k+1}} + M^2 x^{-1} \right) \log x$$

$$\ll K^2 \left(x^{\frac{k+\ell}{2k+2}} + K^2 x^{-1} \right) \log x \ll K^2 \left(x^{\frac{k+\ell}{2k+2}} + 1 \right) \log x$$

and inserting this estimate in (7.49) gives the asserted result.

125

(a) By partial summation

$$\left| \sum_{d \leqslant n} d^2 \psi \left(\frac{m}{d} \right) \right| \ll \max_{D \leqslant n} \left| \sum_{D < d \leqslant 2D} d^2 \psi \left(\frac{m}{d} \right) \right| \log n$$

$$\ll \max_{D \leqslant n} D^2 \max_{D \leqslant D_1 \leqslant 2D} \left| \sum_{D \leqslant d \leqslant D_1} \psi \left(\frac{m}{d} \right) \right| \log n$$

so that, if (k, ℓ) is an exponent pair, we get by Corollary 6.7

$$\left| \sum_{d \leqslant n} d^2 \psi \left(\frac{m}{d} \right) \right| \ll \max_{D \leqslant n} \left\{ \left(m^k D^{k+\ell+2} \right)^{\frac{1}{k+1}} + D^4 m^{-1} \right\} \log n.$$

(b) By Theorem 6.32, we have, if $c\,m^{1/3} \leqslant t \leqslant \sqrt{2m}$ for some $c > 0$

$$\sum_{d \leqslant t} d \psi \left(\frac{m}{d} \right)^2 = \frac{t^2}{24} + O\left(m^{2/3} + (mt)^{1/2} \right)$$

so that, by partial summation

$$\sum_{d \leqslant n} d^2 \psi \left(\frac{m}{d} \right)^2 = n \sum_{d \leqslant n} d \psi \left(\frac{m}{d} \right)^2 - \left(\int_1^{cm^{1/3}} + \int_{cm^{1/3}}^n \right) \left(\sum_{d \leqslant t} d \psi \left(\frac{m}{d} \right)^2 \right) dt$$

$$= \frac{n^3}{24} + O\left(nm^{2/3} + (mn^3)^{1/2} \right) - O\left(\int_1^{cm^{1/3}} t^2\,dt \right)$$

$$- \int_{cm^{1/3}}^n \left\{ \frac{t^2}{24} + O\left(m^{2/3} + (mt)^{1/2} \right) \right\} dt$$

$$= \frac{n^3}{36} + O\left(nm^{2/3} + (mn^3)^{1/2} + m \right)$$

the last term being absorbed by the second one since $m^{1/3} \ll n$.

(c) Since $(m \bmod d) = d \left\{ \frac{m}{d} \right\} = \frac{1}{2}d + d\psi \left(\frac{m}{d} \right)$, we derive

$$\sum_{d=1}^n (m \bmod d)^2 = \sum_{d \leqslant n} \frac{d^2}{4} + \sum_{d \leqslant n} d^2 \left(\psi \left(\frac{m}{d} \right)^2 + \psi \left(\frac{m}{d} \right) \right)$$

and the questions above yield

$$\sum_{d=1}^{n} (m \bmod d)^2 = \frac{n^3}{9}$$

$$+ O\left(\left\{\left(m^k n^{k+\ell+2}\right)^{\frac{1}{k+1}} + n^4 m^{-1}\right\} \log n + nm^{2/3} + (mn^3)^{1/2} + n^2\right)$$

and the exponent pair $(k, \ell) = \left(\frac{1}{2}, \frac{1}{2}\right)$ gives the asserted result, since $m^{1/3} \ll n \leqslant \sqrt{2m}$.

126 Let $r \in \mathbb{Z}_{\geqslant 2}$ fixed. First note that

$$\sum_{n \leqslant x} \left\lfloor \frac{x^r}{n^r} \right\rfloor = \sum_{n \leqslant x^r} \tau(1, r; n).$$

Using Example 4.12, we derive

$$\sum_{n \leqslant x} \left\lfloor \frac{x^r}{n^r} \right\rfloor = x^r \zeta(r) + x \zeta\left(\frac{1}{r}\right) + \Delta(1, r; x^r),$$

where

$$\Delta(1, r; z) = - \sum_{n \leqslant z^{\frac{1}{r+1}}} \left\{ \psi\left(\frac{z}{n^r}\right) + \psi\left(\left(\frac{z}{n}\right)^{1/r}\right) \right\} + O(1).$$

Now, for all $1 \leqslant N \leqslant z^{1/r}$ large, Corollary 6.7 yields

$$\sum_{N < n \leqslant 2N} \psi\left(\frac{z}{n^r}\right) \ll \left(z^k N^{\ell - rk}\right)^{\frac{1}{k+1}} + N^{r+1} z^{-1}$$

$$\sum_{N < n \leqslant 2N} \psi\left(\left(\frac{z}{n}\right)^{1/r}\right) \ll \left(z^k N^{\ell r - k}\right)^{\frac{1}{r(k+1)}} + \left(N^{r+1} z^{-1}\right)^{1/r}.$$

Firstly, for any exposant pair (k, ℓ), we always have

$$\max_{N \leqslant x^{\frac{1}{r+1}}} \sum_{N < n \leqslant 2N} \psi\left(\left(\frac{x}{n}\right)^{1/r}\right) \ll x^{\frac{k+\ell}{(k+1)(r+1)}}.$$

If $\ell \geqslant rk$, then

$$\max_{N \leqslant z^{\frac{1}{r+1}}} \left(\sum_{N < n \leqslant 2N} \psi\left(\frac{z}{n^r}\right) + \sum_{N < n \leqslant 2N} \psi\left(\left(\frac{z}{n}\right)^{1/r}\right) \right) \ll z^{\frac{k+\ell}{(k+1)(r+1)}}.$$

Now assume that $\ell \leqslant rk$. Let $T \in \left[1, z^{\frac{1}{r+1}}\right)$ be a parameter at our disposal, so that the first sum does not exceed

$$\ll \left(\sum_{n \leqslant T} + \sum_{T < N \leqslant z^{\frac{1}{r+1}}} \right) \psi\left(\frac{z}{n^r}\right) \ll T + \left(z^k T^{\ell - rk}\right)^{\frac{1}{k+1}}$$

and choosing $T = z^{\frac{k}{k(r+1)+1-\ell}}$ yields

$$\max_{N \leqslant z^{\frac{1}{r+1}}} \sum_{N < n \leqslant 2N} \psi\left(\frac{z}{n^r}\right) \ll z^{\frac{k}{k(r+1)+1-\ell}}.$$

Noticing that the condition $\ell \leqslant rk$ implies that $\frac{k}{k(r+1)+1-\ell} \geqslant \frac{k+\ell}{(k+1)(r+1)}$ and that $\ell \leqslant 1$ implies that $\frac{k}{k(r+1)+1-\ell} \leqslant \frac{1}{r+1}$, we finally get

$$\max_{N \leqslant z^{\frac{1}{r+1}}} \left(\sum_{N < n \leqslant 2N} \psi\left(\frac{z}{n^r}\right) + \sum_{N < n \leqslant 2N} \psi\left(\left(\frac{z}{n}\right)^{1/r}\right) \right) \ll z^{\frac{k}{k(r+1)+1-\ell}}$$

if $\ell \leqslant rk$, implying the asserted estimate in both cases. Note that using Bourgain's exponent pair $(k, \ell) = BA\left(\frac{13}{84} + \varepsilon, \frac{55}{84} + \varepsilon\right) = \left(\frac{55}{194} + \varepsilon, \frac{55}{97} + \varepsilon\right)$, we get for all $\varepsilon > 0$

$$\sum_{n \leqslant x} \left\lfloor \frac{x^r}{n^r} \right\rfloor = x^r \zeta(r) + x \zeta\left(\frac{1}{r}\right) + O_{\varepsilon, r}\left(x^{\frac{55r}{55r+139} + \varepsilon}\right).$$

127 Using respectively (6.23) and (2.7), we derive for any $0 < \rho < 1$

$$D_{(1,\ldots,1)}(n) = \frac{1}{2\pi i} \oint_{|z| = \rho} \left(\sum_{j=0}^{\infty} z^j\right)^k \frac{dz}{z^{n+1}} = \frac{1}{2\pi i} \oint_{|z| = \rho} \frac{1}{(1-z)^k} \frac{dz}{z^{n+1}}$$

$$= \sum_{j=0}^{\infty} \binom{k+j-1}{j} \frac{1}{2\pi i} \oint_{|z| = \rho} \frac{dz}{z^{n-j+1}}$$

$$= \sum_{j=0}^{\infty} \binom{k+j-1}{j} \times \begin{cases} 1, & \text{if } n = j \\ 0, & \text{otherwise} \end{cases} = \binom{k+n-1}{n}.$$

128 Assume first $0 < \lambda_m < 1$ and let $H \in \mathbb{Z}_{\geqslant 1}$. Using Corollary 6.1 and applying Theorem 6.21 to the function $n \mapsto hf(n)$ for $1 \leqslant h \leqslant H$, we derive for all $\varepsilon > 0$

$$\sum_{N < n \leqslant 2N} \psi(f(n)) \ll \frac{N}{H} + \sum_{h \leqslant H} \frac{N^\varepsilon}{h} \left\{ N(h\lambda_m)^{\frac{1}{m(m-1)}} + N^{1 - \frac{2}{m(m-1)}} + N^{1 - \frac{2}{m(m-1)}} (h\lambda_m)^{-\frac{2}{m^2(m-1)}} \right\}$$

$$\ll \frac{N}{H} + N^\varepsilon \left\{ N(H\lambda_m)^{\frac{1}{m(m-1)}} + N^{1 - \frac{1}{m(m-1)}} + N^{1 - \frac{2}{m(m-1)}} \lambda_m^{-\frac{2}{m^2(m-1)}} \right\}$$

and choosing $H = \left\lfloor \lambda_m^{-\frac{1}{m(m-1)+1}} \right\rfloor$ yields

$$N^{-\varepsilon} \sum_{N < n \leqslant 2N} \psi(f(n)) \ll N\lambda_m^{\frac{1}{m(m-1)+1}} + N^{1 - \frac{1}{m(m-1)}} + N^{1 - \frac{2}{m(m-1)}} \lambda_m^{-\frac{2}{m^2(m-1)}}.$$

If $\lambda_m \geqslant 1$, then the first term above is $\gg N$. Hence we may state the following corollary.

Corollary 7.14 *Assume* $m \geqslant 3$ *and let* $\varepsilon > 0$ *and* $f \in C^m([N, 2N])$ *such that there exist* $\lambda_m > 0$ *and* $c_m \geqslant 1$ *such that, for all* $x \in [N, 2N]$, *we have* $\lambda_m \leqslant f^{(m)}(x) \leqslant c_m\lambda_m$. *Then*

$$\sum_{N < n \leqslant 2N} \psi(f(n)) \ll N^\varepsilon \left(N\lambda_m^{\frac{1}{m(m-1)+1}} + N^{1 - \frac{1}{m(m-1)}} + N^{1 - \frac{2}{m(m-1)}} \lambda_m^{-\frac{2}{m^2(m-1)}} \right).$$

The argument is similar for $\mathcal{R}(f, N, \delta)$, except that we use (6.53) instead of Corollary 6.1. Under the same conditions as in the above corollary, we get for all $\varepsilon > 0$

$$\mathcal{R}(f, N, \delta) \ll N\delta + N^\varepsilon \left(N\lambda_m^{\frac{1}{m(m-1)+1}} + N^{1 - \frac{1}{m(m-1)}} + N^{1 - \frac{2}{m(m-1)}} \lambda_m^{-\frac{2}{m^2(m-1)}} \right),$$

where $0 < \delta < \frac{1}{4}$. We leave the details to the reader.

Chapter 7

129 Let $\mathbb{K} = \mathbb{Q}(\sqrt{-5})$. Suppose that 3 is not an irreducible so that $3 = rs$ with $N_{\mathbb{K}/\mathbb{Q}}(r) \neq 1$ and $N_{\mathbb{K}/\mathbb{Q}}(s) \neq 1$. Since $9 = N_{\mathbb{K}/\mathbb{Q}}(3) = N_{\mathbb{K}/\mathbb{Q}}(r) N_{\mathbb{K}/\mathbb{Q}}(s)$, we must then have $N_{\mathbb{K}/\mathbb{Q}}(r) = N_{\mathbb{K}/\mathbb{Q}}(s) = 3$ and hence $a^2 + 5b^2 = 3$ for some $a, b \in \mathbb{Z}$. This implies that $b = 0$ and thus $a^2 = 3$, which is impossible.

Similarly, if $7 = rs$ where neither r nor s is a unit, then we must have $a^2 + 5b^2 = 7$, implying that either $b = 0$ and $a^2 = 7$ or $b = \pm 1$ and $a^2 = 2$, both cases being impossible.

If $1 \pm 2\sqrt{-5} = rs$ where neither r nor s is a unit, then

$$21 = N_{\mathbb{K}/\mathbb{Q}}\left(1 \pm 2\sqrt{-5}\right) = N_{\mathbb{K}/\mathbb{Q}}(r)\, N_{\mathbb{K}/\mathbb{Q}}(s)$$

and hence either $N_{\mathbb{K}/\mathbb{Q}}(r) = 3$ or $N_{\mathbb{K}/\mathbb{Q}}(s) = 3$, which is impossible as was seen above.

130 θ is algebraic over $\mathbb{Q}$ as the sum of two algebraic numbers over $\mathbb{Q}$. This gives the answer to the exercise, but does not provide the minimal polynomial of θ. To do this, one may use the following lemma, useful for small degrees.[6]

Lemma 7.34 *Let α, β be algebraic over $\mathbb{Q}$ and $P, Q \in \mathbb{Q}[X]$ such that $P(\alpha) = Q(\beta) = 0$. Then the resultant*

$$R = \mathrm{Res}_Y\left(P(X), Q(Y - X)\right)$$

satisfies $R \in \mathbb{Q}[Y]$ and $R\,(\alpha + \beta) = 0$.

Proof We have clearly $R \in \mathbb{Q}[Y]$. Furthermore, R is equal to zero if and only if P and Q have a common root. But $R\,(\alpha + \beta)$ is the resultant of $P(X)$ and $Q(\alpha + \beta - X)$ which have α as a common root, and hence $R\,(\alpha + \beta) = 0$.   □

Applying Lemma 7.34 with $P = X^5 - 2$ and $Q = X^3 - 2$ we get

$$R = \begin{vmatrix}
1 & 0 & 0 & 0 & 0 & -2 & 0 & 0 \\
0 & 1 & 0 & 0 & 0 & 0 & -2 & 0 \\
0 & 0 & 1 & 0 & 0 & 0 & 0 & -2 \\
-1 & 3Y & -3Y^2 & Y^3 - 2 & 0 & 0 & 0 & 0 \\
0 & -1 & 3Y & -3Y^2 & Y^3 - 2 & 0 & 0 & 0 \\
0 & 0 & -1 & 3Y & -3Y^2 & Y^3 - 2 & 0 & 0 \\
0 & 0 & 0 & -1 & 3Y & -3Y^2 & Y^3 - 2 & 0 \\
0 & 0 & 0 & 0 & -1 & 3Y & -3Y^2 & Y^3 - 2
\end{vmatrix}$$

$$= Y^{15} - 10Y^{12} - 6Y^{10} + 40Y^9 - 360Y^7 - 80Y^6 + 12Y^5$$
$$- 1080Y^4 + 80Y^3 - 240Y^2 - 240Y - 40.$$

---

[6]See [5, Proposition 2.1.7].

Furthermore, this polynomial is irreducible over $\mathbb{Z}$ by applying Proposition 7.56 since $|P(m)|$ is prime for

$$m \in \{-653, -579, -532, 459, -447, -429, -427, -367, -337,$$
$$-271, -81, -43, 51, 209, 213, 339, 423, 509, 521, 581\}.$$

Hence $\deg\left(2^{1/3} + 2^{1/5}\right) = 15$.

131 P is irreducible over $\mathbb{Z}$ since $\deg P = 3$ and P has no rational root. Indeed, if P has such a root, then it must be ± 1, and $P(\pm 1) = 1$. This implies that $2\alpha^2 - 3\alpha + 2 \neq 0$ and then β is well-defined. Furthermore, $[\mathbb{Q}(\alpha) : \mathbb{Q}] = 3$ and $\{1, \alpha, \alpha^2\}$ is a $\mathbb{Q}$-base of $\mathbb{Q}(\alpha)/\mathbb{Q}$. Therefore there exist $x, y, z \in \mathbb{Q}$ such that

$$\frac{1}{2\alpha^2 - 3\alpha + 2} = x + y\alpha + z\alpha^2.$$

This may be written as $\left(2\alpha^2 - 3\alpha + 2\right)\left(x + y\alpha + z\alpha^2\right) = 1$ and expanding the product and using the relations $\alpha^3 = \alpha - 1$ and $\alpha^4 = \alpha^2 - \alpha$ we get

$$\alpha^2\left(2x - 3y + 4z\right) + \alpha\left(-3x + 4y - 5z\right) + 2x - 2y + 3z - 1 = 0$$

implying that $x = z = 1$ and $y = 2$, so that

$$\frac{1}{2\alpha^2 - 3\alpha + 2} = 1 + 2\alpha + \alpha^2.$$

This gives $\beta^2 = 7\alpha^2 + 7\alpha - 3$ and $\beta^3 = 25\alpha^2 + 15\alpha - 24$, so that

$$\beta^3 - 5\beta^2 + 10\beta - 1 = 0.$$

One easily checks that the polynomial $Q = X^3 - 5X^2 + 10X - 1$ is irreducible over $\mathbb{Z}$, and hence Q is the minimal polynomial of β.

132 Set $\theta = \sqrt{5} + \sqrt[4]{2}$. We obviously have $\mathbb{Q}(\theta) \subseteq \mathbb{Q}\left(\sqrt{5}, \sqrt[4]{2}\right)$. Conversely, since $\left(\theta - \sqrt{5}\right)^4 = 2$, expanding the product we get

$$\theta^4 + 30\theta^2 + 23 = \sqrt{5}\left(4\theta^3 + 20\theta\right)$$

so that

$$\sqrt{5} = \frac{\theta^4 + 30\theta^2 + 23}{4\theta^3 + 20\theta}.$$

and hence $\sqrt{5} \in \mathbb{Q}(\theta)$. Thus

$$\sqrt[4]{2} = \theta - \sqrt{5} \in \mathbb{Q}\left(\theta, \sqrt{5}\right) \subseteq \mathbb{Q}(\theta).$$

Therefore we get

$$\mathbb{Q}\left(\sqrt{5}, \sqrt[4]{2}\right) \subseteq \mathbb{Q}\left(\sqrt{5} + \sqrt[4]{2}\right)$$

as required. Now using Lemma 7.34, we infer that θ is a root of the polynomial

$$P = X^8 - 20X^6 + 146X^4 - 620X^2 + 529$$

and $|P(m)|$ is prime for $\pm m \in \{6, 12, 18, 60, 66, 120, 132\}$ so that P is irreducible over $\mathbb{Z}$ by Proposition 7.56. Hence

$$\left[\mathbb{Q}\left(\sqrt{5} + \sqrt[4]{2}\right) : \mathbb{Q}\right] = 8.$$

133

(a) First note that, if $A, B \in \mathbb{Z}[X]$ such that $A(X) - B(X) = 2$ and if $a, b \in \mathbb{Z}$ such that $A(a) = B(b) = 0$, then $a - b \mid 2$. Now let $\alpha \in \mathbb{Z}$ be the greatest solution of the equation $(Q(x) + 1)(Q(x) - 1) = 0$. The factor vanished by α has at most d integer roots, and if β is an integer root of the other factor, then $\alpha \geqslant \beta$ and $\alpha - \beta \mid 2$ by the argument above applied with $A(X) = Q(X) + 1$ and $B(X) = Q(X) - 1$, so that $\alpha - \beta = 1$ or 2, implying the asserted statement.

(b) Suppose that P is not irreducible. Since P is not identically 0, 1 or -1, we have $P = QR$ with $Q, R \in \mathbb{Z}[X]$, $Q \neq \pm 1$ and $R \neq \pm 1$. Let $r = \deg Q$ and $s = \deg R$. By the previous question, there are at most $r + 2$ integers m such that $Q(m) = \pm 1$ and there are at most $s + 2$ integers m such that $R(m) = \pm 1$. Now if $|P(m)| = |Q(m)||R(m)|$ is 1 or a prime, then either $Q(m) = \pm 1$ or $R(m) = \pm 1$, so that there are at most $r + s + 4 = n + 4$ integers m such that $|P(m)|$ is 1 or a prime, which proves the proposition by contraposition.

(c) We check that $|P(m)|$ is 1 or a prime for $m \in \{-10, -6, -5, -3, -2, -1, 0, 2, 3, 4, 8\}$ so that P is irreducible[7] over $\mathbb{Z}$.

134
Suppose that $P = QR$ with $Q, R \in \mathbb{Z}[X]$, $Q \neq \pm 1$ and $R \neq \pm 1$. Since P is monic, $\deg Q > 0$ and $\deg R > 0$ and one may suppose that they are also monic. Without loss of generality, suppose that $R(\alpha) = 0$. Since $P(0) \neq 0$, $Q(0)$ is a non-zero integer, so that $|Q(0)| \geqslant 1$ and one may write $Q = (X - \beta_1) \cdots (X - \beta_r)$ with

---

[7]The curve $y^2 = P(x)$ is an example of *hyperelliptic curve* of genus 2 whose jacobian was first treated with a 2-descent method in [9]. Using a refinement of a profound result due to Chabauty and Coleman, it may be shown that this curve has only six rational points, i.e. the two points at infinity and the points $(0, \pm 1)$ and $(-3, \pm 1)$.

$r \in \mathbb{Z}_{\geqslant 1}$ and $\beta_1, \ldots, \beta_r$ are the roots of Q satisfying $|\beta_i| < 1$ for all $1 \leqslant i \leqslant r$. Therefore we get

$$1 \leqslant |Q(0)| = \prod_{i=1}^{r} |\beta_i| < 1$$

giving a contradiction.

135 Let $Q = (2X - 1)P$ so that $Q = 2X^{17} - 17X^{16} - \sum_{i=1}^{12} X^i + 1$, and set $R = 17X^{16}$. Then for all $z \in \mathbb{C}$ such that $|z| = 1$, we have

$$|Q(z) + R(z)| \leqslant 2|z|^{17} + \sum_{i=1}^{12} |z|^i + 1 = 15 < 17 = |R(z)| \leqslant |Q(z)| + |R(z)|$$

so that by Lemma 7.31 we infer that Q and R have the same total number of roots, counting multiplicity, in the disc $\{z \in \mathbb{C} : |z| < 1\}$, and thus Q has exactly 16 roots in this disc. Since $\deg Q = 17$, this implies that Q has exactly one root α with multiplicity 1 such that $|\alpha| \geqslant 1$. We deduce that P has also exactly one root α with multiplicity 1 such that $|\alpha| \geqslant 1$. Furthermore, P is monic and $P(0) = -1$ so that P satisfies all the hypotheses of Exercise 133, and hence P is irreducible over $\mathbb{Z}$. Note also that the condition *monic* cannot be removed in Exercise 133, since the polynomial Q is indeed such that $Q(0) \neq 0$ and has exactly one root α such that $|\alpha| \geqslant 1$, but Q is not irreducible over $\mathbb{Z}$.

136 We first have

$$F_n = n! \, P_n = \sum_{k=0}^{n} \frac{n!}{k!} X^k.$$

1. Let p be a prime factor of $n(n-1)\cdots(n-m+1) = \dfrac{n!}{(n-m)!}$. Therefore for all $0 \leqslant k \leqslant n - m$, p divides $(n!/k!)$ which is the coefficient of X^k in $F_n(X)$, so that $F_n(X) \bmod p$ is divisible by X^{n-m+1}. If $F_n = A_m B$ with $B \in \mathbb{Z}[X]$ is monic such that $\deg B = n - m$, then X^{n-m+1} divides $\overline{A_m} \times \overline{B}$ in $\mathbb{F}_p[X]$. Since $\deg \overline{B} = n - m$, we get $X \mid \overline{A_m}$. This implies that $\overline{A_m}(0) = \overline{0}$ as required.
2. (a) First note that $\theta \in \mathcal{O}_{\mathbb{K}}$ since A_m is monic. By the previous question, we derive

$$N_{\mathbb{K}/\mathbb{Q}}(\theta) = \pm a_0 \equiv 0 \pmod{p}$$

so that $p \mid N_{\mathbb{K}/\mathbb{Q}}((\theta))$.

(b) Since $F_n(\theta) = 0$, we have

$$-n! = \sum_{k=1}^{n} \frac{n!\,\theta^k}{k!}$$

hence there exists an index $k \in \{1, \dots, n\}$ such that

$$v_p\left(\frac{n!\,\theta^k}{k!}\right) \leqslant v_p(n!).$$

Since

$$v_p\left(\frac{n!\,\theta^k}{k!}\right) = v_p(n!) + k\alpha - v_p(k!) = v_p(n!) + k\alpha - ev_p(k!)$$

we get

$$k\alpha - ev_p(k!) \leqslant 0.$$

a. By Exercise 36, we get $k\alpha \leqslant \dfrac{e(k-1)}{p-1}$, and hence

$$(p-1)\alpha \leqslant \frac{e(k-1)}{k} < e \leqslant m$$

so that $p < \frac{m}{\alpha} + 1 \leqslant m + 1$, implying that $p \leqslant m$.

(c) By the first question, all the prime factors of n, $n-1, \dots, n-m+1$ divide a_0 and the previous question shows that each of these prime factors is $\leqslant m$. Thus the numbers n, $n-1, \dots, n-m+1$ form a sequence of m consecutive integers all greater than m which have no prime factor greater than m, contradicting Lemma 7.32.

Remark In [6], the author provided an elegant proof of Schur's result based upon the theory of Newton polygons for polynomials belonging to $\mathbb{Q}_p[X]$. Let us compute the discriminant of P_n. If $\alpha_1, \dots, \alpha_n$ are the roots of P_n in an algebraic closure of $\mathbb{Q}$, we have by Definition 7.2

$$\mathrm{disc}\,(P_n) = (n!)^{2-2n} \prod_{1 \leqslant i < j \leqslant n} (\alpha_i - \alpha_j)^2.$$

We proceed as in the proof of Proposition 7.11. Writing $P_n = (n!)^{-1} \prod_{i=1}^{n}(X - \alpha_i)$, we infer that

$$\frac{P_n'}{P_n} = \sum_{i=1}^{n} \frac{1}{X - \alpha_i}$$

and thus for all $i \in \{1, \ldots, n\}$, we get

$$n! \, P_n' \, (\alpha_i) = \prod_{j \neq i} (\alpha_i - \alpha_j).$$

This implies that

$$\prod_{i=1}^{n} n! \, P_n' \, (\alpha_i) = \prod_{i=1}^{n} \prod_{j \neq i} (\alpha_i - \alpha_j) = \prod_{i=1}^{n} \prod_{i<j} \left(-(\alpha_i - \alpha_j)^2 \right)$$

$$= (-1)^{n(n-1)/2} (n!)^{2n-2} \operatorname{disc} (P_n).$$

Note that $P_n' = P_{n-1}$ and $P_n(X) = P_{n-1}(X) + x^n/n!$ so that

$$P_n' \, (\alpha_i) = P_n \, (\alpha_i) - \frac{\alpha_i^n}{n!} = -\frac{\alpha_i^n}{n!}$$

and hence

$$\operatorname{disc} (P_n) = (-1)^{n(n-1)/2} (n!)^{2-2n} \prod_{i=1}^{n} (-\alpha_i^n)$$

$$= (-1)^{n(n-1)/2+n} (n!)^{2-2n} \left(\prod_{i=1}^{n} \alpha_i \right)^n$$

$$= (-1)^{n(n-1)/2+n} (n!)^{2-2n} \left((-1)^n n! \right)^n$$

$$= (-1)^{n(n-1)/2} (n!)^{2-n}.$$

We deduce that if $n \equiv 2$ or $3 \pmod 4$, then $\operatorname{disc} (P_n) < 0$ and hence $\operatorname{disc} (P_n)$ is not a square in $\mathbb{Q}$. If $n \equiv 0 \pmod 4$, then $n - 2$ is even and thus $\operatorname{disc} (P_n)$ is a square in $\mathbb{Q}$. Now assume that $n \equiv 1 \pmod 4$. By Corollary 3.1, for all $n \geqslant 2$, there exists a prime number p such that $\frac{1}{2}n < p \leqslant n$, so that $v_p(n!) = 1$. This implies that $v_p \left((n!)^n \right) = n$ is odd, and thus $(n!)^n$ is not a square in $\mathbb{Q}$ if $n \equiv 1 \pmod 4$ and $n \geqslant 2$. Therefore $\operatorname{disc} (P_n)$ cannot be a square in $\mathbb{Q}$ in this case. Furthermore, it can be proved that $\operatorname{Gal} (P_n/\mathbb{Q})$ contains a p-cycle for some prime number satisfying $\frac{1}{2}n < p < n - 2$.[8] Using Lemma 7.22, we get the following result due to Schur again.

---

[8] See [6] for instance.

Proposition 7.58 (Schur) *Let $n \in \mathbb{Z}_{\geq 2}$. Then*

$$\mathrm{Gal}\,(P_n/\mathbb{Q}) \simeq \begin{cases} \mathcal{A}_n, & \text{if } 4 \mid n \\ \mathcal{S}_n, & \text{otherwise.} \end{cases}$$

137

(a) Since $\theta^n = -a_{n-1}\theta^{n-1} - \cdots - a_1\theta - a_0$ and $p \mid a_i$, we infer that $\theta^n/p \in M \subseteq \mathcal{O}_{\mathbb{K}}$ and that $N_{\mathbb{K}/\mathbb{Q}}(\theta) = a_0 \not\equiv 0 \pmod{p^2}$ by assumption.

(b) Since $p \mid f$, we deduce that there is an element of order p in $\mathcal{O}_{\mathbb{K}}/M$ by Theorem 7.1, so that there exists $\alpha \in \mathcal{O}_{\mathbb{K}}$ such that $\alpha \notin M$ and $p\alpha \in M$. Hence

$$p\alpha = b_0 + b_1\theta + \ldots + b_{n-1}\theta^{n-1},$$

where not all the b_i are divisible by p, otherwise $\alpha \in M$.

(c) $\beta \in \mathcal{O}_{\mathbb{K}}$ since both α and $b_0 p^{-1} + \cdots + b^{j-1}\theta^{j-1}p^{-1}$ are in $\mathcal{O}_{\mathbb{K}}$. This implies that

$$\beta\theta^{n-j-1} = \frac{b_j\theta^{n-1}}{p} + \frac{\theta^n}{p}\left(b_{j+1} + b_{j+2}\theta + \cdots + b_n\theta^{n-j-2}\right)$$

is also in $\mathcal{O}_{\mathbb{K}}$. Now by the first question $\theta^n/p \in \mathcal{O}_{\mathbb{K}}$ and also $b_{j+1} + b_{j+2}\theta + \cdots + b_n\theta^{n-j-2} \in \mathcal{O}_{\mathbb{K}}$, so that

$$\frac{b_j\theta^{n-1}}{p} \in \mathcal{O}_{\mathbb{K}}.$$

By Proposition 7.9, we infer that the norm of this element must be an integer. But

$$N_{\mathbb{K}/\mathbb{Q}}\left(\frac{b_j\theta^{n-1}}{p}\right) = \left(\frac{b_j}{p}\right)^n N_{\mathbb{K}/\mathbb{Q}}(\theta)^{n-1} = \frac{b_j^n a_0^{n-1}}{p^n}$$

cannot be an integer since $p \nmid b_j$ and $p^2 \nmid a_0$.

138

(a) P is irreducible over $\mathbb{Z}$ by Eisenstein's criterion with $p = 7$.

(b) Since $189 = 3^3 \times 7$ and $756 = 2^2 \times 3^3 \times 7$, we have $\mathbb{K} = \mathbb{Q}(\alpha)$ where $\alpha = \frac{1}{3}\theta$ is a root of $Q = X^3 - 21X + 28$. We have $\mathrm{disc}(Q) = 2^2 \times 3^4 \times 7^2$ and use

Proposition 7.16. The largest square n^2 dividing disc(Q) for which the system of congruences

$$\begin{cases} x^3 - 21x + 28 \equiv 0 \pmod{n^2} \\ 3x^2 - 21 \equiv 0 \pmod{n} \end{cases}$$

is solvable for x is given by $n = 2$ and we get $x = 1$, so that

$$\left\{ 1, \alpha, \tfrac{1}{2}\left(-1 + \alpha + \alpha^2\right) \right\} = \left\{ 1, \tfrac{1}{3}\theta, -\tfrac{1}{2} + \tfrac{1}{6}\theta + \tfrac{1}{18}\theta^2 \right\}$$

is an integral basis for $\mathbb{K}$.

(c) Since disc(P) is a square in $\mathbb{Q}$, we get Gal$(\mathbb{K}/\mathbb{Q}) \simeq \mathcal{A}_3 \simeq C_3$ by Lemma 7.23 or Lemma 7.20. Since $(r_1, r_2) = (3, 0)$, we have $\mathcal{O}_{\mathbb{K}}^* \simeq W_{\mathbb{K}} \times \mathbb{Z}^2$ by Dirichlet's unit Theorem 7.5. Using Theorem 7.11 we get

$$(3) = \mathfrak{p}_3^3$$

with $\mathfrak{p}_3 = \left(3, \tfrac{1}{3}\theta + 1\right)$, and using a computer we obtain

$$(6 - \theta) = \mathfrak{p}_2\mathfrak{p}_3^4 \quad \text{and} \quad (12 - \theta) = \mathfrak{p}_2^3\mathfrak{p}_3^3$$

with $\mathfrak{p}_2 = (2, \theta)$. This implies that

$$(6 - \theta)^3 = (3)^3(12 - \theta)$$

and hence there exists a unit u such that $(6-\theta)^3 = 27u(12-\theta)$. Now expanding $(6 - \theta)^3$ and using $\theta^3 = 189\theta - 756$, we deduce that

$$180\theta^2 - 297\theta + 972 = 27u(12 - \theta)$$

so that

$$9(\theta - 12)(2\theta - 9) = 27u(12 - \theta)$$

and then $u = 3 - 2\theta/3$. Using PARI, the second unit is $u' = \tfrac{1}{9}\theta^2 - \tfrac{5}{3}\theta + 5$ so that

$$\mathcal{R}_{\mathbb{K}} = \left| \det \begin{pmatrix} \log\left|3 - \tfrac{2}{3}\theta\right| & \log\left|\tfrac{1}{9}\theta^2 - \tfrac{5}{3}\theta + 5\right| \\ \log\left|3 - \tfrac{2}{3}\theta'\right| & \log\left|\tfrac{1}{9}(\theta')^2 - \tfrac{5}{3}\theta' + 5\right| \end{pmatrix} \right| \approx 12.594188956\ldots$$

139 Since $-2 \not\equiv 1 \pmod 4$, we have $d_{\mathbb{K}} = -8$.

- Since $-8 \equiv 1 \pmod 3$, we get $(-8/3) = (1/3) = 1$ so that 3 splits completely in $\mathbb{K}$ by Proposition 7.26 and then

$$(3) = \mathfrak{p}_3 \, \overline{\mathfrak{p}_3},$$

  where $\mathfrak{p}_3 = (1 + \sqrt{-2})$ and $\overline{\mathfrak{p}_3} = (1 - \sqrt{-2})$.
- Since $(1 - \sqrt{-2})^2 = -1 - 2\sqrt{-2}$, we get $\mathfrak{a} = \overline{\mathfrak{p}_3}^2$, so that the equality $(1 + 2\sqrt{-2})^n = 3^n$ contradicts Theorem 7.6.
- Now assume that there exists $(p, q) \in \mathbb{Z} \times \mathbb{Z}_{>0}$ such that $(p, q) = 1$ and $\arccos\left(\frac{1}{3}\right) = \frac{p\pi}{q}$. Since $1 + 2\sqrt{-2} = 3e^{i \arccos(1/3)} = 3e^{ip\pi/q}$, we derive $(1 + 2\sqrt{-2})^{2q} = 3^{2q}$, contradicting the previous question. Therefore, $\frac{1}{\pi} \arccos\left(\frac{1}{3}\right) \notin \mathbb{Q}$.

140 First note that, if $\mathbb{k}_1 = \mathbb{Q}\left(\sqrt{a}\right)$ and $\mathbb{k}_2 = \mathbb{Q}\left(\sqrt{b}\right)$, then by assumption on a and b we get $d_{\mathbb{k}_i} \equiv 1 \pmod 3$, so that 3 splits completely in $\mathbb{k}_1$ and in $\mathbb{k}_2$ by Proposition 7.26. This implies that 3 splits completely in $\mathbb{K}$. Now assume that $\mathcal{O}_{\mathbb{K}} = \mathbb{Z}[\theta]$ for some $\theta \in \mathbb{K}$. Then $\mathbb{K} = \mathbb{Q}(\theta)$ and the minimal polynomial μ of θ is of degree 4. By the previous observation, we infer that the reduction $\overline{\mu}$ in $\mathbb{F}_3[X]$ can be expressed as a product of four distinct monic linear polynomials, which is impossible since $\mathbb{F}_3$ has only three distinct elements.

141 We use Proposition 7.29.

- Type I. We have $r_{\mathbb{K}}\left(p^\alpha\right) = \mathcal{D}_{(1,1,1)}(\alpha) = \binom{\alpha+2}{2}$.
- Type II. In this case, p is inert so that by Remark 7.5 we get

$$r_{\mathbb{K}}\left(p^\alpha\right) = \begin{cases} 1, & \text{if } 3 \mid \alpha \\ 0, & \text{otherwise.} \end{cases}$$

- Type III. We have $r_{\mathbb{K}}\left(p^\alpha\right) = \mathcal{D}_{(1,2)}(\alpha)$ which can be computed by using Theorem 2.6 or Popoviciu's result of Exercise 24. For instance, applying Popoviciu's theorem with $a = 1$, $b = 2$ and $n = \alpha$, we get $\overline{a} = 1$ and thus

$$r_{\mathbb{K}}\left(p^\alpha\right) = \tfrac{1}{2}\alpha + 1 - \left\{\tfrac{1}{2}\alpha\right\} = \begin{cases} \tfrac{1}{2}(\alpha + 2) & \text{if } \alpha \equiv 0 \pmod 2 \\ \tfrac{1}{2}(\alpha + 1), & \text{if } \alpha \equiv 1 \pmod 2. \end{cases}$$

- Type VI. We have $r_{\mathbb{K}}\left(p^\alpha\right) = \mathcal{D}_{(1,1)}(\alpha) = \binom{\alpha+1}{1} = \alpha + 1$.
- Type V. We have $g = 1$ and thus we may use Remark 7.5, and since $e = 3$, we get

$$r_{\mathbb{K}}\left(p^\alpha\right) = 1.$$

142

(a) This is done in the proof of Proposition 7.23 using Corollary 7.9.

(b) We have $f_n(\sigma) = \sigma^{n+1}\pi^{-n\sigma/2}\Gamma(\sigma/2)^n$ and hence

$$g_n(\sigma) = \frac{2}{n} - (\log \pi + \gamma)\sigma + \sum_{k=1}^{\infty}\left(\frac{\sigma}{k} - \frac{\sigma}{k + \sigma/2}\right)$$

and thus

$$g_n''(\sigma) = 8\sum_{k=1}^{\infty}\frac{k}{(\sigma + 2k)^3} > 0$$

so that g_n is convex on $(0, +\infty)$.

(c) For all $\sigma \in [1, 2]$, we deduce that

$$g_n(\sigma) \leqslant \max\left(g_n(1),\, g_n(2)\right) = \max\left(2 - \gamma - \log(4\pi) + \tfrac{2}{n},\, 2\left(1 - \gamma - \log\pi + \tfrac{1}{n}\right)\right)$$

and since $n \geqslant 2$ we obtain

$$g_n(\sigma) \leqslant \max\left(3 - \gamma - \log(4\pi),\, 3 - 2\left(\gamma + \log\pi\right)\right) < 0.$$

(d) By above, we infer that f_n is decreasing on $[1, 2]$ so that $f_n(\sigma_0) \leqslant f_n(1) = 1$. Therefore

$$\kappa_{\mathbb{K}} \leqslant \frac{d_{\mathbb{K}}^{(\sigma_0 - 1)/2}}{(\sigma_0 - 1)^{n-1}} = \left(\frac{e\log d_{\mathbb{K}}}{2n - 2}\right)^{n-1}$$

and (7.19) gives then (7.23).

143

Since $\mathbb{K}$ is real, we have $(r_1, r_2) = (n, 0)$, $w_{\mathbb{K}} = 2$ and every character of $X(\mathbb{K}) = X$ is even, so that the class number formula can be written in this case as

$$h_{\mathbb{K}}\mathcal{R}_{\mathbb{K}} = \frac{d_{\mathbb{K}}^{1/2}}{2^{n-1}}\prod_{\substack{\chi \in X \\ \chi \neq \chi_0}}L(1, \chi^{\star}),$$

where $\chi^{\star}$ is the primitive even Dirichlet character that induces χ. Now using (7.41) and the arithmetic–geometric mean inequality, we get

$$\prod_{\substack{\chi \in X \\ \chi \neq \chi_0}}\left|L(1, \chi^{\star})\right|^{\frac{1}{n-1}} \leqslant \frac{1}{2n - 2}\sum_{\substack{\chi \in X \\ \chi \neq \chi_0}}\log f_{\chi^{\star}}$$

and the conductor-discriminant formula, i.e. Theorem 7.34, implies that

$$\prod_{\substack{\chi \in X \\ \chi \neq \chi_0}} |L(1, \chi^\star)| \leqslant \left(\frac{\log d_{\mathbb{K}}}{2n - 2}\right)^{n-1}.$$

144 We proceed as in Example 7.24. We have $15 \equiv 3 \pmod 4$, the order $\mathcal{O} = \mathbb{Z}[\sqrt{-15}]$ has discriminant -60 and conductor $f = \sqrt{60/15} = 2$. By Lemma 7.29, the class number of $\mathcal{O}$ is given by

$$h_{\mathcal{O}} = 2 \times 2 \times \left(1 - \frac{1}{2}\left(\frac{-15}{2}\right)\right) = 2$$

and, by Lemma 7.29, representatives of the two classes of invertible fractional ideals of $\mathcal{O}$ are

$$\left(1, \sqrt{-15}\right) \quad \text{and} \quad \left(3, \sqrt{-15}\right)$$

and hence using PARI/GP we obtain

$$H_{\mathcal{O}} = X^2 - \left(3^3 \times 5^3 \times 10968319\right) X + \left(3^2 \times 5 \times 29 \times 41\right)^3.$$

By Corollary 7.10, we infer that a prime $p \geqslant 7$ can be expressed in the form $p = x^2 + 15y^2$ if and only if $(-15/p) = 1$ and the equation

$$x^2 - \left(3^3 \times 5^3 \times 10968319\right) x + \left(3^2 \times 5 \times 29 \times 41\right)^3 \equiv 0 \pmod p$$

has a solution in $\mathbb{Z}$.

145 Assume $\sigma \in \left[0, \frac{1}{2}\right]$. Using (7.22) with $\sigma = 0$, we derive

$$\zeta_{\mathbb{K}}(it) \ll_{\mathbb{K}} t^{n/2}(\log t)^n \quad (t \geqslant e)$$

and using the bound given in Theorem 7.16, namely $\zeta_{\mathbb{K}}\left(\frac{1}{2} + it\right) \ll_{\mathbb{K},\varepsilon} t^{n/6+\varepsilon}$, along with Lemma 3.14, we get

$$\zeta_{\mathbb{K}}(\sigma + it) \ll_{\mathbb{K},\varepsilon} \left(t^{\frac{n}{2}\left(\frac{1}{2}-\sigma\right)}t^{n\sigma/6}\right)^2 t^\varepsilon \ll t^{n(3-4\sigma)/6+\varepsilon} \quad \left(0 \leqslant \sigma \leqslant \frac{1}{2}, t \geqslant e\right).$$

The case $\sigma \in \left[\frac{1}{2}, 1\right]$ is similar.

146 Let $1 \leqslant T \leqslant e^{-1/n}x^{3/n}$. As in the proof of Theorem 4.63, using (3.41), we first derive

$$\sum_{m \leqslant x} r_{\mathbb{K}}(m) = \frac{1}{2\pi i} \int_{\kappa-iT}^{\kappa+iT} \zeta_{\mathbb{K}}(s) \frac{x^s}{s} \, ds + O_{\varepsilon,\mathbb{K}}\left(\frac{x^{1+\varepsilon}}{T}\right)$$

with $\kappa := 1 + \frac{1}{\log x}$. We shift the line of integration to the line $\sigma = \frac{1}{2} + \varepsilon$ and apply Cauchy's theorem in the rectangle with vertices $\kappa \pm iT$, $\frac{1}{2} + \varepsilon \pm iT$. The residue is given by

$$\operatorname*{Res}_{s=1}\left(\zeta_{\mathbb{K}}(s)x^s s^{-1}\right) = \kappa_{\mathbb{K}} x$$

and using the previous exercise, the contribution of the horizontal segments does not exceed

$$\ll \frac{1}{T} \int_{1/2+\varepsilon}^{1+\kappa} x^\sigma |\zeta_{\mathbb{K}}(\sigma + \varepsilon + iT)| \, d\sigma$$

$$\ll \frac{1}{T} \left\{ \int_{1/2}^{1} x^\sigma T^{n(1-\sigma)/3+\varepsilon} \, d\sigma + x(\log T)^n \right\}$$

$$\ll x^{1/2} T^{n/6-1+\varepsilon} + x^{1+\varepsilon} T^{-1}$$

while the contribution of the vertical segment is bounded by

$$\ll x^{1/2} \left\{ 1 + \int_e^T \left|\zeta_{\mathbb{K}}\left(\frac{1}{2} + \varepsilon + it\right)\right| \frac{dt}{t} \right\}$$

$$\ll x^{1/2} \left(1 + \int_e^T t^{n/6-1+\varepsilon}\right) \ll x^{1/2} T^{n/6+\varepsilon}.$$

Therefore

$$\sum_{m \leqslant x} r_{\mathbb{K}}(m) = \kappa_{\mathbb{K}} x + O_{\varepsilon,\mathbb{K}}\left(x^{1/2} T^{n/6+\varepsilon} + x^{1+\varepsilon} T^{-1}\right)$$

and the result follows from choosing $T = x^{\frac{3}{n+6}}$.

References

1. Andreescu, T., Andrica, D.: An Introduction to Diophantine Equations. GIL Publishing House, Zalau (2002)
2. Axler, C.: A new upper bound for the sum of divisors function. Bull. Aust. Math. Soc. **96**, 374–379 (2017)
3. Bordellès, O.: An inequality for the class number. J. Inequal. Pure Appl. Math. **7**, Art. 87 (2006)
4. Bordellès, O.: The composition of the gcd and certain arithmetic functions. J. Integer Seq. **13**, Art. 10.7.1 (2010)
5. Cohen, H.: Advanced Topics in Computational Algebraic Number Theory. GTM 193. Springer, New York (2000)
6. Coleman, R.: On the Galois groups of the exponential Taylor polynomials. Enseign Math. **33**, 183–189 (1987)
7. de Bruijn, N.G., Lint, J.H.V.: On partial sums of $\sum_{d \mid M} \varphi(d)$. Simon. Stevin **39**, 18–22 (1965)
8. Farissi, A.E.: Simple proof and refinement of Hermite-Hadamard inequality. J. Math. Ineq. **4**(3), 365–369 (2010)
9. Flynn, E.V., Poonen, B., Schaefer, E.F.: Cycles of quadratic polynomials and rational points on a genus-2 curve. Duke Math. J. **90**, 435–463 (1997)
10. Gould, H.W.: Combinatorial Identities. A Standardized Set of Tables Listing 500 Binomial Coefficient Summations. The Author, Morgantown (1972)
11. Graham, S.W., Kolesnik, G.: Van der Corput's Method of Exponential Sums. London Mathematical Society. Lecture Note Series, vol. 126. Cambridge University Press, Cambridge (1991)
12. Hall, R.R., Tenenbaum, G.: Divisors. Cambridge University Press, Cambridge (1988)
13. Hardy, G.H., Wright, E.M.: An Introduction to the Theory of Numbers. Clarendon Press, Oxford (1938)
14. Hobby, D., Silberger, D.M.: Quotients of primes. Amer. Math. Mon. **100**, 50–52 (1993)
15. Huxley, M.N.: Area, Lattice Points and Exponential Sums. Oxford Science, New York (1996)
16. Ivić, A.: Two inequalities for the sum of divisors functions. Zb. Rad. Prir-Mat. Fak. Univ. Novom Sadu **7**, 17–22 (1977)
17. Mincu, G., Panaitopol, L.: On some properties of squarefree and squarefull numbers. Bull. Math. Soc. Sci. Math. Roum. Nouv. Sér. **49**, 63–68 (2006)
18. Ramaré, O.: Explicit estimates on several summatory functions involving the Möbius function. Math. Comput. **84**, 1359–1387 (2015)
19. Robin, G.: Grandes valeurs de la fonction somme des diviseurs et hypothèse de Riemann. J. Math. Pures. Appl. IX Sér. **63**, 187–213 (1984)
20. Trifonov, O.: Lattice points close to a smooth curve and squarefull numbers in short intervals. J. Lond. Math. Soc. **65**, 309–319 (2002)

Index

© Springer Nature Switzerland AG 2020

O. Bordellès, *Arithmetic Tales*, Universitext,

https://doi.org/10.1007/978-3-030-54946-6

Printed in the United States
By Bookmasters